EQUATIONS

- **Ideal gas law**
  $PV = nRT$

- **Nernst equation**
  $EF = 2.3\ RT \log C_1/C_2$

- **Clearance**
  $\dot{V}_1 = C_2\dot{V}_2/C_1$

- **Henderson-Hasselbalch**
  $pH = pK + \log [salt]/[acid]$

- **Ohm's law**
  $E = IR$

- **Cardiac output**
  $C.O. = S.V. \times H.R.$

- **Fick principle**
  $(A - V) \times V =$ Amour

- **Blood pressure**
  $\Delta P = \text{Flow} \times \text{Resistance}$

- **Ventricular work**
  $\text{Work} = PV + wv^2/2g$
  $\text{Work (approximately)} = C.O. \times \text{Mean arterial pressure}$

- **Poiseuille's law**
  $\text{Flow} = \Delta P \times r^4/L \times \text{Viscosity}$

- **Laplace's law**
  $\text{Pressure (P)} = \text{Tension (T)}/\text{radius (r)}$

- **Glomerular filtration rate**
  $GFR = U_x \times \dot{V}/P_x$

TEXTBOOK OF PHYSIOLOGY

SEVENTEENTH EDITION

TEXTBOOK OF PHYSIOLOGY

Byron A. Schottelius, Ph.D.

Department of Physiology and Biophysics,
The University of Iowa College of Medicine,
Iowa City, Iowa

Dorothy D. Schottelius, Ph.D.

Departments of Neurology and Physiology and Biophysics,
The University of Iowa College of Medicine,
Iowa City, Iowa

With 327 illustrations, including 88 in color

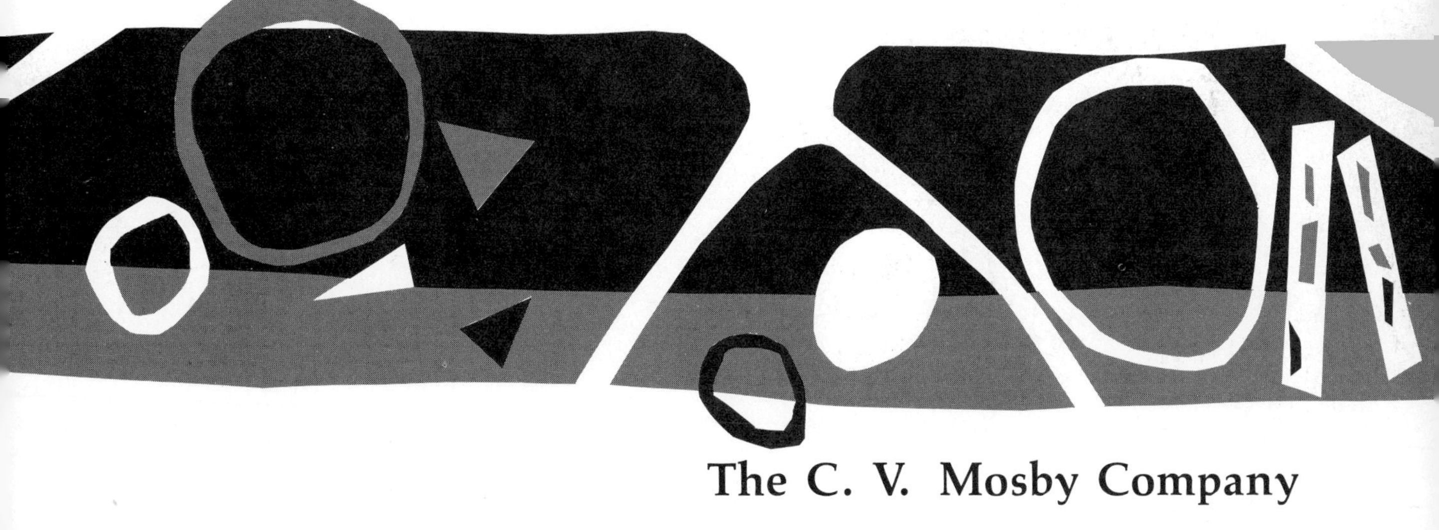

The C. V. Mosby Company

Saint Louis 1973

SEVENTEENTH EDITION

Printed in the United States of America

International Standard Book Number 0-8016-4353-8

Library of Congress Catalog Card Number 72-90106

Distributed in Great Britain by Henry Kimpton, London

in memoriam

WAID W. TUTTLE

1892-1969

PREFACE

We began the planning of this edition by carefully considering the many helpful suggestions provided by readers of previous editions. To the degree that they represented a consensus and were consistent with the scope and aims of the book, these suggestions were incorporated into the text. We herewith thank all those who gave their thoughtful appraisal.

The sequence of topics established in the sixteenth edition has been retained, although it most certainly is not the only approach to the study of physiology. The grouping of topics does not preclude the use of other sequences. To this end, an extra effort has been made to cross reference chapters. Whereas the final chapter, Nutrition, may appear misplaced, its subject matter, in fact, bears upon every preceding chapter and its importance, therefore, obligates us to assign it this position. New material on genetics, muscle energetics, blood, heart, metabolism, endocrinology, renal function, and reproduction has been included. Many new and redrawn illustrations have been provided. Naturally, this has required extensive rearrangement and even deletion of some older material.

Readings have been included at the end of each chapter that have been selected to supplement the text by expanding upon topics, by introducing material not encompassed because of space limitations, or by presenting a different point of view. A special effort has been made to select citations that are not only current but also refer to the "classical" papers in the field. Most of the articles are reviews rather than original research papers, since the latter are generally cited by the authors of the former. Excellence in exposition and illustration has been criterion for selection. The level of presentation in the readings is variable; some of the papers are popularized versions, others are more scholarly treatises. This admixture has been selected to accommodate the individual reader's interest. Nearly all of the readings should be available readily in public or private libraries.

We hope that we have achieved the objectives stated in the preface of the previous edition. If so, it is in no small measure attributable to the assistance of our readers, our students, and colleagues, and our secretary, Mrs. John Malvey.

B. A. S.
D. D. S.

CONTENTS

TEXTBOOK OF PHYSIOLOGY

1

Organization
Protoplasm—chemical composition
Protoplasm—physiological properties
Dynamic state of protoplasm
Life as a stimulus-response phenomenon

LIFE PROCESSES

Even a cursory examination of the human body reveals a most amazing structural complexity. Its chemical constitution is equally complex. Complexity of structure, whether physical or chemical, entails complexity of operation. This complexity and interrelationship in function of its various parts make the study of the living body—physiology—somewhat difficult. The study of any one part (and we can study only one part at a time) demands a knowledge of how its activity is affected by the activity of all the other parts. To overcome this difficulty it may be well at the outset to make a brief survey of the field as a whole.

■ ORGANIZATION

Organs. ■ The human body and all the more highly organized forms of life are composed of various parts; each of these performs a definite function. Such parts are called organs; thus the stomach is spoken of as an organ of digestion and the eyes as sense organs of sight.

Systems. ■ Two or more organs may differ somewhat in their individual functions, but collectively they may serve a definite, ultimate purpose in the body. Such an ensemble of organs is referred to as a system. Thus the mouth, esophagus, stomach, intestines, etc. constitute the digestive system; in this system each organ contributes its part to the more general function of digestion. In a similar manner we speak of the respiratory, the nervous, the muscular, the circulatory, the excretory, and the reproductive systems.

Tissues. ■ By closer examination it can be demonstrated that an organ is made up of two or more kinds of structures known as tissues (Chapter 4), each performing its special duty. In the stomach, muscle tissue and gland tissue are found. The food is moved about by muscle tissue, and the digestive juices are produced by the gland tissue.

Cells. ■ A tissue, in turn, is composed of a countless number of microscopic structures called cells (Chapter 2), which in any given tissue resemble each other closely. Similar to the various parts of a mechanical device, no organ in the body functions independently but only as an integral part of a highly coordinated collection of organs—the living organism.

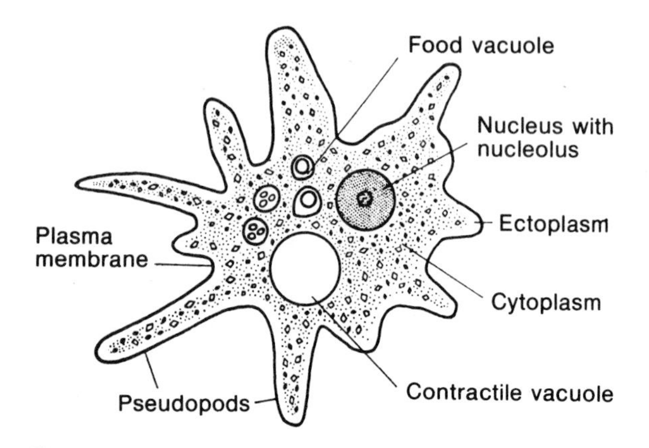

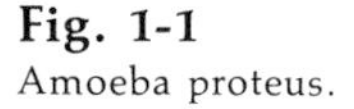

Fig. 1-1
Amoeba proteus.

■ PROTOPLASM—CHEMICAL COMPOSITION

The substance constituting a cell is known as protoplasm. Protoplasm is that particular form of matter exhibiting the properties and activities of life; it is living stuff.

The ameba (Fig. 1-1), an aquatic animal about $^{1}/_{100}$ inch in diameter, appears when viewed through the microscope as an irregular-shaped bit of matter, protoplasm. In this small mass of protoplasm it is nearly always possible to distinguish two parts, a more fluid mass, *cytoplasm,* surrounding a spherical and somewhat denser body, the *nucleus.* Hence we may define a cell as a discrete mass of protoplasm containing a nucleus. The cytoplasm is contained within a very delicate envelope, the *plasma membrane.* Cytoplasm often may appear as a structureless ground substance, but various minute solids in addition to the nucleus can be detected. These will be considered in Chapter 2.

That protoplasm is not a solid in the ordinary meaning of this term is evident from the mechanical changes it can, in many instances, undergo with such great ease. Examined microscopically, protoplasm generally appears as a semifluid transparent substance having a fair degree of viscosity. The physical structure of the semifluid material of cytoplasm is variously described as reticular (network), granular, alveolar (honeycomb), and fibrillar (threadlike).

Protoplasm is not an individual chemical substance such as sodium chloride or sugar. It is composed of a large number of chemical compounds. However, it must not be regarded as merely a mixture of ingredients, for we find practically the same substances in the fluid portion of the blood, to which we do not ascribe the properties of life. Indeed, it is possible to disintegrate cells and to isolate their component parts. Some of these carry out for a time many normal biochemical activities, but these reactions do not constitute life. The cell is an organization of matter and function.

The elements most frequently present in the various compounds found in protoplasm are carbon, oxygen, nitrogen, hydrogen, phosphorus, sulfur, sodium, potassium, calcium, magnesium, chlorine, iodine, iron, and copper; none of these elements is peculiar to protoplasm. A detailed discussion of the compounds in which the aforementioned elements occur belongs more properly in a later chapter; here we must content ourselves with a few cursory, but necessary, observations.

■ Water

From 50% to 90% of protoplasm is water (Chapter 24). Deprived of this water, nearly all protoplasms die quickly; a few forms, notably plant seeds, pass into a condition of latent (inactive) life from which they can be revived by the addition of water. We find some crystalloids dissolved in cell water, but colloids (gel-like substances) are more abundant.

It is difficult to contemplate a piece of machinery such as our muscles being composed of 75% water; it does not suggest the necessary firmness, stability, and rigidity customary in machines. However, the ease with which a contracting muscle changes its shape is evidence of the semifluidity of its protoplasm. The viscosity of protoplasm varies from one type of cell to another. Certain parts of our body exhibit firmness and solidity to a high degree, for example, tendons, ligaments, cartilages, and bones. This, however, does not indicate that the protoplasm lacks a high percentage of water, for these structures are formed of a large amount of solid material deposited between the living cells.

There are a number of specific properties of water that make it uniquely suitable to living systems:

1. It is a liquid at physiological temperatures.
2. It is an excellent solvent for electrolytes.
3. It absorbs large quantities of heat, thus stabilizing the temperature of living organisms.
4. It is immiscible with lipids, thus cellular membranes can limit its movement.
5. The electrolytic dissociation of water itself is extremely low, that is, the pH is neutral.

■ Inorganic materials

Among the most important of the inorganic salts are the soluble chlorides, sulfates, phosphates, and carbonates of sodium, potassium, calcium, and magnesium (Chapter 24). The amount of each salt present in protoplasm varies with the different kinds of cells, but, like water, they are indispensable for life.

■ Organic materials

Except for carbon monoxide, carbon dioxide, and the carbonate salts of metals, any compound containing carbon is known as an organic compound (Chapter 23). Some of the major classes of organic compounds are described below.

Carbohydrates. ■ Carbohydrates are composed of carbon, hydrogen, and oxygen. They include the simple sugars, for example, glucose; the double sugars, of which sucrose and lactose are well-known examples; and the so-called starches, among which are the ordinary starches and a substance known as glycogen. It is especially in the form of glycogen (sometimes called animal starch) that we find the carbohydrates in animal cells. Carbohydrates are essential fuels for living systems.

Lipids. ■ Lipids, a class of compounds that includes the common fats, are composed mainly of carbon, hydrogen, and oxygen. They are present in protoplasm either as simple compounds or in complex combination with other substances (e.g., lipoproteins). Lipids are important constituents of cell membranes (Chapter 2). A large proportion of the energy requirement of living tissues is met by these important fuels. Usually energy production involves the oxidation of a balanced mixture of carbohydrates and lipids; indeed, lipid metabolism is linked to carbohydrate metabolism (Chapter 25).

Proteins. ■ Proteins are one of the most complex substances with which the biochemist deals. They are complex because the molecules contain at least five, and frequently more than five, elements, and the size of the molecule is exceedingly large.

There are many kinds of proteins that differ from each other in composition, solubility, and chemical reactions. Some are soluble in dilute salt solutions (as in milk, blood, and lymph). Only a few are soluble in water. Nearly all proteins are undialyzable; that is, they cannot pass through vegetable or animal membranes (such as the urinary bladder, lungs, and pericardium), and they therefore belong to the class of compounds frequently called *colloids*—the gluelike and jellylike compounds.

Proteins in solution are easily precipitated by certain reagents, such as alcohol, and are coagulated by heat so that the previously soluble protein is rendered insoluble.

Nucleic acids. ■ A primary role in heredity and in the architecture and function of cells is attributable to nucleic acids. These compounds contain carbon, hydrogen, oxygen, nitrogen, and phosphorus. They are complex compounds often bound to protein; hence the term *nucleoprotein*. The name suggests that nucleic acids are found in the nuclei of cells, but they are also present in the cytoplasm.

■ PROTOPLASM—PHYSIOLOGICAL PROPERTIES

In addition to the chemical and physical properties common to all inanimate objects (such as color, cohesiveness, elasticity, acidity, and specific gravity), protoplasm exhibits what are generally termed physiological properties or phenomena. Among these are *active transport, contractility, irritability, conductivity, metabolism, excretion, growth,* and *reproduction*. The physiological properties are the very expression of life. However, this does not necessarily mean that they are independent of the chemical and physical properties of protoplasm.

■ Active transport

A general and essential property of living cells is the capability of selectively concentrating in or excluding from their substance certain materials. This phenomenon is spoken of as *active transport* (Chapter 3). It is the basis for the specialized functions of groups of cells, organs, whereby massive amounts of substances are absorbed or secreted to provide for the constancy of the internal environment of the entire organism.

■ Contractility

One of the most conspicuous characteristics of an animal is its power to move some parts of its body or to change its own position with respect to its surroundings. This is in sharp contrast with the immobility of inanimate objects. Upon observing an ameba for some length of time, it is noticed that at a certain point a projection arises. The protoplasm seems to stream in this direction so that the projection or pseudopod enlarges; at the same time the protoplasm at some other portion of the body can be seen to withdraw. Several of these pseudopods are shown in Fig. 1-1. By thus changing its shape, the ameba moves from place to place. The power of protoplasm to change its form is called *contractility* (Chapter 6).

Not all cells in a highly specialized animal body exhibit the property of contractility. In the human body two structures in particular are endowed with contractility: muscles and certain leukocytes (white blood cells). By the simultaneous contraction of thousands of cells, a muscle shortens in length and increases in thickness. The leukocyte shows its contractility by the formation and withdrawal of pseudopods, similar to the ameba.

■ Irritability or excitability

Another general observation is that slight changes in the environment can induce changes in the activity of a living organism. When an ameba that has been quiescent for a considerable length of time is disturbed by placing a drop of dilute acid near it, a change in its form is soon observed. The animal by its contractility responded to the disturbance.

Irritability is defined as the property of protoplasm and living organisms that permits them to react to stimuli; a change in the environment is called a *stimulus*. The term *excitability* is commonly substituted for the term *irritability*. Among the more common environmental changes are thermal (heat and cold), photic (light), acoustic (sound), chemical, and mechanical (such as impact, pressure, and pull).

A stimulus sets up a change in the protoplasm that is known as the *excitatory state*. This in turn evokes the activity characteristic of the protoplasm that is being stimulated; for example, the protoplasm of a muscle contracts and that of the tear glands forms tears. Excitability is a fundamental property common to all protoplasm, and, in consequence, it is employed as a diagnostic property by which the living is distinguished from the dead. When the form of stimulation to which a particular organ is especially susceptible has been ascertained and when the organ loses its reponsiveness to this stimulation and no treatment can regain it, the organ is dead.

Frequently a quiescent ameba forms a pseudopod, apparently without any stimulus having been applied. This phenomenon is said to show the power of protoplasm to initiate its own activity; that is, it has the property of spontaneity. Although it may be difficult or sometimes impossible to discover any form of stimulation that may have acted upon the ameba, yet for other reasons we cannot grant the existence of this property. One fundamental law of the universe is Newton's law of inertia, which states that a body at rest remains at rest until acted upon by some external force. It is more than likely that the ameba is no exception to this law. It is, however, customary to apply the term *automaticity* to an organ that, after removal from the body, continues its usual activity without any apparent external stimulation. Automatic action is the result of an internal stimulus. When a stimulus calling forth a change in protoplasm ceases, protoplasm returns to its former state. This is regarded as demonstrating the resiliency of protoplasm.

■ Conductivity

The application of a stimulus to a certain part of the body may cause activity in a distant part; for example, the stimulation of the olfactory organs by the odor of fragrant food causes the salivary glands to become more active. The olfactory organs and the salivary glands are interconnected by the nervous system. It must be evident that the stimulation of the olfactory receptors generates nerve impulses, which eventually are conducted to the glands and excite them to activity.

The ability of protoplasm to convey an impulse is known as *conductivity*. Like excitability, conductivity is present in all cells, but both properties find their highest development in nerve tissue (Chapter 5). So far as we know, nerve impulses never originate in the absence of adequate stimulation.

■ Metabolism—energy transformations

The physiological properties of protoplasm are sustained by the expenditure of energy. This constant expenditure of energy by protoplasm *(catabolism)* demands a corresponding intake of energy (Chapter 18) by the cell for restorative purposes *(anabolism)*. These two opposing activities, anabolism and catabolism, are collectively referred to as *metabolism* (Chapter 23). Metabolism includes all the material and energy changes that occur in the body and in its broadest meaning is coextensive with the term *life*. Before continuing our survey with a discussion of the physiological property identified as metabolism, we should review, if only briefly, the subject of energy.

Energy may be defined as the ability to do work or to produce a change. There are many forms of energy, such as heat, light, sound, mechanical, electrical, chemical, and atomic. Two modes of energy, that of *motion* or *kinetic energy* and that of *position* or *potential energy*, are recognized.

Kinetic energy. ■ The energy of motion may be in the form of mechanical energy of a moving body; for example, the energy of wind and waves, of flowing blood, and of a moving part of an animal. Heat is the energy of the movements of individual molecules. The movement of electrons gives rise to the electrical energy of a current.

Potential energy. ■ Potential energy may be regarded as stored energy. No change is produced by it, but there is latent power, which, when released under proper conditions, is capable of doing work. A suspended weight, a coiled spring, and a molecule of glucose all possess potential energy.

Conservation and transformation of energy. ■ The relations between kinetic and potential energy are described by the laws of thermodynamics.* The first law of thermodynamics deals with the conservation of energy; the amount of energy in the universe is constant—energy can be neither created nor destroyed, but it may be altered in form. In a completely *isolated system,* be it the universe, an organism, or a cell, the total energy neither increases nor decreases. However, energy can be changed from one form into another. Thus, electrical potential energy can be transformed into heat, light, or sound. We may also transform kinetic into potential energy; for example, a weight may be raised from the ground to a certain height and placed upon a support. In this action the kinetic energy of motion obtained from the instrument doing the raising is transformed into the potential energy of position of the weight. It is energy of position by virtue of the attraction existing between the earth and the weight. When the support is removed, the energy of position is reconverted into mechanical energy of motion; when the weight strikes the earth, the moving body energy is transformed into sound and heat.

Whereas the first law does not specify the tendency of a process to occur, or its direction, the second law of thermodynamics specifies that left to themselves all physical and chemical processes in an isolated system will proceed in such a direction that the capacity of the system to do work decreases. For example, if two well-insulated cylinders are interconnected by a minute orifice and one, A, is filled with gas at high pressure and the other, B, with the same gas at low pressure, gas molecules will move spontaneously from A to B until the pressures in A and B are equal. At equilibrium of pressures, the system has a decreased capacity to do work, and the original state cannot be restored unless energy from the outside is applied to container B.

We might have arranged to collect the work done in the expansion of the gas from A to B and to use this energy to restore the original state. Our effort would not have been a total success, since there is a natural tendency for all other forms of energy ultimately to be changed into heat. For example, electrical energy is transformed into heat when a current passes through a resistance—a light bulb gets hot. Although work can always be transformed completely into heat, it does not follow that heat can be transformed completely into work. Heat passes spontaneously only from a higher to a lower temperature, and the efficiency with which it may perform work is proportional to the difference between the absolute temperatures.* The relative inefficiency of the conversion of heat into work makes spontaneous processes irreversible; that is, a spontaneous process will not reverse itself spontaneously.

Chemical potential energy. ■ By passing an electric current through water, the water is decomposed into hydrogen and oxygen. The energy of the electric current disappears, and we find it associated with the hydrogen and oxygen atoms. Because of the attraction between the separated atoms, there exists potential energy of position or separation, known as *chemical potential energy.* Chemical potential represents the capacity of a substance to do useful work by undergoing change.

In the elemental state, carbon and oxygen contain chemical potential energy; at the proper temperature (the kindling temperature) they unite, and the energy becomes kinetic, in the form of heat and light. The uniting of oxygen

---

*Thermodynamics is the branch of science that deals with energy and its transformations. A law is a statement of relationship between two or more phenomena. The discovery of such relationships is the goal of scientific research; isolated facts are of little value. According to the relativity theory, matter and energy are two phases of a single principle—energy-matter. Matter may be changed into energy, as occurs in the disintegration of radium and uranium.

*The absolute temperature is stated in degrees Kelvin, where zero is equal to −273 C. The percent efficiency of a heat machine would be $(T_h - T_c)/T_h \times 100$, where T is the absolute temperature of the hotter (h) and cooler (c) bodies.

with another element or with a compound is called oxidation, for example, ordinary burning. All organic compounds that contain carbon and hydrogen or carbon, hydrogen, and oxygen are oxidizable; that is, they have affinity for more oxygen. In consequence, they contain potential energy. Among these compounds are foods such as lipids, carbohydrates, and proteins.

Source of energy for animal life. ■ The potential energy in the food utilized by the animal is in all cases derived directly or indirectly from the plant world. Plants are able to synthesize, or build up, simple inorganic compounds such as water, carbon dioxide, nitrates, sulfates, and phosphates into highly complex organic substances (e.g., sugars, starches, lipids, and proteins). Since these substances contain much potential energy, although the materials from which they are derived are of low energy, energy from some external source must be drawn upon and stored. This production of complex and often stable compounds does not contradict the second law of thermodynamics, since the energy content of the new compounds is always less than the energy used for their synthesis. The external source is the radiant energy of the sun. In chlorophyll, the green pigment of plants, radiant energy produces a series of chemical changes in the inorganic compounds mentioned and thereby transforms them into organic substances such as glucose. The radiant energy is transformed into chemical potential energy and becomes latent in the products formed.

This process of *photosynthesis* may be expressed as follows:

$$CO_2 + H_2O + \text{Radiant energy} \rightarrow$$
$$CH_2O + O_2 + \text{Chemical potential energy}$$

Several CH_2O molecules combine to form $C_6H_{12}O_6$, which is glucose. This simple sugar is regarded as the basic material from which other carbohydrates, fats, and proteins are constructed. It is the storing (conserving) of energy that makes the process of photosynthesis the fundamental process on which the existence of life depends. Being unable to perform this synthesis, the animal takes into its body (directly in herbivorous animals and indirectly in carnivorous animals) the plant-made and energy-rich carbohydrates, fats, and proteins. Green plants are the great living synthetic mechanisms; all animals are predatory.

■ Catabolism

Protoplasm is constantly active. This activity shows itself not only in the changing of its form but also in the production of heat and electrical potentials and in chemical changes. It does not create energy; protoplasm is an energy-transforming mechanism. Hence, to be active it must be supplied with energy. The source of this lies in the chemical potential energy of carbohydrates, lipids, and proteins. These substances are not only found in our food but are also constituents of protoplasm itself. Although it is generally helpful to speak of the food in the protoplasm as fuel and of the protoplasm as the machine, in reality no sharp distinction can be drawn. To be utilized for vital processes (e.g., the contraction of a muscle) the chemical potential energy in the food or protoplasm must be released.

The release of potential energy is known as *catabolism*. Subsequent chapters are devoted to a more detailed account of the energy transformations in our body, but in order to understand any protoplasmic activity it will be necessary to anticipate this by a brief preliminary study.

Liberation of energy. ■ When a large organic molecule is split into two or more smaller molecules, the products formed contain less potential energy than the original molecule; hence some energy must have been set free; for example, during yeast fermentation the large molecule of glucose, $C_6H_{12}O_6$, is broken up into two molecules of carbon dioxide, CO_2, and two molecules of ethyl alcohol, C_2H_5OH. Since this process is accompanied by the production of heat, some (but not all) of the potential energy of the glucose was liberated. The same is true in our body when a molecule of glucose is split into two molecules of lactic acid, $C_3H_6O_3$. But when glucose undergoes complete oxidation, all the latent energy becomes kinetic.

$$C_6H_{12}O_6 + 6O_2 \rightleftharpoons 6CO_2 + 6H_2O + \text{Energy}$$

Utilization of the released energy. ■ The energy liberated during catabolism can be utilized by the protoplasm for its particular functions. In muscle protoplasm some of it eventually appears as mechanical energy, by which a part of the body or the whole body is moved or by which certain substances (blood, food, etc.) are transported from one part of the body to another. In a gland the liberated energy is used for manufacturing secretions. In the mammalian body most of the liberated energy (as much as 80%) takes the form of heat, which is required to maintain the proper body temperature.

Catabolism takes place in every cell without exception, but the amount of catabolism varies from one type of tissue or organ to another and according to the demand for energy. Muscles and certain other organs (e.g., the liver) are the most active structures; in connective tissues, such as bones, catabolism is far less intense.

A purposeful and orderly process. ■ It should be noted that energy transformations are of use in a process only if they are controlled with respect to extent and intensity. Furthermore, provision must be made so that the kinetic energy released can be used. An example of the difference between orderly and disorderly energy transformation can be observed in the burning (oxidation) of gasoline. When simply poured on a surface and ignited, gasoline burns and liberates heat. However, if the gasoline is carefully measured through a carburetor, is mixed with the proper amount of oxygen (derived from the air), is injected into a cylinder, and is ignited, the heat developed can be used to expand the gases formed in the oxidation, and this gas pressure can be used to push a piston and, with proper coupling, to power an automobile. The important point of this analogy is that burning of the fuel is coupled with a total mechanism so that some of the kinetic energy provided through the oxidation can be utilized to do mechanical work. In a similar (but by no means identical) way kinetic energy release in the animal body is ordered, graded, controlled, and coupled.

In biological oxidations a large proportion of the potential energy in the compounds being catabolized also is converted to heat. Since the cells are at a relatively constant temperature, this heat cannot be transformed into work or potential energy. It is, however, useful in maintaining the temperature of the organism and in facilitating its many metabolic processes.

The degree to which heat is produced in the cellular oxidations is a measure of the inefficiency of the organism. Consequently, outside energy must be fed continuously to the cells. Cells are never truly in equilibrium with their environment except at death. Instead, they undergo a continual exchange with their environment—a *steady state.*

■ Anabolism

Organismic activity is associated with the oxidation of food. If a distinction can be made between the food and the protoplasm, we may be allowed to say that, as in all mechanisms, the protoplasmic machinery undergoes wear and tear. For the continued existence and functioning of the protoplasm, the wear must be repaired. This constitutes the process of *anabolism.*

Ingestion. ■ The first and preliminary step in anabolism is supplying the body with the necessary materials for recouping its losses. The ameba ingests its food (small plants or animals) by engulfing it with a pseudopod. The process by which the protoplasm of a cell, devoid of a cell wall, flows around a small body is termed *phagocytosis,* and the cell is called a phagocyte. Certain forms of white blood cells act as phagocytes by engulfing bacteria (Chapter 12). In multicellular animals ingestion usually consists of the introduction of foods into an alimentary tract (Chapter 19).

Digestion. ■ The carbohydrate, lipid, and protein molecules of the ingested food are too large to be used directly in the reconstruction of protoplasm. In addition most of our foods are insoluble; starch, fats, and proteins do not dissolve in water. They are therefore also undialyzable. In this state they cannot enter the majority of the cells of the body, for these cells, in their great differentiation and specialization, have lost the power to engulf solid materials. Hence it is necessary to prepare the food.

On watching an ameba with an engulfed food particle, it is noticed that the food undergoes changes; it seems gradually to be corroded or dissolved. This process is brought about by chemical agents known as *enzymes* that hasten the splitting of the large food molecules into smaller molecules (Chapter 22). By a number of successive cleavages a large starch molecule is split into several hundred smaller molecules of glucose. The successive products thus formed become more soluble and dialyzable and are rendered utilizable by protoplasm. This constitutes digestion.

Absorption. ■ In the human body digestion is carried on in the alimentary canal, which is a long and rather narrow tubelike structure. The soluble products of digestion pass through the walls of the alimentary canal and of the smallest blood vessels found in the walls of this canal and thus into the blood. The transfer of a substance from a free surface (such as the lumen of the alimentary canal) into the blood is called *absorption* (Chapter 20). The absorption of many compounds is accomplished by active transport. Soluble food is carried to all parts of the body by the blood (Chapter 13). As they are needed by the cells, the food molecules pass through the walls of the blood vessels and into the fluid bathing the cells. From this fluid the food enters the cells, commonly as a result of active transport.

Anabolism proper. ■ The food, having entered the cell, is used for various purposes. Of these, only two need concern us at this time: (1) it is used immediately by the protoplasm for the production of kinetic energy or (2) it is built up into protoplasm *(assimilation)*. This power of self-restoration is very impressive when we consider that this synthesis is exceedingly specific.

There are literally thousands of different types of protoplasm. As the function of the protoplasm of a muscle is to contract, that of a nerve is to propagate impulses, and that of a gland is to synthesize secretions, these three protoplasms must differ radically from each other. The body possesses many kinds of glands (e.g., sweat, oil, salivary, and gastric glands); each of these manufactures its particular secretion. The saliva and the gastric juice of a dog are not the same as those made by human glands. This, therefore, demands a large number of distinct glandular protoplasms.

■ Excretion

The oxidation of food has a twofold result: (1) the transformation of potential into kinetic energy and (2) the material change of food into simpler compounds, some of which are waste products. Of these waste products water, carbon dioxide, urea, and uric acid may be mentioned; some of these are removed or eliminated by a process known as *excretion*. The human body has special organs for this purpose. Along with other substances, water, urea, and uric acid are excreted by the kidneys; carbon dioxide and water are eliminated by the lungs. A small amount of waste material derived from the cells of the body is excreted into the alimentary canal and thus is eliminated along with the indigestible or undigested residue of the ingested food.

■ Respiration

Since the greater part of the chemical potential energy in foods is liberated by the process of oxidation, a continual supply of oxygen is needed.

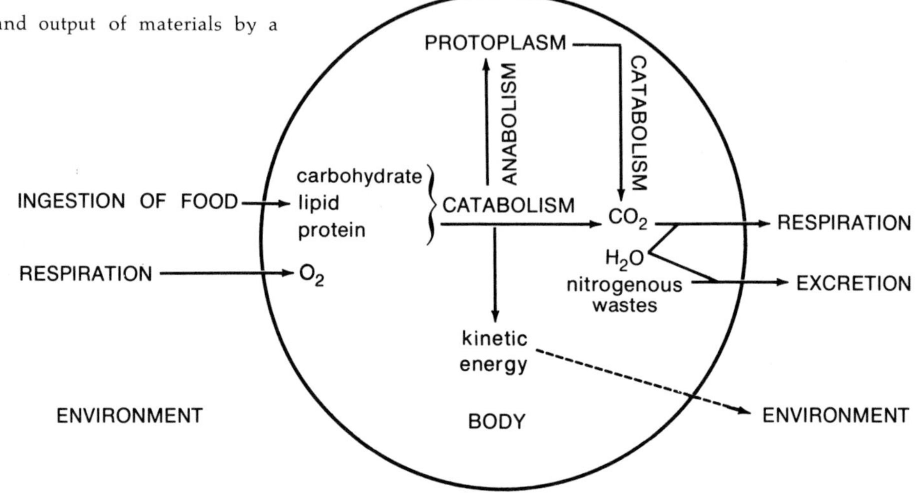

Fig. 1-2
Diagram to illustrate intake and output of materials by a body.

The oxidations result in the formation of carbon dioxide, a gas, which must be removed. Therefore this necessitates an exchange of these gases between the organism and its environment, a process called *respiration* (Fig. 1-2). In man this takes place in the lungs, which have two functions: (1) the transfer of oxygen from the air into the blood and (2) the elimination of carbon dioxide (and to a lesser extent of water) from the blood into the air (Chapter 17).

■ DYNAMIC STATE OF PROTOPLASM

Casual observation of a living structure may under proper conditions reveal a fair degree of permanency and constancy. The apparent resting state and the permanency sometimes attributed to protoplasm are the result of two opposing forces. Life consists of a continual building up and tearing down of protoplasm (Fig. 1-2). When the intake of energy is equal to the outgo, the condition is known as *physiological equilibrium* (Fig. 1-3, *A*), although as we already have noted this is in reality a steady state and not a true equilibrium. This dynamic equilibrium can be maintained only by the constant expenditure of energy. The very existence of the animate machine—protoplasm—depends on it. Life is dynamic; its essence is activity; a *resting cell* is a misnomer.

■ Basal metabolism

In an adult human being in as nearly a completely resting condition as possible, the amount of energy utilized by the body is at a minimum. This condition is known as *basal metabolism.** When the resting body is stimulated, it responds by increasing its catabolism; this enables it to obtain extra energy needed for the increased activity in adjusting itself to the environmental change. Catabolism now exceeds anabolism. However, one of the fundamental characteristics of protoplasm is a tendency to maintain itself in a state of nutritive equilibrium. Consequently, increased catabolism is accompanied or soon followed by increased anabolism.

This situation generally exists under ordinary conditions of life. But since the power of protoplasm to maintain a state of metabolic balance

*Basal metabolism may be compared to the idling of an automobile in which the gas consumption or energy expended is the least compatible with keeping the machine running.

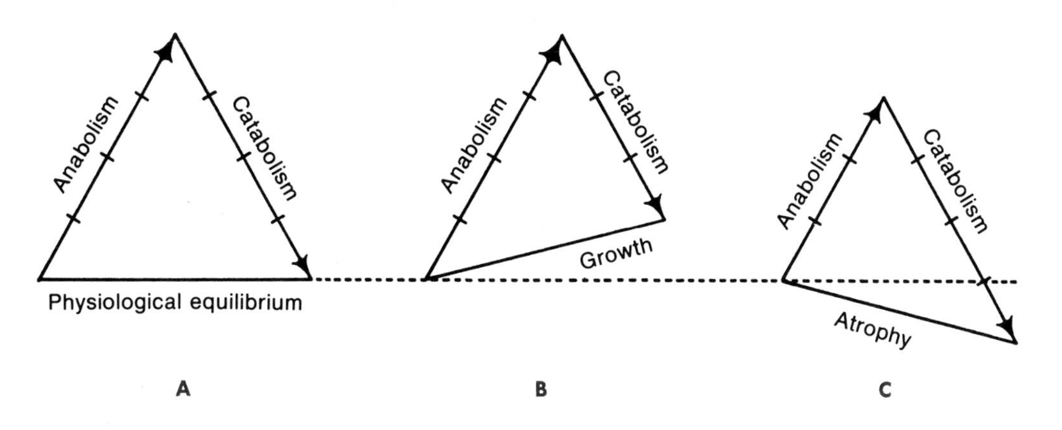

Fig. 1-3
Quantitative relationships between anabolism and catabolism are indicated for the states of **A**, physiological equilibrium (steady state), **B**, growth, and **C**, atrophy.

has its limits, an excessively strong environmental change or a milder form continued for a long time may exceed the ability of protoplasm to rebuild fast enough to compensate for the greatly increased catabolism (Fig. 1-3, *C*). This, of necessity, leads to a diminution in the amount of protoplasm and finally to the cessation of life.

■ Growth

In a child anabolism exceeds catabolism; there is an increase of protoplasm in its body—the condition of *growth* (Fig. 1-3, *B*). Growth may arise from an increase in the number of cells or from an increase in the amount of protoplasm in the already existing cells without an increase in their number.

■ Atrophy and hypertrophy

Almost any severe illness results in a loss not only of the fat present in the body but also of the protoplasm; the patient literally burns up his own flesh. An unused muscle suffers in its anabolism and gradually dwindles away, a condition known as *atrophy* (Fig. 1-3, *C*).

Under certain normal circumstances the anabolism of an organ or tissue in an adult may for a limited length of time exceed catabolism. As a result there is a gain of protoplasm in that particular tissue or organ; this is called *hypertrophy*. The hypertrophy of the muscles of an athlete is a familiar example. Extra work imposed upon the heart may cause it to undergo similar hypertrophy.

■ Differentiation, organization, and integration

In a unicellular organism (e.g., the ameba) the numerous functions common to all animals, such as the ingestion and digestion of food, the intake of oxygen for catabolism of the food, the utilization of the liberated energy for various purposes, and the excretion of waste products, are all performed by just one small mass of protoplasm. This single cell exhibits the properties of irritability, conductivity, and contractility, and it has the power of reproduction.

In a highly evolved animal and in man these many diverse functions are severally performed by a number of discrete structures. These structures differ markedly not only in their physical construction but also in their chemical composition. This *differentiation* is associated with specialization in function. As a result, in the animal economy we find a great division of labor.

Everyday observation teaches us that the various highly specialized functions do not take place at random. To the contrary, they are very closely coordinated with each other. This is called *organization*, and the frequency, duration, and intensity of each individual process are coordinated with those of all other parts.

Frequently the activity of a certain structure determines the activity or the cessation of activity of one or more other structures. In this manner the proper sequence of the functions of an organism's many parts is obtained. As a result of coordination the multitudinous parts act in a unified manner; this constitutes *integration*.

LIFE AS A STIMULUS-RESPONSE PHENOMENON

Adjustment

To the age-old question, What is life? a satisfactory answer has as yet not been found. The great complexity of activity, which varies from one organ to another, makes it virtually impossible to express in one brief statement all that is implied by the word *life*. It is, therefore, not surprising that the various definitions hitherto proposed have taken cognizance of one or two features of this extremely complex phenomenon to the exclusion of other components.

There are negative statements that can be made with some degree of certainty. Although life cannot be separated from matter, life is not matter; the weight of the body immediately after death is the same as that before death. Neither can the individual constituents (water, proteins, salts, etc.) that we find in protoplasm be said to be alive. Although far from satisfactory, it yet may be of some value to regard life as the sum total of the properties and activities of a highly organized aggregate of various chemical compounds that we call protoplasm; it is an organization of materials and functions.

Among these properties we have called attention to irritability as a diagnostic property of a living body. On this property Herbert Spencer based his classic definition: "Life is the continuous *adjustment* of internal relations to external relations." Observation teaches us that this adjustment to environmental changes is possible only within narrow physiological limits. For example, the human body can adjust itself to changes in external temperature only when these changes are very moderate.

The above definition lacks inclusiveness, but the concept embodied in it is of utmost value. It may be epitomized as follows: Life is a stimulus-response phenomenon. Viewed in this way, life is the interplay between the organism and its environment, by which the organism either adjusts itself to the environment or adjusts the environment to itself.

Adaptation

It is of vital importance that the adjustments to the environmental changes are nearly always advantageous to the organism; they are adaptive in that they enable the organism to survive either as an individual or as a race.* Most adaptive reactions fall into one of the following three classes:

1. *Protection*—to protect the organism against injury (defense, avoidance, and escape reactions)
2. *Maintenance*—to procure materials for the growth and maintenance of the protoplasm and to supply energy
3. *Reproduction*—to perpetuate the species

To some environmental changes the body is wholly unable to adjust itself, or the adjustment is incomplete or imperfect. The first results in death and the second in maladjustment. Accidents, diseases, and social infractions are instances of maladjustment. These maladjustments may be due to (1) the suddenness or severity of the environmental change, (2) the absence in a given species of animal of the proper adjustors (e.g., a terrestrial mammal drowning and a fish

*Self-preservation is the first law of nature.

perishing out of the water), or (3) some defect in the physical or chemical constitution of an individual. These defects are frequently a matter of inheritance, as in color blindness. They may reveal themselves in a lack of proper defensive powers against certain external conditions to which a normal person readily adjusts himself, as in hay fever. Quite appropriately they may be called constitutional inadequacies. Even in what may be regarded as a normal person, adjustments are not always perfect. Some organ gradually, almost imperceptibly, fails in its adjustment until finally it is impossible to maintain the physical and chemical conditions necessary for life.

In some instances an organism can acquire an adjusting mechanism with which it was not equipped previously. For example, the average person may succumb to the toxic effects of smallpox of sufficient intensity. But inoculation with vaccine confers upon his body the power to adjust itself adaptively to what would previously have been a lethal dose of the infection.

■ Inhibitors

Everyday experience shows us that certain environmental changes may lessen or even totally suppress the activity in which we happen to be engaged. A clear example is that of a man crossing a street who suddenly stops on hearing the blast of an automobile horn. A decrease in function in response to an environmental change is called *inhibition*. A moment's reflection shows the tremendous importance of inhibition in our physical, mental, and social life.

READINGS

Bernard, C.: An introduction to the study of experimental medicine, New York, 1957, Dover Publications, Inc.

Lehninger, A. L.: Bioenergetics, New York, 1965, W. A. Benjamin, Inc., pp. 1-50.

Schrodinger, E.: What is life? and other scientific essays, New York, 1956, Doubleday & Company, Inc.

Sherrington, C.: Man on his nature, ed. 2, New York, 1953, Doubleday & Company, Inc.

Smith, H. W.: From fish to philosopher. The story of our internal environment, Boston, 1959, Little, Brown and Company.

2

Membranes
Cytoplasm
Nucleus
Cell growth and reproduction

THE CELL

The cell constitutes the structural or morphological unit of the body; it is to the organism what the molecule is to matter. In the preceding chapter a cell was defined as a discrete mass of protoplasm enveloped in a plasma membrane and containing a nucleus. However, this definition offers an inadequate description, for cells are differentiated into many organelles, which are especially adapted to carry on the diverse activities of life. Indeed, the cell is a highly organized molecular factory.

Cells differ largely from each other both in form and composition and therefore also in function. With few exceptions cells are mostly small, that is, on the order of 10 to 100 μ in diameter. Because cells must rely on diffusion for the movement within their protoplasm of molecules of foodstuffs and oxygen, the distances involved must be small, or the inner parts of the cell will be starved. Of course the removal of the waste products of metabolism would be equally hindered. The importance of cell size becomes apparent when it is recalled that the volume of a sphere increases as the cube of the radius, whereas the surface only increases as the square of the radius. Some cells compensate for this by becoming flattened or elongated, thereby bringing all parts nearer the surface. However, a further consideration is that an optimum ratio between nuclear volume and cytoplasmic volume is essential to maintain the important exchange that occurs between the nucleus and cytoplasm. Finally, large size would impose a strain on the plasma membrane that it might not be able to withstand. Cell division with reduction in size is the nearly universal solution adopted to meet these problems.

For all their variety, cells possess a number of common structures and functional capacities. Although a typical cell is nonexistent, the structural details of a composite cell, as illustrated in Fig. 2-1, can profitably be examined, and its functional capacities can briefly be discussed as to *energy transformation, biosynthesis, growth,* and *reproduction.*

■ MEMBRANES

Biological membranes are an important and exciting area of contemporary scientific investigation. The functions of membranes are di-

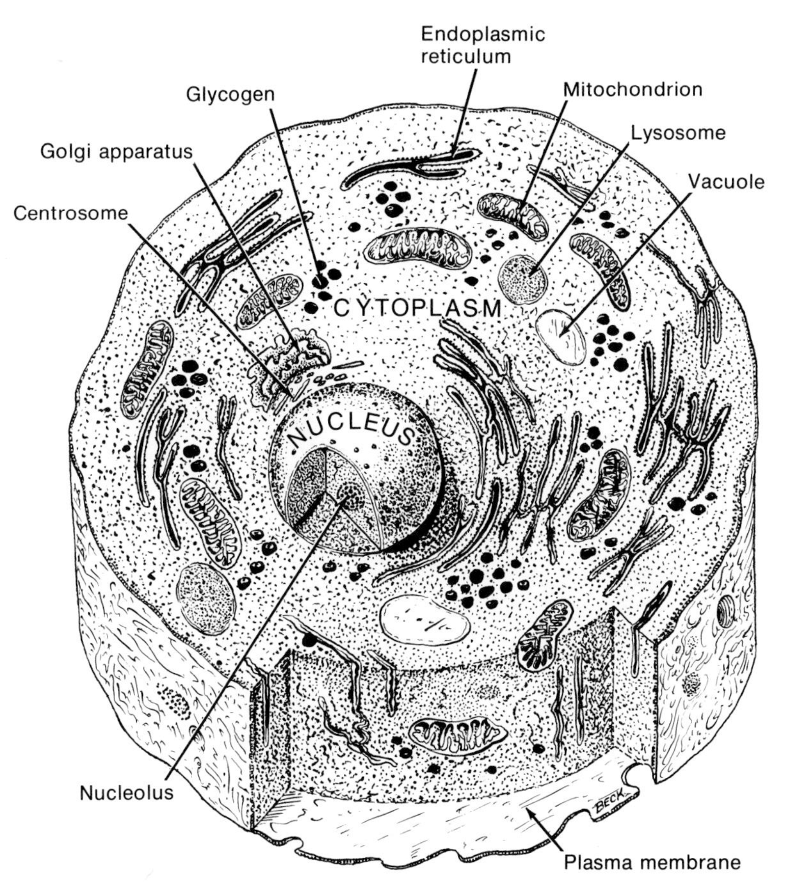

Fig. 2-1
Diagram of a hypothetical composite cell.

verse. First, they serve to define the boundaries between extracellular and intracellular spaces or to separate various intracellular spaces. The plasma membrane and mitochondrial membranes (Fig. 2-1) are examples, respectively, of these functions. In order that substances necessary for growth may be acquired and waste products of metabolism removed, membranes must be *semipermeable* barriers. However, not only matter but also information must be transferred, as exemplified by the influence of hormones (Chapter 28) and nerve impulses (Chapter 5) upon the cell. Exterior membranes enable the cell to recognize similar or different cells. Finally, membranes serve as specific supports for enzymes that participate in a variety of reactions.

■ Composition

All membranes are composed primarily of lipid and protein; carbohydrate residues may be attached to either, or both, of the major constituents. Although the membrane exists for the cell's lifetime, its components are constantly renewed. Three general classes of membranes are recognized, based upon the relative proportions of lipid and protein. Myelin, which functions principally as an insulator for the nervous system, is the simplest type of membrane. It contains about 75% lipid, 20% protein, and 5% carbohydrate. Plasma membranes constitute a second class of membranes in which about 50% of the structure consists of protein. Erythrocyte membranes (Chapter 12), for example, consist of about 43% lipid, 49% protein, and 8% carbohydrate. The third class of membranes is exemplified by the mitochondrial membranes in which up to 75% of the mass of the membrane may be protein.

Lipid. ■ Three types of lipids are present in animal cell membranes. These are phospholipids, glycolipids, and cholesterol. The chemistry of these compounds is discussed in Chapter 23. Plasma membranes contain a large proportion of cholesterol; glycolipids are present exclusively in these structures and not in intracellular membranes. The role of cholesterol in the membrane may be to control the fluidity, a characteristic which largely determines the rate of transfer of matter across the membrane. Some blood group factors (Chapter 12) in red blood cells are glycolipids. It is probable that glycolipids are the antigen components of all tissues and that they are involved in the process of recognition between cells.

Protein. ■ A great deal less is known about membrane proteins than is known about membrane lipids. Plasma membrane proteins, in general, are made up of many polypeptide (Chapter 18) chains in sizes which vary from 15,000 to 280,000 molecular weight. The larger chains comprise the largest fraction of the total protein. In the red blood cell, carbohydrate residues are associated with polypeptides of 80,000 to 100,000 molecular weight. These sugar residues belong to a class called sialic acid; because of their acidic nature they contribute importantly to the charge which exists on the membrane. Glycoproteins have been established as the molecules responsible for the M and N blood groups (Chapter 12).

■ Structure

A single, simple model of the architecture of all biological membranes cannot be provided because of the diversity of membrane types. However, certain general statements are appropriate. The restraining sheath which surrounds the cell, the plasma membrane, is on the order of 100 Angstrom units (A) thick. In electron micrographs at high magnification the membrane appears as two dense lines separated by a less dense area. Despite a number of uncertainties about the fine details of biological membrane structure, a model referred to as the *unit* membrane still provides a reasonable description. The nature of the unit membrane derived from histological (electron microscopy), physiological, and biochemical investigations is that of a bimolecular leaflet of lipid coated with protein. This structure is shown in Fig. 2-2. Phospholipid molecules characteristically have hydrophobic (water insoluble) and hydrophylic (water soluble) ends. The hydrophobic ends consist of fatty acid chains; the hydrophilic ends are phosphate and other groups which are capable of carrying an ionic (electric) charge. Consequently, these ends are described as the *polar* ends. The hydrophobic ends are nonpolar. The polar groups of the phospholipids are oriented toward the protein layers that cover the membrane. Although the bulk of the membrane protein is external to the lipid bilayer, certain protein molecules penetrate and some even extend all the way through the mem-

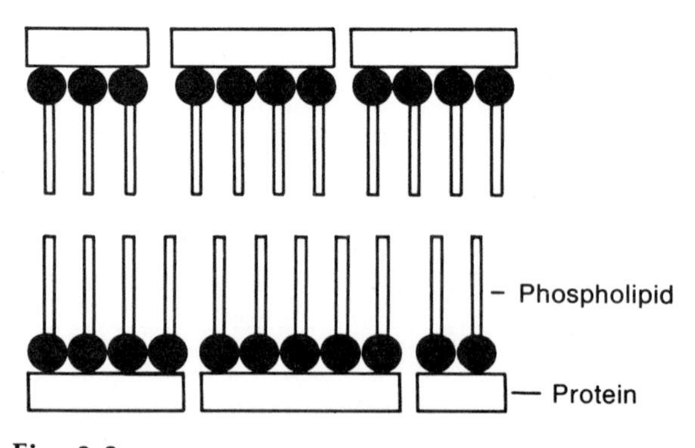

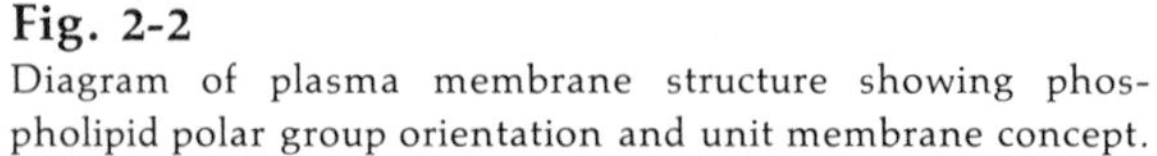

Fig. 2-2
Diagram of plasma membrane structure showing phospholipid polar group orientation and unit membrane concept.

brane. Membrane proteins serve to reduce the surface tension (Chapter 3) and thus to increase the wettability as well as to provide a degree of elasticity to the membrane.

■ Function

The organization of the cell itself is dominated by the cell's membranes, for they are implicated in the storage of information, the influence of drugs, the transport of materials, the scavenging of pathogens, and the organization of a variety of energy transfer systems. The cell membrane is the vital biological interface between the living organism and its external environment. One aspect of life, as noted in Chapter 1, is the continuous adjustment of internal relations to external relations. The maintenance of a steady state, *homeostasis*, is a prerequisite for the organism to achieve independence of its environment. An important level of homeostatic regulation is that of the fluid medium and volume of the cell. The barrier that isolates the cell interior and provides for the constancy on which life depends is the plasma membrane.

Permeation. ■ Molecular traffic across the membrane moves in both directions. A selective and variable permeability exists. The membrane is characterized largely by the rate of permeation of two classes of solutes: lipid-soluble (nonpolar) and water-soluble (polar). Lipid-soluble solutes permeate in rough proportion to their solubility without regard to molecular size, whereas water-soluble solutes move through aqueous channels or pores. Although these pores have not been observed directly, as a result of certain experiments it is estimated that their combined area is $^1/_{100,000}$ to $^1/_{1,000,000}$ that of the total available cell surface. These pores are 3 to 4 A (0.0000003 to 0.0000004 mm) in diameter and consequently strictly limit the size of permeating particles. By and large, particles with a molecular weight greater than 200 are excluded from entering the cell by passive diffusion.

Passive. Penetration of the cell occurs by a number of processes. *Passive diffusion* of substances is due to the presence of a greater concentration of the solute on one side of the permeable membrane—a chemical potential. Often an electrical potential difference develops across the membrane, and solutes bearing a charge of the proper polarity will diffuse under the force of a combined electrochemical potential.

Active. Important instances are recognized when solutes are moved against a chemical, electrical, or electrochemical potential. Such a translocation of material is referred to as *active transport* because the cell must provide the energy to perform this work by its metabolism.

Bulk flow. Several other modes of penetration have been described. If the membrane is quite porous, a bulk flow of solvent may occur that drags along and accelerates solute particles normally diffusing in the direction of the flow, but conversely it hinders those solutes normally proceeding in the opposite direction.

Pinocytosis. Selective or nonselective entrapment of soluble materials of the extracellular environment by infolding of the plasma membrane and subsequent detaching of the closed off pouch to form a free vesicle inside of the cell is referred to as *pinocytosis* or cell drinking. This phenomenon may account for the translocation of relatively large molecules, fat and protein, across the plasma membrane. In certain cells, for example, white blood cells, large particulate bodies such as bacteria are engulfed whole by a similar process known as *phagocytosis* or cell eating.

Recognition. ■ It is a remarkable fact that the cells making up tissues and organs are not firmly held together. Yet, grouping and adhesion are essential if the structure and function of many-celled organisms are to be maintained. The adhesive properties of a cell surface depend on the nature of the proteins accumulated there; consequently, adhesiveness is controlled by genes because these determine the production of specific proteins.

Enzymes. ■ Many of the enzymes (Chapter 22) found in cells are parts of specific membranes. For example, the Na^+-K^+ activated ATP-ase which is associated with the "sodium pump" (Chapter 5) exists in the plasma membrane. The cytochrome enzymes involved in the respiratory chain are all part of the inner membrane of the mitochondrion. On the other hand, the enzyme monoamine oxidase which inactivates catecholamines (Chapter 28) occurs only on the outer mitochondrial membrane.

The localization of enzymes within membranes may subserve several purposes. In the oxidative-phosphorylation processes occurring in the mitochondria (Chapter 23), the most efficient electron transport would be achieved when the proteins (enzymes) were next to each other. The membrane offers the requisite physical support and orientation. Membrane proteins that act as specific binders for a variety of ions, amino acids, and sugars are spoken of as carriers (Chapter 3) and are considered to be the mechanism underlying the phenomenon of active transport. Whether carrier proteins are in fact enzymes is problematical, but they behave much like enzymes in regard to their substrate specificity and reaction kinetics during the process of translocation.

Receptors. ■ It is believed that cell membranes contain special molecular entities which are capable of direct chemical bonding with other molecules present outside of the cell so that a physiological or pharmacological response of the cell ensues. These specific chemical groups in the membrane are called *receptors.* Information is transferred at receptor sites. The information may derive from nerve impulses and the release of neurotransmitters at the nerve terminals, or it may be in the form of a circulating hormone released from an endocrine gland.

Adenylcyclase. A membrane enzyme system called adenylcyclase is the target cell receptor for many circulating hormones. Activation of the adenylcyclase brings about the intracellular conversion of ATP to *cyclic AMP* (adenosine 3′, 5′ monophosphate):

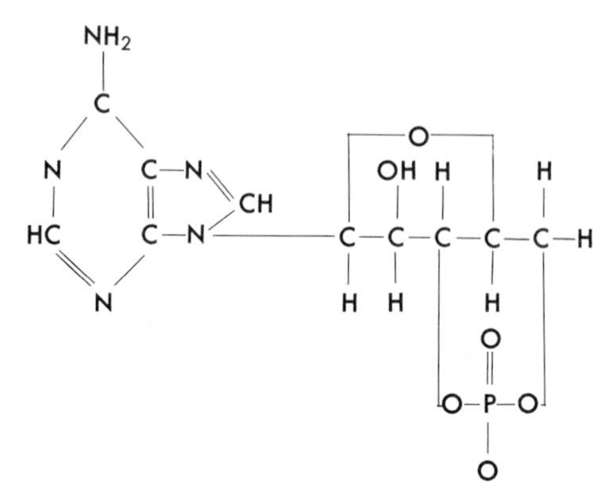

The increased level of intracellular cAMP, in turn, influences the "physiological response" of the cell; i.e., the activities of enzyme systems, the permeability processes, or, in the case of endocrine cells, the synthesis or release of other hormones. This sequence of reactions is summarized in Fig. 2-3.

Because of its intermediate role between the original hormone (first messenger) and the "physiological response," cAMP is often spoken of as the *second messenger.* What cAMP does in a particular cell is determined by the enzyme "profile" in the particular cell; i.e., it depends upon the individuality of the cell. For example, an increase in cAMP in adipose tissue cells activates an enzyme, triglyceride lipase, and evokes an increase in lipolysis, the breakdown of triglycerides to free fatty acids and glycerol. The strength of contraction is augmented in cardiac cells when intracellular cAMP levels are increased. ACTH from the adenohypophysis acting through the adenylcylase system raises the concentration of intracellular cAMP which, in turn, stimulates the production of glucocorticoid hormones (steroidogenesis) by the cells of the adrenal cortex. Hormones produced in response to cAMP stimulation are called third messengers. Other examples of cAMP activity will be pointed out in succeeding chapters.

Adenylcyclase has been found in every mammalian tissue except the mature red blood cell. It also occurs in species lower in the phylo-

A thick and enclose a narrow central space, the cisterna. It is postulated that these sacs and vesicles possess direct continuity with the surface membrane of the cell, and thus their contents are in a sense extracellular. Furthermore, the nuclear membrane appears to be surrounded by the tubules of the reticulum, but with discontinuities that afford areas of contact between nuclear and cytoplasmic matrix. Both a smooth-surfaced and rough-surfaced type of reticulum occur, but the relative amount of each varies with the type of cell. The agranular, or smooth-surfaced, reticulum plays a role in glycogen synthesis and storage in liver cells, in steroid hormone synthesis in the interstitial cells of the testis, and in excitation-contraction coupling in muscle.

■ Ribosomes

Rough-surfaced, or granular, reticulum derives its name from the wealth of small particles, 100 to 200 A in diameter, adhering to the cytoplasmic side of the membranes. These granules are called *ribosomes* because they consist of conjugated *ribonucleic acid* (rRNA) and protein. Ribosomes are also found free in the cytoplasm and a few are located in the mitochondria. Thousands of ribosomes are present in every cell; new ribosomes are formed constantly in the nucleolus to meet the needs of the cell to adapt to changes in its environment or nutrition.

All ribosomes are complex structures consisting of two unequal subparticles, one large and one small. Each subparticle is constructed of one high molecular weight RNA species and many different proteins. The latter are present as single copies in the subunit. Thus, of the approximately twenty proteins associated with the rRNA in the smaller subparticle, no two are alike. The same situation exists in the forty or so proteins of the larger subparticle.

Ribosomes mediate protein synthesis in all organisms. In general, ribosomes attached to the reticulum are involved in the regulation of the synthesis of protein destined to become a cell secretion. The cytoplasmic granules, especially in rapidly growing cells, are associated with protein (commonly enzymes) synthesis for cell growth and repair. The role of ribosomes is discussed again later in this chapter.

■ Centrosome

As its name implies, the centrosome is a cell organelle that lies near the center of the cell and often within the area of the Golgi apparatus. It consists of two bodies, the *centrioles,* each of which has the form of an open cylinder surrounded by nine longitudinally arranged groups of filaments, three to a group. The centrioles are active in organization of fibrillar material of the cell, a function most clearly evident in mitosis (cell division).

■ NUCLEUS

All cells, except mammalian red blood cells and blood platelets, have a nucleus surrounded by cytoplasm. Nuclei vary in size and shape, but the nucleoplasm of a given cell invariably contains three important constituents—the *nucleolus, chromatin granules,* and an amorphous proteinaceous matrix (the *nuclear sap*).

The nuclear membrane appears as two dense lines separated by a less dense zone; the total width is between 250 and 400 A. Ribosomes are sometimes attached to the outer membrane, whereas the innermost membrane is smooth. At intervals the two membranes fuse to form a thin single layer, a structure often regarded as a pore. It is at least a region of special function, and quite likely that function is associated with the translocation of large molecules into and out of the nucleus.

■ Nucleolus

One or more nucleoli are present in the nucleus. They are dense bodies, generally rounded in shape and suspended in the nuclear sap. No membranous envelope surrounds them. The internal structure is made up of a tangled, thick filament or meshwork of fine threads known as the *nucleolonema* and scattered patches of granular material. There is an association between the size of the nucleolus and the extent of protein synthesis the cell is performing. For example, nucleoli in pancreatic cells are especially prominent. The nucleolus is a structure composed of large amounts of RNA. It is the site where ribosomal RNA is synthesized and ribosome subunits are assembled. Because of its central role in regulating protein synthesis, through the production of ribosomes, it has been referred to as the pacemaker of the cell.

■ Chromosomes and genes

Chromatin granules are smaller than nucleoli and of irregular shape and size. They are seen only in the growing nondividing (interphase) cell. Actually they are not granules but areas where chromosome filaments or threads are condensed by being wound into a tight coil, and thus they become dense enough for recognition by microscopy. These chromosomal threads contain the *genes*, the units determining heredity. Chromosomes contain nucleoprotein. The nucleic acid that is conjugated with protein in chromosomes is different than that noted as occurring in ribosomes. In the nucleus the substance is *deoxyribonucleic acid* (DNA). It has been estimated that the human cell nucleus contains an amount of DNA equivalent to five million genes distributed among the twenty-three pairs of chromosomes.

A gene consists of only a segment of the giant DNA molecule. In the thousands of genes of a chromosome the DNA serves to specify the complete chemical structure of the thousands of enzymes required by the living cell to secure energy from nutrients, to repair itself, and to provide a template for the exact copies of itself to be inherited by the cell's offspring. Not all genes are active at the same time, since growth and differentiation of cells requires an orderly sequence of reactions according to the particular requirements of the moment.

■ CELL GROWTH AND REPRODUCTION

The nature and function of any cell is determined by the specific proteins that enter into its structure and the enzymatic activities into which many of these proteins engage.

■ Protein synthesis

Proteins are composed of long chains of amino acids (Chapter 23). Each protein molecule exhibits a specific sequence of amino acids in its chain. Because at least twenty different amino acids are available and because the molecular chains vary in length, the number of possible proteins is enormous. How the cell accomplishes specific protein synthesis for both growth and secretion is the next consideration, although the discussion will be rather cryptic.

Obviously the hereditary material in the gene must contain the information required to organize and direct the development of the cell. This information is coded in the DNA molecule that constitutes the gene. Normally DNA occurs as a giant double-stranded molecule resembling a ladder twisted into a helical shape like a spring (Fig. 2-6). The sides of this structure are composed of alternating units of deoxyribose sugar and phosphate, whereas the rungs are made up of linked pairs of organic bases. A compound made up of an organic base, a five-carbon sugar, and phosphate is known as a *nucleotide*. Only four such bases, sometimes called the letters of the genetic code, are involved in most cells; adenine is always linked to thymine, and guanine is linked only to cytosine. For each type of organism the ratio of adenine-thymine to guanine-cytosine is constant. It is the *sequence* of the linked pairs in the DNA chain that constitutes the genetic code. The synthesis of protein, however, occurs chiefly outside of the nucleus in the ribosomes, as previously noted. Transfer of the necessary information from the nucleus to the ribosomes is accomplished by *messenger RNA* (mRNA). Except that it is very much smaller, consists of a single strand containing ribose instead of deoxyribose, and its side chains substi-

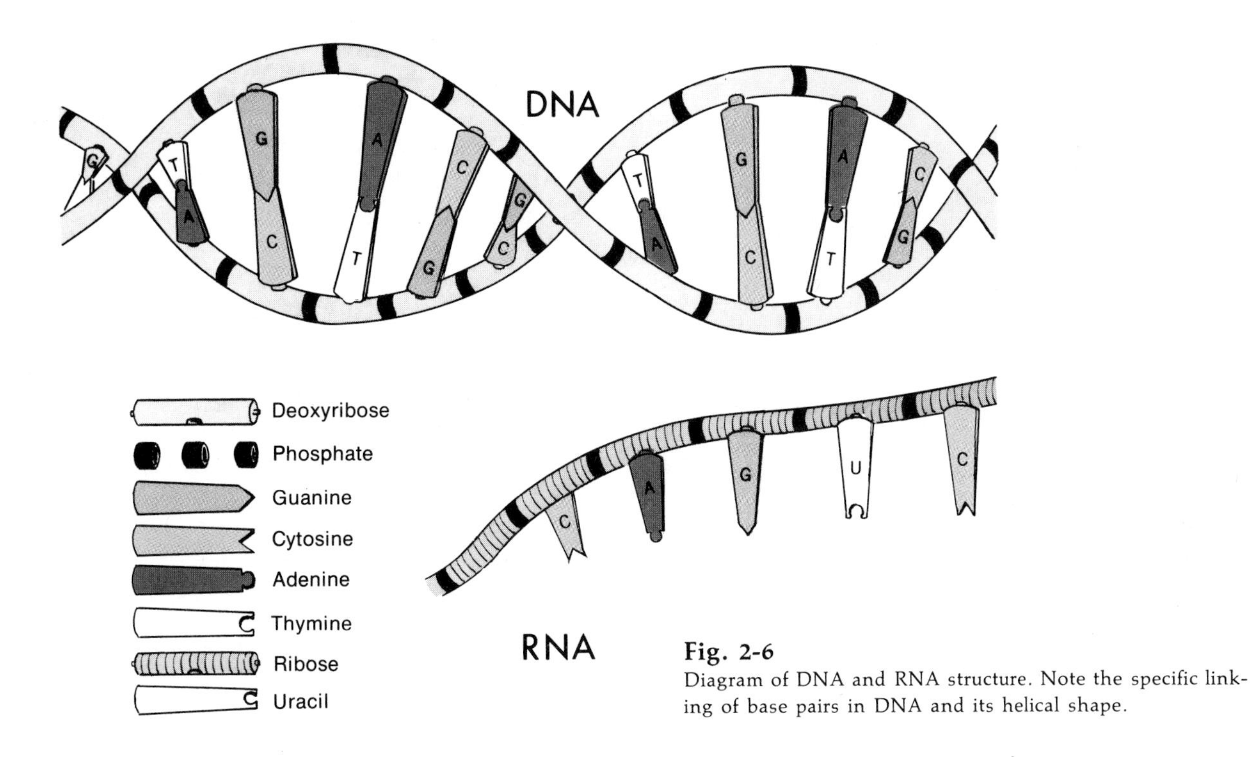

Fig. 2-6
Diagram of DNA and RNA structure. Note the specific linking of base pairs in DNA and its helical shape.

tute uracil for thymine, mRNA is similar to DNA. It is thought that in the formation of mRNA the helix of DNA partially unwinds, and ribonucleotides pair off in a complementary fashion against the deoxyribonucleotides in the now single strand of DNA. Adenine of the DNA pairs with uracil, a compound in RNA with a structure very like that of thymine. Synthesis of a strand of RNA is completed by the action of the enzyme RNA polymerase; the mRNA so formed detaches from the DNA template, which then has its helical structure restored. Messenger RNA leaves the nucleus and attaches to the ribosomes in the cytoplasm (Fig. 2-7), where it forms a template for protein synthesis. The sequence of bases on the mRNA specifies the amino acid sequence in the proteins to be synthesized. Each sequence of three bases on RNA, a *codon*, specifies one amino acid. For example, the message contained in the mRNA sequence uracil-uracil-uracil, which arises from the complementary adenine-adenine-adenine sequence of the DNA strand, is *transcribed* into the amino acid phenylalanine. Since there are four letters (bases) in the code and these are "read" three at a time in the transcription process, the number of possible codons is sixty-four (4^3). Consequently, a particular amino acid among the twenty kinds is usually represented by several codons. For example, the amino acid lysine is expressed in DNA by the linear sequence of the bases thymine-thymine-thymine or thymine-thymine-cytosine. In the complement mRNA codons, the linear sequence becomes either adenine-adenine-adenine or adenine-adenine-guanine.

The proteins turned out by the ribosomes are large molecules, polypeptide chains, whose subunits are the twenty or more amino acids. An amino acid in solution in the cytoplasm is first *activated* by coupling with ATP and is then joined to a soluble transfer RNA (tRNA) molecule, smaller than mRNA, which transports it to

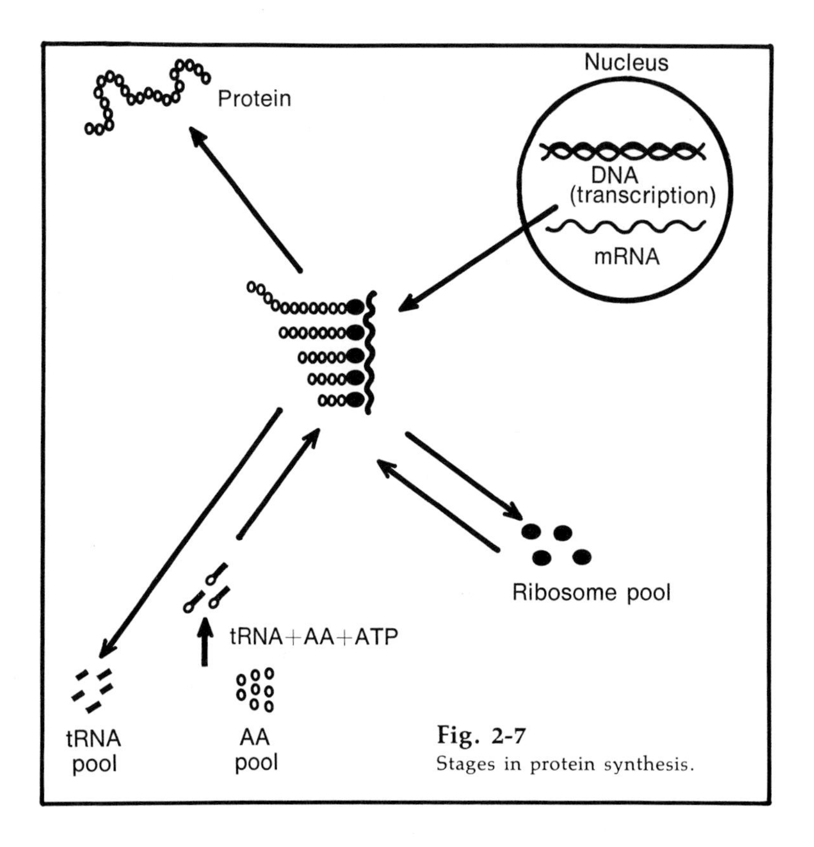

Fig. 2-7
Stages in protein synthesis.

the ribosome. Transfer RNA originates just as does mRNA. There are tRNA's with special affinity for each amino acid.

The initial step in protein synthesis involves the formation of a complex between the smaller subparticle of the ribosome and the mRNA at a codon site which signals the start of the assembly line. Here too a tRNA with a "starter" amino acid (e.g., formyl methionine), which is later discarded, is affixed. The larger subparticle of the ribosome attaches to the starter tRNA, and it participates in the process of peptide bond formation between succeeding amino acids as they are brought to the ribosome by different tRNA's. As quickly as one ribosome begins the translation moves along the mRNA in its production of the polypeptide, another ribosome can start the sequence; thus many ribosomes may be attached to the mRNA simultaneously. This multiplicity of ribosomes of the mRNA gives rise to the structure known as a polysome.

Protein synthesis sometimes goes awry as a result of genetic mutation. Mutations occur due to x-rays, ultraviolet light, radioactive fallout, or certain chemicals to which the cell may be exposed. There are two types of mutation. In the first, one nucleotide (letter) is substituted for another in the DNA chain; in the second, a particular nucleotide in the DNA chain is omitted entirely from the usual sequence. The errors are carried over into the complementary messenger RNA, and consequently its codons specify that different amino acids from the usual be incorporated into the polypeptide chain of the protein.

At least thirty types of abnormal human hemoglobin, the respiratory pigment found in red blood cells, have been identified. The disease known as sickle cell anemia results from a single nucleotide substitution in DNA, whereby, in a sequence of 514 amino acids, valine replaces glutamic acid in the hemoglobin protein of the afflicted individual. This abnormal pigment is largely insoluble when it loses oxygen, and it crystallizes in the red cells, causing the latter to be distorted into a sickle shape. Because this hemoglobin is defective in its role of oxygen transport and the odd-shaped cells obstruct blood flow, this hereditary disease is a severe and often fatal disorder.

■ Cell division

The life-span of a cell is measurable in days, weeks, or decades. Some types of cells, for example, muscle or nerve cells, once differentiated never divide again, for this would be incompatible with their function. However, many types of cells are constantly undergoing growth and division for purposes of restitution and replacement. Liver, intestinal, epidermal, and bone marrow cells are excellent examples. We have already alluded to the necessity of retaining an optimal cell size. Nevertheless, division may occur before growth approaches anywhere near a limiting size. Thus it must be assumed that certain preparatory events occur between divisions—during the *interphase*—which predispose the cell to division regardless of size.

During the late stages of the interphase (Fig. 2-8, *1* and *2*) all of the chromosomes are replicated, or duplicated. The nucleoprotein of which a given chromosome is composed separates into its constituents: protein and DNA. Duplication of the genetic material, the DNA molecule, proceeds as was described previously for the formation of mRNA on a DNA template, except that both strands of the DNA molecule's double helix separate and a new complementary strand of DNA is built up for each original strand. In this way, the number of genes and chromosomes has been doubled. However the two new units are united at one point, the *centromere*. Recombination of the new DNA, now referred to as *chromatids*, with protein completes the replication process. Chromatids, attached at their centromere to the rays of a spindle, appear in frames *4, 5,* and *6* of Fig. 2-8, which illustrates division in a cell with a normal complement of only one pair of chromosomes.

In prophase (Fig. 2-8, *3* and *4*) the nucleolus disappears, the chromatids wind into coils along their entire course, and the *centrioles* move toward the poles of the cell. A spindle with astral rays begins to arise between the poles, and the nuclear membrane disappears. At metaphase the spindle is fully formed and has brought the chromatids into line at the equator of the cell (Fig. 2-8, *5*). Subsequently, in anaphase the centromere divides and the chromatids, now called chromosomes, move along the lines of the spindle toward the poles (Fig. 2-8, *6*). In telophase (Fig. 2-8, *7* and *8*) the separation of the genetic materials is completed. The chromosomes again become thin threads and the nuclear membrane reappears. Finally, *cytokinesis*, the division of the cell into daughters, ensues.

Mitosis, as just described, is the process by which cells perpetuate their own kind and ensure that continuity of structure and function are maintained through transmission of identical sets of genetic material to each daughter cell. *Meiosis*, the division of reproductive cells, or germ cells, is discussed in Chapter 30.

■ Aging

Studies made on cell cultures reveal that normal cells do not divide indefinitely even under optimal conditions; they multiply for only about fifty generations before becoming incapable of division. This is an aging phenomenon. Direct evidence that aging of cells in cultures is compa-

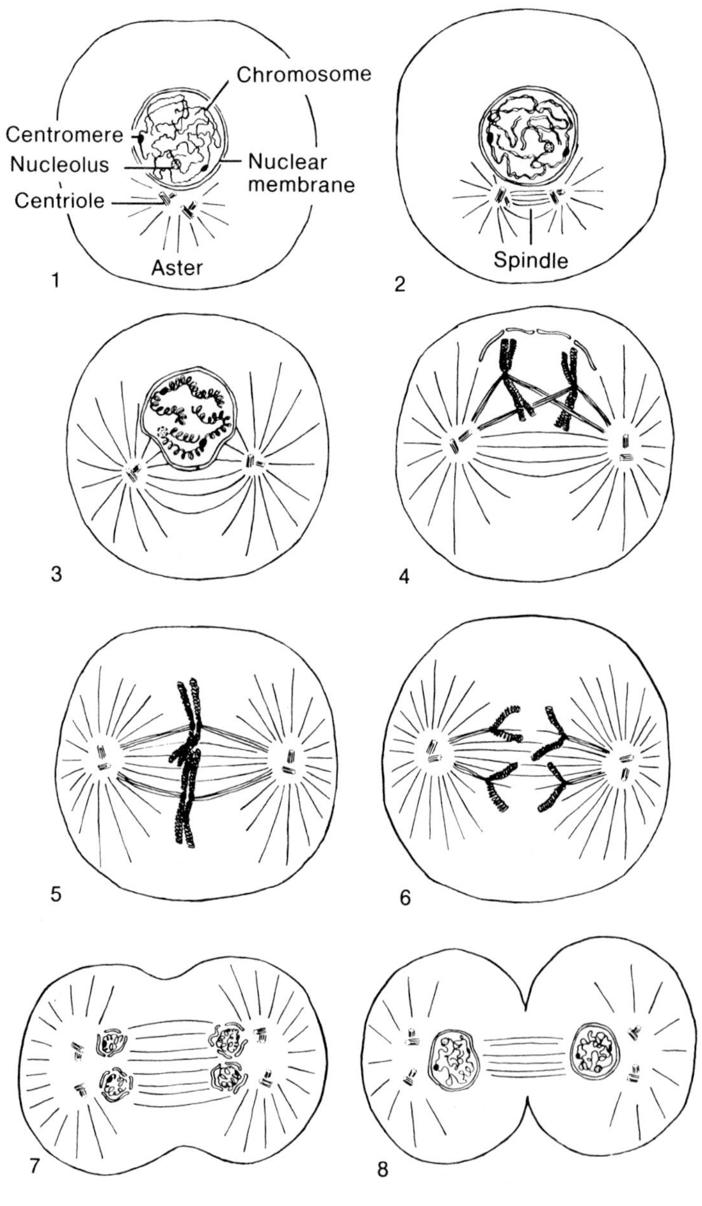

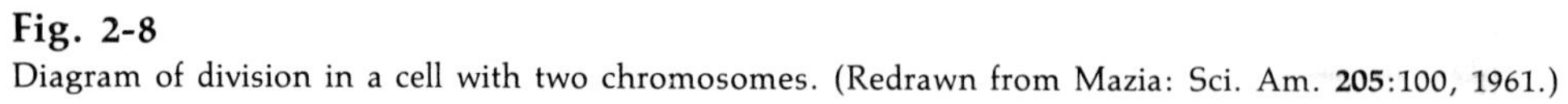
Fig. 2-8
Diagram of division in a cell with two chromosomes. (Redrawn from Mazia: Sci. Am. **205**:100, 1961.)

rable to that in living animals is not available but is suggested by the fact that several organs in man, for example, the brain and kidney, lose weight due to cell loss after middle age. It has been postulated that aging may result from deterioration in the inherited information containing molecules, the DNA. The latter may become faulty and affect the transcription of mRNA for protein synthesis. We have mentioned a type of mutation where one or more nucleotides of the DNA may be missing from the usual sequence. This error might occur during the genetic recombination after cell division. Such a phase-shift mutation would produce abnormal protein in enzymes and disrupt the metabolic pattern of the cells. Although our present knowledge about the genetic code with its implications for human health and happiness is a significant contribution of molecular biology, much, indeed, remains to be discovered.

READINGS

Crick, F. H. C.: The genetic code, Sci. Am. **215**(4):55-63, Oct. 1966.

de Duve, C.: The lysosome, Sci. Am. **208**(5):64-72, May 1963.

Fox, C. F.: The structure of cell membranes, Sci. Am. **226**(2):31-38, Feb. 1972.

Guidotti, G.: The composition of biological membranes, Arch. Intern. Med. **129**:194-201, Feb. 1972.

Kurland, C. G.: Ribosome structure and function emergent, Science **169**:1171-1177, 1970.

Levine, L.: Biology of the gene, St. Louis, 1969, The C. V. Mosby Co., pp. 1-53.

Lowenstein, W. R.: Intercellular communication, Sci. Am. **222**(5):78-86, May 1970.

Neutra, M., and Leblond, C. P.: The golgi apparatus, Sci. Am. **220**(2):100-107, Feb. 1969.

Rasmussen, H.: Cell communication, calcium ion, and cyclic adenosine monophosphate, Science **170**:404-412, 1970.

Singer, S. J., and Nicolson, G. L.: The fluid mosaic model of the structure of cell membranes, Science **175**:720-730, 1972.

Whaley, W. G., Danwalden, M., and Kephart, J. E.: Golgi apparatus: influence on cell surfaces, Science **175**:596-598, 1972.

3

Diffusion
Colloids and crystalloids
Surface tension
Adsorption
Electric constitution of matter
Solutions
Osmosis and osmotic pressure
Cell membrane or plasma membrane
Membrane potentials
Volume distribution and the clearance concept

TRANSLOCATION OF MATERIALS

In Chapter 1 it was stated that protoplasm may be regarded as a solution of crystalloids and colloids. Since the cell lives in water containing various salts, it is pertinent to inquire why the environmental water does not pass into the cell and dissolve it, thereby causing its death, or why the soluble constituents of the protoplasm do not leak out of it. Under such conditions how do the cells, being water solutions, maintain their physical integrity and life?

We have learned that certain substances (e.g., foods) must pass from the environment into the protoplasm and that other materials (waste products) must pass out of the cell. This two-way migration of matter with respect to the cell is encountered everywhere in the study of physiology: the passage of oxygen into and carbon dioxide out of the cell during respiration, the excretion of waste products from the blood through the kidneys, the secretions of gland cells, and the absorption of nutrients from the alimentary canal into the blood and their passage from the blood through the capillary wall into the tissue fluid and from the tissue fluid into the cell. In all these cases *the movement of the material takes place across or through a membrane composed of protoplasm.*

In this chapter some of the many physical and chemical properties of colloidal solutions and some of the more important physicochemical factors concerned in the translocation of materials in the animal body will be discussed.

■ DIFFUSION

Of immense importance to biology is the concept of *diffusion,* the redistribution of material by random movement.

Diffusion of gases. ■ Respiration, the exchange of gases between the organism and the environment, depends on the diffusion of gases. It is well known that gases spread from a place of higher pressure to one of lower pressure. Even gases much heavier than air, such as bromine gas, diffuse upward against the force of gravity. According to the *kinetic theory of gases* the molecules of a gas are in constant, random motion and in consequence spread apart as far as the confines of the receptacle allow. Bombardment of the walls of the container by the molecules of the enclosed gas gives rise to the pressure of the gas. A rise in temperature accelerates the molecules, increases their kinetic energy, and therefore increases the pressure. Increasing the amount of the gas in a container or reducing the size of the container also has this effect. These relationships (Boyle's law, $V = k_1/P$, and Charles' law, $V = k_2T$) can be expressed mathematically in the form of the *ideal gas law*, that is $PV = nRT$, in which P, V, and T are pressure in atmospheres, volume in liters, and absolute temperature, respectively. R is the gas constant (0.082 liter-atmospheres/mole/degree), and n is the number of moles of the gas.* If several different gases are present, the molecules of each gas diffuse and exert pressure independently of those of the other gases (Dalton's law of partial pressure of gases). In respiratory physiology we often refer to the partial pressure as the *tension* of the gas. For instance, in dry atmospheric air the tension of oxygen is 159 mm Hg (Fig. 17-11). Diffusion also takes place when gases are dissolved in a liquid (e.g., oxygen in blood).

Diffusion in liquids. ■ Similar to gas molecules, the molecules of a liquid and of the materials dissolved in a liquid are in constant motion, but, unlike those of a gas, their movements are largely restrained by the attraction exerted by the molecules on each other. The material in solution is known as the *solute;* the medium in which it is dissolved is the *solvent.* The most common solvent is water.

Suppose that a strong sugar solution, Fig. 3-1, *b*, is carefully overlaid with pure water, Fig. 3-1, *a*. Immediately some of the sugar molecules begin to penetrate the pure water. As soon as a number of sugar molecules have penetrated *a*, some of them reenter *b*, but the number doing so is less than that of those going from *b* to *a*. After a sufficient length of time so many sugar molecules have entered *a* that the molecular concentration* of the sugar in *a* is equal to that of *b;* from that time on as many molecules pass from *a* to *b* as in the reverse direction. The system is now in equilibrium, but the motion of the dissolved molecules continues. From this it may be concluded that *when the molecular concentration of a substance in solution is greater in one part of a liquid than in another and when no absolute barrier intervenes, diffusion occurs,* tending thereby to establish an equal concentration of the molecules. This also applies to water molecules. Their concentration at the outset of the experiment is greater in *a* than in *b*, and diffusion takes place as shown in Fig. 3-1. *The velocity of molecules in solution varies inversely as their molecular weight and directly as the temperature of the solution.* Diffusion plays an important part in the passage of materials from the blood to the cells and of cellular products in the reverse direction.

■ Dispersion in water

With respect to the dispersion of materials in water, the resultant mixtures may be grouped into three classes: (1) molecular, or true, solutions, (2) colloidal solutions, and (3) suspensions. These groups differ primarily in the size of the particles concerned, but no sharp lines of demarcation can be drawn between them; whether a substance is in suspension, in colloidal solution, or in true solution is determined by its behavior as well as by the size of its particles.

Molecular solutions. ■ In a molecular solution the dispersed particles are molecules or ions (crystalloids) and have a diameter less than

---

*The product, *PV*, represents energy (or work), since it has the dimensions of force per unit area multiplied by a volume. Atmospheres and liters are more often used dimensions than are gram-weight of force per square centimeter and cubic centimeters of volume (at zero C an atmosphere of pressure exerts 1033.6 grams-weight/cm^2).

*Molecular concentration is the number of molecules per unit volume.

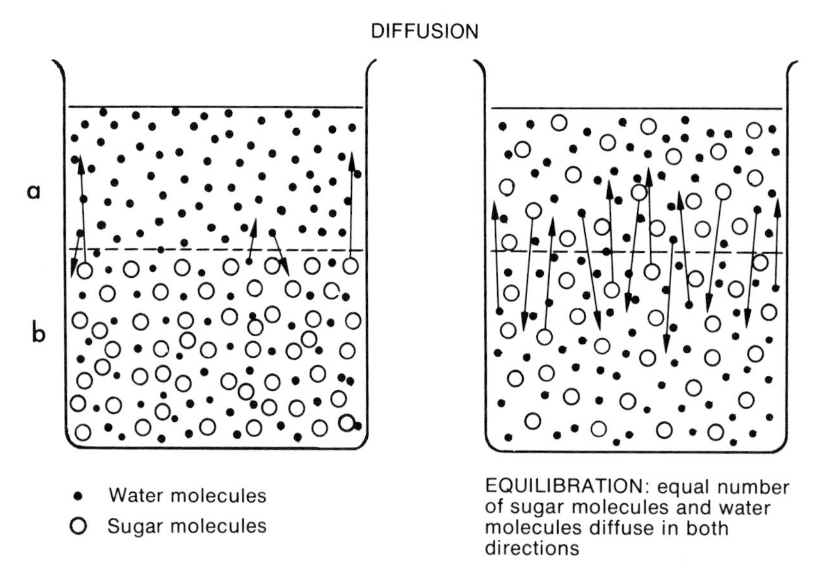

Fig. 3-1
Diagram to illustrate diffusion. Beaker on the left shows state shortly after layering of pure water, **a,** on sugar solution, **b.** Beaker on the right shows equilibrium state.

0.000001 mm. A true solution is stable in that there is no settling of the particles from the medium in which they are found. It is homogeneous. Examples are the solutions of the common salts, acids, bases, and sugars.

Colloidal solutions. ■ Colloidal solutions are those in which the diameters of the particles range between those of particles in a suspension and those in molecular solution, that is, from 0.0001 to 0.000001 mm. Colloidal solutions are fairly stable in that they do not settle out by gravity. However, by very rapid centrifugation (ultracentrifuge) it is possible to cause sedimentation of many colloids (e.g., some proteins) from their solution. Colloidal solutions are heterogeneous; the smaller the particles, the greater is the stability.

Suspensions. ■ In a suspension the particles are greater than 0.0001 mm in diameter. They do not diffuse and can be separated from the liquid by gravity (sedimentation) or by filtration. The cells in the blood are in suspension in the liquid part of the blood.

■ COLLOIDS AND CRYSTALLOIDS

Although crystalloids and colloids gradually merge into each other with respect to the size of the particles in solution and other properties, yet, all in all, we find certain characteristics in which they may differ greatly.

Gelation. ■ Certain colloids are known as *emulsoids;* of these, the protein gelatin will be used as an example. When solid gelatin is stirred in hot water, a colloidal solution is formed; this is known as a *hydrosol.* Sols are fluid because the colloid particles only bind loosely one to another. On cooling, the sol sets or gels to a more or less firm mass called a *hydrogel.* Gels have more firm binding between colloid particles. In the apparently structureless gel there is a more solid or continuous phase that forms a meshwork; in the interstices of this meshwork a more fluid, or dispersed, phase is held. In cream, an emulsion, the droplets of fat constitute a dispersed phase in the continuous water phase; in butter, a colloid, water forms a dispersed phase in the continuous phase of fat. Churning cream into butter is a phase reversal.

The gel of gelatin has a great affinity for the enclosed water; it requires heating at a temperature of 120 C for a considerable length of time or applying great mechanical pressure to drive the water from the gel. Some gels, as that of gelatin, are reversible; other gels are irreversible in that the gel cannot be transformed into a sol. Protoplasm contains proteins in both states, the sol and the gel. The formed bodies in the protoplasm, visible under proper conditions, are regarded as protein gels.

Imbibition. The hydration of dry protein depends on a phenomenon known as imbibition. When a small sheet of gelatin or some other substance (agar or cellulose) that is immiscible in water is placed in cold water, it is soon observed to swell without the substance going into solution. Gelatin may thus take up and hold very tenaciously many times its volume of water; during this process an exceedingly large amount of pressure, known as imbibition pressure, is generated. Imbibition may explain how it is possible for the protoplasmic framework of a jellyfish to hold in an organized manner the 96% water of which it is composed. There is much doubt as to the cause of imbibition. It is influenced by the nature of any salt present in the water; some salts increase, and other salts decrease the amount of water taken up. It is sometimes difficult to distinguish between imbibition and the process of solution.

Diffusibility. A crystalloid, such as NaCl or cane sugar, diffuses rapidly in water. In contrast a piece of soap left in a basin of water dissolves very slowly, and an exceedingly long time is needed for the complete diffusion of the soap throughout the entire volume of water. The same holds true for egg albumin. Sodium chloride diffuses about twenty times faster than albumin; this is due to the larger size of the albumin molecule (about 34,000 molecular weight), as compared with that of sodium chloride (58.5 molecular weight). Although crystalloids in solution diffuse as freely through gels as through water, colloids in solution do not; that is, a gel is impermeable to colloids in solution. This difference enables us to separate crystalloids in solution from colloids.

Separation. ■ We may now discuss how the materials in suspension or in colloidal solution may be separated from those in true solution.

Filtration. When a mixture of sand in a NaCl solution is placed upon a sheet of filter paper, the hydrostatic pressure forces the water and the NaCl dissolved in it through the pores of the filter, but the suspended matter, sand, is retained by the filter. The greater the hydrostatic pressure and the larger the surface of the filter to which the material is exposed, the more fluid is squeezed through in a unit of time. *Filtration means the passage of a substance in solution through a membrane as a result of a mechanical force* (e.g., gravity and blood pressure). By filtration, materials (such as blood cells) can be separated from the fluid in which they are suspended.

Filters differ in the size of their pores. In filter paper the holes are fairly large. Filters with exceedingly small pores are known as *ultrafilters.* In the study of lymph formation (Chapter 16) we shall learn that the capillary walls may be regarded as ultrafilters.

Dialysis. Place a water solution of sugar and egg albumin into a container, the bottom of which (Fig. 3-2, *mn*) is formed by an irreversible gel (such as collodion, parchment, or cellophane). The membrane is freely permeable (allowing passage through) to water and sugar molecules, but it is impermeable to albumin. This vessel is suspended in a large vessel of water (Fig. 3-2, *b*) that is constantly being renewed. After a sufficient length of time practically all the sugar has disappeared from *a*, and only the albu-

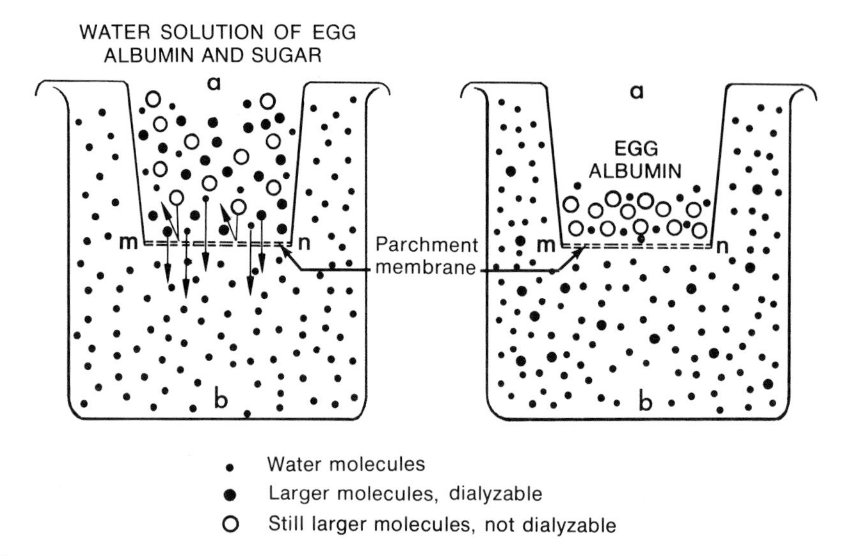

Fig. 3-2
Diagram to illustrate dialysis. Beaker on the left contains sugar and albumin in **a** and water in **b** separated by membrane **mn.** Beaker on the right shows final distribution of substances.

min and some water remain. By this process of dialysis the more diffusible material is separated from the less diffusible material. The former compounds are crystalloids, and the latter are colloids.

The aforementioned property of a membrane is generally designated as *semipermeability.* Some define a semipermeable membrane as one permeable to the solvent and not to the material in solution, the solute. Permeability, however, depends on two factors—the nature of the substance (size of molecule or particle) and the nature of the membrane. For any substance a membrane can be found to which this particular substance is impermeable, and other membranes can be found to which it is permeable. Hence, the term *selective* permeability is more appropriate.

■ SURFACE TENSION

A large array of well-known phenomena are based upon surface tension. Among these are the spherical or nearly spherical form of a falling drop of water, of a soap bubble floating in the air, and of a globule of mercury resting on a flat surface.

Consider a beaker of water. The molecules of the water attract each other; this is shown by the energy (heat) necessary to break them apart, as in evaporation. The molecules in the center are attracted equally by their neighbors from all sides and are therefore free to move in any direction (Fig. 3-3). But the molecules forming the topmost layer are attracted downward only because there are no water molecules above them to pull them upward. This causes the surface of the water to pull itself together, as if the mass of water were surrounded by a stretched elastic skin. The force with which the surface molecules are pulled toward the interior is known as *surface tension.* It is measured in force per unit length, that is, dynes per centimeter. As heat increases the energy of molecules, it enables them to overcome more readily the attraction exerted on

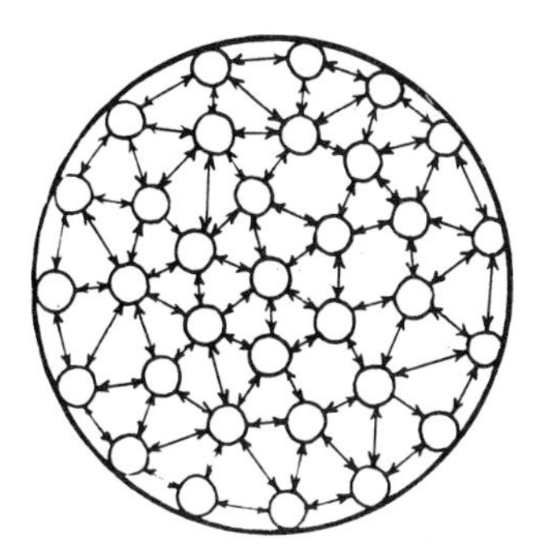

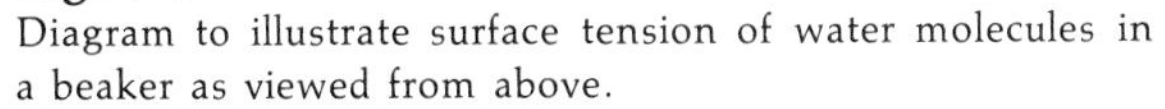

Fig. 3-3
Diagram to illustrate surface tension of water molecules in a beaker as viewed from above.

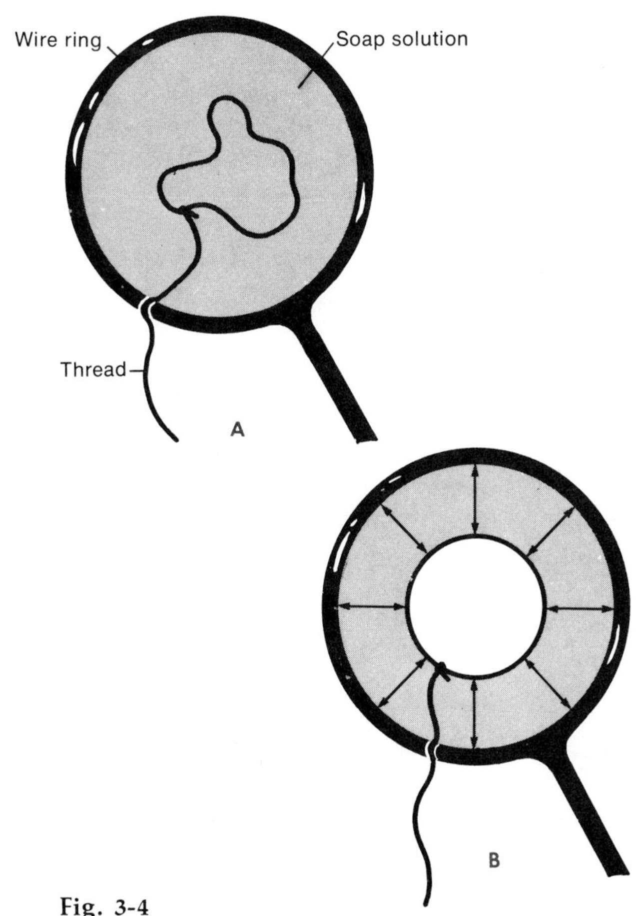

Fig. 3-4
Diagram to illustrate surface tension of a soap film.

them by neighboring molecules; this exhibits itself in a lowered surface tension.

Surface tension is demonstrated by the following experiment: A film of soap solution is caught in a wire ring (Fig. 3-4). Attached to this ring is a loop of thread, which may assume any irregular form as it drifts in the soap film. But when the film enclosed in the loop is broken by touching it with filter paper, the loop immediately takes on a perfectly circular form due to the surface tension of the soap film outside the loop.

When water is allowed to flow out of a narrow tube held vertically, a small amount of water collects at the end of the tube; this water increases in volume until its weight just exceeds the force of the surface tension; the formed drop then breaks away and falls. By noting the size of the drop (or the number of falling drops per milliliter), the relative amount of surface tension can be estimated. A drop of ether is considerably smaller than a drop of water; this indicates that the surface tension of ether is less than that of water. It is the attractive force of the molecules for each other that causes the liquids to assume a mass in which the area of the surface exposed to the air is the least possible, that is, a spherical surface.

Surface tension always exists at the surface of separation, the *interface,* between a liquid and a gas, between two immiscible liquids (e.g., water and oil), and between a liquid and a solid. Surface tension is markedly decreased by nearly all organic substances, such as bile, ether, alcohol, fat, and soap. Oil lowers the surface tension of water; this enables it to quiet waves in a storm. Inorganic salts have a lesser effect. Certain phenomena in physiology perhaps can be best understood in terms of surface tension at the interface separating the two phases of a heterogeneous system. The formation of pseudopods and the process of phagocytosis by white blood cells and the ameba are among the biological activities that have been explained in this manner.

ADSORPTION

Certain solids that present a large amount of surface are able to cause on these interfaces the condensation of gases and substances in solution. When a colored solution of Congo red or methylene blue is allowed to pass slowly through charcoal, glass wool, or absorbent cotton, the fluid loses its color to a greater or lesser extent; the pigment clings to the surface of these substances; this is known as *adsorption.* The adsorption of molecules at an interface causes them to lose their kinetic energy, which then appears as the *heat of adsorption.* In the adsorption of a gas by a liquid, the gas is uniformly distributed throughout the liquid; in surface adsorption there is a local condensation of the adsorbed material upon the surface of the adsorber. Adsorption is a reversible process, but a large amount of energy (heat) must be provided to restore the kinetic energy of the adsorbed molecules.

In addition to the aforementioned separation of dissolved substances from their solvents, other changes may be effected by adsorption. Chemical reactions (such as the union of hydrogen and oxygen passed through spongy platinum) that ordinarily do not take place at room temperature, or only at an infinitely slow rate, are *catalyzed*—made possible or accelerated. Catalysts lower the energy required to initiate chemical reactions. Enzymes are colloidal *biocatalysts.* Since the particles in a colloidal solution have a diameter ranging from 0.0001 to 0.000001 mm, the amount of surface they present is enormous,* and chemical action is enormously accelerated. Heat causes a clumping of molecules (similar to the coagulation of egg albumin by heat), and denaturation greatly reduces the amount of surface and in consequence renders enzymatic activities impossible. Certain substances may become adsorbed on catalysts but not participate in the usual chemical reactions. Indeed, they interfere with or stop these reactions because they reduce the available catalytic surface; they are catalytic poisons. When such substances block the catalytic activity of enzymes, they are referred to as metabolic poisons.

*A sphere of iron having a diameter of 1 mm has a surface area of 0.0314 cm^2. If this bit of iron is pulverized until the particles have a diameter of 0.0001 mm, the total surface area amounts to 314 cm^2.

ELECTRIC CONSTITUTION OF MATTER

The atom is the unit structure of matter. According to the nuclear theory an atom is composed of a nucleus surrounded by *electrons.* The nucleus of the hydrogen atom is composed of 1 *proton;* it is accompanied by 1 electron. The proton is the unit of positive electric charge; the electron is the unit of negative electric charge. Since the mass of the electron may be disregarded (being only $^1/_{1,840}$ the mass of the proton), the mass of the proton is equal to the weight (relative) of the hydrogen atom, which is stated as 1 mass unit.

In all other elements the nucleus contains 2 or more protons and 2 or more *neutrons.* Similar to the proton, the neutron has a mass of 1; however, being devoid of either a positive or a negative charge, it is an electrically neutral particle. In an atom the number of positive protons is always equal to that of the negative electrons; hence the atom is neutral. The number of the electrons in the atom of any element determines the *atomic number* of that element in the periodic table. This is indicated by a subscript placed to the left of the symbol of the element. The combined weight of the protons and neutrons in the nucleus of any element is the *atomic weight* of that element. For example, the nucleus of sodium (Na) contains 11 protons and 12 neutrons; its atomic weight is 23. Since the number of electrons is always the same as that of the protons, the atomic number of sodium is 11. These two facts are indicated as $^{23}_{11}Na$. The symbol of carbon is generally written as $^{12}_{6}C$, indicating that the carbon atom has 6 electrons, 6 protons, and 6 neutrons.

Electrons circle around the nucleus in one or more orbits, or shells, situated at various distances from the nucleus, similar to the orbits of the planets around the sun. The chemical properties and the reactions of one element or compound with another are determined by the number and arrangements of the electrons; the nucleus plays no part in this. The electron or electrons in the *outer shell* constitute the chemical bonds, or *valence*, which enables the atom of one element to combine with that of another element. Hydrogen, sodium, and potassium have but 1 outer electron and are, therefore, *monovalent*. Calcium and magnesium have 2 electrons in the outer sphere and, in consequence, are *divalent*.

■ Isotopes

The atomic weight of hydrogen is stated as 1; this weight is due almost entirely to the mass of the single proton that constitutes the hydrogen atom's nucleus. It is possible to separate from ordinary water, however, samples that on analysis show a hydrogen atom having the weight of 2. This heavy hydrogen is chemically indistinguishable from ordinary hydrogen, which has an atomic weight of 1.* Since the chemical properties of an atom are determined solely by the number of electrons, these two forms of hydrogen must have the same number of electrons, that is, 1. Consequently, the atomic number of both forms of hydrogen is 1, and they occupy the same place in the periodic table. And since the number of protons is always equal to that of the electrons (noting that the atom is electrically neutral), the heavy hydrogen nucleus possesses only 1 proton. The greater weight of the atom of heavy hydrogen is accounted for by the existence in the nucleus of a neutron. The presence of neutrons in the nucleus does not alter the number of positive and negative charges of the atom, nor does it have any influence upon the chemical properties of the element.

Heavy hydrogen is called deuterium, and its nucleus is known as a deuteron. The water formed by the union of heavy hydrogen with oxygen is known as heavy water. Its molecular weight is 20 instead of 18 as in ordinary water. However, these two forms of water are chemically identical. Atoms that have the same number of electrons and protons but differ in the number of neutrons are known as *isotopes*. Isotopes vary in their nuclear mass; that is, although chemically identical, they do not have the same atomic weight.

Isotopes of other elements have been discovered. There are ${}^{12}_{6}C$ and ${}^{13}_{6}C$; in both forms the number of electrons is 6, and, therefore, the atomic number is 6. In the first form the nucleus contains 6 protons and 6 neutrons, but in the second form there are 6 protons and 7 neutrons. Ordinary carbon is a mixture of 98.9% ${}^{12}C$ and 1.1% ${}^{13}C$. Nitrogen exists in 2 isotopes, ${}^{14}_{7}N$ and ${}^{15}_{7}N$; oxygen has 3 isotopes, ${}^{16}_{8}O$, ${}^{17}_{8}O$, and ${}^{18}_{8}O$. By means of the cyclotron or the uranium pile, many isotopes of the known elements have been formed.

■ Radioactive isotopes

Some isotopic atoms are stable. Other isotopes, because of the relative number of protons and neutrons in the nuclei, are unstable, radioactive, and constantly break down into more stable atoms. The atoms of the naturally occurring element radium undergo spontaneous disintegration. In so doing, radium emits *radiations* of particles or rays that can be detected by their power to blacken a photographic plate securely shielded against light or to cause zinc sulfide to glow.* By means of a strong magnetic field it is possible to analyze the radiations emitted by radium into three forms. (1) Positively charged *alpha* parti-

*About 99.98% of hydrogen is ${}^{1}_{1}H$ and 0.02% is ${}^{2}_{1}H$.

*Radium-painted watch dials owed their luminosity to these emanations.

cles are helium atoms, ${}^{4}_{2}He$, that have been stripped of their electrons; therefore, the alpha particles are the nuclei of helium. (2) Negatively charged *beta* particles are electrons set free. (3) *Gamma* rays are similar to x-rays.

In this disintegration of a less stable radioactive element to a more stable condition, energy is set free, and the element is changed into another element; for example, the element radium, ${}^{226}_{88}Ra$, is transformed into radon, ${}^{222}_{86}Rn$, and by a series of transformations radon becomes lead, ${}^{207}_{82}Pb$. In these changes matter is transformed into energy.

The artificially created unstable isotopes undergo a quite similar breaking down into more stable atoms. This is always associated with the emission of one or more of the radiations spoken of previously. These radiations can be detected by various pieces of apparatus, such as the electroscope and the Geiger (or Geiger-Müller) counter, which counts the number of rays emitted in a unit of time.

When a small amount of some radioactive element such as ${}^{14}C$ is incorporated into a carbon-containing compound, this compound is said to be labeled. Compounds labeled in this manner by various elements (Na, K, H, C, P, Fe, etc.) can be traced in their passage through tissue, and the chief areas of deposit can be noted. For this reason such elements are spoken of as *radioactive tracers*. When radioactive sodium, ${}^{24}Na$, is taken (as sodium chloride) into the alimentary canal, within five minutes the Geiger counter shows the presence of the labeled sodium in the fingers. Much has been learned about the action of enzymes, hormones, and vitamins by the aid of tracer elements.

Ionization or electrolytic dissociation

From the atoms of some elements it is possible to remove 1 electron, or more, found in the outer shell; in contrast, the atoms of other elements have the power to incorporate 1 or more additional electrons. In certain instances water has the ability to bring this about.

When a molecule of NaCl dissolves, the Na atom loses 1 of its electrons; this electron is taken up by the Cl atom. By this process the Na atom is rendered more positive, and the Cl atom is rendered more negative. An atom that has lost 1 electron or more is known as a *positive ion* or *cation*, and an atom holding extra electrons is known as a *negative ion* or *anion*. Dibasic acid radicals and bivalent elements give rise to ions that carry two electrical charges.

$$Na_2SO_4 \rightarrow Na^+, Na^+, SO_4^{--}$$
$$CaCl_2 \rightarrow Ca^{++}, Cl^-, Cl^-$$

Although strong, water-soluble, mineral acids, bases, and their salts ionize freely, most organic compounds dissociate feebly or not at all. In many instances chemical reaction takes place between ions.

Electrolytes and nonelectrolytes

As with molecules of a gas, the ions in solution are in constant motion. When a current of electricity passes through a NaCl solution (Fig. 3-5), it *directs* the movement of the ions in such a manner that the positive-charged Na ions move toward the negative pole (connected with the zinc plate of a dry cell), and the Cl ions move toward the positive pole (copper or carbon plate).

From Fig. 3-5 it must be clear that the greater

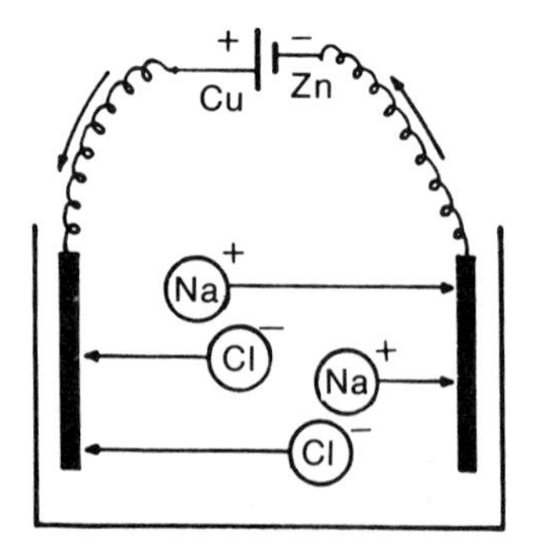

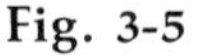

Fig. 3-5
Diagram to illustrate the movement of electrolytes in solution under the influence of an applied potential.

the concentration of ions, the less resistance the electric current encounters in passsing through a fluid and the greater is the electrical conductivity of the solution. Hence, water and water solutions of most organic compounds (e.g., sugar or alcohol) are poor conductors. Such substances are spoken of as *nonelectrolytes.* In contrast, compounds freely ionizing in solution are good conductors or *electrolytes.* An electric current is the flow of electrons through a conductor (e.g., water or copper wire) from a region of greater electron pressure (electrical potential) to one of lower pressure.

SOLUTIONS

Moles and molar solutions

The concentration of a solution (i.e., the amount of the solute in a unit volume of solvent) is frequently expressed in percentage. A more meaningful method is to state it in moles or fraction of a mole. *A mole* (or gram molecular weight) *is the amount of a substance in grams equal to the molecular weight of the substance.* Therefore 1 mole of hydrogen is 2 g; 1 mole of oxygen is 32 g; and 1 mole of carbon dioxide is $12 + (16 \times 2)$ or 44 g. Since the mole is based on the molecular weight, a mole of any substance contains the same number of molecules (Avogadro's number) as a mole of any other substance.

A mole of a gas enclosed under standard conditions (0 C and 760 mm Hg pressure) occupies 22.4 liters of volume. This volume is the gram molecular volume. In other words, enclosed in this space, the impact of the molecules in a mole of any gas exerts on the walls of the container at 0 C a pressure of 1 atmosphere.* If the 22.4 liters of the gas are reduced by an outside force so as to occupy 1 liter of volume, the pressure of the gas will be 22.4 atmospheres (17,024 mm Hg).

A mole of glucose, $C_6H_{12}O_6$, is 180 g. If this amount of glucose could be transformed into a gas and at 0 C could be confined to 1 liter of volume, it also would exert 22.4 atmospheres of pressure. Instead, dissolve 180 g of glucose in a quantity of distilled water sufficient to make 1 liter of solution. This is known as a *molar* solution (M).* This solution placed in an osmometer gives rise to an osmotic pressure of 22.4 atmospheres.† As in gas pressure, the osmotic pressure is independent of the nature of the substance and is determined at a given temperature by the number of particles per unit volume of solution, and therefore a half-molar solution of glucose exerts 11.2 atmospheres of pressure.

Concentrations of solutes in biological fluids are never as large as the units just considered. Therefore it is customary to speak of *millimoles* (mM), that is $^1/_{1,000}$ of a mole.

*Equivalent to 15 lb/in².

Osmolarity and equivalents

Since all molar solutions are equimolecular and since the osmotic pressure of a glucose solution is due to the number of molecules in solution, it might be inferred that a molar solution of NaCl exerts the same osmotic pressure as a molar solution of glucose. However, experiments reveal that in solutions of physiological concentration NaCl exerts about 1.85 times as much osmotic pressure as glucose. The reason for this lies in the ionization of the majority of the NaCl molecules. Because of the difference in the number of particles, even though the number of molecules is equal, a unit that expresses the osmotic effect of a solution has been devised. The *osmol* is the product of the molar concentration of a solute times the number of osmotically active particles it contributes. In biology the usual expression is *milliosmols.*

1 mM glucose × 1 particle = 1 milliosmol
1 mM $CaCl_2$ × 3 particles = 3 milliosmols

*One gram molecular weight dissolved in 1,000 g of water constitutes a molal solution. At low solute concentrations molar and molal solutions are nearly equivalent.
†Because of the tremendous pressure, this experiment can be performed only with more dilute solutions.

The pressure equivalent of osmolarity can be calculated as follows:

1 mole = 1 atmosphere × 22.4 liters
1 osmolar solution = 760 mm Hg × 22.4
1 milliosmol = 17 mm Hg

Osmotic concentrations of solutions are customarily measured by the freezing point depression rather than as osmotic pressure. Pure water freezes at 0 C. A solution containing 1 osmol of undissociated solute (e.g., glucose) in each kilogram of water, a *molal* solution, freezes at −1.86 C. The plasma of man freezes at −0.533 C; therefore, the osmolal concentration is 0.533/1.86 = 0.297. This is 297 milliosmols/kg of body water. Since the osmotic coefficient of NaCl is 1.85 at plasma concentration, the concentration of a NaCl solution required to produce the same osmotic pressure as plasma (297 milliosmol × 17 mm Hg/milliosmol = 5,049 mm Hg) would be 297/1.85 = 0.160 molal.

Many of the important solutes found in the body exist as charged particles, that is, as ions. If the molar concentrations of a monovalent and divalent ion are equal (e.g., Na^+ and Ca^{++}), the total number of electrical charges will be unequal. It is often necessary in biology to consider the electrical equivalence of solutes. One electrical *equivalent* (Eq) of an ionized substance is 1 g molecular weight of the substance divided by its valence; a *milliequivalent* (mEq) is, of course, 1 mg molecular weight divided by the valence. A milliequivalent of Na^+ is 23 mg/1 = 23 mg; however, a milliequivalent of Ca^{++} is 40 mg/2 = 20 mg.

OSMOSIS AND OSMOTIC PRESSURE

Suppose that the walls of the inner vessel, Fig. 3-6, *a,* are permeable to water but not to solutes. The top of the osmometer is supplied with an upright glass tube, Fig. 3-6, *c*. The osmometer is filled with a strong salt or sugar solution and is placed with proper precautions in a large amount of distilled water. It is soon noticed that the solution in *a* enters the glass tube, *c,* and keeps on rising; that is, the volume in *a* is increasing. This increase is due to the distilled water in *b* passing into *a* and forcing the fluid up into the tube, *c,* against the force of gravity. The passage of water (or any solvent) through a selectively permeable membrane from a dilute solution into a more concentrated solution is known as *osmotic flow* or *osmosis.* Osmosis is simply a case of diffusion, but the movement of

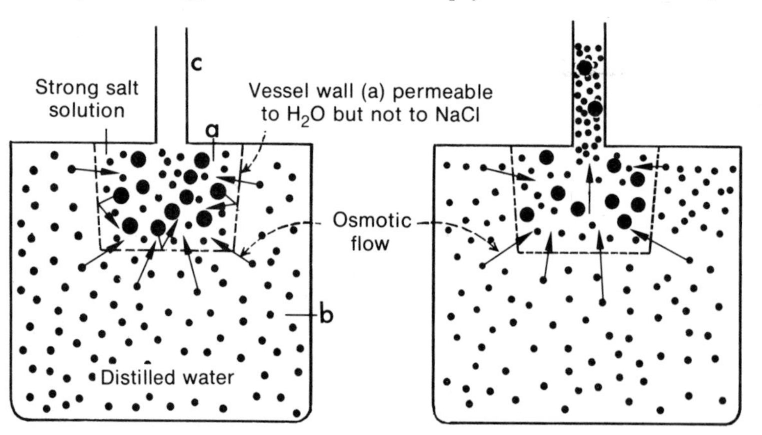

Fig. 3-6
Diagram to illustrate osmosis and osmotic pressure. Large dots represent impermeable ions; small dots, water molecules. Vessel at the left shows initial state; vessel at the right the final state. Note rise of fluid level in tube **c**.

solvent usually causes an obvious volume change, whereas the movement of solute has a negligible effect. The force that the accumulated fluid exerts upon the walls of the osmometer is a measure of the *osmotic pressure* of the solution. Given a sufficient length of time, the column of the solution in the tube, *c*, will attain its maximum height. Measuring the height of this column in millimeters indicates the amount of osmotic pressure. The osmotic pressure across a membrane can be calculated by the following equation: $\pi = RT(C_1 - C_2)$, in which π is pressure in atmospheres, R is the gas constant (0.082 liter-atmospheres/degree/mole), T is the absolute temperature (degrees centigrade + 273), and C_1 and C_2 are the respective concentrations of solute on the two sides of the membrane expressed in moles. For example, at 0 C the osmotic pressure of a 1 M solution of glucose would be equal to $0.082 \times 273 \times 1 = 22.4$ atmospheres. However, the equation for osmotic pressure was derived from the ideal gas law (see p. 31), which assumes that molecules of gas (or solute) occupy no volume and do not attract one another. The solutions to the equations are therefore only approximate answers, being more nearly accurate the less concentrated the solute or gas.

Osmotic pressure may be defined as *the force under which a solvent moves from a solution of lower solute concentration to a solution of higher solute concentration when these solutions are separated by a selectively permeable membrane.*

The origin of the pressure exerted by the osmotic flow may be stated as follows: The water molecules (Fig. 3-6, *a* and *b*) are in constant motion, and the membrane separating *a* and *b* offers equal resistance to their passage in either direction. But as the sugar molecules in *a* displace some of the water molecules, the molecular concentration of the water* is greater in *b* than in *a*. Hence, the diffusion pressure of the water molecules in *b* exceeds that in *a*. In consequence, more water molecules hit the membrane per unit of time and per unit of surface in the direction from *b* to *a* than in the opposite direction; the result is an increase of volume in *a*. Osmotic pressure also may be regarded as evidence of the attraction of a solution for water. A moment's reflection will show that to measure the amount of osmotic pressure accurately it is necessary to prevent any change in volume of the solute solution by application of external pressure.

Osmosis can perform mechanical work; for example, it might have raised a piston placed in tube *c* (Fig. 3-6). This energy exists by virtue of the presence of solute. In effect, the concentration is an expression of energy, and osmosis results from the natural tendency of energy to flow from a higher to a lower potential level. Ultrafiltration, because it concentrates a solute, is the reverse of osmosis and requires an input of energy, which in living systems generally originates from the contractions of the heart.

Osmotic pressure of electrolytes. ■ The amount of osmotic pressure that a solution exerts is determined by the number of particles in solution. In a nonelectrolyte these are the dissolved molecules. But in a solution of an electrolyte the ions formed by electrolytic dissociation exert an osmotic pressure equal to that of an equivalent number of molecules. If, therefore, all the molecules in a KCl solution of a certain molecular concentration were dissociated, the osmotic pressure would be twice as great as that exerted by a nonelectrolyte solution of the same concentration; the osmotic pressure of a $CaCl_2$ or a Na_2SO_4 solution would be three times as great. It is, however, only at infinite dilution that all molecules ionize; the number undergoing ionization depends on the nature of the electrolyte and on its concentration. A 0.01 M solution of sucrose has an osmotic pressure of 0.224 atmosphere; a 0.01 M solution of KCl exerts not 0.448 but 0.435

*The molecular weight of water is 18. A liter of water weighs 1,000 g; thus the concentration of water is about 55.5 M. The addition of solute reduces this concentration.

atmosphere. Similarly a 0.01 M solution of $BaCl_2$ exerts 0.610 (not 0.672) atmosphere.

Colloidal osmotic pressure. ■ Because of the extremely large size of their molecules and therefore the low molecular concentration of their solutions, the colloids (starch, proteins, and so on) exert very little osmotic pressure. For example, the molecular weight of gelatin (a protein) is approximately 36,000. The most concentrated solution of gelatin that can be made contains but 30 g/liter (i.e., a 3% solution and equivalent to approximately 0.001 M); the osmotic pressure of this solution is 0.001 of 22.4 or about 0.02 atmospheres or 15 mm Hg.

Isosmotic solutions. ■ When two sugar solutions of the same molecular concentration of the solute are separated by a semipermeable membrane, no change in volume or pressure occurs in either; they are isosmotic (*iso*, equal). When two solutions having different molecular concentrations are separated by a semipermeable membrane, the water passes from the one having the higher water concentration to the one with the lower concentration. The solution having the lower osmotic pressure is said to be *hyposmotic* with respect to the second solution, whereas the second is said to be *hyperosmotic* to the first solution.

■ CELL MEMBRANE OR PLASMA MEMBRANE

■ Selective permeability of the plasma membrane

To account for the qualitative and quantitative regulation of the exchange of materials between the protoplasm and its environment, we turn to the plasma membrane theory. The immiscibility of the protoplasm of a cell with the fluid surrounding it provides a strong argument in favor of regarding the outer layer of the cell as a special protecting skin. By means of this membrane the cell maintains its individuality. An important characteristic of the cellular membrane is selective permeability. Selective permeability, operative in both directions, determines the kind and amount of the substances passing into and out of a cell. Although this must be regarded as a theory, it is at present perhaps the most acceptable explanation offered for an extensive array of facts, some of which will be briefly considered.

Cytolysis. ■ The plasma membrane of the human red blood cell (RBC) is permeable to water and relatively impermeable to NaCl and certain other cellular constituents. Placed in the proper medium, the red blood cell exerts osmotic pressure. In a highly diluted NaCl solution its volume increases by osmosis, and the membrane may stretch until the contents of the cell escape *(hemolysis)*. In a comparatively strong NaCl solution the greater osmotic pressure of the solution draws water out of the cell which, in consequence, shrinks in volume. In a 0.9% (0.155 M) NaCl solution the red blood cells of man maintain a constant volume. This concentration of NaCl solution has the same osmotic pressure as the contents of the cell; they are, therefore, *isosmotic.*

Isotonic and isosmotic solutions. ■ With respect to the impermeability of the warm-blooded animal cell membrane, a 0.9% NaCl solution is said to be *isotonic* with the cell contents. A 0.32 M urea solution is isosmotic with a 0.155 M solution of NaCl. It is not isotonic because the cell membrane of the red blood cell is permeable to urea in solution; urea enters the cell in a similar manner to distilled water. In consequence, the urea solution is unable to exert any osmotic pressure upon the cell. In no concentration of urea, high or low, does the cell maintain its original volume. The isosmotic NaCl and the urea solutions are, therefore, not equivalent in tonicity.

In spite of the fact that an isosmotic solution contains the same total solute concentration (sum of all solute particles), only an isotonic solution prevents a change in the size of the cell. However, the type of cell under discussion always must be specified. The solutions listed in Table 3-1 are essentially isosmotic solutions but

Table 3-1. Comparison of isosmotic solutions with regard to isotonicity (+) for RBC and muscle cells

Solution	*Red blood cell*	*Muscle*
0.3 M sucrose	+	+
0.15 M NaCl	+	+
0.3 M urea	0	0
0.15 M KCl	+	0

Table 3-2. Intracellular fluid composition of a hypothetical mammalian cell and its extracellular fluid environment

Composition	*Intra-cellular fluid*	*Extra-cellular fluid*
Cations (mEq/liter)		
Na^+	15	150
K^+	150	5
Ca^{++}	—	5
Anions (mEq/liter)		
Cl^-	4	110
HCO_3^-	1	30
Other	160	20
Approximate osmolarity (milliosmols/liter)	300	300

are not necessarily isotonic for the red blood cell or muscle.

A NaCl solution, the concentration of which is greater than 0.9%, is *hypertonic* to the mammalian cell, whereas a NaCl solution, the concentration of which is less than 0.9%, is *hypotonic* to the cell.

Distribution of salts. ■ The difference in the concentration of the inorganic salts in the cells and in the fluid surrounding the cells is most striking. The red blood cells are practically devoid of sodium but contain about 0.6% potassium (in the form of salts); the fluid (plasma) in which these cells float contains 0.35% sodium and only 0.02% potassium. Again, a muscle contains twenty times as much potassium and only one tenth as much sodium as does the fluid that surrounds and nourishes its cells.

For practical purposes the interior of a cell may be considered to be a homogeneous, well-mixed compartment. The *intracellular fluid* (ICF) composition of a hypothetical mammalian cell and its *extracellular fluid* (ECF) environment are approximated in Table 3-2.

■ Variability of permeability

The permeability of the plasma membrane is not the same for all cells, nor is it constant for any one cell. The plasma membrane varies according to the function of the cell. The human red blood cell is relatively impermeable to sodium. Even when their functions appear to be identical, membranes may differ; for example, the permeability of the red blood cell of a rabbit and of a cat is not the same.

According to the theory that selective permeability of the cell membrane is the deciding factor in the exchange of materials between the protoplasm and the environment, it is necessary that the degree of permeability vary with the needs for this exchange as determined by the activity of the protoplasm. It is conceivable that such changes in permeability are brought about by the chemical compounds formed during protoplasmic activity.

The differentiation existing between the outer layer of the protoplasm and the interior is no doubt due to the interaction between this external layer and the environment; consequently, changes in the environment also may cause changes in the permeability. Among the many environmental changes that increase permeability are increases in temperature, radiations (light, ultraviolet light, and x-ray), electric currents, and certain chemical compounds. An increase in the selective permeability (especially to ions) is generally held to be the first result of the stimulation of protoplasmic structure. On the other hand, the loss of excitability by narcosis is believed to be caused by the adsorption of the narcotic upon the cell membrane, thereby reducing its selective permeability.

The influence of chemical compounds is well illustrated by maintaining a cell in an isotonic NaCl solution. In this solution the permeability of the cell membrane steadily increases with time; all materials find ingress into the cell; the cell gradually dies. If a small amount of calcium chloride is added to the sodium chloride solution, the normal permeability is retained for a greater length of time, and the life of the cell is prolonged. In general, *sodium and potassium increase and calcium decreases permeability.*

■ Nature of the cell membrane

How the layer of protoplasm exposed to the exterior acquires the property of selective permeability is still a moot point. The cell membrane is frequently described as a condensation layer of lipid substances and proteins. Lipid substances lower surface tension and always tend to accumulate at the surface of a solution in which they are found, thereby increasing the density of this outer layer.

Perhaps it is best to regard the cell membrane as formed by the gelation of lipoprotein molecules. When an ameba is slightly torn, the protoplasm tends to flow out. If the solution surrounding the organism is a balanced solution (containing the proper concentration of Ca, K, and Na ions), a surface membrane is immediately formed, and the outflow stops. This has been called a surface precipitation reaction. The limiting membrane thus formed is a gelled membrane having a greater density, viscosity, and elasticity than the interior substance. If this experiment is made with a solution deficient in Ca ion or having an excess of Na or K ions, no gelation occurs, and the protoplasm flows out freely to mix with the solution.

■ Osmotic pressure of protoplasm

Protoplasm as a solution of crystalloids and colloids exerts a certain amount of osmotic pressure. The amount of this pressure can be ascertained by methods suggested in our study of cytolysis. A frog's excised muscle retains its weight in 0.125 M NaCl solution; in a solution a little more concentrated (hypertonic) it loses weight; and in a more dilute solution (hypotonic) it gains weight. Granted that the cell membrane is essentially impermeable to NaCl, the osmotic pressure of the muscle must be equal to that of the solution in which it neither gains nor loses weight, that is, 0.125 M NaCl. This solution is approximately 0.7%. For this reason a 0.7% NaCl solution is used as a *physiological salt solution* for cold-blooded animals.* In man the red cells and, therefore, the blood as a whole have an osmotic pressure equal to a 0.9% NaCl solution; such a solution exerts an osmotic pressure of about 7.8 atmospheres or approximately 117 lb/in^2.

■ Modes of penetration

The manner in which the various materials find entrance into cells is an important but often perplexing question. Substances can permeate the plasma membrane either by passing through pores or across the lipoprotein layer. Only small molecules and water can diffuse through pores; lipid-soluble materials diffuse across the membrane by temporarily going into solution in the lipid. Sometimes the membrane is spoken of as a "lipoid sieve." There are, however, many biologically important substances (e.g., amino acids and sugars) that are not lipid-soluble and are too large to penetrate plasma membrane pores. These substances are actively transferred across the membrane by energy-expending mechanisms.

Passive diffusion. ■ The concept of passive diffusion is relatively simple; it involves the redistribution of materials by random movement of

*A physiological salt solution is sometimes erroneously called a normal saline solution.

its particles due to their thermal energy. Movement is from a higher to a lower concentration or electrochemical potential. For example, oxidative metabolism within a cell reduces the tension of oxygen and raises the tension of carbon dioxide. In consequence, more oxygen diffuses in and carbon dioxide diffuses out. Even though metabolic energy is available, there is no need to use this energy to move these gases in or out of the cell, for the translocation is spontaneous.

The net movement of material per unit of time is proportional to the concentration gradient.

$$\dot{Q} = -P\Delta C$$

$\dot{Q}$ is the quantity per unit time, P is the permeability constant of the membrane (the negative sign is appended because the movement is down a concentration gradient), and $\triangle C$ is the difference between the concentrations, that is, the *gradient*. It must be recognized that the permeability constant is not the same from tissue to tissue nor under varied environmental or activity states.

Specific transport. ■ When knowledge about the physical chemistry of two solutions bathing a membrane is not sufficient to account for the movement of a particular molecular or ionic species, mechanisms of translocation other than passive diffusion must be sought. One of these, specific transport, is in part characterized by much more rapid permeation of a solute than is consistent with its water solubility, molecular weight, charge, or structural specificity. Such transport is believed to be mediated by a "carrier molecule" in the membrane, which combines with the solute on one side of the membrane and discharges the solute on the other side. Because the subject of specific transport is so very wide in scope, we will restrict our further discussion to certain essential aspects.

Active transport. Frequently biologically important substances pass across membranes much faster than can be accounted for by simple diffusion. Indeed, the rate of transfer from one side of the membrane to the other is often faster at low or moderate concentrations of the substance than it is at higher concentrations. Such behavior resembles that noted for enzyme activity (Fig. 22-3) whenever the substrate is present in concentrations which "saturate" the enzyme; that is, the enzyme is operating at its optimal capacity and the substrate level exceeds that capacity. The similarity of the transport and enzyme phenomena leads to the hypothesis that transport is brought about by a combination of the water-soluble solute and a "carrier" molecule that shuttles across the membrane. Permeation of a solute that proceeds against a concentration or electrical gradient, if the solute is a charged particle, and that requires metabolic energy for the movement is called *active transport* (Fig. 3-7). If the transport is truly active, then blocking of cellular metabolism or lowering of the environmental temperature, to which metabolism is very sensitive, should depress or stop the solute transfer. Potassium ions, which are commonly more concentrated inside than outside of animal and plant cells (Table 3-2), inevitably leak out due to their concentration gradient whenever the metabolism of these cells is inhibited. The active transport by cells that are dependent on oxidative metabolism for their energy can be blocked by subjecting them to low oxygen tensions or to a metabolic poison such as cyanide. Cells that function well under anaerobic conditions (glycolytic metabolism) can be poisoned with chemicals such as iodoacetate. Human red blood cells maintain their internal K ion concentrations at low oxygen tensions, but not in the presence of iodoacetate or at near-freezing temperatures. Anaerobic metabolism is then their major source of energy. As much as 30% of this metabolism appears to be expended in active transport processes. Another criterion for active transport is that the substance moved should be osmotically active after transfer across the membrane; for, if it is bound or rendered inactive at its new location, then its accumulation could be a consequence of diffusion, that

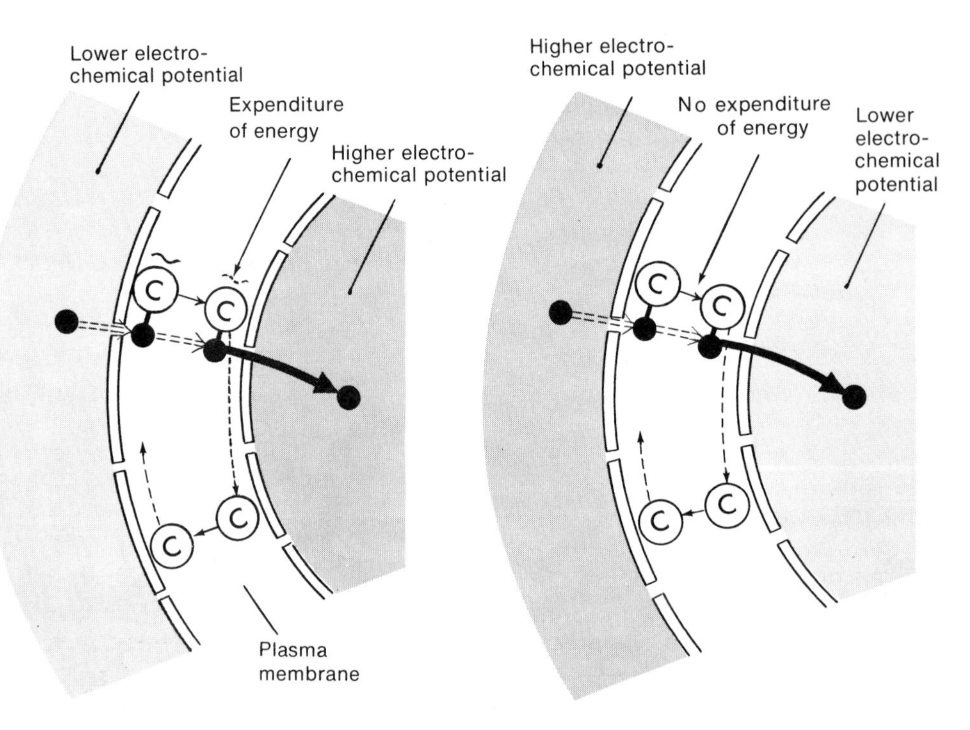

Fig. 3-7
Diagram showing concepts of active transport and facilitated diffusion. Active transport is carrier-mediated transfer, **c**, from lower to higher electrochemical potential with the expenditure, ~, of metabolic energy. Facilitated diffusion is carrier-mediated transfer from higher to lower electrochemical potential without energy expenditure. The same carrier may move other substances in the reverse direction either with or against electrochemical gradients.

is, movement down a concentration gradient. Active transport systems are often called "pumps." Such biological pumps are abundant in living organisms. Many, if not all, cells can pump electrolytes and some of them nonelectrolytes as well. Glucose, for example, can be actively transferred from the gut to the blood through epithelial cells against a concentration gradient.

Facilitated diffusion. Certain lipid-insoluble molecules that are unable to penetrate the membrane pores because of their size do, nevertheless, readily enter the cell. For these substances the direction of movement is with the concentration gradient, and the rate of movement, as in active transport, is faster than can be accounted for by ordinary diffusion. Furthermore, this form of translocation also exhibits saturation kinetics. It is a carrier-mediated, specific transport spoken of as *facilitated diffusion* (Fig. 3-7). This mechanism is insensitive to metabolic poisons (inhibitors) and is only moderately influenced by reduction in the environmental temperature. Facilitated diffusion is therefore distinguished from active transport by the fact that it does not require metabolic energy and that its moving force is a downhill concentration gradient. The sugars, glucose and mannose, appear to cross the red blood cell membrane by facilitated diffusion.

Other modes of penetration exist; hydraulic flow of solvent (filtration), solvent drag upon solute, and pinocytosis are examples. The details of these need not concern us here. Suffice it to say that among the established transport systems are those for Na^+, K^+, Ca^{++}, Fe^{++}, Cl^-, $PO_4^{\equiv}$, $SO_4^{=}$, simple sugars, and amino acids.

MEMBRANE POTENTIALS

It has been noted already in our discussions that the distribution of charged particles between intracellular and extracellular fluid compartments of cells is not uniform. In fact, within most cells the potassium concentration is much greater than outside; the chloride and sodium are distributed in the reverse direction. This inequality is maintained continuously by the normal resting cell. It results from the presence of large indiffusible anions (proteins, amino acids, and organic phosphates) within the cell and from the process of active transport. The presence of the nondiffusible ion ensures that the diffusible ions will not be equally represented in the two compartments, and active transport determines the final distribution of ions.

Let us examine the distribution of substances across a rigid artificial membrane that is permeable to small ions but impermeable to a particular large charged particle. We will assume that this membrane cannot perform active transport and is freely permeable to water. On the left side of the membrane we place a solution of KCl with a relative concentration of 100, on the right side the solution used consists of equal parts of KCl and KA (large anion), each at a relative concentration of 50. The *initial* state is electrically neutral and appears as follows:

Left	Right
100 K^+	100 K^+
100 Cl^-	50 Cl^-
	50 A^-

Diffusion commences immediately with the negatively charged Cl ions moving freely to the right down their concentration gradient. They are accompanied by positively charged K ions, although this movement is increasingly uphill. The process continues as long as the driving force on the Cl ions exceeds the oppositely directed force that progressively develops on the K ions. When the two opposing diffusion forces become balanced, an equilibrium has been reached, and the concentration ratios of K ion and Cl ion across the membrane are reciprocally related; that is:

$$K^+ \text{ (right)}/K^+ \text{ (left)} = Cl^- \text{ (left)}/Cl^- \text{ (right)}$$

The distribution at equilibrium will look approximately as follows:

Left	Right
86 K^+	114 K^+
86 Cl^-	64 Cl^-
	50 A^-

Such a balance of diffusion forces occasioned by the presence of an impermeable charged colloidal particle on one side of the membrane is known as a *Donnan equilibrium.* Further *net* movement of either K or Cl ions cannot take place. The relative numbers of positive and negative charges in a solution on either side of the membrane are equal, so that electroneutrality exists *in bulk.* However, an electrostatic potential difference has developed *just across the membrane,* which exactly balances the concentration gradients of the diffusible ions. In our example, the right side of the membrane is negative with respect to the left side because of the diffusion pressure on the K and Cl ions. Since the membrane is also permeable to water, this substance will move down its concentration gradient. The movement will be from left to right because the osmotic concentration is greater on the right. Net water movement will cease when the hydrostatic pressure, induced by the water movement, is equal to the osmotic pressure difference between the right and left sides.

Examine the hypothetical cell shown in Fig. 3-8. Although the membrane is impermeable to the large anions found in the intracellular compartment, it is moderately to freely permeable to Na^+, K^+, and Cl^-. Because of the high concentra-

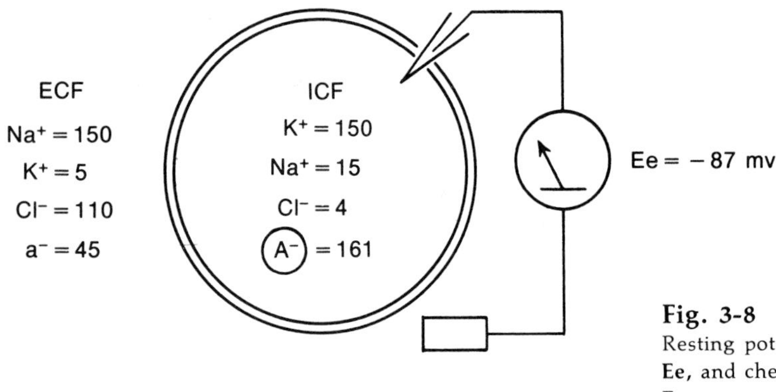

Fig. 3-8
Resting potential of a hypothetical cell. Electrical potential, **Ee**, and chemical potential calculated from Nernst equation, **Ec**, are equal and opposite in polarity. **Ee** is measured by using a glass microelectrode; **Ec** is calculated. The electrochemical potential, **Eec**, is zero; an equilibrium exists.

tion of K ions within a cell a *chemical gradient* (or potential) exists, and some K ions constantly tend to diffuse outward across the membrane. The extent of this egress is limited by the electrostatic attraction developed when indiffusible anions are left behind; an *electrical gradient* that exactly opposes the chemical gradient is produced, and an *electrochemical equilibrium* is established. At equilibrium there is no further *net* movement of ions. With positive charges lined up just outside the cell membrane and negative charges lined up just inside, the membrane is in a *polarized* state; the outside is positive and the inside is negative. This transmembrane potential difference is known as the resting membrane potential or *resting potential.*

The magnitude of the membrane potential can be calculated from the concentration gradient of the diffusible ion by application of a derivation of the Nernst equation:

$$E = 2.3\ RT/F \log C_1/C_2$$

R is the gas constant expressed in appropriate work units (8.3 joules per mole), *T* is the absolute temperature, *F* is the Faraday (96,500 coulombs/mole), and C_1 and C_2 are the diffusing ion's concentrations in intracellular and extracellular fluids. At 25 C (298° absolute) the equation* simplifies to:

$$E_{mv} = -59 \log C_1/C_2$$

For the hypothetical cell shown in Fig. 3-8 the K^+ equilibrium potential, as calculated, would be 87 millivolts (mv).

$$E_{mv} = -59 \log [K_i]/[K_o] = -87$$

*The Nernst equation is one of the most important and frequently used equations in biological literature. It is a statement based on thermodynamic principles, which describes the electrochemical equilibrium existing between the electrical work and the osmotic work needed to move a small quantity of ions in opposite directions across a boundary. Although it is not our purpose or need to explore the derivation of this equation, a few statements may help the reader achieve an intuitive appreciation of the equation. We noted previously that the ideal gas law (PV = nRT) expresses an amount of potential energy (or work). Because n/V represents a concentration (C = moles/liter), the gas law in a modified form can be applied to the calculation of osmotic work (work = 2.3 RT log C_1/C_2). The Nernst equation, in the form EF = 2.3 RT log C_1/C_2, expresses the amount of electrical work required to transfer a small quantity of particles in one direction against an electrical potential (E). At electrochemical equilibrium, the electrical work in one direction equals the osmotic work in the opposite direction.

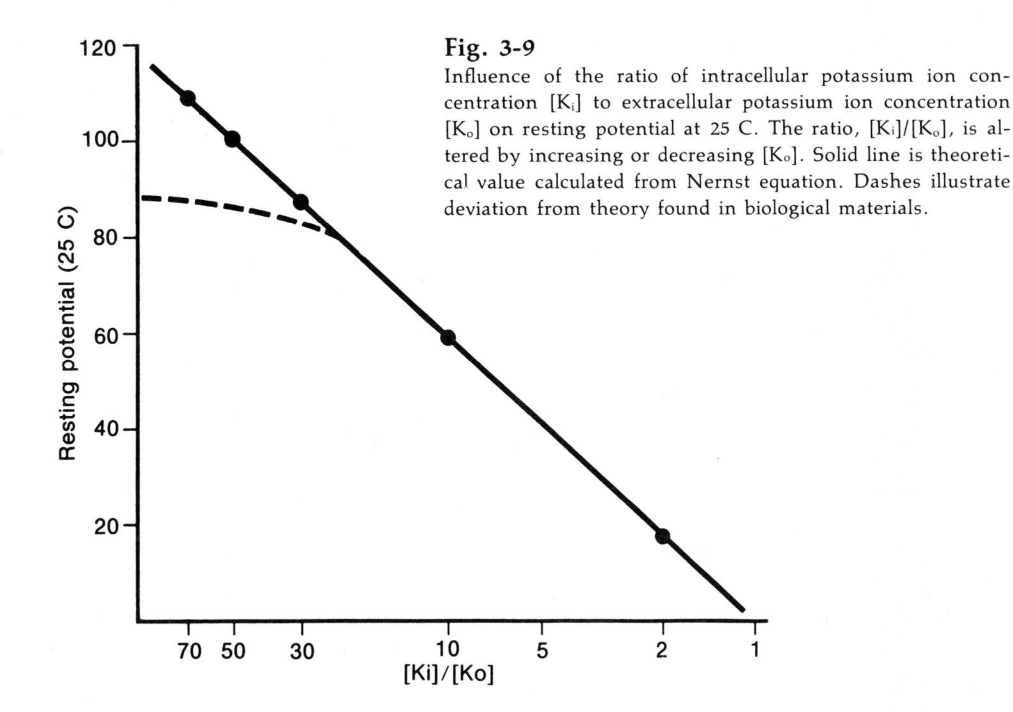

Fig. 3-9
Influence of the ratio of intracellular potassium ion concentration $[K_i]$ to extracellular potassium ion concentration $[K_o]$ on resting potential at 25 C. The ratio, $[K_i]/[K_o]$, is altered by increasing or decreasing $[K_o]$. Solid line is theoretical value calculated from Nernst equation. Dashes illustrate deviation from theory found in biological materials.

When the resting potential is actually measured in the laboratory with a glass microelectrode, 87 mv is the value obtained. The concentration gradient for the Cl ion would favor extensive diffusion into the cell, but the electronegativity inside tends to repel these ions. Application of the Nernst equation reveals that the Cl ion, similar to the K ion, has a distribution across the membrane that produces an electrochemical equilibrium. In the case of many cells, alterations of the chemical gradient for chloride—increase or decrease of the extracellular concentration—have virtually no effect on the resting potential. On the other hand, when the extracellular potassium concentration is altered (Fig. 3-9), the membrane potential does change. For our hypothetical cell in Fig. 3-8 the change would amount to 59 mv for a tenfold change in extracellular concentration. We may, therefore, conclude that in this cell the resting potential is, by and large, a *potassium diffusion potential.* It is especially important to emphasize that the potentials of which we speak arise when only a *very small number* of ions diffuse across the membrane in response to the gradient of the electrochemical potential. If this were not so, the Donnan distribution of the ions would rapidly disappear. That this is so is implicit in the Nernst equation in which the *ratio* of the concentrations and not the absolute values determines the magnitude of the transmembrane potential.

Up to this point another important ion, sodium, which is unequally distributed across the membrane of our hypothetical cell, has been ignored. Indeed, we have regarded the membrane as relatively impermeable to this ion; otherwise why would the cell preferentially accumulate potassium? Studies with radioactive tracers, for example, ^{24}Na, indicate that the membrane actually is moderately permeable to sodium ions. Certainly the electrochemical gradient in the resting cell would favor the entry of these ions into the cell.

The Nernst equation would allow us to predict that the existing distribution of Na ions across the membrane could be produced only if the transmembrane potential (Ee) were 59 mv, inside positive and outside negative:

$$E_{mv} = -59 \log [Na_i]/[Na_o] = -59 \log 0.1 = 59$$

However, the low internal concentration of Na^+ and the observed transmembrane potential are maintained by the so-called "sodium pump." This pump is an active transport mechanism that ejects Na ions from the cell against the electrochemical gradient.

Na ion pumping can be abolished by the addition of metabolic inhibitors to the medium in which the cell is bathed, that is, the extracellular fluid. Interestingly, in many tissues the efflux of Na ions is greatly reduced when the external medium contains no K ions and, conversely, it is increased when the external K ion concentration is increased. These latter findings are strong evidence that in certain tissues (e.g., nerve, muscle, and red blood cells) the same active transport process causes active K ion uptake and Na ion extrusion; that is, Na ion extrusion is coupled to K ion uptake. However, such an exchange is not necessarily a one-to-one event.

Another important role of the "pump" must be noted. Reference to our hypothetical cell (Fig. 3-8) shows that the osmolar concentration of the cell contents is somewhat greater than that of the extracellular fluid. To prevent the continuous diffusion of water into the cell and a potentially damaging increase in volume, a small but significant extrusion of solute must be maintained. This is accomplished by the active transport of sodium.

■ VOLUME DISTRIBUTION AND THE CLEARANCE CONCEPT

The *volume distribution* is a term used to specify the size of a body fluid compartment when a substance is introduced into the body that is distributed uniformly and exclusively within the confines of that compartment. To develop the general concept in the least ambiguous fashion, an elementary principle from general chemistry can be recalled, that is, that the product of the concentration of a substance in a given volume is always equal to the product of a different concentration of the substance in a different volume. This is usually given as follows:

$$\text{Concentration (1)} \times \text{Volume (1)} = \text{Concentration (2)} \times \text{Volume (2)}$$

For example, if we take 1 liter of 0.1 M glucose solution and dilute it with pure water until the concentration is now 0.05 M, the new volume must be 2 liters.

$$C_1 \times V_1 = C_2 \times V_2$$
$$0.1\ \text{M} \times 1\ \text{liter} = 0.05\ \text{M} \times 2\ \text{liters}$$

Of course it would be equally simple to solve for a new concentration if the new volume were specified. This dilution technique can be and is used to measure such fluid compartments as plasma volume and intracellular and extracellular fluid volume. Among the substances used for such measurements are dyes that bind to plasma proteins (Evans blue), radioactive polysaccharides (^{14}C inulin), and heavy water (D_2O). Although simple in principle, the dilution technique used for determining volume distribution requires careful attention to factors such as the toxicity of the substance, the uniformity and speed of its distribution, and its metabolism or excretion during the period of mixing in the body.

In physiology we often are interested in "dynamic" volumes, for example, the cardiac output or blood flow to specific organs, rather than in the more static volumes just considered. To determine these volumes the concept of *clearance* is used. This concept appears in many forms in all areas of physiology. Fundamentally, the concept implies measurement of the rate of transfer of a substance from a given volume, that is, the amount of a substance "cleared" from a volume per unit of time. For measuring clearance the dilution equation is modified to incorporate a statement of volume per unit time (the so-called minute volume), usually milliliters per minute. To indicate this modification it is customary to write the symbol for volume as $\dot{V}$.

$$C_1 \times \dot{V}_1 = C_2 \times \dot{V}_2$$

By way of illustration, suppose that the blood supplied to a given body organ has a specific concentration (C_B), of 0.1 mg/ml, of a substance Z. Furthermore, suppose that substance Z is completely extracted (cleared) from the blood by this organ and subsequently excreted from the body in a concentrated solution that can be collected. The minute volume (V_E) of this collection and its concentration (C_E) can be accurately determined. If this collection had a concentration of 10 mg/ml and a minute volume of 8 ml, the minute volume of the blood, that is, the blood flow to the organ, can be calculated as follows:

$$C_B \times \dot{V}_B = C_E \times \dot{V}_E$$
$$\dot{V}_B = C_E \times \dot{V}_E / C_B$$
$$\dot{V}_B = \frac{10 \text{ mg/ml} \times 8 \text{ ml/min}}{0.1 \text{ mg/ml}}$$
$$\dot{V}_B = 800 \text{ ml/min}$$

It must be emphasized here that the accuracy of the calculated value for the minute volume of blood in this case depends on certain assumptions. These are that the substance Z is neither metabolized nor synthesized in the organ, that it is completely cleared from the blood, and that it is quantitatively excreted. When these criteria are not met fully, use of the clearance concept must be modified to accommodate specific cases.

READINGS

Berlin, R.: Specificities of transport systems and enzymes, Science **168**:1539-1545, 1970.

Diamond, J. M., and Tormay, J. McC.: Studies on the structural basis of water transport across epithelial membranes, Gastroenterology **25**:1458-1462, 1966.

Katz, A. I., and Epstein, F. H.: Physiologic role of sodium-potassium-activated adenosine triphosphatase in the transport of cations across biologic membranes, New Eng. J. Med. **278**:253-261, 1968.

Lehninger, A. L.: Bioenergetics, New York, 1965, W. A. Benjamin, Inc., pp. 152-171.

Orten, J. M., and Neuhaus, O. W.: Biochemistry, ed. 8, St. Louis, 1970, The C. V. Mosby Co., pp. 879-900.

4

Epithelial and connective tissues
Local and communal value of tissues

THE TISSUES

In Chapter 1 it was pointed out that in the more highly developed animals and in man the various functions such as respiration, digestion, and excretion are performed by specialized organs. To facilitate the performance of its function, an organ is constructed of two or more parts known as tissues, which, in turn, are also highly specialized both structurally and functionally. Let us take a skeletal muscle to illustrate this.

Primarily a muscle is composed of cells that excel in the property of contractility. In the various skeletal muscles of our body these cells have a similar appearance, and collectively they constitute muscle tissue. The myriad of muscle cells in a given muscle are bound together so as to form one single piece of machinery; this is accomplished by another type of cell, which collectively is known as connective tissue. Between and penetrating the muscle cells is found still another kind of tissue formed by structurally and functionally specialized cells—nerve tissue. This enables a muscle to receive impulses for initiating its activity. For its nourishment a muscle is supplied with blood vessels. The inner surface of these tubes is lined with flat, scalelike cells (Fig. 13-4), forming what is known as epithelial tissue.

A tissue may, therefore, be defined as a group of more or less similar cells.

■ EPITHELIAL AND CONNECTIVE TISSUES

■ Epithelial tissue

Epithelial tissue is the covering tissue for nearly all the free surfaces within the body, as well as for the exterior of the whole body. In the animal body there are a large number of hollow organs; some are tubular, and others are more saclike. Some hollow organs communicate with the exterior, such as the alimentary canal with its glands (Fig. 19-1), the respiratory tract, and the urogenital tract—kidneys, ureters, bladder, urethra (Figs. 29-1 and 29-2), and the reproductive organs. The free surfaces of all these structures are covered with epithelial tissue.

There are also tubular organs that do not open to the exterior: the vascular system (heart and blood vessels) and the lymph vessels. The covering tissue of these inner surfaces is more generally designated as *endothelial tissue.* The heart, lungs, and abdominal organs are severally surrounded by a double layer of a protective membrane; the surfaces of these two layers are also clothed with epithelial tissue.

In epithelial tissue the cells are arranged in a compact manner, that is, with a minimum of intercellular space. In consequence, the amount of intercellular material (cement substance) is limited. The outlines of the cells are fairly regular, and the nuclei are distinct.

Forms of epithelial tissue. ■ Epithelial tissues line the free surfaces of organs and membranes as a single layer or as many layers of these cells. In the first case it is spoken of as simple epithelium; in the other it is spoken of as stratified epithelium.

Simple pavement (squamous) epithelium. In simple pavement epithelium the cells are flat or scalelike, as shown in Fig. 13-4. In the study of the organs we shall frequently encounter this type of tissue. The squamous cells covering the membranes that line the internal cavities of the body (thoracic and abdominal) and that cover the organs found in these cavities (lungs, heart, stomach, and intestines) secrete a small amount of a serous, or watery, fluid; this fluid lubricates the moving surfaces that come in contact with each other. Serous fluid is also found in the spaces in the joints; here it is known as *synovia* or synovial fluid. When a joint is injured, as by spraining, there may be a great increase in the formation of this fluid, which causes the joint to be swollen.

Simple columnar epithelium. In simple columnar epithelium the cells, forming a single layer, are columnar in shape. In some organs, such as parts of the respiratory tract and of the reproductive organs, the columnar cells are provided with cilia (Fig. 4-1). Cilia may be regarded as hairlike projections of the protoplasm protruding beyond the free border of the cell.

Stratified pavement (squamous) epithelium. Stratified pavement (squamous) epithelium is the most common of all epithelial tissue. It is found in the epidermis of the skin (Figs. 7-3 and 26-1), in the mouth, esophagus, and pharynx, on the anterior surface of the cornea of the eye, and in other regions. In this tissue the lower or deeper situated cells are prismatic, those above these are polyhedral, and, as the cells approach the surface, they become more and more flattened. Epithelial tissue is not supplied with blood vessels, and consequently the outermost layers are not well nourished. In multiplying, the cells of the lower layer crowd the overlying cells upward. The chemical composition as well as the shape of the cells becomes greatly modified; in fact, as they advance toward the surface, they gradually die. The outermost, dead cells are then worn away.

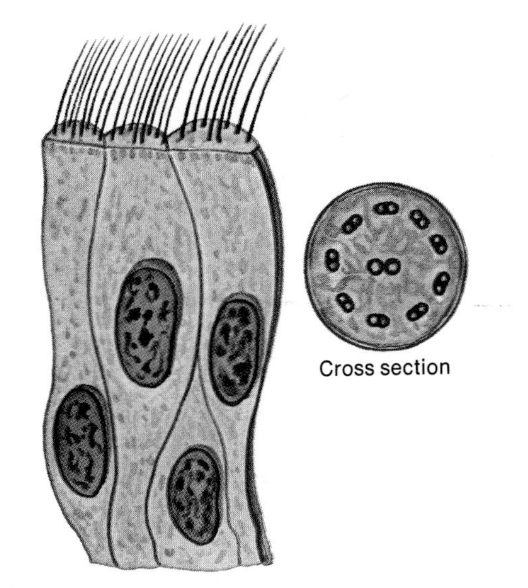

Fig. 4-1
Ciliated epithelium from the trachea.

Stratified columnar epithelium. Stratified columnar epithelium is composed of several layers (Fig. 4-2). The outer layer consists of true columnar cells; the cells of the lower layers are irregular in shape—some are triangular, others polyhedral. This form of epithelial tissue is found in the nose, larynx, trachea, and larger bronchi of the respiratory tract and in certain parts of the male

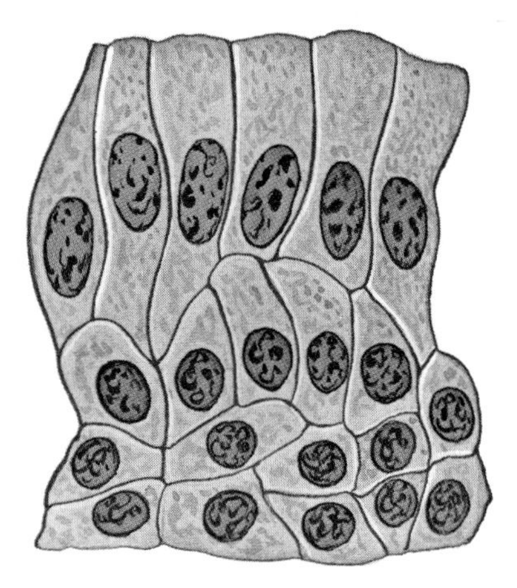

Fig. 4-2
Stratified columnar epithelium.

reproductive organs; in these locations they are supplied with cilia.

Glandular epithelium. In many places in the alimentary canal the columnar epithelium dips down into the underlying tissue, forming little narrow depressions. These depressions constitute the various glands of the stomach and intestine (Fig. 19-6). The mouth of the gland is formed by columnar cells such as those that form the surface layer, but lower down the cells may assume a more cuboidal shape. These cuboidal cells frequently contain granules that are destined to form the ingredients of the secretions made by these glands. The materials manufactured by the gland cells are poured on either a free surface or into the blood.

Functions of epithelial tissue. ■ The functions of epithelial tissue include protection, reception, and secretion.

Protection. As already indicated, epithelial tissue serves as a protection for the more vulnerable structures lying beneath it. How true this is for the hornlike cutaneous epithelium is a matter of almost daily experience.

Epithelial tissue also forms a barrier against material seeking random entrance into the body. Only through a layer of epithelial cells of the lungs, skin, and alimentary canal can a substance gain entrance into the body. Were it not for the great selectivity exercised during the process of absorption by the epithelial cells, life would be impossible. And the way out of the body (excretions via skin, lungs, kidneys, and alimentary canal) is also guarded by epithelial cells.

Reception. In nearly all sense organs (e.g., the retina of the eye) the reception of the stimulus is a function of epithelial cells; these are generally called sensory epithelial or *neuroepithelial* cells.

Secretion. The secretion of most fluids (such as sweat, saliva, and tears) is performed by tubular structures, known as glands, the walls of which are constructed of highly specialized glandular epithelial cells.

■ Connective or supporting tissue

The most widely distributed of all the tissues is one that performs an almost entirely passive part in our body, the connective or supporting tissue.

Forms of connective tissue. ■ Connective tissue includes many diverse structures; the more important forms are elastic, areolar, and adipose tissue, bone, dentin, and cartilage (gristle). Some authors speak of blood as a special tissue; others include it with the connective tissues.

The outstanding structural characteristic of all forms of this tissue is the small number of cells embedded in an exceedingly large amount of intercellular material. This material is manufactured by the cells, and in most types of connective tissue its maintenance appears to be the main, if not the sole, function of the cells. The intercellular material consists of a structureless ground substance, or matrix, in which are deposited various structures that differ largely from one type of tissue to another. The intercellular material bestows on the tissue a certain degree

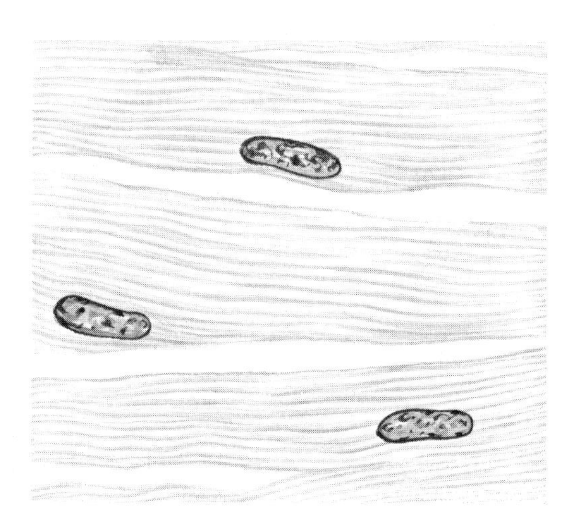

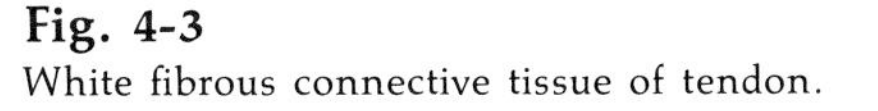

Fig. 4-3
White fibrous connective tissue of tendon.

of firmness. Some connective tissues are pliable, some are distensible, and some are hard and rigid. In all cases it is the intercellular material that enables this tissue to serve the mechanical functions of binding together and of supporting the softer structures of the body. In contrast with muscle and nerve tissues, connective tissue is characterized by its continued ability to multiply. This is illustrated by the healing of a broken bone and by the forming of connective tissue in the healing of a wound (scar tissue).

Elastic tissues. The principal components of elastic tissue are collagen, a fibrous protein, and elastic fibers. The fibrous protein molecules of collagen are linked by derivatives of the amino acid lysine that are present in adjacent polypeptide chains. Collagen provides considerable tensile strength, whereas elastic fibers give the tissue distensibility. White fibrous connective tissue of tendon is illustrated in Fig. 4-3. Elastic tissue occurs in skin, tendon, ligaments, blood vessels, membranes, and other structures. Elastic fibers are often cylindrical and woven in and out among collagen fibers. The components of elastic tissue are synthesized by cells and transported into the intercellular space. In tendons and ligaments these cells of origin are fibroblasts, in arteries the smooth muscle cells produce elastic tissue components.

Elastic fibers consist of microfibrils of 110 A diameter embedded in an apparently amorphous material. The latter is a unusual protein known as *elastin* which contains a unique combination of amino acids. Elastin is synthesized as a precursor, tropoelastin, and, after being carried to the exterior of the cell, is converted to elastin. In this conversion, the enzyme lysyl oxidase brings about the cross-linkage of lysine residues in tropoelastin units. The product of the aggregation, or polymerization, of tropoelastin is elastin. Copper is an essential cofactor for the lysyl oxidase enzyme and several "so-called" collagen diseases are attributable to a copper deficiency. Elastin has a syncytial structure and is not organized into chains. This fused mass structure is the basis for the relative ease with which elastic tissue is deformed.

Areolar tissue. Areolar tissue is composed of a loose network of white and yellow fibers and a ground substance containing cellular elements. It is the most abundant of all the various kinds of connective tissue. It penetrates muscles, nerves, and glands and forms sheaths around them. It also occurs beneath the skin in the dermis (Fig. 26-1) and in the walls of the alimentary, respiratory, and urogenital tracts.

Adipose tissue. In adipose tissue, a modified form of areolar tissue, the cells become charged with fat droplets. Adipose tissue cells are illustrated in Fig. 4-4.

Cartilage. In cartilage the widely dispersed cells and groups of cells lie in small cavities (lacunae), which are found in a dense, resilient matrix. The matrix is chiefly composed of collagen. It is an avascular and nerveless tissue. At the articular surfaces of joints it covers the ends of the bones; because of its exceeding smoothness, it lessens the friction and wear always associated with the action of a joint. Cartilage (gristle)

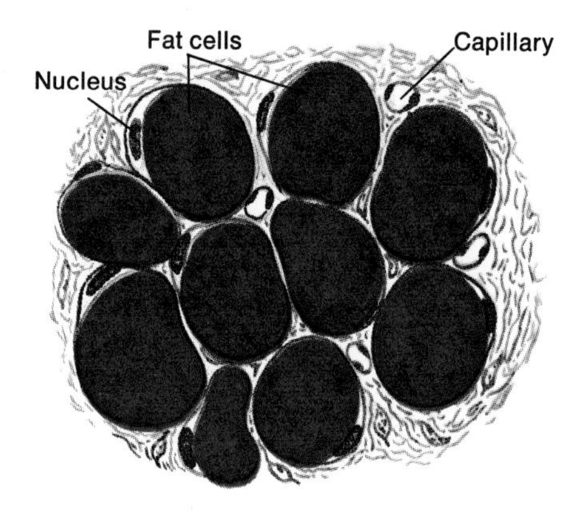

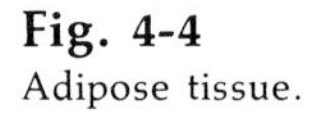
Fig. 4-4
Adipose tissue.

forms the flexible connection between the anterior end of the ribs and the sternum and the rings of the respiratory tree (Fig. 17-5). The intervertebral discs of the spinal column are formed of cartilage, as are the external ear and the distal portion of the nose.

Bone. Bone is the most rigid of all connective tissues due to the large amount of mineral matter deposited in the organic matrix (Fig. 24-3). The structure and formation of bone tissue will be dealt with in Chapter 24.

Functions of connective tissue. ■ The functions of connective tissue are largely passive but from a mechanical point of view indispensable. The most important of these are the following:

1. To give support to the body as a whole (especially performed by bones and cartilages)
2. To connect two distinct organs, as when tendons bind muscles to bones
3. To bind together the various parts of an organ (one of the functions of areolar tissue)
4. To envelop certain organs, for example, the capsule surrounding the kidney
5. To protect delicate organs mechanically, for example, the adipose tissue encasing the eyeball
6. To serve as a storehouse for fat

Other tissues will be discussed in subsequent chapters: nerve in Chapter 5, muscle in Chapter 6, and blood in Chapter 12.

■ LOCAL AND COMMUNAL VALUE OF TISSUES

From this study it is apparent that connective tissue is primarily only of local value. In contrast, the activity of glandular (epithelial) tissue is generally of little, if any, direct value to the tissue itself; its function seems to be performed for the good of the whole body; for example, the sweat glands in the skin are not directly benefited by the secretion of sweat, but, under certain external conditions, it makes possible the continued existence of the whole body. The same is true for muscle and nerve tissues; they serve not themselves but all the various structures constituting the body.

READINGS

Bevelander, G.: Essentials of histology, ed. 6, St. Louis, 1970, The C. V. Mosby Co.

Ross, R., and Bornstein, P.: Elastic fibers in the body, Sci. Am. **224**(6):44-52, June 1971.

Verzar, F.: The aging of collagen, Sci. Am. **208**(4):104-114, Apr. 1963.

Anatomy
Bioelectric phenomena and the nerve impulse
Relation between the stimulus and the generation of a nerve impulse
Propagation of the nerve impulse
Intercellular transmission

NERVE PHYSIOLOGY—CONDUCTIVITY

Many details of the integrated operation of the entire nervous system are as yet unknown; however, in this chapter we will direct our attention to some of the more salient features that are fairly well understood. The physiological properties and function of nerves and nerve fibers, the peripheral extensions of the nervous system, will be described.

■ ANATOMY

Nerve trunk. ■ What is ordinarily spoken of as a nerve is usually an assembly of a large number of fibers, better called a nerve trunk (Fig. 5-1). This trunk is surrounded by a loose connective tissue covering, the *epineurium*. Bundles of individual nerve fibers within the trunk are encased in a relatively strong sheath of connective tissue known as the *perineurium*. Inside the bundles are individual nerve fibers, each surrounded by the network of connective tissue making up the *endoneurium*. Nerve trunks are supplied with a profusion of blood vessels. Small arteries and arterioles are present in the epineurium and perineurium and capillaries in the endoneurium. Most nerve trunks contain a variety of fibers in respect to diameter and to function; that is, *afferent* fibers carrying impulses to the spinal cord and *efferent* fibers carrying impulses from the spinal cord toward the effector organs, muscles, or glands.

Nerve fiber. ■ Nerve cells, or *neurons*, differ greatly from each other in size and outward appearance (Fig. 10-13). In general, the cell body, or *soma*, of a neuron (Fig. 5-2) gives rise to one or more many-branch processes known as dendrites, or *dendrons*, and to an *axon*. Usually the axon is of greater length than the dendron and is referred to as a nerve fiber. The individual nerve fiber is a protoplasmic extension of the nerve cell body; it acquires a rather thick, white covering, known as the *myelin* or *medullary sheath*, a short distance from the cell body and before leaving the spinal cord. This sheath, which varies in thickness in different nerve fibers, is composed of many concentric rings of nonliving lipid and protein formed by the Schwann cells. Such nerve fibers are termed medullated fibers, whereas certain very small-di-

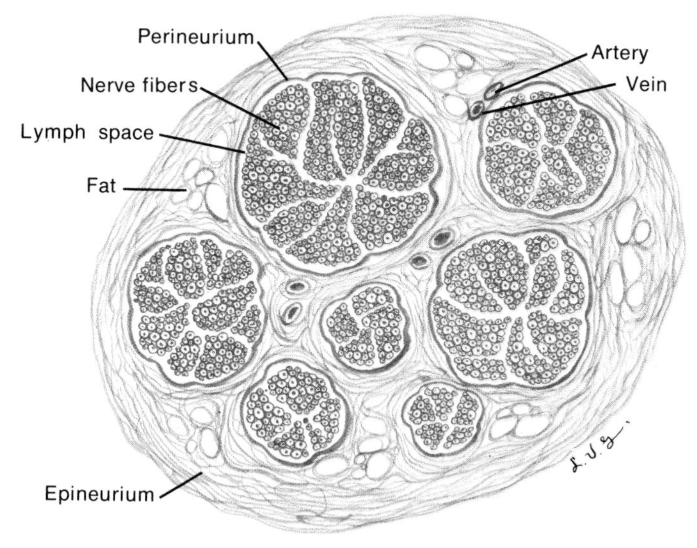

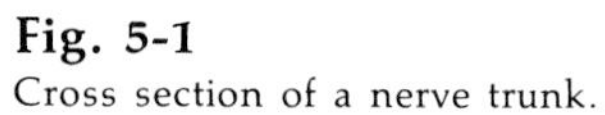
Fig. 5-1
Cross section of a nerve trunk.

ameter fibers—for example, the postganglionic fibers of the autonomic nervous system—being devoid of this sheath, are called nonmedullated fibers. However, the electron microscope has revealed that even these fibers are enveloped in a very thin layer of oriented, fat-containing molecules. The myelin sheath is not continuous but is interrupted periodically by constrictions, the *nodes of Ranvier.* The region between two nodes is termed the internode. All peripheral nerve fibers are separated from the endoneurium by a delicate covering known as the *neurilemma,* or sheath of Schwann. Presumably one Schwann cell provides the sheath for each internode region. The axon usually terminates in multiple small enlargements, the terminal boutons. Many points where the axon terminals of one neuron impinge intimately on the dendrites and soma of another neuron exist in the nervous system. Intercellular transmission of excitation between the individual neural elements occurs at these sites, the *synapses.*

Degeneration and regeneration. A nerve fiber, being an offshoot from a soma, depends on the

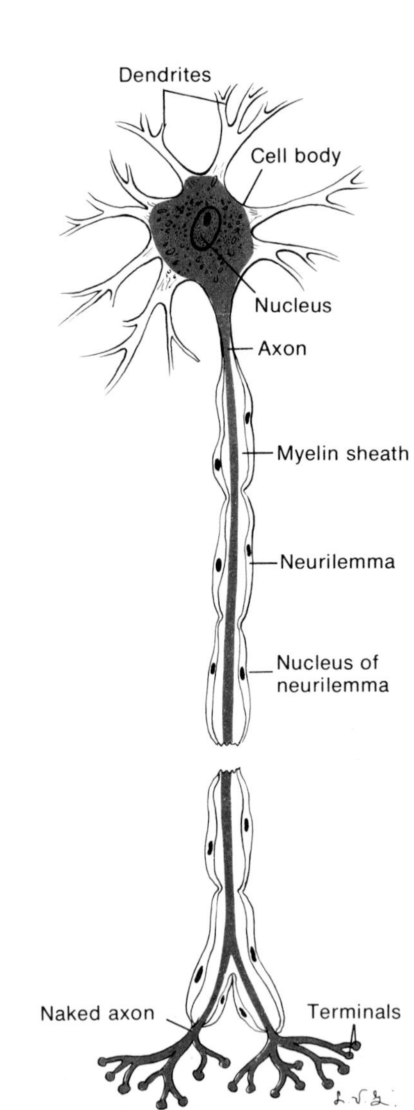

Fig. 5-2
Diagram of a neuron.

cell body for its nutrition. Evidence for flow of materials from the soma down the axon has been obtained. Hence, when a nerve trunk is cut, the central ends of the fibers remain normal, but the peripheral ends (the parts no longer attached to the cell body) undergo degeneration. This process, known as *Wallerian degeneration,* begins about three or four days after the nerve has been cut. After about seven days, regeneration sets in. Regeneration begins with increased activity and growth of the neurilemmal (Schwann) cells; this is followed by an outgrowth of living axons in the central stump, which follow the already laid down neurilemmal cells and thus reach, whenever possible, the proper end organ. The myelin sheath is the last element to be reformed. Since axons in the brain and spinal cord lack neurilemmal sheaths, no regeneration takes place there.

■ BIOELECTRIC PHENOMENA AND THE NERVE IMPULSE

In performing its function as a carrier of messages, the nerve generates at the point of stimulation a disturbance known as a nerve impulse and propagates this impulse along its length. The nerve therefore has the properties of irritability (excitability) and conductivity. Impulses are always propagated to another protoplasmic structure, which may be a muscle, a gland, or a neuron. Activity of a nerve is associated with a number of physical and chemical changes, none of which is directly visible. Electrical changes were the first to be detected and are, no doubt, among the most informative. Excitation of any piece of protoplasm is accompanied by disturbances in its resting electrical state, the transmembrane potential. Except in animals equipped with special electric organs (e.g., the electric torpedo and eel), the electrical changes are minute, and very sensitive apparatus is necessary to detect them.

Resting membrane potential. ■ By use of appropriate apparatus, such as a cathode ray oscilloscope, a number of observations regarding the electrical state of tissues can be made. When the leadoff electrodes of the oscilloscope (Figs. 5-3 and 5-5) are placed on the uninjured surface of a resting nerve or muscle, no difference in potential is revealed. It may therefore be concluded that *all points on the surface of an uninjured resting tissue are isopotential* (equal potential). If, however, the electrodes are placed on a severed nerve or muscle in such a manner that one electrode is on the injured surface while the other is on the uninjured surface, the oscilloscope reveals a difference of potential between these two points; this is known as the injury potential. Furthermore, *the injured part of the protoplasmic structure is electrically negative toward the uninjured part.* In freshly excised frog sciatic nerve the value of the injury potential is about 30 mv.* Over a period of a few hours this potential gradually declines and disappears. Hence in the resting state intact nerve and muscle fibers exhibit a steady difference of electrical potential across their membranes, the exterior being positive relative to the interior (Fig. 5-3). The cause of this polarization of the cells is to be found in the chemistry of the fiber and the extracellular fluids with which they are surrounded.

The origin of membrane potentials has been discussed in Chapter 3 and therefore only a brief recapitulation need be offered here. We should recollect that the positive ions of the extracellular fluid are principally sodium and potassium, whereas the major negative ion is chloride. Each of these electrolytes can diffuse across the membrane. Within the cells are negatively charged anions that cannot leave the cell by diffusion because of their large size. In consequence of this favorable electrical gradient, the cations, sodium and potassium, are attracted into the cell, whereas in large measure the chloride anions are

*The injury potential is much smaller than the true transmembrane potential. Injury provides a low-resistance shunt between outside and inside, and the voltage drop (Ohm's law) is reduced.

excluded. Because in the resting state the concentration of potassium ions is much greater inside than it is outside (the reverse is true for the sodium ions), it is apparent that those sodium ions that do enter are actively extruded. This continuous expulsion is accomplished by the so-called *sodium pump*. In a sense the resting membrane is therefore relatively impermeable to sodium ions; the sodium permeability is about $^{1}/_{100}$ that of potassium; a Donnan distribution exists. It should be noted that as a result of the activity of the sodium pump there exists a strong chemical as well as electrical gradient for the entry of sodium into the cells. The importance of this electrochemical force will be more apparent in the succeeding discussion of the action potential. Accumulation of potassium ions within the cell due to the indiffusible anions creates a chemical concentration gradient favoring the outflow of potassium ions. Consequently, an electrical potential difference develops across the membrane, the magnitude of which, at the equilibrium between electrical and chemical concentration gradients, is the measurable resting potential of the nerve or muscle membrane.

Action potential. ■ Resting potentials of excitable tissues such as nerve and muscle vary from species to species and cell type to cell type, in general falling in the range between 60 and 100 mv. For example, the resting potential of a frog axon is about 71 mv and that of a cortical neuron in the cat brain about 60 mv, but a value near 100 mv is attained in rat skeletal muscle fibers. This variability reflects the differences in tissue ionic concentrations and membrane properties that exist between species and between cells.

An adequate stimulus applied to the tissue reduces the electrical potential difference across the membrane *(depolarization)* and evokes an *action potential* and self-regenerative current flow in the membrane, the action current. To illustrate this point, place lead-off electrodes upon a normal nerve, as shown in Fig. 5-3. Points *b* and *c* are both positively charged (Fig. 5-3, *1*) with respect

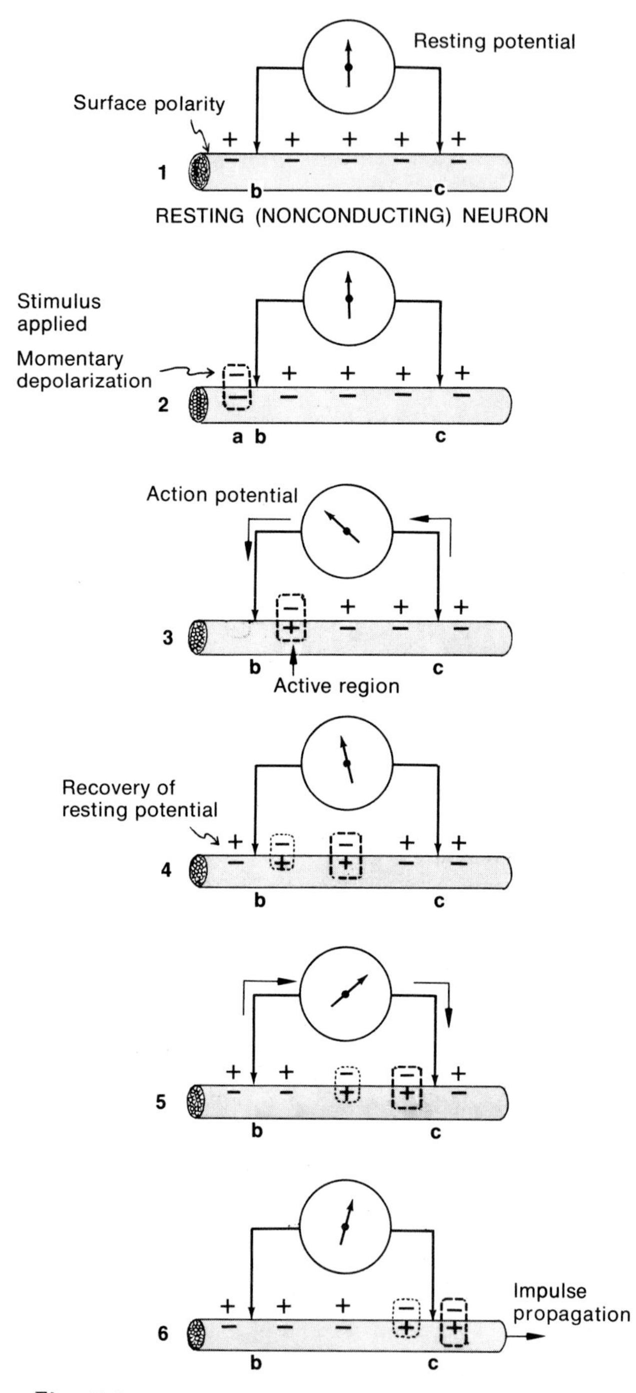

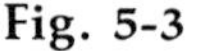

Fig. 5-3
Diagram illustrating resting polarization, **1**; stimulation, **2**; action currents and action potential, **3** to **6**, in nerve fiber.

to the interior of the nerve and are therefore isopotential. The stimulus applied at *a* (Fig. 5-3, *2*) changes the specific permeability of the cell membrane and thereby causes a depolarization due to a reshuffling of the cations and anions. This change in specific permeability and subsequent depolarization spreads along the nerve fiber and constitutes the nerve impulse. The oscillograph reveals a current flow between *c* and *b* (Fig. 5-3, *3*) and also that *b*, the active part of the nerve, is negative toward the inactive part, *c*. As the nerve activity leaves *b*, the original resting permeability is restored, and *b* begins to repolarize (Fig. 5-3, *4*). On reaching *c* (Fig. 5-3, *5*), the nerve activity causes *c* to become negative toward *b*. It must be emphasized that the active region of the protoplasmic *surface* is electrically negative toward the inactive part. These electrical changes, as revealed by the oscillograph (Fig. 5-3, *3* and *5*), constitute a diphasic action potential.

We emphasized that the active region of the membrane surface becomes electrically negative in respect to the inactive surface. The explanation for this reversal of the membrane polarity during activity is provided by the observation that the nerve or muscle membrane becomes highly and *specifically permeable to sodium ions* in response to adequate stimulation. Because of the electrical and chemical concentration gradient for sodium (about ten times greater outside than inside), some of these ions flow inward, thereby reducing the internal negativity. Such reduction in the membrane potential further increases the sodium permeability to many hundredfold times the resting value, and a self-regenerating chain reaction ensues. A deficit of positively charged ions at the active site leaves the membrane *surface* negatively charged with respect to the interior.* It is apparent, then, that

*The actual quantity of sodium ions that enters the nerve or muscle fiber during an action potential is quite small, 1 to 10×10^{-12} moles/cm^2.

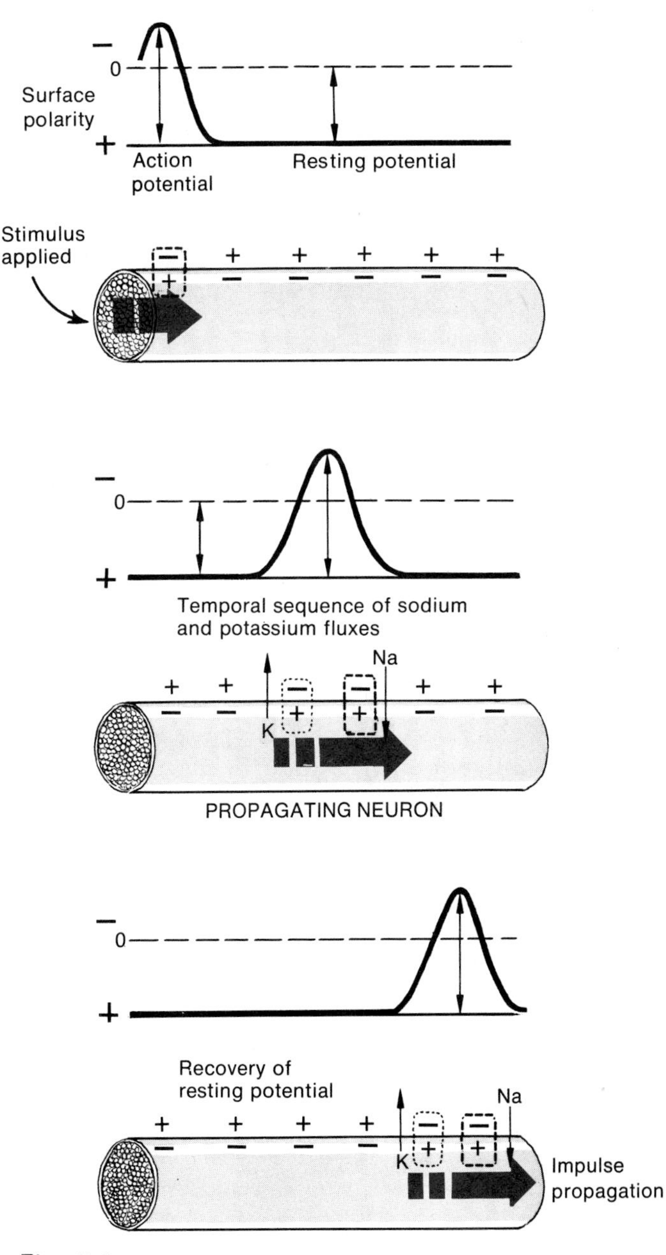

Fig. 5-4
Diagram of a monophasic action potential showing overshoot above zero and temporal sequence of sodium ion influx and potassium ion efflux during propagation of an impulse.

activity not only results in depolarization of the resting membrane but also actually induces a *reversal of the polarity* (Fig. 5-4) and that the action potential exhibits a spike or overshoot; that is, it is larger in magnitude than the resting potential. The measured action potential (from resting level to peak) in a muscle fiber may be as much as 110 to 120 mv. In theory the action potential spike should reach an overshoot value of 59 mv (action potential of 159 mv) as predicted by the Nernst equation from the Na ion concentration ratio:

$$\begin{aligned} E_{mv} &= -59 \log [Na_i]/[Na_o] \\ &= -59 \log 15/150 \\ &= 59 \text{ mv} \end{aligned}$$

Actually, a potential reversal of this magnitude is not attained, since the marked increase in Na ion permeability is of very short duration and reverts to its normal level before the peak of the action potential occurs. The normal membrane polarity, the resting potential, is restored *(repolarization)* by a slightly delayed efflux of potassium ions. This entire sequence of events at the active site occurs within a few milliseconds (Fig. 6-6). During a relatively long period of inactivity the sodium pump restores the original sodium-potassium distribution of the resting membrane. Although experiments have confirmed the theory of the action potential already outlined, little knowledge exists concerning the molecular mechanism that underlies the all-important changes in permeability of the membrane.

To be of value as a means of relaying messages, the change in permeability, depolarization, and resulting action potential must be propagated along the nerve fiber. The evidence for such propagation is obtained by recording diphasic action potentials. Once the stimulus has induced an adequate depolarization at the site of stimulation, the action potential is carried along the fiber quite independent of the original stimulus. This is accomplished by a local current flowing between the active point and immediately adjacent points on the membrane that excites the region just ahead of the impulse and releases local energy resources capable of renewing the signal. It is important not to confuse the nerve or muscle action potential with the original stimulus signal *(stimulus artifact)* when electrical stimuli are employed. The latter travels along the tissue surface at the velocity of light; the former travels much more slowly because of the self-regenerative membrane processes just described. Thus the action potential progresses along the nerve fiber in the form of a wave of activity that constitutes the nerve impulse. These phenomena have been found to be basically similar in all types of nerve fibers, and therefore it is highly probable that the impulses in various fibers are of the same general nature.

Recording the action potential. The brief duration and minute amplitude of action potentials are not faithfully reproduced by mechanical recording apparatus. It is customary, therefore, to amplify and then to display action potentials on an oscilloscope (Fig. 5-5), which can record potentials lasting only a few millionths of a second. A cathode ray tube (CRT), which is similar to the cathode ray tube familiar to most people as the picture tube in a television set, is the principal component of the instrument. In the neck of the cathode ray tube there is an electron gun that projects a concentrated high-velocity beam of electrons toward the phosphor-coated face of the tube. This thin coating glows (fluoresces) under the influence of the electron bombardment. Horizontal movement of the beam is controlled by the application of electrical potentials to the deflection plates through the sweep circuit; a positive charge draws the beam toward the plate, whereas a negative charge will electrostatically repel the beam. In operation the beam is made to move at a slow, measurable velocity across the face of the tube from left to right and return from right to left almost instantaneously. The glow of the moving spot describes what appears to be a horizontal line or trace across the tube face. Application of an electrical

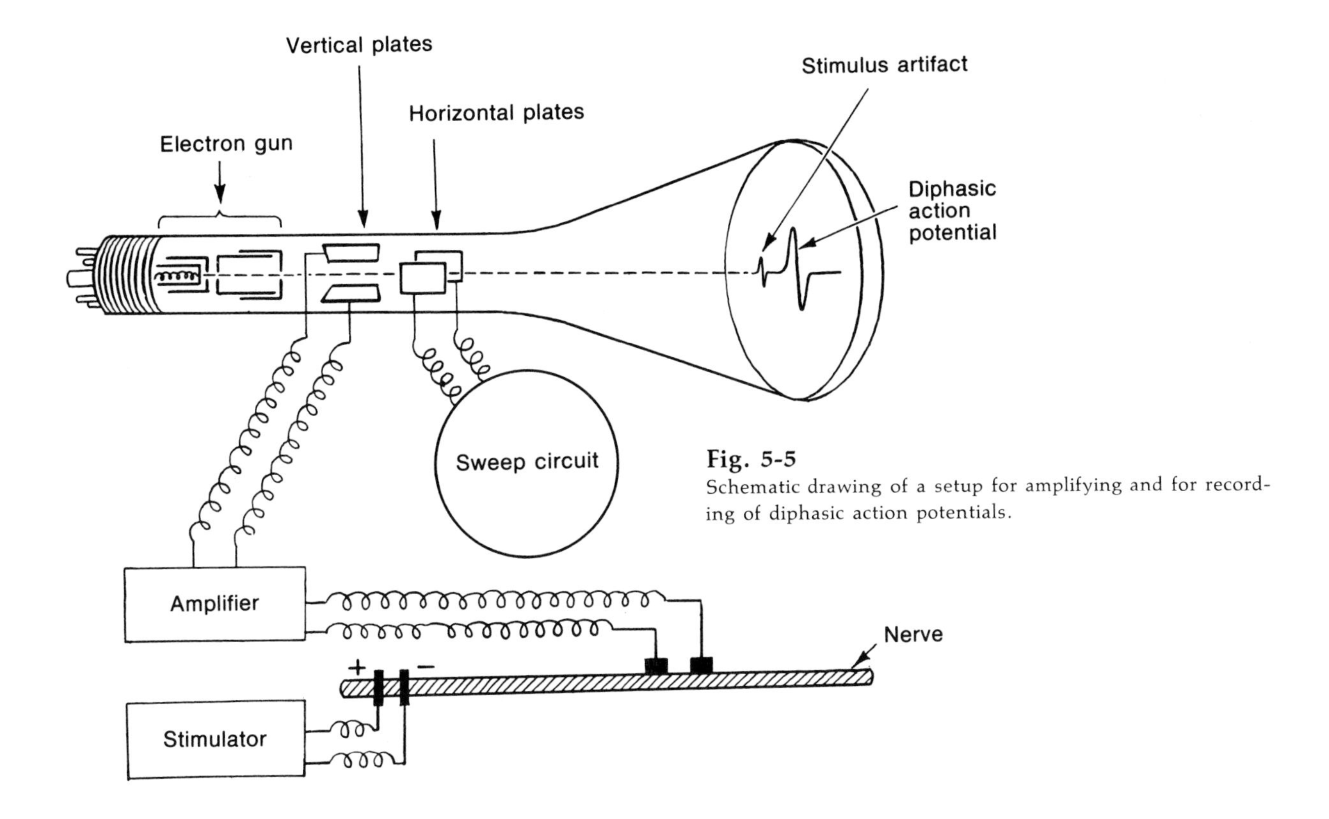

Fig. 5-5
Schematic drawing of a setup for amplifying and for recording of diphasic action potentials.

potential to the vertical plates during the passage of the trace from left to right causes the beam to move either up or down.

When the nerve in Fig. 5-5 is electrically stimulated, the rapidly conducted stimulus artifact appears on the scope face as the first deviation of the beam. Due to its much slower velocity the action potential reaches the recording electrodes and is displayed on the scope after an interval during which the beam has returned to its original base line. The presence of the action potential beneath the first electrode produces an upward deflection indicating surface negativity of the tissue. Negativity beneath the second electrode causes the beam to move downward. Together the upward and downward deflections make up the diphasic action potential record. The tissue beneath the second electrode often is crushed deliberately so that the action potential will be recorded only by the first electrode. Such a single deflection record is called a monophasic action potential (Fig. 5-4). Monophasic action potentials are less complicated and more accurate representations of the activity at a given site, but they are not evidence for the propagation of the impulse because activity is recorded beneath a single electrode.

■ RELATION BETWEEN THE STIMULUS AND THE GENERATION OF A NERVE IMPULSE

Electrical stimulation. ■ Many forms of externally applied stimuli (such as thermal, mechanical, and chemical) may be employed to cause excitation of nerve or muscle tissue. However, because it is easily applied, is readily controlled as to strength and duration, closely simulates the physiological process of excitation, and causes minimal damage to the tissue, an electric current is by far the most commonly used form of stimulation. Electronic stimulators are instruments of precision and versatility. Their use derives from the need to regulate more accurately the three fundamental characteristics of a stimulus—*strength* (intensity), *duration*, and *rate of change of intensity*.

Characteristics of the stimulus. ■ As in all stimulus-response phenomena, the response of a nerve to a stimulus is determined by the characteristics of the stimulus and by the condition of the tissue.

Intensity; law of all-or-none. It is a common laboratory observation that stimulation of a sciatic nerve trunk of a frog by an electrical shock of very low intensity produces no propagated action potential but that by increasing the intensity of individual shocks a stimulus strength which just elicits a nerve impulse can be discovered. This intensity of stimulus is called the *minimal, liminal,* or *threshold stimulus; it is a measure of the irritability of the tissue.* A stimulus below the liminal in strength is termed a *subliminal stimulus.* As the intensity of the shocks is gradually increased from threshold value, the action-potential amplitude increases. At a particular value of stimulus strength the amplitude of the response becomes fixed; it no longer increases with greater intensities of stimulation. This value is called the *maximal stimulus.* Intensities greater than maximal are termed *supramaximal.*

The greater response following stronger stimulation can be attributed to a difference in the degree of irritability of the various fibers constituting the nerve trunk. Feeble stimulation excites only the nerve fibers that have a lower threshold of irritability. Individual nerve fibers follow the principle of the *law of all-or-none;* that is, any stimulus of threshold, or greater, intensity always produces a nerve impulse of constant amplitude, but one appropriate to the existing environmental state of the nerve. The absolute amplitude of the response and the threshold stimulus level can be changed by altering the environment, for example, by reducing the temperature. Once the threshold level is achieved, the action potential develops independently because the energy required for the nerve impulse is contained in the axon, and the stimulus serves only to trigger the response. A maximal stimulus applied to a nerve trunk excites all of the nerve fibers, and the response represents the summed activity of all.

Influence of duration. To be effective, that is, to evoke a response, a stimulus must act for a certain length of time; the minimal time required for this is called its *excitation time.* Within limits, the stronger the stimulus, the shorter is its excitation time. The relations between strength and duration of a threshold stimulus are illustrated in Fig. 5-6, where the curve denotes combinations of stimulus intensity and duration that just produce a response. A strong stimulus, *a,* requires but a brief length of time; for progressively weaker stimuli, *b, c,* and *d,* the duration gradually increases. A stimulus of strength *e,* known as the *rheobase* (*rheos,* to flow), must be applied for an indefinite length of time; a stimulus of lesser intensity than this is ineffective no

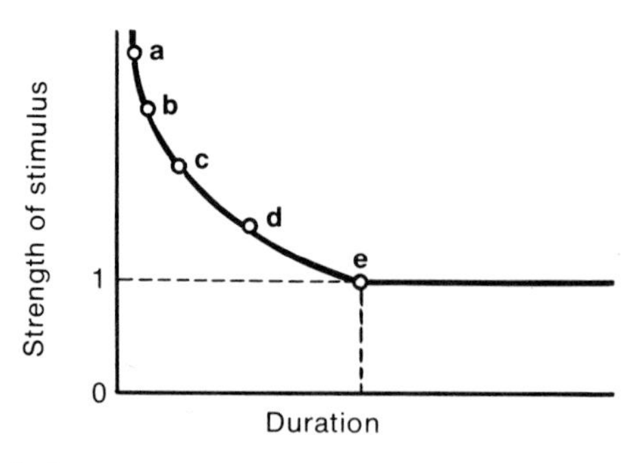

Fig. 5-6
Strength-duration curve illustrating relationship between strength and duration of stimuli. Horizontal line, **1**, indicates minimal threshold.

matter how long it acts. At the other extreme, if the time allowed is too brief, no matter how intense the stimulus, no nerve impulse is generated. This is seen in the passage of high-frequency alternating currents through the body without causing any muscular contraction.

Rate of change; accommodation. Suppose that a nerve is stimulated by a constant (galvanic), rather than interrupted (faradic), current of uniform and medium intensity for one minute or more; a single response occurs at the moment of application of the current. During the continued flow of current no additional impulses are generated. This is not due to fatigue, since the nerve will respond for hours to faradic shocks of the same intensity. It is said that the *nerve accommodates itself to the presence of constantly flowing current.* Accommodation represents a stabilization of the resting membrane potential, an increase in the threshold for excitation probably associated with a further reduction of Na ion permeability. If the intensity of the constant current flowing through the nerve is suddenly increased or decreased, stimulation occurs. In general terms it may be stated that *a stimulus becomes adequate only if the rate of change exceeds a certain lower limit. Accommodation to all forms of stimuli seems to be a universal property of protoplasm.*

Cable properties and excitation. ■ To understand the stimulatory effects of a current applied to a tissue, it is useful to examine certain properties of excitable membranes. Resting nerve and muscle fibers are in many respects like insulated electrical cables; they are long and cylindrical, and both the membrane and axoplasm are resistive; that is, they impede the flow of current. Because a potential difference exists across the membrane, it also behaves as a capacitor. Capacitance, the ability to separate unlike charges, is also a property of cables.

In Fig. 5-5 the stimulating electrodes are designated as + and −; the former is properly called the *anode* (anion collector) and the latter, the *cathode* (cation collector). When an electrical potential difference is applied to the two electrodes in contact with the nerve, current flows between them. Much of this current flows directly from anode to cathode along the surface of the nerve because of the low resistance of the external fluid surrounding the tissue. A portion of the current, however, flows inward through the membrane at the anode, longitudinally through the axoplasm, and outward through the membrane at the cathode. Current flow through a resistance produces a potential difference between the two ends of the resistive material, the end nearest the anode being more positive than the end nearest the cathode. In the case of the nerve membrane, the current moving inward through the membrane at the anode makes the membrane more positive on the outside; the membrane becomes *hyperpolarized.* The current flowing outward through the membrane at the cathode reduces the positivity of the surface at that point; that is, it *hypopolarizes* or *depolarizes* the membrane. It is the reduction of the membrane potential to a critical (threshold) level that effects the membrane permeability changes, which lead to the sudden increase in Na ion influx and the action potential. Since outward flow of current depolarizes the membrane, it is *cathodal* stimulation that normally triggers the response.

The capacitive properties of the membrane do not alter the direction of current flow through the membrane, but because they slow down the rate of flow, they influence the time course and spatial distribution of the membrane potential changes in an exponential fashion. The alterations in potential of adjacent regions of the membrane and between the electodes are smaller than those occurring immediately beneath the electrodes. Membrane capacities determine the shape of the strength-duration curve. A stimulus of low intensity, and consequently lesser current, will require a longer time to reverse the charge on the membrane capacities in the region of the cathode than will one of greater intensity.

Local excitatory state; summation. ■ The immediate result of the stimulation of a nerve fiber is the setting up of a chemical and physical change in the membrane at the site of stimulation, which eventually leads to the formation of a nerve impulse; this change is sometimes spoken of as the local excitatory state (l.e.s.). Although a sufficiently strong stimulus results in a local excitatory state capable of initiating the series of events culminating in a nerve impulse, the local excitatory state following a subliminal (or subminimal) stimulus is too weak to have this result. Oscilloscope recordings made at the site of these subthreshold stimuli reveal minute membrane potential depolarizations lasting for a few milliseconds, which are not much greater than one tenth of the magnitude of the action potential and are restricted to a region only a few millimeters to either side of the cathode; that is, they are not propagated. On the other hand, should the local excitatory state initiate a nerve impulse, its own existence is obscured by the rapidly rising propagated action potential. The local excitatory state undoubtedly reflects the linkage of sodium permeability and membrane potential changes. If the depolarization is small, the increase in sodium permeability is likewise small, and the membrane is readily and rapidly stabilized by a net outward flow of potassium ions. Thus excitable tissues have an inherent ability to reestablish their normal equilibrium.

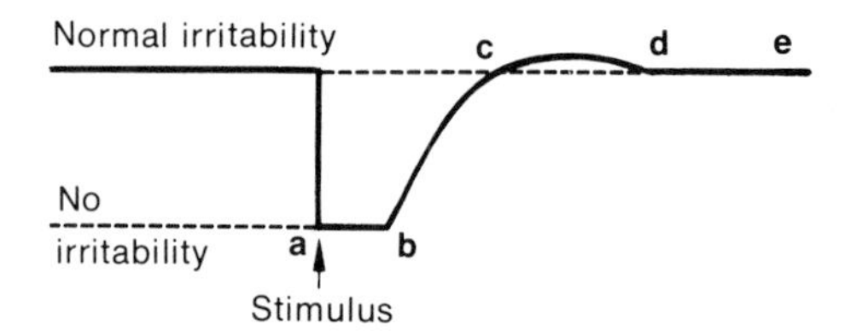

Fig. 5-7
Diagram showing relative irritability of a stimulated nerve. **a**, Time of stimulation; **ab**, absolute refractory period; **bc**, relative refractory period; **cd**, period of supernormal excitability; **de**, recovery of normal irritability.

The local excitatory state subsides within a few milliseconds. If, before the expiration of this time, a second or a series of subliminal stimuli are applied, the lingering effect of the first stimulus is reinforced, and a nerve impulse may be generated. This reinforcement is known as the *summation of subliminal stimuli* or *temporal summation.* Since a state of dynamic equilibrium is the basis of irritability, it may be said that every normal excitatory stimulus, adequate or inadequate, always increases the excitability of protoplasm.

Frequency of stimulation; the refractory period. ■ The onset of an action potential in a nerve fiber is associated with a great reduction in excitability—the *refractory period.* If a second adequate stimulus is sent into a nerve within, say, 0.4 msec after the application of the first stimulus, it is ineffective; only one action potential is generated. The duration of the refractory period varies inversely with the diameter of the nerve fiber. As shown in Fig. 5-7, immediately after stimulation (at *a*) excitability is completely lost. No strength of stimulus is now able to evoke a response; this is the *absolute refractory period.* Soon, however, the irritability is gradually regained (from *b* to *c*). During this *relatively refractory period* a stimulus of greater than liminal intensity is required to generate an impulse. In recovery, from *c* to *d*, the degree of excitability may go a little beyond normal *(hyperexcitability)* for a short period of time. Since the refractory period for the membrane follows the moving action potential, it limits the number of impulses a nerve fiber can carry in a unit of time; the number varies with the duration of the refractory period. For example, if the total refractory period were 1 msec in duration, the maximal frequency of nerve impulses in response to liminal stimuli would be 1,000/sec.

The existence of the refractory period depends on the fact that the irritability of a nerve fiber lies in its polarized state and the generation of an impulse in its depolarization. If another stimulus is presented during the depolarized state of the fiber, further depolarization is impossible, and therefore no impulse can be formed. Since the threshold stimulus is followed by an absolute refractory period, the reversal of the membrane potential is complete, and therefore a stronger stimulus can have no greater effect than that produced by the threshold stimulus; this establishes a basis for the all-or-none principle.

■ PROPAGATION OF THE NERVE IMPULSE

Local current. ■ The disturbance in equilibrium created in a nerve fiber by adequate stimulation is self-propagated along the fiber. We use the term *propagation*, rather than conduction, to emphasize the fact that the nerve impulse is not identical with the stimulus artifact but is an independent phenomenon. Once a specific area of the membrane has been critically depolarized by the stimulus, a pronounced inward flow of Na ions (current) ensues (Figs. 5-3 and 5-4), in effect reversing the flow caused by the stimulating current. The area undergoing this change, the active region, is sometimes spoken of as the "sink" for the current. Although the active area is now depolarized, the adjacent regions still are normally polarized, and current is drawn from these "sources" toward the "sink." This current flows inward through the membrane at the active sites and spreads longitudinally through the axoplasm. The *local circuit* is completed when this current flows outward through the membrane in the adjacent inactive regions. Because the outward flow depolarizes the inactive regions, exactly as the stimulator current at the cathode had done previously for the active region, these now become "sinks" for current from even more distant regions. Thus, the nerve impulse is propagated as a wave of activity by local current flow producing successive depolarizations of the membrane immediately ahead of the active region. It is important to separate in our thinking the effects of a current provided by an external source, such as a stimulator, and those currents that are associated with the action potential.

Double propagation. ■ When the stimulation of a nerve occurs somewhere in its course (Fig. 5-8, *S*), the impulse travels in both directions, as can be verified by detecting the action potential at points *A* and *D*. Although two-way propagation is possible, in the animal body this seldom occurs, since normally a nerve fiber is stimulated only at its end. Stimulation that causes impulses

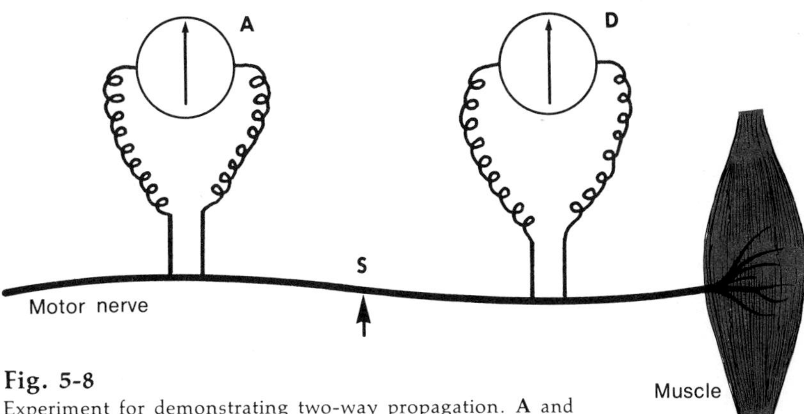

Fig. 5-8
Experiment for demonstrating two-way propagation. **A** and **D**, Galvanometers, electrodes of which are resting on the motor nerve; **S**, stimulating electrode position.

Table 5-1. Some properties of peripheral nerve fibers

Fiber type	*Diameter* (μ)	*Velocity* (*m/sec*)	*Action potential spike duration* (*msec*)	*Absolute refractory period* (*msec*)
A	22-1	120-5	0.5-0.4	1.0-0.4
B	3 or less	15-3	1.2	1.2
C	1.3-0.3	2.3-0.6	2	2

to be propagated in a direction opposite to that normally expected, for example, away from the muscle in a motor nerve, is known as *antidromic stimulation;* that stimulation which causes impulses to travel in the normal direction, for example, toward the muscle in a motor nerve, is termed *dromic stimulation.*

Isolated propagation. ■ A nerve trunk is composed of thousands of nerve fibers distributed to many organs. When an impulse is generated in a certain fiber which makes connections with a given effector, the impulse in passing down this fiber does not spread to the adjacent fibers in the nerve trunk and thereby activate other effectors. Without isolation the coordinated activity of structures depending upon nerve impulses would be an impossibility, e.g., the graded contraction of skeletal muscles and the tactile discrimination in the sense organs found in the skin. A nerve is comparable to an electrical cable composed of many wires in which some insulating material, in this case myelin, separates the individual wires and prevents an electrical current that is passing through any one wire from escaping to the others.

Peripheral nerve fiber types. ■ A peripheral nerve, as we have suggested already, consists of a multitude of individual fibers with a common function—that of carrying messages. Some, the afferent fibers, carry information about the organism's environment to the central nervous system, whereas others, the efferent fibers, relay coded orders from the central nervous system to muscles and glands. There is a continuous spectrum of fiber diameters, ranging from a few tenths to over 20 μ, in a typical nerve. However, by means of propagation velocity studies, as described in the following discussion, it is possible to classify nerve fibers into at least three major types known as A, B, and C fibers. Fibers of the A type are the largest in diameter and are myelinated afferent and efferent fibers. Subdivisions of this type, known as alpha, beta, and gamma fibers, have been demonstrated. The type B fibers are smaller myelinated efferent axons of the autonomic nerves. Unmyelinated fibers of very small diameter, both afferent and efferent in function, are classified as C fibers. Table 5-1 summarizes some of the properties of peripheral nerve fiber types.

Velocity of the nerve impulse. ■ The velocity of a nerve impulse can be determined by finding the velocity of the passage of the action potential along a stretch of nerve (Fig. 5-5). If a nerve trunk is used, the electrical response is termed the *compound action potential.* On the scope face both the application of the stimulus and the response of the tissue are recorded by the moving electron beam. Fig. 5-9 is a reproduction of such a record. By noting the time lapse between the stimulus artifact and the separate peaks of the compound action potential recording and by knowing the distance between the cathode and first recording electrode (Fig. 5-5), the rate of travel of the impulse can be calculated. Coarse medullated fibers in the sciatic nerve of a frog

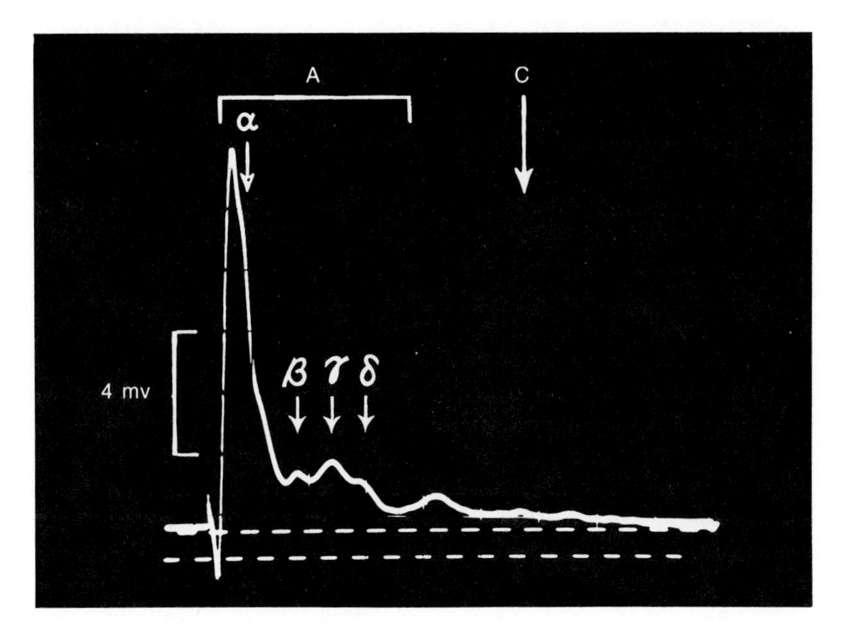

Fig. 5-9
Oscilloscope trace showing compound action potential from frog sciatic nerve. Letters indicate fiber types with different diameters and velocities. Stimulus artifact precedes response of fibers. Time scale is given in lower trace by square wave pulses at 1,000/sec.

conduct at a velocity from 30 to 43 m/sec; in thin fibers the velocity is about 16 m.*

Nerve impulse velocity usually varies directly as the diameter of the fiber. In mammalian myelinated nerve fibers the velocity generally increases by 6 m/sec for each micron increment in overall diameter. The velocity for a given fiber does not vary with the strength of the stimulus, but the larger diameter fibers have a lower stimulus threshold intensity. Despite the fact that the compound action potential peaks are of varying amplitude, with the larger diameter fibers exhibiting the largest values, there is no reason to believe that individual nerve fiber–action potential magnitudes vary with diameter. Indeed, individual action potentials are probably reasonably uniform in amplitude; the variation observed in the compound action potential is a consequence of recording from outside of the nerve trunk and in part due to the numbers of fibers of a particular type in the nerve. The magnitude of a compound action potential recorded with externally applied electrodes is even smaller than is the action potential obtained from a single fiber (Fig. 5-9). After reviewing the number of insulative layers present in a peripheral nerve trunk (Fig. 5-1), one can recognize what a limited amount of the total action current actually reaches the recording electrodes under these conditions.

Saltatory conduction. ■ The presence of a myelin sheath around most types of vertebrate nerve fibers has a pronounced effect upon propagation velocity. Since this material is a good insulator and is applied rather tightly around the axon in the internodal portions of the nerve fiber (Fig. 5-2), effective contact of the electrically ex-

*For comparison: The velocity of sound (at zero C) = 321 m/sec; the velocity of light = 186,000 miles/sec.

citable axon membrane with the exterior is restricted to the nodes of Ranvier. Hence local circuit current flow during activity appears to be restricted to the region of the nodes; it flows inward at the active node (sink) and outward at adjacent nodes (sources). Recalling the internodal distance, from 1 to 2 mm, and the role of local circuit current flow in excitation, one can appreciate how the activity in myelinated nerve fibers dances *(saltare)* from node to node in contrast to the continuous, wavelike progression in unmyelinated nerve. This form of impulse propagation is known as *saltatory conduction.* As a consequence of the current flow being restricted to nodal regions, the propagation velocity in myelinated fibers is many times greater than is that of unmyelinated fibers of equal diameter. Furthermore, saltatory conduction provides for an economy in the metabolic activity of recovery, that is, ion pumping, because only a fraction of the axon membrane presumably undergoes depolarization during activity.

Blocking of a nerve impulse. ■ In a nerve fiber, blocking may be caused by cooling a stretch of the nerve; on warming, the nerve activity is restored. The same results can be obtained by applying narcotics, such as chloroform, ether, and cocaine; the activity returns on removal of the drugs. In blocking a mixed nerve, the sensory fibers, especially those conveying the impressions of pain, are affected before motor fibers. Pressure applied to a nerve may also cause a block, as is seen in the falling "asleep" of a limb when its nerve is pressed upon for a sufficient length of time. The removal of the pressure may stimulate the nerve, producing, in the case of a "sleeping" limb, the prickling sensation.

Propagation without decrement. ■ When it is stated that a nerve fiber follows the all-or-none law, it must not be inferred that an impulse traversing a particular fiber is always of the same strength.* Let us enclose a stretch of nerve (Fig.

*The all-or-none law means that although the magnitude of the response to stimulation may be influenced by many environmental conditions, the strength of the stimulus does not have this effect. "All-or-none" does not mean "always-the-same."

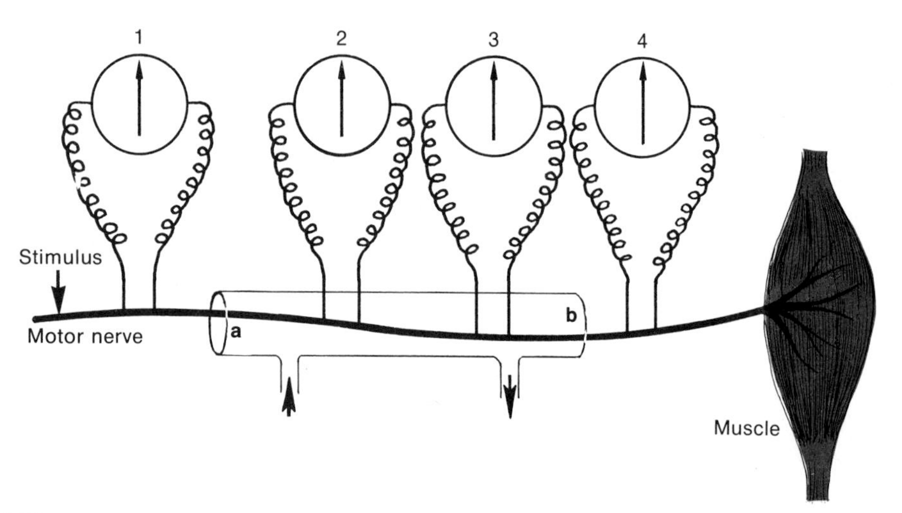

Fig. 5-10
Arrangement used to demonstrate propagation without decrement. Recording instruments **1** and **4** are placed on normal nerve; **2** and **3** are on nerve in region to be narcotized. Compare with Fig. 5-11.

5-10, *a* to *b*) in a glass tube and place upon the nerve the leads of four recording instruments *(1* to *4)*. On stimulating the nerve, all four instruments will show, in sequence, the same degree of potential difference; that is, the action potential in traveling down the nerve suffers no decrement (waning). If now the nerve inside the tube is subjected to the fumes of ether, the ability to propagate an impulse gradually decreases. If the nerve is again stimulated before irritability is totally lost, the action potential at *1* is of the usual strength, but that shown at *2* and *3* is greatly reduced. However, the size of potential at *3* is the same as that at *2*, showing that the action potential, although greatly reduced, is propagated without further decrement. When the weakened impulse arrives at *b*, it regains, as shown by *4*, its original magnitude. These effects are illustrated in Fig. 5-11. What was said concerning the size of the impulse is also true for its velocity. The lowest intensity of stimulus able to excite a maximal contraction of the muscle before the nerve is narcotized is also the strength necessary when the propagation amplitude and velocity between *a* and *b* have been reduced almost to the vanishing point. This affords additional proof of the all-or-none phenomenon and emphasizes the fact that the energy for the action potential is derived, not from the stimulus, but from sources within the nerve.

Fatigue. ■ The chemical changes that take place in an active nerve fiber are exceedingly small, and there is a speedy and almost complete recovery, since *under the usual conditions of stimulation a nerve fiber cannot be fatigued.* Indefatigability of a nerve may be demonstrated in the following manner. A sciatic nerve is stimulated at a certain point, and a stretch of nerve between this point and the muscle is cooled in order to block the impulses. The stimulation, at the rate of from 10 to 20/sec, is maintained for an hour or more. At the expiration of this time, without interrupting the stimulation, the nerve is allowed to warm up; when the proper temperature has been reached, the muscle contracts. This indefatigability has also been demonstrated by using curare to block the transmission of the impulse to the muscle and, after applying a stimulus for

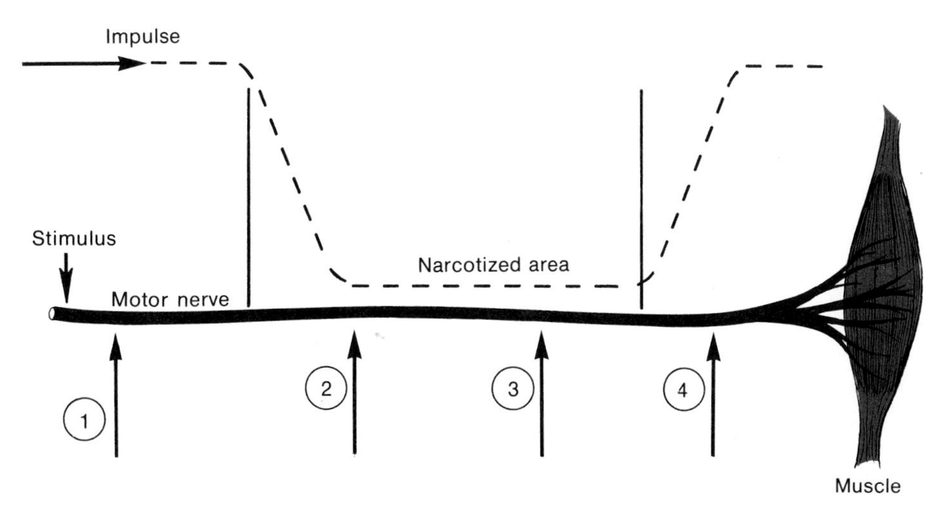

Fig. 5-11
Diagram illustrating propagation without decrement both in normal and in narcotized area. Dotted line indicates relative magnitude of action potential. Numerals correspond to those used in Fig. 5-10.

a sufficient length of time, by removing the block with an application of eserine.

Chemical and thermal changes. ■ Oxygen is necessary to maintain the activity of the nerve fiber, since in an atmosphere of nitrogen it gradually disappears, to be restored on the readmission of oxygen. Both at rest and during the activity of a nerve there is a consumption of oxygen, a production of CO_2, and the generation of heat. In contrast to muscle in which maximum activity may increase the metabolic rate 1,000 times, in nerve the increase is slightly less than twofold. That the energy expended during neural activity is not derived from the stimulus follows from the observation that an impulse which has been greatly weakened by passing through a partially narcotized stretch of nerve (Fig. 5-11) is immediately restored to its original strength on reaching the normal segment. And as the impulse is always propagated without decrement, it must be concluded that as an impulse travels along a nerve fiber it gathers energy from the nerve fiber itself. Adenosine triphosphate (ATP) is the immediate source of energy for the sodium pump.

It has been calculated that a single impulse causes an initial rise of temperature of 0.000001 C; this is followed by a prolonged recovery phase, as long as thirty minutes, in which up to thirty times as much heat is liberated. Hence the generation and propagation of the impulse is associated with a very limited energy transformation; it is during the restoration of the polarized state that most of the energy is required.

Nerve impulse and electrical current. ■ Although the nerve impulse is a bioelectric phenomenon, it must not be supposed that an impulse is similar to an electrical current passing through a wire. From our previous discussion, four notable differences between the propagation of an impulse by a nerve and that of conduction of an electrical current by a wire reveal themselves.

1. A wire is a passive conductor; a nerve fiber is an active agent.

2. The energy of an electrical current is derived directly from the battery or some other source of electricity outside the conducting wire. The energy of a nerve impulse is not obtained from the stimulus but from the nerve itself. A nerve cooled to below zero C is an electrical conductor but is incapable of propagating a nerve impulse.

3. As an electrical current passes along a stretch of wire, its potential gradually declines due to the resistance encountered.The strength of a nerve impulse is maintained at a constant level from an internal energy source. This is known as propagation without decrement.

4. Although an electrical impulse travels at the rate of 300,000,000 m/sec, the velocity of a nerve impulse in man is 100 m/sec or less.

■ INTERCELLULAR TRANSMISSION

Physical continuity of protoplasm between two nerve cells or between a nerve cell and an effector cell does not exist, but contiguity in the form of a synapse is present. As we have seen, the nerve impulse is a wave of activity propagated from one part to the neighboring part of the fiber. When this bioelectric disturbance reaches the end of the axon, how is the next structure (muscle, gland, or other neuron) influenced by it? Two methods of intercellular transmission are possible—electrical and chemical.

Electrical transmission. ■ Although chemical transmission is the only form of intercellular transmission in the mammalian nervous system for which there is adequate evidence, at selected sites in some vertebrate and in many invertebrate species, electrical synapses have been identified. Transmission at the junction between two excitable structures is accomplished by current flow across the plasma membranes and between the *presynaptic* and *postsynaptic* elements, much like that described for the local current flow along the axon. Effective depolarization of the postsynaptic cell by this current initiates a new, propagated action potential. Electrical synapses are structurally much different from typical chemical synapses; the membranes of cells at electrical synapses are generally in much closer proximity, and even in some instances the presynaptic and postsynaptic membranes are fused together. Such anatomical modifications reduce the intercellular resistance and help to assure an adequate flow of current for depolarization of the postsynaptic cell.

Chemical transmission. ■ At chemical synapses, the nerve impulse as an electrical phenomenon ceases when it reaches the end of the presynaptic axon. On arriving at the synapse (or neuromuscular junction), it brings about the release of a chemical compound *(neurohumor)*, which is known as a *transmitter;* this substance excites the postsynaptic cell.

A narrow cleft, several hundred Angstrom units wide, separates the presynaptic and postsynaptic elements. In the nerve endings of the presynaptic cell there are numerous mitochondria and *synaptic vesicles*. Presence of mitochondria suggests that these terminals are metabolically active. Synaptic vesicles, spheres about 300 to 500 A in diameter, are usually clustered close to the membrane. From available evidence it is concluded that these vesicles are stores of the chemical transmitter that were synthesized by the nerve. Nerve impulses in the presynaptic cell bring about the release into the synaptic cleft of the neurohumor contained in the vesicles. In turn, the transmitter diffuses across the cleft and increases the permeability of the postsynaptic cell membrane. The increased permeability is nonspecific, that is, not limited to sodium ions, but occurs for all small cations; consequently, the postsynaptic membrane becomes depolarized, and an action potential is initiated. Obviously, not all the vesicles are released by a single impulse, but *synaptic fatigue* may occur after continuous activity when the number of vesicles is seriously depleted. Two potent transmitters are *acetylcholine* and *norepinephrine*, although evidence indicates these are probably not the only neurohumors in the nervous system. Further discussion of transmitters will be postponed until we deal with skeletal muscle and the autonomic nervous system, especially in relation to the heart.

READINGS

Baker, P. F.: The nerve axon, Sci. Am. **214**(3):74-82, Mar. 1966.

DeRobertis, E.: Molecular biology of synaptic receptors, Science **171**:963-971, 1971.

Hille, B.: Ionic permeability changes in active axon membranes, Arch. Intern. Med. **129**:293-298, Feb. 1972.

Hodgkin, A. L.: The conduction of the nervous impulse, Springfield, Ill., 1964, Charles C Thomas, Publisher.

Katz, B.: Nerve, muscle and synapse, New York, 1966, McGraw-Hill Book Company, pp. 1-141.

Katz, B.: Quantal mechanism of neural transmitter release, Science **173**:123-126, 1971.

Nachmansohn, D.: Proteins in excitable membranes, Science **168**:1059-1066, 1970.

Tasaki, I.: Nerve excitation, Springfield, Ill., 1968, Charles C Thomas, Publisher, pp. 3-26.

6

Types of muscle tissue
Structural properties of skeletal muscle
Chemical properties of skeletal muscle
Biophysics of skeletal muscle contraction
Effects of muscular work and training
Visceral or smooth muscle
Cardiac muscle
Cilia

MUSCLE PHYSIOLOGY—CONTRACTILITY

Because muscles are essentially machines for converting chemical energy into mechanical work, they play a major part as responding organs in adjusting the body to environmental changes. Adjustments may include the purposeful movement of the whole body from one point in space to another or the movement of a limited part of the body in respect to the body itself or to the environment. The maintenance of posture against the effects of gravity is especially significant. In addition, vital processes (e.g., contraction of the heart, constriction of blood vessels, breathing, and peristalsis of the intestine) are dependent on muscular activity. These various functions are carried out by several types of muscles, but, despite their many characteristic differences, they possess in common the property of contractility. The importance of the body musculature is even more readily appreciated when it is learned that it constitutes 43% of the body weight, contains more than one third of all the body proteins, and contributes about one half of the metabolic activity of the resting body.

■ TYPES OF MUSCLE TISSUE

Muscles consist of thousands of fibers or cells, which are the structural units of a muscle (Figs. 6-1 and 14-8). There are three broad classes of muscle; they differ in structure (histologically), in location (anatomically), in function (physiologically), and in their manner of innervation (neurologically).

Smooth muscle. ■ Smooth (unstriated, visceral, or involuntary) muscle fibers are devoid of cross striations, although they do exhibit faint longitudinal striations. They possess a single, centrally located nucleus. Found in the walls of internal or visceral organs (generally hollow organs), these fibers are usually not subject to the will and receive their innervation via the autonomic nervous system.

Cardiac muscle. ■ Cardiac (or heart) muscle is involuntary and innervated by the autonomic nervous system. Its cells show both longitudinal and imperfect cross striations and possess a single nucleus.

Skeletal muscle. ■ Skeletal (striated or volun-

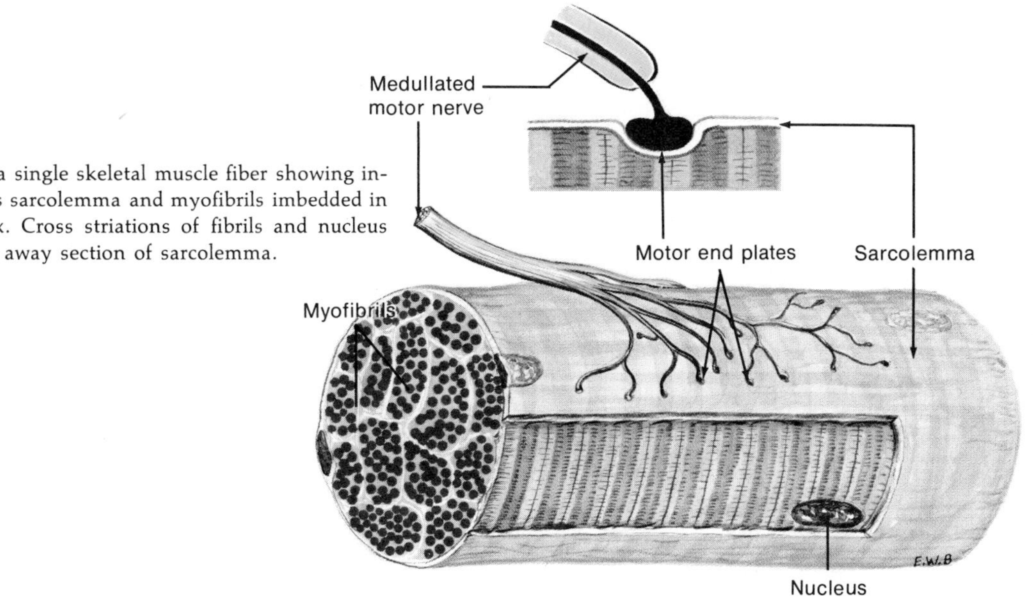

Fig. 6-1
A small portion of a single skeletal muscle fiber showing innervation as well as sarcolemma and myofibrils imbedded in sarcoplasmic matrix. Cross striations of fibrils and nucleus appear beneath cut away section of sarcolemma.

tary) muscle fibers are multinucleate and clearly display longitudinal and cross striations. These muscles in nearly all instances are attached to bones. Their innervation is derived from somatic nerves, and, to a large extent, they are under volitional control.

■ STRUCTURAL PROPERTIES OF SKELETAL MUSCLE

Tendons. ■ Nearly all striated muscles are attached to bones by means of tendons (Fig. 6-2). A tendon is composed of dense, white, fibrous (inelastic) connective tissue. These fibers, extending in the direction of the length of the tendon, are grouped into small bundles or fasciculi in which the fibers are held together by an interfibrillar cement substance. The fasciculi are grouped into larger bundles surrounded by areolar connective tissue, and between these bundles the tendon cells are found. A number of such larger bundles are held together by a layer of areolar tissue, the peritendineum, to form the tendon. The blood vessels, lymph vessels, and nerves pass into a tendon along the areolar tissue. Where the muscle and the tendon meet, the fibers of the tendon are affixed to the sarcolemma of the muscle fibers. To further strengthen this union between a muscle and its tendons, the areolar tissue surrounding the tendon bundles forms a continuation of the tissue enveloping the muscle bundles. Because of its great strength, a tendon is less likely to rupture or break than is a muscle or a bone.

Levers. ■ In the contraction of a muscle, the two ends approach each other, and since these ends are attached to two articulated bones, the result is a movement of one or of both of the bones. The tendon attaching a muscle to an immovable or less movable bone is called its origin; the tendon attaching a muscle to a more readily movable bone is known as its insertion. As shown in Fig. 6-2, the biceps muscle (*bi-*, twice; *caput*, head) of the arm has its two upper tendons (origin) affixed to the scapula (shoulder blade) and its lower tendon (insertion) affixed to the radius of the forearm. The radius makes a

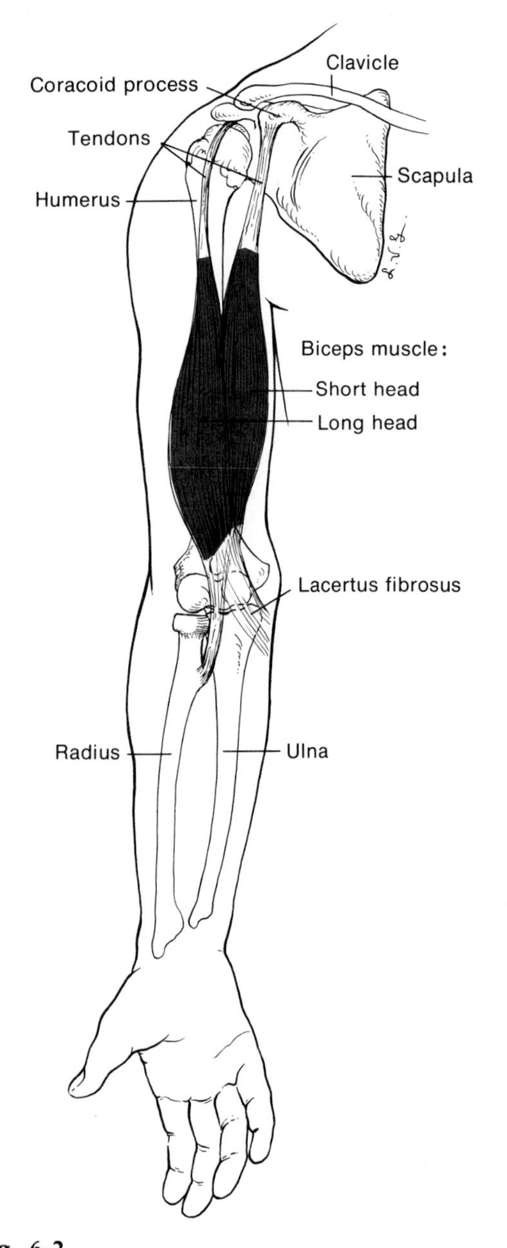

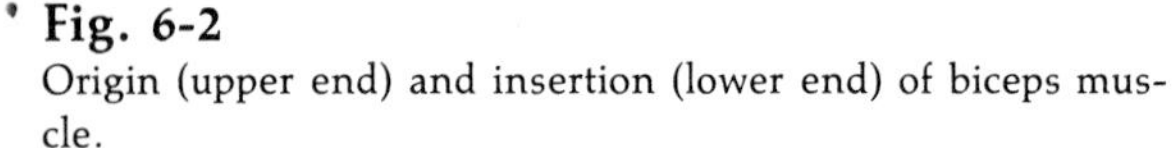
Fig. 6-2
Origin (upper end) and insertion (lower end) of biceps muscle.

hinged joint with the humerus at the elbow. By the contraction of the biceps muscle, the forearm is moved through an arc of a circle; that is, the arm is bent or flexed. Muscles by which the various parts of a limb are bent are known as *flexors;* those by which a limb is straightened or extended (e.g., the triceps brachii of the forearm) are called *extensors.* Flexors and extensors are antagonistic muscles.

The bones of the limbs and the ribs act as levers. In most cases (as exemplified by the biceps muscle) the muscle is attached to the lever so as to enhance the extent and the speed of the movement produced by a certain amount of contraction at the expense of force (levers of the third order). Depending on the position of the tendons upon the two bones and on the manner of the articulation of the bones, various modes of motion such as flexion, extension, circumduction, rotation, and gliding are possible.

Muscles. ■ Skeletal muscle fibers are elongated cylindrical cells, large numbers of which are bound together by areolar connective tissue into bundles or fasciculi. The fasciculi are, in turn, surrounded by connective tissue sheaths and grouped together into still larger bundles. Finally, the whole muscle is enveloped by a connective tissue sheath known as the *epimysium.* Muscles are abundantly supplied with blood vessels; these and the afferent (sensory) and efferent (motor) nerves enter the muscle along the areolar tissue.

Motor end plates. ■ On entering a muscle, the large-diameter fibers of an efferent nerve separate to distribute themselves among the thousands of muscle fibers (Fig. 6-3). Since the number of fibers in the muscle greatly exceeds the number of fibers in the motor nerve, the individual nerve fibers branch repeatedly so that a single nerve fiber innervates from 5 to as many as 200 muscle fibers. The small terminal branches, devoid of myelin sheaths (Chapter 5), are invaginated into the sarcolemma and form a special structure known as the *motor end plate*

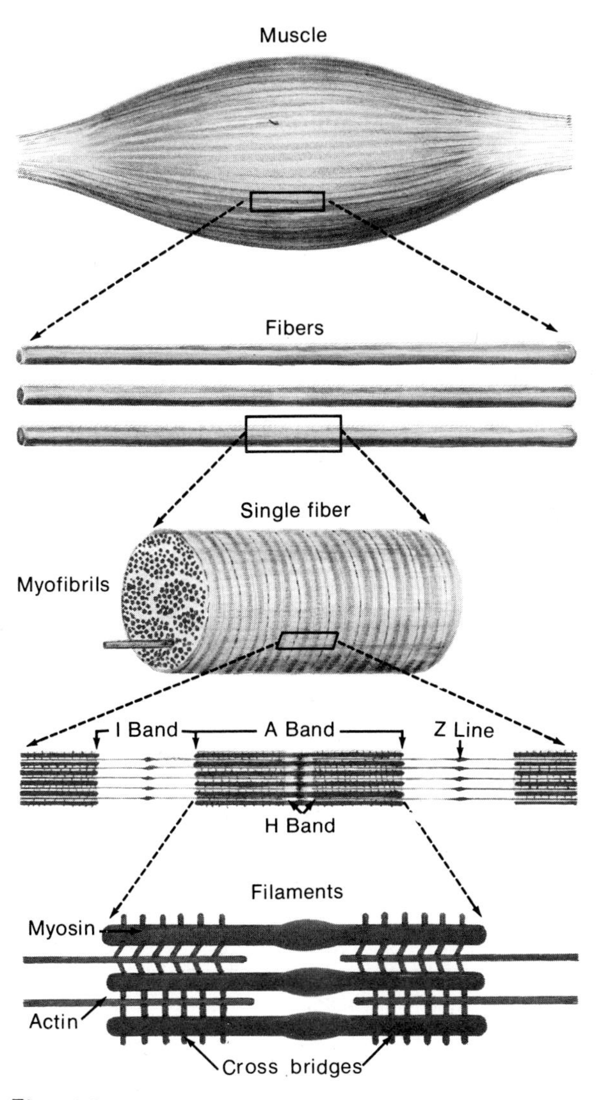

Fig. 6-3
Striated muscle dissected in schematic fashion. The muscle is made up of many fibers, which appear cross striated. Single fiber consists of myofibrils, which show alternating dark and light bands. A single sarcomere is the region between two Z lines containing I and A bands. Bands derive from presence of two sets of interdigitating filaments, thick filaments (myosin) with cross bridges and thin filaments (actin).

(Fig. 6-1). Each muscle cell may have one or more motor end plates. An individual nerve fiber with all the muscle fibers it innervates, the functional neuromuscular unit, is termed the *motor unit.*

Muscle fibers. ■ The structural unit of a muscle is the single fiber. In striated muscles these are from 0.01 to 0.1 mm in diameter and from 1 to 40 mm (rarely 120 mm) in length. Size and amount of tissue, especially elastic tissue, between fibers increase with age of the individual. Strength-building exercises tend to increase the fiber bulk and cross-sectional area. Each fiber is surrounded by a tough, exceedingly thin elastic sheath, the *sarcolemma,* beneath which are located the numerous nuclei. Muscle fiber protoplasm is a semifluid substance appropriately known as *sarcoplasm.* Within this matrix are embedded a large number of delicate strands, about 0.001 mm in diameter, termed *myofibrils* (Figs. 6-1 and 6-3).

When observed by means of a microscope, myofibrils appear to be transversely or cross striated due to alternating light and dark bands (Figs. 6-3 and 6-4). The dark bands are called anisotropic or A bands, whereas the light bands are spoken of as isotropic or I bands. Dense Z lines cross the center of each I band, thereby dividing the myofibril into smaller units known as *sarcomeres.* Adjacent fibrils are aligned so that the striations are in register; that is, they appear as a continuous band across the muscle fiber.

At the center of the resting sarcomere and naturally of the A band, there is a less dense region. This H zone (Fig. 6-4) occurs because of the definite arrangement of *filaments* within the sarcomere. As shown in Figs. 6-3 and 6-4, these filaments are of two sizes. The thicker ones are present only in the A band, whereas the thinner ones occupy the I band and part of the A band. Their absence from the central portion of the A band accounts for the less dense H zone. Living fibrils are made up of two sets of overlapping and interdigitating filaments surrounded by a solu-

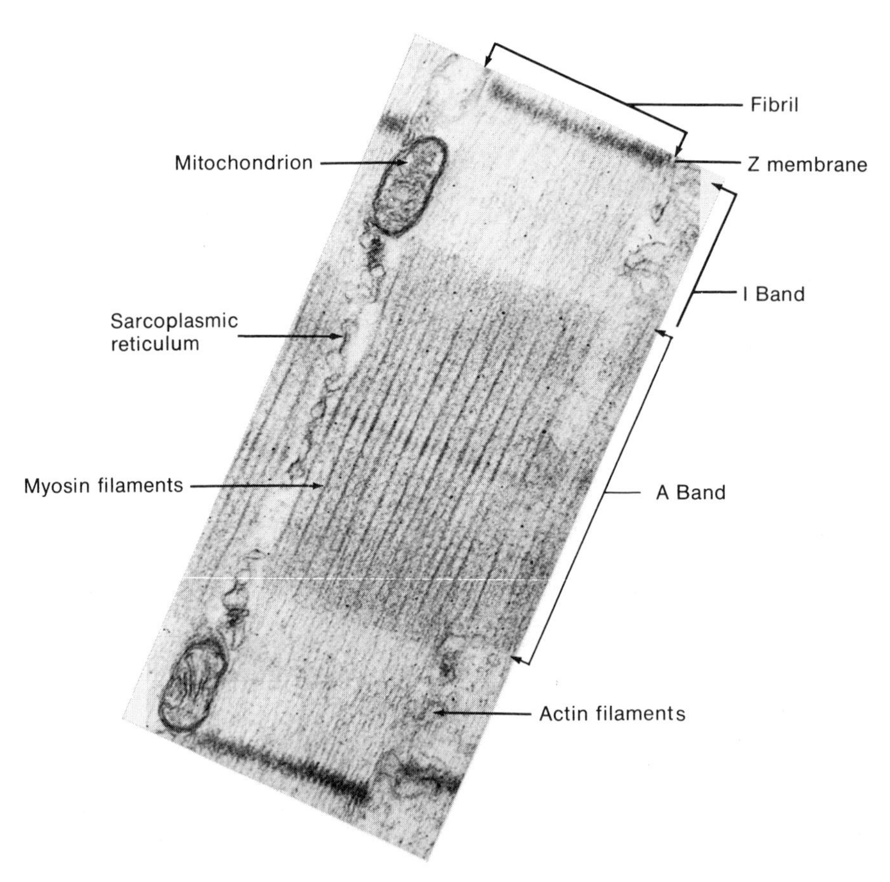

Fig. 6-4
Electron micrograph of mouse skeletal muscle fibrils. Actin and myosin filaments are clearly discernible. Mitochondria, sarcoplasm, and sarcoplasmic reticulum also are evident between fibrils. Single sarcomere occurs between Z lines of fibril. Less dense region in center of A band is H zone. (×43,000.) (Courtesy Dr. W. E. Scott and Dr. B. A. Schottelius.)

tion of salts and soluble proteins. The importance of this anatomic specialization will be discussed in a subsequent section on the chemistry of muscle contraction.

Sarcoplasmic reticulum. ■ An organized network of sacs and tubules, known as the *sarcoplasmic reticulum,* is distributed throughout the sarcoplasm (Figs. 6-4 and 6-5). It is in some respects similar to the endoplasmic reticulum found in other kinds of cells. This sarcoplasmic reticulum consists of tubules lying parallel to the myofibrils; at the Z line and H zone these tend to fuse and form sacs that completely surround the individual fibrils. By coalescence with contiguous sacs, a transverse channel across the entire muscle fiber is produced. At the level of the Z line the sacs of the sarcoplasmic reticulum appear to enclose a smaller tubule, distinctly separated from the sacs by its own membrane. Together these vesicular structures are referred to as a *triad,* or, in some species, as a *diad.* The enclosed tubule belongs to the *transverse tubule system,* or T system, which constitutes a deep infolding of the sarcolemma into the interior of the

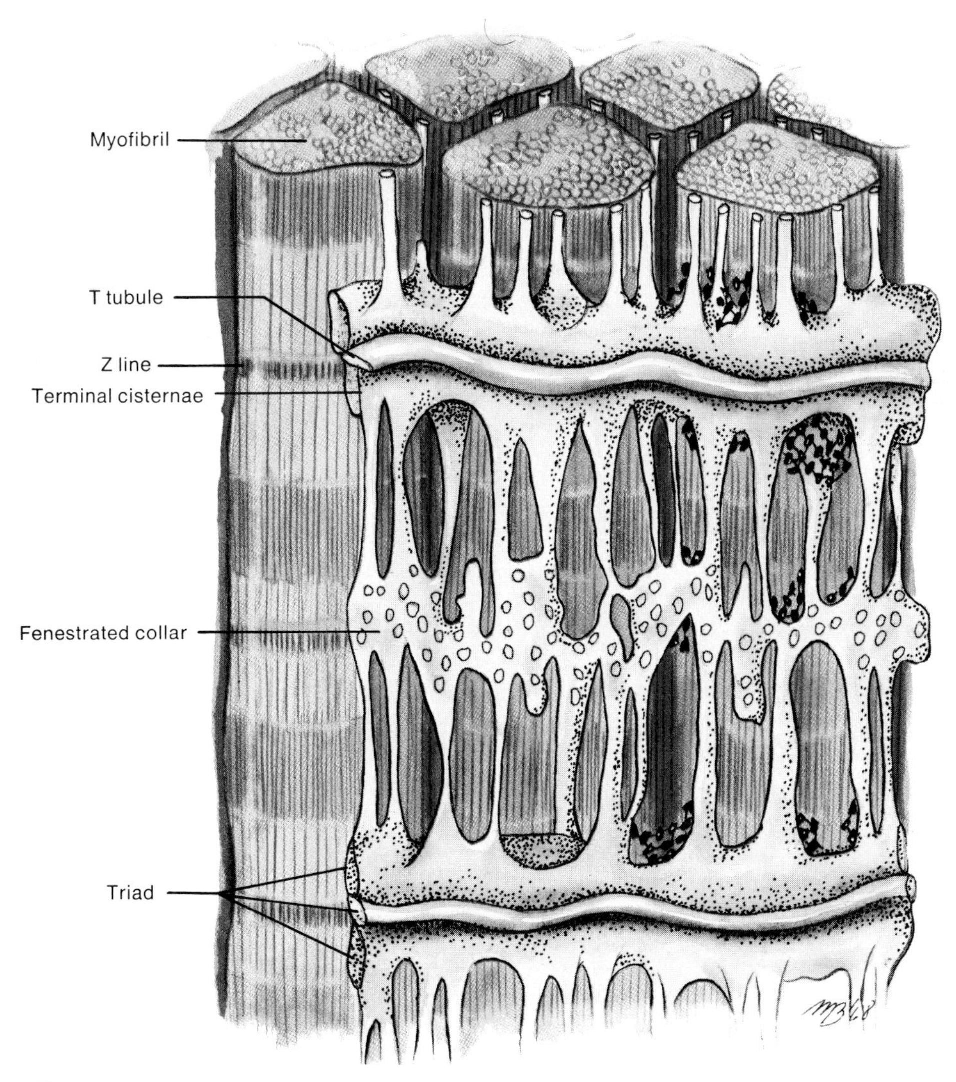

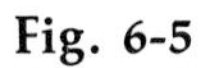
Fig. 6-5
Relationship of sarcoplasmic reticulum and T tubule system to individual skeletal muscle fibrils.

muscle fiber. Surface area of the fiber's outer membrane is increased manyfold by the presence of the T system.

When one remembers that the radius of a single skeletal muscle fiber may vary from 5 to 50 μ, one can appreciate the significant role that the T system and the sarcoplasmic reticulum play in the metabolism, excitation, and contraction of muscle. The T system serves as a duct system for the movement of fluid containing ions and other substances into and out of the cell. Longitudinal distribution of these substances in the spaces between the myofibrils, as well as of those substances synthesized within the sarcoplasm or mitochondria, is a function of the reticulum. The specific role of the T system and sarcoplasmic reticulum in the phenomenon of excitation-contraction coupling will be described later.

■ CHEMICAL PROPERTIES OF SKELETAL MUSCLE

The approximate chemical composition of skeletal muscle may be summarized as shown in the accompanying flow diagram (below).

Pale and red muscles. ■ A muscle fiber is composed, as already noted, of a number of delicate fibrils surrounded by a more fluid sarcoplasm. Many mitochondria and the sarcoplasmic reticulum are present. A respiratory pigment, *myoglobin* (muscle hemoglobin), that functions in the transport of oxygen from the blood vessels (capillaries) in the extracellular space to the sites of oxidation (mitochondria) is also present within the sarcoplasm. Some fibers have a greater content of myoglobin than others; since this protein resembles the red pigmented protein, hemoglobin, of the red blood cell in that it contains iron, it gives the fiber a red appearance. Other fibers deficient in myoglobin are pale. Most muscles include both kinds of fibers. When the dark fibers predominate (e.g., in the soleus muscle), the whole muscle is red in color; the white or pale muscles (e.g., the pectoral muscles of a chicken) are to a very large degree composed of pale fibers. Muscle fibers rich in myoglobin have a high capacity for oxidative metabolism with a strong activity of the Krebs cycle and electron transport enzymes (Chapter 23). White muscle fibers have a high rate of anaerobic glycolysis with intense activity of glycolytic enzymes and phosphorylase.

Red fibers are considerably slower in their contractile action but undergo fatigue less rapidly than do white fibers. In consequence of these two characteristics, the red fibers are well adapted for static or postural contractions. Postural contractions are sustained, tonic contractions chiefly concerned in maintaining the position of the body in space (as in standing) for a consider-

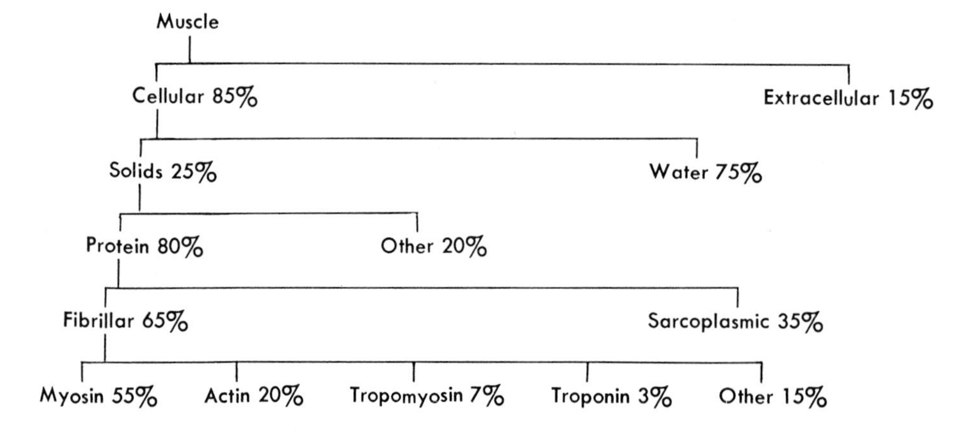

able length of time or in maintaining the position of a part of the body. Body posture is largely achieved by extensor (antigravity) muscles well supplied with red fibers. On the other hand, phasic contractions, by which changes in the position of the body or a limb are brought about, are better served by the more rapidly acting white fibers predominant in flexor muscles.

Extractives. ■ Many of the nonprotein constituents of muscle are soluble in water and are usually spoken of as extractives. These include components of the glycolytic cycle (Chapter 23) —creatine, *creatine phosphate, adenosine triphosphate,* adenosine diphosphate, amino acids, lactic acid, and so on. Adenosine triphosphate (ATP) is one of the most important substances in muscle cells. The terminal phosphate group of this compound is linked to the structure by an "energy-rich" bond. Rupture of this bond produces inorganic phosphate and adenosine diphosphate (ADP), with the simultaneous release of a large amount of energy.

$$ATP \rightarrow ADP + H_3PO_4 + Energy$$

Energy released by the breakdown of ATP is used in the performance of work.

Carbohydrate in muscle is present mostly as glycogen in amounts up to 1%; lipids and salts constitute about 2% and 1%, respectively. Among the salts, the principal ion is potassium; only small quantities of sodium, calcium, chloride, and magnesium are present. The specific role of potassium ions in the origin of the resting potential was described in Chapter 3; the role of sodium and potassium in excitation was considered in Chapter 5, although we will make further reference to it in this chapter. The role of other extractives is also discussed later in this chapter and in Chapter 23.

Proteins. ■ The enzyme components of muscle that catalyze the various steps in glycolysis (the breakdown of glycogen to CO_2 and H_2O) are proteins. Other proteins of muscle include different enzymes and the structural proteins, of which *myosin, actin, tropomyosin,* and *troponin,* comprising together well over one half of the total protein, are the most important.

■ BIOPHYSICS OF SKELETAL MUSCLE CONTRACTION

The visible shortening during contraction of an active muscle constitutes only the mechanical phenomena; the process of contraction also involves other more subtle phenomena. Our purpose in this section will be to describe and to integrate the several aspects of contraction at the different levels of anatomical complexity and to a degree, at least, in the temporal sequence of their occurrence. For the latter purpose, we will consider contraction as exhibiting four major aspects in sequence: electrical phenomena, chemical phenomena, mechanical phenomena, and thermal phenomena.

■ Electrical phenomena and muscle stimulation

Biopotentials. ■ When skeletal muscle is removed from the body, it remains quiescent until it is stimulated from an external source; that is, it possesses no automaticity. However, the resting muscle fiber exhibits a steady difference of electrical potential across its membrane, the resting potential. This potential, its origin and maintenance, have been discussed in previous chapters; we can refer to these to refresh our memory. Our problem is to answer the question: What normally furnishes the stimulus to initiate muscle activity? Recall that all skeletal muscles receive efferent or motor nerves from the central nervous system and that these nerve fibers make intimate contact with the muscle fibers (Fig. 6-1). When the nerve fibers in the central nervous system are excited, an electrical disturbance, the nerve action potential, ensues. This action potential is propagated to the nerve terminals where intercellular transmission across the neuromuscular junction (synapse) occurs. Depolarization of the terminal nerve membrane releases the *acetylcholine* contained within the synaptic vesicles, and this transmitter diffuses with a finite delay—*synaptic delay*—across the gap—*synaptic gutter*—between the nerve membrane and the underlying specialized region of the muscle fiber membrane. This specialized region consists of many accordion-like pleats, the junctional folds of the sarcolemma. In addition to its histological specialization, this subsynaptic region has a functional specialization; it becomes highly permeable to all small cations when exposed to acetylcholine, and, thereby, its normal transmembrane potential is reduced almost to zero; that is, it becomes depolarized. It undergoes no reversal of potential, but, nevertheless, it serves as a "sink" for local current flow from adjacent areas of the muscle membrane. Such local circuits, as we have previously learned, excite and evoke propagated action potentials in muscle membranes. The muscle action potential (Fig. 6-6), because of quantitatively different cable properties, is of slightly longer duration and is more slowly propagated (about 3 m/sec) than is that of large nerve fibers.

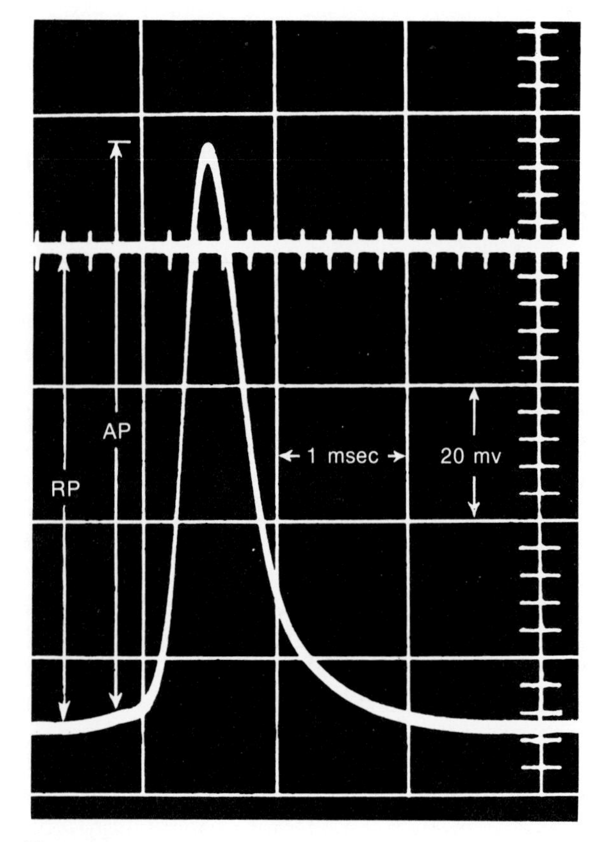

Fig. 6-6
Photograph of intracellular microelectrode recording of action potential, **AP,** from excised mouse skeletal muscle. **RP** is the resting potential. Oscilloscope grid lines indicate time on horizontal axis and potential on vertical axis. Compare with Fig. 14-4.

Excitation-contraction coupling. ■ Effective contraction of a muscle fiber is the end result of collective and coordinated shortening in its assembly of myofibrils. Because of the relatively large diameter of a skeletal muscle fiber, there is scant probability that the propagating action potential would influence all of the myofibrils equally and essentially simultaneously. Current would reach the deeper embedded fibrils much reduced and delayed, if at all. The T system and the sarcoplasmic reticulum provide the means for synchronizing myofibril activity. Since the T system membrane is continuous with the sarcolemma, the electrical activity on the surface is conducted inward to the depths of the fiber at each Z line. Its influence is transferred to the sarcoplasmic reticulum at the triad. Although all the details of this *excitation-contraction coupling* phenomenon are not available, it is quite certain that the result of electrical activity is the release of free calcium ions from the reticulum into the neighborhood of the contractile filaments. The probable interaction of calcium with contractile proteins will be considered later.

Independent muscle irritability: curare. ■ We have indicated already that the normal means of stimulating a muscle is *indirect* through its motor nerve. *Direct* stimulation of a muscle may elicit a contraction either by excitation of the muscle fibers or the nerve fibers within the muscle. Under these circumstances, it cannot be stated whether the muscle fiber possesses independent irritability. To resolve this, recourse is taken to the action of a drug, *curare.* When a muscle-nerve preparation is placed in a dilute solution of curare for a sufficient length of time, stimulation of the muscle directly causes a contraction, but stimulation of the nerve is without effect. This latter occurrence is not due to the action of curare upon the nerve fibers, for when the motor nerve is placed in the solution (the muscle not being in the solution), the stimulation of the nerve is capable of causing the muscle to contract. Evidently neither the muscle nor the motor nerve is affected; hence curare must paralyze the neuromuscular junctions, rendering it impossible for the nerve impulse to activate the muscle.* By this and other experiments the *independent irritability* of the muscle has been established.

The injection of a curare solution into a frog (with a destroyed brain) produces the same results as those already outlined. If when the effects of curare are well developed, a few drops of an *eserine* solution are injected, the paralyzing action of the curare is abolished.

From these and related experiments three points of interest may be noted. (1) The myoneural junction is a specialized structure about a thousand times more sensitive to the action of curare and certain other chemical compounds than is either nerve or muscle. (2) Some drugs are extremely specific in their action; that is, they act on certain body structures without affecting others. The greater the specificity of the chemicals employed in treating diseases, the greater their value. (3) One chemical compound may be able to counteract the physiological effect of another; this is known as antagonism. As we pursue our study, abundant evidence of the specific action and of the physiological antagonism of compounds normally present in the body will be found.

*A preparation of curare has been used to lessen the severity of convulsions and to produce muscular relaxation as during abdominal operations and the setting of fractured bones. Curarine is the active principle of curare.

Electrical stimulation. ■ Much of our knowledge of muscle activity has been gained by the study of excised muscles of frogs. For investigative purposes the sartorius muscle is often employed, since it is easily accessible, thin, and survives for relatively long periods of time when supplied with adequate nutrients. The preparation frequently used in a teaching situation is the gastrocnemius muscle and its motor nerve; although its survival period is shorter, it is simple to dissect and produces very strong contractions. Fig. 6-7 illustrates a typical arrangement of the apparatus used in a student laboratory for direct stimulation of excised muscle and permanent recording of the contractile responses.

Since considerable space was devoted to a description of the characteristics of a stimulus in the previous chapter, here we need only to review them briefly in regard to muscle contraction. Under a given set of environmental conditions, the *strength* of a stimulus delivered to a whole muscle determines the extent of contraction. In Fig. 6-8 the *threshold* intensity of the stimulation is approximately 0.8 v, values less than this are of *subliminal* intensity. As the voltage is increased from 0.8 to 5, the magnitude of the contraction also increases; this response is known as *multiple fiber* or *quantal* summation. A strength of stimulation is attained (at 5 v) beyond which no further increase in contraction occurs because all of the muscle fibers have been brought into action; this is called the *maximal stimulus.* A stimulus of a strength between the threshold *(liminal)* and the maximal is known as a *submaximal stimulus.* The effects of strength of a stimulus on contractile activity are also depicted in Fig. 6-9, *A.*

Although under special circumstances graded contractions of single muscle fibers can be demonstrated, under normal conditions the contraction of a single muscle fiber is always maximal; in other words, a minimal stimulus causes the in-

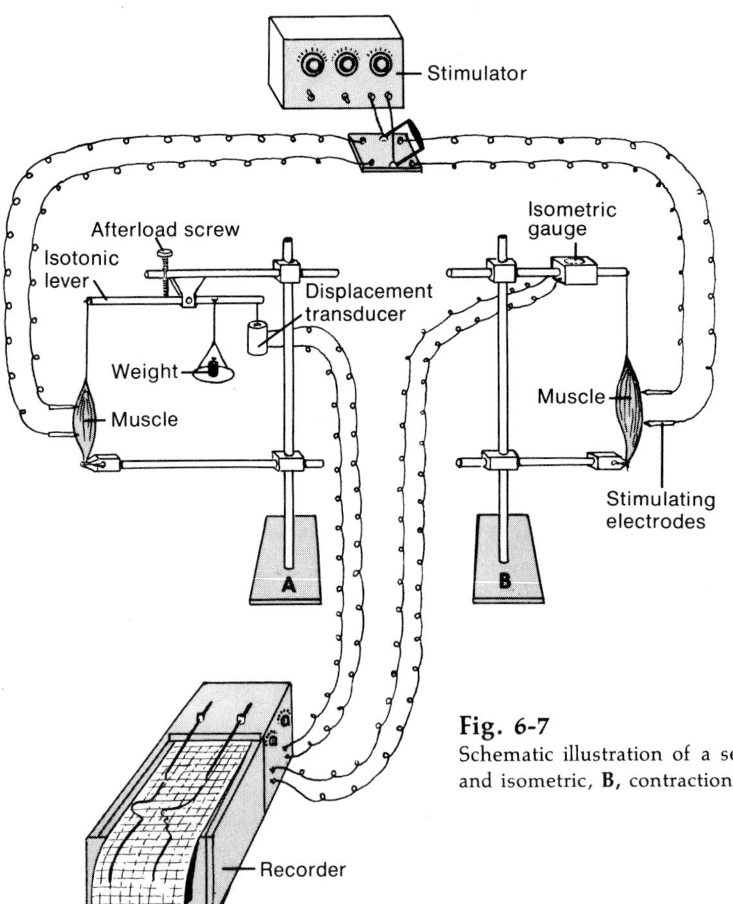

Fig. 6-7
Schematic illustration of a setup used to study isotonic, **A**, and isometric, **B**, contractions.

dividual muscle fiber to contract to the same extent as a stronger stimulus does. This phenomenon, known as the *all-or-none* law, is easily demonstrated in heart muscle, as will be discussed in Chapter 14. The greater contraction of a whole skeletal muscle following a stronger stimulation may be attributed to a larger number of muscle fibers participating in the contraction. It is possible that some fibers are less irritable or, being buried deep in the muscle, are less accessible to the stimulus and exhibit a higher threshold. In any event the maximal stimulus is just sufficiently strong to cause all the fibers to contract. Although the individual skeletal muscle fiber and motor unit follow the all-or-none law, the muscle as a whole does not because of the variation in number of contributing fibers, and therefore *it is possible for the muscle as a whole to execute contractions of graded extent or force.* A moment's reflection shows the tremendous importance of this characteristic in everyday life. It must be emphatically stated that, with varying conditions of nutrition, exercise, initial length, and fatigue, the extent of the contraction of a muscle fiber

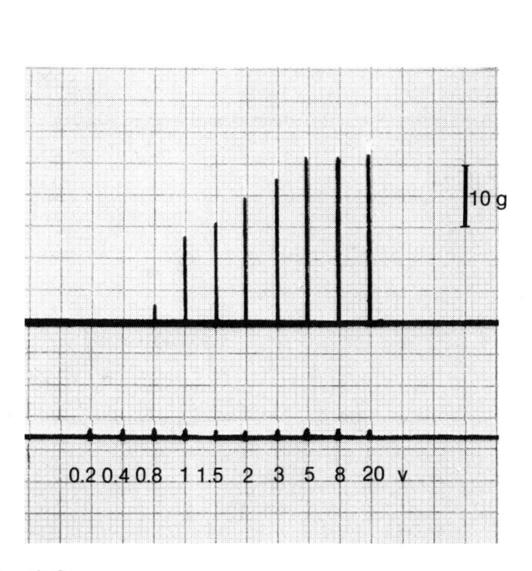

Fig. 6-8
Effect of increasing strength of stimulus on isometric contraction of frog sartorius muscle. Stimulus strength is shown in volts; contraction tension is in grams.

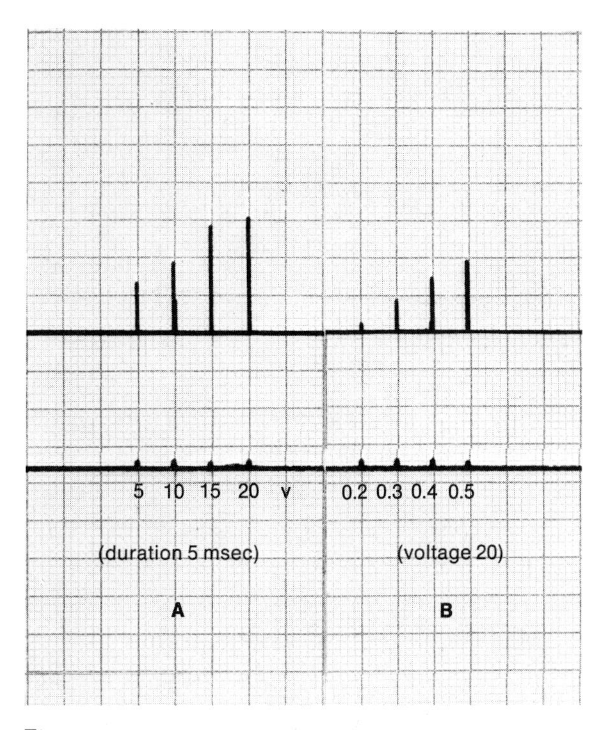

Fig. 6-9
Effect of stimulus strength and duration on isometric contraction of frog sartorius muscle. In **A**, duration is constant and strength of stimulus varies from 5 to 20 v; in **B**, voltage is constant and duration varies from 0.2 to 0.5 msec.

and the tension developed will vary, but the all-or-none law holds good for any given condition of the fiber or the motor unit.

Duration of the stimulus is equally as important as the strength (intensity). If the length of time during which the stimulus is applied is shortened sufficiently, no contraction results, no matter how strong the stimulus. This is illustrated in Fig. 6-9, *B*.

Rate of change of intensity is the third important characteristic of a stimulus. The efficiency of a stimulus varies with the rapidity of change in strength from zero to maximum (DuBois-Reymond's law). At *M* in Fig. 6-10, a supramaximal stimulus is delivered to the muscle by pressing the switch on the stimulator, and the muscle responds vigorously; the sudden withdrawal of the current from the muscle on release of the switch also produces a contraction, *B*. At *a* the current is gradually introduced into the muscle by means of a rheostat and reaches its full intensity at point *b* without having any effect on the contractile activity of the muscle. From point *b* to *c* the current is just as gradually withdrawn,

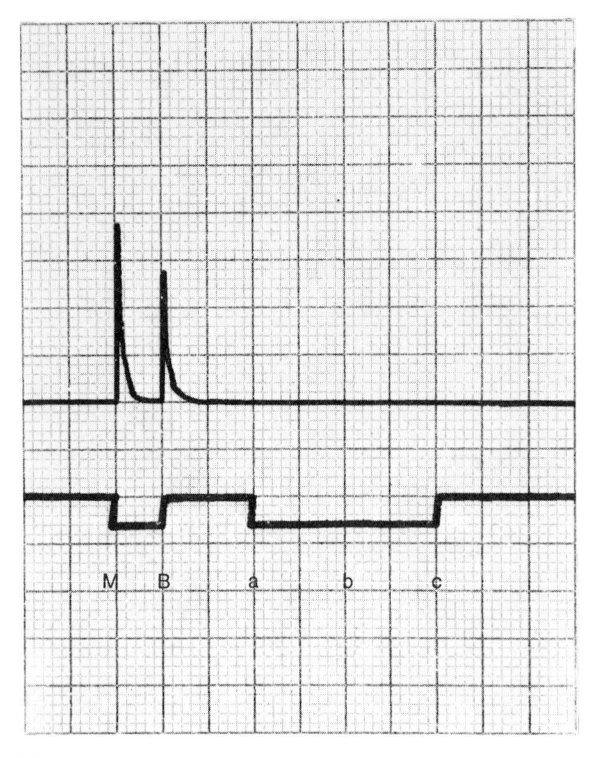

Fig. 6-10
Tracing showing the effect of sudden application, **M**, and removal, **B**, of stimulus current. From **a** through **b** current was slowly applied to attain the same density as in **M**; from **b** to **c** it was slowly withdrawn.

also without stimulating the muscle. *The muscle accommodates itself to a gradually increasing strength of stimulus.*

From the foregoing discussion it may be concluded that the effectiveness of a stimulus is for muscle, as for nerve, a function of its strength, duration, and rate of change in intensity.

■ Chemistry of contraction and relaxation

It has been said in relationship to muscular contraction that: "Facts and theories are natural enemies."* Each new discovery in the areas of chemistry and ultrastructure (electron microscopy) of muscle has produced its share of new theories. Fundamentally, muscle is a machine that converts chemical energy into mechanical work. Consequently, any investigation of muscle contraction ultimately becomes a study in molecular biology, for muscle fulfills its physiological function by means of the interaction between molecules. Despite a number of still unresolved problems, the general outline of a currently acceptable interpretation of the facts can be described.

Interdigitation theory. ■ From the changes evident in the myofibrils due to differential extraction of actin and myosin from them by strong and weak salt solutions, it has been concluded that the A band filaments are, indeed, made up of *myosin* molecules; the I band filaments are made up of *actin* molecules (Figs. 6-3 and 6-4). However, at least two other structural proteins, *tropomyosin* and *troponin*, are associated with the actin filaments. The actin molecule is a globular protein (G-actin) that joins with like units to form a chainlike structure (F-actin). Tropomyosin, which is present throughout the extent of the thin filaments, seems to serve as an agent for maintaining the F-actin structure. The thin filament comprises two such F-actin chains twisted together; each has a helical arrangement of 13 G-actin units per revolution; the diameter of the filament is on the order of 50 A, its length about 1.5 μ. Troponin molecules are incorporated in the filament at intervals of about 400 A. In the longitudinal section, as shown in Figs. 6-3 and 6-4, only one (sometimes two) actin filament appears between each myosin filament; a transverse section reveals that each myosin filament is actually surrounded by six actin filaments spaced evenly in 60° increments around the periphery.

The myosin molecule is an asymmetrical linear structure in which one end is thickened or globular in form. It is this region of the molecule that exhibits the ability to combine with actin and to

*From Wilkie, D. R.: Facts and theories about muscle. In Progress in biophysics, London, 1954, Pergamon Press.

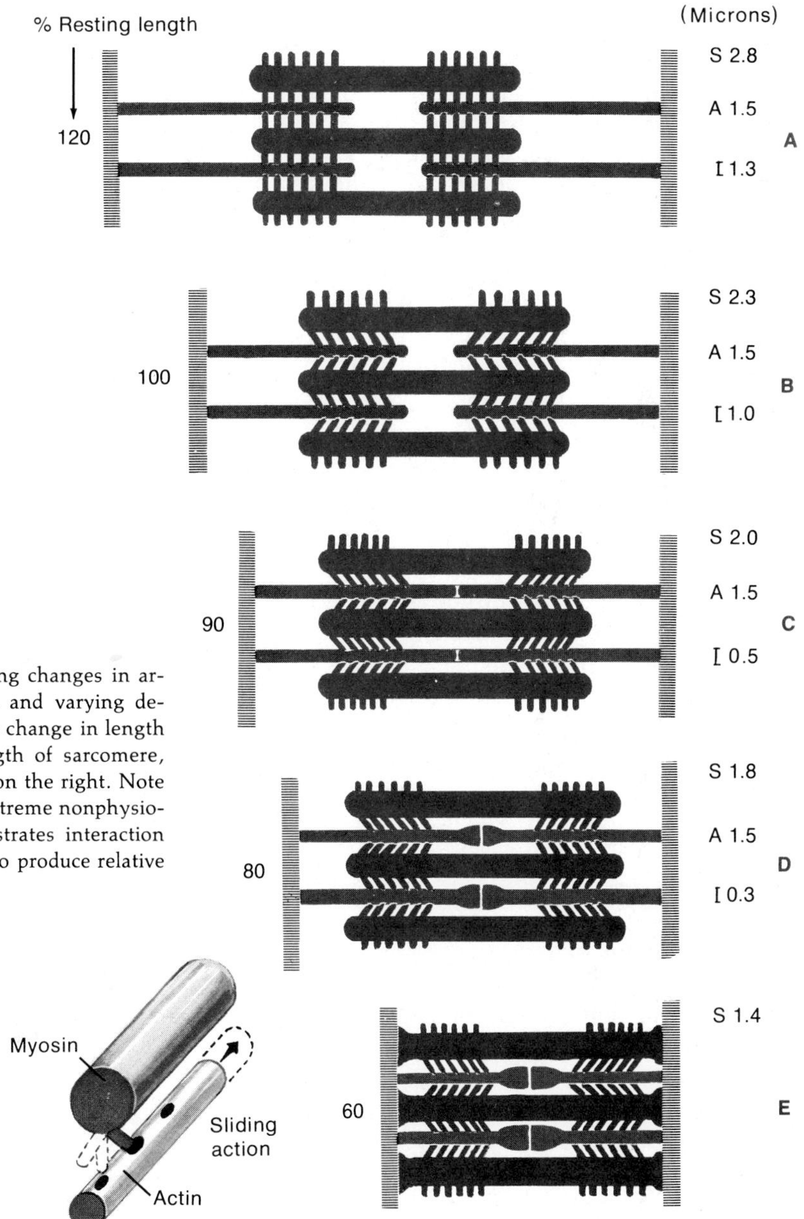

Fig. 6-11
Diagram of a normal sarcomere, **B,** showing changes in arrangement of filaments under stretch, **A,** and varying degrees of contraction, **C** through **E.** Relative change in length appears on the left, and approximate length of sarcomere, A band, and I band, in microns, is shown on the right. Note constancy of A band length except under extreme nonphysiological shortening. Inset in lower left illustrates interaction of myosin cross bridge and actin filament to produce relative movement of filaments in contraction.

hydrolyze ATP. We will have occasion to refer to this activity again. Myosin molecules aggregate longitudinally with their thickened ends oriented toward the lateral aspects of the A band; that is, by analogy they may be likened to a group of clubs, or hammers, arranged to form a bundle with their handles parallel and overlapping and with one half of the heads lying to the right, the remainder lying to the left. The diameter of the thick filament so constituted is approximately 120 A, its length 1.5 μ. Protoplasmic extensions known as *cross bridges* project from the myosin filament at either end; a bridge-free area occurs in the center of the filament. Six cross bridges exist in each approximately 400 A segment of the myosin filament. They are, however, distributed in a helical manner, with a 60° radial pattern, so that one revolution around the periphery is completed by each set of six bridges. Thus, each bridge in a set points toward a separate actin filament; bridges occurring at approximately 400 A intervals along the myosin filament point toward identical actin filaments.

According to older theories, muscle contraction resulted from the folding or coiling of molecules and filaments of the contractile proteins. New observations, however, show that during contraction and stretching over a large range (Fig. 6-11) the length of the A bands remains constant. Since the A band is equal to the length of the thick myosin filaments, it is evident that the length of these filaments is also constant. Because the width of the H zone increases or decreases with the length of the I band and because the distance from the end of one H zone to the beginning of the next in an adjacent sarcomere is nearly constant, it can only be concluded that the length of the thin actin filaments does not change an appreciable amount. When a muscle contracts, as presented in the schema of Fig. 6-11, the two sets of filaments slide past one another; this is sometimes spoken of as the *interdigitation* (or *sliding filament*) *theory of contraction.*

Role of calcium ions. ■ Let us now focus our attention on the essential role of the calcium ion in contraction. It was pointed out previously that the excitatory wave on the muscle fiber membrane is conducted inward by the T system and in an as yet undetermined manner causes the release of calcium ions from the sarcoplasmic reticulum. Free calcium ions activate the interaction of myosin and actin, but apparently only through the intervention of tropomyosin and troponin. We noted before that tropomyosin is bound uniformly to actin and that troponin complexes with tropomyosin at specific, repeating sites about 400 A apart on the actin filament. It is probable that troponin in the nonactive muscle serves as a regulatory protein to prevent the interaction of myosin and actin. However, troponin is also a calcium receptive protein. Under the influence of an increase in free calcium ion concentration, troponin is thought to undergo a structural change. This abolishes the previous repression of the myosin-actin interaction, and contraction ensues.

Relaxation is brought about when the free calcium ion concentration is reduced. Withdrawal of calcium ions from the vicinity of the contractile filaments is a function of the *calcium pump* located within the sarcoplasmic reticulum membrane. This sequestering of calcium ions may require a marked fraction of the energy consumption associated with contractile activity.

Myosin-actin interaction. ■ The cross bridges are the sites of the actual interaction between myosin and actin. Activated by the release of calcium ions from the sarcoplasmic reticulum, the myosin projections bond to the thin filaments and by their ratchetlike motion (Fig. 6-11) pull the latter inward toward the center of the sarcomere. The arrangement of the myosin molecules in the thick filament, which we described earlier, assures that the thin filaments on either side of the A band are moved in opposite directions, that is, toward each other. Consequently, the Z lines are drawn closer together, and as each sarcomere is reduced in length, the muscle shortens. By its movement a single cross bridge displaces the actin filament about 100 A. Thereafter the bond between the bridge and the thin filament is ruptured, and the bridge returns to its initial position. Repetitive to-and-fro movement of a cross bridge provides for a greater total displacement. The action of the individual cross bridges is not synchronized; each acts independently. At any given instant only a fraction of the bridges actively generates force and displacement; when these bridges release their load, others take up the task so that tension or shortening is maintained.

Sources of muscle energy. ■ Myosin, in addition to its structural role, is an enzyme that catalyzes the breakdown of ATP to ADP. The energy required for the cross-bridge movement is obtained, whether directly or indirectly is uncertain, from the breakdown of ATP. Since the rates of anaerobic and aerobic metabolism usually are too slow to maintain adequate levels of ATP in a muscle during sustained contractions, a supplemental source of energy is available in the form of *creatine phosphate* (phosphocreatine), which through the action of a transferase enzyme restores the depleted ATP. The reactions may be summarized as follows:

$$\text{ATP} \rightarrow \text{ADP} + \text{Pi} + \text{Energy}$$
$$\text{CP} + \text{ADP} \rightarrow \text{ATP} + \text{C}$$

During periods of inactivity, or less intense activity, the creatine is rephosphorylated by ATP (enzymatic reactions are reversible) produced in intermediary metabolism. Another means of obtaining additional energy is also available to the muscle under certain circumstances. The enzyme *myokinase,* which is present in the sarcoplasm, catalyzes the following reaction:

$$2\ \text{ADP} \rightarrow \text{ATP} + \text{AMP}$$

Although the ultimate source of energy for muscle contraction is carbohydrate, there is strong evidence that lipids are utilized at times. These lipids are in the form of free fatty acids (FFA) that can diffuse rapidly from the blood stream into the muscle or derive from the muscle's fat stores. All the enzymes necessary for the oxidative metabolism of FFA to CO_2 and H_2O are present in muscle. The availability of oxygen determines the degree to which FFA constitute a source of energy. Under optimal circumstances, these substances may be the most important source. Certainly under anaerobic conditions, as during sustained activity, glucose is utilized and lactic acid produced. The series of chemical events concerned in the liberation and utilization of energy is long and involved. More detailed discussion of these events appears in Chapter 23.

Rigor mortis. ■ Most of us are familiar with the fact that soon after death the body passes into rigor mortis, the stiffening of death. An excised frog muscle appears translucent and is flaccid and extensible. When dead, the muscle loses its irritability, shortens, and becomes opaque, rigid, and inextensible; these changes are attributable to the depletion of ATP and the accumulation of lactic acid. Normally, as the muscle passes into rigor mortis, its glycogen disappears, and, *pari passu,* lactic acid forms.

■ Mechanical phenomena

Form curve. ■ In order to study some of the characteristics of a muscle twitch (the response to a single stimulus), the form curve is obtained by recording the twitch of a frog's sartorius muscle on an electronic recorder as shown in Fig. 6-7. Since it is desirable to know the time relationships of the various events taking place during the contraction, two pens are used. One records the stimulation; the other depicts the muscle response. The speed with which the chart paper moves beneath the pens can be adjusted according to need. When the writing points of the pens are aligned vertically, the correct time relationship is maintained. The tracing in Fig. 6-12 was obtained in this manner.

It should be noted that a short lapse of time

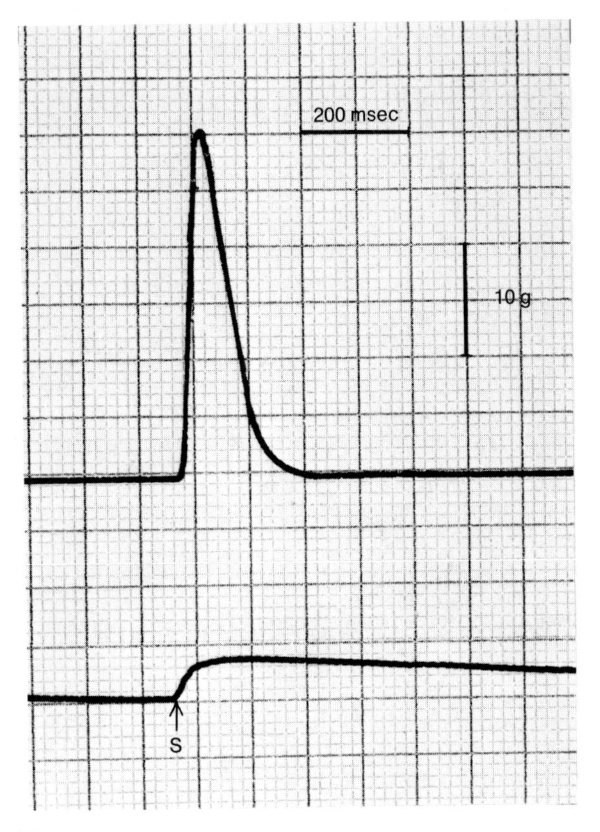

Fig. 6-12
Form curve of a frog sartorius muscle twitch. Stimulation of the muscle occurred at **S**. The latent period was 20 msec; the contraction period lasted 40 msec, and relaxation required 160 msec.

occurs between the application of the stimulus at *S* and the beginning of the response; this is known as the *latent period.* The duration of this interval is determined by noting the number of chart spaces between *S* and the beginning of contraction. In the curve given in Fig. 6-12 this amounts to about 0.02 sec, whereas, by means of more precise apparatus, the latency of mammalian muscle has been demonstrated to be about 0.001 sec. The length of the phase of increased tension (contraction proper) is about 0.04 sec, of the relaxation phase 0.16 sec, and of the whole twitch about 0.2 sec. The length of these periods varies greatly with the specific muscle, with its condition or state (e.g., temperature), and with the load to be lifted. This is illustrated in Figs. 6-13 and 6-14. Variation between different muscles of the same animal can be appreciated from the following contraction times recorded from the cat: internal rectus of the eye, 0.008 sec; gastrocnemius, 0.04 sec; soleus, 0.1 sec.

Isometric and isotonic contraction. ■ A contracting muscle unable to move a load retains its original length; for this reason, its contraction is called *isometric* (*iso,* same; *metric,* length). Although it develops tension, it performs no mechanical work; eventually all the expended energy appears as heat (physiological work). On the other hand, when the resistance offered by the load is less than the tension developed, the muscle shortens and performs mechanical work. This is known as an *isotonic* contraction (*iso,* same; *tonus,* tension). The upper traces in Fig. 6-14 illustrate isotonic contractions of a frog sartorius muscle subjected to 10 g *(A)* and 20 g *(B)* loads. In this experiment, the muscle is affixed at one of its ends to an isotonic lever and at the other end to an isometric gauge, rather than to a stationary clamp as seen in Fig. 6-7. When stimulated, the muscle first develops tension (isometric phase) equal to the load it is expected to lift, as shown in the lower traces; when the tension equals the load, the muscle proceeds to shorten and lift the load. As the load is increased, the start of the isotonic phase is delayed in time, and the magnitude and velocity of shortening are reduced. All so-called isotonic contractions involve an isometric phase, for even an unloaded muscle must lift its own weight. If the load is one that the stimulated muscle neither lifts nor drops, the contraction is truly isometric; if the load is greater than the tension that the muscle can develop, the muscle lengthens during contraction; that is, negative shortening occurs.

The muscles of the body sometimes contract

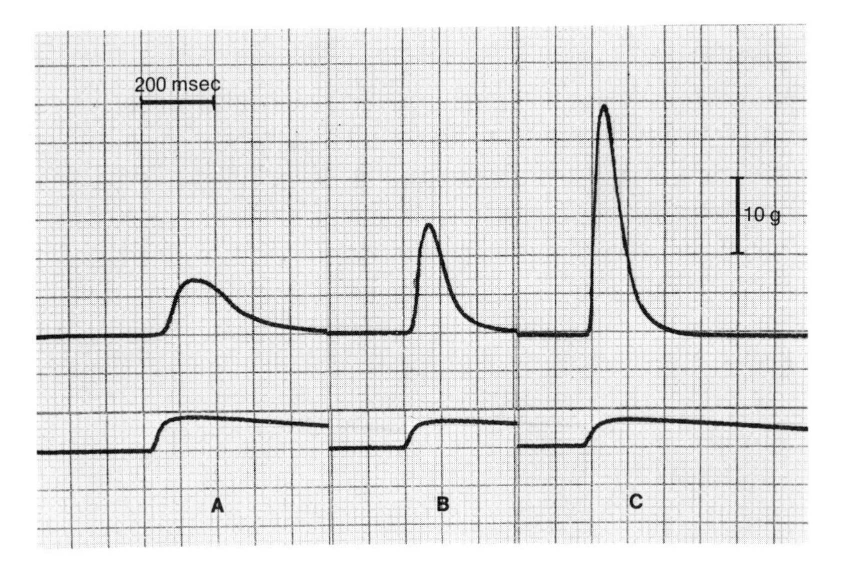

Fig. 6-13
Influence of temperature on the form curve of frog sartorius muscle. Temperatures in **A, B,** and **C** were 0, 15, and 25 C. The respective latencies were 40, 20, and 10 msec. The total twitch times, **A** through **C,** were 440, 280, and 230 msec. Intensity of stimulation was constant.

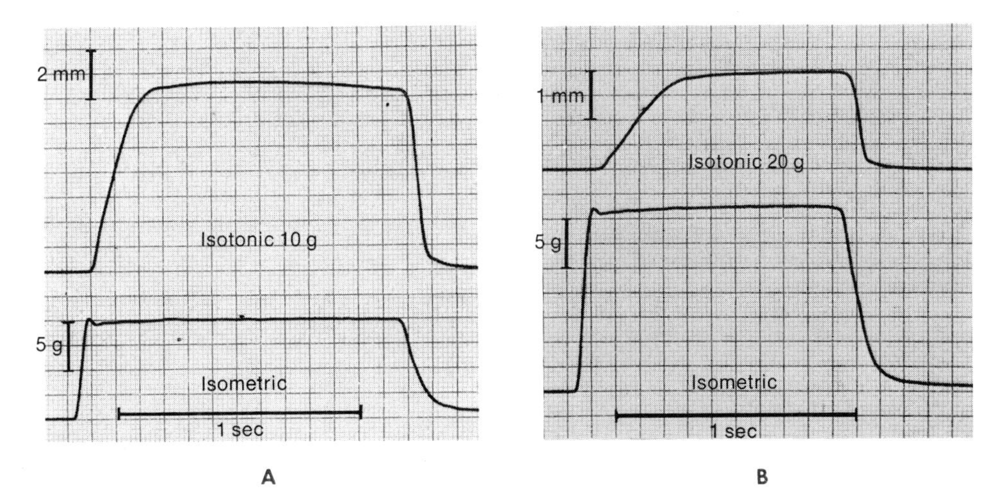

Fig. 6-14
Influence of load on velocity and magnitude of shortening of frog sartorius muscle. Muscle was attached to isometric gauge at one end and isotonic gauge at the other end. Upper records in **A** and **B** are isotonic; lower records are isometric. Shortening and tension calibrations appear on the left; size of the load is indicated on the right. In **A,** with 10 g load the isometric contraction starts 70 msec before isotonic contraction. In **B,** with 20 g load the isometric phase precedes the isotonic phase by 100 msec. The velocity of shortening is about 39 mm/sec in **A** and about 6.3 mm/sec in **B.**

purely isometrically and sometimes isotonically. For example, the muscles of the hands and arms in holding an object and the muscles of the trunk and legs in supporting the body in its erect position against the force of gravity are contracting isometrically. However, when the last-mentioned muscles propel the body by alternate contraction and relaxation, as in walking, the contractions are isotonic.

Summation. ■ The muscle twitch is the fundamental unit of recordable muscular activity. Previously, it was noted how an increase from minimal to maximal strength of a stimulus brought about a greater response of the tissue. It is equally as important to study the effect of *repetitive stimulation.*

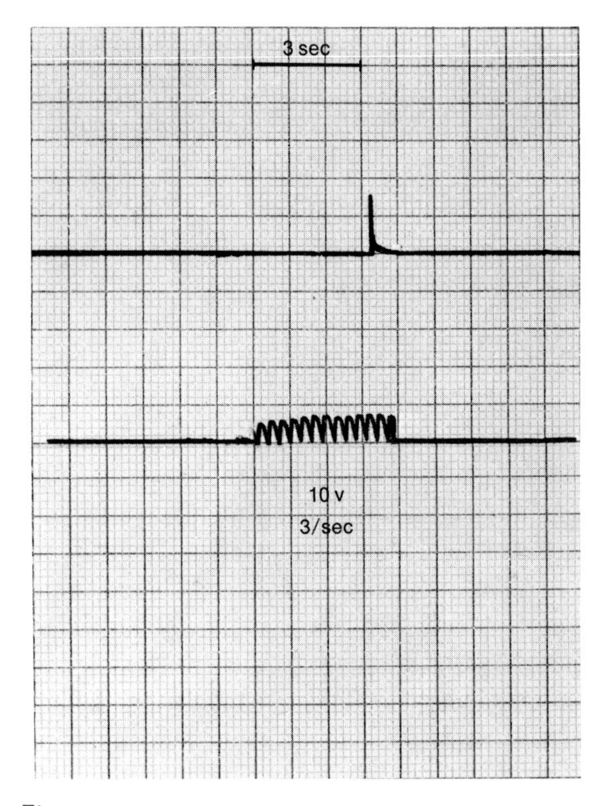

Fig. 6-15
Summation of subliminal stimuli applied to frog sartorius muscle. Stimuli were of 3 v intensity and delivered at the rate of 3/sec. Contractile response occurred on eleventh pulse.

When a muscle is stimulated by a single subliminal stimulus, no contraction takes place. Yet, if two or more of these inadequate stimuli that are just below threshold intensity are delivered in rapid succession, a muscle contraction is evoked. In Fig. 6-15 the muscle was stimulated eleven times (as shown by the row of short vertical lines) with subthreshold shocks of constant intensity before a contraction was obtained. This is referred to as *summation of subliminal stimuli.* It is evident that, although the first few stimuli were inadequate to produce those chemical and physical changes that culminate in muscle activity, *they nevertheless increased the irritability of the tissue* so that a subsequent stimulus was able to complete the process. The height of the contraction is not of necessity extremely small, for the magnitude of response will depend entirely on the number of muscle fibers that possess approximately equal excitabilities.

Summation of twitches (wave summation) occurs when two stimuli, each capable of causing a muscle to contract, follow each other in rapid sequence so that the second stimulation is delivered before the first twitch is completed. The second twitch is superimposed on the first twitch and generally is somewhat greater in extent. The result is more striking the more closely the second stimulus follows the first. Even when the first twitch is maximal, that is, when no increase in the intensity of the first stimulus (supramaximal to begin with) can elicit a stronger contraction, the second stimulus is nevertheless followed by a higher contraction. This apparently runs counter to the all-or-none law and requires an explanation. Since the greater response cannot be due to an increase in numbers of responding fibers, it must be due to a difference in the state of the contractile mechanism when the second stimulus is applied. The activities responsible for the first twitch are not complete before the second twitch begins. Obviously the duration of the *active state* of the first response makes possible the augmentation of the second response. Of course the increased response depends on measurement from the original starting position (base line); of itself, the second twitch in a summated pair is never twice as great as the first. The reasonableness of the latter statement, if not immediately apparent, should be appreciated after study of a succeeding discussion on muscle elasticity.

When a freshly excised frog muscle is stimulated with single shocks of constant strength at a frequency of about 1/sec, a series of contractions is obtained in which each of the first few twitches is a little greater than is the preceding one (Fig. 6-20, *A*). This is known as the *staircase phenomenon (treppe).* It is possible that the accumulation of metabolic products formed during activity and slight increase in temperature of the muscle act to create more favorable conditions for excitation-contraction coupling and work. In the intact animal these factors may contribute to the "warming up" by increasing local blood flow through the muscle because of dilation of small blood vessels.

Tetanus or tetanic contraction results when a number of stimuli are applied to a muscle in rapid succession so that little time is offered for relaxation between the successive contractions (Fig. 6-16). There is more or less fusion of the twitches; the greater the frequency of the stimuli, as shown progressively from *A* to *C*, the more nearly complete is the fusion (incomplete tetanus), until at *D*, the line drawn by the pen shows no evidence of individual twitches. Such a steady state of contraction is known as *complete tetanus.* It may be defined as *a sustained contraction of a muscle due to the fusion of many twitches* following each other in rapid succession; the external cause lies in the large number of stimuli presented to the muscle in a unit of time. Due to summation, the height of the tetanic contraction is usually considerably greater than is that of a twitch produced by a single stimulus of the same intensity (compare *D* with *A* in Fig. 6-16). The tension developed during tetnaus may be four or more times as great as that developed during a single twitch.

The frequency of stimulation required to induce complete tetanus (fusion) varies with the

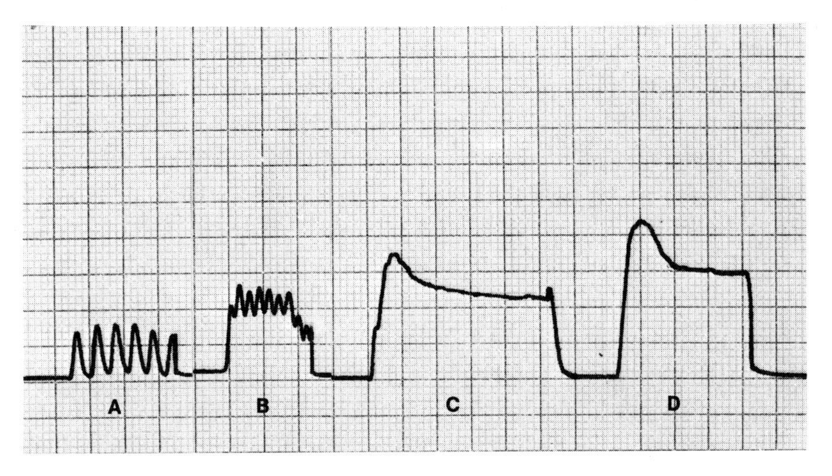

Fig. 6-16
Record showing summation of twitches in isometrically stimulated frog sartorius muscle. In **A, B, C,** and **D** the frequency of pulses was 8, 15, 25, and 40/sec, respectively. A smooth tetanic contraction appears in **D.**

nature of the muscle; for the fast-acting external eye muscles, this is about 350 shocks/sec; for the frog sartorius, about 40; for human gastrocnemius, about 100. The fusion frequency is inversely proportional to the twitch contraction time. It is also true that the condition of the muscle (temperature and fatigue) influences the fusion frequency. For example, the frequency is diminished in a cooled or fatigued muscle because of prolongation of the individual twitch response.

A complete, or smooth, tetanic contraction is not necessarily a maximal contraction in which all the muscle fibers participate. If even a relatively small number of motor units act asynchronously, the result is a smooth, sustained contraction. Although the pen record gives no evidence of the repetitive nature of muscle activity, we know from our study of action potentials that the excitatory processes responsible for the completely fused twitches are discontinuous.

Voluntary contractions of a muscle in our body, even when of extremely short duration, as the winking of an eye, are seldom, if ever, simple muscle twitches but are more or less prolonged contractions due to the fusion of many asynchronous twitches; voluntary contractions are tetanic in nature. These twitches are the result of a large number of distinct impulses (from 5 to 50/sec, depending on the intensity of the volitional effort) coming from the brain cells and relayed by motor nerves.

When two adequate stimuli are applied to a muscle in succession, say $^{1}/_{30}$ to $^{1}/_{50}$ sec apart, both elicit a twitch. But if the second stimulation occurs within $^{3}/_{1000}$ sec after the first stimulation, no response to the second stimulation takes place. This failure to respond indicates a lack of irritability. It seems that the activity of all protoplasm is associated with a loss of irritability during a certain phase of the activity. This is known as the *refractory period.* Although in a frog muscle the refractory period is very short (about $^{5}/_{1000}$ sec) and has passed before the muscle begins to contract, in the heart muscle it is relatively long; therefore, this phenomenon will be studied more closely in dealing with that organ.

Other factors determining the extent of contraction. ■ The two physical properties of *extensibility* and *elasticity** may be demonstrated by fastening one end of an excised muscle of a frog in a clamp and attaching progressively heavier weights to the free end. The muscle elongates and, within limits, the greater the weight, the more the muscle stretches. Provided muscle damage (rupture) has not occurred, on removal of the weights the muscle by virtue of its elasticity shortens almost to its original length. Since most muscles in the body are generally somewhat stretched, these two properties are of value in that they tend to keep the muscle in continuous readiness for contraction. They prevent or lessen the danger of rupturing when excessive strain is placed upon a muscle, and they render smooth the work of various parts of the body.

Initial length (or tension) of a muscle has a strong influence on the contraction. To demonstrate the effect of initial length on tension a setup similar to that illustrated in Fig. 6-7, *B*, may be employed. With the clamp at the lower end of the muscle adjusted upward a few millimeters on the rod, the muscle hangs loosely and produces a small deflection of the pen when tetanically stimulated; that is, it develops only a minimal amount of tension. If the clamp is now moved downward, millimeter by millimeter, and the muscle restimulated at each new length, the recorded isometric tension increases markedly. When at a length approximately equal to that it had in the body before dissection—the resting or optimal length—the muscle develops its maximum tension. This is illustrated for skeletal muscle by the isometric *length-tension diagram* in Fig.

*Elasticity is that property by which a body resists deformation by an external force, but by which it regains its former shape when the force is removed. Extensibility and elasticity are antagonistic.

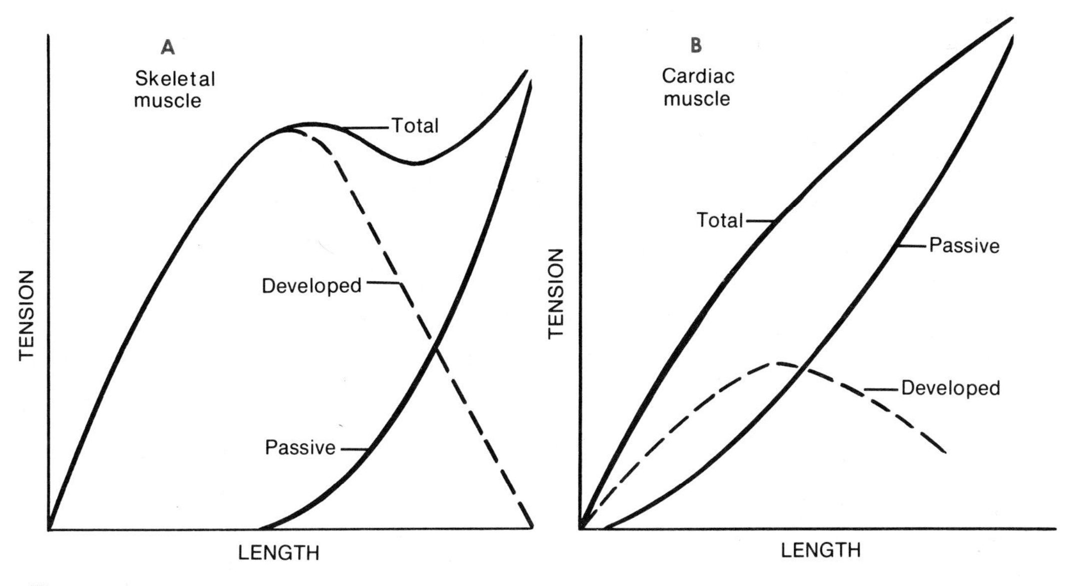

Fig. 6-17
Isometric length-tension diagrams of skeletal, **A**, and cardiac, **B**, muscle. Length of the muscle increases from left to right; tension increases in an upward direction. The total tension is the sum of the passive tension, resistance of the muscle to stretch, and the developed (or active) tension due to stimulation.

6-17, *A*. As the stretch and stimulate procedure is continued, two changes in the response occur. First, the stretch alone produces recordable tension, the so-called *passive tension* curve. Second, the *developed* or *active tension* declines as the muscle length is further increased. Total tension, being the sum of passive and developed tension, increases, but at a slower rate than does that observed at the shorter lengths. Eventually, a stretched length is attained when no contractile tension is developed. In a whole muscle this length is often found to be about 130% of the optimal length; single muscle fibers often develop some tension up to 200% of the resting length. At the level of the individual sarcomere, the capacity to generate tension disappears when the stretching withdraws the thin filaments from their customary position between the thick filaments. Slack in the muscle also reduces the capacity to develop tension, since all or most of the energy released by the contraction is used in sarcomere shortening, that is, no external load or resistance opposes the myosin-actin interaction.

Under isotonic conditions, as illustrated in Fig. 6-18, *A, the velocity of contraction depends on the load.* To demonstrate this phenomenon best, the muscle is brought to its resting length with an appropriate load; however, the afterload screw is adjusted so that the load cannot further stretch the muscle (and so that the muscle cannot spontaneously assume a shorter length). Tetanic stimulation is delivered to the muscle, and the shortening is recorded with the chart paper moving rapidly past the pen. This is repeated with a series of loads, each heavier than its predecessor. Records similar to those in the upper traces of Fig. 6-14, *A* and *B*, are obtained. The velocity of shortening (cm/sec) is calculated from the initial slope of the records and plotted against the

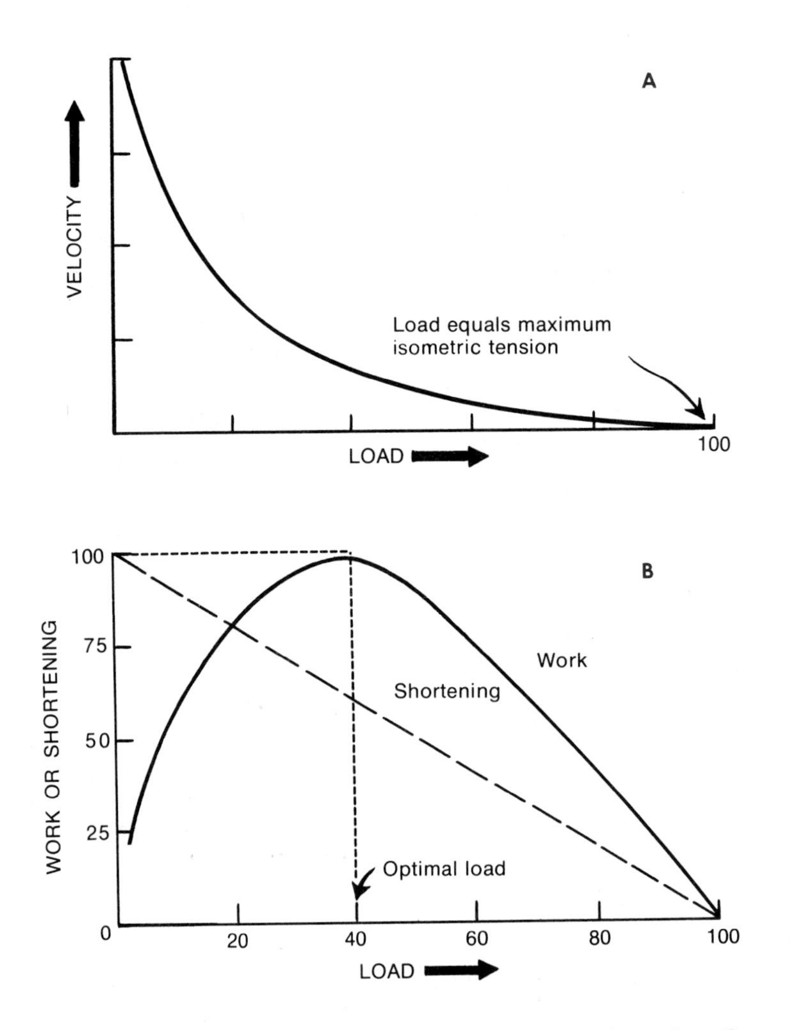

Fig. 6-18
A is a typical load-velocity diagram for skeletal muscle in relative units. **B** shows the influence of load on shortening (dotted line) and work (solid line) in the same relative units appearing in **A.**

value of the load, as in Fig. 6-18, *A*. The diagram obtained is known as a *load* (or *force*)-*velocity curve;* it is hyperbolic in nature.

Analysis of its mechanical properties indicates that muscle is a two-component system, a contractile component and an elastic component. The latter occurs both in *series* (partly in the end tendons and partly within the sarcomeres) and in *parallel* (connective tissue of the whole muscle and sarcolemma of the individual fiber) with the former. The contractile component can develop tension only by shortening and stretching the elastic component. Under certain circumstances, the contraction response is usually greater when the elastic component is somewhat prestretched.

The development of tension enables a muscle to overcome resistance and, under proper conditions, to perform mechanical work such as lifting

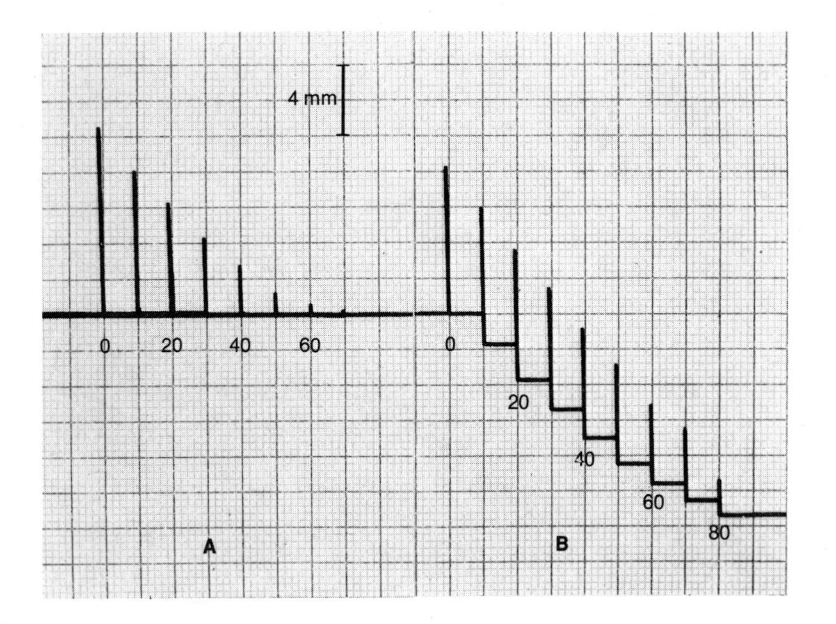

Fig. 6-19
Effect of initial length on the shortening and work performed by a frog sartorius muscle. In **A,** the muscle was after-loaded; that is, the muscle did not support the load prior to stimulation. In **B,** the muscle was stretched by each increment of load; that is, it was preloaded. The amount of stretch is evident from the change in baseline observed with each new load. Values for shortening and calculations of work performed by the muscle under after-loaded and preloaded conditions are given in Table 6-1. In **B** the muscle performed the greatest amount of work after being stretched about 8 mm. Strength and duration of stimulation were the same in **A** and **B.**

Table 6-1. Effect of zero initial tension (after-loaded) and positive initial tension (direct-loaded) on the capacity of a frog sartorius muscle to perform work*

Weight in grams	*Height in centimeters*		*Work in gram-centimeters*	
	After-loaded	*Direct-loaded*	*After-loaded*	*Direct-loaded*
0	1.02	0.81	0	0
10	0.80	0.78	8.0	7.8
20	0.62	0.56	12.4	11.2
30	0.42	0.53	12.6	15.9
40	0.28	0.52	11.2	20.8
50	0.12	0.52	6.0	26.0
60	0.03	0.42	1.8	25.2
70	0.01	0.28	0.7	19.6
80	0.00	0.10	0	8.0

*Data in this table are based on Fig. 6-19.

a weight. When a muscle contracts without a load or when the load is too heavy to be lifted, no mechanical work is done. The amount of mechanical work done by a muscle is determined by multiplying the weight of the load, in grams or kilograms, by the height, in centimeters or meters, to which it is lifted; the result expresses the work in gram-centimeters or kilogram-meters. Results obtained from an experiment on an after-loaded frog sartorius muscle, which was not brought to its resting (optimal) length before setting the afterload screw, are presented in Fig. 6-19 and Table 6-1.

It can be seen from Table 6-1 (after-loaded work column) that up to a certain point an increase in the load is followed by an increase in the amount of work done, although the height of contraction decreases. The load (30 g) at which maximum work is done per contraction is known as the *optimum load.* Fig. 6-18 illustrates a typical *work* (or *shortening*)*-load curve.* The optimal load is generally found to be one approximately 40% of that which the muscle just cannot lift. As a rule, with the same strength of stimulus, a fairly well-loaded muscle does more work than one underloaded or overloaded. In the body the mechanical leverages of the skeleton are arranged so that the muscle will usually be confronted with optimal conditions of load and velocity.

When a contracting muscle can barely lift the load (70 g), the tension developed is for all practical purposes the greatest tension the muscle can generate. This expresses the *maximum strength* of the muscle under the conditions of the experiment. In the human body, however, it is generally held that no muscle can comfortably do sustained work that exceeds one half to one third its total capacity.

Aside from factors that influence the nutrition and the general state of the muscle, the maximum strength is determined by the number of muscle fibers. For example, due to the feather-shaped (pennate) arrangement of its fibers with respect to its central tendon, the gastrocnemius muscle is composed of a large number of short fibers, and in consequence the extent of its contraction is small, but the maximum strength is great. The tension developed by the gastrocnemius muscle during running may be six times the weight of the runner. The sartorius muscle has long fibers, arranged in parallel; it can shorten a great deal but is not overly strong.

That the initial length (or tension) of a muscle has a strong influence on its contraction already has been noted. The effect upon isotonic contractions is also pronounced, as shown in Fig. 6-19, *B*. The greater amount of work done by a muscle lifting a medium-sized load over that done by a muscle with a lesser load is generally attributed to the stretching of the muscle. That a direct-loaded muscle (prestretched) performs better than does an after-loaded muscle (not prestretched) is evidenced by the data in Table 6-1. It is probable that the prestretching exerts its effect by taking up "slack" of the elastic elements and leaving a greater proportion of the energy released in activity available for movement of a load or tension development. Of course, this process cannot go on indefinitely, and soon the tension or work will diminish rapidly.

Effect of temperature on muscle contraction is illustrated in Fig. 6-13. Cooling of a muscle increases the duration of all three phases (latent period, contraction, and relaxation), but the du-

Table 6-2. Effect of temperature on the duration of periods of muscle contractions

Length of cooling period	*Increase in duration above normal (%)*		
	Latent period	*Contraction*	*Relaxation*
5 min	12.5	21	62
15 min	25.0	46	115
20 min	37.5	82	172

ration of the relaxation is disproportionately increased (Table 6-2). This sluggishness may be responsible for poor athletic performances in short bouts of exercise when there is an improper "warming up." An increase in temperature of a muscle within physiological limits speeds up metabolic processes and decreases the "viscosity" of the protoplasm, thereby reducing the duration of the various phases of activity.

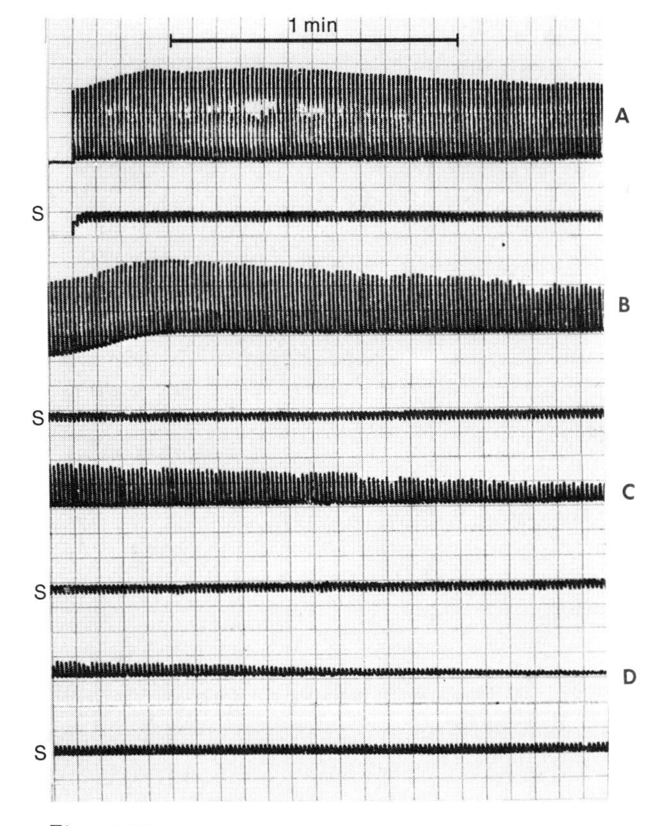

Fig. 6-20
Effect of prolonged stimulation on contractile response of frog sartorius muscle. Record is continuous from **A** through **D.** Stimulation indicated by lines marked **S.** Staircase phenomenon appears at beginning of record in **A.** Contracture appears in **B** when muscle fails to relax to original baseline between stimuli. Gradual decrease in contractions signifies fatigue of muscle.

*Fatigue** develops when a muscle is subjected to a prolonged period of stimulation (Fig. 6-20). In this experiment, stimuli of constant intensity were applied at a rate of about 1/sec. The contractions may show a constant height for some time, but soon they gradually decrease in height if the work is at all strenuous. Finally, no more twitches can be elicited, even if the intensity of the stimulus is increased. This is complete fatigue. It is characterized by decreased irritability, conductivity, and contractility; however, since recovery is possible under proper conditions, it must be recognized as a transitory loss of these properties.

As fatigue develops, the shortening and relaxation processes become slower; this is especially true for the latter. Due to the slowness of relaxation, the muscle fails to regain its original length before the next stimulus arrives, and it remains in what is known as a state of *contracture* (Fig. 6-20, *B*). This state is an early sign of impending fatigue.

If a fatigued excised muscle is perfused† with physiological salt solution, the fatigue diminishes somewhat. This is largely due to the washing away of intermediate and final waste products, chiefly acid in nature, such as CO_2, lactic acid, pyruvic acid, and acid phosphates. These are sometimes spoken of as *fatigue substances.* If glucose is supplied to the muscle, fatigue is re-

*Few subjects are of greater economic importance than fatigue. Fatigue is the antithesis of industry. But all fatigue is not caused by overexertion; many of us on occasion have had a "tired feeling" after a night's rest. Other easily recognized causes of fatigue are lack of proper food; lack of carriers of oxygen to the brain and muscles (anemia); lack of proper distribution of food and oxygen (circulatory disturbances); bacterial infection (toxins); and, last but not least, in the industrial world boredom or lack of interest created especially by monotony.

†An excised organ, such as a skeletal muscle, is perfused with a certain solution by inserting a tube into the artery of the organ and, under pressure, forcing the solution through the arteries, capillaries, and veins of the organs. As the fluid issues from the vein, it may be caught and used over and over, if so desired.

moved more speedily. It appears, therefore, that the cause of muscle fatigue may be twofold: (1) the accumulation of intermediate or final waste products and (2) the deficiency in energy-furnishing materials.

When the nerve of a muscle-nerve preparation is stimulated for a sufficiently long time, the muscle eventually fails to contract; that is, fatigue has set in. If now the muscle is stimulated directly, contractions are obtained; hence the fatigue shown when the nerve was stimulated cannot be referred to the muscle and must, therefore, be attributed to either the nerve fibers or the myoneural junctions. A nerve fiber is practically incapable of experiencing fatigue as this term is generally understood. It must be concluded, therefore, that the fatigued structures are the neuromuscular junctions. Junctional fatigue results from a depletion of transmitter substance. That a muscle itself can also experience fatigue is readily shown by further stimulating the muscle directly.

■ Thermal changes

Heat production. ■ Muscular activity is accompanied by the production of heat, since only a small part of the chemical potential energy set free is transformed into mechanical work. Even at rest there is a considerable heat production, which in man can account for over one fourth of his basal metabolic rate. Resting heat is a manifestation of the metabolism essential for keeping the muscle in a state of readiness for contraction, that is, to assure the integrity of the excitable membrane and the transport of substances used in the repair and restoration of cell constituents. During the contraction and relaxation phases of activity the heat production increases greatly. This *initial heat* is independent of the presence of oxygen and is associated with the breakdown of ATP. In the isometric phase of an isotonic twitch (Fig. 6-14) and in an isometric twitch, the initial heat released in known as the *activation heat;* the summed activation heat set free in an isometric tetanus is termed *maintenance heat.* If shortening occurs, an additional initial heat, the *shortening heat,* is produced, which in quantity is directly proportional to the amount of shortening.* Recovery of the muscle is associated with evolution of *delayed heat* (or *recovery heat*). Delayed heat can be further subdivided into an anaerobic phase and an aerobic phase. Heat in the anaerobic phase arises from the breakdown of glycogen to lactic acid but is of small magnitude compared to the aerobic phase, perhaps as little as one twentieth. Aerobic delayed heat is produced for an extended period, up to thirty minutes, and represents the oxidation of the fuels to CO_2 and H_2O. Heat, of course, is energy wasted through the inefficiencies in chemical processes. Nevertheless, this heat may be of great value in the maintenance of body temperature.

Fenn effect. ■ A muscle that contracts isotonically and performs work, i.e., it shortens against a load, generates more energy than the identical muscle contracting isometrically at the *same initial length.* This phenomenon of *extra energy* release in an isotonically contracting muscle, when compared to one that is contracting isometrically, is known as the *Fenn effect,* so named for its discoverer. How the muscle "senses" the need for extra energy expenditure is an unknown, but important, aspect of muscle energetics. Currently it is subject to speculation only and to intensive investigation.

An idealized diagram illustrating the Fenn effect appears in Fig. 6-21, which is a graph of the initial energy plotted against the load to which the muscle is subjected. In an isometric contraction (load P_o), the energy expended is recorded as heat; it consists of a stable activation heat (maintenance heat in a tetanic contraction), the

*The statement that the shortening heat is directly proportional to the amount of shortening has been challenged. According to some investigators the extra heat is proportional to the work. We prefer to reserve judgment until further evidence is presented.

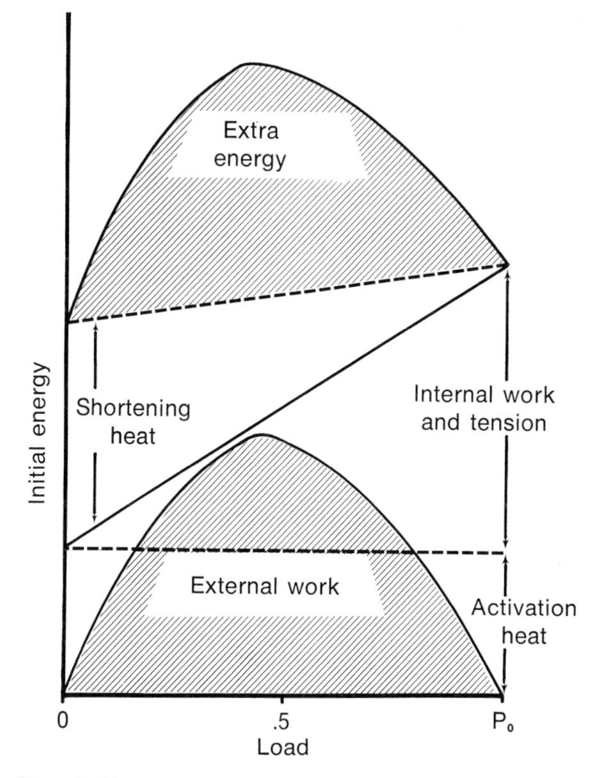

Fig. 6-21
Idealized diagram showing the initial energy released by muscle contracting against loads varying from zero to maximum isometric (P_o).

lower dashed line in the figure, and a heat associated with internal work and tension generation. Activation heat in all likelihood represents the energy cost of the calcium pump. With the application of decreasing loads which the muscle can shorten against, extra initial energy is released to perform external work and also as shortening heat. Heat produced by internal shortening and tension generation declines along with the load, as would be expected. Thus, in isometric contractions at lengths shorter or longer than optimal the tension would be some fraction of P_o (Fig. 6-17, *A*) and the total heat would follow the solid diagonal line in Fig. 6-21. Shortening heat in isotonic contractions is inversely proportional to the load, again as expected. The uppermost curve in Fig. 6-21 represents the total initial energy released as heat measured in a contraction-relaxation cycle; delayed heat is not depicted. The external work which the muscle performed in the shortening phase of the contraction is very nearly equal to the extra energy degraded into heat during the relaxation of the muscle (shaded areas in Fig. 6-21). The net effect, then, is that the external work which the muscle does and the shortening heat which it generates represent an extra energy expenditure that is rather precisely regulated, despite the fact that at the time of stimulation the muscle is "unaware" of the need.

■ EFFECTS OF MUSCULAR WORK AND TRAINING

No other factor affects practically all the organs of the body more profoundly and more frequently than does muscle work. Its influence on the heart, respiration, perspiration, appetite, and heat production is a matter of common observation. As even a brief discussion of these effects necessitates some knowledge of the bodily functions concerned, such discussion will be undertaken with the study of these individual functions. At this time attention will be focused on the changes in the neuromuscular system itself. The effects of training or exercise depend on their intensity and duration and are either of an immediate (acute) or of a lasting (chronic) nature. The best-recognized chronic effects are an increase in the size and strength of the muscles and a greater efficiency and endurance in the performance of work.

Size.* ■ It is a well-known fact that a muscle which is not used (as in casts and splints) undergoes atrophy. Under proper working conditions, muscles may show a considerable increase in size, as shown by the arm muscles of a weight lifter

*At birth the musculature constitutes about 25% of the body weight; this increases to 43% at maturity.

or the general musculature of other athletes. The greater size of a trained muscle is *caused by hypertrophy of the individual muscle fibers* and is not due to the appearance of new fibers. An increase in the toughness of the connective tissue that binds the fibers together makes the muscle better able to withstand any additional mechanical demands placed upon it.*

Strength. ■ An increase in the size of a muscle increases its strength. But the strength gained by exercise is frequently greater than can be accounted for by the increase in the size of the muscle. More than one explanation has been offered for this; it may be pictured in the following manner.

Voluntary impulses are carried by efferent nerves to the motor units. As the lifting of objects of different weight shows us, we are able to gauge the number of impulses necessary to innervate the number of motor units required for any particular piece of work. To discharge the same number of impulses for lifting a light object as is needed for a heavier one might work disastrously. Some investigators maintain that under the ordinary conditions of life we can voluntarily, even with our best endeavor, innervate only a part of the motor units of any skeletal muscle. We have heard of the marvelous strength of people under great mental stress, such as intense fear or religious mania, and also during delirious conditions. It is conceivable that in these unusual states of mind more nerve pathways are activated or the inhibition normally exercised by the cerebral cortex is materially reduced, and hence more motor units are brought into action. Perhaps in training, a person learns to utilize nerve pathways hitherto lying idle.

Efficiency. ■ By increased efficiency is meant that a trained person accomplishes a given amount of work with a smaller expenditure of energy than is possible for an untrained individual. The ratio of the amount of work done to the total amount of energy expended (work done/work done + heat produced) expresses the mechanical efficiency. In bicycling the efficiency ranges from 20% to 25%; in a few athletes it approaches 40%. This compares favorably with the modern steam engine in which the maximum efficiency is stated as about 20% (25% for the gas engine and 40% for the diesel engine). Muscle efficiency varies slightly from one person to another and considerably with the nature of the work and other external conditions. As common experience teaches us, unaccustomed work is done inefficiently. The causes of greater efficiency in the trained individual lie in the following:

1. The optimal rate for doing the work has been established.
2. The greater economy on the part of a trained man is largely due to better neuromuscular coordination—a matter of skill. In performing a piece of work that is entirely new, a person is likely to contract many muscles that are not concerned in the work. By practice, he gradually eliminates these useless and extravagant contractions; it is then said that he does the work with greater ease or that the work is done more efficiently.
3. By training, the body may rid itself of undesirable fat that impedes the movements of muscles or parts of the body or, as in walking and climbing, adds an unnecessary load and thus causes a greater expenditure of energy.
4. In a trained subject there is a better utilization of O_2 and a smaller oxygen debt after cessation of the work.

Rapidity with which work is done. In climbing stairs (Fig. 6-22) the greatest efficiency occurred when the work was done in 100 sec. Doing this same amount of work in 25 or 250 sec markedly reduced the efficiency. In this experiment the efficiency was highest when each step occupied 1.3 sec.

Fatigue. As fatigue sets in, efficiency decreases; therefore, suitable rest periods are desirable to maintain efficiency.

*Incidentally, it also increases the toughness of a piece of meat.

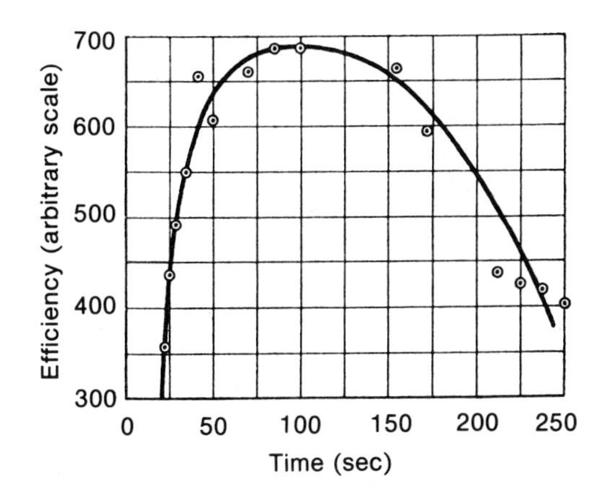

Fig. 6-22
Graph showing efficiency of stair climbing (ordinate) at different speeds (abscissa).

In the human body the feeling of fatigue must not be confused with the actual loss of physiological properties due to activity. The former, also known as subjective fatigue, is often merely a case of boredom; objective fatigue is a physiological loss of power to continue work at the previous rate. The prevention or recovery from objective fatigue is a matter of the disposal of waste products and the supplying of oxygen and nutrients. Intermediate waste products, lactic acid and pyruvic acid, are either oxidized still further into CO_2 and H_2O or, by the expenditure of energy, built up into higher compounds, such as glycogen. (Incidentally, lactic acid formed in muscles may be removed by the circulating blood and carried to the heart, there to be used as such in the production of energy for cardiac activity.) The CO_2 is carried to the lungs and the acid salts to the kidneys for elimination. To prevent fatigue, not only oxygen but also nutrients must be brought to the muscles from the organs of supply. All these activities depend on the circulation of the blood and on respiration. Any interference with the blood supply hastens the onset of fatigue. An example of this is seen in the tetanically contracted leg muscles during motionless standing. Because of the high degree of tension in the tonically contracting muscles, the capillaries and small veins are compressed and circulation is made difficult; as a consequence, this form of work, in distinction from rhythmic muscular activity (e.g., walking), fatigues an individual rapidly.

Although in the excised frog muscle fatigue is due to a steady, but slow, decline in the release of acetylcholine at the neuromuscular junction, in the human subject true mechanical fatigue appears as a failure of the contractile mechanism. This conclusion is implicit in the earlier discussion of the need for adjustments in the cardiovascular system in order to prevent or recover from objective fatigue. More direct evidence is available from studies wherein the subject contracted a muscle group by maximum voluntary effort until total unresponsiveness was achieved. At this time, an electric shock was applied to the motor nerve of the muscle group and recordings made of the nerve and the muscle action potentials. Both were found to be present. The experiment demonstrates that in the intact subject, under physiological limits (maximum voluntary effort), propagation and transmission are relatively indefatigable at levels of neural activity where the contractile mechanism fails.

The length of time necessary for recovery from fatigue varies from one muscle to another and also with the degree of fatigue experienced. It has been found that 80% recovery occurs in ten minutes and 95% occurs in twenty minutes from fatigue brought on by ergographic* work. But after severe and prolonged athletic contests, an hour or even a day may be needed for complete restoration. If volitional impulses are sent to the muscle after fatigue has set in, that is, if a person strains to continue the work even though no

*An ergograph records the amount of work done by a group of muscles or by a single muscle (such as the flexor or abductor of the index finger). Several kinds of ergographs have been devised, for example, the finger ergograph.

Table 6-3. Effect of work rate on the development of fatigue

Number of contractions	*Fatigued by*	*Work done*
1 per sec	14 contractions	0.912 kg-m
1 per 2 sec	18 contractions	1.080 kg-m
1 per 4 sec	31 contractions	1.842 kg-m
1 per 10 sec	No fatigue	Almost indefinite

visible contractions result, the time necessary for recovery is increased.

The amount of work a muscle can do before complete fatigue sets in depends largely on the number of contractions executed in a unit of time, as is shown in Table 6-3. This work was done with a finger ergograph.

For every load there is a certain rate at which the most work in a given length of time can be accomplished. Too rapid a rate of work fatigues a muscle quickly and therefore cuts down the total amount of work that the muscle can do; too slow a rate, although preventing fatigue, also decreases the output.

A variety of drugs influence fatigue. Caffeine, an alkaloid found in tea and coffee, is a mild cerebral stimulant and is generally said to increase the work output and to delay the onset of fatigue. Benzedrine is claimed to have considerable power to delay fatigue of the nervous system, but some investigators attribute this to a dulling of the sense of fatigue. Tobacco smoking affects most people adversely, whether they are habituated to its use or not. It reduces physical fitness and the capacity for work in all individuals.

Rhythm. Most persons work with a certain rhythm; for example, some persons walk with longer and slower steps, others walk with shorter and faster steps. It is a familiar fact that adopting a new rhythm brings on fatigue more speedily.

Body weight. Increase in body weight due to the accumulation of fat lessens efficiency.

Warming up. It has been demonstrated that preliminary warming up improves performance in athletic contests.

Age. Efficiency gradually lessens with age.

The muscles in our body work at a great *mechanical disadvantage.* For example, the biceps muscle that flexes (raises) the forearm has its two origins on the scapula and its insertion on the radius a short distance below the bend of the elbow (Fig. 6-2). The forearm is a lever of the third order; the fulcrum is placed at the elbow (joint of radius and humerus), the arm and hand constitute the weight, and the power is applied, as already stated, a short distance from the fulcrum. This causes the power arm to be very short, an arrangement that does not lend itself to the development of great power. However, because of the long weight arm, this anatomical structure does admit great speed of movement of the hand.

Endurance. ■ Delaying the onset of fatigue in a trained person is largely a matter of acquiring better adjustment of circulation, respiration, and other functions to the increased needs of the body. Hence the speed with which fatigue develops is frequently used as a measure of a person's ability to carry on a certain piece of work; it may also be an indication of the state of health.

One aim of physical training is the proper and harmonious development of the body during the period of growth. In training, opposing muscle groups should be properly balanced as to length and strength. Inequality in this respect leads to faulty position of various parts of the body with respect to each other and to inefficiency in action. Training should be of a nature so as to be useful to the individual not only during the period of training but also for the entire span of his life.

Posture. ■ Lack of training frequently leads to an improper body posture. A slumped posture displaces the center of gravity of the body and thereby imposes work upon the muscles to counteract gravity; this leads to greater fatigue than is experienced when the correct position in standing and walking is maintined. Obesity, relaxed abdominal muscles (protruding abdomen), and weak back muscles are important factors in creating a poor posture.

Muscle tonus. ■ The so-called resting muscles in the animal body are frequently not in a perfectly relaxed condition. Although showing no outward signs of activity, they are in *a state of mild contraction, which causes them to resist being stretched;* this is known as muscle tonus. Tonus is well exemplified by the masseter and temporal muscles (Fig. 20-2), which raise the lower jaw. Although we may be unconscious of the position of the jaw and are not volitionally sending impulses to the muscles, yet they are, as long as we are awake, constantly holding the jaw up in opposition to the force of gravity. The cause of muscle tonus lies in a continuous, but low frequency, stream of impulses descending from the central nervous system; therefore, cutting the motor nerve abolishes the tonus of a skeletal muscle; in consequence the muscle is flaccid. Tonic contraction is a reflex phenomenon.

In some body structures the nutrition and therefore the ability for continued functioning depend to a certain extent on the effect of their own activity. This is particularly true for skeletal muscles in which *nutrition and function are inseparable.* No matter how well supplied with blood and food, a muscle permanently inactive undergoes atrophy.

Muscle cramp. ■ A spasmodic, painful, involuntary contraction of muscle is known as a muscle cramp. This definition, however, is not wholly accurate, since some forms of cramp are painless. There are numerous ways in which muscle cramps can originate, and what is known about the mechanisms cannot be claimed to be clear or complete knowledge. Some of the causes are overactivity of motor neurons due to excessive facilitatory impulses from higher levels of the central nervous system (Chapter 8), spontaneous discharge of action potentials on the muscle fiber membrane, failure of the sarcoplasmic reticulum to reaccumulate calcium after its release, and abnormal interactions between actin and myosin.

Exercising muscle uses glycogen in preference to blood glucose (Fig. 23-2). In a genetic abnormality of muscle metabolism, McArdle's disease, the skeletal muscles have a deficiency of the enzyme phosphorylase (Chapter 23) and are therefore incapable of adequately using glycogen during exertion. When a victim undertakes exercise, his muscles fail to relax properly and soon become hard, shortened, and painful; i.e., a cramp or contracture ensues. It has been proposed that enough ATP is present to satisfy the contraction effort, but not enough is available for the activity of the calcium pump.

Myotonias are genetically determined disorders wherein the defect is in the sarcolemma, the excitable membrane of the muscle. The muscles of myotonia victims exhibit increased irritability, generating action potentials even from simple mechanical stimulation. Indeed, repetitive discharges follow stimulation, whether normal (neural) or abnormal (mechanical), and delayed relaxation (cramp) is characteristic of these muscles. The biochemical defect in the myotonic muscle membrane is not known; in fact, there are a number of myotonias, each differing in certain muscle membrane properties, but each exhibiting the characteristic spontaneous contraction which is not accompanied by pain.

Ordinary muscle cramp often occurs after hard exercise, in untrained muscles caused to contract suddenly and forcefully, and in the feet or legs during sleep. It is quite painful. The spasmodic contraction is associated with irregular, high-frequency, and high-voltage muscle action potentials. The cause of this phenomenon is unknown; it is not due to an abnormality of the muscle membrane but is associated with the motor nerves and neurons.

The painful nature of some muscle cramps has been attributed to an unidentified pain substance which escapes from the contracting muscle and accumulates in the interstitial fluid, where it stimulates nerve fibers subserving pain (Chapter

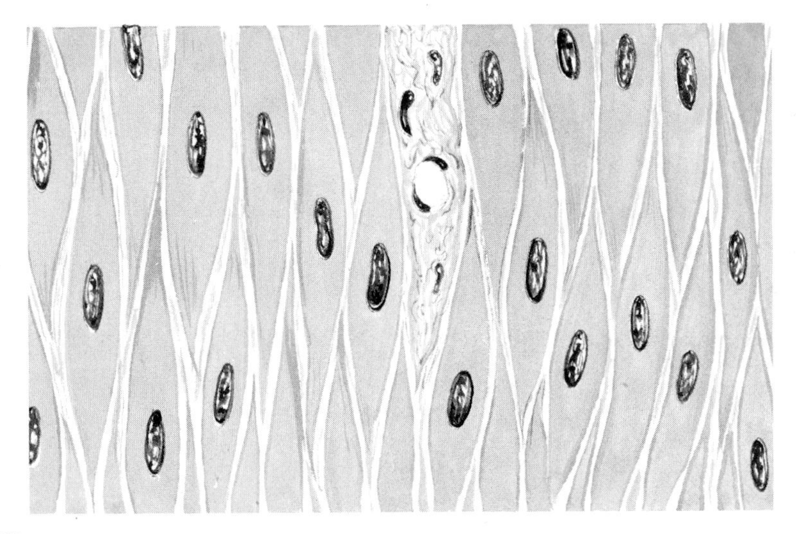

Fig. 6-23
Smooth muscle fibers.

7). This pain substance has been described as lactic acid and as potassium ions that have leaked from the muscle; there is no direct evidence for either, although the substance is most logically a metabolite produced by contracting muscle.

■ VISCERAL OR SMOOTH MUSCLE

Visceral muscles are found chiefly in the walls of the four great tracts of hollow organs—the circulatory, the respiratory, the alimentary, and the urogenital (Figs. 13-4 and 19-10). They are also found in the interior of the eye (the ciliary muscles and the muscle in the iris), in the skin (the pilomotor, contraction of which erects the hair and produces goose bumps), and in the ducts of glands.

Ultrastructure. ■ Visceral* muscles not only exhibit a sharp contrast with skeletal muscles but differ from each other in their structure, location in the body, mode of action, innervation, and general functions in the animal economy. Their spindle-shaped fibers are 50 to 100 μ long and 2 to 5 μ in diameter and contain a single elongate nucleus (Fig. 6-23). Each is encased in a sarcolemma surrounded, except at special sites, by a heavy layer of glycoprotein. The special sites are points where two adjacent cells come into close approximation, perhaps less than 100 A separation, or even fuse their membranes. These points of contact are termed "tight" junctions and probably are areas of low electrical resistance between the cells, which serve in intercellular transmission and propagation of action potentials. A poorly developed sarcoplasmic reticulum has been detected in electron micrographs. No sarcomeres are evident in the smooth muscle cell, although faint longitudinal striations appear, which are myofilaments of 50 to 80 A in diameter. Although myosin is not readily demonstrated, the filaments do seem to be composed of F-actin. Both total contractile protein content and energy-rich compound concentration are much lower than they are in skeletal muscle.

*Visceral—the adjective of the noun *viscera,* which is the plural of viscus. A viscus is an organ in any one of the body cavities (cranial, thoracic, abdominal, or pelvic), but especially the abdominal cavity.

Physiological classification. ■ Smooth muscles exhibit considerable diversity in their arrangement and position in the body as well as in their mode of behavior. Two major groupings, based on physiological function, are recognized. The *unitary* muscles occur in sheets or layers and are characterized by their ability to contract spontaneously. This activity originates within the muscle, since any smooth muscle cell may act as a pacemaker to initiate contraction, and spreads throughout the whole muscle as if it were, indeed, a single unit. Unitary muscles are especially responsive to the mechanical stimulus of stretch, but their activity is modulated and coordinated by neural influences. Muscles of the gastrointestinal tract and the ureters are especially good examples of unitary muscles.

Smooth muscles that do not contract spontaneously, that are unresponsive to stretch, and that have a multiple innervation by motor nerves so that each cell is innervated (in contrast to unitary muscles) are known as *multi-unit* muscles. They may occur in sheets (the larger blood vessels; aorta, carotid arteries), in small bundles of fibers (pilomotor), or even in single cells (spleen). In a diffuse way they are organized into motor units. The previously described "tight" junctions are neither present nor required by this class of smooth muscle.

Certain smooth muscles, for example, the urinary bladder, resemble both unitary and multi-unit muscles in that they may react spontaneously as well as reacting to motor nerve impulses.

Although their diversity restricts the possibility of generalized statements about smooth muscle, at least the following can be said:

1. Smooth muscle motor innervation is exclusively autonomic.
2. All smooth muscles undergo slow contractions, lasting in some instances for several seconds, which can be sustained with a minimum expenditure of energy.
3. Smooth muscles exhibit a basal resting tension known as *intrinsic tone.*

Functions. ■ Irrespective of its form or its position in the body, the function of a muscle is merely that of contraction. However, the result of this activity is determined by a muscle's arrangement and position in the body; frequently the final result is spoken of as the function of the muscle. By their rhythmic contraction and relaxation, the muscles in the walls of the alimentary canal and in the urinary and reproductive organs propel the contents of these organs forward *(peristalsis).* In the walls of blood vessels they govern the amount of blood passing through a vessel and thus its distribution. Smooth muscles in the form of a strong circular band or ring *(sphincter)* control the opening and closing of a tube or orifice.

In conjunction with the receptors and the nervous system, *skeletal muscles adjust the organism to its environment; the visceral muscles are concerned with those processes necessary for the maintenance of the receptors, the skeletal muscles, and the nervous system.* These processes, collectively known as the *vegetative functions,* include the digestion and absorption of food, circulation of blood, respiration, and excretion.

Action of visceral muscles. ■ Because of the differences in smooth muscles, descriptions of their actions are not necessarily applicable to all. Provided we recognize the likelihood of exceptions to any general statement, it seems profitable to examine certain aspects of the electrical, chemical, and mechanical phenomena of visceral muscle.

In unitary muscles the resting potential can only be defined as the maximum level of polarization between periods of electrical activity. At this point the interior of the fiber is negative by 55 to 60 mv, relative to the extracellular medium. In some smooth muscles the resting potential shows a rhythmic oscillation, like a sine wave, and action potentials arise at any time during the cycle. Other smooth muscles produce action potentials only after a slow depolarization (the *prepotential*) to a threshold value. The development of a prepotential resembles the behavior noted in cardiac muscle (Fig. 14-16), except that it may occur even more slowly. Intercellular transmission in unitary muscles probably occurs at "tight" junctions as an electrical depolarization of one cell by another. Cells that are innervated directly respond to the release of neurohumors from the motor nerves. Neurohumors act on the permeability of the membrane and induce depolarization with subsequent firing off of action potentials.

The ionic basis for the resting potential is similar to the ionic basis of nerve and skeletal muscle in that it is predominantly a potassium diffusion potential. However, other ions make a contribution. For example, smooth muscle fibers immersed in a medium that is low in chloride ions have a smaller resting potential than normal. A sharp increase in external calcium ion produces an increase in the resting potential and consequently a decrease in excitability; this probably accounts for the inhibition due to calcium ion, which is seen in Fig. 6-23. Although excess potassium ions stimulate smooth muscle cells of the aorta, they tend to exert a plasticizing effect on arteriolar smooth muscle. The permeability of resting smooth muscle to sodium ions is very high; therefore, the sodium pump must be quite efficient, since the internal concentration of sodium ions is much lower than that found in the extracellular fluid. The origin of the action potential is attributable to sodium ion entry. Stretch of unitary smooth muscle causes the membrane to become depolarized, action potentials to be propagated, and contraction to be initiated. In this respect, smooth muscle cells behave much like the stretch receptor cells found in skeletal muscle (Chapter 8).

Excitation-contraction coupling in smooth muscle is not well understood. There is convincing evidence that calcium ions are required, as with skeletal and cardiac muscle, since action potentials will not evoke contractions in the absence of calcium. The effects of the excessive calcium concentration described before are, of course, exerted on the membrane (perhaps calcium competes with sodium and thus prevents depolarization) and not on the contractile process; thus no conflict arises between these observations. A long latency between the excitation wave and the mechanical response is uniformly observed in smooth muscle. The duration of this latency can be twenty times that seen in skeletal muscle. This phenomenon correlates with the primitive state of the sarcoplasmic reticulum; apparently, the bulk of the calcium ion diffusion occurs through the sarcolemma, and consequently the coupling process is delayed.

Since smooth muscle contains enzyme systems similar to those found in skeletal muscle, the chemical processes involved in smooth muscle contraction are probably like those found in skeletal muscle. In contraction, the oxygen consumption of smooth muscle is not increased significantly beyond that used at the resting level, which requires only one fourth the oxygen consumption of comparable skeletal muscle tissue. Visceral muscle can continue active for a considerable period of time in the absence of oxygen, and for this reason it is difficult to demonstrate fatigue in this tissue.

Tonus. ■ The tension developed by smooth muscles is a function of the frequency with which action potentials are discharged. This is true of both unitary and multi-unit muscles. Smooth muscles exhibit tonic activity; however, the extent of the contraction of the muscle fibers may vary greatly. The influence of temperature, pH, and ions is especially noteworthy (Fig. 6-24). Because of smooth muscle *tonus* the hollow organs more or less resist stretching. Thus, when the stomach is empty, its cavity is virtually obliterated by the tonus of the gastric muscles. As food is put into the stomach, the wall yields and stretches; for this reason, intragastric pressure remains fairly constant as the stomach is being filled. The same principle applies to the urinary bladder. This very important property may be stated as follows:

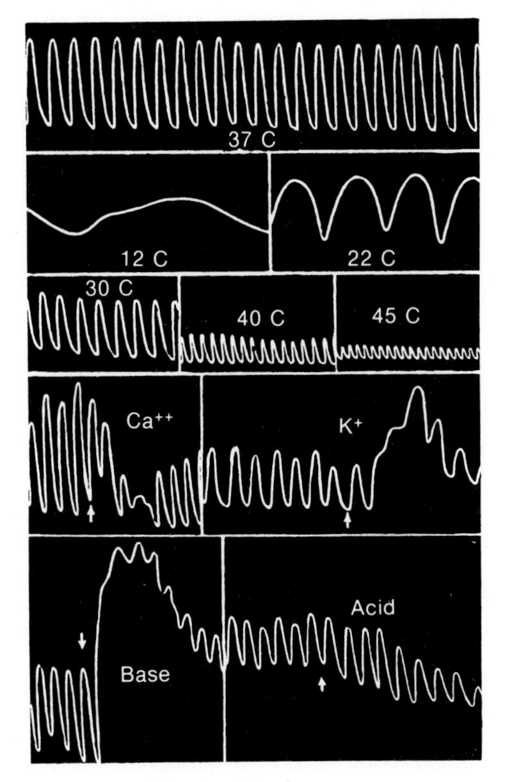

Fig. 6-24
Effects of environmental changes (temperature, increased ion concentration, and pH) on activity of smooth muscle. Muscle (rabbit intestine) was bathed in Locke's solution, which was appropriately altered to effect the change. (Modified from Francis, Knowlton, and Tuttle: Textbook of anatomy and physiology for nurses, The C. V. Mosby Co.)

Within limits the walls of the hollow organs accommodate themselves by their tonus to the volume of the contents without any marked alteration in the internal pressure or in the tension of the wall. The tonus of visceral muscles is influenced by the nervous system; however, in many organs the cutting of the nerve supply does not permanently abolish the tone. Hence, in contrast with skeletal muscles, the origin of the tonus of these organs must reside in the organ itself or, alternatively, be due to chemical excitatory substances in the blood.

Innervation. ■ The visceral muscles differ from striated muscles in their innervation in that the latter receive their impulses directly, without any relaying, from the spinal cord or brain, whereas the nerves for the smooth muscles run in the autonomic nervous system (see Fig. 11-2). Motor end plates, as described for skeletal muscle, do not occur in smooth muscle; the nerve fibers lie in more or less intimate contact with the muscle cells and quite probably release transmitter at many points along the bare axon. In contrast to the absolute dependence of skeletal muscles on neural impulses for their activity, the visceral muscles, in general, show a large amount of independent activity. Another striking difference between these two types of muscles is that many visceral muscles have a double nerve supply, one for excitation and one for inhibition.

Smooth muscle in the gastrointestinal tract is stimulated by acetylcholine, either coming from its motor nerves or through the circulation. This neurohumor depolarizes the muscle cell membranes and increases the force of contraction by increasing both the frequency and propagation velocity of action potentials. The neurohumors, epinephrine and norepinephrine, either excite or inhibit smooth muscle. In the human, for example, epinephrine inhibits intestinal smooth muscle but stimulates uterine smooth muscle. Inhibition occurs because the cell membranes become hyperpolarized and action potentials are abolished. Excitation by epinephrine transpires through the same mechanisms as those described for acetylcholine. Although the aorta and other large arteries are typical multi-unit muscles, the smaller vessels and veins behave like unitary muscles in that they are responsive to stretch; this may reflect differences in the degree of innervation. The response of vessels from different regions of the body to neurohumors varies; for example, cerebral blood vessels are insensitive to epinephrine and norepinephrine; coronary vessels are made to relax by these agents; and renal and mesenteric vessels contract. It should be obvious that generalizations about smooth muscle can be misleading; it is probably always safest to specify the structure being discussed.

■ CARDIAC MUSCLE

Cardiac muscle is cross striated (see Fig. 14-8), as is skeletal muscle. Its fibers exhibit branching and contain myofibrils and filaments of actin and myosin arrayed similarly to those in skeletal muscle. Likewise, the mechanisms of contraction are essentially the same as those in skeletal muscle. The physiological characteristics peculiar to cardiac muscle are discussed in Chapter 14.

■ CILIA

Location. ■ Cilia are protoplasmic filaments projecting beyond the free surface of epithelial cells, which are generally columnar in form; each cell gives rise to from ten to thirty cilia. Ciliated epithelium occurs in many locations, for example, nearly the whole respiratory tract, the Eustachian tube (which connects the middle ear with the pharynx—Fig. 9-29), the tear duct and sac, the cavities of brain and cord, and certain parts of the reproductive organs.

Action. ■ Because of their contractility, the cilia sway forward and backward. During the forward motion (the active phase) the cilia are rigid, but they are limp during the backward stroke (the recovery phase). Their action may be conveniently studied in a pithed frog whose lower jaw has been removed and the animal placed in a supine position. When a small particle, blood clot, is placed on the roof of the mouth, the swift forward motion of the cilia causes the clot to move downward toward the esophagus. In our body dust and mucus are in this manner moved upward in the respiratory tract. In general, the function of cilia is to move suspended matter and the fluid bathing the epithelium. The activity of cilia is greatly influenced by changes in temperature. Cooling decreases this activity, and warming increases it. They are rendered inactive by the application of anesthetics such as ether.

READINGS

Bendall, J. R.: Muscles, molecules and movement, New York, 1969, American Elsevier Publishing Co.

Davies, R. E.: A molecular theory of muscle contraction: calcium dependent contractions with hydrogen bond formation plus ATP-dependent extensions of part of the myosin-actin cross-bridges, Nature **199**:1068-1074, 1963.

Ebashi, W., and Endo, M.: Calcium ions and muscle contraction, Progr. Biophys. **18**:123-183, 1968.

Hoyle, G.: How is muscle turned on and off? Sci. Am. **222**(4):84-93, Apr. 1970.

Huxley, H. E.: The mechanism of muscular contraction, Science **164**:1356-1364, 1969.

Layzer, R. B., and Rowland, L. P.: Cramps, New Eng. J. Med. **285**(1):31-40, 1971.

Margaria, R.: The sources of muscular energy, Sci. Am. **226**(3):84-91, Mar. 1972.

Mommaerts, W. H. F. M.: Energetics of muscular contraction, Physiol. Rev. **49**(3):427-508, 1969.

Porter, K. R., and Franzini-Armstrong, C.: The sarcoplasmic reticulum, Sci. Am. **212**(3):72-80, Mar. 1965.

Ruegg, J. C.: Smooth muscle tone, Physiol. Rev. **51**(1):201-248, 1971.

Sandow, A.: Excitation-contraction coupling in skeletal muscle, Pharmacol. Rev. **17**:265-320, 1965.

Satir, P.: Cilia, Sci. Am. **204**(2):103-116, Feb. 1961.

7

General features
Somatic sensations
Visceral sensations

RECEPTOR PHYSIOLOGY—SOMATIC AND VISCERAL SENSATIONS

■ GENERAL FEATURES

The stimulation of a receptor or sense organ informs the nervous system of environmental changes to which adaptive adjustment must be made for the preservation of life. That the nerve impulse generated by the stimulation of a receptor may evoke a reflex or a sensation is a matter of common knowledge. However, since impulses are frequently inhibited in some parts of the nervous system, it is possible that the stimulation of a sense organ may elicit neither a sensation nor an overt response. It is customary to say that our sensations are generated by the activity of the cortical cells of the cerebral hemispheres. As we shall learn later on, certain areas of the cerebral cortex are neurally closely associated with the several sense organs, and the artificial or pathological stimulation of these cortical areas evokes a sensation or hallucination (Chapter 10). Whatever the relation between brain and mind may be and whatever a sensation may be, it is safe to say that, without a corresponding physical or chemical activity on the part of certain brain cells, sensations never originate.

In this chapter we will restrict our study to what are commonly spoken of as general sensations—touch, pain, temperature, pressure, visceral, and kinesthetic. The special senses will be discussed in Chapter 9.

■ Excitation and biopotentials

Irritability. ■ A sense organ or receptor possesses a high degree of excitability so that its threshold stimulus is much below that of the afferent nerve fiber itself. To what an almost unbelievable extent this development has taken place may be illustrated by the following examples. The pacinian corpuscle, a receptor specialized for detection of pressure changes, responds to mechanical compression of the capsule that displaces it by as little as 0.2 μ. Threshold detection of temperature changes is achieved by cold receptors when the decline in temperature is about 0.004 C/sec; warmth is detected when the temperature rises 0.001 C/sec.

Adequate stimulus. ■ This high degree of sensitivity of a sense organ is limited to only one kind of stimulus; as the organ gained in sensitivity to one form, its ability to respond to other forms of stimuli decreased. For example, a commonly occurring receptor in man responds primarily and strongly to pressure changes but gives a weak response to cold stimuli. Naturally, one must distinguish between receptor sensitivity and the psychological effect in the brain, the sensation. The reason for this receptor specificity must be sought in the physical and chemical structure of the sensory epithelium. Depending on their selectivity, the sense organs may be spoken of as thermoreceptors (cold or warmth), pressoreceptors (pressure), or nociceptors (pain).

Receptor potential. ■ Receptors are *transducers,* since they convert one form of energy into another. Although adequate natural stimuli may occur in a variety of energy forms (pressure, heat, mechanical, chemical), none are, except by accident, of electrical nature. Yet each produces in the receptor a depolarization of its membrane. This depolarization is spoken of as a *receptor potential* or *generator potential.* The receptor potential is a graded response; that is, the magnitude of the depolarization is a function of the intensity of the stimulus, within limits. Depolarization of the receptor site causes it to become a sink for local current flow from the adjacent nerve fiber membrane. When the receptor potential attains a specific level of depolarization, the threshold level, the local current flow initiates an action potential in the nerve fiber. The receptor potentials, as we noted above, are graded; they are also stationary, that is, they are nonpropagated responses. Their duration varies according to the type of receptor involved, but it reflects the duration of the applied stimulus and is considerably longer than an action potential. This type of response contrasts with that of nerve fibers where the action potentials are all-or-none and propagated.

Available evidence suggests that receptor depolarization is brought about when the stimulus increases the receptor membrane permeability. The increased permeability is nonselective; that is, the receptor membrane, unlike the nerve membrane, becomes permeable to all small ions. Since this is so, the maximum depolarization would be to zero potential (no separation of charges across the membrane), and an overshoot or spike, as seen in action potentials, would not exist.

Adaptation. Receptor potentials show a maximum level of depolarization on application of the stimulus. This is known as the "on" response of the receptor. Thereafter, with maintained stimulation the receptor potential declines either to a new steady state, or static state of depolarization, or to its original resting potential level. This phenomenon is called *adaptation* of the receptor. On cessation of the stimulus the receptor gives an "off" response, which is commonly a brief hyperpolarization. Quickly adapting receptors are known as *phasic receptors,* and those that adapt slowly are *tonic receptors.* For example, touch receptors at the base of hair follicles are phasic, and tendon receptors are tonic. The precise mechanism of adaptation is unknown. It may in part be associated with viscous or other mechanical properties of the nonneural tissues associated with the receptor or its membrane. That adaptation is not due to fatigue is shown by the fact that when the stimulus is momentarily released and then reapplied, the dynamic "on" response and adaptation to the static response level reoccur.

Coding. ■ The stimulus energy, after transduction to electrical energy, is *amplitude modulated by the receptor* and *frequency modulated by the nerve,* since the number of action potentials elicited per unit time is commonly a function of the generator potential amplitude. Adaptation of the receptor causes a decrease in the frequency of propagated action potentials. In phasic receptors, often only one action potential arises when adaptation is particularly rapid. However, within limits, the magnitude of the static phase of the generator potential in tonic receptors and the frequency of the nerve impulses are proportional to the logarithm of the stimulus intensity. We speak of the frequency pattern as the *coding* of the information contained in the specific stimulus. Different receptors for the same sensation have varied thresholds; hence when a larger area is stimulated by a more intense stimulus, more receptors respond. This manner of obtaining additional information is known as *recruitment.* In phasic receptors, intensity is signaled by an increase in the number of receptors activated.

■ Classification of sense organs

Many classifications of sense organs are employed. It is convenient to group the sense organs according to the source of stimulation and to the location of the receptor into two classes: *exteroceptors* and *interoceptors.*

Exteroceptors. ■ Exteroceptors are stimulated by changes occurring outside the body. In some of these, contact receptors (skin and tongue), the source of the stimulation is in direct contact with the sense organ. In provoking a sensation in a distant receptor (ears), the source of the stimulus, situated at some remote point, affects an intermediate agent (air) by which the disturbance (sound) is brought to the receptor.

Interoceptors. ■ Interoceptors are stimulated by changes occurring within the body. These may be subdivided into proprioceptors, labyrinthine sense organs, and visceroceptors.

Proprioceptors. Proprioceptors are the sense organs embedded in muscles, tendons, and ligaments (joints). By them we sense the tension (stretch) to which these structures are subjected. The impressions gained are also known as *kinesthetic* sensations.

Labyrinthine sense organs. Labyrinthine sense organs inform us of the position and changes in the position of the head. These will be discussed in Chapter 9.

Visceroceptors. Visceroceptors are found in the viscera (in the walls of the respiratory and digestive organs, bladder, etc.) and are stimulated by chemical and mechanical changes. The sensations resulting from their stimulation include pain, hunger, thirst, flushing, suffocation, nausea, sexual, and distention sensations (stomach and urinary bladder). By means of impulses from the visceroceptors, the activities of the various internal organs are regulated to the needs of the body.

■ Attributes of sensations

Sensations differ from one another in many respects—modality, quality, intensity, adaptation, duration, and projection—which will be briefly described.

Modality. ■ Sensations aroused by stimulation of various sense organs, such as the eye or ear, bear no resemblance to each other; to most people it is inconceivable that the stimulation of the eyes could evoke the sensation of taste or hearing. This characteristic of a sensation by which we distinguish it clearly from all other sensations is known as its *modality.* The modality of a sensation aroused by the stimulation of a given sense organ is fixed—Mueller's law or the *law of specific nerve energies.* The nature of modality is not determined by the manner in which the sense organ is stimulated. For instance, when the eye is turned strongly nasalward and gentle pressure with a fingertip is exerted upon the eyeball, a luminous circle (known as a phosphene) is seen.* By analogy it should be recalled that whether a motor nerve is stimulated chemically, mechanically, or electrically, the result is always the same—a muscle contraction.

*This is exceedingly brilliant if the experiment is made in complete darkness. Incidentally, we may note that visual sensations can be had without any physical light being present.

Sensations may be classified roughly according to the nerve tracts that convey the stimulus information to the brain. The *special senses* that are served by cranial nerves include vision, audition, taste, olfaction, and labyrinthine (or vestibular) sensations. These are considered in Chapter 9. The *cutaneous* or *superficial sensations* are touch-pressure, temperature, and pain. They are served by certain cranial nerves and cutaneous branches of spinal nerves. Muscular branches of spinal nerves and certain cranial nerves serve the *deep sensations* that include, in addition to kinesthetic sensations, deep pain and deep pressure sensations. Finally, the *visceral sensations* such as hunger, nausea, and visceral pain are served by afferent fibers in the autonomic nervous system (Chapter 11). The cutaneous and deep sensations are known collectively as the *somatic sensations.*

Quality. ■ Sensations of the same modality may differ in quality. Thus we speak of a warm or cold breeze and of a red or green light. Many individuals lack the power of discriminating between certain qualities of a sensation. For example, some individuals cannot distinguish between red and green (color blindness); and there is also tone deafness.

Intensity. ■ Two sensations of the same quality may differ in intensity. We speak of a warm or hot object. The former delivers little energy and the latter delivers much energy to the receptor. Sensations cannot be measured quantitatively, but sensations of the same modality and quality may be compared as to their intensity; even this comparison is qualitative, for we cannot say that one sensation is twice as great as another. Although we cannot measure the intensity of a sensation, the strength of the stimulus evoking the sensation (e.g., heat, stretch, or pressure) can generally be stated quantitatively. Weber discovered a relationship between our ability to detect an increase or decrease in the intensity of a sensation and the increase or decrease in the strength of the stimulus. This is known as *Weber's law.* This law states that *the least perceptible increase in the intensity of a stimulus is a constant fraction of the original stimulus.* For the various sensations these ratios differ; for hearing the ratio is 1/10 to 1/20, and for weight (pressure) it is 1/20 to 1/40. Weber's law does not apply to weak or very strong stimulation.

Adaptation. ■ From our study of muscles and nerves it will be recalled that these structures adapt to a constant stimulus. For example, in Fig. 8-12 the frequency of impulses in a nerve serving a stretch receptor is plotted versus time. The degree of stretch was constant. It will be noticed that the frequency of the impulses gradually decreased due to adaptation. It was stated earlier that the speed with which adaptation sets in differs in the various receptors. We are familiar with the fact that a slight bending of a hair causes a distinct sensation, but, if the hair is kept in this new position, the sensation quickly disappears. This is true for touch in general, as everyday experience teaches us. On the other hand, impulses generated by the stretching of a muscle spindle or tendon (muscle sense or proprioceptive impressions) show far less adaptation (Fig. 7-1). Fortunately, the adaptation to pain sensation is exceedingly slow or absent.

Duration. ■ That a sensation occupies a certain length of time is self-evident. However, the sensation is not necessarily co-existent with the stimulus. We have already seen that because of adaptation the duration of the sensation may be shorter than that of the stimulation. On the other hand, sensations may outlast the period of stimulation and thereby give rise to aftersensations.

Projection or localization. ■ We consider sensations as more intimately concerned with the brain than with any other part of the body. Yet when a sensation is experienced, we are never conscious of our brain. *Sensations are invariably projected or referred to one of two distinct locations: either to some part of our own body or to some part of the external world.* The sensation

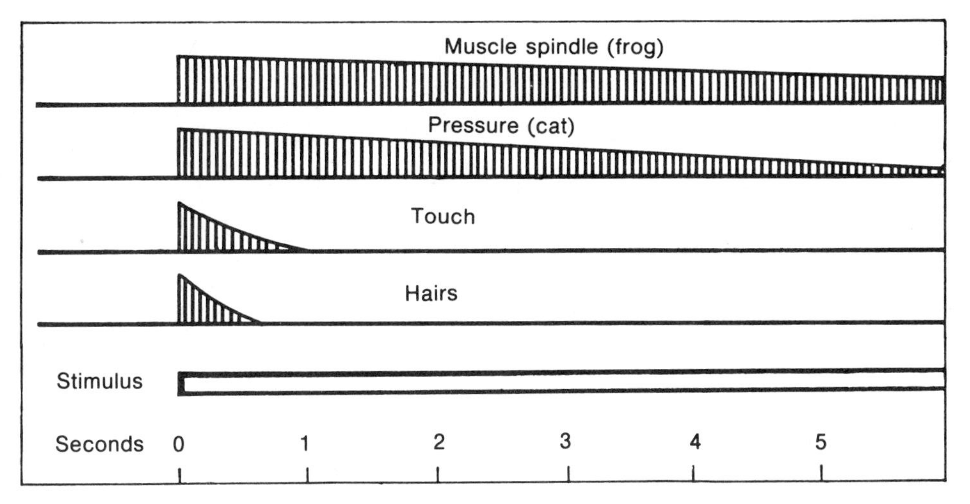

Fig. 7-1
Diagram showing adaptation to continued stimulation in different types of receptors. (Modified from Adrian: The basis of sensation, W. W. Norton & Co.)

of hunger is thought of as situated in a certain region of our own body and is never projected externally. We associate the sensation of heat with the hot object, but the pain on touching this object is always referred to that part of the body affected by the heat. Other sensations are projected either to a definite part of the body or diffusely to the whole body; among these are thirst, nausea, sex, equilibrium, muscle sense (proprioceptive impressions from muscles and joints), fatigue, tickling, itching, and the sense of well-being or its reverse. The sensations of heat and cold are in this respect on the border line. On touching a warm or cold object, the sensations are projected externally, but if the air around us cools our feet or if the face burns as in a fever, we think of our feet or face and not of the outer world.

A moment's reflection shows us the vital importance of properly projecting our sensations; indeed, most of our sensations, particularly those that are sometimes designated as the special sensations, would be of no value to us unless proper projection accompanied them.

■ SOMATIC SENSATIONS

■ Cutaneous sensations

From the skin we obtain impulses that give rise to four sensations—*heat, cold, touch-pressure,* and *pain.* The first three sensations are limited almost entirely to the skin and to the ends of the alimentary canal.

Receptive fields. ■ The four cutaneous sensations are distributed in a punctiform manner over the skin. By means of a stiff bristle or hair, it is possible to find points on the skin that are insensible to touch; such areas pricked with a needle may give rise to pain. By the use of warm or cold thin metal rods, it is possible to find spots or areas that are sensitive either to heat or cold and to find others that are insensitive to both these forms of stimulation (Fig. 7-2). A square centimeter of the palmar surface of the fingertip contains about 60 pain and 100 touch spots; the back of the finger has about 100 pain and 9 touch spots. A single afferent axon may serve an area of a few square millimeters to several square centimeters, depending on the region of the body; for example, the lips are more densely innervat-

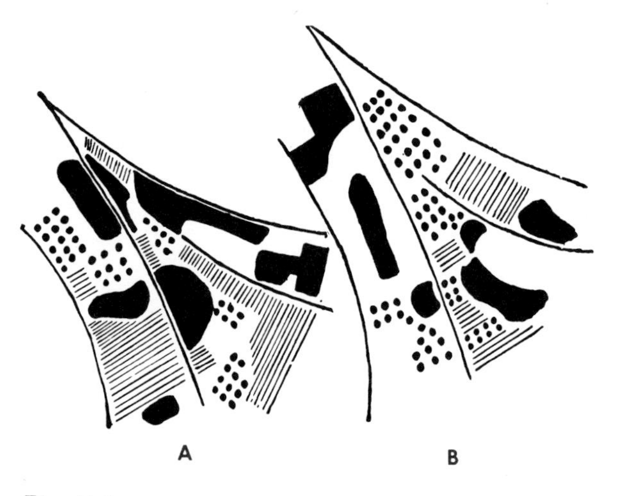

Fig. 7-2
Cold spots, **A,** and heat spots, **B,** of an area of skin of the right hand. In each case the most intense sensations were experienced in the black areas, less intense in the lined areas, and least intense in the dotted areas. Blank areas represent parts in which no special sensation of either kind was experienced. (Goldscheider.)

ed than is the back. Although a single axon may innervate as many as 100 hair follicles *(divergence),* each follicle may be innervated by from 2 to 20 different axons *(convergence).*

Receptor specificity. ■ According to the theory of specific nerve energies, it is generally thought that in the skin and adjacent tissues there are several distinct sense organs, each capable of being most efficiently stimulated by one form of stimulus and always giving rise to one definite sensation. Many kinds of cutaneous receptors have been described. That these variously described specialized sense organs are specific receptors is questionable, since they are absent in many areas of the body where all of the typical sensations can be elicited. For example, hairy skin shows sensitivity to cold, warmth, touch, and pain; yet only free nerve endings (Fig. 7-3) and nerve baskets around hair follicles are found on close examination. Even where the so-called specialized receptors are evident, as in glabrous

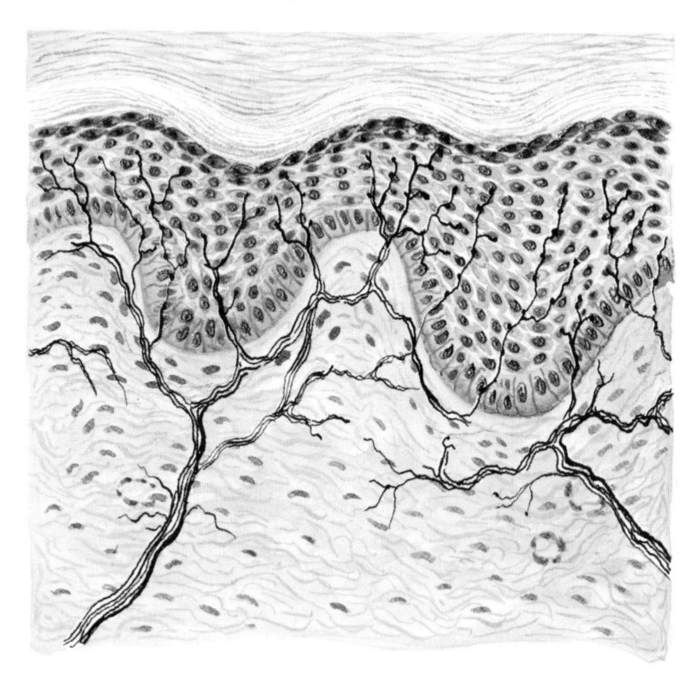

Fig. 7-3
Free nerve endings of sensory nerve in skin.

skin, some doubt exists as to the nature of the sensation produced by their stimulation. By and large, encapsulation of a nerve terminal may offer protection from strong stimuli rather than confer on the terminal any specificity of sensation. The latter phenomenon is best attributed to the nerve terminal itself.

That distinct nerves mediate the various cutaneous sensations is rendered likely from the following observations:

1. The punctiform manner in which the sense organs are distributed in the skin.

2. In certain pathological disturbances of the central nervous system there results what is called the dissociation of the cutaneous sensations. For example, the fingers of a physician who suffered from a spinal disease were completely analgesic (incapable of perceiving pain), but he was still able to feel the pulse of a patient. Sometimes it happens that the mucosa of the mouth loses its sense of temperature, whereas that of touch remains normal.

3. Pressure upon a nerve trunk may also bring about dissociation. We have all experienced the peculiar effect of a limb going "asleep"; in this state the sensations of pressure and cold are almost completely lost, but that of heat remains.

4. The application of cocaine, an anesthetic, to a cutaneous nerve causes the disappearance of the sensations in the following order: pain, cold, heat, and touch. From these and other observations it has been concluded that the nerve fibers having the smallest diameter are the first to be affected by cocaine and mediate the sense of pain; the largest fibers are the last to be influenced and convey impulses that arouse the sensation of touch.

Touch-pressure. ■ The stimulus for touch sensation is the deformation of the skin by unequal pressure.* Touch verges imperceptibly into the sensation of pressure so that the two modalities are commonly considered together. The degree of this sensitivity may be found by means of von Frey's esthesiometer. From a number of horse hairs of various thicknesses, the thinnest one is found, which upon being pressed onto the skin until it bends, is able to elicit a touch sensation; the amount of pressure, in grams, equivalent to this stimulation can be determined by means of a pair of scales. This pressure gives the value of the stimulus. The fingertips and the lips possess the highest degree of tactile sensitivity; the middle of the back the least.

There are at least two different mechanoreceptors in the skin. The first adapts rapidly, often giving only one action potential, and has a low threshold; that is, stimulation is achieved by only 5 to 20 mg of pressure with a von Frey hair. The second undergoes adaptation but even in the static phase maintains relatively high firing rates, up to 300 impulses per second.

Tactile localization. By the tactile sense we can tell more or less accurately which part of the skin is touched. In our brain there is a field corresponding to the field of the periphery of our body exposed to the environment. Normally if two widely separated areas of the skin are simultaneously stimulated by two discrete objects, two sensations are experienced. When these two areas are brought into close proximity to each other so that they can be simultaneously stimulated by one object, two distinct objects are felt. This is done in Aristotle's experiment (Fig. 7-4), in which the middle finger is crossed over the

*When a person sticks his finger in a cup of mercury, a sensation is experienced only at the line where the surface of the mercury and the air touches the skin.

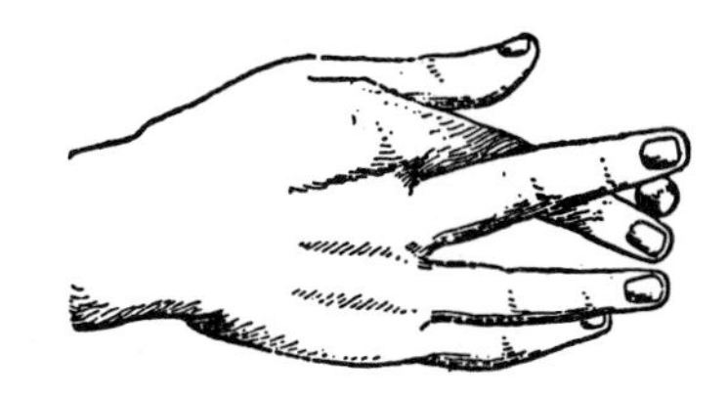

Fig. 7-4
Aristotle's experiment.

index finger and the point of a blunt pencil is moved along the crotch thus formed.

Tactile discrimination. By tactile discrimination is meant the power to discern two discrete sensations when the skin is stimulated at two points. If the stimulated points are close enough together, only one sensation is experienced. To determine the degree of tactile discrimination for various regions of the skin, we find the least distance by which the two blunted points of a compass must be separated in order to be felt as two (the compass test). Tested in this manner the sensitivity of the tip of the tongue and that of the index finger is about 1.1 mm; that of the palm, 8 mm; and that of the middle of the neck, 67 mm.

Temperature. ■ Although no special end organs can be identified with temperature sensitivity, both cold and warmth receptors occur in the skin. They are in the dermis near blood vessels. Cold receptors increase their firing rate upon a decrease in temperature; warmth receptors increase their firing rate with an increase of temperature. The energy of the stimulus is, of course, heat, since temperature itself is only an expression of the relative amount of this energy present. Temperature has a punctate distribution (Fig. 7-2). On the human forearm, cold spots are more numerous (average 13 to $15/cm^2$) than are warm spots (average 1 to $2/cm^2$). Temperature sensors record the temperature at the depth at which they occur in the skin; consequently, they are also activated by the internal heat content.

■ Pain

The sensory modality known as pain is common to cutaneous, deep, and visceral sensations; that is, it is detected in virtually all areas of the body. It can be defined as an unpleasant sensation that has a considerable affect (subjective response) and is poorly localized.

Adequate stimulus. ■ There is no adequate stimulus for pain; any form of stimulus energy evokes pain if sufficiently strong. For example, a forehead skin temperature of 44.9 C in the average individual elicits pain; tissue damage occurs between 44 and 45 C. Pain is the most primitive and one of the most important of all sensations. It is generally held that the free nerve endings (Fig. 7-3) of small myelinated fibers constitute the receptors for pain. They are the most widely distributed receptors in the body, being found in the skin, cornea, blood vessels, muscles, tendons, joints, and most viscera. The threshold of irritability of these receptors is high; consequently the stimulations befalling them are of such intensity as to threaten health and life. For this reason they are frequently called *nociceptors* (*nocere,* to injure). They elicit protective and defensive reflexes. Because of their vital importance, the impulses from nociceptors always take precedence in the reflex activity of the nervous system.

Many internal organs are insensitive to what are generally regarded as painful stimuli. Tumors have been removed from the brains of patients who remained conscious during the entire operation, without causing any pain. Handling, cutting, or cauterizing the intestine never gives rise to expressions of pain. However, pain can be elicited from most hollow organs by an increase in the tension of their walls; this is brought about (1) by great distention (e.g., accumulation of gas in the intestine, or the passage of a gallstone through the bile ducts) or (2) by excessive contraction of their musculature (i.e., colic).

Pain can be relieved by reducing the irritability of the nerve, as by compression, cold, drugs (analgesics and narcotics, e.g., aspirin, morphine, codeine, Novocain), or by reducing the sensitivity of the cerebral cells, as in general anesthesia.

Characteristics. ■ Pain may variously be described as burning, throbbing, pricking, or stabbing. Many times these descriptions identify the sensation by its duration or causative agent. Prolonged pain is often said to be a burning sensation despite the absence of heat energy. Sharp pain is called bright and diffuse; aching pain is usually dull.

One explanation of the diverse origins of pain

sensations is that the receptors are excited by noxious chemicals released from the surrounding tissues in response to the applied stimulus. Such a concept would account for the poor localization, the apparent spreading of the sensation in the affected area, and the close relationship between pain and tissue damage. Experimental evidence suggests that this chemical may be a vasodilator substance, a polypeptide. Because of uncertainties as to its exact composition, it resembles oxytocin, bradykinin, and other polypeptides; it is usually referred to as *neurokinin.*

Referred pain. ■ Pain can be elicited by stimulating a nerve fiber at any point along its course, but the sensation is always *referred* or projected to the endings of the nerve; this is made evident when a person hits his crazy bone, by which the ulnar nerve is stimulated in its course. Pain cannot always be definitely localized, especially when it is severe and of long duration; the sensation then seems to spread to neighboring parts, as is felt in toothache.

Pain experienced in internal organs is sometimes referred to another and, generally, external part of the body. In diseases of some internal organs, certain cutaneous areas may become *hyperalgesic* (excessively sensitive to pain) so that the least stimulation, such a gentle touch or even a breath of air, applied to this part of the skin gives rise to pain. It seems that when an internal organ receives its nerve supply from a certain segment of the cord that also supplies fibers to an external area, the painful stimulation of the internal organ (which is generally less sensitive than the external area) is referred to the external area. How this anatomical arrangement furnishes a basis for referred pain is not entirely clear, and the postulated mechanisms are not essential to our study. However, it can be readily understood that a knowledge of referred pain may be of great diagnostic value to a physician.

Recently a new theory of pain perception known as the "gate control theory" has been proposed. This theory has many commendable aspects, especially in that it explains many observations about pain that have been heretofore at variance with the generally accepted specific modality theory. The details of and evidence for this new theory are too extensive for our consideration.

■ Deep sensations

General observation shows that we are able to judge the position or movements of our limbs without the aid of our eyes. We "feel" the degree to which the fingers are bent or the arm is extended. This is well illustrated in our ability to lay our hand, in the dark, upon a certain object whose position in space is known to us.

The deep sensations arise from receptors in the muscles, tendons, and joints. The importance of these receptors is attested to by the fact that at least one half of the nerve fibers innervating muscle are afferent. Several types of sense organs are recognized—*muscle spindles, tendon end organs, joint organs, pacinian corpuscles,* and *pain endings.* Muscle spindles (Fig. 8-11), the best known and most widely studied of the proprioceptors, are embedded between muscle fibers of all antigravity muscles. They are stimulated whenever the muscle is stretched and are, therefore, length recorders. The greater the stretch applied to the muscle, the greater is the frequency of the impulses from the spindles; this information delivered to the spinal cord causes reflex contraction of the muscle in which the spindle is located; the contraction reduces the stretch, and the spindle afferent activity ceases. In Chapter 8 the stretch reflex is considered in more detail.

Tendon end organs, or Golgi tendon organs, are encapsulated structures located in the tendinous ends of muscles. They are sensitive to stretch of the tendons, whether caused by the contraction of muscle with which they are associated (active stretch) or by an antagonistic muscle (passive stretch). They are tension recorders. Impulses from tendon organs on reaching the spinal cord bring about inhibition of the stretch producing contraction. Their participation in the stretch reflex is also a subject of Chapter 8.

Joint organs are, as the name implies, sense organs found in the joints. They are recorders of position and movement. Adaptation of joint receptors either does not occur or is extremely slow. The mechanoreceptor known as a pacinian corpuscle consist of layers (lamellae) of connective tissue surrounding, like the skin of an onion, the bare nerve terminal of a large myelinated nerve. This structure is a pressure receptor of exceeding sensitivity. Besides being found in muscle tissue, they are common in the connective tissue of the abdomen (mesenteries) and at branching points of small arteries. The free nerve endings serving the sensation of pain are, as we indicated before, ubiquitous in deep tissues.

Although we are but dimly conscious of the kinesthetic impressions (muscle sense), yet it is difficult to overestimate their importance. They acquaint us with the position of our limbs in respect to the whole body; as we shall learn in Chapter 8, they are indispensable in establishing muscle tonus and thereby in maintaining the proper posture of the body in standing. In *tabes dorsalis* certain fibers of the dorsal (sensory) roots of the spinal cord are destroyed, and the sense of position is largely lost, especially that of the legs. Deprived of these afferent impulses, the patient, with eyes closed, is unable to maintain an erect position, and there is generally a marked disturbance in locomotion.

Free nerve endings that respond to distention are found in the walls of many visceral organs. There are complex endings in the aorta, at the bifurcations of the carotid artery, and in the right atrium of the heart (Chapter 14). These structures record pressure changes in the circulatory system. Certain receptors in the lungs are sensitive to stretch induced by the expansion of the lungs in inspiration (Chapter 17).

Hunger is a sensation projected to the region in the immediate neighborhood of the stomach and is due, as shown by Cannon, to the contraction of the musculature of the stomach (Chapter 19).

Thirst sensation is generally associated with dryness of the mucosa of the mouth and pharynx. As to the causative agent, it is perhaps twofold, local and general. The dryness of the mucosa may be due to strictly local conditions, such as breathing dry, hot air or dusty air. Or it may be due to the loss of water from the body, which results in a diminished secretion of saliva. Loss of body water with an increase in the osmotic pressure of the blood (as in profuse sweating) may be the more general stimulus exciting the osmoreceptors, or thirst sense organs. Thirst and drinking are discussed further in Chapter 24.

■ VISCERAL SENSATIONS

We have no sense of touch, heat, or cold in the internal organs, as is indicated by the fact that we do nor feel the gliding of the various organs, such as the lungs, diaphragm, and heart, along the adjacent structures; nor normally do we feel the movements of such organs as the stomach and intestines. The structures immediately below the skin may give rise to a sensation of pressure if the stimulation is sufficiently strong; and a sensation of fullness (akin to pressure) is obtained from the bladder and rectum. Pain, on the other hand, can be elicited from most organs.

READINGS

Lowenstein, W. R.: Biological transducers, Sci. Am. **203**(2):99-108, Aug. 1960.

Melzack, R.: The perception of pain, Sci. Am. **204**(2):41-49, Feb. 1961.

Melzack, R., and Wall, P. D.: Gate control theory of pain. In Soulairac, A., Cahn, J., and Charpentier, J., editors: Pain, New York, 1968, Academic Press, Inc., pp. 11-31.

Mountcastle, V. B.: Physiology of sensory receptors: introduction to sensory processes. In Mountcastle, V. B., editor: Medical physiology, ed. 12, St. Louis, 1968, The C. V. Mosby Co., vol. 2, pp. 1345-1371.

Anatomy
Reflex arc
Reflex characteristics
Reflex actions
Spinal cord as a conductor

REFLEX ACTIONS—THE SPINAL CORD

Stimulation of the receptors studied in the preceding chapter is of no avail unless the impulse generated finds expression in the activity of an effector. To transmit the impulse from the receptor to the effector is the primary function of the nervous system. Supplementing this primary function of conduction, the nervous system in its development acquired the power to coordinate the multitude of impulses received from the receptors and thus integrate the activities of the various parts of the body according to its needs.

■ ANATOMY

The nervous system may be conveniently divided into (1) the *central nervous system*, comprising the brain and spinal cord, and (2) the *peripheral nervous system*, which includes the cranial and spinal nerves and the *autonomic nervous system*. The central nervous system is composed of many millions of neurons. As in all other organs, these cells are held together by connective tissue. In addition to usual forms, we find in the central nervous system a special kind of connective tissue known as *neuroglia*.

The brain and spinal cord, being vital organs and extremely delicate, are well protected. First of all, they are enveloped in three membranes, or *meninges:* the *dura mater* (a strong fibrous membrane), the *arachnoid*, and the *pia mater*. The pia mater, closely applied to the brain and cord, is a vascular coat and is therefore the nutritive membrane for these organs. Inflammation of these coverings is known as meningitis. In addition to these membranes, the brain is protected by the rigid cranium; the spinal cord is protected by the flexibly joined thirty-three vertebrae of the spinal column. In the bones forming the skull and in between the vertebrae we find orifices (foramina) for the exit of the cranial and spinal nerves (Fig. 8-1).

Cerebrospinal fluid. ■ Through the entire length of the cord extends a narrow central canal that is continued anteriorly into the brain, where in four locations it widens out to form the ventri-

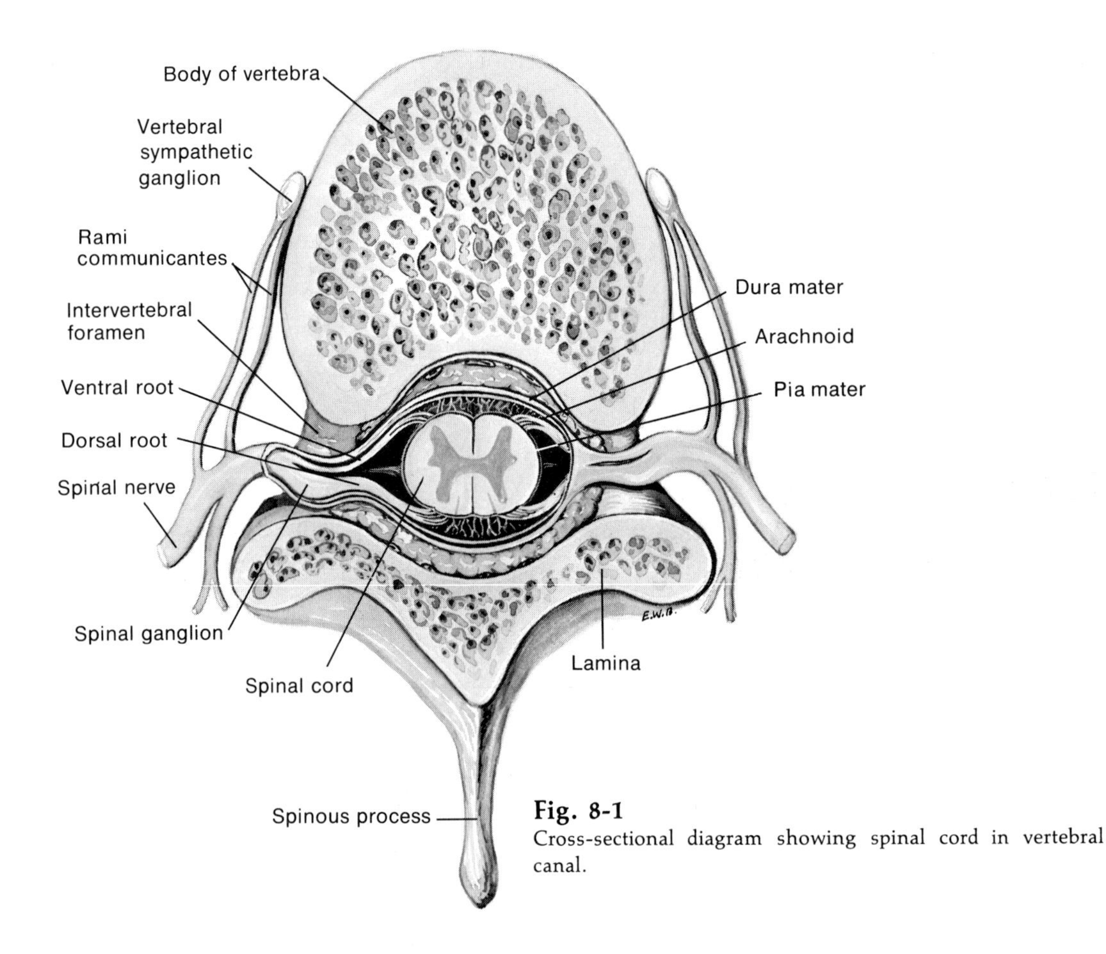

Fig. 8-1
Cross-sectional diagram showing spinal cord in vertebral canal.

cles of the brain. The central canal and ventricles, as well as the space between the arachnoid and the pia mater, are filled with *cerebrospinal fluid* (CSF). It serves as a support and a cushion. The composition of the CSF is different than that of the blood plasma or extracellular fluid. It contains less K^+, Ca^{++}, HCO_3^-, glucose, and protein than plasma, but slightly increased amounts of Na^+, Cl^-, and Mg^{++}. The pH of CSF is from 0.1 to 0.2 unit lower than that of plasma.

Lumbar puncture. In certain diseases it is desirable to learn the composition and pressure of the cerebrospinal fluid. The spinal cord extends from the base of the skull to the second lumbar vertebra; here it ends in a large number of nerves which comprise the lower ten pairs of spinal nerves and are collectively known as the *cauda equina.* These nerves are enclosed in a sac formed by the meninges and extend to the third sacral vertebra. It is therefore possible to insert a hypodermic needle between the third and the fourth lumbar vertebrae and to introduce it into the sac. In this manner some of the cerebrospinal fluid can be drawn into the syringe either for diagnostic purposes or to relieve excessive pressure (as in encephalitis or inflammation of the brain).

Spinal cord. ■ The spinal cord, situated in the

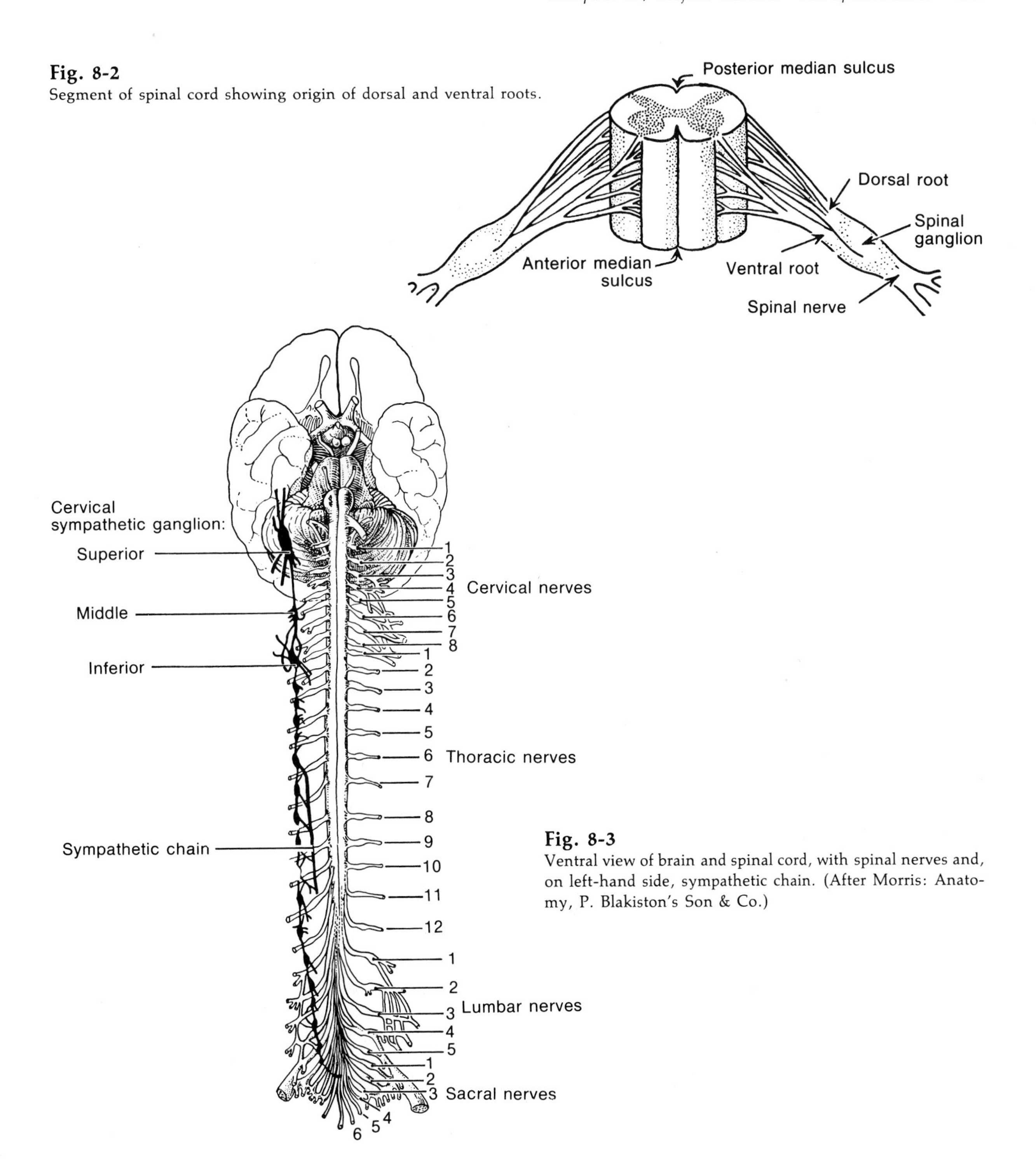

Fig. 8-2
Segment of spinal cord showing origin of dorsal and ventral roots.

Fig. 8-3
Ventral view of brain and spinal cord, with spinal nerves and, on left-hand side, sympathetic chain. (After Morris: Anatomy, P. Blakiston's Son & Co.)

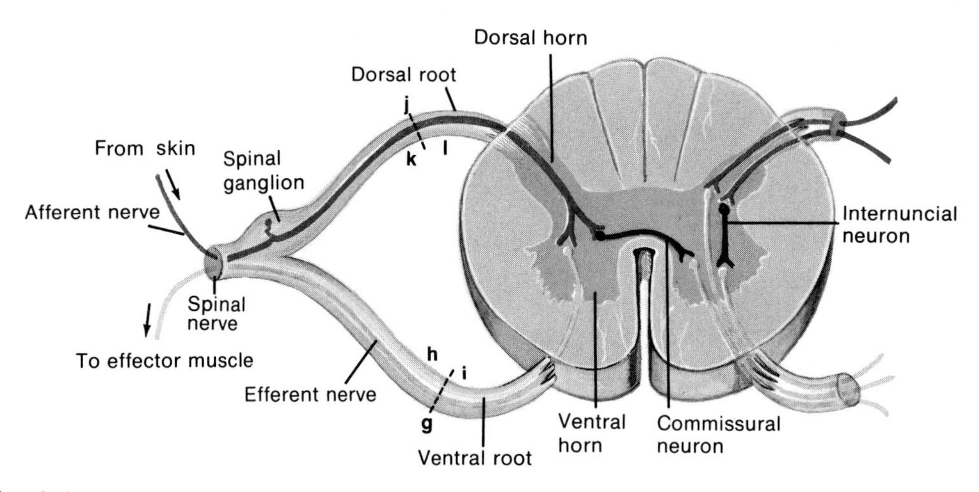

Fig. 8-4
Cross-sectional diagram illustrating some of simpler connections between afferent and efferent nerve fibers. For ascending and descending branches of these fibers, see Fig. 8-5. Letters on diagram are discussed in text.

vertebral canal of the vertebral column, is cylindrical in form (Fig. 8-1) and about 18 inches long and $^3/_4$ inch in diameter. As shown in Fig. 8-2, the cord is almost completely divided into two lateral halves by the ventral (anterior) and the dorsal (posterior) median sulci. From the ventral aspect of the cord spring thirty-one pairs of ventral roots (Fig. 8-3); there are the same number of dorsal roots. Each root contains a large number of nerve fibers. The fibers of the dorsal and ventral roots of one lateral half of a segment of the cord commingle to form the spinal nerve (Fig. 8-4); this divides into many branches to supply various regions of the body. The thirty-one pairs of spinal nerves are grouped as in Table 8-1.

Table 8-1. Spinal nerves

Cervical	8 pairs
Thoracic	12 pairs
Lumbar	5 pairs
Sacral	5 pairs
Coccygeal	1 pair

Gray and white matter. ■ In Chapter 5 we learned that a typical neuron is composed of a cell body, an axon, and generally two or more dendrites (Fig. 5-2). In the cell body and its dendrites (not in the axon) are found many angular bodies or granules that color deeply with certain stains and are for this reason known as *chromatophil granules* (Nissl bodies). They are composed chiefly of a nucleoprotein and are involved in the nutrition of the cell. In the central nervous system, cell bodies are massed to form the gray matter; the axons, surrounded by myelin sheaths, constitute the columns or tracts of white matter. The cell body plays no special part in the propagation of nerve impulses; on it depends the nutrition of the entire neuron. Gray matter has roughly an H shape in the center of the spinal cord, as shown in Figs. 8-2 and 8-4.

Degeneration and regeneration of nerve cells. ■ The cutting of a nerve fiber has two results: it causes degeneration of the part of the fiber separated from its cell body, and the staining material in the Nissl bodies decreases (chromatolysis). As nerve fibers in the central nervous system are devoid of a neurilemma, regeneration of a severed fiber does not take place here. Outside of the central nervous system there is a close relation between the time of myelin restoration and the resumption of function by the regenerated fiber. A mature neuron is not capable of reproducing itself; similar to muscle cells (fibers), they are irreplaceable. After the age of 1 year, the number of neurons is not increased; indeed, there is a loss with aging.

■ REFLEX ARC

Neurons and reflex arcs. ■ The reflex arc is the nerve chain between a receptor and an effector (muscles or glands). The afferent neuron of this arc passes into the spinal cord by way of a dorsal root; the efferent nerve leaves the cord by a ventral root. Between these two neurons we nearly always find one or more interneurons (Fig. 8-4). *The reflex arc constitutes the functional unit of the central nervous system.* Cutting any ventral root (Fig. 8-4, *g*) causes paralysis of the muscle or muscles supplied with motor fibers by this root. Stimulating the peripheral end, *h,* activates its effector organs; stimulation of the central end, *i,* has no effect. Cutting a dorsal root, *j,* brings about the loss of sensation (touch, pain, etc.) from some part of the body. The stimulation of the peripheral end, *k,* has no effect; stimulating the central end, *l,* may have two distinct results: (1) the impulse arriving in the brain may elicit a sensation and (2) it may produce a reflex action, e.g., the contraction of a muscle.

Short and long reflex arcs. ■ In its passage through the spinal cord several pathways are open to an afferent impulse. In its simplest form, an impulse coming in over the afferent neuron (Fig. 8-5, *a*) passes without the intervention of an internuncial neuron into the efferent motor neuron, *1,* on the same side of the cord. An afferent impulse in neuron *b* traverses an internuncial neuron to reach the motor fiber, *2.* In *c* the impulse passes over to the opposite side of the cord and leaves by neuron *3.* In these examples the impulses leave by the same segment of the cord as they entered; these are short reflex arcs. The majority of afferent fibers, *d* or *j,* soon after entering the cord split into an ascending, *ad* or *aj,* and a descending branch, *dd* or *dj,* which, respectively, travel up and down the cord for varying distances. Both of these may give rise to one or more lateral branches, *m* and *n,* which make connections with a number of internuncial neurons. These interneurons can discharge their impulses into a number of efferent neurons, *4, 5,* and *6* at synapses *g, h, i, l,* and *k.* These constitute the long reflex paths serving to connect receptors of, for example, the feet with the muscles of the hand. They also enable an impulse over a single afferent fiber, *d* or *j,* to activate many effectors. On the other hand, a given motor fiber, *5* or *6,* may receive impulses from two or more afferent fibers, *d* and *j.* It is this vast array of interneurons that enables the central nervous system to function as a coordinating and integrating mechanism. By means of the ascending fiber, *f,* the impulse may travel up the cord and finally reach the cerebrum and there evoke a sensation.

■ The synapse

Structure. ■ At a synapse the apposed fibrils of the neurons do not continue one into the other but only make close contact with each other; that is, there is neural contiguity but not continuity. Evidence of this is the fact that degeneration of an afferent neuron does not extend into the efferent neuron but stops at the synapse. Axons tend to divide repeatedly at their terminal ends, and these fine presynaptic branches diverge to make functional contact with a number of postsynaptic cells. The presynaptic branches may be distributed to any or all of three locations on the postsynaptic cell: the dendrites (axodendritic synapses), the body or soma (axosomatic synapses), or the axon hillock (axoaxonic synapses). By and large, the structure of the synapse is much the same regardless of the location.

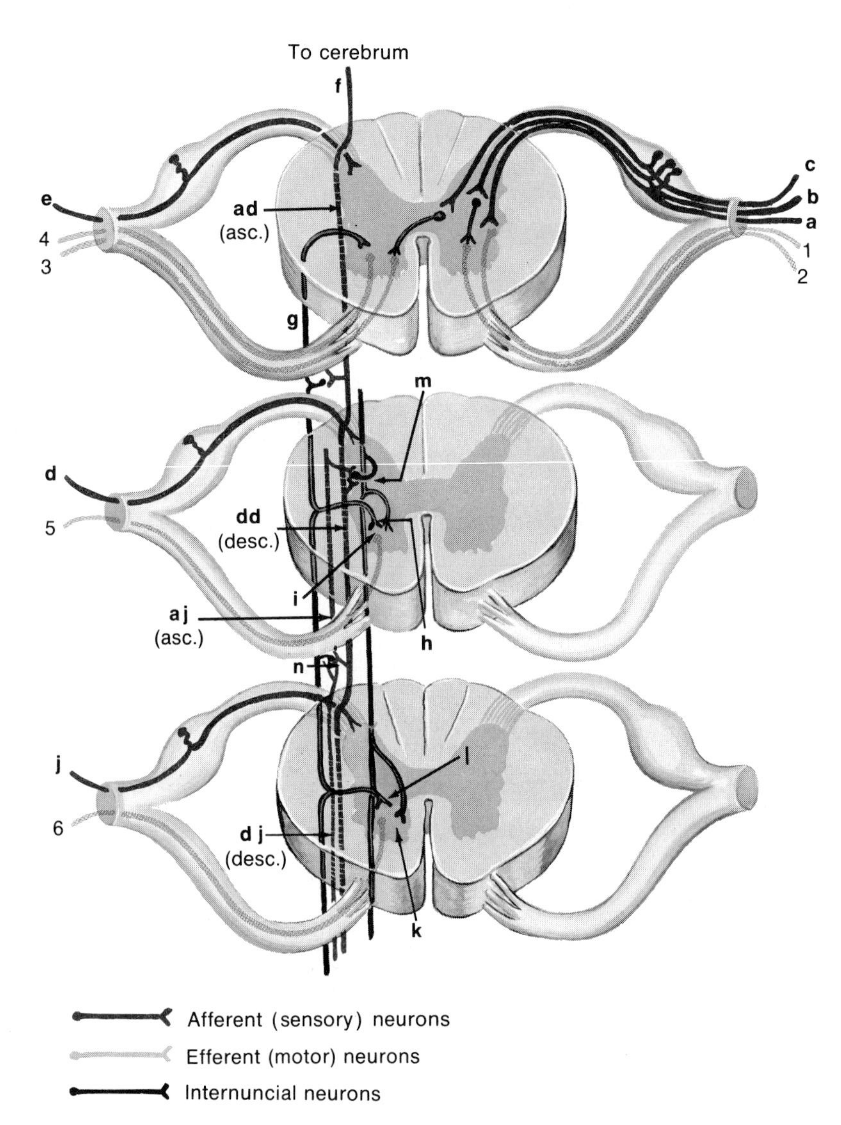

Fig. 8-5
Diagram indicating some neural connections between afferent and efferent fibers in the spinal cord. Both ascending, **ad** or **aj**, and descending, **dd** or **dj**, pathways are shown.

The presynaptic terminal end is a bulblike structure (bouton) separated from the postsynaptic cell by a synaptic cleft of several hundred Angstroms. Both the presynaptic and postsynaptic membranes in electronmicrographs show a thickening of their membranes at the synaptic region. At some synapses in the central nervous system the terminal bulb may appear invaginated within the postsynaptic cell. The dendrites of certain brain cells bear fine spiny projections, and here the presynaptic bulb folds over the spine much like a cap. A large number of spheroidal vesicles usually is present in the presynaptic terminal, especially near the presynaptic membrane; these contain the chemical transmitter agents. Some evidence indicates that the synaptic vesicles of inhibitory transmitters are different from those of excitatory transmitters; that is, inhibitory vesicles are more flattened. Large numbers of mitochondria occur in the presynaptic terminal and are assumed to be there because of the metabolic energy required for transmitter synthesis.

Function. ■ The synapse is characterized in several important respects; in some of these it differs quantitatively, if not qualitatively, from the nerve fiber. It is these peculiarities of the synapse that largely determine the coordinated responses of an animal to environmental changes. Before continuing our study, we should note the following functional characteristics of the synapse:

1. It can transmit impulses only in one direction.
2. It retards the passage of an impulse more so than the nerve fiber does. Consequently, there is a synaptic delay in the passage of an impulse from the presynaptic to the postsynaptic fiber.
3. It is a site where repetitive discharge of a neuron may occur.
4. It is very vulnerable to the action of many chemical compounds, some of which increase its responsiveness (strychnine), whereas others decrease it (anesthetics).
5. It has the characteristics of summating and of inhibiting impulses.

Unidirectional transmission. Experiments have shown that in the central nervous system an impulse can pass along a certain route in only one direction. If, in Fig. 8-4, the normal afferent nerve is stimulated, the impulse passes up to and across the synapse and thence down the efferent nerve to the muscle. But when the efferent nerve is stimulated, the impulse not only reaches the muscle but also passes up the nerve to the synapse but does not reach the afferent nerve, as shown by the absence of an action potential in this nerve. This function of the synapse, which allows an impulse to pass over it in one direction but never in the opposite direction, *is known as the law of forward conduction.*

Central synaptic transmission. An action potential in an axon causes the discharge of transmitter agents from its presynaptic terminals. Synaptic delay, approximately 0.5 msec in the central nervous system, is occasioned by the time required for diffusion of the transmitter across the synaptic cleft. The action of the transmitter on the postsynaptic membrane may be either excitatory or inhibitory; in either case it involves changes in the electrical potential of the postsynaptic membrane. Excitatory impulses bring about a local, graded depolarization of the postsynaptic membrane known as the *excitatory postsynaptic potential* (EPSP). The EPSP is a transient potential; it lasts for only a few milliseconds before the membrane resting potential is reestablished. Gradation of the EPSP is a consequence of the convergence of presynaptic terminals from many afferent nerve fibers upon the postsynaptic cell. Each presynaptic fiber impulse produces an increment of the depolarization, the action of all fibers summates. At the threshold level of depolarization, about 10 mv decline in the resting potential, the postsynaptic cell generates a propagated action potential. We have discussed the ionic basis for these electrical changes elsewhere in the text. Only rarely, if ever, can an impulse in a single presynaptic fiber bring the EPSP to threshold level. This is accomplished either by a train of action potentials in the single presynaptic fiber *(temporal summation)* or by single action potentials carried by many presynaptic fibers (*spatial summation*). The EPSP is of longer duration than the action potential; the importance of this becomes more obvious when it is recognized that temporal summation is possible because of the lingering effect of previous action potentials and that spatial summation is achieved from asynchronous trains of impulses in afferent fibers.

Inhibition at the synapse is achieved by two means, *postsynaptic inhibition* and *presynaptic inhibition*. In postsynaptic inhibition, the afferent impulse releases a transmitter that hyperpolarizes the postsynaptic membrane; that is, the interior of the cell becomes even more negative with respect to extracellular fluid. This potential change, a kind of mirror image of the EPSP, is termed the *inhibitory postsynaptic potential* or IPSP. Like the EPSP, the IPSP is a local graded response. Inhibitory transmitters appear to exert their hyperpolarizing effect by altering the membrane permeability to K and Cl ions, but not to Na ions. Increased movement of the K and Cl ions down their concentration gradients would lead to increased internal negativity. Presynaptic inhibition has its origin in the effect of certain interneurons, previously excited by one set of afferent fibers, upon the presynaptic terminals of another set of afferent fibers. The interneurons do not produce IPSPs in the presynaptic terminals; instead, they partially depolarize them. Consequently, impulses in these presynaptic fibers are reduced in magnitude, or are blocked, and are far less effective in releasing excitatory transmitter from the terminals. Presynaptic inhibition is of long duration, probably due to repetitive firing of the interneurons. The ionic mechanism and nature of the interneuron transmitter in presynaptic inhibition is uncertain.

Transmitters of the CNS. Two compounds, *acetylcholine* and *norepinephrine* (Chapters 22 and 28), are generally accepted as transmitters in the CNS; they have been proved to be so in the peripheral nervous system. Acetylcholine is widely, but unevenly, distributed in the CNS. In general, the concentration and distribution of the enzymes necessary for the synthesis (choline acetylase) and degradation (acetylcholinesterase) of acetylcholine parallel its concentration and distribution. The results of a variety of experiments indicate that some CNS synapses are *cholinergic*; acetylcholine is the transmitter; it is true that many are not. The catecholamine norepinephrine is likewise widely and variously distributed in the CNS. Its highest concentrations appear in the hypothalamus, the cranial nerve nuclei, and certain fiber tracts that descend in the spinal cord to innervate the cells of the preganglionic sympathetic nervous system. The enzymes for synthesis (tyrosine hydroxylase, dopa decarboxylase, and β-hydroxylase) and degradation (monoamine oxidase and catechol-O-methyl transferase) of norepinephrine occur in the same CNS regions as does the catecholamine. It is probably stored in synaptic vesicles, but some is free in the cytoplasm. Many of the effects of norepinephrine in the CNS are inhibitory, albeit not all. Other compounds with possible transmitter function are present in the CNS. Among these are serotonin (5-hydroxytryptamine), histamine, and glutamate. An especially interesting compound, *γ-aminobutyric acid* (GABA), is synthesized only in the CNS. GABA's role in the mammalian CNS is uncertain; however, it serves as an inhibitory transmitter in certain invertebrate peripheral nerves, and such a function is, at least, probable in the mammalian CNS. Its inhibitory action is expressed by stabilizing membrane potentials at or near the resting level, thus preventing excitation from another source.

■ REFLEX CHARACTERISTICS

Reflex actions differ from certain other activities in that they are not volitionally performed;

indeed, reflexes may take place without the individual being conscious of them, as, for example, the constriction of the pupil in a bright light.

Specificity and predictability of reflexes. ■ It is desirable in studying a particular reflex to eliminate, as far as possible, all other neural activity to avoid the complexities suggested above. In studying the spinal cord, the brain is frequently removed. In a frog this is done by pithing the brain with a needle thrust through the foramen magnum; the animal is known as a spinal animal. Immediately after this operation the frog is perfectly flaccid, appears lifeless, and responds to no form of stimulation. This condition, called spinal or neural shock, promptly disappears,* and proper stimulation now evokes reflexes. For example: the application of a *nociceptive* stimulus, such as dipping the toes in dilute acid or pinching them, is speedily followed by the withdrawal of the foot from the harmful stimulus. In this all the flexor muscles of the leg participate and every joint is flexed. This very evidently is an escape reaction. In our body, flexor responses serve not only this purpose but are also necessary for rhythmical locomotion.

In a spinal animal under stated conditions a given stimulus produces a certain reflex action; we can predict with reasonable certainty that the next application of this stimulus under the same circumstances will call forth the same response. In contrast with this, in a normal animal, especially in a more highly organized animal such as a dog, the result of a given stimulus may vary and cannot always be foretold. Human behavior, especially, may be indeterminate and unpredictable.

Purposefulness. ■ Reflex actions are generally characterized by being useful; they seem to be executed for a purpose. When the paw of a spinal dog is pricked with a pin, a flexion response takes place. On the other hand, if pressure is applied with a flat object to the pad of the foot, the leg is straightened and pressed down by the extensor muscles. It enables the animal to support its body and to prevent falling. It will be noticed that these actions, taking place in the spinal animal and, therefore, in the absence of consciousness, are of the same nature as those occurring daily in the life of the conscious dog. In summary: *a reflex action is purposive and adaptive in that it is conducive to the well-being of the organism.* Although a reflex is adaptive, it is not adaptable; it is always of a stereotyped nature.

Susceptibility to chemical changes. ■ The susceptibility of the synapse to chemical changes is illustrated by the pronounced influence of general anesthesia (e.g., ether) on this structure, as compared with that on nerve fibers and muscles. Likewise, the reflexes in a frog are abolished in about thirty minutes because of lack of oxygen; nerve trunks retain their irritability for three or more hours under the same condition.*

Reflex time. ■ The time elapsing between the stimulation of a receptor and the beginning of a response by an effector (in which the spinal cord is involved chiefly) is called reflex time. In an experiment, the flexor reflex in a cat had a total reflex time of 10.4 msec. The conduction of the impulse over the afferent and the efferent nerves required 6.5 msec. The balance of 3.9 msec was used by the impulse to pass through the polysynaptic pathways in the spinal cord. This has been called *nuclear delay.*

Integration. ■ Because of the intricacy of the nervous system and the large number of various stimuli constantly befalling the animal body, the activity of one part of the nervous system frequently modifies another. Although the events at individual synapses are, so to speak, basic functions, it is clear that they do not have great sig-

*In man, spinal shock, caused by severing the spinal cord, for example, in the thoracic or lumbar region, may last for many weeks. Spinal shock is a condition of muscular paralysis accompanied by absence of sweating, low blood pressure, and retention of urine.

*In cold Ringer's solution an excised nerve may retain its irritability for several days.

nificance in the overall operation of the nervous system. Any study of reflex activity requires that populations of cells, or *neuron pools,* and their interactions be considered.

An afferent volley* in a peripheral nerve will, when it reaches the spinal cord, bring about either excitation or inhibition of the neurons with which it makes synaptic contact. Because of the number of afferent fibers involved and the high degree of divergence of individual fiber terminals, a rather extensive pool of postsynaptic neurons will be affected. If the volley is an excitatory one, the effect is to create a *central excitatory state;* if it is inhibitory, a *central inhibitory state* is produced. However, in this excitatory state not all the postsynaptic neurons in the pool will necessarily respond by generating action potentials; probably the greater portion will generate only subthreshold EPSPs. The neuron pool is described as being fractionated into a *discharge zone* and a *subliminal fringe zone.* To increase the response of the pool, that is, to increase the discharge zone and decrease the subliminal fringe zone, either additional volleys must arrive at the synapses while the EPSPs are still present (temporal summation), or a volley in another nerve must reach a portion of the previously subliminally excited neurons through convergence and add its depolarizing effect to that of the first nerve (spatial summation).

An indirect form of inhibition, known as *occlusion,* sometimes occurs in neuron pools that are in large measure commonly excited by two different nerves. Suppose that stimulation of a nerve, *A,* brings about discharge of 100 neurons in a given pool and that stimulation of a nerve, *B,* also excites 100 neurons, of which 50 were the same as those responding to the volley in nerve *A*. Clearly, observing this from outside, we would anticipate that simultaneous activity in *A* and *B* would yield a 200 neuron response. In fact, the response would involve only 150 neurons. The total is not equal to the sum of the parts; occlusion has occurred.

Often, when two volleys in a single nerve fiber are separated in time so that temporal summation is avoided, the response to the second volley is, nevertheless, greater than to the first. The EPSPs did not summate, but *central facilitation* transpired. A logical explanation offered for this is that previous activity in the afferent nerve terminals increases the quantity of transmitter released during succeeding bouts of activity. The same kind of phenomenon underlies what is known as *posttetanic potentiation.* In this the response to an initial volley is compared to that of a volley following a period of rapid repetitive stimulation. The second volley elicits a greater response; the effect is one of prolonged facilitation. In contrast to central facilitation, inactivity of a synapse for a period of time reduces the effectiveness of transmitter release, a phenomenon termed *central depression.*

The large numbers of interneurons and their complex interconnections also provide a means for change so that the input-output relationship of reflexes is not simply one to one. If an afferent nerve volley activates a neuron pool through several chains of interneurons, because of differences in propagation time and synaptic delays in the several chains, the pool will be stimulated not once, but each time the original volley makes its way through any chain. The response of the neuron pool is prolonged through repetitive firing of the discharge zone and recruitment from the subliminal fringe zone. *Afterdischarge,* as it is termed, is characteristic of polysynaptic reflex arcs.

Final common pathway. ■ It will be noticed in Fig. 8-6 that the impulses from both afferent fibers, *a* and *b,* leave the cord by the same efferent fiber, *d,* to reach the muscle, *e.* For this reason the efferent nerve is spoken of as the *final common pathway.* The final common pathway is a necessity, for there are fully three times as

*A volley is the response of a nerve, not a nerve fiber, to a single stimulus. Repetitive responses are termed volleys.

many afferent as efferent spinal nerve fibers. It must be evident that only those afferent impulses that lead to the same final common pathway can reinforce each other.

Spatial summation of impulses. ■ Not only may an impulse in an afferent nerve be inhibited (blocked) at the synapse by an impulse in a second nerve, but also impulses may cause summation. In Fig. 8-6 both afferent fibers, *a* and *b,* make synaptic connection with the efferent fiber, *d.* The stimulation of *a* or *b* by itself may not produce a response in *d.* But if both *a* and *b* are stimulated simultaneously, muscle *e* responds. The *synaptic potentials* created by *a* and *b* at the synapse, *c,* are summated. The insufficient concentration of the transmitter substance released

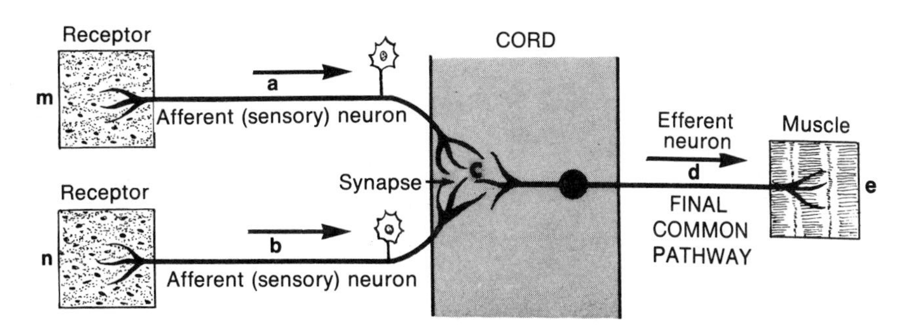

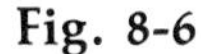
Fig. 8-6
Diagram illustrating concept of spatial summation at synapse, **c**.

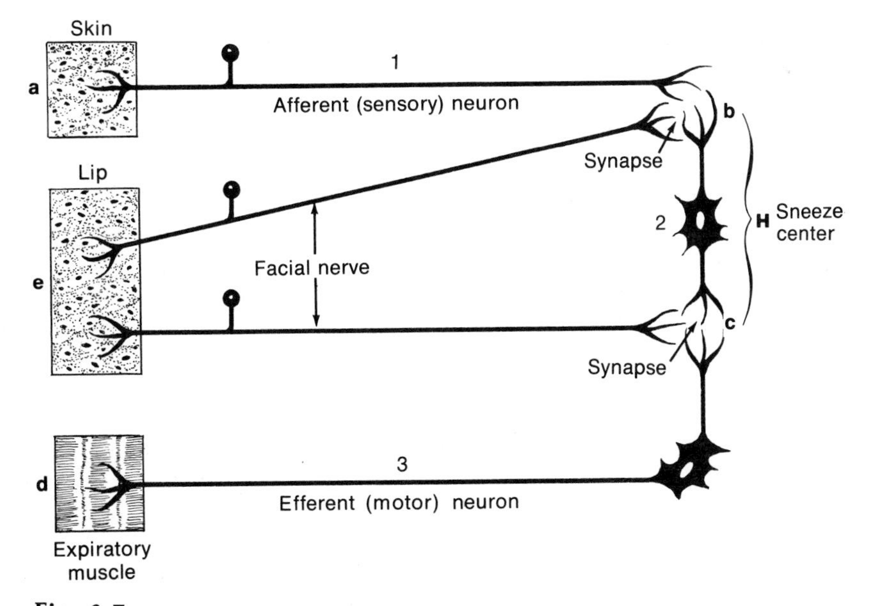

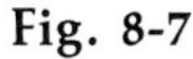
Fig. 8-7
Diagram illustrating concept of reflex inhibition at either synapse **b** or **c**, or both.

at *c* by the ineffective impulse from *a* is augmented by that formed by the impulse arriving from *b*.

Inhibition of reflexes. ■ Most of us are familiar with the fact that an oncoming sneeze can be stopped by pressure applied to the upper lip. The afferent impulse for the sneeze was formed, let us say, by a cold draft stimulating the skin at *a* in Fig. 8-7. The impulse is conveyed to the sneeze center, *H*; from here it normally travels through the neurons, *2* and *3*, to the expiratory muscles, *d*. In so doing it crosses two synapses, *b* and *c*. When simultaneously with the cold draft the facial nerve in the lip, *e*, is stimulated, this impulse reaches the synapse, *b* or *c*, and interferes with the generation of an impulse in either neuron *2* or *3*. In our daily life one impulse blocks another if the first is of greater physiological importance than the second. Pain is a great inhibitor. This holds true also for the emotion of fear.

We may call attention to a marked difference between the inhibitory mechanism of skeletal muscle and that of cardiac and smooth muscle. Although cardiac muscle and intestinal muscle are supplied directly with efferent inhibitory nerves (the vagus and splanchnic nerves, respectively), *no inhibitory nerves to skeletal muscles of vertebrates are known*. The inhibition of the last-named muscles is a central phenomenon involving an inhibition of their motor neurons.

■ REFLEX ACTIONS

Heretofore we have discussed certain simple reflexes in order to define their characteristics and to analyze aspects of synaptic activity. In this section we examine the functional role of specific reflexes in postural and phasic activity. Although our earlier discussion often spoke of a volley or volleys of afferent inputs in which the stimulus was applied maximally and synchronously, it behooves us to remember that in natural reflex activity, stimulation originates in receptors and that afferent input is rarely synchronous.

Reciprocal innervation. ■ When the paw of a spinal dog is pricked with a pin (Fig. 8-8), the *ipsilateral* (same side) flexor muscles are caused to contract. This *flexion* reflex has a broad receptive field; generally any site on the distal part of the limb will cause widespread withdrawal effects. However, the site of stimulation does influence the withdrawal pattern in that different sites bring about different strengths of contraction in the various muscles; to this extent the flexion reflex is not stereotyped (although under identical condition it is). The dependence of the response of the site stimulated is known as *local sign*. Flexion reflexes are polysynaptic (involve interneurons) and show afterdischarge; that is, the contraction outlasts the application of the stimulus in time. They are served by a spectrum of afferent fiber diameters ranging downward from 12 μ and are generally elicited by noxious stimulation.

If the dog's paw is subjected to pressure stimulation (Fig. 8-8), the leg becomes extended like a pillar. This polysynaptic *extensor thrust* reflex plays an important role in locomotion. By rhythmical alternation of extension and flexion, stepping is achieved.

Contraction of flexor muscles is not the only response observed when a dog's paw is pricked with a pin. Examination of the antagonistic extensor muscle reveals that it is relaxed. The relaxation is due to inhibition of the extensor motoneurons through the mechanism of *reciprocal innervation*. Fig. 8-9 illustrates this; the afferent fiber excites the ipsilateral flexor neurons at *b* and inhibits the ipsilateral extensor neurons at *a*. Furthermore, the afferent fiber makes connections (not necessarily so directly as indicated in the diagram) with neurons supplying the extensors and flexors of the *contralateral* limb. Although the flexor neurons of the left limb are activated by afferent stimulation, those in the right limb, at *d*, are inhibited. Extensor neurons of the right limb, at *c*, are stimulated, and a *crossed extension* reflex obtains as a result of the *double reciprocal innervation*. The crossed exten-

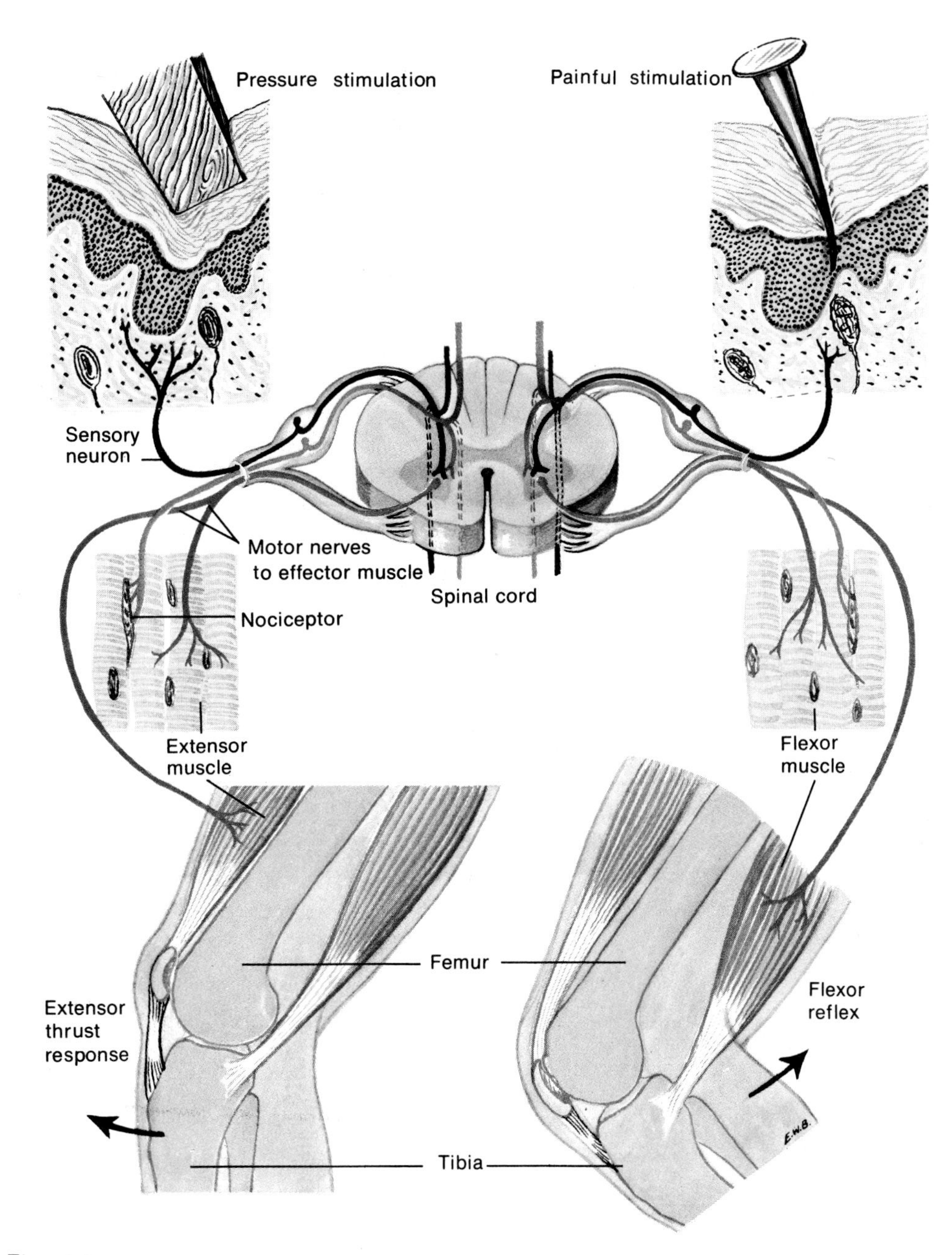

Fig. 8-8
Diagram illustrating flexor reflex withdrawal from painful stimulation (pin prick) and extensor thrust reflex due to pressure stimulation. Internuncial neurons have been omitted.

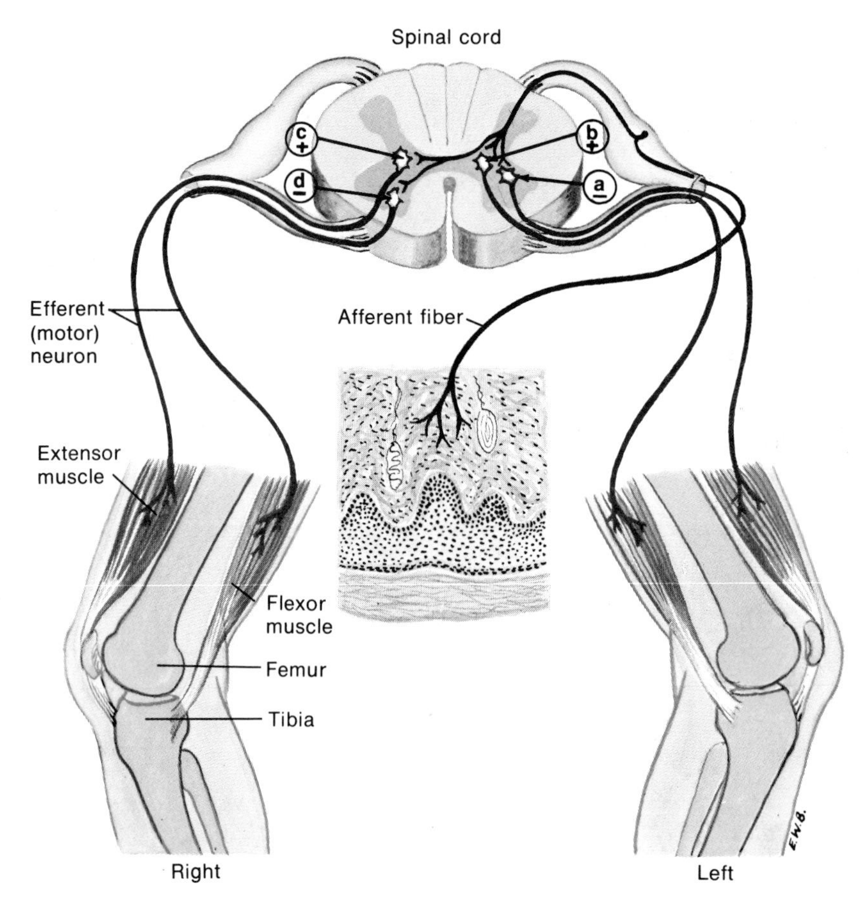

Fig. 8-9
Diagram illustrating reciprocal innervation of antagonistic muscles at **a** and **b** and double reciprocal innervation at **c** and **d.** Excitation is indicated by plus signs, inhibition by minus signs.

sion reflex is not a separate response but is a part of the flexion reflex pattern that serves to support the animal when one limb is raised.

The results of the lack of reciprocal innervation of antagonistic muscles are well illustrated in the effects of strychnine or tetanus toxin. Strychnine reduces synaptic "resistance" throughout the entire central nervous system so that the slightest stimulation of a sensory surface causes widespread muscular activity, and the inhibitory effects that normally accompany a reflex action seem to be converted into excitation. As the extensors in a frog's limbs are more powerful than the flexors, a frog under the influence of strychnine will become rigid in a fully extended position; as a result the frog may be held horizontally by holding it by the toes of the hind legs. When a patient with lockjaw wishes to open his mouth, instead of inhibiting the masseter and temporal muscles (by which the jaw is raised), these muscles, as well as the digastric and mylohyoid (the muscles that lower the jaw), are excit-

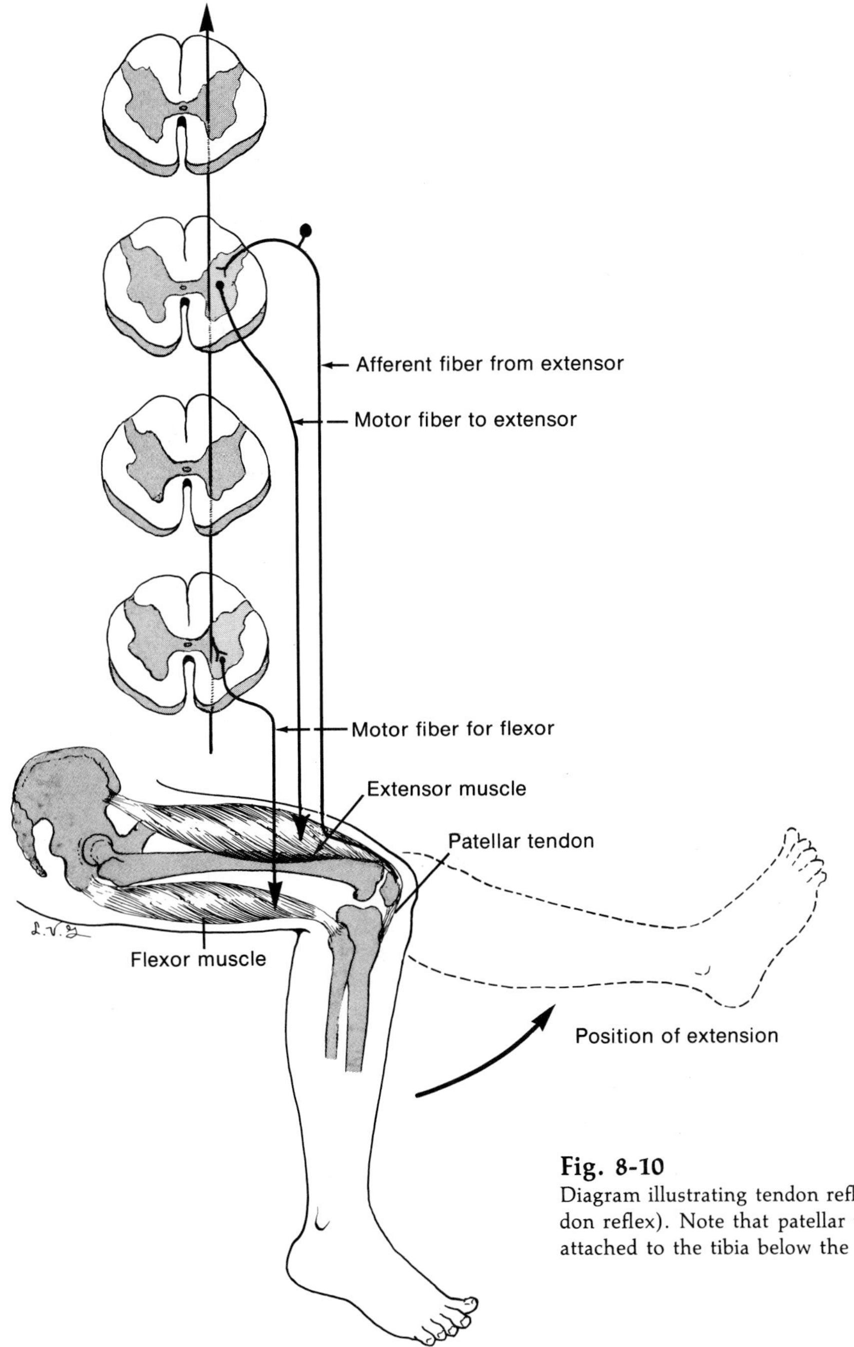

Fig. 8-10
Diagram illustrating tendon reflex (knee jerk or patellar tendon reflex). Note that patellar tendon of extensor muscle is attached to the tibia below the knee.

ed; as a result, the stronger closers overbalance the openers. Antagonistic muscles are frequently called on to contract simultaneously. For example, the forearm may be rigidly extended so that a considerable amount of external force is needed either to raise or to lower it; this is the result of the *cocontraction* of both the flexors and the extensors. Whenever any part of the body or the whole body must be held in a rigid state, as in standing, antagonistic muscles are brought into action at the same time. What are generally called the antagonistic muscles now work *synergistically*.

Tendon reflexes. ■ A distinction is frequently made between the reflex muscular contractions by which a newly acquired position of the body is maintained and those contractions by which movements of the body or a part of the body are brought about (and which are generally of short duration). The former (often of considerable duration) usually are spoken of as *tonic, static,* or *postural contractions;* the latter are called *phasic contractions.* Frequently it happens that the sudden stretching of a muscle in our body results in a marked phasic myotatic reflex. To illustrate: When a person lets his lower legs hang free, as by sitting on the edge of a table, striking the patellar ligament just below the kneecap causes the lower leg to be extended and the foot to be kicked forward (Fig. 8-10). This phasic contraction is known as the *knee jerk,* or *patellar reflex,* and is an example of a class of reflexes variously called *tendon, myotatic* or *stretch* reflexes. The tendon jerk reflexes have a short latency and do not exhibit afterdischarge. These facts suggest that they are served by monosynaptic pathways. They can be inhibited by coincident, strong stimulation of excitatory pathways to antagonistic muscles. Our discussion of the muscle spindle in Chapter 7 and Fig. 8-10 helps explain the origin of the knee jerk. The patellar ligament may be regarded as a continuation of the tendon of the quadriceps extensor by which the lower leg is extended. This reflex occupies the third and fourth lumbar segments of the cord. The great sensitivity of the sensory end organs in a muscle may be appreciated from the fact that the stretching of a muscle by as little as 0.05 mm and for as short a length of time as $^1/_{20}$ sec is sufficient to evoke the reflex. Another example of tendon reflexes is the extension of the foot by the contraction of the gastrocnemius muscle when the tendon of Achilles is tapped (ankle jerk).

The magnitude of the reflex varies directly as the degree of tone in the responding muscle. Conditions that increase muscle tone, such as mental excitement or hyperexcitability of the nervous system (as in hysteria and strychnine poisoning), also increase the extent of the knee jerk; any decrease in the irritability of the nervous system (as by sleep or a restful state of mind) is associated with a decrease in the extent of this reflex. The threshold stimulus for the knee jerk has been found higher after physical or mental work. The knee jerk may be exaggerated (reinforced) by a second stimulus (e.g., a sudden, loud noise) when this is properly timed with respect to the tapping of the patellar tendon. The patellar reflex is absent if the dorsal fibers in the spinal cord or the motor neurons, chiefly of the spinal cord or brainstem, are destroyed.

We may profitably ask: What is the functional significance of the phasic response of a muscle to blows applied to its tendon? The brevity of the jerk response is, of course, due to the synchronous excitation of stretch receptors by the suddenly applied stretch and to the reasonably uniform diameters of the large afferent and efferent fibers, so that propagation velocities are about the same. Although the tendon jerk responses are of diagnostic significance to the physician, the natural stimulus for the stretch reflex is gravity. For example, the quadriceps muscle that we used to illustrate the patellar tendon reflex is, during standing, stretched by bending of the knee in response to gravitational force. Maintained pull upon the muscle induces repetitive, asynchronous firing of the stretch receptors and results in a sustained, smooth contraction (tonic) of significance to posture.

■ Stretch or myotatic reflex and muscle tone

When a muscle in the body is pulled upon by gravity or by the contraction of its antagonistic muscle, the muscle contracts and thereby resists further stretching. The extent of this contraction depends on the degree of stretching. The stretching stimulates the stretch receptors (muscle spindles), and the proprioceptive impulses set up thereby travel to the cord and return by the efferent (motor) fibers *to the muscle in which they originated.** The resulting contraction is referred to as a *stretch,* or *myotatic, reflex.* When due to maintained (e.g., gravity) stretch stimuli, in contrast to tendon jerk reflexes, this activity is tonic.

The influence of gravity can readily be demonstrated by the following experiment: A brain-pithed frog is suspended by means of a hook passed through the lower jaw; this frog will hang motionless for a long time, and the muscles of the hind legs appear, at first sight, to be fully relaxed. It will be noticed that the hind legs have certain characteristic bends at the hip, knee, ankle, and toes. Now open the abdomen and cut the roots of the left sciatic nerve, which innervates the musculature of that leg. It will be seen that the left leg hangs a little more nearly straight, the characteristic bends are not so marked as in the opposite leg, and the toes hang a little lower; the muscles of this leg are flaccid to the touch. Instead of cutting the sciatic trunk, which contains both afferent and efferent fibers, the severing of only the dorsal (afferent) roots produces the same results, showing that afferent impulses are necessary for the generation of tonic contractions. Of all muscles, the *antigravity* muscles exhibit the greatest amount of tonic contractural activity.

*This distinguishes the myotatic from all other reflexes.

As an example of an antigravity muscle, none is better than the musculature that elevates the inferior maxilla. Due to gravity, the lower jaw has a constant tendency to drop, but the proprioceptive impulses created by the stretching of the masseter and temporal muscles (by which the jaw is raised) keep these muscles in constant tone, and as a result the mouth is kept closed.

A simplified diagram of the neural basis for the stretch reflex appears in Fig. 8-11. The muscle spindle is actually a fusiform (cigar shaped) capsule containing two to ten specialized muscle cells known as *intrafusal fibers.* Spindles are distributed widely in skeletal muscles, especially antigravity muscles. Because they are attached to neighboring muscle cells, the *extrafusal fibers,* by connective tissue, they are anatomically in parallel with their neighbors. Intrafusal fibers are long and slender; their *poles* (ends) are composed of striated contractile tissue; the central region may vary but for our purposes can be described as nonstriated, swollen, and nucleated. It is referred to as the *nuclear bag.* Lymph separates the nuclear bag from the capsular connective tissue.

Although three major types of nerve fibers have been observed to penetrate the capsule, a description of only two is necessary for a basic understanding of spindle activity. A large-diameter, 8 to 12 μ, fiber penetrates the capsule and distributes endings over the nuclear bag region. These fibers degenerate when appropriate dorsal roots are transected, and therefore, they are afferent; they are the so-called *primary afferent* fibers. The second type of fiber is small, 3 to 7 μ in diameter, and degenerates on appropriate ventral root transection. These small motor fibers, known as *fusimotor fibers,* or *gamma efferents* (gamma for size), terminate in end plates on the contractile polar regions of the spindle. A number of very small fibers enter the spindle; these are probably pain afferents and sympathetic fibers that innervate the vasculature of the spindle. The extrafusal muscle fibers are inner-

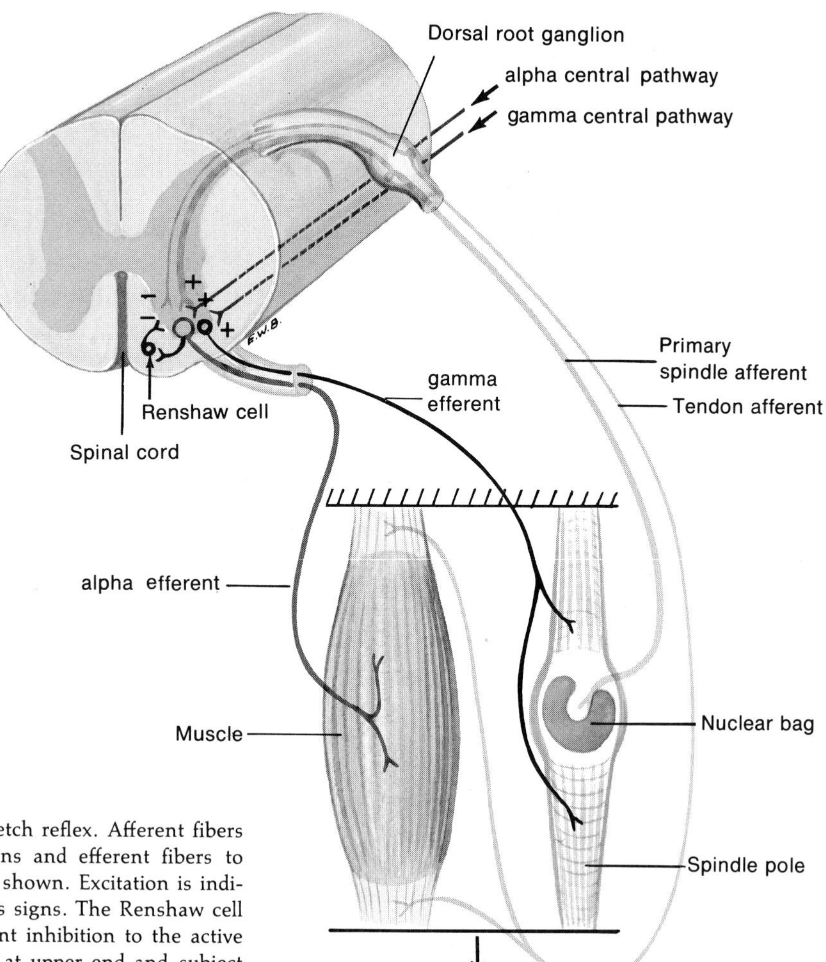

Fig. 8-11
Diagram of the neural basis for the stretch reflex. Afferent fibers from muscle spindle and tendon organs and efferent fibers to muscle and spindle (gamma fibers) are shown. Excitation is indicated by plus signs, inhibition by minus signs. The Renshaw cell is an interneuron that provides recurrent inhibition to the active motoneuron pool. Muscle rigidly fixed at upper end and subject to stretch in direction of arrow at lower end.

vated, as we know, by large-diameter *alpha efferent* nerve fibers.

Stretch of the nuclear bag region, as by pulling down on the muscle (Fig. 8-11), distorts the primary afferent endings and causes a discharge of impulses in the primary afferent. These action potentials on reaching the spinal cord activate alpha motoneurons and bring about contraction of the extrafusal fibers of the muscle. Because the spindle is in parallel with the extrafusal fibers, contraction of the latter reduces the tension on the spindle, which now ceases to fire. Firing of stretch receptors under tension is regular and rhythmic. The rate is somewhat higher on application of the stretch, but it soon adapts to a steady rate that can be sustained even for hours, if the stretch is maintained. Fig. 8-12 illustrates the effect of constant stretch on the response of a single spindle from a frog's toe muscle. Frequency of firing is plotted against time; after an initial high frequency phasic burst, the spindle adapts and a tonic level of activity is maintained

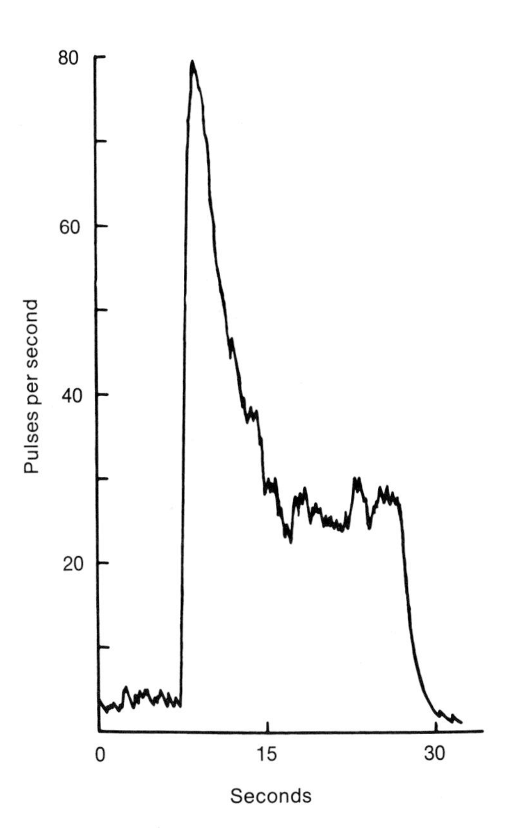

Fig. 8-12
Graph showing frequency of impulses in afferent nerve of a frog's toe muscle stretch receptor. Muscle was both stretched and released suddenly. Adaptation occurred within about 7 seconds. Compare with Fig. 7-1.

until the stretch is released. Individual spindles have different thresholds of stretch stimulation; each signals the above-threshold intensity of the stimulus by frequency of discharge. Together, frequency and number of active spindles assure the graded contraction of the muscle appropriate to the stimulus.

In the tendinous ends of the muscle there are fibrous capsules enclosing the terminals of one or two large-diameter fibers (Fig. 8-11). These nerve fibers are *tendon afferents,* and their endings in the fibrous capsules are the *Golgi tendon organs* or *tendon end organs.* Stretch applied to the muscle displaces the tendon organs and initiates action potentials in the afferent nerve. The Golgi tendon organs are in series with the extrafusal fibers; therefore, the organs do not distinguish between passive stretch of the muscle or active contraction of the muscle, either of which would elongate the tendons. Golgi tendon organs have a much higher threshold than do muscle spindles, perhaps two hundred times higher. As the intensity of the stretch upon the tendon exceeds threshold, the firing rate increases. The tendon afferent ends in the spinal cord upon the alpha motoneuron, which it inhibits. Thus, strong passive stretch or active contraction of a muscle results in *autoinhibition* of that muscle's stretch reflex response. Autoinhibition can serve as a protective mechanism against potentially injurious stretch, whether induced passively or actively; it also can be a means of adjusting, through the tendon end organ feedback loop, the degree of active contraction that can be maintained.

When the activity in a ventral root is fairly strong and repetitive, contraction of the muscle does not reduce spindle firing. In fact, it may increase. Resolution of this apparent paradox lies in understanding the function of the gamma efferent system. Impulses in this system cause the polar regions of the spindle to shorten. Since the intrafusal fibers are in a fixed position relative to the extrafusal fibers, shortening in each polar region must proceed in the direction away from the nuclear bag and stimulation of the primary endings by this distortion increases the firing rate of the receptor. No significant contribution is made to the muscle's total tension by intrafusal fiber contraction, since their numbers are small compared with extrafusal fibers and since they are not anatomically suited to transmission of tension to their ends. The importance of the gamma system arises from its ability to regulate the sensitivity of the spindle. By shortening the spindle relative to the extrafusal fiber length, the gamma system resets, as it were, the level of activity in the primary endings.

The gamma system is involved in other reflex patterns as well. For example, in the crossed extension reflex the afferent fibers from the side ipsilateral to the noxious stimulus excite not only the contralateral extensor alpha motoneurons but also the gamma system to the extensor muscles, thereby reinforcing the extension. Central pathways from the brainstem and cerebellum are important contributors to the stretch reflex pattern, since activity in them has marked effects, inhibitory and excitatory, on both alpha and gamma motoneurons.

Inhibition of the stretch reflex is not solely attributable to tendon end organ activity, as previously described. In the ventral horns of the spinal cord, collateral branches from axons of motoneurons synapse with small interneurons (Fig. 8-11), known as *Renshaw cells.* The axons of the latter make synaptic contact with numbers of alpha motoneurons; the effect that Renshaw cells have on motoneurons is inhibitory and, consequently, it is often referred to as *recurrent inhibition.* Such inhibition represents a negative feedback because the degree of inhibition is directly related to the intensity of activity in the motoneurons. Despite extensive investigation, the functional role of recurrent inhibition in normal reflex activity is, at best, obscure.

Muscle tone, or *tonus,* is a term descriptive of the degree of resistance that a muscle exhibits in response to passive stretch. It is a consequence of the myotatic reflex, for, as we have already described it in the frog, the cutting of dorsal roots makes the muscles flaccid. This *hypotonic* muscle offers little resistance to stretch. On the other hand, destruction of central pathways from the vestibular nuclei and reticular formation that normally inhibit the stretch reflex leads to increased tone, *hypertonia,* with increased resistance to stretch. It is likely that in a muscle with a certain amount of tonus a few fibers are contracted maximally at any given time; as these fibers undergo fatigue, others are brought into activity. This may explain why a muscle can exhibit tonus for an almost unlimited time without experiencing fatigue. Tonus is diminished by sleep* and anesthetics.

■ SPINAL CORD AS A CONDUCTOR

Since it may be desirable that a given afferent nerve be in communication with two or more efferent nerves distributed to as many responding organs, and since it may be necessary that a single responding organ be accessible to two or more afferent inputs, a place for these manifold neural connections must be supplied. To furnish this and to regulate the passage of impulses through these connections (synapses) are the primary and fundamental functions of the central nervous system. The gray matter of the brain and spinal cord may be compared with the central exchange of a telephone system in which the wire carrying an incoming message is plugged into one provided for carrying the message to the desired party. How the manner of the "plugging in" may be modified, by delay, inhibition, or facilitation, we have considered in the foregoing pages.

Ascending and descending fibers. ■ Some fibers constituting the white matter of the cord carry impulses from the lower to the higher levels of the cord and even to the brain (Fig. 10-9); these are ascending fibers. Other fibers have their cell bodies in the brain (Fig. 10-10) or in the upper parts of the spinal cord and conduct impulses to various lower levels; these are descending fibers. If the cord is transected at a certain level, the ascending fibers, which have their cell bodies below the section, degenerate above the cut; the descending fibers degenerate below the section. By making sections at various levels of the cord or brain and by noting the position

*This is well illustrated in the dropping of the jaw by an individual asleep in a sitting position.

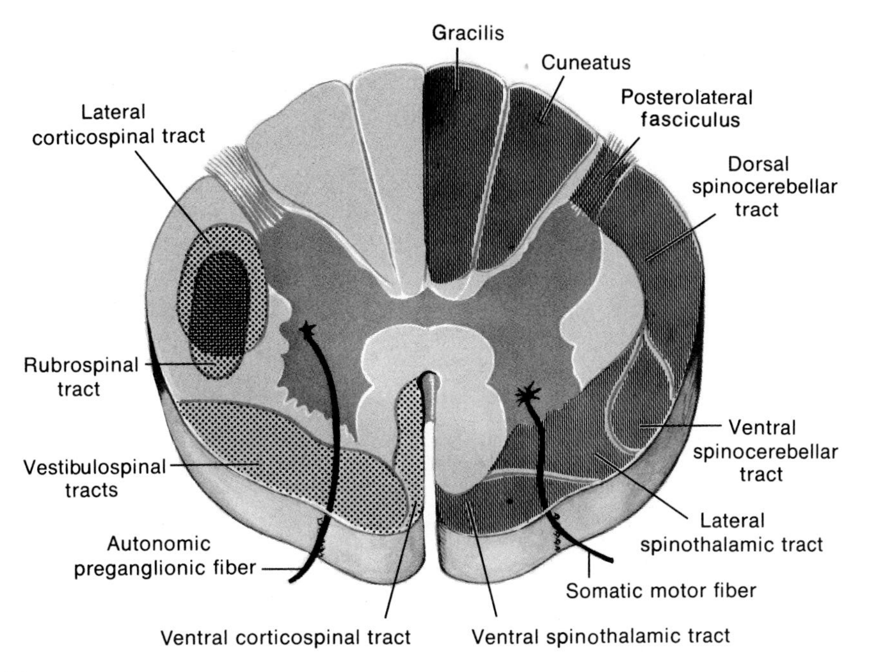

Fig. 8-13
Diagram of a cross section of the spinal cord. Hatched areas on right depict ascending tracts, or columns, of fibers; stippled areas on left indicate descending tracts, or columns.

of the ascending and descending degeneration and the chromatolysis in the cell bodies by histological stains, the various fibers have been traced from their points of origin to the next relaying neuron. In this manner and by observing pathological conditions in man, it has been determined that the fibers conveying impulses up or down the cord are arranged in an orderly manner.

In Fig. 8-13 the most important ascending and descending tracts or columns are indicated. To avoid confusion, the ascending tracts are indicated only in the right-hand half of the cord, and the descending tracts are indicated only in the left-hand half. Of course, both ascending and descending tracts are present in both halves.

We may mention four important ascending tracts: the fasciculus gracilis, the fasciculus cuneatus, the spinocerebellar, and the spinothalamic. The fibers of the first two mentioned tracts carry proprioceptive impulses from muscles, tendons, and joints and tactile impressions from the skin to nuclei located in the medulla. The spinothalamic fibers carry impulses of pain, temperature, and sexual sensations to the thalamus. The spinocerebellar fibers (or indirect cerebellar tract) shown in Fig. 10-15 carry proprioceptive impulses directly to the cerebellum. The further transmission of these various impulses to the higher brain levels will receive attention in Chapter 10.

READINGS

Barnes, C. D., and Kircher, C.: Readings in neurophysiology, New York, 1968, John Wiley & Sons, Inc.

Davson, H., and Welch, K.: The relations of blood, brain and cerebrospinal fluid. In Siesjö, B.K., and Sørenson, S.

C., editors: Ion homeostasis of the brain, New York, 1971, Academic Press, Inc.
DeRobertis, E.: Molecular biology of synaptic receptors, Science **171**:963-971, 1971.
Granit, R.: The basis of motor control, New York, 1970, Academic Press, Inc.
Lippold, O.: Physiological tremor, Sci. Am. **224**(3):65-73, Mar. 1971.
Merton, P. A.: How we control the contraction of our muscles, Sci. Am. **226**(5):30-37, May 1972.
Rubin, R. P.: The role of calcium in the release of neurotransmitter substance and hormones, Pharmacol. Rev. **22**:389-428, 1970.
Wilson, V. J.: Inhibition in the central nervous system, Sci. Am. **214**(5):102-110, May 1966.

9

Vision
Hearing
Labyrinthine sensations
Chemical senses—taste and smell

RECEPTOR PHYSIOLOGY— SPECIAL SENSES

The general features of receptor physiology were discussed in Chapter 7. In the pages that follow we will extend our study of receptor physiology to those modalities known collectively as the special senses.

■ VISION

■ Optics

The sensory neuroepithelium of the eye is located in the retina, which forms the innermost coat of the three tunics of the eyeball. Its stimulation by light provides the basis of three sensations: light, color, and form or shape. The extent of its sensitivity is readily apparent from the fact that under proper conditions, a ray of green light stimulating the eye for only 1/8,000,000 sec is perceptible. This is an amount of energy equal to that liberated by 1/300,000,000,000 of an ounce falling the distance of 1/25 inch. To obtain this sensitivity and the discrimination of form, a picture of the object sending the light into the eye must fall upon the retina. To bring this about is the object of nearly the whole structure that we call the eye; before treating this subject, we must review a few points concerning the action of light.

Spectrum of light. ■ The word *light* may be used in a subjective or in an objective sense. In the first sense we denote the sensation experienced when the retina is stimulated; in the objective sense is meant the light waves that are capable of stimulating the retina. Light travels at a velocity of about 186,000 miles per second. Sunlight is heterogeneous, being composed of wavelengths ranging from 0.00076 mm (red) to 0.00038 mm (violet).* When a beam of sunlight passes through a prism (Fig. 9-1), these wavelengths are dispersed and a spectrum (band of colors) is seen. The violet light, having the shortest wavelength, is bent (refracted) farthest from its original course; the red is refracted least. Wavelengths shorter than those of violet light

*In violet light the frequency is 757,000,000,000,000 and in red light it is 392,000,000,000,000 vibrations per second. All light waves travel with the same velocity.

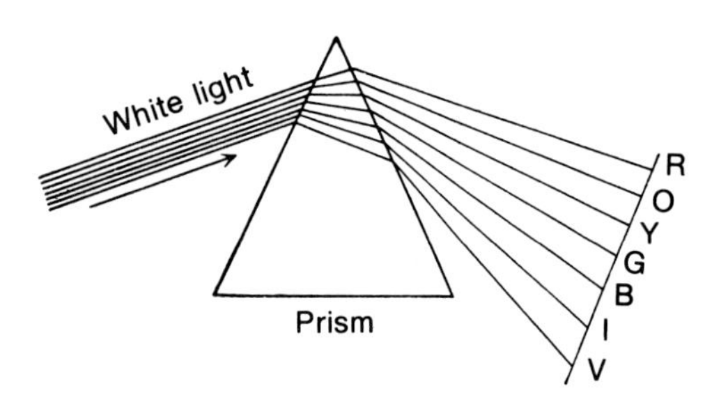

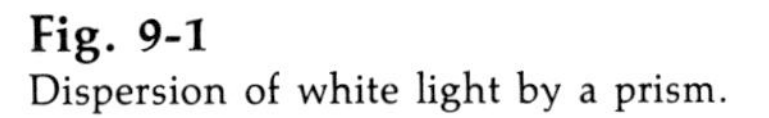

Fig. 9-1
Dispersion of white light by a prism.

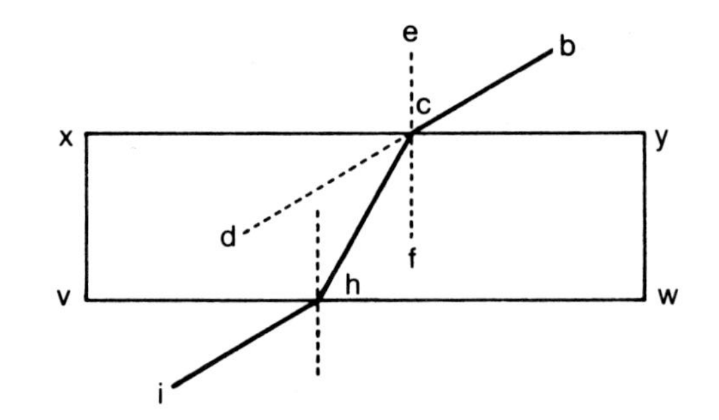

Fig. 9-2
Diagram illustrating the refraction of light by a plane surface.

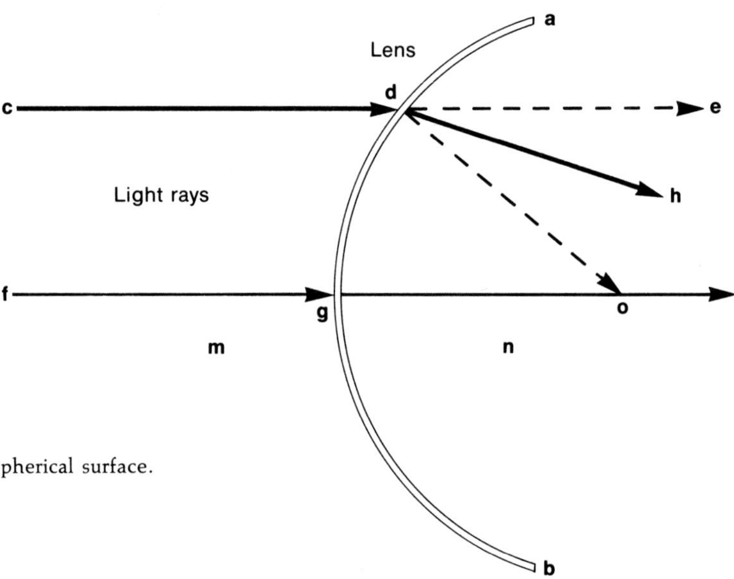

Fig. 9-3
Diagram illustrating the refraction of light by a spherical surface.

(ultraviolet) and longer than those of red light (infrared) do not stimulate the retina and are, therefore, invisible (Fig. 24-4).

Refraction of light. ■ The velocity with which light travels is uniform as long as the medium through which it passes remains the same and as long as the waves proceed in straight lines. Suppose that a flat piece of glass (Fig. 9-2) is placed in the path of the ray, *bc*. If the velocity of light in glass were the same as that in air, the ray would proceed straight onward to *d*. But, because of the physical and chemical differences between air and glass, the velocity is reduced, and, as a result of this, the ray is bent out of its straight course in such a manner that it is refracted toward the normal (*ef*, a line perpendicular to the surface, *xy*). The refracted ray describes the course, *ch*. At point *h* the ray strikes the other surface of the glass, *vw*, and reenters the air. As the velocity of the light is now in-

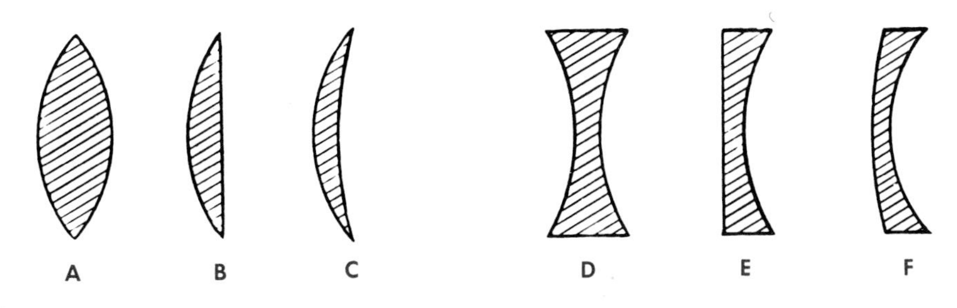

Fig. 9-4
Various types of lenses. **A,** Biconvex; **B,** plano-convex; **C,** concavo-convex; **D,** biconcave; **E,** plano-concave; **F,** convexo-concave.

creased, the ray is refracted away from the normal erected at *h* upon the surface, *vw*, and describes the course, *hi*. The velocity of the light in glass is less than that in air; consequently we say the *optical density* or *index of refraction* of glass is greater than that of air. We may generalize: *light in passing from a rarer into a denser medium is refracted toward the normal,* and in passing from a denser into a rarer medium it is bent away from the normal.* A ray of light normal to the surface is retarded in speed but undergoes no refraction.

It is immaterial whether the refracting surface is a plane or a spherical surface. Suppose that *ab* (Fig. 9-3) is a segment of a glass sphere, *n*, to the left of which there is air, *m*. The ray of light *cd* strikes the spherical surface of the glass at *d*. As glass has a greater optical density than air, the ray is refracted toward the normal *do*. The normal to the surface, *ab*, at the point *d* is the radius of the sphere from its center of curvature, *o*, to the point *d*. The ray would have proceeded to *e* if the glass had not been placed in its path, but now it is bent toward the normal *do* and describes the course *dh*. The ray *fg*, being normal to the surface, undergoes no refraction and continues through in the direction of *o*.

Lenses. ■ We can apply the previous discussion to spherical and cylindrical lenses. Spherical lenses are those whose surfaces are segments of spheres. There are two classes of spherical lenses: *convex*, or plus, lenses and *concave*, or minus, lenses. Various types of lenses are shown in Fig. 9-4.

Convex lenses. The passage of light through a convex lens may be described as follows: In Fig. 9-5, *ab* represents a biconvex lens, and *cd*, *ef*, and *gh* represent three parallel rays of light. In reality, all rays of light are divergent, but if the light comes from a great distance, the divergence is so small that it may be ignored, and the rays may be considered parallel. The ray *cd*, striking the anterior surface of the glass at *d*, is refracted toward the normal *op* (*o* being the center of curvature of the anterior surface of the lens); hence, it will describe the course *di*. At *i* the ray strikes the posterior surface of the lens and, on issuing from the glass into the air, is bent away from the normal *qr* and describes the course *ij*. Applying the same process of reasoning to the ray *gh*, we see that its refracted ray is *kl*. The ray *ef* is normal to both the anterior

*This is a qualitative statement. The amount of refraction that a beam of light undergoes is determined by the index of refraction of the two transparent media and by the angle that the incident ray makes with the normal. In Fig. 9-2, *ecb* is the angle of incidence and *hcf* is the angle of refraction. The law of refraction states: angle of incidence: angle of refraction : : index of second medium : index of first medium.

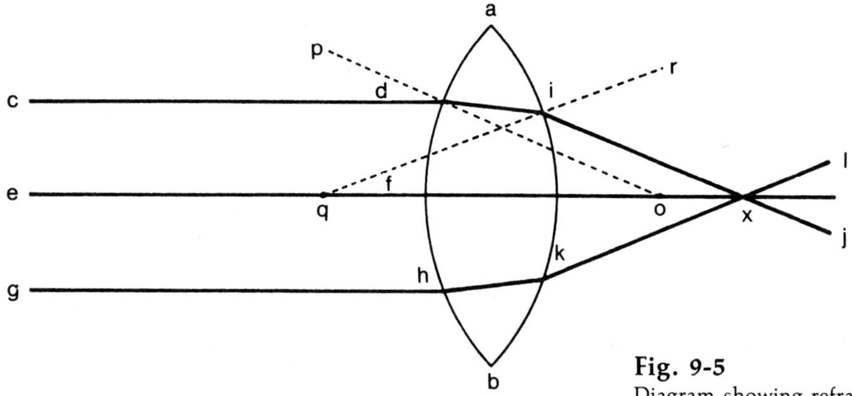

Fig. 9-5
Diagram showing refraction of light by a biconvex lens.

and posterior surfaces (on it are located the centers of the two curvatures), and hence this ray will pass through without undergoing any refraction. From the diagram it will be noticed that these three rays after refraction cross at a common point, *x;* this point is called the *principal focus.*

Conjugate foci. In Fig. 9-6, *a* is a luminous point situated near the lens. The principal focus is *x.* The rays of light, *ab* and *ad,* are divergent, and for this reason their focusing is delayed beyond *x;* the emergent rays cross at point *f.* The luminous point, *a,* and its focus, *f,* are said to be *conjugate foci.* The nearer the luminous point lies to the lens, the farther back of the lens lies its conjugate focus. It will be noticed that a convex lens is a gathering, or *converging, system;* it brings or tends to bring the rays of light to a focus.

The diopter. Two factors determine the focusing power of a lens: (1) the index of refraction of the lens and (2) the radius of curvature of the surfaces. As already stated, refraction occurs when the light passes from one medium into another having a different optical density. The greater the difference between the two media, the greater is the bending of the light. Suppose that the optical density of lens *A* in Fig. 9-7 is 1.33, as compared with the density of air (1.00),

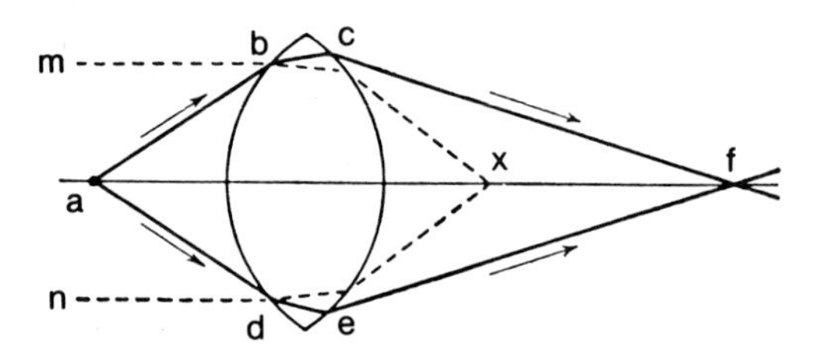

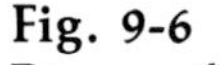
Fig. 9-6
Diagram illustrating conjugate foci, that is, the luminous point **a** and its focus **f.**

Fig. 9-7
Diagram showing effect of optical density of a lens on its refracting power.

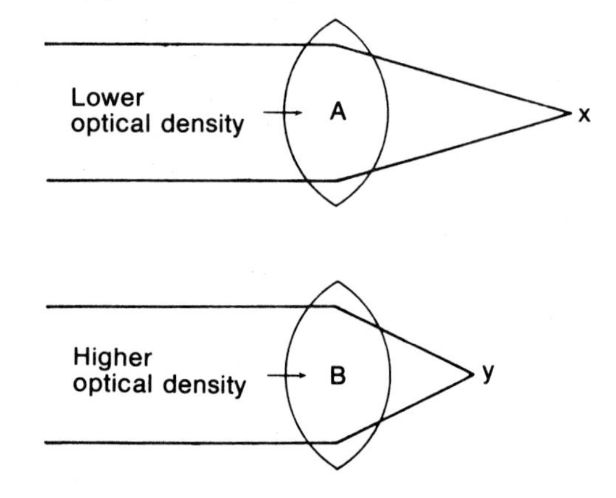

and that its principal focus lies at *x,* for example, 2 inches back of the lens. Lens *B* is similar to lens *A* in shape and size, but, being made of a different kind of glass, its optical density is 1.43. The parrallel rays falling upon it will be bent more, and consequently the principal focus, *y,* will lie closer to the lens, for example, 1 inch back of the lens. Hence, lens *B* is *stronger,* that is, has a greater focusing power (and also magnifying power) than *A. The focusing power of a lens varies directly as the optical density.*

Again, suppose we have two lenses made of the same kind of glass and, therefore, having the same optical density, but one of them (Fig. 9-8, *A*) has surfaces that are segments of larger spheres than the surfaces of the other lens, *B.* More tersely stated, the radii of the curvatures of *B* are shorter than those of *A.* When parallel rays fall upon these lenses, the principal focus of *B* lies nearer to the lens than does that of *A;* that is, the focusing power of *B* is greater than that of *A.* Evidently, *the focusing power of a lens varies inversely as the radius of curvature.*

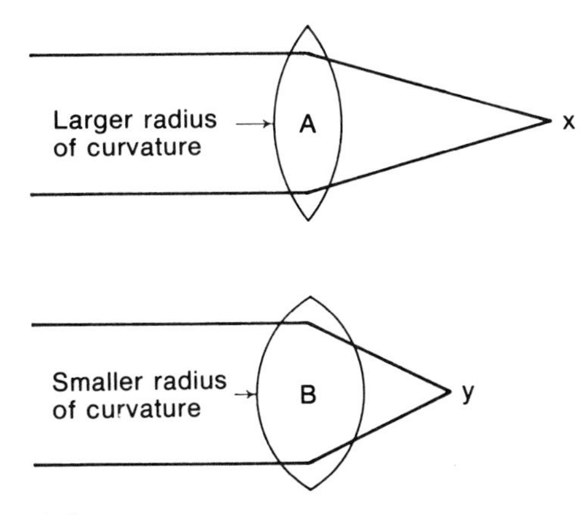

Fig. 9-8
Illustration showing effect of radius of curvature of a lens on its refracting power.

The strength of a lens is measured in terms of the focal distance. If a lens has the power to focus parallel rays at a point 100 cm back of the lens, its focal length (distance between the principal focus and the lens) is 100 cm; such a lens is said to have *1 diopter* refractive power. A lens with a focal distance of 25 cm has 100/25, or 4, diopters refractive power; the power in diopters may, therefore, be found by dividing 100 by the focal length in centimeters.

Images. When light from an object falls upon a convex lens and a screen is placed at the proper distance on the other side of the lens, a picture of the object will be thrown upon the screen; this image, as illustrated in Fig. 9-9, is real (behind the lens) and inverted. Notice that we have drawn rays of light from the two extremities of the object (*a* and *b*) and passed them straight through the center of the lens, *n;* the point *n* is called the *nodal point,* which may be defined as a point in an optical system of such a nature that

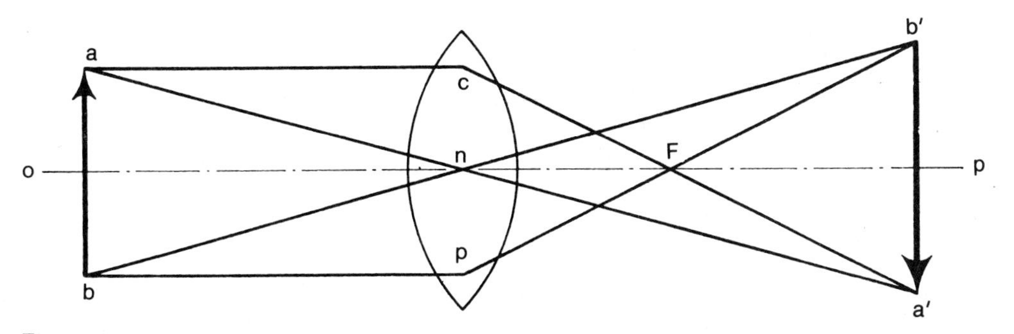

Fig. 9-9
Diagram illustrating formation of an image by a convex lens.

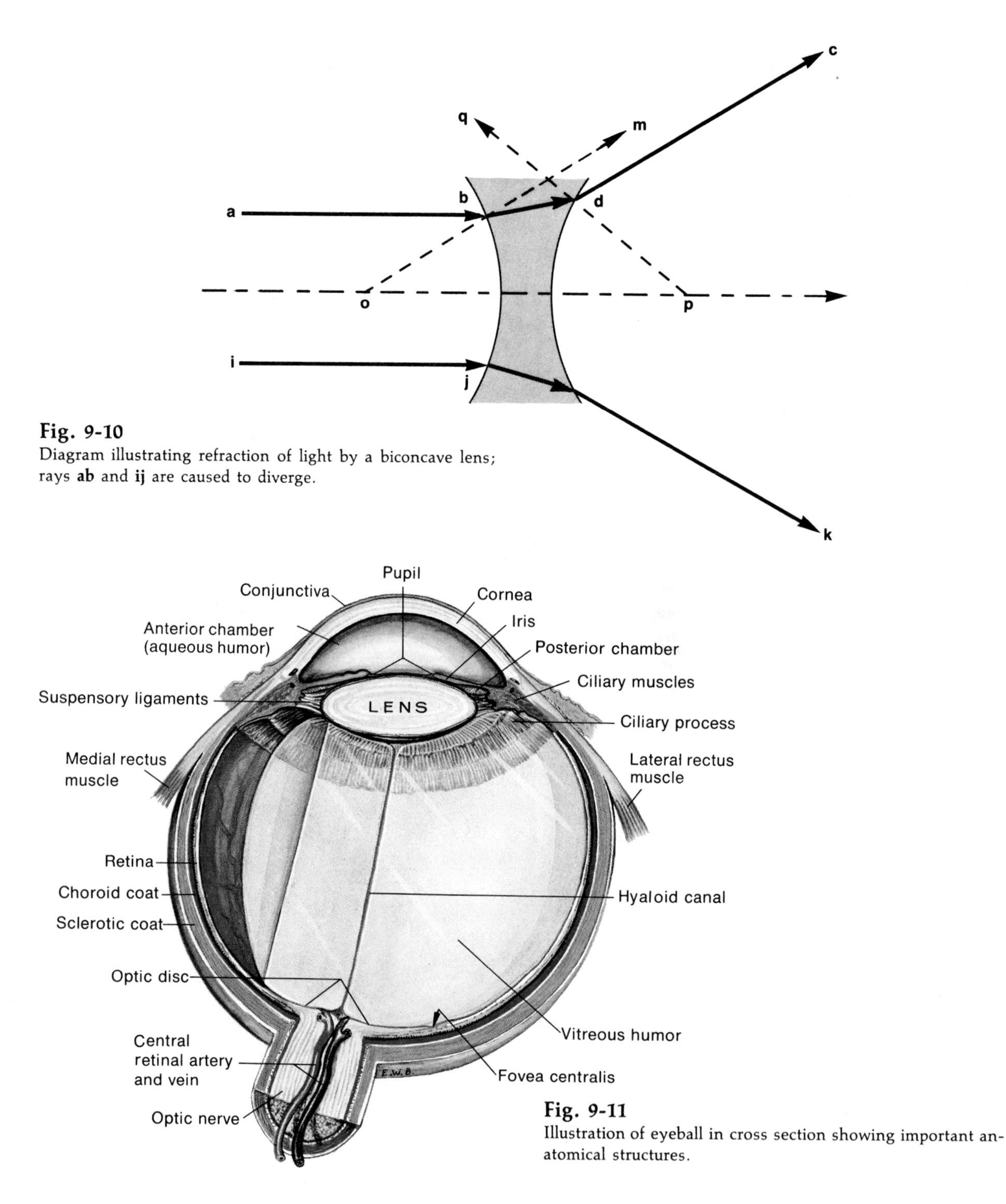

Fig. 9-10
Diagram illustrating refraction of light by a biconcave lens; rays **ab** and **ij** are caused to diverge.

Fig. 9-11
Illustration of eyeball in cross section showing important anatomical structures.

a ray of light may be considered as passing through it without having undergone refraction.*

Concave lenses. Concave lenses also are frequently used in spectacles. The ray *ab,* in Fig. 9-10, striking the surface of the glass at *b,* is bent toward the normal *om* and therefore describes the course *bd.* On entering the air at *d,* the ray is bent away from the normal *pq* and describes the course *dc.* In a similar manner, the ray *ij* is refracted so that it travels the course *jk.* From this it is evident that *a concave lens is a dispersing system;* it scatters the light, and no real images can be formed.

■ The eye as an optical instrument

Anatomy. ■ The eyeball, a nearly spherical structure about 1 inch in diameter, is composed of three coats, which surround the transparent media through which the light travels to reach the sensory epithelium (retina). The outer coat (Fig. 9-11) is the sclerocorneal, or sclerotic, coat. The *sclera* is the white of the eye. It consists of dense connective tissue and serves well in protecting the inner structures and in giving a definite, constant shape to the eyeball. The anterior portion of the sclera is modified so as to be transparent; it forms the glassy part, the *cornea.* The *choroid* forms the middle coat; it is a pigmented, vascular coat. In the anterior chamber of the eye, the choroid forms the ciliary body and the pigmented iris, which are seen through the cornea. The innermost coat of the eye, the *retina,* is the neural layer in which is found the sensory epithelium and in which the optic nerve fibers (second cranial nerves) originate. It is the functional, sensitive coat upon which the images of the objects to be seen must be cast; it is comparable to the film of a camera. Enclosed within these coats are found (1) a transparent, fluidlike medium, the *aqueous humor,* (2) the *crystalline lens,* which lies posterior to the anterior chamber, and (3) a transparent, jellylike substance known as the *vitreous humor,* which lies between the lens and the retina. The cornea, lens, and humors are transparent, and, as their optical densities are greater than is that of air, the light on passing into the eye is refracted by the anterior surface of the cornea and by the anterior and posterior surfaces of the lens. When these media lose their transparency, as in cataract of the lens, vision is impaired or altogether lost.

*In reality there are two nodal points, but they are situated so close together that, for practical purposes, they may be regarded as one. This is not necessarily situated in the center of the lens.

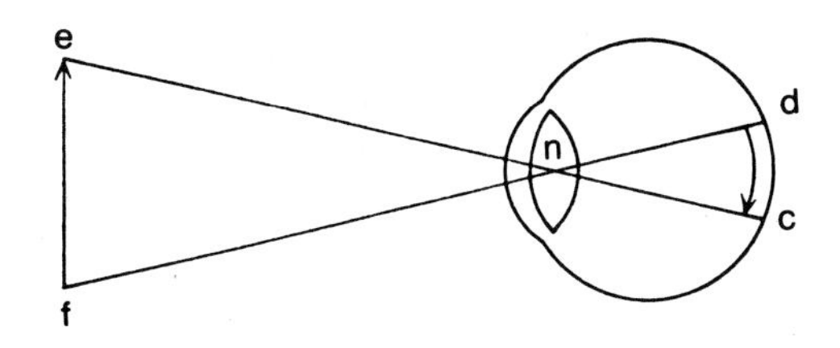

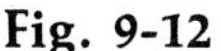

Fig. 9-12
Diagram showing formation of a retinal image, real and inverted.

Retinal images. ■ In Fig. 9-12, rays of light from the extremities of the object, *ef,* pass through the nodal point, *n* (which actually is located fairly close to the posterior surface of the lens); under the proper conditions, the refracted rays focus upon the retina at *cd,* thereby forming a real and inverted image. If the focus is situated in front of or behind the retina, the image is blurred and vision is indistinct. Objects situated at different distances from a lens have their foci at different distances back of the lens; the nearer an object is situated toward the lens, the farther back of the lens lies the focus. The question is, what objects are focused upon the retina? The answer will depend on the condition of the eye, since not all eyes are the same in this respect. Let us first study the eye in what we may call the normal condition—*emmetropia.*

Emmetropia. ■ Suppose an individual having so-called normal eyes stands, with eyes closed, near a window; the eyes are then at rest. On opening them he immediately and without any effort sees the buildings across the street clearly and distinctly but not the specks on the window pane. From this we must conclude that distant objects (20 feet or more removed) are focused upon the retina of the normal eye when this eye is in the resting condition. Such an eye is called an *emmetropic eye;* this may be defined as *the eye in which the posterior principal focus* (Fig. 9-13, *c*) *lies upon the retina when the eye is at rest.* Under these conditions an object, *a,* nearer than 20 feet sends divergent rays and is focused at *b,* back of the retina, and cannot be seen distinctly.

Accommodation. An individual with emmetropic eyes can see near objects if he makes an effort; that is, by some process he can bring the focus (Fig. 9-13) of the near object, *a,* upon his retina. The process by which this is accomplished is called *accommodation.* To understand how this is brought about, we must refer to another point in the anatomy of the eye. Notice in Fig. 9-11 that near the lens the choroid coat is thickened and gives rise to a number of thumblike processes, the *ciliary processes,* which encircle the lens. From the ciliary processes spring very delicate cords, the *suspensory ligaments,* which span the space between the processes and the lens and are attached to the periphery of the lens. With this arrangement the lens is held in position in the more or less fluid contents of the eye. The intraocular pressure, which bears upon the retina and choroid, forces the ciliary processes apart in a radial direction. This pressure causes tension to be exerted upon the suspensory ligaments which, in turn, pull upon the periphery of the lens. Since the lens is plastic, the traction exerted by the ligaments flattens it so that the radius of curvature of the anterior surface is increased and its refractive power decreased; such is the condition of the lens when the eye is at rest, and, if the eye is emmetropic, refraction is as indicated in Fig. 9-13.

From the sclerotic coat near its junction with the cornea there proceed delicate, smooth muscle fibers, the *ciliary muscles* (Fig. 9-11); the other ends of these fibers are embedded in the choroid coat. When the ciliary muscles contract, the choroid coat is drawn forward. This allows the ciliary processes to approach each other centrally and thereby relaxes the suspensory ligaments; then the lens, by virtue of its elasticity, assumes a more spherical form; that is, the radius of curvature of the anterior surface is decreased and the lens' refractive power is correspondingly increased.

The change in focusing power is readily understood by comparing Figs. 9-13 and 9-14. In Fig. 9-13 the eye is at rest, and parallel rays from a distant object focus upon the retina, *c,* while di-

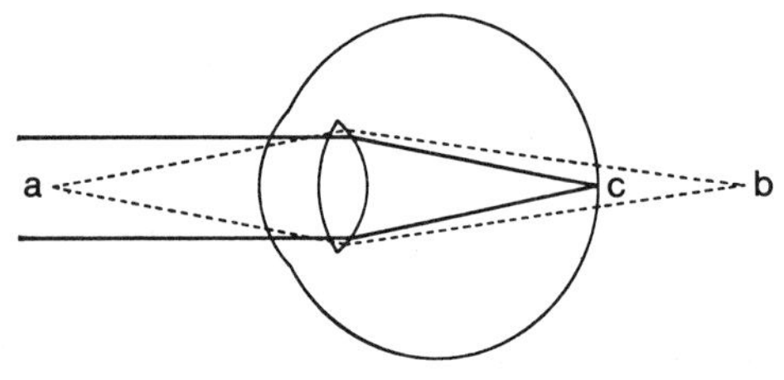

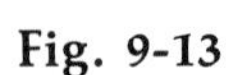
Fig. 9-13
Diagram of emmetropic eye at rest. **c,** Posterior principal focus of parallel rays lies on retina, **b.** Focus (hypothetical) of near object, **a,** lies in back of retina.

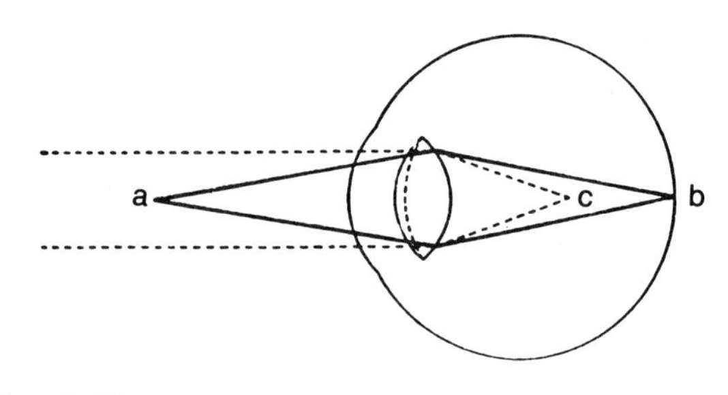

Fig. 9-14
Diagram of emmetropic eye in accommodation. Near object, **a,** is in focus on retina at **b.**

vergent rays from a near object, *a,* focus back of the retina, *b.* When the eye is accommodated (Fig. 9-14), the anterior surface of the lens has a shorter radius of curvature and a greater refractive power; hence the near object, *a,* has its focus upon the retina, *b,* and is seen distinctly, but the far object is focused in front of the retina, at *c,* and is therefore not seen distinctly. By means of the parasympathetic (autonomic) outflow in the oculomotor or third cranial nerve, we are able to govern the extent of the contraction of the ciliary muscles (and therefore the amount of accommodation) according to the distance of the object from the eye. In addition, the complete reflex act of accommodation involves constriction of the pupils (miosis) by the sphincter muscles in each iris and the convergence of the eyes due to activity in the extraocular muscles (Fig. 9-25). The accommodation reflex requires about 5/6 sec.

Far point and near point. The accommodative power of the eye has its limit; when this limit is reached, the nearest object (point) upon which the retina can focus is called the near point of vision. The far point is the point that the eye sees without accommodation. For the emmetropic eye the far point is infinity (any object beyond 20 feet); the near point is approximately 6 inches.

The ciliary muscles are therefore constantly at work during near vision. In people engaged in near work, few muscles of the body are used as constantly as the ciliary muscles.

■ Defects of vision

Intraocular pressure. ■ Good vision can be obtained only if the images are cast upon a smooth screen. Any wrinkles or creases in the retina distort the images. Distortion is avoided by two factors: (1) a fairly inelastic scleral housing and (2) the more or less fluid media within the eye are under a constant amount of pressure—*intraocular pressure*—which keeps the retina smoothly applied to the choroid coat. Fluid volume changes occur chiefly in the anterior chamber (Fig. 9-11). The 5 to 6 ml/day produced by the ciliary bodies are normally drained away by the canal of Schlemm and eventually reenter the bloodstream. Loss of humor due to punctures may lower the intraocular pressure. In abnormal elevation of the pressure (glaucoma), the retinal blood vessels may be squeezed shut, thereby interfering with the nutrition and function of the retina, and, unless soon relieved, sight may be lost. Glaucoma causes about one half of the blindness in adults. The normal intraocular pressure is about 24 mm Hg.

Presbyopia. ■ The degree of accommodation depends on the elasticity of the lens; the greater the elasticity, the greater is the power of accommodation, and the closer the near point of vision is situated to the eye. With advancing years, gradual loss of elasticity reduces the accommodative power and causes the near point to recede. In consequence a book must be held progressively farther away, until finally the images of the letters on the retina become too small to be recognized. To this condition the term *presbyopia* (sight of old age) is applied. The loss of elasticity of the lens *(lenticular sclerosis)* begins early in life, for it can be detected at the age of twelve or fourteen years, but in an emmetropic eye it gives little or no trouble until the age of forty to forty five years. The waning accommodative power can be supplemented by properly fitting *convex* lenses.

Myopia. ■ Not all eyes are emmetropic; many are either myopic or hyperopic. *Myopia* or nearsightedness is that condition of the eye in which the posterior principal focus falls in front of the retina (Fig. 9-15). This is, in most instances, due to an elongated condition of the eyeball. As a result, in myopia, when the eye is at rest, not the far but the relatively near object, *d,* is focused at *d′* upon the retina. The far point of vision is therefore nearer than infinity (20 feet), and the near point is nearer than 6 inches. Weakness of the coats of the eyeball is regarded as a cause of elongation of the eyeball. Myopia generally

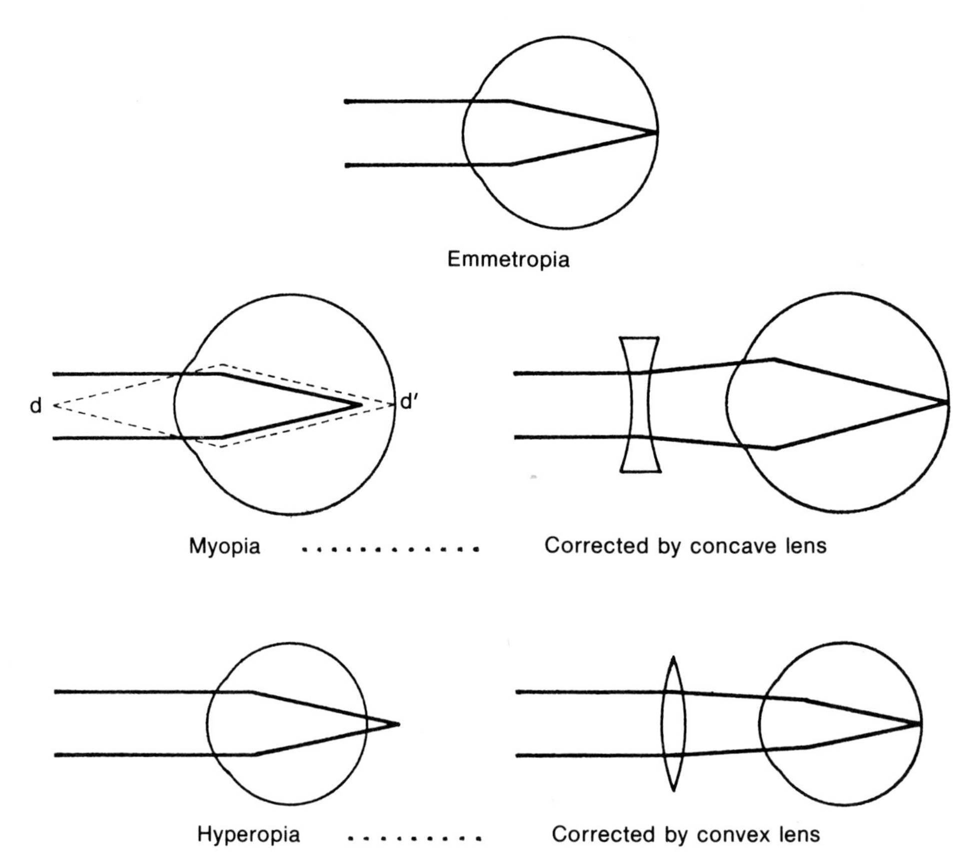

Fig. 9-15
Diagram of emmetropic eye at rest and of defects in myopia and hyperopia. Correction of defects by concave (myopia) and convex (hyperopia) lenses is shown.

becomes manifest during the teen years; there is a genetic influence.

From Fig. 9-15 it can be seen that in myopia the refractive power of the eye is too great with respect to the position occupied by the retina. Hence, to correct this defect, the refractive power must be decreased or the focusing of the light must be delayed. This can be brought about by causing the light entering the eye to be more divergent; *concave* lenses have this effect.

Hyperopia. ■ *Hyperopia* (hypermetropia or farsightedness) is generally due to a shortening of the eyeball; consequently the posterior principal focus lies back of the retina (Fig. 9-15) when the eye is at rest. In this condition no rays of light are focused on the retina, and the individual sees nothing distinctly. By accommodation, parallel rays can be brought to a focus upon the retina; in fact, to see anything at all, far or near, the hyperopic person must always accommodate. The far point of vision for the hyperopic eye does not exist, and the near point lies farther than 6 inches. To correct this defect, the refractive power of the eye must be increased

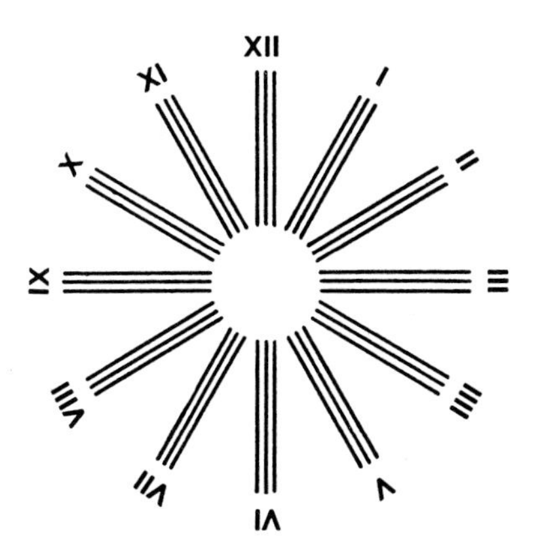

Fig. 9-16
Typical chart used for detection of astigmatism.

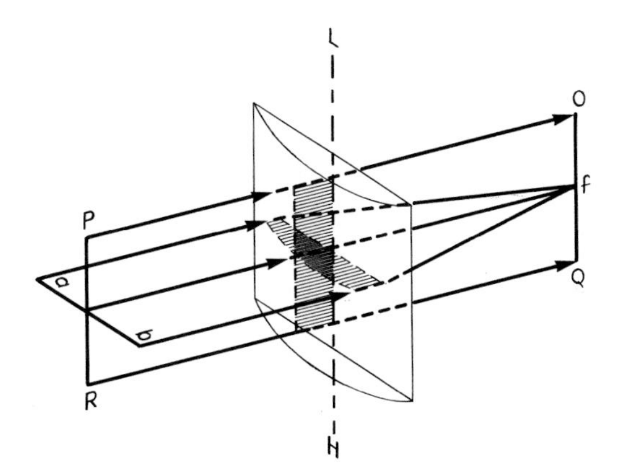

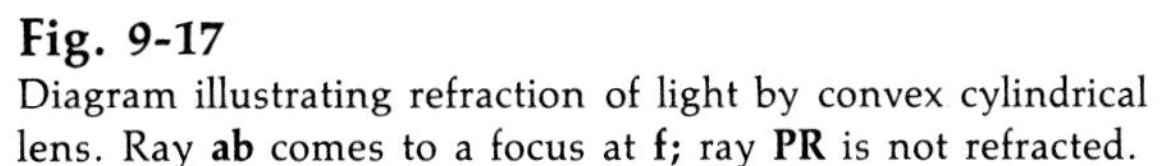

Fig. 9-17
Diagram illustrating refraction of light by convex cylindrical lens. Ray **ab** comes to a focus at **f**; ray **PR** is not refracted.

by a *convex* lens, which causes the parallel rays to converge before entering the eye and therefore to come to a focus sooner.

Astigmatism. ■ A defect more common and more trying than myopia or hyperopia is *astigmatism.* Until now we have regarded the three refracting surfaces of the eye as segments of spheres, but this condition seldom exists. In very many eyes one of the surfaces, perhaps the anterior surface of the cornea, has a shape like the back of a teaspoon; that is, the various meridians do not have the same radii of curvature. Suppose the cornea having this shape is placed so as to correspond in position to the spoon with the handle held horizontally. In this position the vertical meridian (edge to edge of the spoon) has a shorter radius of curvature and hence a greater refractive power than the horizontal meridian (tip to base). A vertical beam of light falling upon such a cornea is focused sooner than is a horizontal ray. If an individual having such a cornea looks at radiating lines (Fig. 9-16), one set may be sharply focused and seen clearly, whereas the other set is focused either in front or back of the retina and is seen indistinctly. In this case, the individual is emmetropic in one plane and myopic or hyperopic in the plane at right angles to this. All kinds of combinations are possible.

In case the difference between the radii of curvature of the meridians is very slight, it does not trouble the individual; this is called physiological astigmatism. But if the error is great, it seriously interferes with good vision and the defect must be corrected, especially if the person is engaged in near work. This is accomplished by *cylindrical* lenses. Suppose from a cylinder of glass we cut a section parallel with the axis of the cylinder; we obtain a convex cylindrical lens (*LH* in Fig. 9-17). A ray of light, *PR,* parallel with the axis, *LH,* passes straight through without any refraction, *OQ;* but a ray, *ab,* at right angles to the axis is focused at *f.*

Aberrations. ■ Ability to discern details is greatly aided by the iris. The iris (Fig. 9-11) is a very thin membrane with a central aperture, the *pupil.* The pigment granules deposited in it give color to the eye and render the iris opaque.

The iris functions like the diaphragm, or stop, in an optical instrument. All the rays entering the eye from the object looked at do not focus properly upon the retina. In Fig. 9-18, *A,* parallel rays fall upon the lens; it will be noticed that the

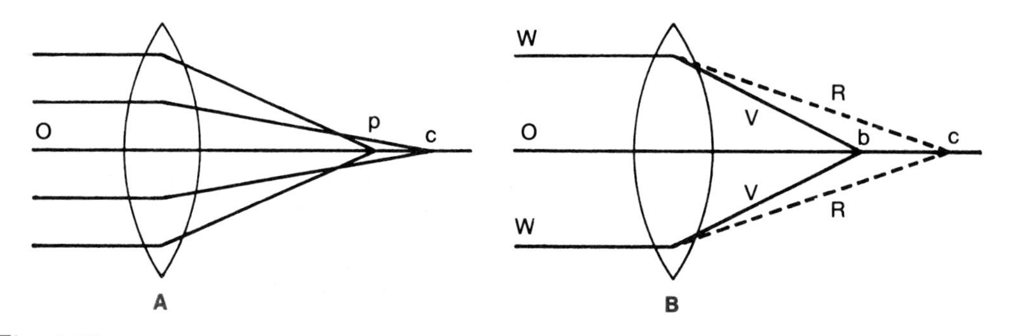

Fig. 9-18
Diagrams illustrating aberration. **A,** Spherical and, **B,** chromatic. Violet, **V,** and red, **R,** components of white, **W,** light are refracted unequally.

peripheral rays (those nearest the edge of the lens) focus at *p* before the more central rays focus at *c*. This defect, known as *spherical aberration,* causes a blurring of the image upon the retina and thereby leads to indistinct vision. By its opacity the iris cuts off the peripheral rays and thus lessens the aberration. It can be demonstrated readily that the more divergent are the rays of light entering the eye, the greater is the spherical aberration; hence, it is necessary that during near vision the pupil be reduced in size (accommodation reflex).

A second defect, *chromatic aberration,* arises because light rays of short wavelength (violet) when passing through a lens are refracted more than are rays of longer wavelengths (red) and come to a focus sooner (Fig. 9-18, *B*). This defect is not of great importance in the eye, since the brain apparently compensates for the error.

There are two sets of smooth muscles in the iris, the circular or sphincter muscles (constrictors) and the radial muscles (dilators). The circular muscle fibers run parallel with the rim of the iris, and by their contraction the pupil is constricted. By observation, we can notice that a person's pupils decrease in size during accommodation. The neuromuscular machinery of the eye is so constituted that whenever an impulse is sent to the ciliary muscles for accommodation, an impulse is also brought to the sphincter muscles of the iris for pupil constriction.

Pupillary reflex. ■ Another condition affecting the size of the pupil is the amount of light entering the eye. This is seen readily by observing the pupil of an individual passing from a dark to a light room. Constriction brought about in this manner is called the photo-pupil reflex. It requires about two-tenths second. The object of this reflex is to shield the eye from too great and sudden illumination that results in overstimulation of the retina. The center for the light reflex is located in the midbrain, and the motor nerve fibers for the sphincter muscles leave the midbrain with the parasympathetic fibers in the third cranial nerve.

The radial or dilator muscles of the iris are innervated by sympathetic fibers originating at the thoracic levels of the spinal cord. Pupillary dilation is termed *mydriasis.* It is a common response to fright and pain.

Atropine (belladonna) paralyzes the endings of the third cranial nerve in the iris and in the ciliary muscles; it therefore causes pupillary dilation and abolishes accommodation. Another class of drugs, by stimulating the iris, causes great pupil constriction (pinhole pupil); among these are opium and morphine. The pupils are influenced by many conditions, among which are pain, fear, muscular exertion, and anoxia (as in dyspnea and asphyxiation). This last is taken advantage of in gauging the depth of anesthesia, for when a patient is not getting a sufficient amount of air or when circulatory failure threatens, the excessive carbon dioxide in the blood causes the pupils to dilate.

■ The eye as a sense organ

The exceedingly large part played by vision in modern civilization is apparent. It is estimated that approximately two thirds of all afferent nerve fibers conveying impulses to the central nervous system receive their stimuli from the eyes.

Anatomy of the retina. ■ The retina can be made visible by means of an ophthalmoscope. In its simplest form this instrument consists of a concave mirror with a central aperture. The mirror serves to reflect light from a given source upon the retina of the observed eye; the central aperture enables the observer to see the retina thus illuminated. The retina constitutes the sensitive coat of the eye; it is in the retina that the stimulation by light takes place. It forms the innermost of the three eyeball coats and borders on the vitreous humor; however, it occupies, as a hemispheric shell, only the posterior half of the interior of the eyeball (Fig. 9-11). The retina is composed of three principal layers of neurons: the neuroepithelium, the bipolar cell layer, and the ganglionic cell layer (Fig. 9-19). Each chain, consisting of three neurons, extends from the choroid coat, through the thickness of the retina, to the vitreous humor; collectively, lying side by side, the chains occupy almost the entire compass of the retina.

The neuroepithelial layer is the outermost layer; it is separated from the choroid by a layer of pigmented cells. The outer ends of the neuroepithelial cells are highly modified to form the *rods* and *cones;* the other end makes connection with the bipolar cells. These make synaptic connection with the ganglion cells. The axons of the ganglionic cells, bending at right angles, converge toward a point on the inner surface of the retina known as the *optic disc* (Figs. 9-11 and 9-19) and form the *optic nerve.* Having pierced all three layers of the retina, they leave the eyeball and proceed to the brain (Figs. 9-20 and 9-25). The central retinal artery provides the blood supply for the inner half-thickness of the retina. Rods and cones receive their nourishment from capillaries of the choroid coat; no capillaries are in direct contact with the photoreceptors.

Light, having passed through the transparent media, penetrates the retina and in the rods and cones sets up physical and chemical changes that cause the generation of impulses. The pigmented epithelium and choroid prevent the reflection of light not absorbed by the photoreceotors and thus improve the quality of the optical image. Impulses are propagated through the three neurons previously described to certain nerve structures lying in the diencephalon, known as the lateral geniculate bodies (Fig. 9-20). A fourth relay of neurons conveys the impulses to the occipital lobes of the cerebrum (Fig. 10-11). That the rods and cones are the visual cells in which the nerve impulses are formed is rendered likely from the following observations concerning the optic disc or blind spot and the fovea.

The optic disc or blind spot. Where the optic nerve fibers leave the eyeball (Fig. 9-19), the retina is devoid of rods and cones, and, consequently, sight is absent. The existence of a blind spot can be demonstrated by means of the cross and circle shown below.

\+ ●

Hold the page about 12 inches from the right eye, the left eye being closed. Look at the cross; the circle will be seen by indirect vision. On bringing the book a little closer to or a little farther from the face, at a certain distance the circle disappears from view because its image now falls upon the blind spot. Incidentally, we may call at-

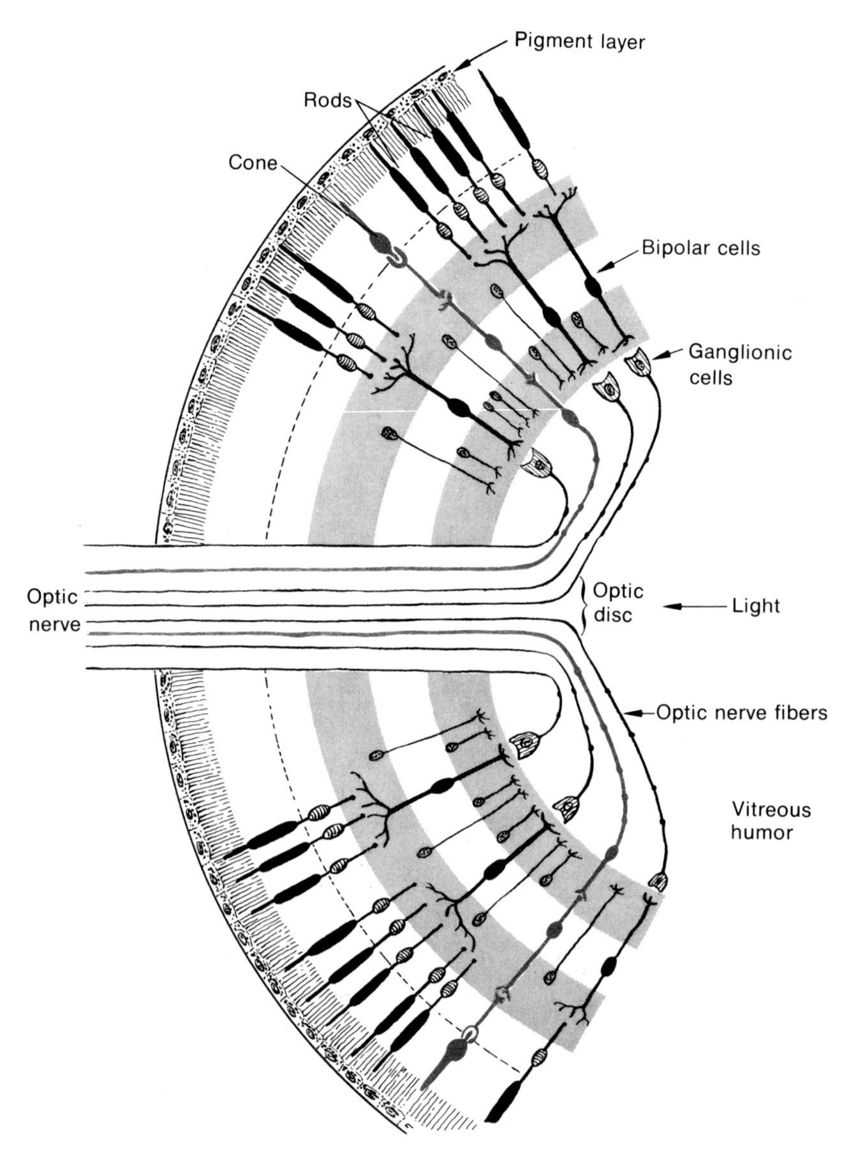

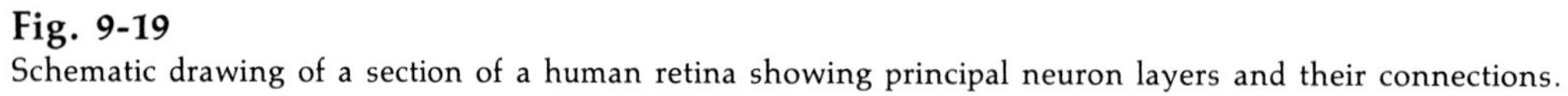
Fig. 9-19
Schematic drawing of a section of a human retina showing principal neuron layers and their connections.

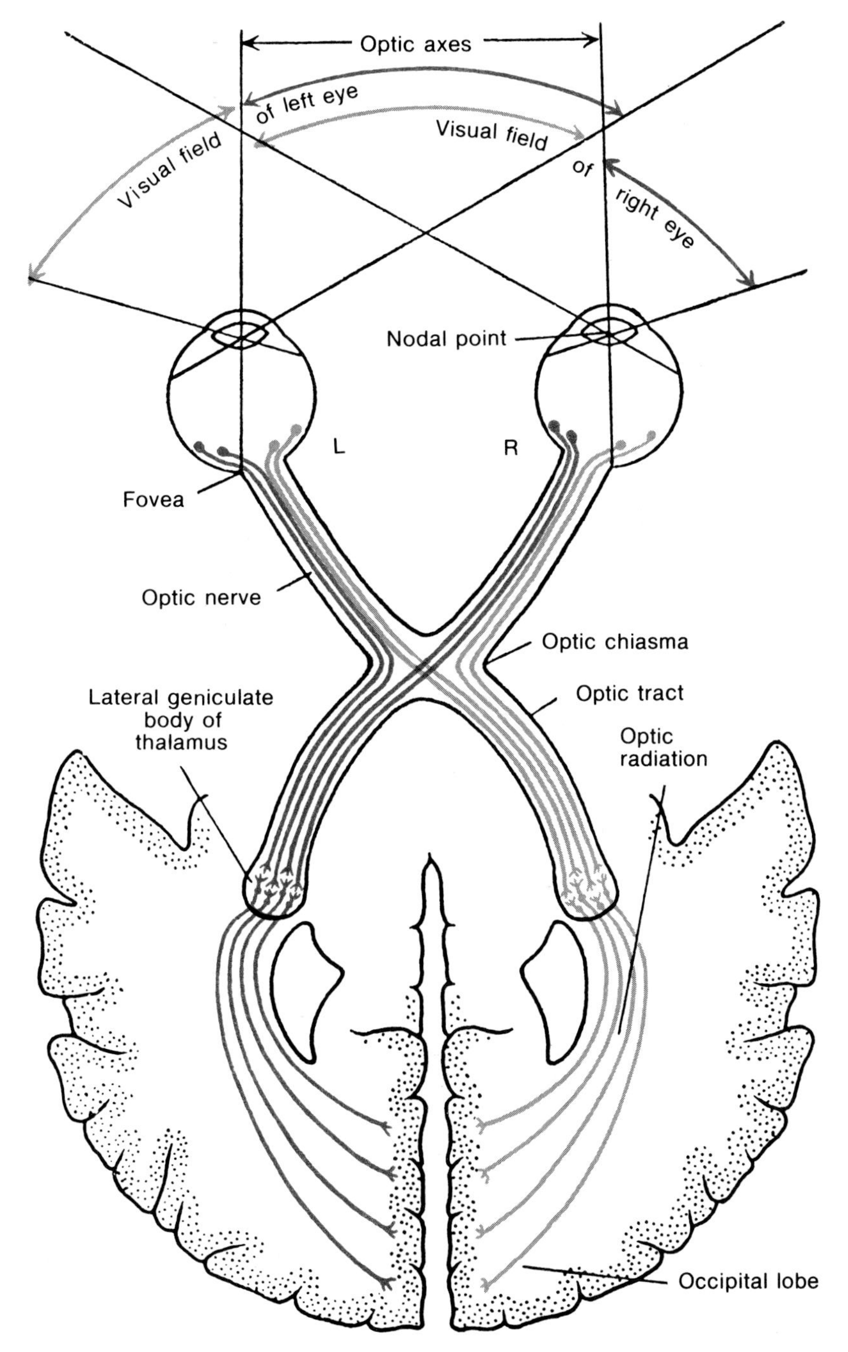

Fig. 9-20
Illustration of neural connections between retina and occipital lobe (visual cortex). Note the crossing of optic nerve fibers from the nasal portions of the retinas of the left and right eyes at the optic chiasma. Visual fields and corresponding nerve tracts are shown by the same color.

tention to the great difference between nerve fibers and the receptors. There are more nerve fibers in the blind spot than anywhere else in the retina, but the light falling upon them is incapable of stimulating them; it is only when the light falls upon the specially constructed neuroepithelium, the rods and cones, that a nerve impulse is generated.

The fovea. There is one small area of the retina in which the rods and cones, especially the cones, are more numerous than in any other part of the retina. This is called the *yellow spot* (macula lutea); its central depression, known as the *fovea centralis* (Fig. 9-11), is formed by slender and closely packed cones, the rods being absent. This part of the retina has the keenest vision for detail.

Electrical activity of retina. ■ Some foveal cones connect with a single bipolar cell and this with a single ganglion cell. Thus, a private pathway in the optic nerve is provided. In other areas of the retina, many rods and cones converge upon a single bipolar cell, which, in turn, may show divergence by contacting several ganglion cells. Although we have mentioned only three cell layers in the retina, in fact, other cell layers are present, and some of these provide complex interconnections between rod and cone neurons. Interactions between retinal elements through convergence, divergence, and intraretinal association provide the neural basis for the facilitation and inhibition that takes place in retinal function. Facilitation (summation) is demonstrated by the fact that the threshold for visual stimulation is decreased when the area of stimulation is increased; that is, when more receptors are exposed, the intensity need not be so great. Under certain conditions, stimulation of two adjacent retinal areas by separate light sources causes the sensitivity of one area to be reduced, a form of inhibition.

The electrical activity of the retina is complex because of the considerable number of interacting cells. However, the initial event appears to be a depolarization of the photoreceptors by the stimulus. At the ganglion cell level three distinct behavior patterns are evident. First, some ganglion cell axons increase their rate of discharge on application of the stimulus to the retina and show a progressive adaptation. At cessation of stimulation, these cells are briefly silent before resuming their normally slow discharge frequency. These axons are referred to as ON fibers. Some ganglion cell axons cease firing at the onset of light; these are OFF fibers and give a short burst of spikes at a frequency greater than normal when the light stimulus ends. A third behavior pattern, known as ON-OFF, occurs in axons that respond to both the application and removal of the light stimulus with a brief burst of increased activity. These patterns of activity are the neural substrate of visual function and are of great importance to visual acuity and perception of movement.

Metabolism of the retina. ■ The retina is an extension of the cerebral cortex, and its metabolism resembles that of the cortex, although the magnitude of this activity is quite low. All neural cells are highly susceptible to a curtailment in their oxygen supply and to their inability to contract an oxygen debt. Anoxia lowers visual acuity (especially in dark adaptation), decreases the visual field, impairs color vision, slows the speed of dark adaptation, and, if there is any tendency toward imbalance of the extraocular muscles, it exaggerates this and may thus cause *diplopia*—double vision. Therefore, in high-altitude flying, without the benefit of extra oxygen, vision and judgement may be seriously impaired. The rapid drainage of blood from the head in certain aerial maneuvers deprives the retina of its required oxygen and results in *blackout*.

Resolving power. ■ By resolving power is meant the ability to discern two discrete luminous bodies as two. On a black surface two white dots, 2 mm apart, are readily recognized as two points when the observer stands 3 or 4 feet away. But when this distance is increased to 10

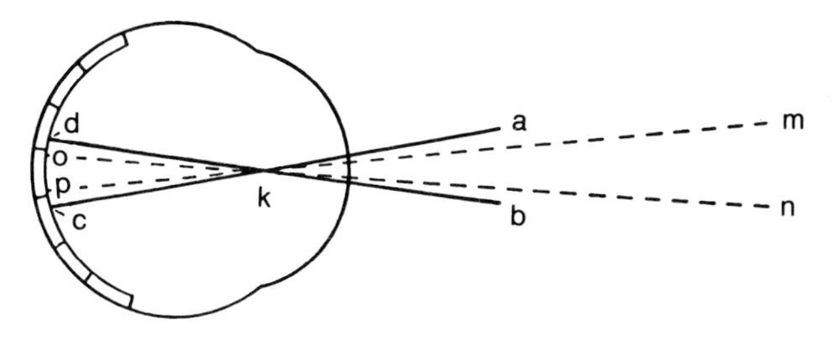

Fig. 9-21
Diagram illustrating the resolving power of the retina.

feet, the two points fuse into one. The basis for this resolving power lies in the dimensions of the visual elements, the rods and cones; to appreciate this, refer to Fig. 9-21. It will be recalled that a ray of light going through the nodal point, *k,* of the eye undergoes no refraction. In Fig. 9-21, *a* and *b* are two luminous points, which have their foci at *c* and *d*, respectively. It is thought that when these two foci or images fall most sharply or intensely upon two discrete cones separated by a third cone, which is not so intensely stimulated, the two retinal images call forth in the brain two discrete sensations; that is, the individual sees *a* and *b* as two separate objects. If now the points are placed farther from the eye, at *m* and *n*, the foci, *o* and *p*, fall either upon two neighboring cones or upon a single cone; in these cases the two points are seen as one object. Since the cones are more slender and more densely packed in the fovea than anywhere else, to see the detail of an object we always turn our eyes in such a manner that the images of the object fall upon the two foveas.

Visual acuity. ■ The ability to distinguish the detail of an object, for instance, printed letters placed 20 feet (infinity) from the observer, is a measure of visual acuity. At a distance of 20 feet an emmetropic person can read letters of a certain size; if they are smaller, he is not able to recognize them. At this distance the images of the letter on the retina of a person afflicted with myopia, hyperopia, or astigmatism are too blurred to be read. Let us suppose that the smallest letters that a person with defective eyes can read are those that a normal eye reads at 40 feet; his visual acuity is then said to be 20/40. In this fraction the numerator expresses the distance between the observer and the letters; the denomination is the distance at which a normal eye can read them. Obviously the visual acuity of a normal eye is expressed as 20/20.

Field of vision. ■ A considerable part of the periphery of the retina is not utilized because, due to the configuration of the face (nose and eyebrows), the images of external objects cannot be cast upon it. This is shown by a difference in the angles of the lines passing through the nodal point of either retina in Fig. 9-20. The total area of the retina upon which images can be projected is known as the *visual field;* within this field the visual acuity progressively decreases as we pass from the fovea to the periphery. Objects having their foci outside of the fovea are said to be seen indirectly and appear indistinct.

Daylight and twilight vision. ■ On stepping from bright sunlight into a dimly illuminated room we are unable at first to see distinctly. On remaining in the dim light for a few minutes, we gradually begin to discern the objects about us. On reentering the brightly lighted area we are dazzled, and for a few seconds vision is poor, but recovery occurs rapidly.

The eye is a double organ; one for seeing in bright light and the other for seeing in dim light. The vision in bright light is called *photopia* or daylight vision; the other is known as *scotopia* (*scotos,* darkness) or twilight vision.

In photopic vision the greatest sensitivity of the retina is found in the fovea centralis; in scotopic vision the fovea is blind for all practical purposes, and the retinal area of greatest irritability lies some distance from the fovea. The reader can readily verify this by looking at a very faint star. On looking directly at it (foveal vision) the star is invisible, but on shifting the gaze a little to one side of the star so that its image falls upon an extrafoveal portion of the retina, it is readily perceived.

In photopia all colors are readily discernible; in scotopia the eye is totally color blind. This may be proved by observing a flower garden at early dawn. The objects and patterns are recognized, but only as grays of various intensities; no colors are seen. In scotopia red objects are not seen at all, not even as gray; the scotopic eye is completely blind to this light.

Duplicity theory. ■ These notable differences between photopic and scotopic vision are explicable only on the assumption that in the retina there are two distinct neural systems, one of which operates in bright light and the other in dim light. According to the *duplicity theory of vision,* the apparatus for photopia is found in the cones and that for scotopia in the rods. The initial change set up in the rods and cones by photic stimulation is, in all probability, a photochemical change.* In the outer portions of the rods there is present a pigment known as *rhodopsin* or *visual purple.* Rhodopsin is a protein (opsin) to which is bound a chromophore (retinene) that gives the compound color and sensitivity to light. When exposed to light, rhodopsin is bleached; that is, it is split into its two components. The retinene may undergo chemical reduction to vitamin A and be stored in the pigmented epithelium. Oxidation of vitamin A yields retinene. Restoration of rhodopsin occurs in the dark by the action of an enzyme that facilitates the recombination of retinene and opsin. In its bleaching response the retina is quite similar to photographic film. Indeed, it has been possible to produce permanent retinal pictures called optograms (Fig. 9-22). However, the retina excels photographic film in that rhodopsin is restored in the dark. The bleaching of the rhodopsin is an indispensable condition for the stimulation of the rods. Let us see how the behavior of rhodopsin can explain the differences between daylight and twilight vision.

*A photochemical change is a change in the chemical composition of a material produced by light. Among the well-known instances are the changes in a photographic film and in a delicately colored fabric or flower, in the tanning and freckles of the skin, and, more important, in the transformation of CO_2 and H_2O into glucose in green plants by chlorophyll.

Fig. 9-22
Optogram, or retinal picture, of a window caused by the bleaching of rhodopsin.

1. As the fovea contains no rods and therefore no rhodopsin, scotopic vision at the fovea is an impossibility. At a certain distance from the fovea, the retina has the greatest concentration of rods and therefore the greatest sensitivity in dim light.

2. Color vision is mediated by the cones, not by the rods. This fact is based on the observation that our ability to distinguish the different colors is most acute in the fovea where the cones are most abundant and on studies made of function in individual photoreceptors from isolated retinas. When in photopic vision lights of various colors are thrown upon the outlying portions of the retina, they are seen as colorless; that is, we are color blind with the peripheral retina; here the cones are scarce or entirely absent, although rods are present. According to the duplicity theory, scotopia is rod vision and therefore is colorless.

3. Rhodopsin, we have learned, is bleached by light. But, like some photographic film, it is not affected by red light and hence, according to the theory, the longest wavelengths of light cannot be seen by the scotopic eye.

Dark adaptation. ■ Rhodopsin is bleached by bright light, and it takes time to restore it;

hence, it can readily be understood why we are temporarily blind after going from a brightly to a very dimly illuminated room. Dark adaptation provides for regeneration of visual purple by which the irritability of the retina is increased. This enables a feeble light, which previously made no impression, to generate retinal impulses again.

Although man has both photopic and scotopic vision, some animals have only one or the other. It is for this reason that chickens go to roost at sundown and that owls fly at night. In certain pathological conditions a person loses the power of dark adaptation; he is then more or less unable to find his way in fairly dim light. This condition is known as night blindness or *nyctalopia.* It occurs frequently in persons subsisting on an insufficient and monotonous diet.

Positive afterimage. ■ The length of time a photic stimulus must act in order to call forth a sensation is exceedingly short. An electric spark of sufficient intensity, lasting only 1/8,000,000 sec, is visible. But the sensation experienced lasts a great deal longer than this, as the following experiment shows. After his eyes have been dark adapted for several minutes, have a person turn on an electric light for a fraction of a second, while his gaze is directed toward the frosted lamp shade. In the succeeding darkness he experiences a reproduction of the visual impression in which many of the details can be recognized. This is known as the *positive afterimage* and demonstrates that our visual sensations continue to exist after the cessation of the stimulus.

The afterimage effect is the basis for the fact that we may experience a uniform (steady) sensation although the photic stimulation of the retina is intermittent. When in the dark a live coal is twirled through the air, we see a series of glowing lines. If no positive afterimage existed, the live coal would be seen only in the place it actually occupied at any given moment, and the light would not appear as a line but as a moving point. The positive afterimage ordinarily lasts only a fraction of a second, and, therefore, in order to produce the results described, the live coal has to be moved with a certain velocity. Many other phenomena, for example, motion pictures, are based on this principle. The positive afterimage is always seen in the same color as the original.

Colors. ■ Colors are subjective phenomena to which we assign specific terminology. They are changes in our consciousness and have no objective existence. That which in the outer world corresponds to the color sensations, and by which the sensations are generally produced, are the electromagnetic radiations of various wavelengths.* When light rays having a wavelength of about 0.00076 mm strike the retina, they set up definite changes, which create nerve impulses; these impulses stimulate certain brain cells whose activity we interpret as red. This is also true for the other wavelengths; each one produces its own specific result in our consciousness. By the mixing of two or more colors (wavelengths of light), other color sensations may be produced. For example, if a color wheel, composed of a red and a green sector, is rapidly rotated, because of the positive afterimage, there is a physiological fusion of the red and green, and as a result we experience an entirely new sensation, that of yellow.

Complementary colors. It is also possible to mix two colors so as to produce a sensation of white. Such colors are called complementary—red and greenish blue, yellow and deep blue, and violet and greenish yellow.

Negative afterimage. When a person looks intently for several seconds at a small yellow card and then at a white sheet of paper, he sees a reproduction of the card in the color complementary to yellow, i.e., blue. This is the *negative afterimage.* If in this experiment a sheet of yellow

*The word *color* used in the subjective (or psychological) sense designates a sensation; in its objective (or physical) sense it refers to a particular wavelength (or a combination of wavelengths) of visible radiation.

paper is substituted for the white, the individual discovers that the area of the retina previously stimulated by the small yellow square is now yellow blind. By similar experiments it readily can be shown that black is the negative afterimage of white, and white of black. This phenomenon is also known as *successive contrast.*

Color blindness. It was stated that by mixing two or more colors we are able to call forth new sensations. All the color sensations that a normal individual can experience may be produced by the proper mixing of the three primary colors —red, green, and blue. For this reason normal color vision is called *trichromatic color vision.* For most color-blind individuals all the color sensations they are capable of experiencing can be produced by the proper mixing of two colors, and hence, they have *dichromatic color vision.* The most common form of color blindness is the red-green blindness. These individuals have more or less difficulty in distinguishing between red and green. As these are the colors used for danger and safety signals, respectively, in traffic on land and sea or in the air, it is of great importance that the color vision of sailors, pilots, and drivers be tested. Color blindness may be acquired (retinal disorders), but it is mostly an inherited defect. About 8% of men and less than 0.5% of women are red-green color blind. Blue and total (monochromatic) color blindness is rare.

Helmholtz's theory of color vision. According to Helmholtz's theory of color vision, the retina contains three fairly distinct elements or structures. Each retinal element responds to light of any wavelength but not to the same extent to all lengths; the various retinal elements also differ from each other in this respect. The three elements are generally spoken of as the red, the green, and the blue.

The red element responds most strongly to the longest wavelengths (red), less to the intermediate wavelengths (green), and still less to the shortest (violet or blue). The green element shows strongest reaction with the intermediate wavelengths and reacts but slightly either with those that are longer or shorter. The blue element is most affected by the shortest wavelength of light, and its reaction gradually decreases as the wavelengths become longer.

These elements of which we speak actually represent inherent differences in the spectral sensitivities of the cones. However, since color sensation is a psychological phenomenon, other elements (synapses) in the visual system also must have a role in color vision.

Projection. ■ We have become acquainted with the fact that our visual sensations are projected beyond the body. This projection takes place in a definite manner, as the following experiment indicates. When the tip of the finger is pressed upon the sclerotic coat on the nasal side of the pupil, a circular luminous field, called a *phosphene,* is seen; this phosphene seems to be situated on the temporal side. When the pressure is applied to the temporal side of the eyeball, the phosphene is situated on the nasal side. The pressure of the finger mechanically stimulates the retina at, for example, *a* in Fig. 9-23, and the sensation produced is projected in the direction of the line joining *a* with *n,* the nodal point; hence, the object causing the stimulation appears

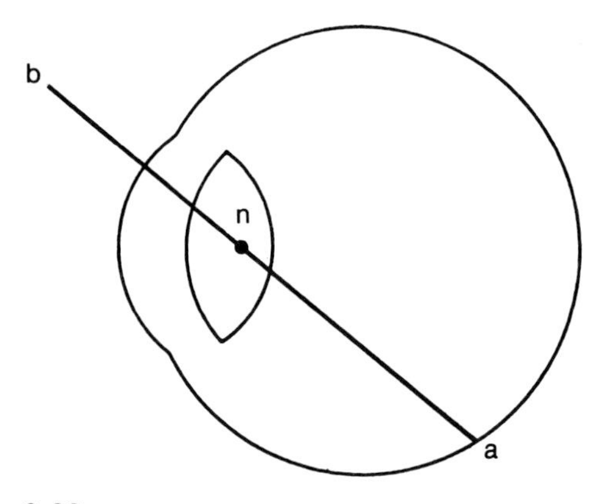

Fig. 9-23
Diagram illustrating the law of projection. **a,** Stimulated point on retina; **n,** nodal point; **ab,** line of projection.

to be situated at *b*. By this conscious projection, the inverted images on the retina are reinverted and we see things right side up.

In viewing an object with both eyes, we project the sensation from the bridge of the nose. To demonstrate this, prick two pinholes in a card as far apart as the distance between the pupils of the eyes. On placing the card close to the face and looking at the bright sky or lamp shade, only one luminous circle is seen and this appears to be situated in the median plane of the body.

Single binocular vision. ■ Since we have two eyes, each with its own image, and consequently both optic nerves carrying impulses, we rightfully may ask why, under ordinary circumstances, we see the object looked at as a single and not as a double object. We said "under ordinary circumstances," for that we can see double—diplopia—is easily demonstrated. While looking at an object with both eyes and seeing it single, with a finger press one eyeball out of its normal position; double vision results. But even without thus influencing the eyes, in the ordinary manner of looking we constantly see certain things double. Hold a finger about 8 inches and a pencil 21 inches from the face. Look at the pencil; the finger is seen double.

Evidently there are many objects that we normally see double, and there are some that we see single. The explanation for this comes from the theory of *identical* or *corresponding points.* In Fig. 9-24, point *a* of one retina and *b* of the other

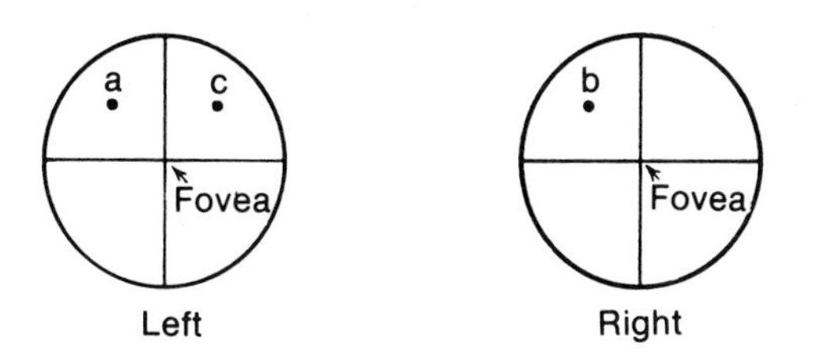

Fig. 9-24
Diagram illustrating identical points on right and left retinas. **a** and **b**, Identical, or corresponding, points; **c** and **b**, unidentical points.

may be regarded as corresponding points, for they lie in the same direction and at the same distance from their respective fovea. If the image of an object falls in the left eye upon *a* and in the right eye upon *b,* the object is seen as a single object, but if one image falls upon *c* and the other upon *b,* the object is seen double; *b* and *c* are said to be unidentical or noncorresponding points. The two foveas are identical, and hence their simultaneous stimulation by the two images of an object leads to single vision. Since the images of the object of regard fall upon the two foveas, the object looked at with attention is seen single.

An anatomical basis for corresponding points appears in the semidecussation (half-crossing over) of the optic nerve fibers in the chiasma (Fig. 9-25). The fibers from the nasal halves of the two retinas (the gray fibers from the left eye —Fig. 9-20—and the red fibers from the right eye) cross over at the optic chiasma; they leave the chiasma with the fibers from the, respectively, opposite temporal halves. In the optic tract the temporal fibers of one retina and the nasal fibers from the other retina proceed to the lateral geniculate body where they make synaptic connection with the fibers of the optic radiation that end in the cortex of the occipital lobe. An object situated to the left of the observer (in the gray part of the visual field of the left eye in Fig. 9-20) has its focus on the nasal half of the left and on the temporal half of the right retina. By the fibers from these two retinal halves (colored gray in Fig. 9-20), the impulses are sent to the right occipital lobe of the cerebral cortex. The particular fibers stimulated (in both retinas) report to the same group of cells in the cortex, and one sensation is experienced.

The individual always strives to obtain single and to avoid double vision. To obtain this single vision when the eyes move from one object to another, the movements must be so coordinated that the images of the object of regard shall fall on the two foveas.

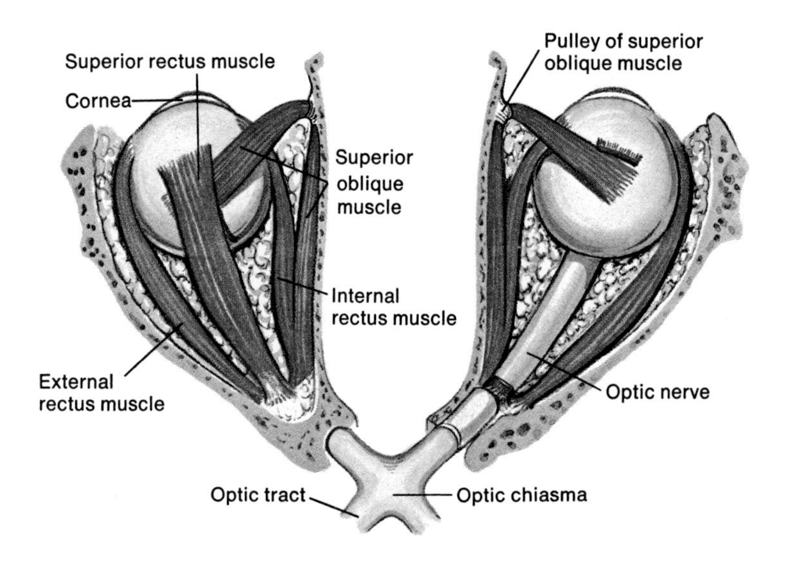

Fig. 9-25
Illustration of relationship of four extrinsic eye muscles of left and right eyeballs. Inferior rectus and inferior oblique muscles of each eye are situated below eyeball and are not visible.

Table 9-1. Muscles of the eyeball

Muscle	*Direction*	*Innervation*
Internal rectus	Inward	Third cranial nerve (oculomotor)
External rectus	Outward	Sixth cranial (abducens)
Superior rectus	Up and in	Third cranial (oculomotor)
Inferior rectus	Down and in	Third cranial (oculomotor)
Superior oblique	Down and out	Fourth cranial (trochlear)
Inferior oblique	Up and out	Third cranial (oculomotor)

■ Ocular movements

Eye muscles. ■ The great motility of the eyes is a matter of common observation. Eye movements are brought about by the six, striated, extrinsic eye muscles whose origins are fixed upon the bones of the orbit* and with insertions upon the connective tissue of the eyeball. The position of these muscles may be seen in Fig. 9-25. The direction in which the individual muscles turn the eyeball and their innervation are given in Table 9-1.

The turning of the eyeball outward (toward the temporal side) is called abduction; turning it inward (toward the nose) is called adduction; to bring about these two movements, the external rectus and the internal rectus, respectively, are sufficient. But to raise the gaze straight upward the superior rectus and the inferior oblique are necessary (to neutralize the inward movement caused by the superior rectus by the outward

*The bone socket in which the eyeball is placed.

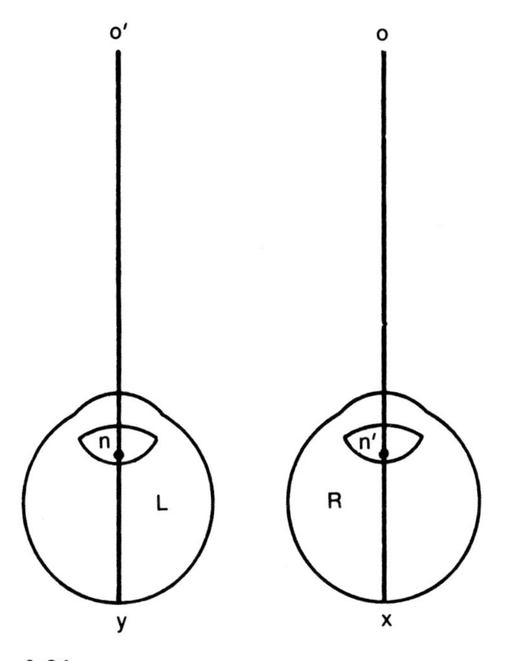

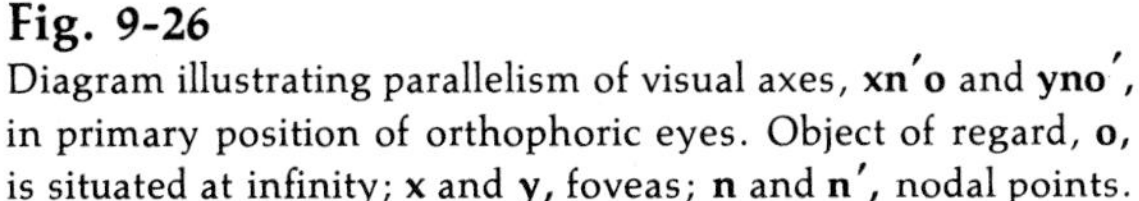

Fig. 9-26
Diagram illustrating parallelism of visual axes, **xn′o** and **yno′**, in primary position of orthophoric eyes. Object of regard, **o**, is situated at infinity; **x** and **y**, foveas; **n** and **n′**, nodal points.

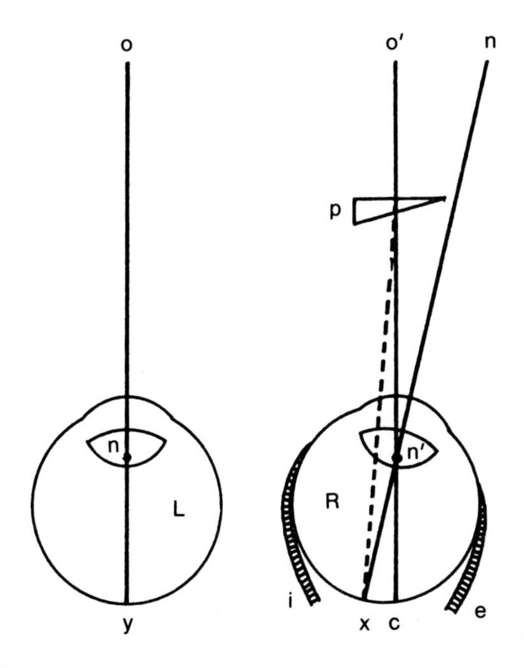

Fig. 9-27
Diagram illustrating strabismus (heterophoria) and its correction with prism, **p.**

movement produced by the inferior oblique). For the same reason, the downward motion requires the inferior rectus and the superior oblique. For the oblique movement three muscles are necessary. For example, if our eyes fixate a point lying straight ahead and on the horizon (primary position of the eyes), and if we wish to look at an object 2 feet distant and situated in the median plane of the body and about on a level with the top of the head, both eyes must be moved upward and inward; for this the superior rectus, inferior oblique, and internal rectus of each eye are used. To secure single binocular vision in near work, the process of accommodation is always associated with *convergence* of the eyeballs brought about by the internal recti muscles.

All these coordinated muscular activities are controlled by a most intricate nervous mechanism, and it is not surprising that the coordinating mechanism by which the two eyes are made to work in harmony sometimes fails. This occurs normally when we become drowsy. Since noncorresponding points are stimulated, *diplopia* results. The diplopia caused by alcoholic intoxication is too well known to need more than mention.

Heterophoria. When the eyes are in the primary position (looking straight forward at an object on the horizon), under normal conditions the visual axes of the eyes may be said to be parallel. The visual axis is the line joining the object of regard, the nodal point, and the fovea. Thus, in Fig. 9-26, the visual axes *o′ny* and *on′x* are parallel and must be understood to meet at the point of regard, *o,* situated at infinity. Under these conditions the twelve extrinsic muscles may be said to be at rest in so far as without extra innervation to any one muscle in particular they hold the eyes properly in the primary position. Without any effort on the part of the individual, the object of regard, *o,* is seen single. This condition is known as *orthophoria.* Let us assume that the external rectus muscle *e* (Fig. 9-27) of the right eye is a little stronger than, or has a mechanical advantage over, its antagonist, the internal rectus, *i.* Suppose a person with this defect is looking at a point on the horizon and that all the extrinsic muscles of both eyes receive equal impulses from the central nervous system; because of the greater effect produced by the external rectus muscle, *e,* the visual axis of the right eye will be turned outward (temporalward) to a greater or lesser extent. This causes the fovea, *x,* to be moved toward the nose, and the object, *o′*, will have its focus at *c* and not on the fovea. Unidentical points (*y* and *c*) are stimulated, and the individual sees double. But there is always a striving for single vision. In the case under consideration, the central nervous system sends supplementary impulses to the weaker internal rectus muscle, *i,* of the right eye so as to offset the greater tension of the external rectus. The eye is now properly placed, and the person has single binocular vision; however, to obtain it, the eye muscles are taxed by extra work; this results in eyestrain.

The condition just discussed in which the extrinsic muscles of the eyes are not properly balanced, but in which the individual can, by extra innervation of a certain muscle or group of muscles, remedy the defect, is called *heterophoria.* In many cases the eyestrain may be relieved by prismatic lenses, whereby the light entering the eye is so deflected from its previous course as to focus upon the fovea. Let us place in front of the right eye, in Fig. 9-27, a prism, *p,* with its base toward the nose; the ray from *o′* is bent so as to focus upon the fovea, *x.* By the prism we accomplish what the individual previously did by extra innervation; eyestrain is relieved.

Strabismus. When the imbalance between the eye muscles is so great that the individual cannot by extra innervation remedy the defect, it leads to a permanent deviation of the visual axes, and the person is then said to have *strabismus* (cross-eye or squint). This results in double vision. Fortunately, the person so afflicted learns to neglect one of the images; he obtains single vision by using only one eye. Strabismus and heterophoria are frequently not due to any muscle defect but to refractive errors (myopia and hyperopia); therefore properly fitted glasses may correct the fault. In children from three to six years of age corrective orthoptic exercises have in many instances proved to be highly beneficial.

Eyestrain. Eyestrain can be caused by almost any error of refraction and by muscle imbalance; especially do we find it in hyperopia and astigmatism. For hyperopic children, close work at school, especially when small print is used or the light is poor, is almost an impossibility; any attempt to read or write is at the expense of all the accommodation of which they are capable.

All conditions in which a constant excessive innervation of the muscles for accommodation and convergence and a constant close observation of indistinct retinal images are necessary not only cause fatigue but may involve pain, constant watering of the eyes, chronic inflammation, compression of the eyeballs, increasing nearsightedness, and distention of the ocular vein and other disturbances. It is therefore of prime importance that the eyes be used correctly and not abused and that, because of the large amount of close work demanded, the eyes of school children be examined carefully.

■ Protective mechanisms

The eyes are protected in two ways, by their eyelids and by their tears.

Eyelids. ■ Reflex closure of the eyelids protects the eyes from the entrance of foreign bodies and from sudden bright lights. This reflex has its afferent limb in fibers of the fifth cranial

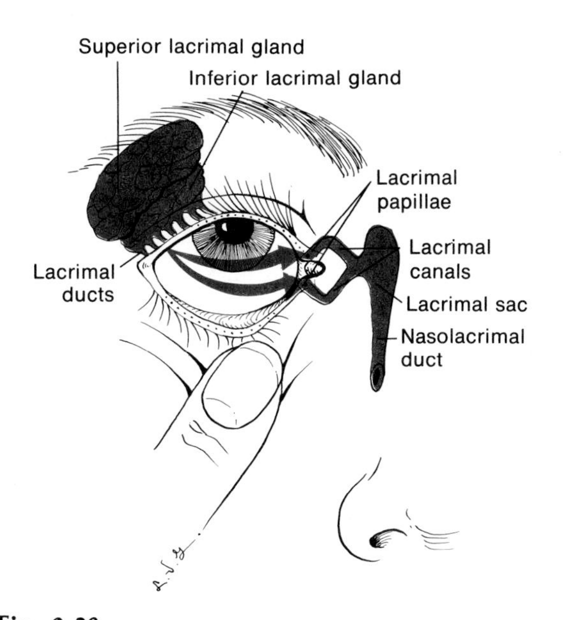

Fig. 9-28
Illustration of structures involved in tear formation and removal.

nerve; its efferent limb, in fibers of the seventh cranial nerve, innervates the orbicularis oculi muscles of the eyelid. Spontaneous blinking occurs throughout waking life on the average of twenty-five times per minute and each blink lasts about $^1/_5$ sec. Reflex closure of the eyelid serves to keep the corneal surface moist and to clear it of mucus. Voluntary closure can be used to interrupt the visual process, for only about 1% of the incident light penetrates the closed eyelid. The eyelashes, about 200 in number per lid, also serve to prevent the entrance of foreign objects. Sensitive mechanoreceptor nerve endings at the base of the follicle discharge when a lash is bent; this discharge evokes a blink reflex.

Tears. ■ Lacrimal glands at the upper and outer portion of each orbit (Fig. 9-28) produce tears. Besides psychic production, secretion of tears is induced by stimulation of receptors in the eyelids and conjunctivae. The lacrimal glands are innervated by autonomic fibers from the midbrain traveling successively in the fifth and seventh cranial nerves. Perhaps less than 1 ml of tear secretion is produced per day. It is isosmotic with normal saline solution (0.9% NaCl) and mildly alkaline (pH 7.4). An enzyme that hydrolyzes mucus and has bacteriocidal action is present in the secretion. Tear fluid prevents the drying of the cornea and conjunctiva, thereby serving as a lubricant for lid movement; it washes away foreign bodies and improves the optical surface of the cornea by forming a thin film over it. Most tear fluid evaporates, but any excess flows to the inner canthus (angle) of the eye and is collected by the nasolacrimal duct and brought to the inferior meatus of the nose.

■ HEARING

The ear is constructed so that the endings of the auditory (eighth cranial) nerve are most effectively stimulated by sound.

Characteristics of sound. ■ Sound, in its objective sense, consists of air vibrations. When a violin string is plucked, it vibrates forward and backward. In so doing it condenses the air in front of it and rarefies the air behind it. These condensations and rarefactions travel through the air at a rate of 1,090 feet per second. The vibrations of a shorter string produce a greater number of waves per second, and the pitch of the sound is higher; but whether the string is plucked gently or violently does not affect the number of vibrations per second, and consequently the pitch will remain the same. When the string is pulled violently, the excursion of the string will be greater, and what is called the amplitude of the wave will be larger; this causes the sound to be louder. The *intensity* or loudness of a sound therefore depends on the amplitude of the waves; the *pitch* is determined by the number of vibrations per second. The human ear can respond to sound waves ranging from 16 to about 20,000 double vibrations per second. Within these limits about 11,000 different pitches can be distinguished.

Another characteristic of sound is its *color.* Suppose a violin string is caused to vibrate two thousand times per second, and a similar string in a piano is also caused to vibrate with the same force as the violin string; the two sounds have the same pitch and the same intensity, but no one mistakes one for the other. The *color* or *timbre* is produced in the following manner: When a string vibrates, it vibrates as a whole, forward and backward; the waves thus formed constitute the fundamental sound and determine the pitch of the sound. But the string, at the same time, vibrates in its parts, that is, each half, third, or fourth of the string vibrates independently of the whole string. This causes other waves and other sounds, the *overtones,* to accompany and to fuse with the fundamental wave. These overtones vary with different sorts of instruments; in consequence, each distinct source of sound impresses its overtones on the fundamental tone; this timbre enables us to recognize the source of the sound.

Resonance. ■ When a sound wave falls upon an object, such as a windowpane, it causes the latter to vibrate. When a person sings a certain note near a piano, the piano continues for a short time to send out the same note after the singer has stopped, because the string that corresponds in number of vibrations to that of the note sung is thrown into sympathetic vibration or *resonance.*

■ The ear

The ear is divided into three parts: the external, middle, and internal ear (Fig. 9-29). The external ear includes the auricle *(pinna)* and the external auditory *meatus,* which is a canal leading to the eardrum *(tympanic membrane).* On the

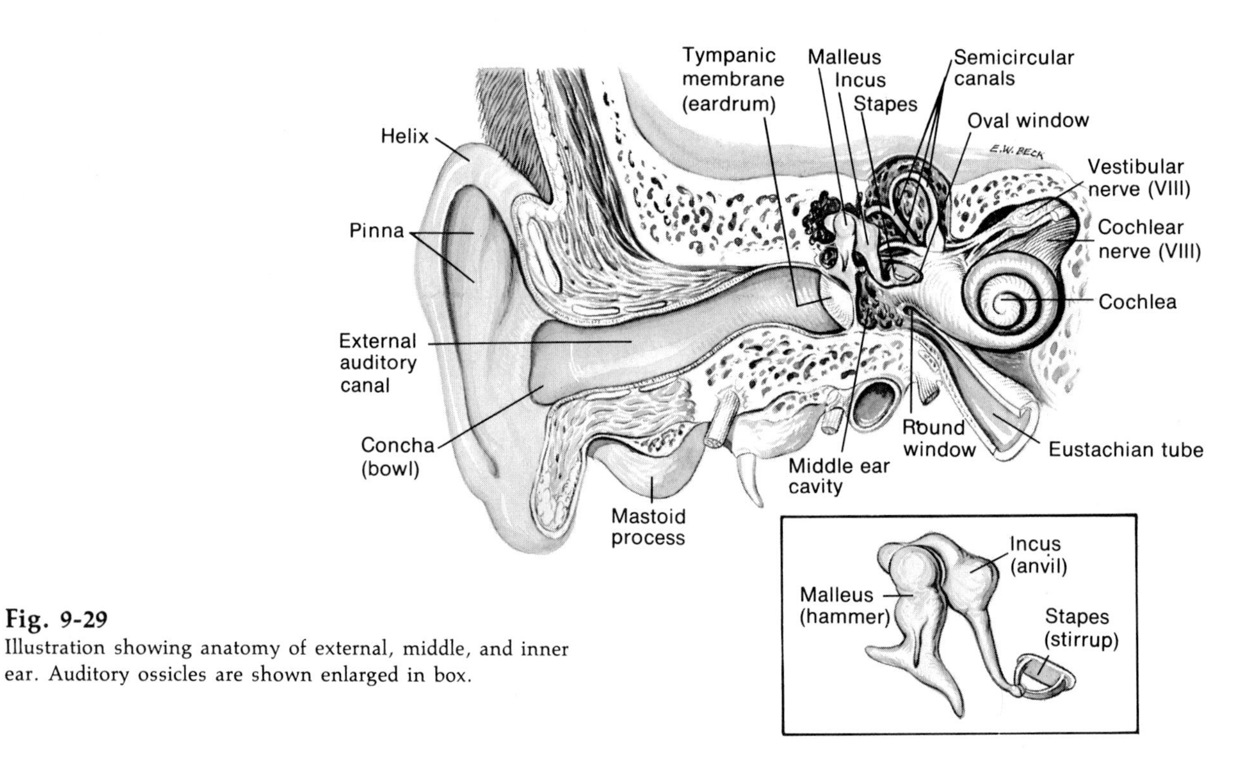

Fig. 9-29
Illustration showing anatomy of external, middle, and inner ear. Auditory ossicles are shown enlarged in box.

inner side of the eardrum lies the middle ear. The middle ear is a cavity, the tympanic cavity, in the temporal bone; it is lined with mucosa and contains a chain of three small bones or auditory *ossicles,* extending from the eardrum to the oval window or membrane. These ossicles are called the *malleus* (hammer), *incus* (anvil), and *stapes* (stirrup). The handle of the hammer (Fig. 9-29, inset) is embedded in the eardrum, its head forming a joint with the body of the incus. The long process of the incus articulates with the head of the stapes, and the foot of the stapes is fixed by a membrane into the oval window, which opens into the *scala vestibuli* of the cochlea.

The tympanic cavity is closed to the outside air by the eardrum; it is shut off from the inner ear by the *round* and the *oval* windows (membranes); from its floor originates the Eustachian tube that connects the tympanic cavity with the pharynx (Figs. 9-29 and 17-1).

The inner ear, lying on the other side of the oval window, consists of a series of winding cavities cut in the petrous portion of the temporal bone. These cavities are known as the *osseous* or *bony labyrinth.* Most of these cavities are tubelike, constituting what is known as the *cochlea* or snail and the *semicircular canals;* between these two sets of winding tubes there is a dilated connecting part, the *vestibule.*

In these winding canals of the osseous labyrinth there is placed a *membranous canal* or *labyrinth* (Fig. 9-30), which in general has the shape of the bony canal, except that in the vestibule the membranous canal is composed of two sacs, the *saccule* and *utricle.* From the utricle spring the three membranous *semicircular canals;* from the saccule originates the membranous canal found in the cochlea of the ossesous canal and known as the *duct of the cochlea.*

The snail or cochlea is a spiral-shaped tube making two and a half turns (Fig. 9-30, *A*). Fig. 9-30, *B,* shows a cross section of one of the coils. A bony shelf (lamina spiralis ossea) reaches out from the wall into the canal. From the free edge of this bony shelf the basilar membrane extends to the opposite wall; by this, the tube is divided into two parts, the *scala tympani* and the *scala vestibuli.* Across the latter stretches Reissner's membrane, thereby forming the duct of the cochlea or the *scala media.* The scala media is filled with *endolymph;* the other two scalae contain *perilymph.* Endolymph is similar to intracellular fluid but has a lower concentration of protein. It is not present in the *tunnel of Corti.* Perilymph, in contrast, has a high concentration of Na ion and is high in protein. The basilar membrane (Fig. 9-30, *C*), 30 mm long, is constructed of fibers that run transversely; as its shape is that of a truncated triangle, these fibers are of varying lengths.It is estimated that there are 24,000 fibers in this membrane. On the basilar membrane is found the organ of Corti, the most important parts of which are the hair cells resting on the fibers of the basilar membrane and around which end the fibers of the cochlear division of the eighth cranial nerve. As many as 120 hairs arise from each cell and make contact with the overlying *tectorial membrane.* Two sets of hair cells are present; the inner and outer, which are separated by the arches of the tunnel of Corti.

Conduction of sound to the nerve ending. ■ Air vibrations enter the external auditory meatus. The auricle in man is rudimentary, and it is doubtful whether it plays any great part in collecting and reflecting sound waves. The meatus serves to protect the eardrum from mechanical and, to a lesser extent, from thermal injuries.

Sound waves entering the meatus cause the eardrum to vibrate at a corresponding frequency. This sets the auditory ossicles in motion and thus causes the foot of the stapes to move into and out of the oval window. By the vibration of the stapes, the perilymph in the vestibule and in the scala vestibuli is set in motion. Acting as levers, the ossicles magnify the pressure waves

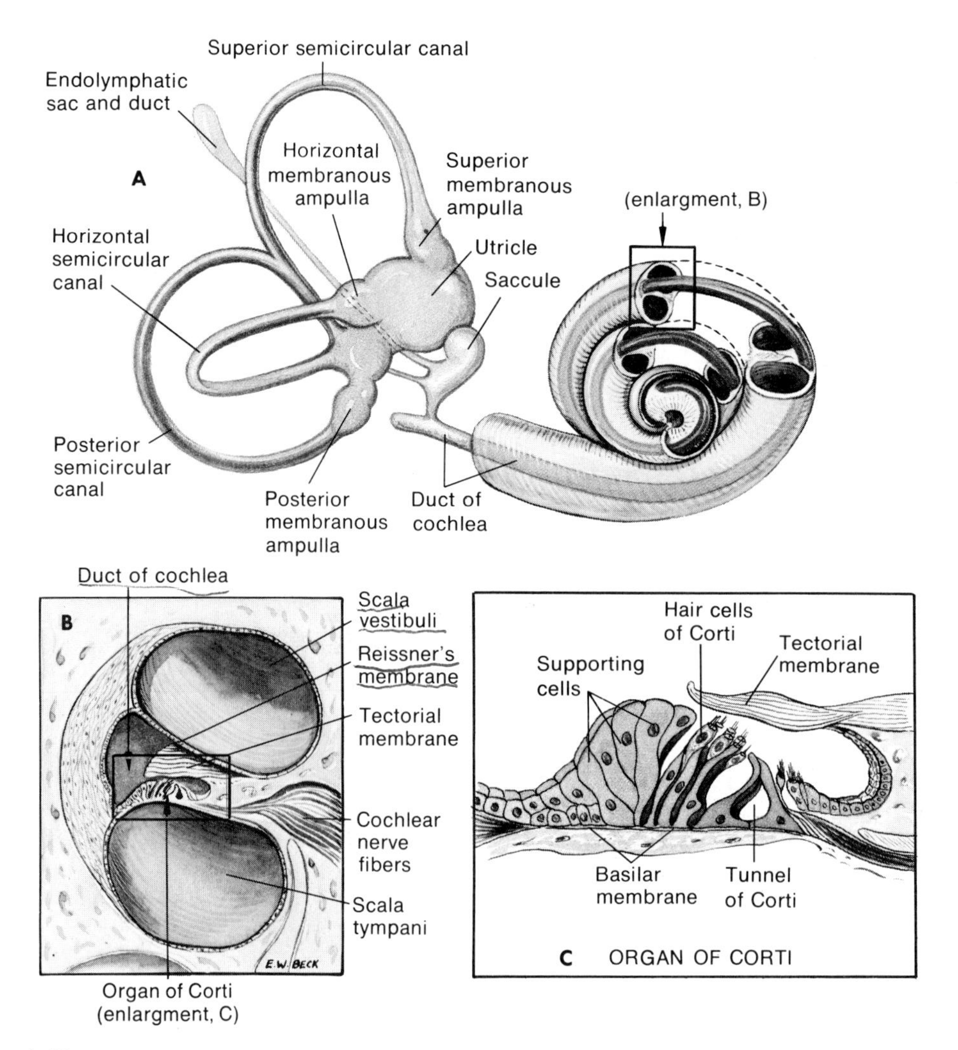

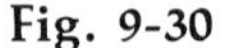

Fig. 9-30
Illustration showing relationships of anatomical structures of the inner ear. **B** and **C,** Successive enlargements of areas shown in **A** and **B.**

about twenty-two–fold. As the perilymph bathes the basilar membrane, the waves in this lymph cause the membrane to vibrate, and this, in turn, disturbs the hair cells resting on the basilar membrane. The hair cells are the auditory sensory cells, in the same manner as the rods and cones are the visual sensory cells; by their activity the nerve endings that surround them are stimulated and nerve impulses are generated.

Electrical potentials of the cochlea. ■ The scala media even in the unstimulated condition supports a potential of +80 mv. Hair cells show typical resting potentials of −80 mv; thus, the total potential between endolymph and hair cells is of the order of 160 mv. This *endocochlear potential* depends on an adequate supply of oxygen. It is also sensitive to displacement of the basilar membrane, so that when a sound wave is applied to the eardrum, a potential shift of a few millivolts, the *cochlear microphonic,* is detected. The microphonic appears to be associated with radial displacement of the membrane and movement of the outer hair cells. To what extent the cochlear microphonic excites the auditory nerve endings is unknown.

Excitation of auditory nerve fibers occurs when the endocochlear potential goes negative, as it does when the basilar membrane rises toward the tectorial membrane and causes bending of the cilia of the outer hair cells. Auditory information is coded in the frequency of all-or-none action potentials in the individual nerve fibers and in the numbers of fibers brought into activity.

Resonance and traveling wave theories. ■ Many theories exist regarding the response of the basilar membrane to different sound frequencies. Among these are the Helmholtz resonance (place) and the traveling wave (frequency) theories. Helmholtz maintained that the transverse fibers of the basilar membrane serve as a series of resonators. The fibers vary in length from the longest (about 0.36 mm) at the apex of the cochlea to the shortest (about 0.21 mm) at the base. Similar to the strings of a harp or piano, these fibers, even though embedded in a common matrix, are thought to vibrate individually. Suppose that three fundamental sound waves strike the eardrum simultaneously; the complex vibration set up in the conducting mechanism of the external, middle, and inner ear causes three fibers of the basilar membrane to be thrown into sympathetic vibration, each fiber being attuned to one of the three fundamental sounds (air waves). As a result, three nerve fibers carry impulses to the brain and three sensations are created, although, as our experience shows us, there is considerable commingling. Since the human ear is able to distinguish only 11,000 tones of different pitch, the number of hair cells (estimated at 16,000) is sufficiently great to act as receptors for the perception of all audible wavelengths. The resonance theory assumes that pitch is determined by the location of active fibers and that intensity is determined by the frequency of firing and by the numbers of active fibers.

The traveling wave theory proposes that the entire basilar membrane responds to pressure changes in the cochlea induced by ossicle movement; the shorter stiffer fibers near the base respond in phase with the applied signal, but the longer, more elastic fibers at the apex respond somewhat sluggishly and thus become out of phase. This, then, results in a wave of displacement that moves from base to apex. However, the maximum displacement, a peak in the wave, will occur at a specific distance from the base in accord with the frequency of the pressure change, that is, the frequency of the sound stimulus. Low frequencies produce a wave peak at the apex of the basilar membrane, high frequencies peak at the base and damp out before reaching the apex. This theory assumes that pitch is represented by the frequency of nerve impulses and intensity by the numbers of active nerve fibers.

Both theories, resonance and traveling wave, are supported by experimental evidence. Although the resonance theory cannot adequately explain all facets of auditory sensitivity, its simplicity makes it attractive.

In favor of the resonance theory we may state that, in individuals who are able to perceive only the lower pitched tones (as in boilermakers' disease), the shorter fibers of the basilar membrane have been found to be defective; the apical end of the membrane, with its longer fibers, has been shown to be degenerated in those persons not able to hear the lower tones. This has also been found to be true in animals in which either one or the other end of the membrane was experimentally destroyed.

■ Acuity of hearing

Localization. ■ Our ability to locate the source of a sound is not very impressive. If the source lies to the right or left of us, we have no difficulty in determining it, but if the sound comes from in front, behind, above, or below us, we have no idea in which direction the source lies; it is only by turning the head so that the sound strikes one ear more than the other that we can inform ourselves of its location.

Acuity. ■ The relative acuity of hearing can be determined by the distance at which a person can hear a given intensity of sound, for example, the ticking of a watch. Scientifically constructed audiometers have been devised by which the acuity can be accurately measured for sounds ranging from 16 to 33,000 vibrations per second. Audibility curves obtained by audiometry reveal that acuity is a function of the frequency of the sound. Hearing is most acute between 1,000 and 4,000 cycles/sec. Efficient hearing depends on (1) the proper transmission of the sound waves by the conducting mechanism to the inner ear, (2) the generation of nerve impulses by the receptor (organ of Corti), (3) the propagation of impulses by the auditory nerve to the cerebral cortex, and (4) the setting up of the sensation in the auditory area (Chapter 10) of the brain.

Eustachian tube. ■ For the eardrum to vibrate freely and thereby set in motion the other conduction mechanisms, the air pressure on the two sides of the drum must be equal. A gas enclosed in a body cavity is speedily absorbed by the blood and surrounding tissues; hence, the pressure of the gas is gradually reduced. To prevent this from occurring in the tympanic cavity, the Eustachian tube permits air to flow from the pharynx into the middle ear; thus an equality of pressure on the two sides of the eardrum is assured. The tubes are open, however, only during swallowing, yawning, and while blowing the nose. In a head cold the Eustachian tubes may be occluded by the swollen condition of the mucosa; the air in the middle ear is then absorbed and hearing impaired.

Acoustic reflex. ■ Two small muscles in the middle ear by their contraction are able to reduce the response of the cochlea through operation of what is known as the *acoustic reflex.* The *tensor tympani* muscle attaches to the malleus and the *stapedius* muscle to the stapes; when both are contracted, the ossicle system has an increased stiffness and the transmission of sound energy to the oval window is markedly diminished. Reflex contraction of these muscles is induced by irritation of the meatus and face, noises, and swallowing or yawning. The acoustic reflex serves as a protective measure against prolonged intense sound for periods of minutes before the muscles gradually relax. Because of the latency inherent in reflex activity, this reflex cannot offer protection against explosive sounds. Besides its obvious protective function in respect to prolonged sound, the reflex seems to be operative intermittently under normal conditions. In certain disease states this reflex capability is lost, and the intensity of sounds is a frequent complaint of the victim.

Bone conduction. ■ If the foot of a vibrating tuning fork is pressed upon the bones of the head, the ears being plugged, the sound is very noticeable; the bones conduct the sound waves to the inner ear and thereby set the lymph in the cochlea in vibration. Normally, air conduction is more effective than is bone conduction.

Deafness. ■ Deafness may be either conduction deafness or nerve deafness. In the first the conduction apparatus is at fault. This may be due to a number of abnormal conditions, such as thickening of the eardrum, occlusion of the Eustachian tubes, ankylosis* of the auditory ossicles, or the immobilization of the stapes in the oval window. Impacted cerumen (ear wax) impedes the tympanic vibration. Certain infectious diseases (e.g., mumps, scarlet fever, measles, diphtheria, and meningitis) may cause grave disturbances in hearing. In all of these conditions

*Ankylosis is the growing together of two neighboring bones in a joint and rendering the joint immobile.

hearing aids may be of value; these either reinforce the air conduction, or they depend on bone conduction.

Nerve deafness, due to deterioration of the cochlear structure or degeneration of the auditory nerve, frequently occurs. Hearing aids are of little value; this is certainly true in extensive nerve destruction. Tone (or tune) deafness is in hearing what color blindness is in vision. It is familial.

Efferent nerves. ■ Although the vast majority of nerve fibers in the auditory nerve are afferent, a small number are efferent. These motor fibers, originating in the olivary nuclei of the medulla, exert an important influence on cochlear response to sound. Motor activity reduces or suppresses the formation of action potentials in the afferent fibers. Details of the mechanism are unknown, but the fact that the central nervous system can modulate its own acoustic input by this arrangement is significant.

■ LABYRINTHINE SENSATIONS

Anatomy. ■ Of the membranous labyrinth previously described, only the duct of the cochlea (Fig. 9-30) is concerned with hearing. The other parts of this labyrinth include three distant structures: the utricle, the saccule (the otolith organs), and the three semicircular canals, or ducts. All these parts are filled with endolymph. The three membranous semicircular ducts spring from the utricle and are so placed that the three ducts in each ear are all at right angles to each other. For each of the three ducts in one ear there is a duct parallel to it in the other ear, and each duct has at one end a dilated part, termed the *ampulla.* In the ampulla is found a structure called the *crista acoustica* (Fig. 9-31), which may be described as a hillock in which the epithelial cells lining the wall of the duct assume a columnar shape and are provided with cilia. Covering the top of the crista is a gelatinous mass into which the cilia project.

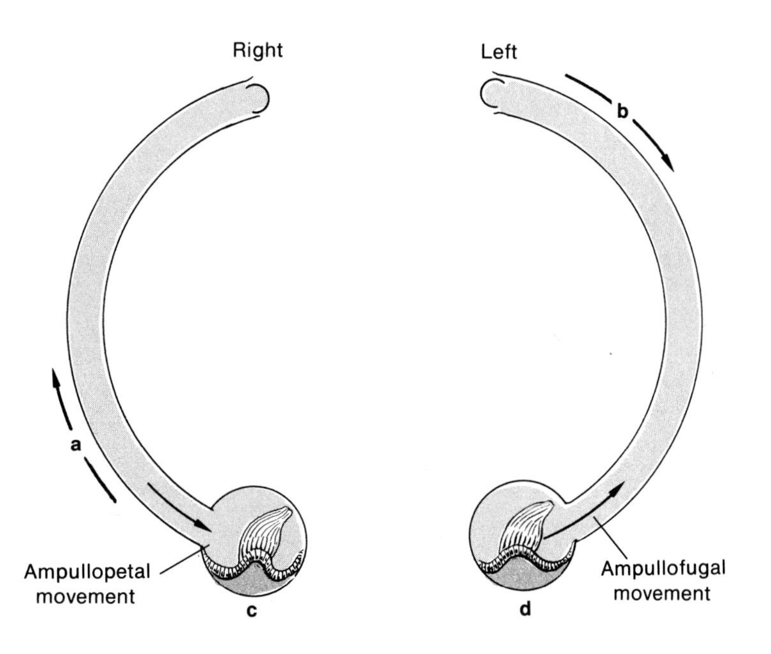

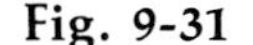

Fig. 9-31
Diagram illustrating effect of rotation of the head, shown by arrows **a** and **b**, on endolymph flow in ampullae, **c** and **d**, of horizontal semicircular ducts in right and left ear, respectively.

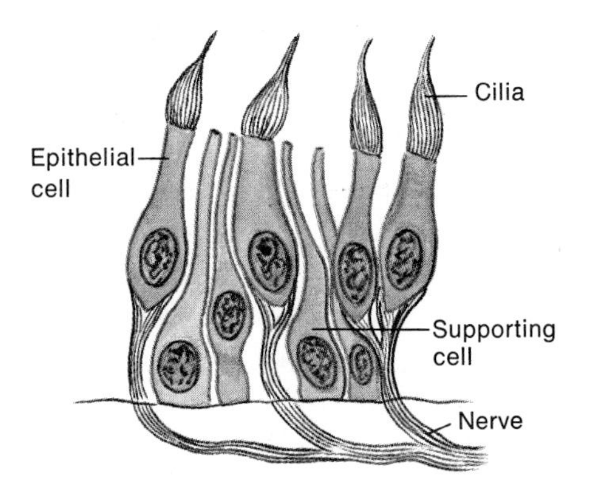

Fig. 9-32
Illustration showing hair cells of macula acoustica. (Halliburton.)

In the utricle and saccule is found the sensory epithelium known as the *macula acoustica.* This resembles the crista just described except that lying in among the hairs of the epithelial cells there occur a number of small concretions of calcium carbonate, known as *otoliths.* Fibers of the vestibular division of the eighth cranial nerve end in the neighborhood of the ciliated cells in the crista acoustica and macula acoustica (Fig. 9-32).

The fine structure of the cilia of the macula acoustica of a frog's saccule is shown in Fig. 9-33. Ciliary tufts on receptor cells exhibit two distinct structures; these are the *kinocilium,* which has a bulbous terminal, and the *stereocilia,* which are rodlike. The kinocilium is attached at the bulbous end to adjacent stereocilia by thin filaments and at its base to a notch in the cuticular membrane of the receptor cell. At the notch, the membrane is deformable, but the remainder of the cuticular membrane is relatively rigid. Stereocilia are affixed at their bases to the rigid membrane. Through this organization, pressure applied at the stereociliary side of the tuft causes the membrane beneath the kinocilium to be depressed; pressure applied at the kinociliary side of the tuft elevates the membrane at the notch. The membrane at the notch is in close contact with the cytoplasm of the receptor cell. It is postulated that the dimpling of the membrane brings about depolarization of the cell and increased action potentials in the nerve, whereas a reduction in action potentials would occur when the membrane was elevated.

Ciliary tufts with kinocilium and stereocilia are found in the macula, saccule, and ampulla of the semicircular canals. Although the kinocilium is missing in the ciliary tufts of the mammalian organ of Corti, a basal remnant is present and similarly placed in a cuticular notch. The mechanism for action potential generation proposed above could be envisaged as operating in the mammalian auditory system.

The labyrinth is a system of sense organs that are stimulated by changes in the position of the head or by changes in the velocity of motion of the head, either angular or linear. This is especially evident in pilots, and perhaps in astronauts, when subjected to intense accelerative forces. Under these circumstances they can become completely disoriented in space.

Compensatory eye movements. ■ When a person seated on a revolving chair is rotated, the eyes execute peculiar movements. During the rotation the eyes fix upon, and keep in view, a certain object; the eyes, therefore, move in the direction opposite to that of the body. When the eyes have turned as far as possible and the object can no longer be seen, the eyes swiftly move in the direction of the body rotation and fixate another object, and thus the process is repeated. The slow movement of the eyes opposite to the direction of rotation and the swift motion in the direction of rotation is known *rotatory nystagmus.*

These movements of the eyes are in response to *changes in the velocity of angular motion* to which the body is subjected. When the body rotation is maintained steadily for some time, the nystagmus ceases, to be resumed (but in the opposite direction to the original—*postrotatory nystagmus*) on stopping the rotation. Angular

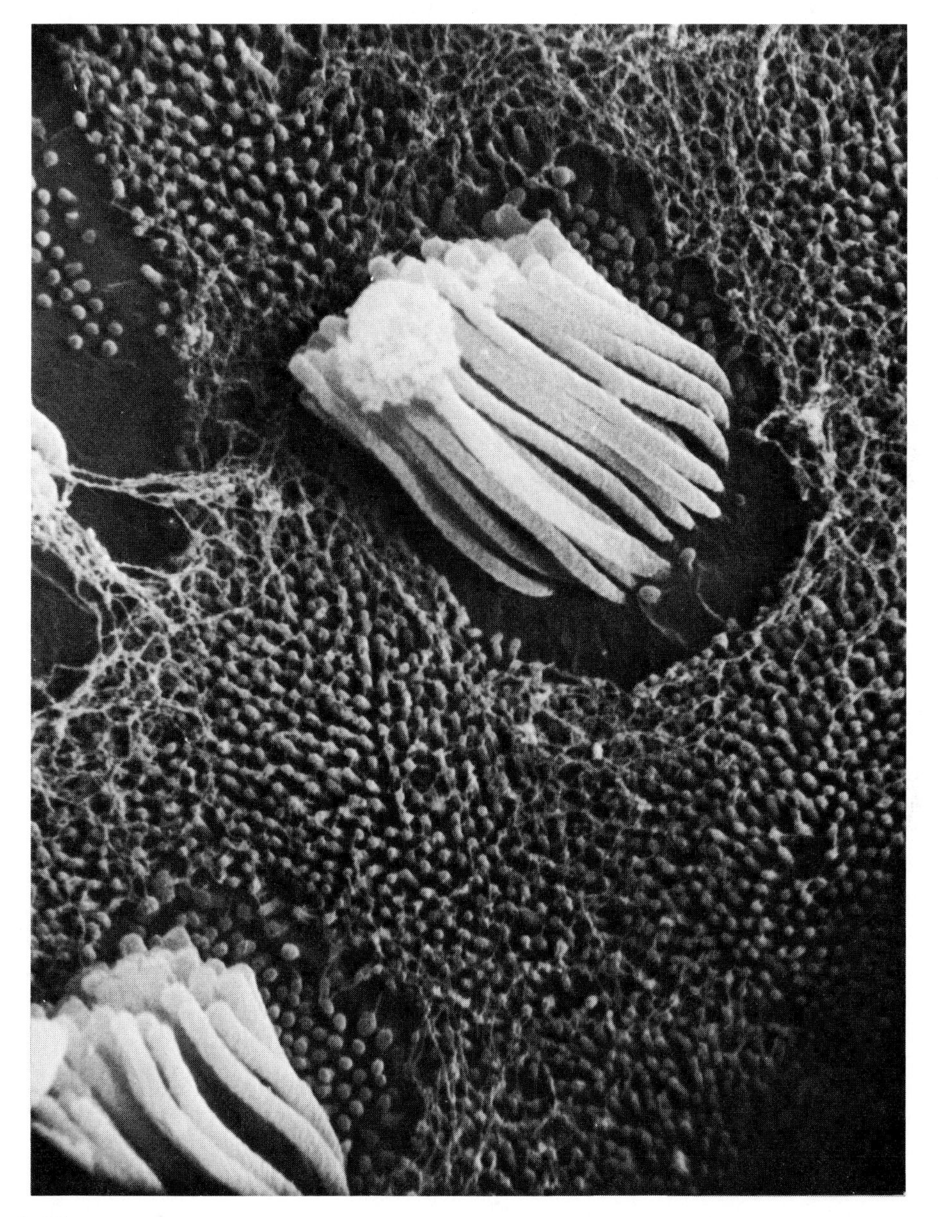

Fig. 9-33
A scanning electron micrograph of the macula acoustica in a frog's saccule. The structure with a large bulbous end is the kinocilium; the rod-like structures are stereocilia. Length of the cilia is about 5 μ. (Courtesy Drs. D. E. Hillman and E. R. Lewis.)

movement of the head in a given direction causes the endolymph in the semicircular canals in the plane of the movement to be forced, by virtue of its inertia, either toward (ampullopetal) or away from (ampullofugal) the ampullae (Fig. 9-31); for example, rapid acceleration of the movement of the body around the vertical axis stimulates the horizontal canals. This change in pressure bends the cilia of the crista and thereby stimulates or inhibits the fibers of the vestibular nerve. In Fig. 9-31, where rotation of the horizontal canals is clockwise, the crista in *c* undergoes ampullopetal movement; the vestibular nerve increases it firing rate. Ampullofugal movement occurs in *d* and the vestibular nerve ceases to fire. The difference in activity between the paired ampullae provides the cortical sensation of movement and determines the direction of nystagmus. When body rotation stops, the cristae reverse their deflections and the vestibular nerve action potentials halt in *c* but increase in *d*. Impulses are conveyed to the vestibular nuclei in the medulla, from where they are relayed, eventually, to the extrinsic muscles of the eyes.

Motion sickness. ■ Motion sickness (e.g., seasickness) is caused by changes in the velocity of either angular or linear movements. In the latter, the sense organs in the utricle and saccule are stimulated. Removal of the labyrinth renders a dog immune to motion sickness (no vomiting). The impulses from the labyrinth are sent to the cerebellum (Figs. 10-1 and 10-2); from there they are relayed to the vomiting center in the medulla. The utricle and saccule are also stimulated by abnormal positions of the head (positional reflexes).

Righting reflexes. ■ A cat held upside down and then allowed to fall rights itself in midair and lands on its feet. Vision is not necessary for these righting reflexes; a blindfolded cat behaves in the same manner. But removal of the otoliths or destruction of the utricle abolishes them. Righting reflexes are in response to *changes in the position of the head* and are initiated by the pulling of the otoliths upon the hair cells, due to gravity. The impulses are conveyed by the vestibular nerve to the cerebellum and to the vestibular nuclei in the medulla. These neural structures are concerned with maintaining the equilibrium of the body. From these centers efferent impulses are sent to the muscles of the neck by which the position of the head is readjusted with respect to gravity. The further righting of the body is mediated by the proprioceptors of the neck and body muscles. The sensory impressions from the labyrinth are of much more importance in some of the lower animals than in man, where destruction of these organs causes but a temporary disturbance.

We may conclude that the proprioceptive impressions from the labyrinth, together with visual and muscle proprioceptive impressions, are of the greatest value in equilibration, those from the labyrinth being especially concerned with maintaining the normal position of the head.

■ CHEMICAL SENSES—TASTE AND SMELL

Gustatory or taste sensations. ■ The chemical senses—taste and smell—are mediated by sense

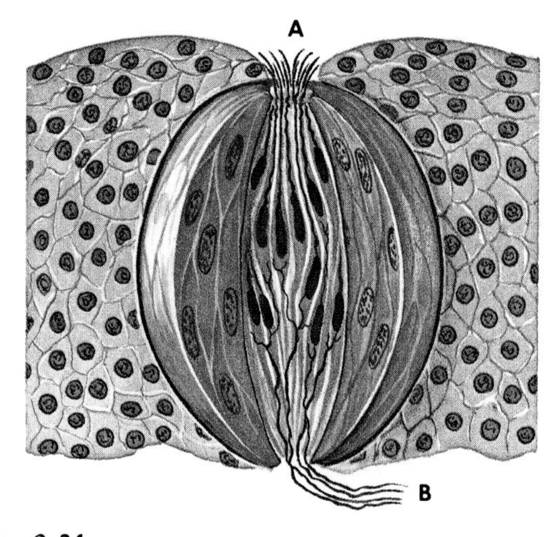

Fig. 9-34
Illustration of a taste bud showing pore, **A,** and nerve fibers, **B.**

organs located in the tongue and nose, respectively. In the mucosa covering the tongue, especially on the tip, edges, and posterior third, are found taste buds consisting of barrel- or bottle-shaped groups of epithelial cells. Some of these cells terminate at their external end (Fig. 9-34, *A*) in a hairlike process that projects into a pore or depression of the bud so that the stimulating agencies may come in direct contact with them. The other end of the taste cells is surrounded by the terminal filaments of the gustatory nerves (Fig. 9-34, *B*). Taste buds are constantly degenerating and being regenerated. They have a short life-span. The receptor cells have a resting potential of −50 to −95 mv at rest; when stimulated they respond with a generator potential that excites the nerve fibers. Intensity of the stimulus is coded in the magnitude of the generator potential, in the frequency of impulses in the nerve fibers, and in the number of active fibers. Adaptation is usually rapid, a matter of seconds.

To be tasted, the substance must be in solution. If the tongue is dried with a towel and a lump of dry sugar is placed upon it, there is no sensation until some sugar is dissolved by the saliva that has collected between the tongue and the sugar. Insoluble substances have no taste.

The importance of the sense of taste in the nutrition of the body is evident in the role it plays in the secretion of digestive juices. The instinctive food selection (specific appetite) shown by many animals is abolished by severing the nerves for taste sensation.

Tastes have been classified as sweet, sour, salt, and bitter; some authorities add to this an akaline and a metallic taste. The various tastes are not experienced equally well at all parts of the tongue and mouth; sweet and salt are best, but not exclusively, appreciated at the tip of the tongue, sour at the sides, and bitter at the back. A solution of magnesium sulfate stimulating the tip of the tongue creates a salty taste, but on being swallowed it tastes bitter.

The glossopharyngeal (ninth cranial) is the nerve of taste for the posterior third of the tongue; the facial (seventh cranial) nerve innervates the anterior two thirds. The trigeminal (fifth cranial) is the nerve of common sensation (touch and temperature) for the tongue and for the mucosa of the mouth and nose.

Sensation of smell. ■ The sensation of smell is mediated by the olfactory (first cranial) nerve, which has its endings in the nasal mucosa just above the superior turbinate bone of the nose (Fig. 17-1). The receptors are stimulated chemically by inhaled gases. They are bipolar neurons, ciliated at the dendritic end, that face the nasal cavity. On stimulation a generator potential develops, and bursts of spikes originate from the receptor cells. How the olfactory stimulus is coded for relay to the brain is unknown.

The sense of smell is closely allied to the sense of taste. In fact, many of the so-called tastes (flavors) of certain foods are in reality odors. A piece of raw onion chewed with the nose shut cannot be distinguished from a piece of raw potato. For this reason a cold in the head interferes with appreciation of food, thus causing loss of appetite. The value of the sense of smell is that by creating appetite it favors the flow of the digestive juices. It also informs us of injurious substances, such as decaying food; however, this is by no means an infallible guide. The sense of smell warns of the presence of odorous gases, the inhalation of which may be harmful.

READINGS

Amoore, J. E., Johnston, J. W., Jr., and Rubin, M.: The stereochemical theory of odor, Sci. Am. **210**(2):42-49, Feb. 1964.

Botelhs, S. Y.: Tears and the lacrimal gland, Sci. Am. **211**(4):78-86, Oct. 1964.

Davson, H.: Physiology of the eye, ed. 3, Boston, 1972, Little, Brown and Company.

Dunphy, E. B.: The biology of myopia, New Eng. J. Med. **283**:796-800, 1970.

Hillman, D. E., and Lewis, E. R.: Morphological basis for a mechanical linkage in otolithic receptor transduction in the frog, Science **174**:416-419, 1971.

Lerman, S.: Cataracts, Sci. Am. **206**(3):106-114, Mar. 1962.

Michael, C. R.: Retinal processing of visual images, Sci. Am. **220**(5):105-114, May 1969.

Miller, W. H., Gorman, R. E., and Bittensky, M. W.: Cyclic adenosine monophosphate: function in photoreceptors, Science **174**:295-297, 1971.

Moulton, D. G., and Beidler, L. M.: Structure and function in the peripheral olfactory system, Physiol. Rev. **47**:1-52, 1967.

Oakley, B., and Benjamin, R. M.: Neural mechanisms of taste, Physiol. Rev. **46**:173-211, 1966.

Rosenzweig, M. R.: Auditory localization, Sci. Am. **205**(4):132-142, Oct. 1961.

Stevens, S.S.: Neural events and the psychophysical law, Science **170**:1043-1050, 1970.

Young, R. W.: Visual cells, Sci. Am. **223**(4):81-91, Oct. 1970.

10

Encephalization
Brainstem
Cerebrum
Cerebellum

THE BRAIN

The brain is divided into the following parts (Figs. 10-1, 10-2, and 10-7):

1. Prosencephalon (forebrain)
    - Telencephalon
        - Cerebral hemispheres
        - Corpus striatum
        - Internal capsule
        - Corpus callosum
    - Diencephalon
        - Thalamus
        - Hypothalamus
        - Geniculate bodies
2. Mesencephalon (midbrain)
    - Corpora quadrigemina
    - Cerebral peduncles
3. Rhombencephalon (hindbrain)
    - Metencephalon
        - Pons
        - Cerebellar peduncles
        - Cerebellum
    - Myelencephalon
        - Medulla oblongata

It is frequently convenient to regard the brain as made up of three parts: the cerebrum, the cerebellum, and the brainstem. This latter part includes the diencephalon, mesencephalon, metencephalon (except the cerebellum), and myelencephalon. The brainstem may be regarded as a much modified prolongation of the spinal cord; the cerebrum and the cerebellum are developed as two large expansions of the stem. In the lower vertebrates, the fishes, the brainstem and the cerebellum are fairly well developed, but the cerebrum is altogether lacking or but poorly represented. With advances in the scale of animal life, through the amphibians, reptiles, birds, and mammals, the cerebrum gradually not only becomes anatomically larger in size, as compared with the rest of the brain, but acquires, especially in the highest mammals and more particularly in man, a functionally dominant place. We may therefore speak of the brainstem and cerebellum as being phylogenetically* the old brain; the cerebrum is of much later origin.

ENCEPHALIZATION

In the central nervous system we speak of lower and of higher levels of function. The spinal

*Phylogeny, the evolution of a race; ancestral development. Ontogeny, the development of the individual.

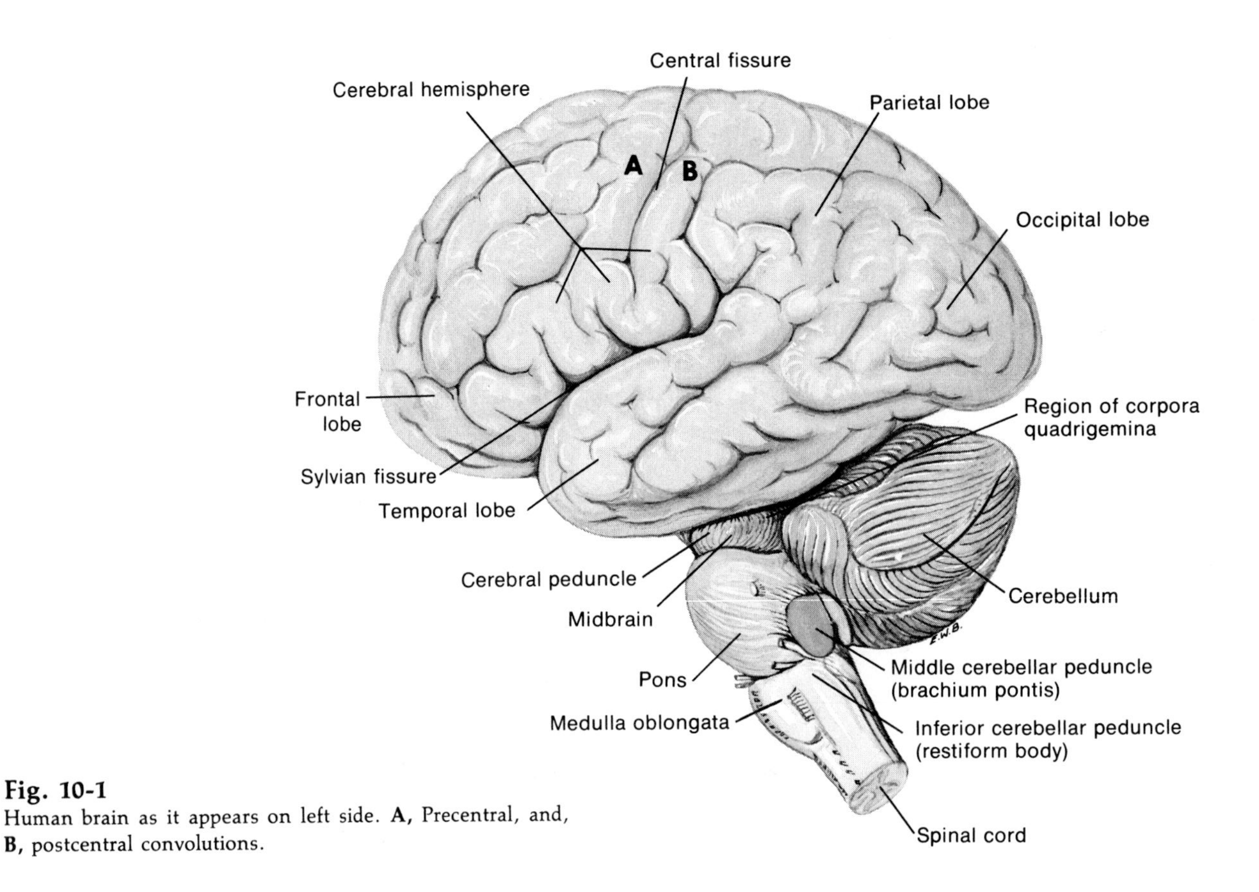

Fig. 10-1
Human brain as it appears on left side. **A,** Precentral, and, **B,** postcentral convolutions.

cord, for example, forms the lowest level; above this we progressively advance to the medullary, the midbrain, the hypothalamic, the thalamic, and the cortical levels. In the relative degree of development of the neural levels of integration lies the important and conspicuous difference between animals placed higher or lower in the scale of life. A few basic facts as to these levels of neural functions are as follows:

1. In a more lowly organized vertebrate (fish or amphibian), the spinal cord exhibits a great deal of independent activity; that is, in its response to environmental changes it is but little influenced by neural centers of the higher levels.

2. The higher an animal stands in the scale of life, the greater the control exercised by the higher neural levels over the lower levels and the less the ability of the lower levels to act independently.

3. The highest level of neural integration is found in the cerebral cortex. The greater its development, the greater is its dominance over all subcortical levels. This is the outstanding difference between the human brain and that of all other mammals.

4. Muscle and gland responses following the discharge of impulses from the higher levels are more varied and are able to meet more efficiently a greater diversity of environmental changes. They have lost much of the stereotyped and predictable characteristics of spinal reflexes.

5. The mechanisms of the upper neural levels

are of later phylogenic development than are those of lower levels. The upper neurons, in influencing the responses to environmental changes, do not supplant the lower levels but act through them.

6. As a corollary, the greater the development of the upper levels (especially of the cerebral cortex), the greater is the disturbance caused by their ablation or malfunctioning.

The gradual moving forward (i.e., toward the cerebrum) of the integrating mechanism is designated as *encephalization.* This is well illustrated in the development of the visual mechanism.

In all vertebrates the optic impulses from the retinas are sent to some nuclei in the brainstem; in man these are the lateral geniculate bodies (Fig. 9-20). In a frog all the retinal fibers discharge their impulses to the optic lobes. From there the impulses are carried by motor nerves to various skeletal muscles. No visual fibers are sent to the rudimentary cortex. Consequently, the removal of the frog cortex leaves the subcortical station in full control; the photic impulses from the retina share in determining the behavior of the animal—we say the animal sees. In slightly more highly developed animals, a few neural connections are established between the visual subcortical nuclei in the central gray matter and the cortex; this enables the cortex (by means of efferent fibers) to control muscular activity to a limited extent, but the subcortical station retains most of its reflex control. Even in birds, the removal of the cortex causes scarcely any disturbance in vision. Still higher in the animal scale, the number of visual fibers from the subcortical station to the cortex (projection fibers) becomes progressively greater; the functional connections between the subcortical station and the muscles become less. In man the optic impulses received by the lateral geniculate bodies find access to the efferent nerve fibers by way of the cortex; hence, removal of the cortex (occipital lobes) renders a person completely blind.

■ BRAINSTEM

The central gray matter in the brainstem does not form a solid, continuous column as in the spinal cord. Instead, it is broken up into a number of separate masses of various sizes known as ganglia and nuclei. The fibers constituting the white matter describe devious pathways, winding over, under, and between the ganglia and nuclei. This makes the anatomy of the brainstem complicated.

■ Reticular formation

The cells and nuclei scattered throughout the brainstem together constitute what is called the *reticular formation.* Although this region is not well defined anatomically, it has an important role in both visceral and somatic functions. It occupies a large part of the medulla oblongata, and the red nuclei can be thought of as the upper termination (Fig. 10-10). The vestibular nuclei are included within the formation, as are the cell groups serving the special visceral efferent system, that is, the autonomic nerves. The reticular formation receives impulses from the spinal cord by way of collateral fibers from the ascending sensory pathways. It also has input from the cerebellum and cerebral hemispheres and sends impulses to these structures. The diffuse connections with the cortex are made through a number of relays in subcortical nuclei.

Reticular formation motoneurons are both excitatory and inhibitory. Some of these cells appear to play a significant role in consciousness and attention because of the complex interactions of afferent and efferent activities that occur in the formation. For this reason they are referred to as the *reticular activating system.* It is suggested that this system is especially sensitive to anesthetics and its destruction results in unconsciousness. Stimulation of the activating system produces "arousal" and changes in the electroencephalogram (Fig. 10-12). Certain cells located principally in the medulla exert an inhibitory effect on antigravity (extensor) muscles but

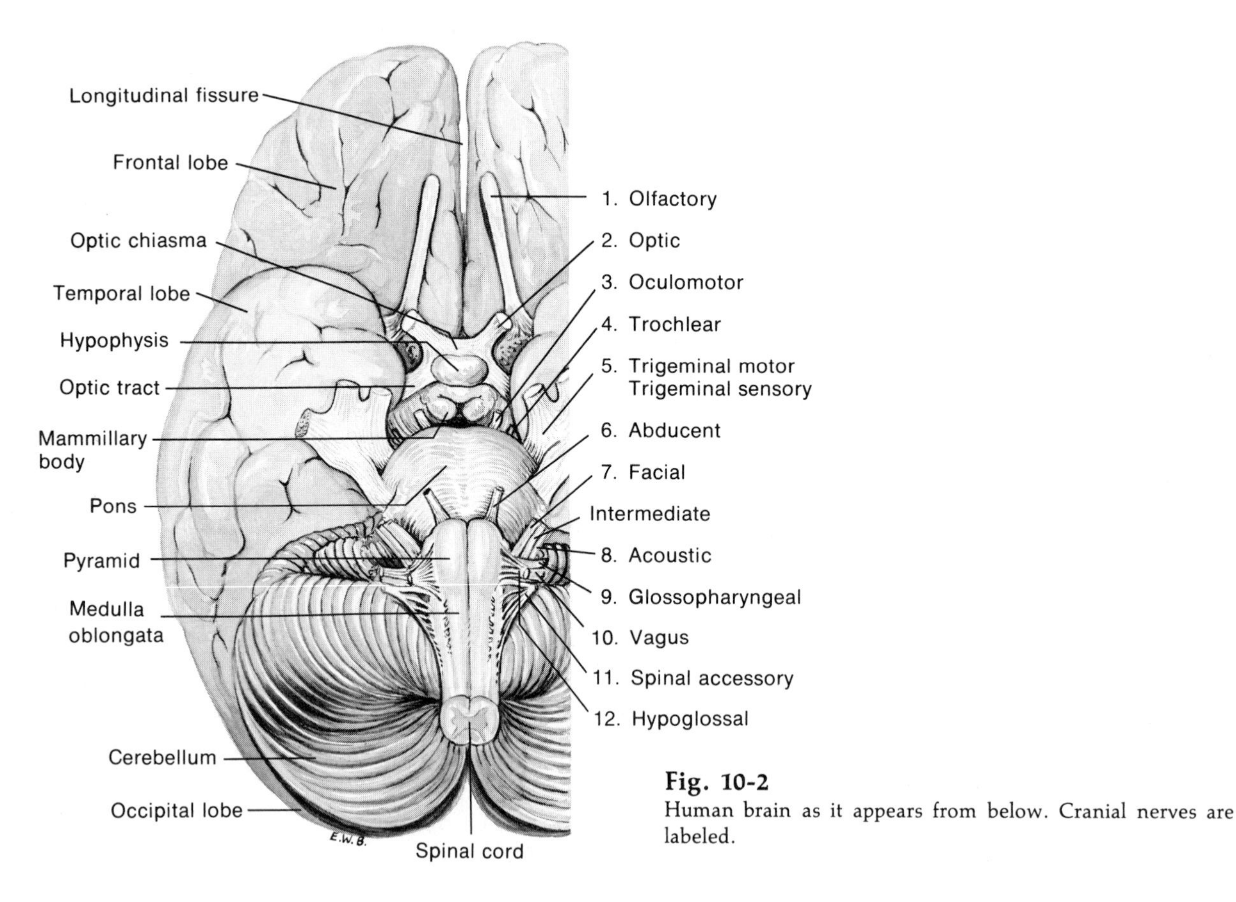

Fig. 10-2
Human brain as it appears from below. Cranial nerves are labeled.

facilitate flexor motoneurons in the spinal cord. Higher levels of the nervous system can cause inhibition or excitation at the spinal cord level by their influence on reticular formation cells. Neurons that excite the antigravity muscles are usually more laterally and rostrally (forward) located. Isolation of the reticular formation from the cerebral cortex by high transection of the midbrain yields a state known as *decerebrate rigidity,* in which there is exaggerated tone in the antigravity muscles due to an imbalance of excitatory and inhibitory activity in the formation.

Many cells of the reticular formation are involved in visceral functions. They constitute the centers for regulation of respiration, blood pressure, cardiac activity, and alimentary tract secretion and movement. The reticular formation is also connected to the auditory and visual systems through projections from the central relay nuclei of these systems.

■ Cranial nerves

From the brain issue twelve pairs of cranial nerves (Fig. 10-2). Some of these are afferent, some are efferent, and some are mixed nerves. All except the first pair, the olfactory nerves, arise from the brainstem. The afferent and efferent fibers sustain the same relationship to the gray matter as do the spinal nerves. One of each pair is discussed.

First cranial or olfactory nerve. ■ The olfactory nerve carries the impulses for the sensation of smell. The cell bodies lie in the upper part of the nasal mucosa, and the fibers from these cell bodies pass to the olfactory bulb. From there a new relay takes the impulses to the sensory area of smell in the cerbral cortex.

Second cranial or optic nerve (Figs. 9-20, 9-25, and 10-2). ■ The course of the optic nerve to the lateral geniculate bodies and the relaying by the optic radiation fibers of the projection system to the occipital lobes has been discussed in Chapter 9. A few of the fibers of the optic nerve end in the midbrain and there make synaptic connections with the third, fourth, and sixth cranial nerves.

Third cranial or oculomotor nerve (Fig. 10-2). ■ Originating in the midbrain the oculomotor nerve is the motor nerve for four of the six extrinsic eye muscles and for the elevator of the upper eyelid.

Fourth cranial or trochlear nerve. ■ The trochlear nerve originates in the midbrain and innervates the superior oblique muscle of the eyeball.

Fifth cranial or trigeminal nerve. ■ The trigeminal nerve issues from the pons of the brainstem and resembles the spinal nerves in being composed of a sensory and a much smaller motor root. The cell bodies of the sensory fibers lie in the Gasserian ganglion outside of the brain; from this ganglion arise three large roots by which their fibers are distributed to the skin of the face, to the eyeball, to the mucosa of the mouth and nose, and to the teeth. The motor fibers innervate the muscles of mastication.

Sixth cranial or abducens nerve. ■ The abducens nerve is the motor and sensory nerve for the external rectus muscle of the eyeball.

Seventh cranial or facial nerve. ■ The facial nerve is the motor nerve for the muscles of the face, ears, and scalp. This nerve also contains gustatory fibers.

Eighth cranial, auditory, or statoacoustic nerve (Figs. 9-29 and 10-2). ■ The auditory nerve is a sensory nerve composed of two parts, cochlear and vestibular; these were considered in the discussion of the ear (Chapter 9).

Ninth cranial or glossopharyngeal nerve. ■ The glossopharyngeal is a mixed nerve. Its motor branches supply the muscles of the pharynx and the base of the tongue. It supplies secretory fibers to the parotid (salivary) gland. Its sensory fibers are supplied to the tongue and pharynx and, together with the seventh cranial nerve, constitute the nerves of taste.

Tenth cranial, vagus, or pneumogastric nerve. ■ The vagus nerve is a mixed nerve. Its motor fibers are supplied to the muscles of the larynx and of the alimentary tract (extending from the esophagus to the large intestines), and its inhibitory fibers supply the heart. The glands of the stomach and the pancreas are innervated by this nerve. Its sensory fibers end in the heart and in the mucous membranes of larynx, trachea, lungs, esophagus, stomach, gallbladder, and intestines.

Eleventh cranial or spinal accessory nerve. ■ The spinal accessory nerve arises in the medulla, is a motor nerve for sternomastoid and trapezius muscles, and sends many other motor fibers directly into the vagus nerve.

Twelfth cranial or hypoglossal nerve. ■ The hypoglossal nerve arises in the medulla and is probably a mixed nerve for the muscles of the tongue and larynx.

■ Medulla oblongata

The medulla oblongata (Figs. 10-1 and 10-2) forms the connection between the spinal cord and all the brain anterior to the medulla. In our discussion of the reticular formation we became acquainted with its many important and even vital nerve centers: the centers for respiration, phonation, vasoconstriction, vasodilation, cardiac inhibition and acceleration, mastication, deglutition, salivary and gastric secretions, and perspiration.

Some nerve fibers found in its white matter

merely pass through the medulla, being bound for either higher or lower parts of the nervous system. This is true for the ascending fibers in the spinocerebellar tracts of the cord (Fig. 8-13) and for the descending fibers in the corticospinal tract; these will be discussed presently in considering the pathways into and out of the cerebral cortex. From the gray matter of the medulla originate the ninth, tenth, eleventh, and twelfth cranial nerves.

■ Hypothalamus

The hypothalamus, a portion of the diencephalon, lies just below the thalamus and above the pituitary body (hypophysis). It consists of several distinct masses of gray matter or nuclei. The positions of some are indicated in Fig. 10-3, and its relation to the pituitary gland is shown in Fig. 10-2. In our study we shall mention only an anterior and a posterior portion of this structure. The hypothalamus receives nerve fibers from the thalamus; as this latter structure is closely associated with the cerebral cortex, impulses from the cortex can reach the hypothalamus indirectly by this route. The medulla oblongata and the spinal cord also send impulses to the hypothalamus. The hypothalamus is efferently connected with the pituitary body and in some measure controls the hormonal activity of this gland. Efferent fibers are also sent to the thalamus (thus indirect contact is made with the cerebral cortex of the frontal lobes) and to the reticular formation. In a subsequent discussion in this chapter, the part the hypothalamus plays in

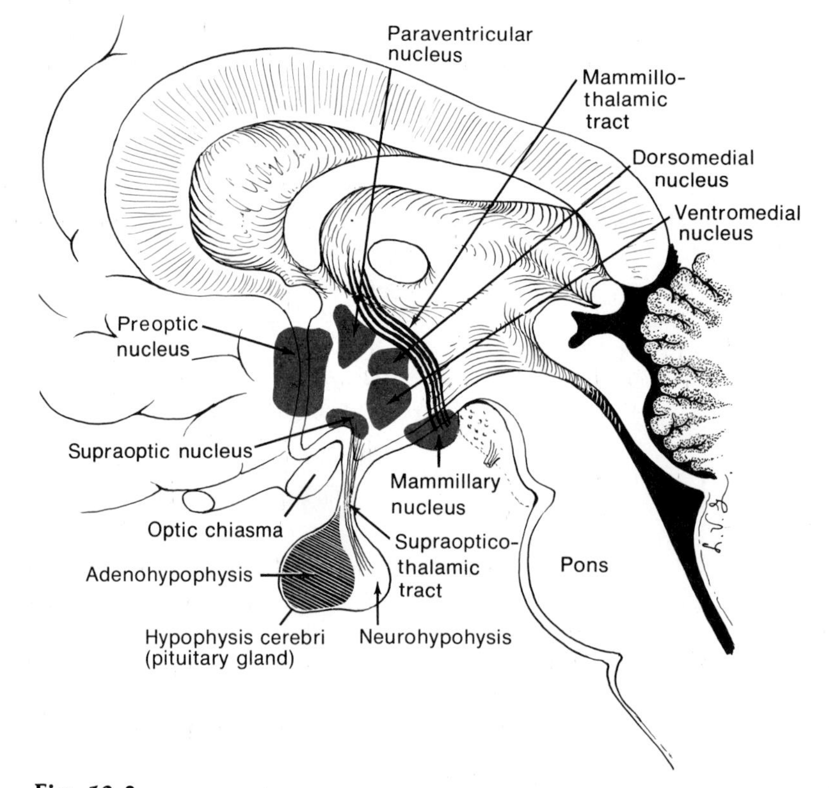

Fig. 10-3
The hypothalamus showing location of nuclei and mammillary body.

the objective manifestation of emotions will be considered.

■ The thalamus

Another mass of gray matter in the diencephalon is the thalamus (Figs. 10-4; 10-6, *OT;* and 10-7). Its neural connections with the other parts of the brain and its functions can best be considered when studying the cerebral cortex.

■ CEREBRUM

■ Gray matter

In contrast to the spinal cord and brainstem, the gray matter of the cerebrum (or cerebral hemispheres) is deposited as a mantle on the exterior and covers the white matter lying in the interior of the hemispheres. For this reason it is called the *cortical gray matter,* or *cortex* (Latin, bark). In Figs. 10-5 to 10-7 it is represented by

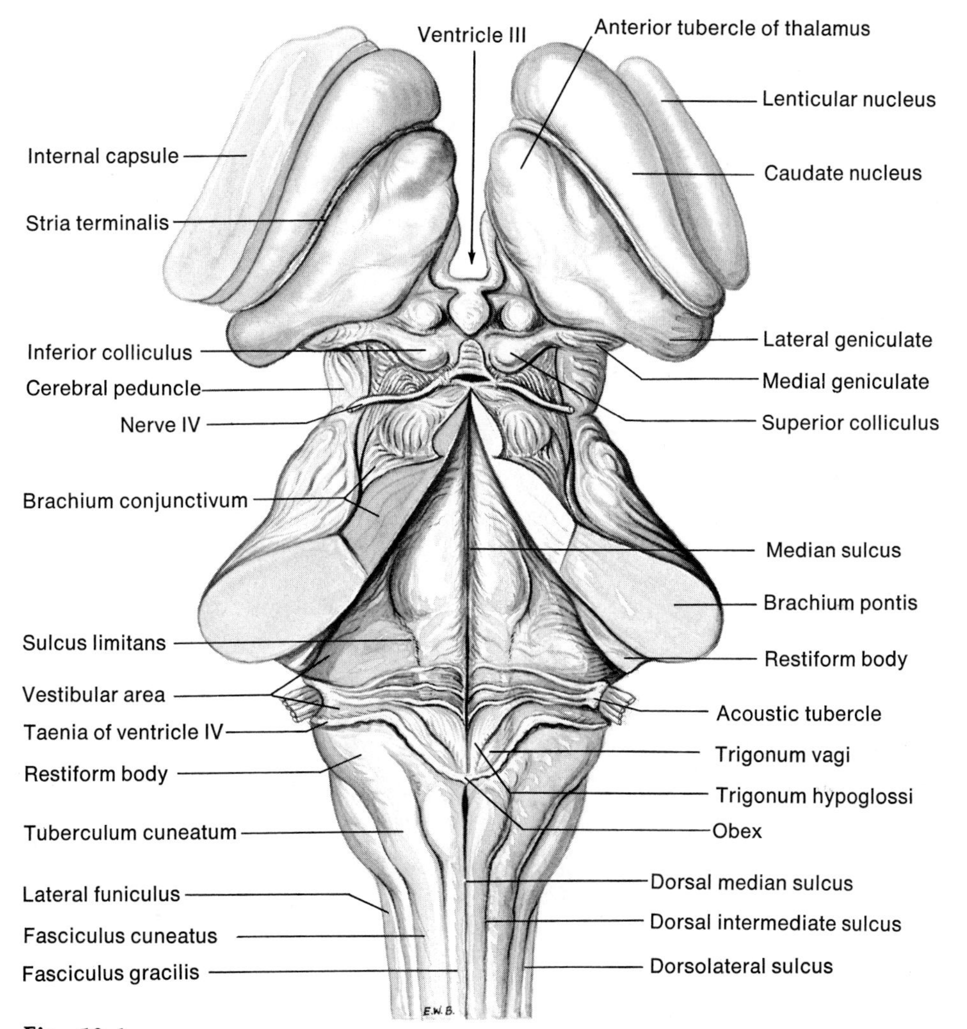

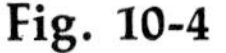
Fig. 10-4
Human brainstem as it appears from above.

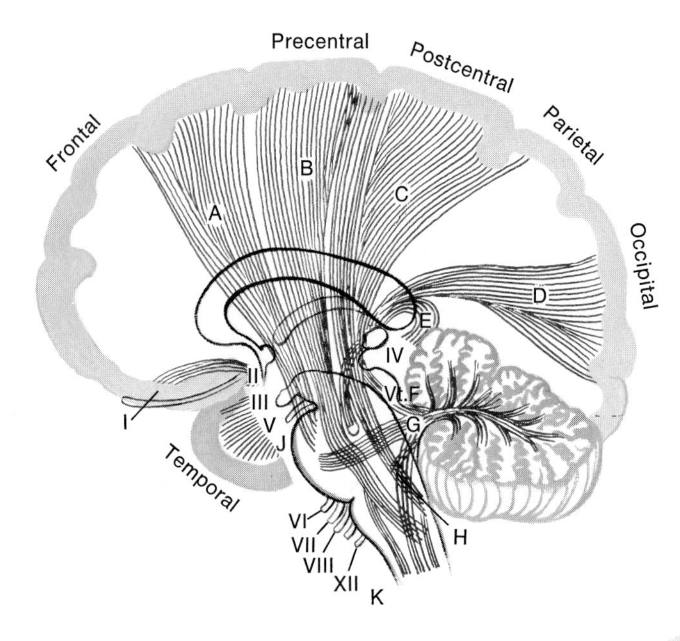

Fig. 10-5
Diagram showing projection fibers of cerebral cortex and fibers of cerebellar peduncles. **A,** Tract connecting frontal convolutions with cells lying in pons; these cells are connected with cerebellum by means of middle cerebellar peduncle, **G; B,** motor fibers of pyramidal tracts, which, after crossing over to opposite side in medulla, **K,** descend cord; **C,** sensory fibers carrying impulses for tactile and motorial sensations to postcentral gyrus; **D,** visual tract; **E,** auditory tract; **F,** superior, and, **H,** inferior cerebellar peduncles. Roman numerals indicate cranial nerves.

Fig. 10-6
Diagram illustrating association fibers of cerebral hemispheres. **A,** Between adjacent convolutions; **B,** between frontal and occipital areas; **C** and **D,** between frontal and temporal areas; **E,** between occipital and temporal areas; **CN,** caudate nucleus; **OT,** optic thalamus.

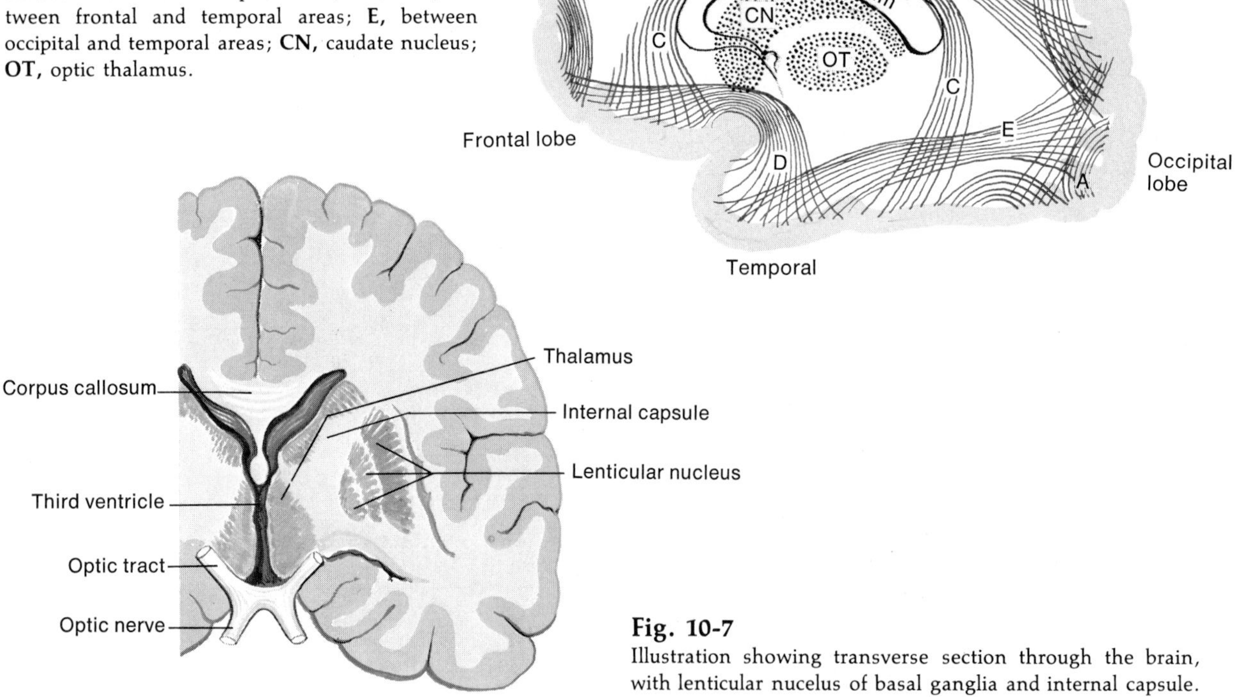

Fig. 10-7
Illustration showing transverse section through the brain, with lenticular nucelus of basal ganglia and internal capsule.

the stippled areas at the periphery. In the cortex (from 2 to 4 mm thick) the millions upon millions of neurons are arranged in six layers. The layers differ from each other in the size, shape, distribution, and density of population of the cell bodies and in the arrangement of the cell processes.

Convolutions. ■ At many places the gray matter dips down into the brain so that folds (Fig. 10-1) are formed which give the hemispheres viewed from above much the appearance of the kernel of an English walnut. By this means the amount of cortical gray matter is much increased. In general, the higher animals have a more convoluted cortex than do the lower animals. The grooves are known as *fissures,* or *sulci* (singular, sulcus), and the folds, as *convolutions,* or *gyri.* We may call attention to three fissures: the lateral (Sylvian) fissure, the central fissure (Fig. 10-1), and the longitudinal fissure, separating the two hemispheres (Fig. 10-2).

Lobes. ■ The cerebral cortex is divided into four major divisions: the frontal, the parietal, the occipital, and the temporal lobes (Figs. 10-1 and 10-2). Each lobe has two or more convolutions.

■ Basal ganglia

The interior of the hemispheres contains clearly defined masses of gray matter within the central white matter. These are the thalamus, hypothalamus, and *basal ganglia.* The last consists of a medial nucleus dorsal to the thalamus, the *caudate;* and a lateral nucleus, the *lenticular,* which is made up of the globus pallidus and putamen (Fig. 10-7). The medial and lateral nuclei are separated by a band of white matter, the *internal capsule.* Connections of the basal ganglia with the reticular formation are made via the red nucleus (Figs. 10-10 and 10-15) and substantia nigra. The basal ganglia receive fibers from and send fibers to the thalamus, the cerebral cortex, and one another. Function of the basal ganglia is not well understood, since they are difficult to get to in living animals, a necessity for functional studies. However, certain animal experiments and careful observations of human disorders have shown that the basal ganglia are of great importance in the inhibitory functions of the extrapyramidal system, to which we will address our attention later. Together with the cerebellum the basal ganglia serve to integrate locomotor and postural reflexes through connections with the reticular formation. The mechanisms subserving the integration are virtually unknown. These reflex patterns are not greatly disturbed in decorticate (removal of the hemispheres above the level of the thalamus—a so-called midbrain animal) animals but are severely disturbed in decorticate man. Diseases of the basal ganglia in man lead to disorganization of integrated motor functions and, depending on the nuclei involved, the appearance of two classes of spontaneous abnormal movements. Parkinsonian victims (palsy) exhibit fine tremors associated with a degree of muscle rigidity, including the facial muscles. Writhing movements (chorea), even speech disorders and dementia, may occur in Huntington's chorea.

■ White matter

The nerve fibers, constituting the white matter of the interior of the hemispheres, are processes either of the cortical cells or of cells located in the central gray matter of the brainstem (Figs. 10-5, 10-6, and 10-9). These may be divided into three classes: projection fibers, association fibers, and commissural fibers.

Projection fibers. ■ Projection fibers are those that carry impulses (1) afferently from the brainstem to the cortex and (2) efferently from the cortex to the lower parts of the central nervous system (Fig. 10-5).

Association fibers. ■ Association fibers are those that originate in cortical cells and that carry impulses to other areas on the same side of the cortex (Fig. 10-6).

Commissural fibers. ■ Commissural fibers connect the two cerebral hemispheres. The

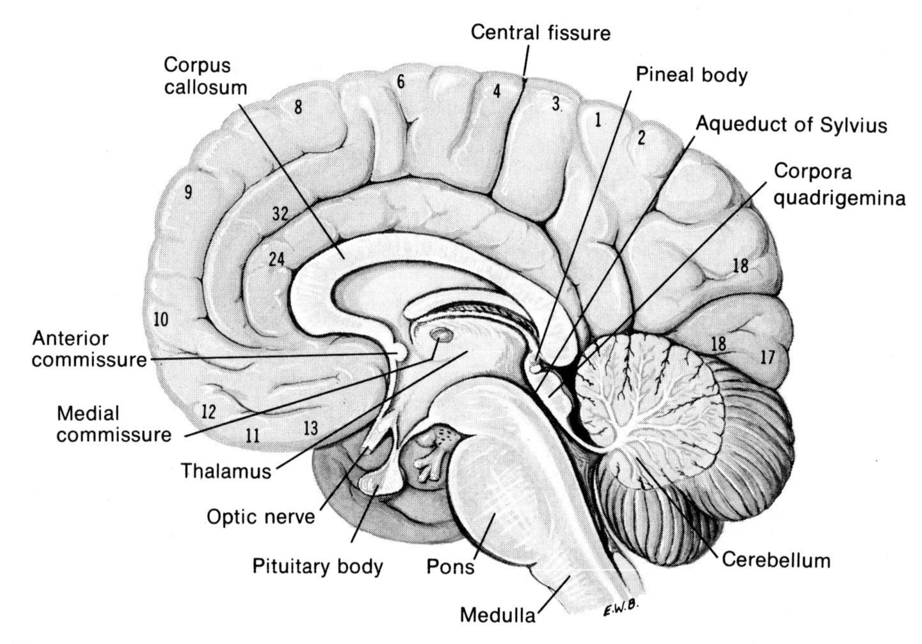

Fig. 10-8
Human brain as it appears on the medial surface. Numbers refer to cortical areas. Compare with Fig. 10-11.

corpus callosum, shown in longitudinal section of the cerebrum in Figs. 10-6 and 10-8, is composed of such fibers. By this and other commissures the parts of one hemisphere are connected with corresponding parts in the other hemisphere.

■ Neural connections

The cerebrum has two important functions. It constitutes the highest integrative center of the nervous system, and it is the seat of psychic functions, such as sensation, perception, memory, judgment, volition, and consciousness. To serve as the chief integration center, the cerebrum must be connected with the subcenters in the brainstem and spinal cord, which in turn are neurally connected with the receptors and effectors throughout the body. Before tracing some of the important afferent and efferent impulses into or out of the cerebral cortex, five important general observations should be made.

1. Receiving organs located in the skin, in the muscle spindles, and in the tendon end organs make *direct* (without relay) connections with the central gray matter of spinal cord and medulla (Fig. 10-9). However, to reach the cortical gray matter, an impulse from these receptors must pass through a chain of three neurons: the primary, secondary, and tertiary neurons.

2. All afferent impulses (except those of the olfactory or first cranial nerve) pass through the thalamus. The thalamus may appropriately be called the port of entry to the cerebrum. It will be noticed from Fig. 10-9 that the secondary neuron makes connection with the tertiary neuron in the thalamus.

3. The efferent connection between the gray matter of the spinal cord and the muscles is also *direct* (Fig. 10-10), but an efferent impulse from the motor cells in the cortex must traverse at least two consecutive neurons to reach a muscle.

4. Impulses originating from stimulation to

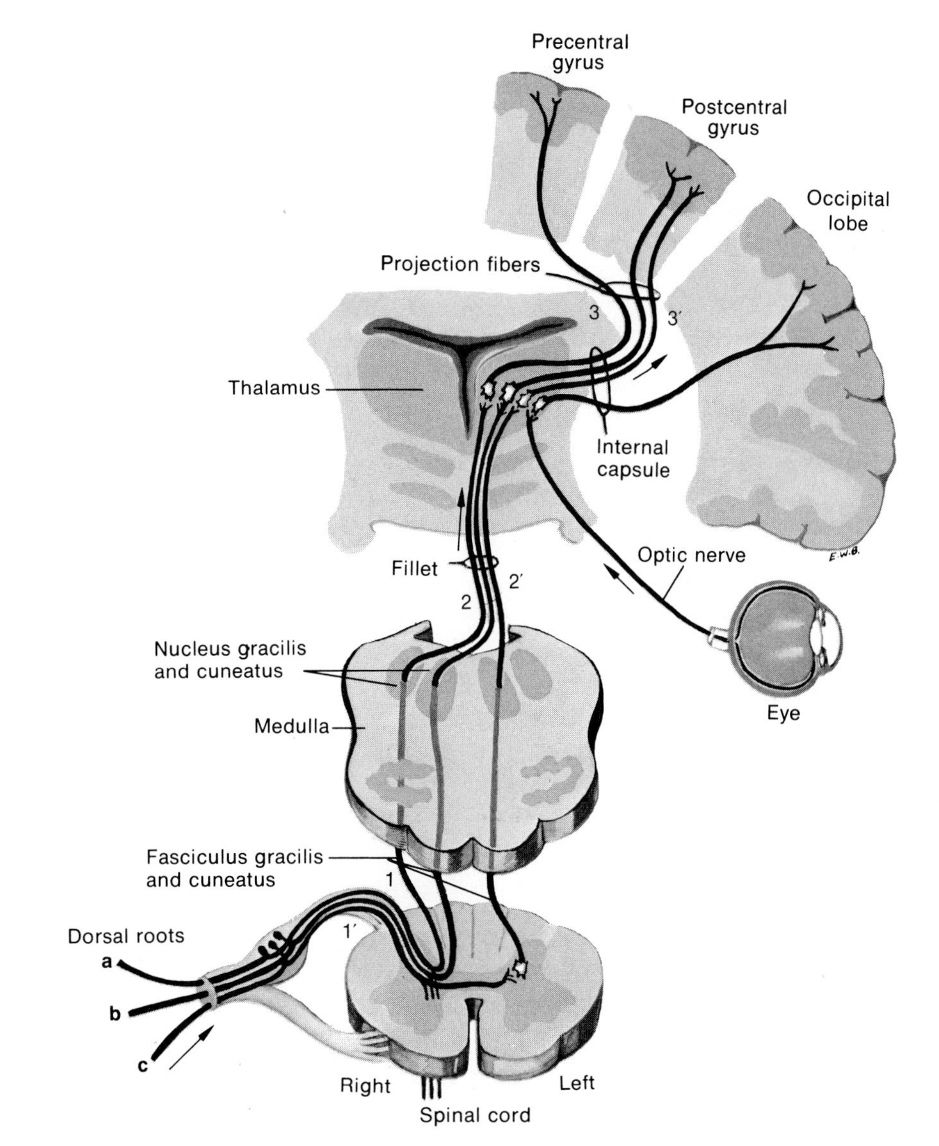

Fig. 10-9
Illustration of the most important ascending neural pathways between the left side of the spinal cord and the cerebral cortex.

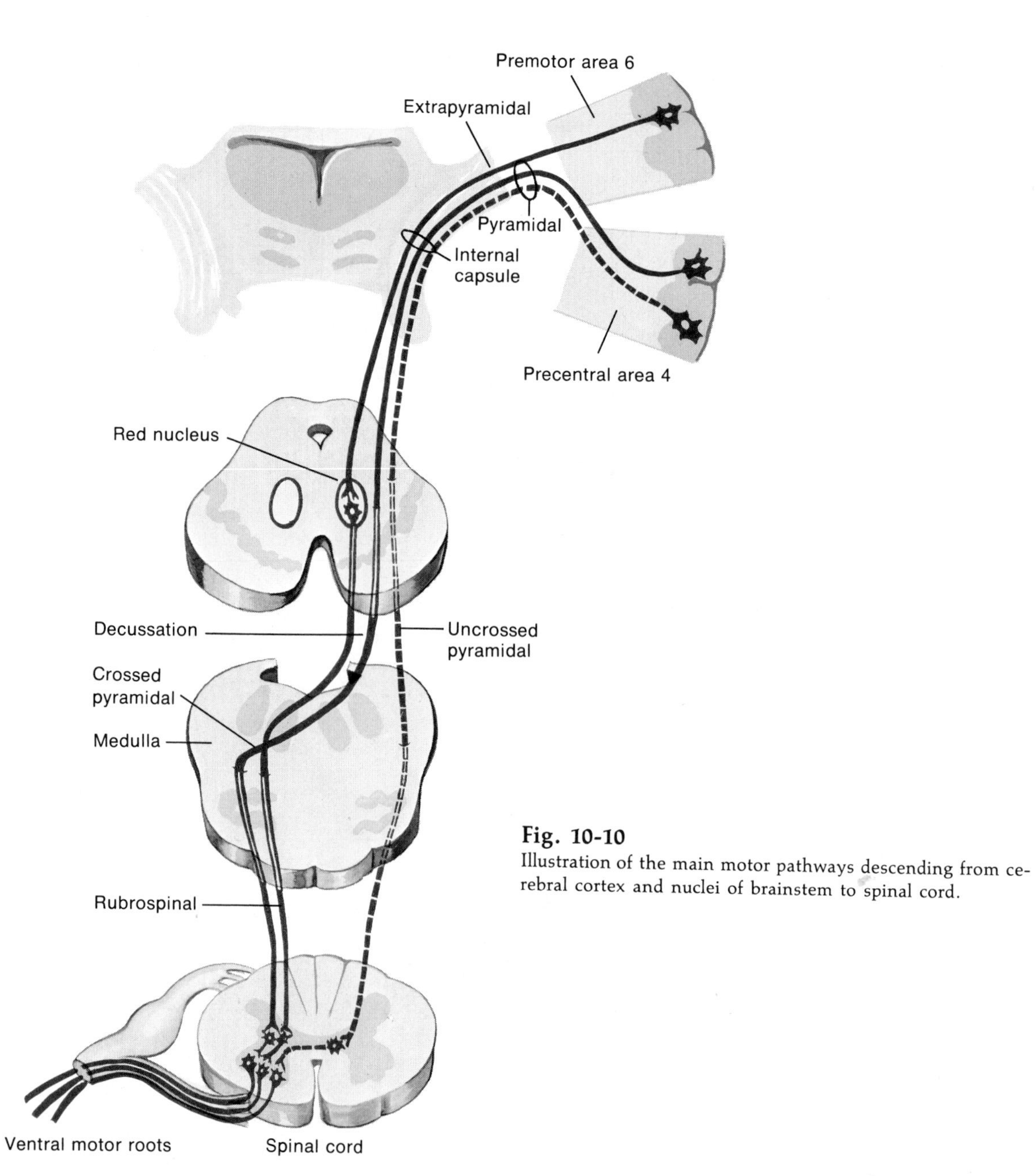

Fig. 10-10
Illustration of the main motor pathways descending from cerebral cortex and nuclei of brainstem to spinal cord.

the left of the body are usually received by the right cerebral cortex; impulses leaving the left hemisphere are generally dispatched to the right side of the body *(crossed pyramidal tract),* although an uncrossed tract exists.

5. The cerebral cortex can be influenced by a peripheral organ and can influence this organ only by way of the central gray matter. The more recently acquired cerebral cortex does not supplant, but supplements, the older central gray matter.

Pathways into the cortex. ■ Impulses from the skin, *a* (Fig. 10-9) (pressure, touch, and those leading to tactile localization and discrimination), and impulses, *b*, from the muscles, tendons, and joints (kinesthetic sensations of position and movements of limbs) enter the cord by the afferent fibers in the dorsal roots. These fibers of the first order extend upward in the fasciculus gracilis and fasciculus cuneatus (Fig. 8-13) of the cord; they terminate in two masses of gray matter in the medulla, the *nucleus gracilis* and the *nucleus cuneatus.* Synaptic connections are made with cells, the axons of which (Fig. 10-9, *2*), after crossing over to the opposite side of the medulla, proceed upward through the medulla, pons, and midbrain and end in the thalamus. This bundle of fibers of the sensory neurons of the second order is known as the *fillet* or *lemniscus.* On its way through the brainstem, the fillet is joined by sensory fibers of the fifth, seventh, and ninth cranial nerves and distributes collaterals to the reticular formation. The fibers of the third order originate in the thalamus and convey the impulses to various parts of the cerebral cortex, mainly to the postcentral gyrus. These are part of the sensory projection fibers (Fig. 10-5, *C*).

The dorsal root fibers (Fig. 10-9, *c*) carrying impulses of temperature and pain (and perhaps to a limited extent, of touch) soon after entering the cord cross over, *1′*, to the opposite side. Here synaptic connections are made with the neurons of the second order, *2′*, which ascend the cord in the spinothalamic tract (Fig. 8-13), join the fillet, and end in the thalamus. As previously described, the third order of neurons conveys the impulses to the cortex.

The optic nerve fibers (Fig. 10-9) end in the lateral geniculate body, which is physiologically part of the thalamus. From here the next relay, the visual radiation, takes the impulses to the occipital lobe (Figs. 9-20 and 10-5, *D*). Impulses from the cochlea of the ear are received by the medial geniculate body and are brought to the superior temporal convolution by another relay.

The afferent fibers of the third order on their way to the cortex go through a narrow passage, the internal capsule, between masses of gray matter (Fig. 10-7). On their way from the capsule to the cortex these fibers spread out, like a funnel, and form the afferent projection fibers shown in Fig. 10-5. Collectively this mass of fibers is sometimes spoken of as the *corona radiata.*

Pathways out of the cortex. ■ The fibers that carry impulses out of the cerebrum fall into two classes: the *pyramidal* and the *extrapyramidal.*

Pyramidal system. The pyramidal, or corticospinal, fibers originate chiefly in the precentral convolution (Figs. 10-1, *A;* 10-5, *B;* and 10-10). This region is also designated as area 4 (Fig. 10-8). Leaving the cortex, these fibers pass through the internal capsule. They then proceed down the brainstem to the medulla. On the way, some fibers synapse with motor nuclei of the efferent cranial nerves. The remainder of the fibers proceed to the medulla where the majority of them cross over to the other side (decussation of the pyramids—Fig. 10-10). They continue down the cord in the crossed pyramidal or lateral corticospinal tract (Figs. 8-13 and 10-10) and end at various levels. By means of internuncial neurons, their impulses are discharged to the spinal cells in the anterior horn and are brought, by the fibers in the ventral roots, to the skeletal muscles. Most, but not all, impulses from the cerebral cortex innervate contralateral muscles. The pyramidal fibers from the cortex that do not cross over in the medulla pass down in the direct pyramidal or ventral corticospinal tract (Figs. 8-13 and 10-10) and cross in the cord before making synaptic connections with the cells of the anterior horn.

The pyramidal system is phylogenetically new and prominent in primates. In its passage through the medulla, the fibers are collected into pyramid-shaped tracts, thus the name pyramidal tract. About 75% of the fibers cross at the decussation. The terminals of corticospinal fibers are primarily excitatory for spinal cord motoneurons, especially those concerned with precise and delicate movements of the limbs and hands. Destruction of the pyramidal tracts is seldom a pure achievement, for the fibers lie close to or are mingled with those of other systems. However, lesions in the medullary pyramids can be precise and cause weakness (paresis), but not paralysis, of the limbs and hands of the opposite side of the body. Because other systems support the final common pathway neurons in the spinal cord, recovery to a marked degree occurs, and after several months the functional defect may be virtually undetectable.

Extrapyramidal system. The fibers of the extrapyramidal system originate largely (but not exclusively) in the premotor area (area 6, Fig. 10-8). Having passed through the internal capsule, they travel to the basal ganglia and several levels of the brainstem. We will use the corticorubral and rubrospinal fibers as an example. (For other tracts, consult special texts.) The corticorubral fibers (Fig. 10-10) end in the red nucleus in the midbrain. The fibers originating here proceed down the brainstem, decussate, and travel down the cord in the rubrospinal tract (Fig. 8-13). At various levels, the fibers synapse with motor cells.

The connections of the extrapyramidal system in the brainstem reticular formation are with excitatory and inhibitory nuclei. Since the fiber relays from the brainstem do not enter the medullary pyramids, these fibers are termed extrapyramidal; however, in the spinal cord there is considerable intermingling of the two systems. The commonest lesions of the higher levels of the CNS involve both pyramidal and extrapyramidal fibers, for example, in the internal capsule. Effects of the extrapyramidal defect are most evident, but how the defect is expressed depends on the extent of the lesion and on the relative loss of the inhibitory and excitatory components of the system. Lesions may result, because the fibers cross the cord, in *hemiparesis* (weakness on one side) or *hemiparalysis* (paralysis on one side). Although the lower facial muscles are affected, this is not true of upper facial muscles, which have bilateral representation in the hemisphere. Sometimes lesions produce a large deficit in the brainstem inhibitory regulation of reflex arcs. Consequently, the stretch reflex is overactive, and a degree of hypertonia known as *spasticity* appears in the muscles. Naturally, a lesion in the spinal cord, after the fibers have crossed, leads to flaccid paralysis, for all excitatory or inhibitory inputs to motoneurons disappear.

■ Localization of functions in cortex

We have learned that, after proper relaying, the impulses from the sense organs are brought to certain quite circumscribed areas of the cortex; these are collectively known as *sensory areas.* On the other hand, in other cortical areas are located the neural cells that discharge impulses to the muscles; these are *motor areas.* Stimulation of these two areas in a conscious person (under local anesthesia) gives rise, respectively, to a sensation or to muscular activity. Disease or removal of any of these areas results, respectively, in a disturbance in sensation or in the inability to execute voluntary movements. Both these defects occur chiefly on the side of the body opposite to the lesion.

Sensory areas. ■ Sensations are unusual experiences in man, but they do not occur if the cere-

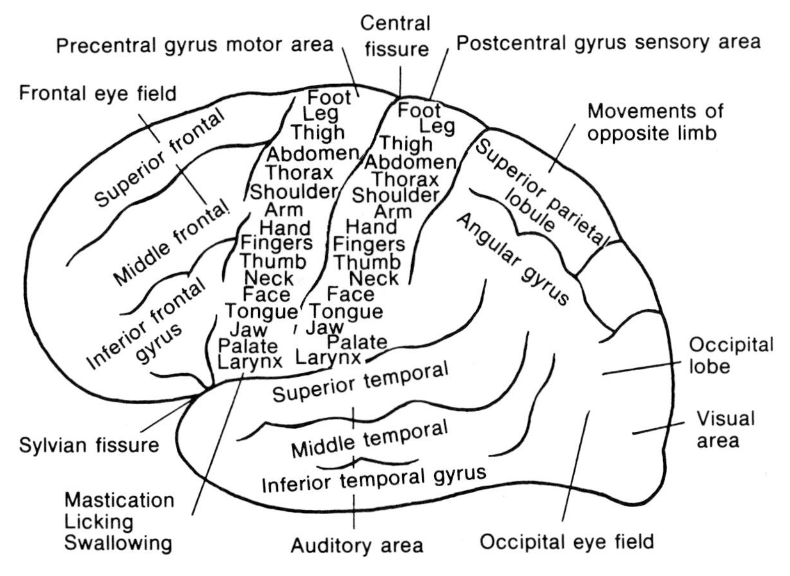

Fig. 10-11
Lateral view of cerebral hemisphere showing localization of cortical motor and sensory functions.

bral cortex is temporarily interfered with (anesthesia) or permanently destroyed. The same difficulties are encountered in explaining sensations, as in explaining other psychic and many physical phenomena. We will consider six sensory areas.

Somesthetic or body-sense area. Proprioceptive impulses from muscles, tendons, and joints, the impulses from the touch, heat, and cold receptors of the skin, and the deep sensibilities from the underlying tissues are brought by previously described pathways to the contralateral postcentral convolution of the parietal lobe (Figs. 10-5, 10-9, and 10-11). This area is therefore known as the somesthetic (*soma,* body; *aisthesis,* sensation) or body-sense area. From Fig. 10-11 it will be seen that a point-to-point representation of the various sensory parts of the body surface exists in the postcentral convolution. Neighboring cortical areas (most of the parietal lobe) are also concerned in this function. A conscious patient, in whom the postcentral areas were stimulated, experienced tingling and tactile sensations that he projected to specific parts of the opposite side of his body; he also felt sensations of movements of parts of his body, but no actual movements occurred. On the other hand, stimulation of a certain part of the skin evoked electrical changes in the corresponding area of the postcentral convolution.

The somesthetic area receives impulses by which we become aware (1) of the position of our limbs in space, (2) of the particular area of the skin stimulated (tactile localization), (3) of the number of discrete points of the skin stimulated (tactile discrimination), and (4) of the relative weight of objects placed upon the body. By the proper evaluation of these various items of information and by the proper synthesis of the necessary factors, we arrive at a conclusion as to the weight, texture, size, and shape of an object; that is, we have three-dimensional knowledge or stereognosis (*stereo,* solid; *gnosis,* knowledge).

Destruction of a certain sensory area may not

completely abolish the sensation normally attributed to this area. A crude, dull sensation perhaps remains; its modality can be recognized. It lacks, however, the characteristics of quality, intensity, and extensity by which it is differentiated from other sensations of the same modality. These residual sensations are generally credited to the thalamus. Normally, it is believed the thalamus transmits the impulses it has received from the receptors to the sensory cortex. Here these crude sensations are brought under the influence of impulses simultaneously received from other, but relevant, sources, and of memories of past impressions; this transforms it into a finished sensation, having the characteristics previously described.

Visual area. The visual area (Fig. 10-11) occupies that part of the occipital lobe known as the cuneus and receives impulses from the retina by way of the diencephalon and optic radiation (Fig. 10-5, *D*). When the upper part of this cortical area is stimulated, the eyes are turned downward; stimulation of the lower part causes the eyes to be turned upward. Destruction of one occipital lobe results in blindness of the corresponding halves of both retinas and therefore the field of vision of the opposite side. The reason for this lies in the semidecussation of the optic nerves, as illustrated in Fig. 9-20. From this figure it will be seen that destruction of the left occipital lobe (fibers in red) must cause blindness of the left half of both eyes and blot out the right field of vision for each eye.

Surrounding the visuosensory center (Fig. 10-8, *17*) lies a field, *18*, called the visual association area. Destruction of this area does not cause blindness, but the person loses the power to recognize what he sees; this condition is called mind or psychic blindness.

Auditory area. The impulses from the cochlea of the ear are brought by a circuitous route to the superior temporal lobe (Fig. 10-5, *E*). Each ear is bilaterally represented, but most of the impulses are carried to the contralateral lobe (Fig. 10-11). Stimulation of the lobe on one side causes the animal to prick up its ears and turn its eyes and head in the opposite direction. The cochlea is represented in this area in a point-to-point manner.

Taste. The receptive area for taste lies in the lower part of the postcentral convolution (Fig. 10-11); it is situated close to the sensory area for the tongue and near the motor centers for the muscles of mastication and deglutition.

Smell. There is uncertainty as to which part of the cortex receives the impulses from the olfactory sense organ. There is no olfactory representation in the thalamus as there is for other sensory modalities.

Pain. Pain impulses are said to terminate in the postcentral convolutions. However pain sensations have not been evoked by stimulation of the cortex in a conscious person, and neither does removal of the cortex abolish them. Decorticate animals show exaggerated pain reaction. This seems to be brought about by the thalamus, which, in conjunction with the cerebral cortex, may be regarded as the normal primary receptive center for pain sensations.

Motor areas. ■ The skeletal muscles are represented in the cortex chiefly by two areas: the primary and the secondary motor areas. The area in which the pyramidal (corticospinal) fibers primarily have their origin is located in the precentral convolution (area 4) of the frontal lobe (Figs. 10-8 and 10-11). The secondary area occupies the premotor area (area 6); most of the extrapyramidal fibers issue from this part.

Precentral motor area. The precentral motor area (area 4) is divided into many smaller areas (Fig. 10-11) in which the various parts of the body are represented in a definite order,* chiefly contralaterally. It is here that the pyramidal system originates. Each subarea is associated with a definite and restricted group of muscles so that

*An upside-down map of the body.

the contractions resulting from the stimulation of any one area are limited to a few or to only one muscle. The higher the animal stands in the phylogenetic scale, the more minute are these subdivisions; this concerns especially the movements of the distal parts of the limbs and the mechanism of phonation. To illustrate: the cortical area allotted to the fingers (and more especially in man to the highly opposable thumb) is larger than that for the entire trunk. This makes possible the more refined and delicate voluntary movements and actions acquired later in life, such as writing and speaking.

Premotor area. The premotor area (Fig. 10-8, area 6) is phylogenetically an older system than is the precentral area. The extrapyramidal system begins in this area. It differs from the latter in two respects. It is considerably less excitable (having a higher threshold), and each center innervates a larger group of muscles. It therefore controls the grosser movements of the body, for example, postural adjustments. It lacks the high degree of discreteness exhibited by the precentral areas.

In man ablation or pathologic lesion of the precentral convolution or of its projection fibers results in the contralateral loss of finer movements (e.g., of the fingers or face); the grosser movements (as those of a limb) are retained. If the lesion is more extensive, the premotor area may be involved, and the paralysis affects also the grosser movements. Such lesions may be the result of a thrombus or embolus in a blood vessel (nutritional disturbances) or of the rupturing of a vessel (mechanical disintegration of brain tissue); this is known as apoplexy or *stroke.** Since the afferent and efferent spinal nerves are unimpaired, reflex actions are possible. The neuromuscular disturbances may gradually subside somewhat, other parts of the motor areas (e.g., in the opposite hemisphere) taking over the lost functions. In this condition, the skilled movements recover least.

Broca's area. The third (inferior) frontal convolution (Fig. 10-11), also known as Broca's convolution, is concerned with speech. Although lesions in this area on the left side in a right-handed person cause impairment in his speech (motor *aphasia*),* this area must not be looked on as a speech center. Communicating one's thought by speech is an exceedingly complex action in which many parts of the cortex are concerned; Broca's convolution is but one of these. These lesions are generally accompanied by some mental disturbances. An impairment in the ability to express one's thoughts in writing is known as *agraphia.*

Prefrontal lobe. ■ A large amount of frontal cortex, areas 9 to 13 (Fig. 10-8), does not function as a sensory area; and it does not discharge impulses to skeletal muscles. It is difficult to assign specific functions to these areas (collectively termed the *orbitofrontal cortex*). They are not silent areas, as was believed until recently, and they are not the seat of intelligence as is commonly supposed. Two phases of their activity may engage our attention.

From recent investigations it appears that the orbitofrontal cortex is in close neural connection, both afferently and efferently, with the thalamus and, through it (and perhaps directly also), with the hypothalamus. Since the hypothalamus is intimately associated with visceral functions, this neural linkage enables the cerebral cortex to exert influence on the autonomic functions. For example, stimulation of areas 8 and 9 increases gastric secretion and gastrointestinal movements. Stimulation of area 13 has been observed in man to cause noticeable changes in respiration and blood pressure.

Limbic system. ■ To this point we have discussed the brainstem and some of its nuclei and the cortical regions involved in sensory and

*Most often, however, this takes place in the internal capsule and is the result of arteriosclerosis.

*The reverse is true in left-handed persons.

motor functions. These cortical regions, *neocortex,* are the highest expression of phylogenetic development of the nervous system, but between them and the brainstem lies the older more primitive cortex, the *archicortex,* which is a part of the system known as the *limbic system.* This system consists of the hippocampus, parts of the orbitofrontal and temporal lobes surrounding the Sylvian fissure (Fig. 10-1), the cingulate gyrus (Fig. 10-8, *24*), some of the thalamic and hypothalamic nuclei (Fig. 10-3), and parts of the basal ganglia, including the amygdaloid nucleus. Many of the specific structures named above are not labeled in the illustrations, since it serves our purpose to regard the brain as consisting, in some respects, of three concentric circles expanding outward from the upper end of the spinal cord. The core is the brainstem; it is surrounded by the limbic system (the central ring) and most of the cerebral cortex, or the *neocortex* (the outer ring).

The limbic system has extensive two-way interconnections with the cerebral cortex via the thalamus and with the reticular formation of the brainstem. Its own components are intraconnected. Retention of recent memories appears to be one function of the hippocampus; permanent memories are stored in the neocortex, particularly perhaps in the temporal lobe. The hippocampus also has roles in control of attention through its relationship to the reticular formation and the latter's arousal-attention function, and it exerts a tonic influence on the pituitary-adrenal stress mechanism. Since stimulation of the amygdala in experimental studies gives rise to a variety of visceral sensations and autonomic reactions, this part of the limbic system is considered to modulate the activity of the hypothalamus with respect to such behavior and emotional patterns as aggression, sexual, eating, and drinking. With the hippocampus supplying coded and analyzed data concerning the environment and past experience and the amygdala providing information about visceral events, the limbic system as a whole may serve to select the appropriate behavior to any given set of stimuli, thus forming a substrate for conditioned reflexes. Psychosomatic reactions, visceral responses to emotional states, are components of limbic system activity.

■ Electroencephalogram—brain waves

The electrical activity of the brain was first described in 1875 by Caton, an English physiologist. The electrical potentials were similar to the electrical activity seen in muscle and nerve during activity. In 1929 Hans Berger, a German neuropsychiatrist, reported that rhythmical electrical activity could be recorded through the intact skull of man. Subsequent work has shown that a wide spectrum of frequencies is present in the electroencephalogram (EEG). These have been divided into *alpha* (8 to 12/sec), *beta* (18/sec and faster), *theta* (4 to 7/sec), and *delta* (1 to 3.5/sec). The exact pattern of waves and the dominance of frequencies depend on many variables. Age is a factor; stable adult patterns in the predominantly *alpha* range are not achieved until late adolescence. The area of the brain studied is a crucial variable (Fig. 10-12), *alpha* activity being predominantly posterior (occipital) in origin and the anterior areas usually showing low voltage *beta* activity. *Theta* and *delta* activity are associated with younger subjects, with sleep or states of impaired consciousness, and with various types of brain pathology or pathophysiology. The EEG is used clinically to investigate epilepsy, infectious diseases, trauma, tumors, and other pathologic cerebral states.

■ Metabolism of the brain

The metabolism of the brain shows several marked peculiarities. Although constituting only 2.2% of the body weight, the adult brain seems to contribute from 8% to 10% of the basal metabolism. The brain obtains its energy almost exclusively from carbohydrates; its respiratory quotient is therefore 1. Since its reserve of car-

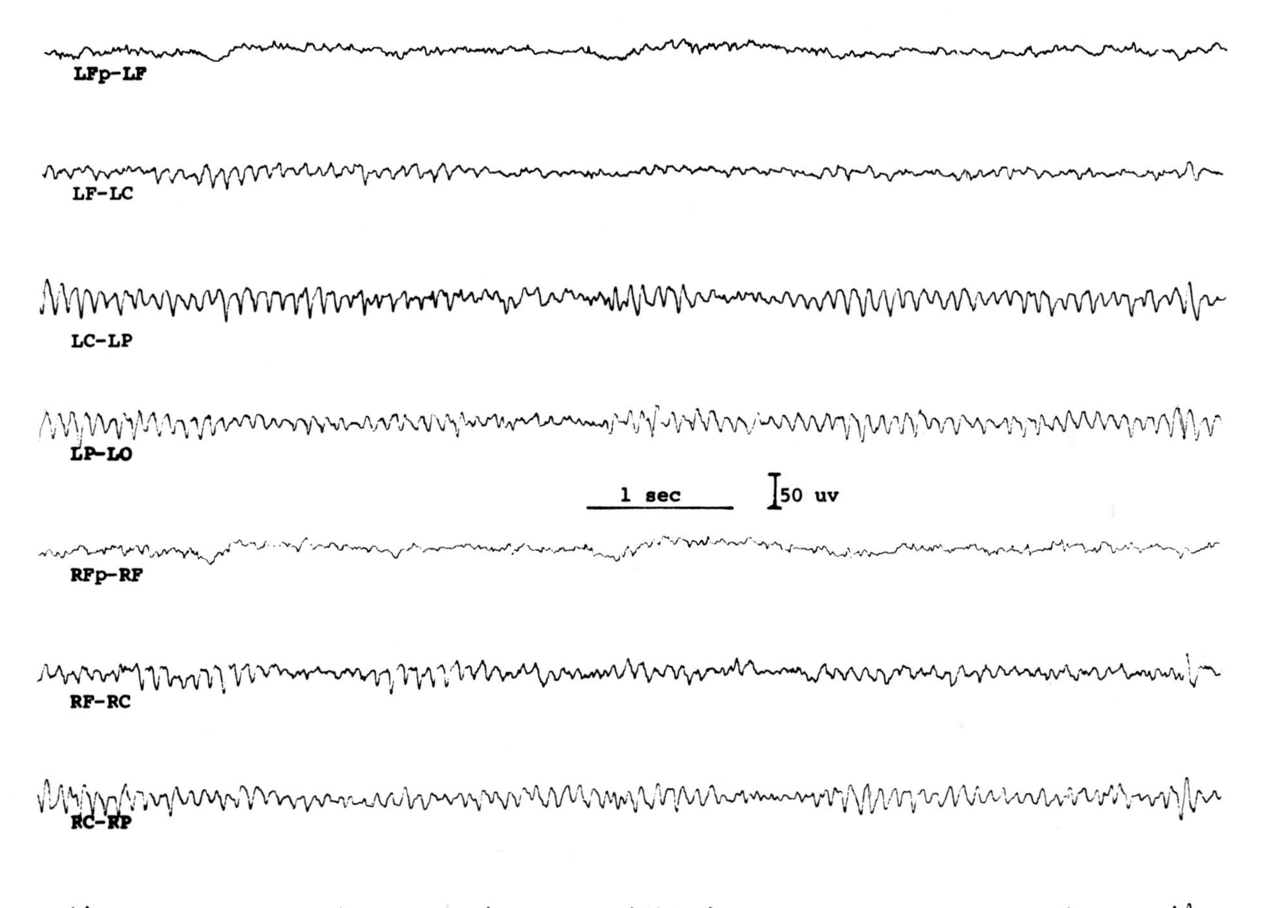

Fig. 10-12
Normal electroencephalogram from young adult male. Traces from top to bottom are referred to as channels 1 through 8: Ch. 1, left frontal polar to frontal (precentral); Ch. 2, left frontal to central; Ch. 3, left central to parietal; Ch. 4, left parietal to occipital; Ch. 5, right frontal polar to frontal (precentral); Ch. 6, right frontal to central; Ch. 7, right central to parietal; Ch. 8, right parietal to occipital. Vertical bar = 50 μv; horizontal bar = 1 second. Ten to $10^1/_2$/sec *alpha* rhythm is dominant in the posterior areas (3 and 4, 7 and 8). Lower voltage *beta* is seen in the frontal polar to frontal derivations (1 and 5) and is interspersed with *alpha* in other channels. No dominant *theta* or *delta* activity is seen in the waking electroencephalograms of normal adults.

bohydrates is very limited, a decrease in the supply of glucose causes mental disturbances; a blood sugar below 40 mg/100 ml (as in overdose of insulin) induces mental confusion, dizziness, convulsions, and even loss of consciousness. Cortical activity is greatly dependent on vitamin B_1; without it, the oxidation of carbohydrates does not go beyond pyruvic acid.

The amount of oxygen consumed by the adult human brain varies from 40 to 50 ml/min; the centers in the medulla and spinal cord consume less than the cortex. A few moments' interruption in the oxygen supply brings about unconsciousness, and complete deprival causes irremediable damage to most cerebral cells in about four minutes.

With the exception of the spleen, the brain shows a higher concentration of the amino acid, glutamic acid, than does any other organ. That glutamic acid plays an important part in the metabolism of human brain tissue is shown in that the human brain keeps up its consumption of oxygen in the absence of glucose if glutamic acid is available.

Of all mammalian tissues, the brain has the highest level of adenylcyclase; it is especially concentrated in synaptic membranes. Stimulation of brain cells causes a rapid elevation in cAMP. Some recent experiments indicate that "learning" is associated with increased cAMP and RNA synthesis.

■ Sleep

The most significant feature of sleep is a loss, to a greater or lesser extent, of consciousness, a function generally associated with cortical activity. From it the person can be aroused by ordinary, harmless stimuli; in this it differs from those conditions that simulate sleep; for example, coma, hypnosis, and syncope. Sleep is composed of two different states: (1) slow wave sleep (SWS), characterized by slow synchronized EEG patterns, and (2) paradoxical sleep (PS), in which low-voltage fast waves are associated with rapid eye movements and reduced skeletal muscle tonus.

That during sleep the irritability of the sense organs is decreased and that the individual reacts less readily and less critically to changes in his environment is common knowledge. In light sleep (SWS) the kinesthetic reflexes may be increased due to loss of cerebral restraint, as shown by twitchings and jerkings of the limbs; but in deep sleep (PS) they are greatly reduced. The patellar reflex is greatly reduced, if not entirely abolished. Other reflex centers (e.g., the photopupil constricting center in the midbrain and the vital centers in the medulla) are little affected. All parts of the cortex may not be under the influence of sleep to the same extent at all times, for although a comparatively loud noise of no importance to the sleeper may pass unnoticed, the slight noise of guarded footsteps may awaken him instantly.

Cause of sleep. ■ Many theories concerning the cause of sleep have been formulated. One of these deals with the lessening or cessation of afferent impulses streaming into the cerebral cortex. It is, of course, well known that darkness, the absence of sound, and a comfortable bed (one by which no part of the body is excessively stimulated by pressure) are external conditions highly favorable to sleep and that to fight off sleep the muscles must be kept in action. Investigators concluded that the lessening of proprioceptive impressions from the muscles, either by fatigue or voluntarily, was a large factor in precipitating sleep.

The theory that sleep is induced by a general inhibition exercised by the hypothalamus has considerable support. In the pathological condition of epidemic encephalitis (inflammation of the brain), the patient is in a state of deep and prolonged sleep—sleeping sickness. In this disease the posterior nucleus of the hypothalamus of the diencephalon is involved. Electrical stimulation of this part in experimental studies produces sleep and suggests that in the hypothalamus there exists a *sleep center.* Transverse section of the hypothalamus in rats causes a loss of the normal rhythm of sleep. There is probably a *waking center* also in the brainstem. Somnolence, whether normal or abnormal, would be caused by the withdrawal of impulses from a waking center in the brainstem. If both a sleeping and a waking center exist, the relation between them is somewhat hazy. Nevertheless, sleep is more than just a passive damping of the reticular formation activating system; it is an active phenomenon. Biochemical mechanisms participate; an increase in brain serotonin accompanies SWS. A decrease in serotonin and norepinephrine leads to the disappearance of both SWS and PS.

Syncope. ■ In syncope there is a complete and sudden loss of consciousness. The causes of fainting are varied and the condition is not fully understood. Most frequently it is due to cerebral anemia caused by cardiac or vasomotor disturbances. The retarding influence of gravity on venous circulation or hemorrhage may induce syncope.

■ Integration

In general, the activities of the living body are adaptive; that is, the responses to environmental changes (both internal and external) are of a purposive nature, the purpose being self-preservation and the continuation of the race.

Selecting from a number of conflicting responses that single response or combination of responses leading to adaptive or purposive action is designated as integration.

Importance of afferent impulses. ■ As in the activity of the lower parts of the nervous system, we have no concrete evidence that the cerebrum has the power to discharge impulses without having been stimulated. This is well illustrated by the inability of a monkey to use its arm when the afferent nerves are completely severed. Since the stimulation of the proper motor area in the cortex causes the usual response, the paralysis of the arm is due to lack of afferent impulses from skin and from muscle spindles and tendon end organs. As far as neuromuscular activity is concerned, it is initiated by a stimulus. This is also true for psychic life.

Point-to-point representation: association fibers. ■ The point-to-point relation between the sensory surfaces of the body and the receiving cortex and a similar relation between the motor cortex and the effectors make possible the discrete and narrowly circumscribed activities (e.g., of a single finger) following the stimulation of a limited area of a sensory surface. This, however, demands an almost countless number of association (internuncial) neurons by which almost any part of the cortex is directly or indirectly brought into contact with every other part. *It is in the number of these cells and in the more varied connections they make that the human brain excells that of lower animals.* It is this, among other important factors, that determines man's greater manual skill, his power of speech, and, no doubt, his greater intelligence.

Reinforcement of impulses. ■ The reinforcement of impulses from different receptors is an essential factor in many complex, highly integrated activities. This is well demonstrated by the act of standing. In assuming and maintaining the erect position of the body, at least four distinct afferent impulses are involved: visual, labyrinthine, kinesthetic, and, to a lesser extent, tactile. A normal person, closing his eyes, finds no difficulty in maintaining his position because the remaining afferent impulses are sufficient to enable the cortex to discharge adequate motor impulses. But in the pathological condition of tabes dorsalis, the kinesthetic sensations are lost. The patient on closing his eyes is unable to stand, no matter how desirous he may be of doing so. The inability to perform this motor act under these conditions is not due to paralysis of the motor cortex or its efferent fibers but to the lack of adequate incoming impulses.

In view of these facts, it is hardly correct to

speak of the cortical motor areas as the seat of volitional impulses. The impulse for performing a certain muscular act, such as the closing of one's hand, is not formulated in that small cortical area, the artificial stimulation of which results in this action. A motor area in the precentral convolution merely represents a point in the afferent-efferent nerve pathway of a reflex arc and must not be regarded as endowed with any mysterious function not possessed by other parts of the arc.

Cortical inhibition. ■ An indispensable factor in integration is the restraint or inhibition that the cerebral cortex constantly exercises over its own activity and over that of lower levels of the nervous system. The greater the number of possible connections (by association fibers) between the afferent and the efferent cortical pathways, the greater is the need of this faculty. In the cortex are areas especially concerned with inhibition. Of these we may mention band 4s (a special, narrow strip in area 4), area 8, and area 24 (Fig. 10-8). These areas are known as suppressor areas. When movements are evoked by the stimulation of some part of area 4, subsequent stimulation of a suppressor area inhibits the muscular action. On the other hand, extirpation or lesion of a limited amount of a suppressor area is associated with spasticity (i.e., hypertonicity) of some muscle or muscles on the opposite side of the body.

The power of the cortex to inhibit the flow of certain impulses while others are allowed free play is the basis of the phenomenon of attention. In mental concentration all afferent impulses that threaten to interfere with the prosecution of the thing we are after must be repressed, and the passage of all helpful impulses must be facilitated. Extraneous information reaching the reticular formation is probably, for example, suppressed by hippocampal action. The role inhibition plays in our physical, mental, and social life can scarcely be overestimated. However, too great a general inhibition may work disastrously. Fear may exercise more powerful inhibition on cerebral activity than does any other single factor; it thereby restrains the individual from doing well, or at all, that for which he may be naturally fitted (inferiority complex). Being set free during anoxia from the control exercised normally by the higher centers allows unrestricted activity of the lower centers. In this respect anoxia acts like alcoholic intoxication. When severe anoxia is recovered from, the only organ that has been seriously damaged is the central nervous system. The phylogenetically and ontogenetically* late development of the cerebral cortex, especially in its function of inhibition, readily explains much of the physiological and mental behavior of an infant or child, notably its lack of sustained interest.

Plasticity and fixity of the nervous system. ■ Another factor responsible for the vastly expanded activity of the cerebral cortex and for its high degree of integration lies in the great plasticity or flexibility of the cortical gray matter as compared with the fixity of the central gray matter. The neural pathway for afferent and efferent impulses through the central gray matter of spinal cord and brainstem is fixed. Although impulses may be reinforced or inhibited, new ways of responding to any given sensory stimulation are never acquired during the life of the individual animal except by way of the cerebral cortex.

In a newborn child the growth and development of the cortical gray matter and its neural connections lag far behind that of the central gray matter. The activities mediated by the spinal cord and the brainstem, such as suckling, swallowing, breathing, crying, urination, and the photopupil reflex, are exhibited immediately after, or soon after, birth. The reflex patterns for these activities are complete. Like the so-called instinctive actions of lower animals, these activities, fortunately, are not dependent on previous experiences of the individual. In contrast with

*Ontogeny is a brief recapitulation of phylogeny.

the fixity of the central gray matter, the cortical gray matter can acquire an entirely new method of response to a given stimulus, as is illustrated in the formation of what Pavlov called *conditioned reflexes.*

■ Reaction time

Reaction time is the length of time required to respond voluntarily to the stimulation of a sense organ.

The average reaction time for sight is 0.25 sec; for hearing, 0.17 sec; for touch, 0.15 sec. It is increased by fatigue, mental excitement, and worry. It may be somewhat decreased by an increase in the intensity of the stimulus and to a slight extent by practice. The length of an individual's reaction time is of great importance in certain industries. In a dilemma, in which a choice must be made between two or more forms of stimulation, the reaction time *(choice reaction time)* is prolonged.

■ Conditioned reflexes

A hungry dog on being shown a piece of meat has an increased flow of saliva; the hearing of a musical sound has no such effect. But Pavlov caused a certain tone to be produced while a hungry dog was shown and subsequently given a piece of meat. After a number of repetitions, the responsiveness of the dog's nervous system was so altered that the hearing of the sound was followed, when the dog was hungry, by a flow of saliva, even though the meat was not shown or given. The dog learned to associate the hearing of this sound with the obtaining of a piece of meat; the flow of saliva is in anticipation, one might say, of a future stimulus—the receiving of meat.

The stimulus of eating meat (taste) which is immediately followed by a flow of saliva is known as an *unconditioning stimulus* (US), and the reflex evoked by it is an *unconditioned reflex* (UR); the sound stimulus to which the animal finally learns to respond by salivation Pavlov called a *conditioning stimulus* (CS) and the reflex following it, a *conditioned reflex* (CR). In establishing a CR, stimulation of various sense organs, such as skin, nose, ears, and eyes, has been used. Conditioned reflexes, employing the salivary gland, muscles, heart, blood vessels, alimentary canal, and so on, have been formed. In the developing fetus, infant, and child, a particular UR makes its appearance at a time when the general development of the various structures and functions is sufficiently advanced to make this new reflex possible and to render it necessary and profitable. It is possible for an adult person to outgrow a certain UR, and yet he transmits the potential for that particular reflex to his children.

Relation between conditioned and unconditioned reflexes. ■ The fundamental distinction between these two reflexes lies in the fact that the UR is inborn; the CR is never acquired spontaneously; a CS is always associated with an US in the formation of a CR. The CS may either occur simultaneously with, or it may precede, the US, but no CR is ever formed by a CS following an US. The shorter the lapse of time between the CS and the US, the more quickly the CR is established. A strong CS leads more rapidly to the formation of a new reflex, but if excessively strong, it hinders it. Again, an intermittent CS has greater effect than a continuous stimulus.

Weakening of conditioned reflexes. ■ When the CS that created the new reflex is not applied for some time, the CR becomes weakened but not entirely lost; it can be revived by a few applications of the conditioning and unconditioning stimuli. The oftener the application of both the CS and the US the more firmly established becomes the CR and the less rapidly it weakens or becomes extinct. If a CR has been established and the CS is then applied several times without the application of the US, the CR decreases and finally becomes extinct.

Specificity and discrimination. ■ When a musical sound of a certain pitch, for example, that of a tuning fork having 800 vibrations per second, is the CS, the animal responds not only to sounds of this pitch but also to those considerably higher or lower. But if, while the animal is being conditioned to the sound of 800 vibrations, another sound of either slightly higher or lower pitch is also used but is never accompanied with or followed by the giving of food, the animal is finally able to discriminate between a sound of 800 and one of 812 vibrations per second.

Summation. ■ It is possible also to cause summation of conditioned reflexes. A dog secreted 60 drops of saliva to the CS of smelling camphor; to the stimulation of the skin it responded with 30 drops; when the two stimulations were applied simultaneously, the yield was 90 drops in the same length of time.

Inhibition. ■ Conditioned reflexes are very susceptible to inhibition. When in a dog a CS (e.g., the stimulation of the skin) has been given and an unexpected sound occurs before the US is applied, the CR does not take place or does so only feebly (distraction). By repeating this sequence (CS ⟶ Noise ⟶ Food), the noise loses its inhibitory power; in common language, the dog learns to ignore the noise. Here the limbic system is exerting its suppressive power on the reticular formation alerting function.

Association. ■ *Learning always proceeds from the known to the unknown.* Let us suppose an animal is conditioned to a flash of light. If now simultaneously with this a second stimulus, say, a musical note, is applied, the animal associates the second stimulus with the first CS, so that, after a time, when only the note is heard without the light stimulus, the reaction still takes place (memory function of the limbic system). In this manner a secondary CR may be built on a previously acquired reflex. Dogs are not able to form CRs beyond the tertiary; in man there is no known limit.

Conditioned reflexes not inherited. ■ Although thousands of generations have acquired conditioned reflexes, yet each new generation must learn them for itself. *Acquired reflexes are not transmitted from parents to their offspring.* The acquired knowledge, skill, and education of a large number of ancestors have never given the newborn offspring a better start. Education must start "from scratch" with each generation.

Conditioned reflexes and cerebral cortex. ■ The cerebral cortex is necessary for the establishment of conditioned reflexes. As we may regard the stereotype reflex as the unit of spinal cord activity, so the CR is the unit of cortical activity. It appears that, at least in lower animals, the physiological and psychological neural activity in the formation and in the retention of a CR may be a more or less diffuse process in which a large part of the cortex, rather than a sharply localized area, is concerned.

The more complicated the problem that has been learned, the greater the loss of function following the extirpation of a given amount of cortex. To illustrate: In a dog conditioned to a sound stimulus, the removal of both temporal lobes (generally regarded as the site of hearing) had the following results:

1. The unconditioned sound reflexes, such as pricking up the ears, remained.
2. The CR based on sound stimulation of a simple nature, for example, the striking of a gong, was temporarily lost but did recover to a limited extent. It seems possible, in some instances, for neighboring areas to take over, by proper training, the function of the removed cortical area.
3. The CR based on more complex sound stimulation (e.g., the calling of the dog's name), which requires a greater power of analysis and discrimination, were completely and permanently lost. The results of these lesions are more drastic and are followed by less recovery in adult than in young animals.

■ Associative memory

To make possible the acquisition of a conditioned reflex, it is necessary to assume that the repeated CS, coupled in proper sequence with the US and its response, produces a more or less permanent change in a plastic neural mechanism. This residual imprint is called associative memory. As to the manner of the formation of these retained impressions and as to the physical or chemical changes the nervous system experienced in their formation, we have little knowledge.

At birth, the cerebral cortex is still largely undeveloped; the infant, as we have seen, is a reflex organism. For some time after birth the motor areas are not excitable. Many fibers to and from the brain lack myelin sheaths, and it is generally conceded that the function of a neuron does not appear until this structure has been completed. The pyramidal fibers are not fully myelinated until well along in the second year. When these and other connections between the cerebral cortex and the spinal cord have been completed, the child is ready to walk. Prior to this, the infant was able to use all these muscles, but the activities were random; they lacked cortical direction. The growth of the brain also finds expression in an increase in the number, length, and distribution of the dendritic processes by which the number of possible neural connections is greatly multiplied (Fig. 10-13). The outward expression of this is a gradual expansion of the range of cerebral activity, both sensory and motor. For how long a time during a person's life the possibility of opening new neural pathways (learning) continues is not known; there is reason to believe it never ceases.

Association areas. ■ The ability to modify its reactions and behavior seems to be common to most, if not to all, protoplasm. But this property finds its highest development in the cortical gray matter. Since man is the most teachable of all animate creation, in what respect does the human brain differ from that of the lower animals? When motor and sensory areas have been properly located, the greater part of the human cortex remains unaccounted for. Stimulation of

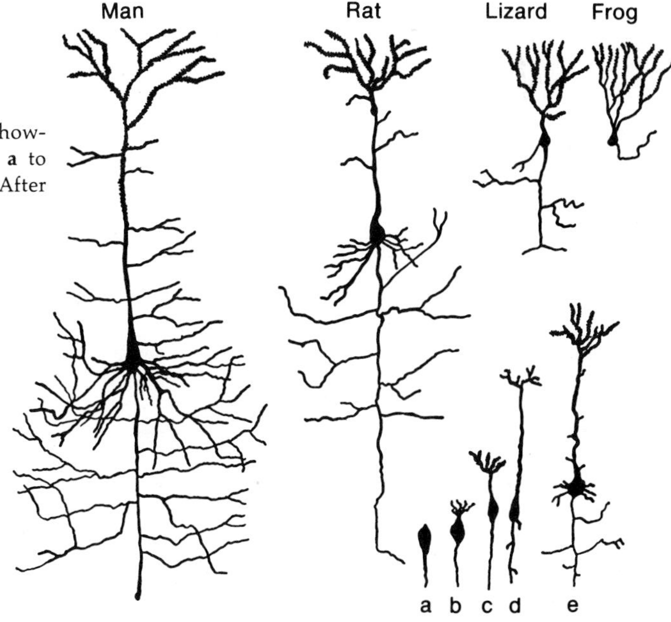

Fig. 10-13
Cells from cerebral cortex of frog, lizard, rat, and man showing progressive increase in size and complexity. Cells **a** to **e** show development of pyramidal cells in embryo. (After Ramon y Cajal.)

these parts evokes no muscular response, and their removal leads to no sensory paralysis. They are generally referred to as association areas; each lobe, the frontal, the parietooccipital, and the temporal, contains an association area. In the cortex of the association areas are located the cell bodies of the association fibers (Fig. 10-6), by which the sensory and the motor areas are linked together in every conceivable way. As animals rise in the scale of life and as they progressively display greater intelligence, the complexity of the cerebral neurons increases (Fig. 10-13), and the association areas become larger and the intricacy of the neural network more bewildering. This development finds its climax in man. It is therefore frequently thought that the neural machinery for the higher mental operations is to be found in the association areas. However, emotional and the various other phases of intellectual activities cannot be localized in circumscribed cortical areas. In pathological conditions (e.g., tumors) large areas can be destroyed without any great mental disturbance, and the extent of the lesion is of greater consequence than is its precise location. The prefrontal lobes (i.e., the area lying anterior to the premotor areas, Figs. 10-8 and 10-11) have been specially investigated, both experimentally and clinically.

In the brain operation known as lobotomy, no part of the cortex is removed, but the white matter within the prefrontal lobes, connecting the cortex with underlying structures (e.g., thalamus, hypothalamus, and limbic structures), is severed. This operation, resorted to for the relief of excessive anxiety or chronic depression and for intractable pain (e.g., of carcinoma), has in many cases proved highly beneficial. Freed of the excessive emotional impulses from these underlying structures, the previous state of fear, delusion, or melancholia gives way to a feeling of euphoria. Unilateral lobotomy does not adversely affect intelligence in most cases, but, in addition to the emotional changes already mentioned, the patient may be highly distractible,

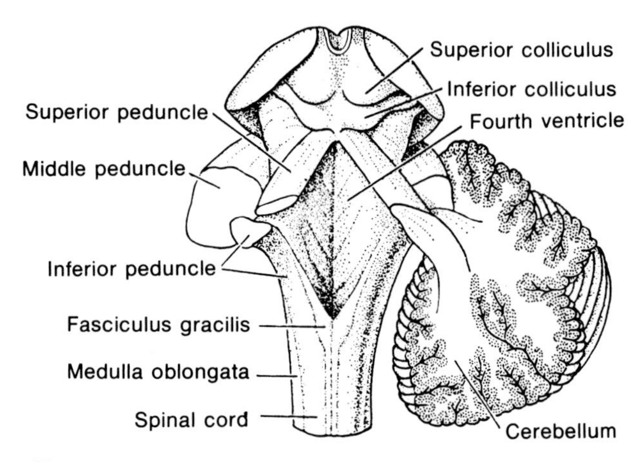

Fig. 10-14
Dorsal view of brainstem with part of right cerebellar hemisphere. Three cerebellar peduncles on the left side have been cut across.

may lack foresight, cannot critically evaluate himself, and shows altered social behavior; in short, he has a changed personality.

■ CEREBELLUM

The cerebellum forms a conspicuous part of the brain (Figs. 10-1, 10-2, and 10-14). Certain parts are phylogenetically very old; it finds its greatest development in man. Like the cerebrum, the cerebellum has a mantle of gray matter covering the internal white fibrous structure.

Neural connections. ■ The nerve fibers entering or leaving the cerebellum pass through three large nerve tracts: the superior, middle, and inferior cerebellar peduncles (Figs. 10-2 and 10-14). The cerebrum is concerned with the adjustment of the body to impulses received from the exteroceptors (e.g., eyes, ears); impulses to the cerebellum are chiefly derived from the proprioceptors, that is, the labyrinthine impressions from the inner ear and the kinesthetic impressions from muscles and tendons (Fig. 10-15). The efferent fibers are distributed to skeletal muscles through brainstem nuclei. It will be noted that

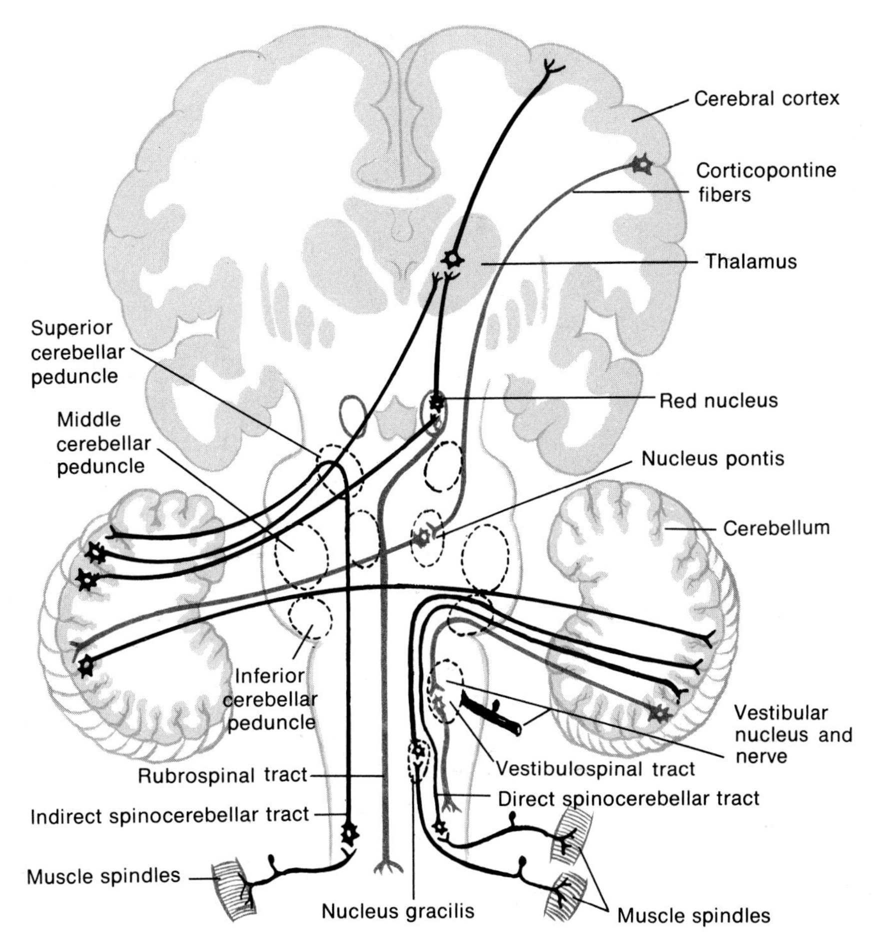

Fig. 10-15
Illustration of pathways in superior, middle, and inferior cerebellar peduncles and the neural connections of cerebellum with the spinal cord and cerebrum.

all these neural connections are made with the homolateral side of the body. With the cerebral cortex, the cerebellum makes afferent and efferent connections contralaterally; these connections are by way of the red nucleus and the thalamus. In this manner, both halves of the cerebellum are connected with both sides of the body.

Functions. ■ In contrast with the cerebrum, the cerebellum is not concerned with exteroceptive sensations, and its total extirpation in animals causes no paralysis. Its functions may be treated under three headings.

Equilibration. The removal or pathological condition of that part receiving the impulses from the vestibular mechanism causes disturbances in the muscular coordinations needed for equilibration. A person thus afflicted has a staggering, reeling gait *(cerebellar ataxia)* similar to that of an inebriated person. The flocculonodular area seems to be involved in motion sickness.

Postural reflexes. The function of the anterior

lobe is largely that of inhibition. When a certain area in the cerebral motor cortex is stimulated and simultaneously a definite area in the anterior cerebellar cortex is also stimulated at a slow rate, the muscle contraction initiated by the cerebral stimulation is inhibited. On the other hand, removal of the anterior lobe area increases the stretch or tendon reflexes and causes spastic contractions of the muscles. In a human being, lesions of this part cause voluntary movements to be jerky and associated with tremors.

Synergic function. Voluntary muscular activity is characterized by a high degree of coordination of several muscles (agonists, antagonists, and synergists) participating in a purposive action. There is a proper timing of the contraction of the individual muscles and their antagonists; the action is carried out with proper speed, force, and amplitude and results in a smooth and efficient action. The synchronous and orderly activity of groups of muscles is known as *synergia.* Synergia is more or less affected by lesions of the posterior lobe of the cerebellum; for example, in the finger-to-nose test, miscalculation of the rate, force, and direction of the contracting muscles causes the person's aim to be wide off the mark and to overshoot it—*asynergia.*

READINGS

Atkinson, R. C., and Shiffrin, R. M.: The control of short-term memory, Sci. Am. **225**(2):82-90, Aug. 1971.

Brazier, M. A. B.: The analysis of brain waves, Sci. Am. **206**(6):142-153, June 1962.

Geschwind, N.: The organization of language and the brain, Science **170**:940-944, 1970.

Heimer, L.: Pathways in the brain, Sci. Am. **225**(1):48-60, July 1971.

Hertz, L. The biochemistry of brain tissue. In Bitlar, E. E., and Bitlar, N., editors: The biological basis of medicine, vol. 5, New York, 1969, Academic Press, Inc., pp. 3-37.

Hobson, J. A.: Sleep: physiologic aspects, New Eng. J. Med. **281**:1343-1345, 1969.

Hubel, D. H.: The visual cortex of the brain, Sci. Am. **209**(5):54-62, Nov. 1963.

Kandel, E. R.: Nerve cells and behavior, Sci. Am. **223**(1):57-70, July 1970.

Llinas, R., and Hillman, D. E.: Physiological and morphological organization of the cerebellar circuits in various vertebrates. In Institute for Biomedical Research: Neurobiology of cerebellar evolution and development, Chicago, 1969, AMA Education and Research Foundation.

Luria, A. R.: The functional organization of the brain, Sci. Am. **222**(3):66-78, Mar. 1970.

Rosenzweig, M. R., Bennett, E. L., and Diamond, M. C.: Brain changes in response to experience, Sci. Am. **226**(2):22-29, Feb. 1972.

Sperry, R. W.: The great cerebral commissure, Sci. Am. **210**(1):42-52, Jan. 1964.

Anatomy
Characteristics
Synaptic transmission

THE AUTONOMIC NERVOUS SYSTEM

The adjustment of the body as a whole to external environmental changes is mediated largely by the skeletal muscles. These are innervated by somatic nerves. The visceral muscles and the glands of the vegetative organs whose activities are concerned with maintaining the constancy of the internal medium are regulated by a division of the nervous system known as the autonomic nervous system.

■ ANATOMY

The somatic nerves innervating the skeletal muscles make direct connection between these organs and the central nervous system. They are not interrupted or relayed in their passage from the central nervous system to the peripheral organs. But the gap between central nervous system and the vegetative or visceral organs is bridged by two neurons. With the aid of the right-hand half of Fig. 11-1, this can be readily understood.

Certain cells in the lateral horn of the cord send their medullated (white) fibers out by the ventral root. These fibers, the *preganglionic* fibers of the autonomic nervous system (ANS), leave the mixed spinal nerve by what is known as the white ramus and may end (Fig. 8-1) in a *lateral ganglion* (paravertebral) or continue uninterrupted to a *collateral ganglion.* Here they make synaptic connection with neurons that send their nonmedullated (gray) fibers out by the gray ramus to rejoin the spinal nerve and, with the afferent and efferent somatic fibers, they proceed to the various parts of the body or proceed from the collateral ganglia to the target organs. The structures innervated by the postganglionic fibers are smooth muscles, cardiac muscle, and glands.

Autonomic neuromuscular junction. ■ Three types of smooth muscle innervations occur at the terminals of autonomic nerves. First, individual innervation of cells through one or more 200 A wide neuromuscular junctions occurs in ciliary muscles and circular smooth muscle coats. These muscles perform relatively fast, simultaneous contractions. A second innervation pattern exists

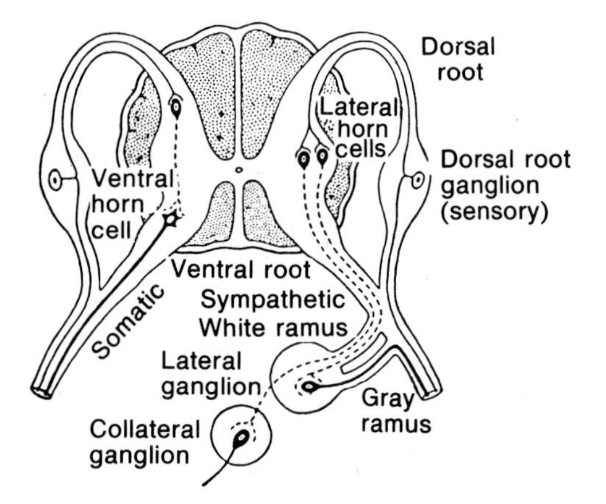

Fig. 11-1
Diagram illustrating somatic (at left) and autonomic nerves (at right). Preganglionic fibers are represented by broken lines and postganglionic fibers by heavy black lines. (From Pearce and Macleod: Physiology, The C. V. Mosby Co.)

in longitudinal coats of the alimentary canal and in most vascular smooth muscles. In these, only a few smooth muscle cells are directly innervated and affected by transmitter. Noninnervated cells within 1 to 2 mm are excited by electrotonic coupling (a local current flow between adjacent cells), rather than by transmitter. Such muscles are geared for slow, graded, often spontaneous contractions in rather circumscribed areas. The third pattern exemplified by the nictitating membrane, pupillary constrictors, and urinary bladder has 20% to 50% of the cells directly innervated; the remainder are either excited by electrotonic coupling or by transmitter diffusing from nerve terminals 0.1 μ distant.

Sympathetic and parasympathetic nervous system. ■ The cell bodies that give rise to preganglionic fibers lie in four divisons of the central gray matter: (1) in the midbrain, (2) in the medulla oblongata, (3) in the thoracolumbar region of the cord, and (4) in the second, third, and fourth segments of the sacral cord (Fig. 11-2). That part springing from the thoracolumbar cord is called the *sympathetic* nervous system, and the part originating in the other three divisions is called the *parasympathetic* nervous system.

Lateral and collateral ganglia. ■ The ganglia of the autonomic nervous system may be grouped, according to their location, into three classes: *lateral* (paravertebral), *collateral,* and *terminal.* Many of the preganglionic fibers of the sympathetic system end in one of the twenty-two pairs of ganglia lying along the vertebral column. Some of the fibers issuing from a certain segment of the cord end in the corresponding ganglion. Other fibers do not make synaptic contact in this ganglion but pass through it on their way to some paravertebral ganglion above or below. In this manner the lateral or paravertebral ganglia are connected by intervening fibers, thus forming the sympathetic chain lying on each side of the spinal column (Figs. 8-3 and 11-2).

Other neurons in the lateral horn of the thoracolumbar cord send their preganglionic fibers to the collateral ganglia that lie a short distance from the spinal column: the celiac, the superior mesenteric, and the inferior mesenteric ganglia. In Fig. 11-2 these are shown as situated to the right of the sympathetic chain. The *splanchnic nerves* are composed of the preganglionic fibers to these collateral ganglia. From the lateral and collateral ganglia, the postganglionic fibers supply many organs, as can be seen in Fig. 11-2: dilators for the pupil; vasoconstrictors for the blood vessels of skin, stomach, intestines, kidneys, and other organs; secretory nerves for the sweat glands; cardiac accelerator fibers; motor fibers for the pyloric, ileocolic, and internal anal sphincter; inhibitory fibers for the muscles of the stomach, intestine, and bladder (Fig. 19-2); and pilomotor nerves for erectors of the hairs.

Terminal ganglia. ■ The ganglia of the parasympathetic nervous system differ from the ganglia of the sympathetic system in that they lie (Fig. 11-2) in or close to the organ they innervate and, in consequence, the *postganglionic*

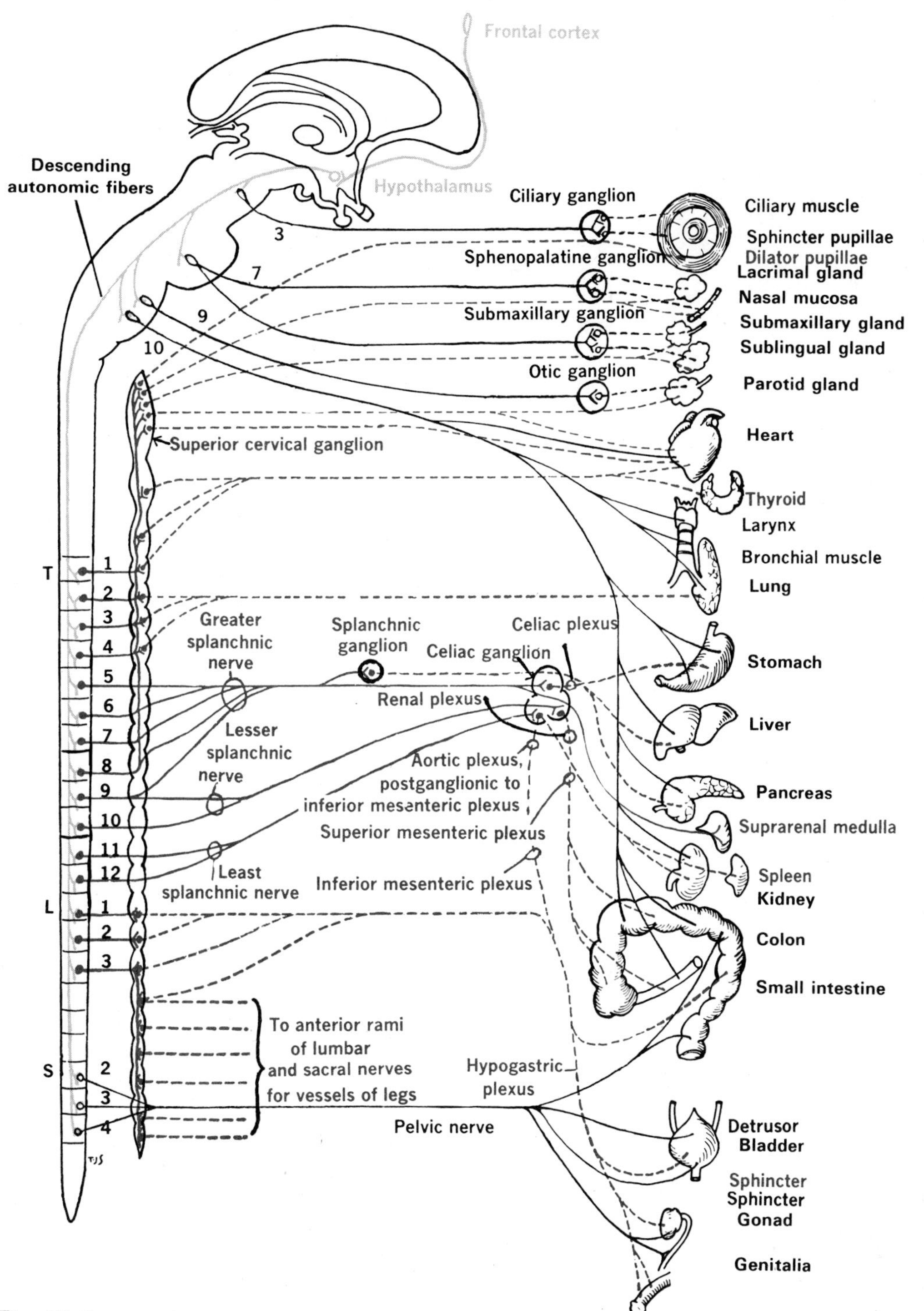

Fig. 11-2
Diagram of the autonomic nervous system. The sympathetic fibers are in red; the parasympathetic fibers are in black. The preganglionic pathways are represented by solid lines, and the postganglionic fibers are represented by broken lines. (From Mettler: Neuroanatomy, The C. V. Mosby Co.)

fibers are short. The preganglionic fibers from the second, third, and fourth sacral nerves unite to form a single nerve trunk, known as the *pelvic nerve,* and terminate in the pelvic ganglion. From there the postganglionic fibers supply the lower part of the colon, the rectum, the internal anal sphincter, the bladder and its sphincter, and the blood vessels of the genital organs.

Preganglionic fibers from the midbrain leave by the third cranial nerve and end in the *ciliary ganglion* just back of the eyeball (Fig. 11-2); the postganglionic fibers innervate the muscles of accommodation (Chapter 9) and the sphincter of the iris. Preganglionic fibers leave the medulla oblongata by the seventh (facial), ninth (glossopharyngeal), and tenth (vagus) cranial nerves (Fig. 10-2). The fibers in the seventh and ninth cranial nerves end in ganglia lying in or near the salivary glands and innervate the secretory cells and the blood vessels. Preganglionic fibers in the vagi are distributed to the heart (inhibitory), various parts of the digestive tract, gallbladder, liver, pancreas, and kidneys.

■ CHARACTERISTICS

The autonomic nervous system does not possess any autonomy in the sense that, independent of the other parts of the nervous system, it exercises control over the organs innervated by it. The term *autonomic* is a misnomer. It does not act as an independent reflex mechanism. Afferent fibers are found in the autonomic nerves (as in the vagi from the heart or in the splanchnics from the intestines); these have, like all other afferent nerves, their cells of origin in the dorsal root ganglia and therefore do not belong to the autonomic system proper; they are simply *visceral afferents.* Impulses, carried by them, in order to return to the visceral organs must pass through the central nervous system. The afferent limbs of vital visceral reflexes (such as cardiac, aortic, urinary, and respiratory) are always located in parasympathetic nerves. Splanchnic nerve afferents are concerned with hunger sensations and pain in the abdominal viscera.

Cerebrospinal *versus* autonomic system. ■ By means of the cerebrospinal, or somatic, system volitional control is exercised over the skeletal muscles. Generally speaking, we cannot voluntarily send impulses to the heart and to the smooth muscles and glands of the vegetative organs; neither do the autonomic reflexes affect our consciousness. For these reasons the autonomic nervous system is also spoken of as the *involuntary* or the vegetative nervous system. These two systems, however, must not be thought of as two distinct, unconnected parts of the entire nervous system; they mutually influence each other. Activity of the cerebral cortex frequently has a powerful effect on the vegetative organs, as shown by the influence of emotions on cardiac, vascular, respiratory, and digestive functions. The voluntarily induced activity of the skeletal muscles would fail to attain the desired goal were it not for the assistance rendered by the autonomic nervous system in increasing cardiac activity, in redistributing the blood, and in stimulating the sweat glands.

A sharp contrast between these two systems lies in the degree of automaticity of the respective organs that they innervate. Severing the motor nerve to a skeletal muscle causes complete paralysis and atrophy of the muscle. Although cutting an autonomic nerve may modify the activity of an organ, it does not result in permanent, complete cessation of activity or in atrophy.

Localization. ■ The two divisions of the autonomic nervous system differ in the distribution of their impulses in the peripheral organs. Since the ratio of preganglionic to postganglionic fibers in the *sympathetic system* is 1 to 20 or more, the impulses reaching a peripheral organ or organs are *diffuse;* there can be no sharp localization of the responses produced. This is illustrated in the innervation of blood vessels and sweat glands. In the *parasympathetic system* there are in some instances only two postganglionic fibers for every preganglionic fiber; the activity is therefore more *localized.*

Double nerve supply. ■ As can be noticed from Fig. 11-2, nearly all organs supplied with autonomic nerves receive innervation from two sources; one from the sympathetic and one from the parasympathetic.In many instances this double nerve supply is mutually antagonistic; for example, the vagi (parasympathetic) inhibit the heart; the cervical sympathetics accelerate it. In the intestinal musculature this relationship of the two systems is reversed. Some structures are supplied with two mutually antagonistic muscles; in such instances the nerve supply is from the sympathetic system for the one muscle and from the parasympathetic for the other muscle. For example: the sphincter (circular) muscle fibers of the iris are innervated by the parasympathetic fibers in the third cranial nerve; the radial muscle (dilator) fibers are innervated by the cervical sympathetic. By this arrangement the two nerves exercise a mutual check on each other.

In most instances the relationship between the nerves from this double supply should be looked on as complemental rather than antagonistic. As an illustration of this, consider the urinary bladder. The filling of this organ is made possible by the sympathetic; the emptying is controlled by the parasympathetic.

Sympathetic system not indispensable. ■ A cat has been kept alive for three and one-half years after the entire sympathetic nervous system was removed. Although there was no change in muscle tone, the capacity for muscular work was reduced by 35%. This great curtailment naturally follows the abolition of the circulatory adjustments necessary for sustained muscular activity. For the same reason the animal was unable to defend itself against cold by the erection of its hair, the vasoconstriction of its cutaneous blood vessels, and the secretion of epinephrine to increase its metabolic rate. Unless kept in a warm room the animal shivered constantly and was in danger of experiencing a fall in body temperature. Neither was it able to tolerate hemorrhages, anoxemia, asphyxia, or emotional excitement. To meet these conditions, extra effort must be made for homeostasis; this is brought about by organs largely controlled by the sympathetic nervous system. In these emergency functions the sympathetic nervous system is greatly aided by adrenal epinephrine release.

Hypothalamus and autonomic nervous system. ■ In order that the activities of the diverse vegetative organs may maintain a nearly constant state of the internal medium *(homeostasis),* these activities (individually under the control of the autonomic nervous system) must be properly coordinated. To illustrate: Maintenance of the normal body temperature is a proper balancing of heat production and heat loss. Heat production is the result of tissue (chiefly muscular) activity. The loss of heat is brought about by the radiation of heat (due to cutaneous vasodilator action) and to the evaporation of sweat. Each of these three processes is controlled by a distinct part of the nervous system. As these activities vary in extent with ever-changing internal and external conditions, it is evident that a unifying and coordinating mechanism is necessary; this is the function of the hypothalamus. Similarly, the establishment of a water and an energy balance and the maintenance of a constant body weight are under hypothalamic control. *The hypothalamus is an integrative center for the visceral functions governed by the autonomic nervous system.* It is generally thought that in many instances the hypothalamus exercises these controls by influencing the hormone secretions of the adrenal glands, pituitary, and perhaps other endocrine organs.

■ SYNAPTIC TRANSMISSION

Transmitters in the autonomic nervous system are classified into two types: *cholinergic* and *adrenergic.* The occurrence of two transmitters and a number of different receptors provides the

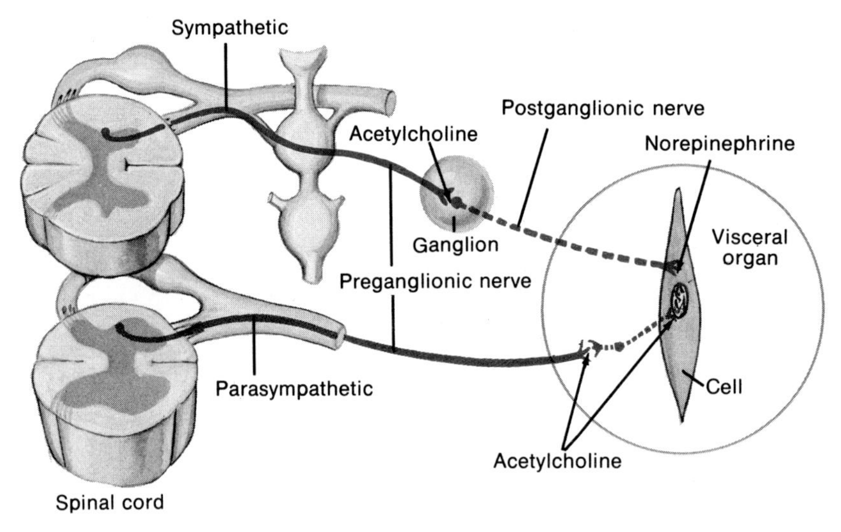

Fig. 11-3
Illustration of sympathetic and parasympathetic outflow from different levels of spinal cord. Transmitters at ganglia and target organ are indicated.

basis for the variability in response observed upon stimulation of parasympathetic and sympathetic nerves.

Acetylcholine. ■ The transmitter released at the terminals of all preganglionic fibers, both sympathetic and parasympathetic, and at the endings of the postganglionic fibers of the parasympathetic system (Fig. 11-3) is acetylcholine. For this reason, these nerve fibers are called cholinergic. Evidence derived from pharmacological studies indicates that at least three types of acetylcholine receptors must exist in the various postsynaptic structures. First, there is the motor end plate of skeletal muscle (Chapter 6) that can be blocked by curare. Atropine blocks the effect of acetylcholine on smooth and cardiac muscle and on exocrine glands (salivary, sweat, etc.), parasympathetically innervated structures, but has no effect upon transmission at the neuromuscular junction of skeletal muscle. Finally, synaptic transmission in the autonomic ganglia is blocked by the drug hexamethonium, which is without effect on skeletal muscle transmission. Sometimes, the receptors for skeletal muscle and ganglia are spoken of as *nicotinic* receptors and those for smooth muscle, cardiac muscle, and glands as *muscarinic* receptors, since the naturally occurring compounds nicotine and muscarine mimic the action of acetylcholine at these respective sites.

Norepinephrine and epinephrine. ■ The transmitter released at the terminals of the postganglionic fibers of the sympathetic nervous system, with two major exceptions, is norepinephrine (Fig. 11-3). Epinephrine is the major secretory product of the adrenal medulla (Chapter 28) and is a hormone rather than a neurotransmitter. However, both of these catecholamines are specified as adrenergic substances. In general, norepinephrine is involved in the maintenance of sympathetic tone and the adjustment of circulatory dynamics through its vasconstrictor effects. Epinephrine which is released in emergency states stimulates metabolism and promotes increased blood flow to skeletal and coronary (heart) blood vessels; it constricts blood vessels in the skin and mucous membranes as well as the renal arteries. Furthermore, it increases the rate and force of contraction of the heart and dilates the bronchioles. Although norepinephrine does accelerate the heartbeat, its effect is masked by a reflex slowing (Chapter 14) induced by the marked elevation of the blood pressure that vasoconstriction causes. Pharmacological evidence indicates that two types of adrenergic receptors must exist; they are termed *alpha* and *beta.* Norepinephrine is most effective at alpha receptors, the activation of which produces vasoconstriction and pupillary dilation, for example. Epinephrine acts upon both alpha and beta receptors, but action upon the latter is responsible for the vasodilation in skeletal muscle, bronchial relaxation, and the increased strength and rate of the heartbeat. Stimulation of alpha receptors, then, generally causes contraction of smooth muscle whereas stimulation of beta receptors initiates relaxation.

The exceptions to the adrenergic transmission at sympathetic postganglionic nerve terminals are (1) the sweat glands and a few blood vessels and (2) the adrenal medulla. It is not surprising that this gland should respond to cholinergic stimulation, for it is embryologically a modified sympathetic ganglia.

READINGS

Axelrod, J.: Noradrenaline; fate and control of its biosynthesis, Science **173**:598-606, 1971.

Burn, J. H.: The autonomic nervous system, ed. 3, Philadelphia, 1968, F. A. Davis Co.

DiCara, L. V.: Learning in the autonomic nervous system, Sci. Am. **222**(1):30-39, Jan. 1970.

Rubin, R. P.: The role of calcium in the release of neurotransmitter substances and hormones, Pharmacol. Rev. **22**:389-428, 1970.

von Euler, U. S.: Adrenergic neurotransmitter functions, Science **173**:202-206, 1971.

12

THE BLOOD

■ INTERNAL MEDIUM

The cells of our body are highly specialized; that is, they have acquired a high degree of efficiency for a particular function. Specialization is generally accompanied by a loss of other, frequently vital, functions. Because of this, the more highly specialized a cell becomes, the less it is capable of carrying on an independent existence. In consequence, any organ, tissue, or cell when separated from the body perishes in a brief span of time. To continue their existence in the body, they must be protected against adverse environmental changes, for example, changes in temperature. Indeed, so helpless are the individual cells that nature has supplied each cell with its own diminutive *internal environment.* The internal environment bears the same relationship to the cell that the external environment bears to the body as a whole.

The internal environment is formed by a thin film of liquid known as tissue fluid, which bathes the cell (Fig. 12-1). The formation and composition of tissue fluid are subjects considered in Chapter 16. To maintain a constant supply of food and oxygen and to prevent an undue accumulation of waste in the tissue fluid are the prime functions of a constantly moving common carrier, the blood. In its passage through the capillaries of the intestinal wall the blood acquires a supply of nutrients. In the capillaries of the lung, oxygen is acquired, and these materials are carried to the tissue fluid and thus to the cells. It picks up the waste products and transports them to the kidneys and lungs for excretion. The blood is, therefore, the *middleman* between the external and the internal environment. Together the blood and the tissue fluid are known as the internal medium, or extracellular fluid.

Homeostasis. ■ Because of the inability of the cells to defend themselves against any noxious changes in the internal environment, it is necessary to maintain as nearly constant as possible the physical and chemical constitution of the tissue fluid. Two factors tend to upset this constancy: (1) external environmental changes (e.g., temperature changes) and (2) internal changes. Body activities tend to produce internal changes in (1) the concentration of various nutrients, (2) the concentration of oxygen, (3) the

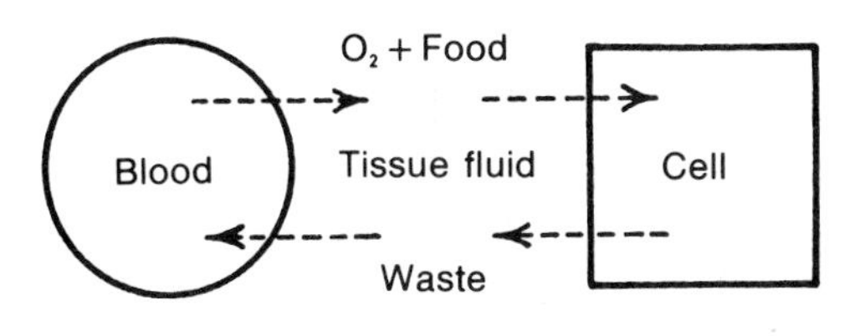

Fig. 12-1
Diagram showing relationships between blood, tissue fluid, and cells.

concentration of the H and OH ions, (4) the temperature, (5) the osmotic pressure, (6) the concentration of waste products, (7) the volume, and (8) the colloidal content. Since its constituents and, therefore, its properties fluctuate continually, *the constancy of the internal medium is a dynamic balance,* varying only within the narrow limits compatible with the existence and functioning of the cells.

Maintenance of the constancy of the internal medium is described as *homeostasis.* Homeostasis is the result of the integrated cooperation of nearly all the vital organs of the body—the digestive apparatus, skin, and circulatory, respiratory, and excretory systems; collectively these are known as the *vegetative organs.* As stated by Claude Bernard: *Vital mechanisms have one objective—preservation of a constant internal environment.* Homeostasis constitutes the main theme of the study of *physiology.*

■ AMOUNT, FUNCTIONS, AND COMPOSITION OF BLOOD

Amount of blood. ■ The amount of blood in the human body is approximately 8% of the body weight.* Normal human adult blood volume is 4.5 to 5 liters. The amount is constant and is not increased or decreased for any length of time by drinking fluid, by injections, or by hemorrhage (Fig. 24-3). When a person drinks a large quantity of water, the water is speedily taken from the blood by the tissues or eliminated by the kidneys. The amount of blood in the human body is of great importance; without a minimum amount, circulation stops and renders life impossible.

Functions of blood. ■ The functions of blood are discussed in many subsequent chapters; here they may be summarized under three headings:

1. Transport functions
   a. Nutrients from the alimentary canal to the cells
   b. Oxygen from the lungs to the cells
   c. Waste products from the cells to the organs of excretion
   d. Heat formed in the more active tissues to all parts of the body, thereby aiding in the regulation of the body temperature
   e. Hormones transported to all parts of the body
2. Acid-base balance mechanisms
3. Immunologic reactions

Composition of blood. ■ From these functions it can be gathered that the blood is a complex fluid containing the nutrients absorbed from the alimentary canal, the oxygen taken up in the lungs, the waste products produced by cellular activity, the hormones, the antibodies, and other substances. It also must be apparent that the composition of the blood varies with place and time.

By microscopic investigation it can be seen that what appears to the naked eye as a perfectly homogeneous fluid is in reality one in which an immense number of small bodies or corpuscles float. By means of a centrifuge corpuscles can be separated from the liquid, for they are considerably heavier (Fig. 12-2). The composition of blood is as follows:

Plasma, a light yellow liquid—about 55% of the blood volume
1. Water—90%
2. Dissolved solids:
   a. Three proteins, forming from 6% to 8% of the plasma, are serum albumin (approximately 4.5%), serum globulin (2%), and fibrinogen (0.3%). They

*The blood and the tissue fluid constitute almost 20% of the total body weight.

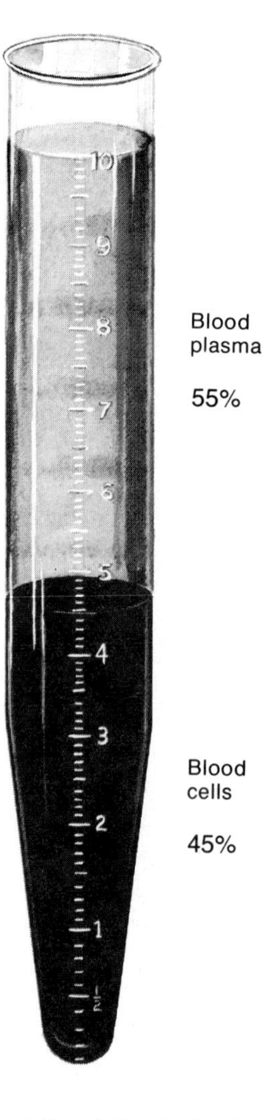

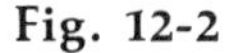
Fig. 12-2
Tube containing centrifuged blood, showing relative proportion of cells to plasma.

contribute to the viscosity and, to a lesser extent, to the osmotic pressure of the blood. Fibrinogen plays a major part in coagulation. Of the three forms of globulin (alpha, beta, and gamma), gamma globulin is concerned with the immunization of the body against foreign cells and substances.

b. Supplies for cells: glucose (about 0.1%), fat and fatlike substances, amino acids, and salts
c. Cellular products: enzymes, antibodies, and hormones
d. Cellular waste products (nitrogenous): for example, urea and uric acid

3. Gases: oxygen, carbon dioxide, and nitrogen

Formed elements—about 45% of the blood volume

1. Red blood cells (RBC, or erythrocytes); carriers of O_2 and CO_2
2. White blood cells (WBC, or leukocytes); scavengers and immunizing agents
3. Platelets; blood coagulation

PROPERTIES OF BLOOD

Specific gravity

Specific gravity is the ratio of the weight of a certain volume of a substance to the weight of an equal volume of water. The specific gravity of the blood varies from 1.052 to 1.061. The specific gravity of the blood depends largely on the number of red blood cells. When freshly drawn blood to which an anticoagulating agent has been added is placed in a vertically suspended glass tube, the red cells settle to the bottom because of their greater specific gravity. The *rate of sedimentation* is expressed in millimeters per hour; normally it varies from 4 to 10 mm/hr.

Specific gravity may be determined by letting drops of blood fall into mixtures of varying proportions of chloroform and benzene. If a drop does not sink or rise, it has the same specific gravity as that mixture. A hydrometer placed in the mixture will indicate the specific gravity by the distance to which it sinks into the fluid.

Osmotic pressure

The osmotic pressure of human blood averages about 5,100 mm Hg or 6.7 atmospheres (about that of a 0.9% NaCl solution). Osmotic pressure is due chiefly to the various salts, waste products, sugar, and other crystalloids dissolved in the plasma. The proteins found in the plasma also exert a small amount of osmotic pressure. Since the composition of the blood undergoes continual, although small, changes because of the passage of water, dissolved foods, and waste products into and out of the blood, the osmotic pressure also varies slightly. Such variations, however, are speedily corrected by the kidneys (Chapter 29).

■ Viscosity

The viscosity of a liquid is due to the mutual attraction of its molecules, which thereby offer resistance to flow. The viscosity of the blood is about five times as great as that of water. Whatever increases or decreases the number of blood cells or the amount of protein affects the viscosity in the same sense. The degree of viscosity is of importance because the greater the viscosity, the more slowly a fluid flows through a tube and the greater the force necessary for its propulsion.

■ Hydrogen ion concentration

The maintenance of the chemical environment in which a cell normally functions is of prime importance; therefore the hydrogen ion content of the blood is held within very narrow limits. We will discuss briefly some of the concepts necessary for an understanding of this phenomenon.

Acids and bases. ■ An *acid* is defined as any substance that liberates H ions or protons when it dissociates in solution; a *base* is any substance that binds or removes H ions or protons from solution. In simplified form the ionization of an acid may be thought of as:

$$HA \rightleftharpoons H^+ + A^-$$

As the reaction proceeds to the right, HA yields hydrogen ions; therefore, it is an acid. In the reverse direction the anion, A^-, accepts an H ion; therefore, A^- is a base. A strong acid such as hydrochloric completely dissociates in dilute aqueous solutions to form:

$$HCl \rightarrow H^+ + Cl^-$$

A weak acid such as acetic only partly dissociates as:

$$CH_3COOH \rightleftharpoons H^+ + CH_3COO^-$$

In the first case Cl^- can only be considered a very weak base, since it will not react with the proton; on the other hand, the acetate ion is a strong base, since it readily accepts the H^+. The conjugate base of a strong acid is a weak base; the conjugate base of a weak acid is a strong base. Hydroxyl ions (OH^-) represent strong bases, and usually we refer to sodium or potassium hydroxides as strong bases; in aqueous solution they dissociate completely into Na^+ or K^+ and OH^-.

Neutralization. ■ A molar solution of HCl is made by adding 36.5 g to an amount of water sufficient to make 1 liter. If all the acid molecules ionized, 1 liter of this solution would contain 1 g of H ions. Similarly, 1 liter of a molar solution of NaOH (23 + 16 + 1 = 40 molecular weight) has 17 g of OH ions. As we have taken the gram molecular equivalent of H^+ (1) and of OH^- (17), the number of H^+ in 1 liter of a molar HCl solution is the same as the number of the OH^- in 1 liter of a molar NaOH solution. When these two solutions are mixed, the H and the OH ions unite to form H_2O and NaCl, a neutral salt; this constitutes neutralization.

Normal solutions. ■ The molar solutions of HCl and of NaOH just mentioned are known as normal solutions. A normal solution of an acid contains in 1 liter of the solution as many grams of the acid as its molecular weight divided by the number of replaceable hydrogen atoms in the molecules. Since this number is 1 in the case of HCl, the molar and the normal solutions are the same. For 1 N sulfuric acid (H_2SO_4) we take (2 + 32 + 64) ÷ 2, or 49 g, of the acid. A liter of a normal solution of any acid contains therefore 1 g of replaceable hydrogen ions; 1 liter of a normal solution of any base contains 17 g of replaceable hydroxyl ions.

Actual acidity. ■ A normal solution of any acid will neutralize an equal volume of a normal solution of any base. But all normal acids are not equally strong. For example, a fourth normal (0.25 N) HCl is very destructive to living tissue, but 0.25 N acetic acid may be safely used. This marked difference depends on the degree of *ionization* the acid undergoes. In a given concentration of a strong acid, such as hydrochloric, nitric, or sulfuric, ionization takes place to a greater extent (i.e., more ions are set free per unit of volume) than in weak acids, such as lactic or carbonic, of the same concentration. For example, in 0.1 N HCl, 91% of the molecules are ionized; in the same concentration, acetic acid is ionized only 1.3%. This degree of ionization can be expressed as a *dissociation constant* and is derived from the law of mass action. When we add acetic acid to water, a reversible reaction occurs as indicated before; this reaction proceeds rapidly until an equilibrium is reached, which can be expressed as follows:

$$\frac{[H^+]\,[CH_3COO^-]}{[CH_3COOH]} = K_a$$

K_a is an equilibrium constant and over a very narrow range of conditions is the dissociation constant. Since these numbers are small and awkward to handle, it is often expressed in the form of its negative logarithm; in this form it is called the pK. Some examples are:

$$\begin{aligned}
\text{acetic acid } K_a &= 1.86 \times 10^{-5} \\
\log K_a &= \log 1.86 + \log 10^{-5} \\
-\log K_a &= 4.73 \\
pK &= 4.73 \\
\text{carbonic acid } K_{a_1} &= 7.9 \times 10^{-7} \\
pK_1 &= 6.1 \\
K_{a_2} &= 6.0 \times 10^{-11} \\
pK_2 &= 10.4
\end{aligned}$$

Since carbonic acid (H_2CO_3) has two H ions, it has two ionization constants; the first H ion is more readily exchanged.

The same course of reasoning holds for bases. Although normal solutions of all bases have the same power to neutralize acids, the strength of a base depends on the degree of ionization; this causes sodium hydroxide, NaOH, and potassium hydroxide, KOH, to be much more powerful than calcium hydroxide (limewater), $Ca(OH)_2$.

The number of H or OH ions present gives the actual or true degree of acidity or alkalinity; the number of replaceable H or OH ions (whether in the ionic form or lodged in the undissociated molecule) give the neutralizing power, which is sometimes called *titratable* acidity or alkalinity.

Degree of acidity or alkalinity—pH. ■ According to the theory of electrolytic dissociation, some of the molecules of pure water ionize to form an equal number of positive hydrogen and negative hydroxyl ions. The dissociation of water is represented by the following equation:

$$HOH \rightleftharpoons [H^+] + [OH^-]$$

Since the amount of dissociation is so small, the undissociated water may be considered as constant; therefore:

$$[H^+]\,[OH^-] = K_w$$

This value has been determined to be equal to 1×10^{-14} at 25 C. As with the ionization constant of weak acids, this can be referred to as $pK_w = 14$. In 1,000 g of water there is 0.0000001 g of H^+, or 1×10^{-7}; since there must be an equal number of OH^-, its value also would be 1×10^{-7}. These numbers are difficult to handle so the degree of alkalinity or acidity is designated by the symbol pH, which can be defined as the logarithm* of the hydrogen ion concentration with sign reversed. Consequently, the pH of water is found as follows:

$$\begin{aligned}
[H^+] &= 1 \times 10^{-7} \\
&= \log_{10} 10^{-7} \\
&= 7
\end{aligned}$$

Fig. 12-3 indicates the relationship between normality of an acid or alkali and the pH. The following points should be noted:

1. A pH 7 stands for neutrality.
2. A pH less than 7 indicates acidity. As the

*Log of 1 = 0
Log of 10 or 10^1 = 1
Log of 100 or 10^2 = 2
Log of 1,000 or 10^3 = 3
Log of $^1/_{10}$ or 0.1 or 10^{-1} = −1
Log of $^1/_{100}$ or 0.01 or 10^{-2} = −2
Log of $^1/_{1,000}$ or 0.001 or 10^{-3} = −3, etc.

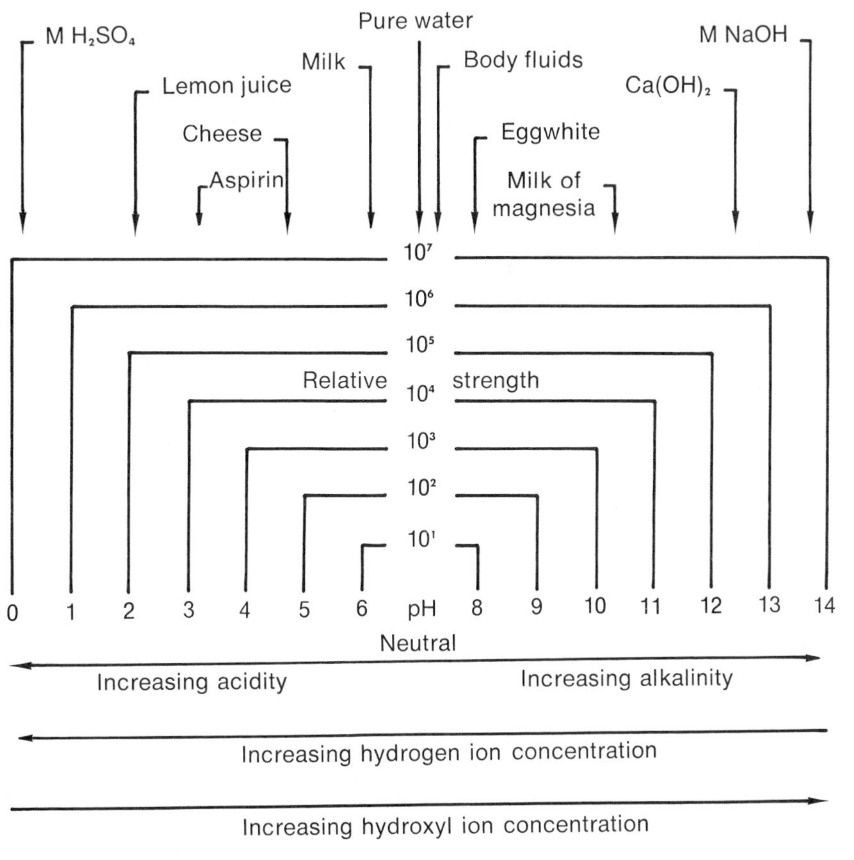

Fig. 12-3
The pH scale.

degree of acidity *increases,* the pH value *decreases.*

3. A pH greater than 7 indicates alkalinity. The greater the alkalinity, the greater is the pH value.

4. Since the pH expresses the logarithm, a pH 3 represents ten times the hydrogen ion concentration of pH 4 and one hundred times that of pH 5.

pH of the blood. ■ The pH of the blood is approximately 7.4, with a normal range of 7.3 to 7.5. Various mechanisms are involved in holding the pH of blood at a constant value. These mechanisms all involve maintaining the buffer capacity of blood.

Buffer action. ■ A solution of a weak or poorly ionized acid and its almost completely dissociated salt is able to prevent or minimize a drastic change in pH when either a strong acid or base is added. This solution is called a buffer mixture, or pair; carbonic acid–sodium bicarbonate and sodium dihydrogen phosphate-disodium hydrogen phosphate are examples of such buffer pairs. When a weak acid is in a mixture with its salt, the hydrogen ion concentration will be considerably lower than that for the weak acid itself and is expressed as

$$[H^+] = K_a \frac{[acid]}{[salt]}$$

If we now take the negative logarithm of this

$$-\log\ [H^+] = -\log\ K_a + \left[-\log \frac{[acid]}{[salt]}\right]$$

by substitution:

$$pH = pK + \log \frac{[salt]}{[acid]}$$

This is the *Henderson-Hasselbalch equation;* it can be used to calculate the pH of a buffer system and the effect of the addition of a strong acid or strong base. As an example, we shall use the carbonic acid–sodium bicarbonate system, since it is present in plasma. Approximate plasma concentrations of these two substances are $NaHCO_3$ = 0.026 M; H_2CO_3 = 0.0013 M, which, substituted into the equation, yields:

$$pH = 6.1 + \log \frac{0.026}{0.0013}$$

$$= 6.1 + \log 20$$
$$= 6.1 + 1.3$$
$$= 7.4$$

If we add a quantity of strong acid so that the hydrogen ion concentration in the solution is increased by 0.001 M (or N), the ratio would become $\frac{0.026}{0.0023}$ and:

$$pH = 6.1 + \log \frac{0.026}{0.0023}$$

$$= 6.1 + \log{\sim}11$$
$$= 6.1 + 1.04$$
$$= 7.14$$

The same quantity of acid added to water would have given a pH = 3. The capacity of buffers to resist drastic change is well demonstrated. Other buffer systems present in the blood include the proteins and phosphate salts.

Hemoglobin is the most important protein buffer in the blood. The total amount of substances in the blood available for neutralizing acids is known as the *alkali reserve* and is equivalent to about a 0.5% sodium bicarbonate solution.

Acidosis and alkalosis. ■ The ratio of buffer systems in the blood and body is spoken of as the *acid-base balance.* An increase or decrease in the amount of H_2CO_3 without a corresponding increase or decrease in $NaHCO_3$ changes the pH value. The tissues are highly susceptible to these changes. A decrease in pH much below 7.3 is known as *acidosis;* it causes grave disturbances and may result in coma and even in death. An increase beyond 7.5 constitutes *alkalosis;* it may cause convulsions and is incompatible with life. Metabolic acidosis results from an excess production of acid (e.g., lactic and acetoacetic) by the tissues or from failure of the kidneys to excrete H^+ as NH_4^+ or NaH_2PO_4. Metabolic alkalosis may arise from ingestion of unusual amounts of bicarbonate or from loss of gastric secretion, as by vomiting. Failure to adequately discharge CO_2 from the body (e.g., in pulmonary disease) leads to an accumulation of H_2CO_3 and respiratory acidosis. Prolonged hyperventilation will induce respiratory alkalosis.

The body has several mechanisms for maintaining the proper ratio of the buffer systems and thus the proper acid-base balance; these are functions of the lungs and kidneys (Chapters 17 and 29).

■ RED BLOOD CELLS OR ERYTHROCYTES

Size. ■ The erythrocytes in man and nearly all other mammals are biconcave circular discs (Fig. 12-4). There is no nucleus, Golgi apparatus, mitochondria, RNA, or centriole present; consequently they are often called corpuscles. The size varies in different animals and conditions. In man, the average diameter is 7.5 μ; the average volume is 87 μ^3. Because of the similarity in shape and size we cannot distinguish morphologically between the red blood cells (RBC) of man and those of some common animals, such as the dog, rat, rabbit, mouse, or ox.

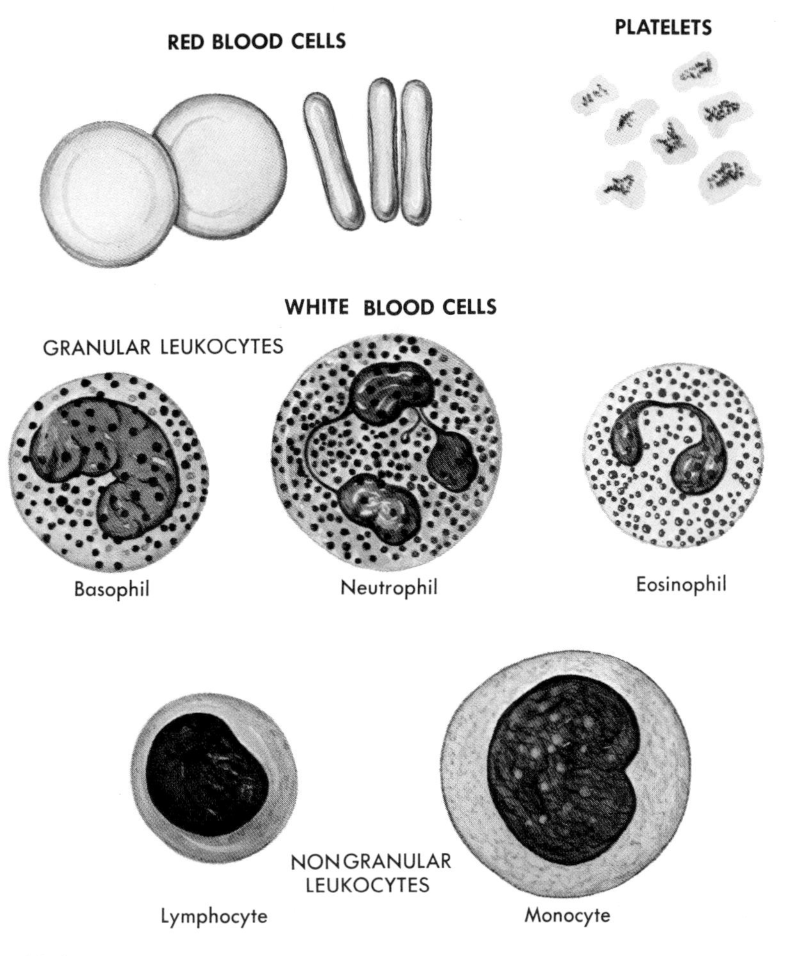

Fig. 12-4
Various formed elements of the blood. (From Anthony: Basic concepts in anatomy and physiology, The C. V. Mosby Co.)

Number. ■ The counting of the cells is done by means of the hemocytometer. The average number of the RBC is about 5.4 million/mm^3 of blood in men; in women the number is about 4.8 million; in the newborn infant the count may be considerably higher.* Individual variations are by no means small. In healthy men between the ages of nineteen and thirty years, the number varies from 4.7 to 7.1 million/mm^3; in women the number varies from 4.3 to 5.3 million.

By means of the hematocrit (Fig. 12-2), it can be shown that the RBC generally occupy about 45% and the plasma 55% of the volume of the blood. The number of RBC is increased when the blood is concentrated by a great loss of fluid from the body through the skin (as after muscular exercise), by the intestines (in diarrhea), or by the kidneys; on the other hand, dilution of the blood

*It must be noted that these figures give the concentration (or population, as it is sometimes called) and not the actual number in the body.

by a great influx of liquid (as occurs after a considerable hemorrhage) is associated with a decrease in the number. Whenever there is a chronic interference with the oxygenation of the blood, as in high altitude, the erythrocytes become more numerous.* The total number in the body is about 25 trillion, with an area estimated at 3,500 m^2, which is about sixteen hundred times the surface area of the entire body. This enormous amount of surface is of extreme importance in the function of the erythrocytes.

Metabolism. ■ Despite the absence of a nucleus, erythrocytes are not metabolically inert. The RBC has an active glycolytic metabolism which is important to the viability of the cell and to the regulation of oxygen transport (Chapter 17). Relatively high levels of ATP are found in red blood cells even though the mechanisms for oxidative phosphorylation (Chapter 23) associated with mitochondria are not present. Energy derived from glycolysis is used to maintain the ionic distribution across the red blood cell membrane and to convert methemoglobin to hemoglobin. An especially significant product of erythrocyte metabolism is the phosphate compound known as 2,3 diphosphoglycerate (DPG). This compound occurs only in trace amounts in most tissues but is present in red cells in concentrations about four times that of ATP. DPG modulates the affinity of hemoglobin for oxygen.

■ Hemoglobin

Composition. ■ The characteristic and most important constituent of the red blood cell is a colored protein, hemoglobin (Hb). It is a conjugated protein composed of a protein, globin, united to a prosthetic group called *heme.* The Hb molecule is roughly spherical in shape, consists of two pairs of subunits known as α and β, possesses a water-filled cavity down the center, and has a molecular weight of about 67,000. In the synthesis of heme, Fe^{++} is added to protoporphyrin IX (Chapter 23) to yield:

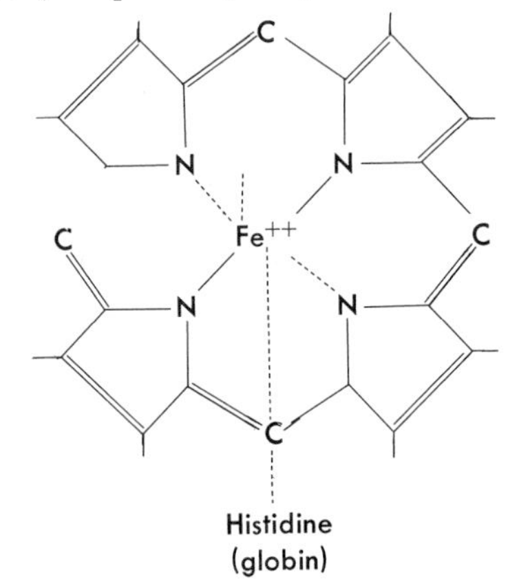

The ferrous iron is linked to the four nitrogen atoms of the porphyrin and to the histidine of the globin. On the opposite side from the histidine linkage is an open space for the oxygen. Four molecules of heme are attached to each globin moiety to form hemoglobin.

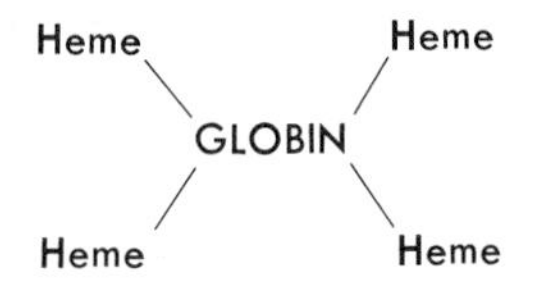

The iron content of the Hb molecule is about 0.34%; iron is the functional part of hemoglobin and essential for life. Hemoglobin constitutes about one third of the red blood cell volume. It has been estimated that a single RBC can carry as many as 10^9 molecules of oxygen.

Function. ■ If blood (or a solution of Hb) is shaken with air, the oxygen in the air dissolves in the water of the blood and unites with the iron of the hemoglobin in the cells, forming *oxyhemoglobin* (HbO_2). Hemoglobin has the unique ability to bind oxygen loosely and reversibly, with the iron atom remaining in the *ferrous state* (Fe^{++}). Each Hb molecule may combine with from 0 to 4 oxygen molecules, one per heme. An increase in acidity causes a dissociation of oxyhe-

*If at sea level the usual population of red cells is taken as 5.2 million, at an elevation of 1 km (3,280 ft) this becomes 5.4, and at an elevation of 7 km (22,960 ft) it becomes 8.1.

moglobin, as does a reduction in oxygen pressure and an increase in tissue carbon dioxide. This reversible change in oxyhemoglobin will be discussed with respiration (Chapter 17).

Other gases can combine with hemoglobin. Carbon dioxide combines with hemoglobin, but the combination is with the globin rather than with the heme. The resultant compound is *carbaminohemoglobin,* which aids in the transport of CO_2 in the blood. The affinity of hemoglobin for carbon monoxide (CO) is about two hundred and ten times that for oxygen. Carboxyhemoglobin is found in the blood of smokers and those individuals exposed to internal combustion engine fumes.

When the iron in hemoglobin is in the ferric state (Fe^{+++}), the compound is called *methemoglobin.* Neither O_2 or CO can combine with methemoglobin, although it combines readily with cyanide and a number of other anions. Methemoglobin is produced by a number of drugs. Hereditary methemoglobinemia has been described; it results from a defect in the enzyme controlling the reduction of iron ($Fe^{+++} \longrightarrow Fe^{++}$) in the RBC. Sickle cell hemoglobin and other abnormal hemoglobins were mentioned in Chapter 2.

Amount of hemoglobin. ■ Since hemoglobin supplies the tissues with oxygen, it is very important that a sufficient quantity of this protein be present in the blood. The average amount is about 16 g/100 ml of blood in men and about 14 g/100 ml in women. At birth the hemoglobin content may be very high, 23 g/100 ml; this drops to a lower level of from 10 to 12 g at the end of the first year. The amount present in the blood frequently is measured by an apparatus called a hemoglobinometer, of which several kinds have been devised. In these instruments the color of a small sample of blood or of blood diluted with a certain volume of a liquid is compared with a standard color scale; the depth of the color indicates the amount of hemoglobin. The hemoglobin content can also be measured spectrophotometrically after conversion of the hemoglobin to cyanomethemoglobin.

■ Life history of erythrocytes

Destruction of red blood cells. ■ Being devoid of a nucleus, the RBC age rapidly; the average life-span is estimated at about four months. The jostling and squeezing that the older cells undergo in the circulating blood causes a certain amount of fragmentation. The spleen is especially effective in trapping old cells and fragments. After entering the blood, a red cell may travel as much as 700 miles before it is destroyed. The membrane of the RBC becomes less flexible as the cell ages and this loss of resiliency contributes to its eventual fragmentation. Even a small increase in environmental temperature, as in a fever, alters the membrane structure and increases the fragility. Humoral agents found in the blood, e.g., prostaglandin (PGE_2), epinephrine, and isoproterenol, decrease the deformability of the red cells and consequently play a role in their destruction.

The whole cell or its fragments are disposed of by large cells known as macrophages. This phagocytosis of the RBC or its fragments takes place in the *reticuloendothelial system,* which consists of phagocytic cells lining the vascular and lymph channels. Those cells which eliminate the worn out RBC are located in the liver, bone marrow, and spleen (Fig. 29-1). The macrophages chemically digest the red blood cells and set free the hemoglobin. The Hb is usually further broken down into heme and globin, although in pathological states some may escape in the urine as hemoglobin subunits (α and β). Normally, the iron is liberated from the heme and the latter is converted through a series of steps into the compound *bilirubin.* Any heme which does enter the plasma is bound by a protein, *hemopexin,* and carried to the liver where the heme is converted to bilirubin and iron. Hemopexin is also discussed in Chapter 21. The bilirubin formed from heme is excreted from the

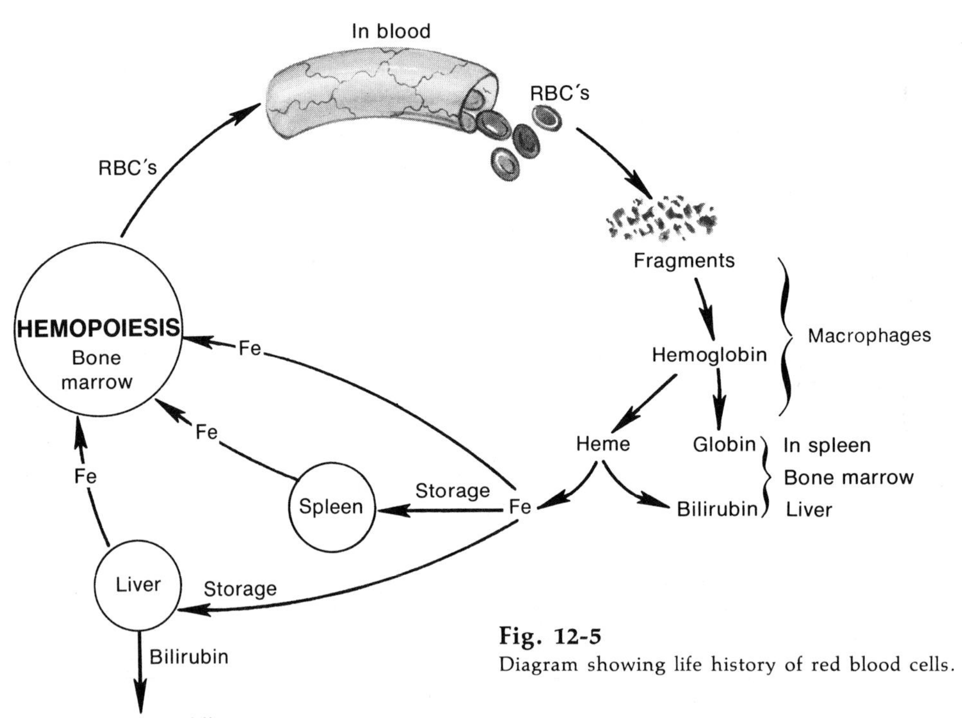

Fig. 12-5
Diagram showing life history of red blood cells.

liver with the bile.* The liberated iron may be used immediately in the formation of new red blood cells, but if the liberated iron plus the iron in the diet exceeds the demand, it is stored principally in the liver but also in all tissues including those of the bone marrow and spleen (Chapter 24). Storage is in the form of *ferritin* which is an iron-protein complex. The life history of red blood cells is illustrated in Fig. 12-5.

Formation of RBC—erythropoiesis. ■ To offset destruction, cells are made continuously. In the fetus the liver, red bone marrow, and spleen are centers for the formation of the red cells *erythropoiesis*. After birth the red blood cells are made exclusively in the red bone marrow found in spongy bones such as the ribs, sternum, and certain bones of the skull, and in the ends* (epiphyses) of the long bones (e.g., femur and humerus). Throughout the blood-forming tissues there are located stem cells that continually give rise to precursors of the erythrocyte. These large nucleated cells, known as *erythroblasts,* synthesize hemoglobin and undergo division. Finally, when sufficient hemoglobin has been formed, the cell loses its nucleus and is released into the bloodstream as an erythrocyte.

Control of the production of red blood cells is very precise so that sufficient numbers are avail-

*It is the chemical breaking up of the hemoglobin that gives the well-known coloration (blue, green, and yellow) to a bruise in which the blood escapes from the ruptured blood vessels into the subcutaneous tissue.

*The shaft (diaphysis) of long bones is formed of compact bone and contains in its central cavity fatty or yellow marrow.

able for tissue oxygenation, but not so many as to impede circulation. Three components are essential to this response. First, there must be adequate stimulation of the stem cells. Second, there must be adequate substrate iron. Finally, there must be normal functioning of the erythroid generating tissue. Any condition that decreases the amount of oxygen reaching the tissues increases the rate of erythrocyte production. Anoxia in tissues, especially in the kidneys, causes the formation of *erythropoietin* (Chapters 28 and 29). This hormone acts on the stem cells and increases the number of erythroblasts formed. Erythropoiesis may be increased by four to five times basal production under erythropoietin stimulation.

Two vitamins of the B complex, vitamin B_{12} and folic acid, are especially necessary for erythrocyte production. Both of these vitamins are important in nucleic acid synthesis and thereby influence cellular protein synthesis. They would, of course, be necessary for hemoglobin synthesis.

The rapidity with which new hemoglobin and red blood cells are made is well illustrated by the fact that radioactive iron fed to an anemic dog is detected in the red cells within four hours, and within four to seven days all the iron fed is transformed into hemoglobin. It is estimated that per kilogram of body weight 3,500 million red blood cells are formed per day. Iron supply sets the upper limit of production.

Anemia. ■ A marked deficiency in the number of cells or in the amount of normal hemoglobin,* regardless of the number of cells, is known as *anemia.* The blood is then unable to unite with an adequate amount of oxygen; this condition is called *anoxemia.* As a result active muscles do not receive all the oxygen needed in their catabolism; this decreases the liberation of energy. Anemic persons are for this reason deficient in physical strength, and they fatigue rapidly; during physical exertion they exhibit great shortness of breath and heart palpitation; their power of resistance to adverse conditions is lessened; mental work is difficult.

It is evident that anemia may be caused either by an excessive loss or destruction of the cells or by diminution in their production. The size and Hb content of the cells found in the blood are influenced by the origin of the anemia. If the cells are of normal size and contain a normal amount of Hb, they are called *normocytic* and *normochromic.* Cells with above normal size and Hb content are said to be *macrocytic* and *hyperchromic.* Conversely, the terms *microcytic* and *hypochromic* describe cells that are small and without the normal Hb concentration. Several types of anemia are known, but only a few examples will be considered.

Hemorrhagic anemia is the result of the loss of a large volume of blood and is a serious threat to the circulation. To prevent this, a great influx of water from the tissue spaces into the blood occurs. However, the blood is diluted by this inpouring of water, the red blood cell count is reduced, and anemia exists. Hemorrhagic anemia is an example of normocytic-normochromic anemia. Incidentally, it may be noted that the withdrawal of water from the body tissues explains the intense thirst experienced after an extensive hemorrhage and the necessity of copiously supplying the conscious patient with water. Soon after a large hemorrhage the activity of the red bone marrow is greatly increased; the amount of iron normally stored diminishes. The amount of red marrow in the epiphyses of the long bones increases and may encroach upon the fatty marrow in the shaft.

Nutritional anemias are the result of iron deficiency. This deficiency comes about by an increase in iron requirement combined with a diet low in this compound. In iron deficiency the anemia is microcytic and hypochromic.

---

*When the hemoglobin content of the blood becomes less than 10 g/100 ml, a person is generally regarded as anemic. However, no great distress is experienced until the amount is decreased to almost one-half normal.

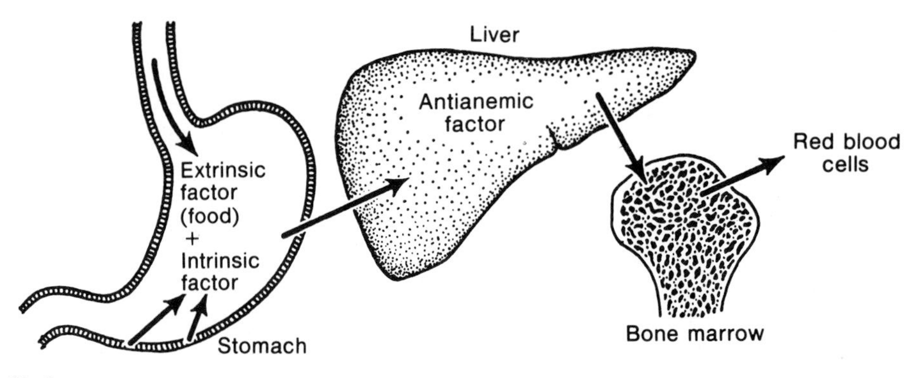

Fig. 12-6
Vitamin B_{12} (extrinsic factor) is absorbed from the small intestine, provided the intrinsic factor from the stomach is present. The vitamin is stored in the liver. It is necessary for the proper maturation of red blood cells produced in the bone marrow. Without the intrinsic or extrinsic factor a condition known as pernicious anemia results.

Pernicious anemia is due to a defect in the formation of erythrocytes and to the fact that their number in the blood is greatly reduced. Blood and bone marrow examination indicate that the blood cells do not mature correctly and that the erythrocytes formed are larger than normal and are very fragile. Administration of iron proves of no value in treatment of this anemia, but feeding large quantities of liver ($^1/_4$ to $^1/_2$ lb/day) causes the blood count to return to normal. Dried hog stomach or muscle treated with normal gastric juice has the same beneficial effect as liver. Therefore, two factors are concerned (Fig. 12-6): an *extrinsic factor* found in food and an *intrinsic factor* present in normal gastric juice. The intrinsic factor is secreted by the normal gastric mucosa and is discussed in Chapter 20. It is required for the absorption of vitamin B_{12}, the extrinsic factor; this vitamin is essential for the normal maturation of red blood cells. Pernicious anemia is macrocytic and hyperchromic.

Sickle cell anemia is caused by blood destruction resulting from the presence of an abnormal type of hemoglobin. The abnormality is a defect in the globin portion and stems from the inheritance of a gene that produces an alteration in the globin structure, a difference in amino acid content (Chapter 2). Sickle cell anemia is one type of hemolytic anemia.

■ Hemolysis

Hemoglobin is readily soluble in dilute salt solutions and therefore in plasma; but notwithstanding this solubility, the hemoglobin normally remains within the erythrocyte. By various means the pigment can be released from the cell. When hemoglobin is released and dissolves in the surrounding liquid, the process is called *hemolysis.*

Osmotic hemolysis. ■ Two factors are concerned in osmotic hemolysis: the selective permeability of the cellular membrane and osmotic pressure. The highly selective membrane of the red blood cell is permeable to H, OH, and HCO_3 ions, urea, and ammonium salts and slightly to glucose and amino acids. It is impermeable, or nearly so, to Na, K, and Ca ions, hemoglobin, and disaccharides. The contents of the RBC (such as water, salts, and hemoglobin) exert osmotic pressure equal to that of the surrounding plasma, which is equal to that of a 0.9% NaCl solution.

On placing red blood cells into dilute salt solution, for example, 0.6% NaCl, some water enters them. This causes the biconcave disc to become more nearly spherical, thereby making room for the absorbed water. The osmotic pressure is lowered to that of the salt solution, and no further change takes place. By using progressively lower concentrations of NaCl solutions, more and more water enters the cell until, with a solution of, for example, 0.5%, it has become spherical, and its volume has increased considerably. Since the cell wall is not very elastic, placing cells in a still lower concentration of NaCl (0.48%) causes the entering water to exert a hydrostatic pressure sufficient to stretch the walls of some cells, and hemolysis takes place. In 0.48% NaCl only a few cells are hemolyzed. As the concentration of the solution is progressively decreased, more and more cells are hemolyzed, until in a 0.32% solution all hemoglobin is released, and the blood is said to be completely hemolyzed or *laked.* This difference in the behavior of the individual cells depends on their age. The membrane becomes weaker with age; consequently the fragility is increased. From the foregoing it is evident that in supplementing a reduced volume of circulating blood, as after a hemorrhage, a solution isosmotic with the blood must be used.

Chemical hemolysis. ■ It may be recalled that a cell membrane contains materials, such as lecithin which are dissolved by fat solvents. Ether and chloroform therefore, cause hemolysis by destroying the integrity of the cell membrane. Bile salts, saponin, and snake and spider venoms also have hemolytic action. The toxins produced by certain bacteria and the parasitic organism of malaria (which enters and destroys the erythrocytes) are also able to cause hemolysis. When hemolysis occurs within the blood vessels, most of the liberated hemoglobin is broken down to bilirubin and eliminated by the liver, or it is excreted by the kidneys.

■ Crenation

When a drop of blood is placed in a solution of greater osmotic pressure than that of a 0.9% NaCl solution, water leaves the cell. As a result, it shrinks and shrivels, a condition known as *crenation.*

Cells placed in a solution of urea or ammonium chloride having an osmotic pressure greater than that of a 0.9% NaCl solution do not undergo crenation but are hemolyzed. Since these substances do not dissolve the cell membrane, it must be concluded that the membrane is permeable to the molecules of these substances; water follows the molecules into the cell and the volume increases. These solutions behave like distilled water.

■ WHITE BLOOD CELLS OR LEUKOCYTES

Classification. ■ According to size, granules, staining reactions, and number and shape of nuclei, leukocytes or white blood cells (WBC) are divided into classes: granulocytes and nongranular leukocytes (lymphocytes and monocytes). WBC are true cells as they contain both nucleus and cytoplasm. They are colorless and more resistant than RBC to changes in the external medium.

Granulocytes, or granular leukocytes. Granulocytes are subdivided into neutrophils, eosinophils, and basophils (Fig. 12-4). The neutrophils are the most numerous, constituting approximately 65% to 75% of all leukocytes; their number increases in many infectious diseases.

Lymphocytes. These nongranular leukocytes (Fig. 12-4) are the next most abundant, constituting about 20% to 25% of all leukocytes. They are about the same size as the red blood cells.

Monocytes. Monocytes (Fig. 12-4) are considerably larger than lymphocytes, are also nongranular, and constitute about 7% of the total number of leukocytes. Monocytes are also called mononuclear macrophages.

Number of leukocytes. ■ The number of leukocytes is generally from 5,000 to 9,000/mm^3 of blood, but the number is quite variable. It is greatly increased by strenuous physical exercise (to as high as 35,000), in certain emotional stresses, and in pain. This increase, known as physiological or activity leukocytosis, is only an apparent increase; it is due to the washing of the cells out of stagnant pools into the active circulation. A real increase, called pathological or inflammatory leukocytosis, occurs during infectious diseases.

Life history of leukocytes. ■ Lymphocytes and monocytes are made in lymphatic tissue of lymph nodes, tonsils, spleen, thymus, and mucosa of the intestine. Granulocytes are formed in red bone marrow. Leukemia is an excessive formation (or a lack of removal) of white blood cells and may interfere with the production of red blood cells and platelets (thrombocytes). Hence certain forms of leukemia are associated with anemia and retarded blood coagulation. It is not known where these cells are destroyed, and their life-spans vary from a few hours to 200 days.

Functions of leukocytes. ■ The primary function is to provide a mobile system of protection for the body. Although all WBC are said to be more or less capable of ameboid movements by means of pseudopods, this property, as well as that of phagocytosis, is especially developed in the neutrophil group of granulocytes and in the monocytes. Their amebic capability enables them to pass through the walls of the capillaries and into the tissue spaces, for which reason they are sometimes called the wandering cells (Fig. 16-4). This process is known as *diapedesis.*

In normal blood or plasma, granulocytes and monocytes move at random. But when a clump of bacteria is near, their course is directed toward the bacteria. On arriving there, they cease to move and begin to engulf the bacteria, a process called *phagocytosis.*The substance exerting this attractive force is a product of bacterial metabolism or a substance derived from the body tissues when subjected to bacterial invasion. The response of a motile cell to the directional influence of a chemical substance is known as *chemotaxis.* The value of chemotaxis lies in the speedier engulfment of the bacteria and their subsequent disposal. Generally the neutrophils and monocytes perform their functions only after they have left the blood, that is, in the tissues.

The granulocytes and monocytes engulf not only bacteria but also almost any other particulate matter of small dimensions. When a tissue has been broken down to a certain extent by some pathological process, these cells engulf and remove the cellular detritus. Blood clots, the worn-out red blood cells or the fragments of their disintegration, and useless organs or parts of organs (e.g., the tail of a metamorphosing tadpole) are thus disposed of.

Leukocytes contain many enzymes. Some, quite similar to those in the alimentary canal, are able to digest and liquefy the protein material (such as bacteria and cellular debris) ingested by the leukocytes. The body normally possesses a reserve of phagocytes stored in the liver, spleen, bone marrow, and lymph nodes, and in case of an infection it immediately sets to manufacturing more at an accelerated rate. It is evident that conditions which lessen the activity of the phagocytes or interfere with the rate of their formation must weaken this line of defense. Among such conditions are exposure to extreme heat or cold, worry, loss of sleep, alcoholism, wasting diseases (e.g., diabetes, cancer), and malnutrition, especially the lack of essential amino acids and vitamins.

Lymphocytes are important in the production of serum globulins, both beta and gamma. Immune substances *(antibodies)* are associated with gamma globulins, so that the lymphocytes occupy a position in defense reactions of an immunological nature.

■ PLATELETS

The platelets are spherical or oval bodies about 3μ in diameter (Fig. 12-4). They number about $250{,}000/mm^3$ of blood and are formed in the red bone marrow. Platelets play a significant part in hemostasis, but they also have an important role in body response to injury. Thus they are involved in immunological and inflammatory reactions due to injury of vascular epithelial linings and to entry of foreign substances into the bloodstream. They are believed to be phagocytic of antigen-antibody complexes, viruses, and bacteria.

■ HEMOSTASIS

■ Hemostatic mechanism

Several elements participate in *hemostasis*—the stanching of bleeding. These include the blood vessels, the plasma (and especially the platelets contained therein), the vessel wall, and neural and humoral activity. The process of hemostasis is not completely understood, for many specific details about the interaction of contributing mechanisms are unknown. Therefore, we cannot provide a unified theory that would receive the approval of all investigators. However, we will discuss the various elements in a traditional way, emphasizing those aspects for which there is substantial evidence.

Blood vessels. ■ The role of the vessels in hemostasis is two-fold. They prevent leakage of the blood into the tissues under normal circumstances and they undergo constriction when the vessel wall is severed. This constriction is very effective in reducing blood flow and thereby blood loss. It is a reflex response initiated by the trauma and mediated by sympathetic nerves. It may be prolonged by humoral agents (e.g., epinephrine) in the bloodstream or by myogenic spasm. The importance of vessel constriction to hemostasis varies with vessel size and type. It is more successful in arteries with strong muscular coats than in veins.

Platelets. ■ A dominant role in hemostasis is attributable to the platelets. They are also believed to contribute to the integrity and continuous repair ("nutrition") of the endothelial cell lining of the vessels along with vitamin C. The role of platelets in hemostasis involves three stages: adhesion, release (secretion), and aggregation. When a vessel is injured, the endothelial cell lining loses its nonwettability and smoothness. Exposure of subendothelial layers, the basement membrane, and collagen to the bloodstream induces the platelets to adhere to these structures. Platelets which adhere to the collagen fibers at the edge of the wound change their shape and begin selectively to release their substance. The most important secretions are ADP, platelet factor 3, which is a phospholipid (PF-3), and serotonin. ADP promotes the adhesion of additional platelets one to another and aggregation, layer by layer, ensues until a platelet plug is formed that fills the rupture in the vessel wall. Such platelet plugs are formed even when there is a deficient clotting system in the individual. Aggregation of platelets is not simply a surface phenomenon, for aggregated platelets have undergone intracellular changes. It is known that agents which stimulate aggregation, e.g., epinephrine, inhibit adenylcyclase and consequently lower the platelet level of cAMP. Prostaglandin PGE_1 and caffeine depress aggregation and increase adenylcyclase activity and cAMP levels. The PF-3 which is released during platelet aggregation derives from the platelet membranes and is an important factor in the intrinsic clotting mechanism. Serotonin release probably aids in promoting or maintaining vessel constriction.

Plasma. ■ The plasma participates in hemostasis in several ways. Plasma contains two complex enzyme systems; one is concerned with blood coagulation, the other involves fibrinolysis. The end product of the system concerned with coagulation is a protein known as fibrin. This substance may serve as the substrate for the growth of fibroblasts and thus be essential for the formation of scar tissue and the final repair of injured vessels.

Vessel walls, neural and humoral activity. ■ We have stated that the subendothelial structures induce platelet adhesion by contact. Another contribution of the vessel wall to hemostasis is the release of an intracellular, particulate lipoprotein known as *thromboplastin.* Tissue thromboplastin is a trigger for the extrinsic clotting system to be described shortly. Besides promoting constriction of vessels, neural and humoral activity brought about by traumatic bleeding acts to mobilize platelets from storage sites, especially the spleen, and for the release of coagulation accelerators and inhibitors into the bloodstream from sites of production or storage.

■ Coagulation

It is a familiar fact that shortly after the blood leaves the vessels it loses its fluidity and becomes jellylike. This change is known as *coagulation* or *clotting.* On learning that no less than thirty-five compounds take part in the formation of a firm clot, it is readily understandable that blood coagulation is one of the most complicated chemical processes in our body. Maintaining the fluidity of the blood while it flows through normal vessels and yet preserving its coagulability is not only one of the marvels of our body but is also of vital importance.

Formation of a clot. ■ When a beaker of freshly drawn blood stands for a few minutes, a meshwork of microscopic threads or fibrils is formed throughout the entire mass of blood. This is an example of the intrinsic clotting mechanism in operation. The reaction is due to traumatization of the blood, in this case removal from its natural environment. These fibrils entangle the cells and transform the liquid blood into a gel. Soon the fibrils begin to shrink (clot retraction) and to press out of the gel a slightly yellowish fluid; the shrinkage continues for some time and progressively more fluid is formed, and the gel grows firmer to form a clot. In this fluid, called *serum,* floats the solid clot or coagulum composed of the fibrils and entangled blood cells.

When fresh blood is whipped with some foreign object (e.g., a small coil of wire), the usual solid clot is not formed, but the beater becomes covered with a mass of red, stringy material; the vessel contains a liquid known as *defibrinated* blood. Washing the stringy material reveals a white, fibrous, and highly elastic substance; it is *fibrin,* a protein derived from a precursor, *fibrinogen,* normally present in the plasma.

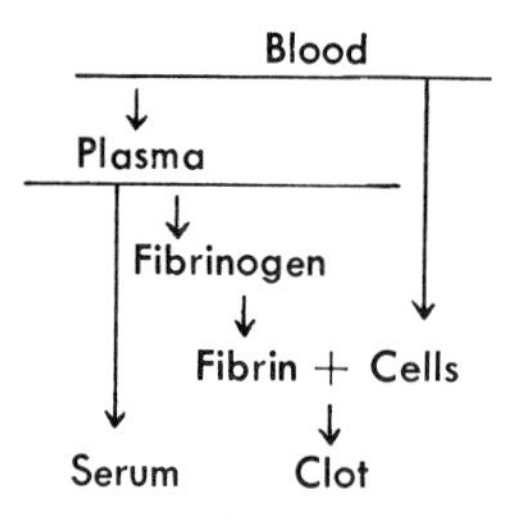

When a blood vessel is damaged, the extrinsic clotting mechanism operates and fibrils of fibrin adhere to the injured vessel walls to form a clot as described above. This clot seals the opening and prevents the escape of blood. As clot retraction progresses, a process requiring thirty to sixty minutes, the damaged edges of the vessels are drawn together to reinforce hemostasis. Clot retraction is dependent upon the presence of platelets. It is believed that they may provide the energy for the shortening or syneresis of the fibrin threads that occurs.

Production of thrombin. ■ The final, or fourth, step in clotting involves the conversion of a soluble plasma protein, fibrinogen, into an insoluble protein, fibrin, by an enzymatic process. The enzyme responsible is *thrombin,* a protein of 34,000 molecular weight (MW). Thrombin is normally absent from the circulating blood and, once formed, is inactivated progressively by *antithrombins.* Thrombin cleaves fibrinogen, a protein of 340,000 MW, into fibrin monomers and fibrinopeptides. The fibrin monomers subsequently polymerize to form fibrin threads for clotting.

The production of thrombin from its inactive precursor, *prothrombin,* is the third step in coagulation. The conversion of prothrombin to thrombin is the result of a complex series of enzymatic processes. The rate of thrombin production determines the time required for clotting. Prothrombin is a glycoprotein of 68,000 MW synthesized in the liver. Its concentration in the plasma is normally about 15 mg/100 ml, but this depends upon an adequate supply of vitamin K for its formation.

Prothrombin
↓ (3)
Thrombin
↓
Fibrinogen —(4)→ Fibrin

Activation of prothrombin. ■ In the blood there are a number of different factors or procoagulants that, when activated, are responsible in turn for the activation of the next factor in the sequence. These factors are proteins, mostly beta globulins, that are designated by Roman numerals according to international convention. All of the complex reactions are necessary for the rapid conversion of prothrombin to thrombin. Calcium and phospholipid are necessary cofactors in the conversions. The phospholipid, probably phosphatidyl choline (Chapter 23), derives from platelets, PF-3, in the intrinsic clotting mechanism and is the same as tissue *thromboplastin.*

We can schematize the processes of complex formation which are responsible for the activation of prothrombin in the flow diagram which follows:

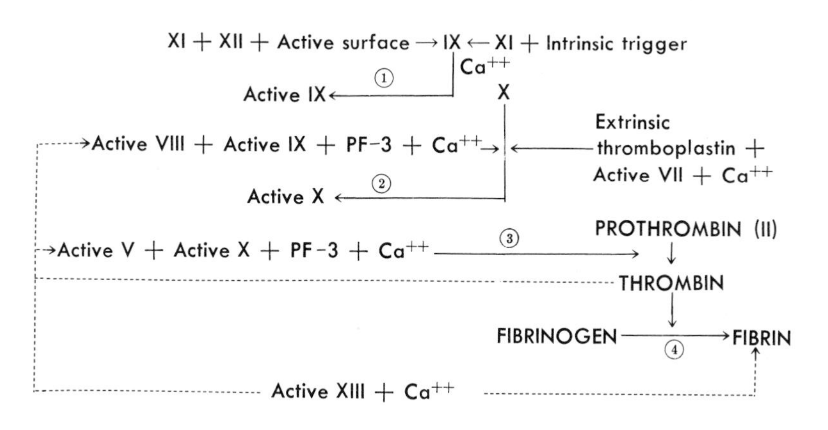

It should be noted that some investigators consider factors VII, IX, and X to be degradation products of prothrombin and not original clotting factors. Despite this, the schema below is a reasonably accurate and current description of the stages of complex formation involved in coagulation. The active surface in the intrinsic mechanism has been postulated to arise from platelets. Since blood lacking factor XII will clot, an intrinsic "trigger" that reacts with factor XI and calcium to activate factor IX is presumed to exist. Its source is unknown. After the formation of active factor X, both intrinsic and extrinsic clotting systems proceed identically. There is evidence that small quantities of thrombin are required to activate factors V, VII, and XIII. Factor XIII is known as the fibrin stabilizing factor. It is required in the formation of the clot. A deficiency of this factor leads to prolonged bleeding, delayed wound healing, and the formation of excessive scar tissue.

Clotting abnormalities. ■ The protein factors just discussed are often referred to by a more descriptive nomenclature. Factors I and II are fibrinogen and prothrombin, respectively. Proaccelerin and proconvertin are the names often given to factors V and VII, respectively.

Factor VIII is also called AHG (antihemophilic globulin). The absence of this factor in the plasma is the most frequent and serious cause of genetically determined clotting defects (Chapter 30). It is designated as *hemophilia A* and has been recognized for over 2,000 years. A trivial cut or bruise may result in prolonged bleeding. The abnormality is a sex-linked characteristic, with the disease transmitted from a female carrier to a son. The incidence of hemophilia A is about 1 in 15,000. The spleen is a rich source of factor VIII. Factor IX is otherwise known as the plasma thromboplastin cofactor (PTC), or Christmas factor. A deficiency in plasma concentration of factor IX is also a genetically determined defect and is responsible for *hemophilia B.* This disease is only one fifth as common as hemophilia A.

Hemorrhagic disease results from deficiencies in factors X (Stuart-Prower factor) and XI (plasma thromboplastin antecedent, or PTA factor). Factor XII (Hageman trait) deficiency induces long clotting times, but the victims are not truly bleeders.

There are other sources of bleeding abnormalities, as one would expect from our discussion of hemostasis to this point. Since the liver (Chapter 21) is concerned with the production of most clotting factors (II, VII, IX, X), disease of that organ or vitamin K deficiency leads to clotting disorders. The estrogen component of oral contraceptives alters the hemostatic balance which leads to an increased incidence of thrombosis. A reduced platelet count, *thrombocytopenia,* is the most frequent cause of bleeding states. It may occur because of depressed production as happens in pernicious anemia, certain drug therapies, and irradiation; or it may result from increased peripheral destruction by antibody. Thrombocytopenia causes bleeding in capillaries rather than large vessels and the presence of minute, discolored blotches on the skin is referred to as *thrombocytopenic purpura.*

Anticoagulants. ■ There appears to be a specific plasma inhibitor for most of the active factors in the coagulation mechanism. The most common is an inhibitor of factor VIII, although the best understood are the antithrombins. Of the six antithrombins recognized, only II and III are of significance. Antithrombin II is a plasma cofactor of heparin and antithrombin III is a heterogeneous globulin which is capable of inactivating thrombin. Heparin is present in the tissues of normal individuals in small quantities. Sufficient quantities for clinical use are prepared from the lung tissue of animals. Heparin in combination with its plasma cofactor, antithrombin II, inhibits the conversion of fibrinogen to fibrin. It also appears to facilitate the reaction of thrombin with antithrombins and the adsorption of thrombin on fibrin which, in turn, leads to inactivation.

Dicumarol is a drug of value in preventing intravascular clotting; it interferes with the synthesis of prothrombin by the liver. Both aspirin and prostaglandins (Chapter 29) inhibit platelet activity, thus they have anticoagulant effects.

Coagulation time. ■ Coagulation time is the length of time required for coagulation to occur after the blood has been shed, i.e., it is a measure of the effectiveness of the intrinsic clotting system. The time varies somewhat in different people; from five to eight minutes may be regarded as the normal range. It sometimes happens that the blood clots in the usual length of time but that the clot is too lacking in firmness to adhere to the bleeding surface. It is then necessary to determine the *bleeding time* by puncturing the skin and applying to it at regular intervals a piece of filter paper and noting the length of time elapsing before the paper is no longer stained with blood. Bleeding time concerns the interaction of blood and the injured tissue, i.e., it measures the effectiveness of the extrinsic clotting system. This time averages from two to six minutes.

Retarding coagulation is sometimes desirable either for practical or experimental purposes. Clotting within blood vessels may be retarded by heparin and by dicumarol. Clotting of blood outside the body can be retarded by cooling, or it can be prevented by the addition of sodium oxalate, which removes the free calcium from the solution. Any rough surface hastens clotting; on the other hand, a nonwettable surface such as a paraffin- or silicone-lined container retards coagulation.

There are numerous methods by which coagulation may be hastened. Warming or agitation of a container of blood will speed coagulation, as will the addition of tissue extract or thrombin. Bleeding can be controlled by application of pressure; a tourniquet is used if large vessels are involved. Local application of epinephrine aids in the control of small hemorrhages by causing a constriction of the blood vessels. The application of a solution of thrombin is of great assistance in controlling certain hemorrhages.

Intravascular coagulation or thrombosis. ■ That coagulation can take place in the body during life is demonstrated by introducing a foreign body, for example, a rough needle thrust through a blood vessel; it becomes covered with a coagulum. Thrombosis may occur when a blood vessel is crushed or its inner lining (endothelium) is injured, in arteriosclerosis, in varicose veins, and by bacterial infection of a vessel, the heart, or a valve.

A clot remaining attached at the site of formation is known as a *thrombus;* when it breaks loose and floats in the blood, it is called an *embolus* and may then be carried to distant parts of the body. Both the thrombus and the embolus may seriously interfere with blood circulation by occluding the vessels. In the brain this may give rise to apoplexy; in the heart it may cause coronary thrombosis and angina pectoris; in a restricted area it may cause death of the tissues (gangrene or necrosis).

A thrombus or embolus gradually undergoes liquefaction by the digestion of its protein material. This is due to a proteolytic enzyme (plasmin) present in its precursor form in the blood.

■ Fibrinolysis

We have stated that the plasma contains two complex enzyme systems that are involved in the hemostatic phenomenon. In addition to the coagulation system, there is the *fibrinolytic system.* In this system a plasma protein, *plasmin,* of about 100,000 MW cleaves fibrin into fibrin degradation products (FDP).

Production of plasmin. ■ The plasma contains *plasminogen,* the inactive precursor of plasmin. It is a protein of about 143,000 MW. Plasminogen can be activated in either of two ways, one direct, the other indirect. The direct activation is known to result from the proteolytic activity of a naturally occurring proteolytic enzyme, *urokinase.* Other proteolytic enzymes such as trypsin are capable of activation. The source of urokinase in the body is uncertain; it and other direct activators are released, however, by direct tissue injury and, in the absence of cellular damage, by mechanisms unknown.

The indirect activation of plasminogen appears to be a two-step process. In this, proactivators in the plasma are first activated by other activators found in various tissues such as veins. Activation of the proactivators also is accomplished by bacterial products such as *streptokinase* from *Streptococcus* bacteria.

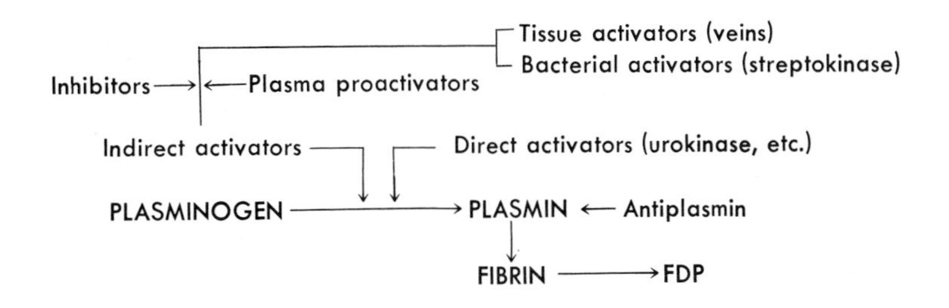

As in the case of the coagulation system, the fibrinolytic system is carefully balanced by inhibitors. These act upon direct and indirect activators as well as upon plasmin.

The available evidence indicates that there is a system present in the plasma for removal of intravascular clots or thrombi. This is in accord with the observation that sterile blood clots are dissolved within a few hours or days. Platelets contain antiplasmin, perhaps to protect the platelet plug from an undesirable hemostatic disturbance caused by fibrinolysis. Pathological activation of fibrinolysis causes severe bleeding problems and the presence of high levels of inhibitors increases the probability of thrombosis.

■ ANTIBODIES AND IMMUNITY

Bacteria, molds, viruses, and protozoa may enter the body by way of the skin, alimentary canal, or respiratory tract. Having gained entrance, they may cause great harm either by the formation and liberation of toxins or by the actual destruction of the tissues of the body; in a word, disease. Diphtheria and smallpox are examples of the former; of the latter, leprosy is an example.

Lines of defense. ■ The body is equipped with several lines of defense. The first is the unbroken skin and the mucosa of the digestive tract that bar entrance to most microorganisms. Other lines of defense lie within the body itself.

It has been noted already that leukocytes constitute a line of defense by virtue of their phagocytic and digestive functions. This line of defense is called *cellular mediated immunity* (CMI). In addition, there exists another line of defense known as *humoral mediated immunity* (HMI). Each of these lines of defense depends upon lymphocytes. Because there is not universal agreement among authorities about all the aspects of immunity, we will consider only a brief outline of a current and satisfactory hypothesis.

Cellular mediated immunity. ■ During fetal life, certain "stem" cells are set free from the bone marrow and some lodge in the thymus gland. Here they become lymphocyte producing cells, lymphoblasts. In a way that is, as yet, obscure the thymus "programs" some of these cells to become "immunocompetent," to be able to resist disease. Numbers of these programmed cells enter the bloodstream and populate other lymphoid tissue, notably the spleen and lymph nodes. In these sites, the immunocompetent lymphocytes form germinal centers that can produce large numbers of specifically sensitized lymphocytes by clonal growth when appropriate stimulation by foreign substances occurs. These small, sensitized lymphocytes enter the circulation. They are relatively long lived and may persist for years in humans. Unique receptor sites are present on the lymphocyte surface that react to particular invading foreign cells or substances. The latter are spoken of as *antigens*. The sensitized, immunocompetent lymphocytes respond to the presence of antigen in a variety of ways; they may affix themselves to the invader and destroy it directly or they may release substances (lymphotoxins) that are toxic to the invaders. Certainly, they produce a chemotoxic substance that attracts and holds macrophages (monocytes) to the antigenic site. The combined action of lymphocytes and macrophages in destroying or neutralizing the antigen constitutes what is called CMI. Although the thymus gland regresses after adulthood is attained, some lymphoid tissue persists in the gland. Consequently, "stem" cells from the bone marrow can repopulate this tissue, and additional immunocompetent lymphocytes can be made available to the other lymphoid tissues. Furthermore, the germinal centers in the spleen and lymph nodes retain numbers of lymphoblasts that possess a "memory" for the antigen. CMI is also responsible for graft rejection and delayed hypersensitivity.

Humoral mediated immunity. ■ The bone marrow and lymphoid tissues also produce other lymphocytes that are somewhat less mobile than the small, long-lived lymphocytes discussed above. These cells have a life-span of only a few days. They are capable, upon stimulation by antigen, of being transformed into larger cells known as plasma cells or plasmacytes. Plasma cells synthesize immunoglobulins, the *antibodies* which are found in the plasma. Antibodies are the vehicle for HMI. Production of specific antibody in response to specific antigen stimulation is not immediate but exhibits a lag period of several days to a week, after which there occurs an abrupt rise in the plasma concentration (titer) of specific antibody. Catabolic processes remove circulating antibody so that the titer declines, although the "memory" of the antigen is retained by lymphocytes in the lymphoid tissues. Reexposure to the same antigen produces a rapid and marked increase in specfic antibody titer. These facts explain the need for frequent "booster" shots in order to maintain optimal humoral mediated immunity.

There are many kinds of antibodies. The nature of the antibody is determined by the nature of the antigen. When the antigen is a *toxin*, the antibody is an *antitoxin*. The body responds to foreign cells by the production of materials causing agglutination and lysis of the invaders. Inoculation with a foreign protein calls forth the formation of an antibody causing precipitation of the protein.

Active immunity. ■ The inoculation of an animal or the infection of a human being with bacillus or virus stimulates the production of antitoxin. If the animal or human being recovers, natural *active immunity* is said to have been acquired. For certain diseases (e.g., smallpox, typhoid fever, and rabies) artificial active immunity can be established by vaccines and toxoids. A vaccine consists of bacteria or viruses that have been highly attenuated by treatment with heat, ultraviolet rays, drying, and so on. They have lost their power to set up any serious pathological disturbances in the host but are capable of inducing the production of antibodies. Toxoids are chemically altered toxins that have antigenic properties but are not toxic.

Passive immunity. ■ The injection of the serum of an animal that has been immunized by the injection of a toxin (or the serum of a human being who has recently recovered from the disease) into another animal confers on the host a *passive immunity*. The recipient is not manufacturing antibodies; he is merely using the antitoxins borrowed from the immune animal. Antitoxins can be isolated from bacterial cultures or from immune serums. They have been obtained against diphtheria, scarlet fever, tetanus, botulism, and the venoms of spiders and snakes.

Passive immunization by means of antiserums (immune serums) and of antitoxins provides protection that usually lasts but a few weeks.

Tests for immunity. ■ The injection of a very minute dose of a certain toxin is frequently employed to determine the susceptibility of an individual to a particular disease. Thus in the Schick test for diphtheria, one fiftieth of the minimal lethal dose in 1 ml of a 0.9% NaCl solution is injected subcutaneously; susceptibility to diphtheria is indicated by an area of inflammation at the point of injection within twenty-four hours. An individual not having active immunity to diphtheria, as shown by the Schick test, may acquire immunity by two or three injections of toxin-antitoxin. In this preparation the antitoxin is just a little less than that necessary to neutralize the toxin, and the latter thereby stimulates the production of antitoxin in the body.

Allergy. ■ Some individuals are hypersensitive to certain foreign substances and react abnormally to them, a condition known as *allergy*. The exciting agents, known as *allergens*, may be organic or inorganic, but most frequently they belong to the proteins, especially the proteins of egg, milk, fish, meat, and wheat. Young people are more susceptible. The disturbances are of various kinds, such as the hives and rashes (eczema) some individuals suffer after eating a certain food for which they have an idiosyncrasy; others are afflicted with dyspepsia, diarrhea, or cardiac disturbance. And still more common are the asthma and hay fever caused by the proteins of pollen, dandruff, feathers, and dust entering the body by way of the respiratory passages. Mere contact may also be a causative factor, for example, with poison ivy. It is frequently possible to determine the particular protein (from pollen or other sources) to which the victim is hypersensitive from the results of a skin or patch test.

Reactions. ■ Antigen-antibody reactions can take place in several different ways. The antibody has been given a name indicative of the type of reaction that occurs.

Agglutinins. The addition of serum from a person recovering from typhoid to a suspension of typhoid bacilli causes clumping of the bacilli; the germs are said to agglutinate. This is brought about by the presence in the immune blood of an antibody known as an agglutinin. Agglutinins are specific, and the clumping observed is a means of diagnosing typhoid fever and of detecting typhoid carriers.

The serum of one individual may cause agglutination of red blood cells of another (Fig. 12-7). This has important bearing on the transfusion of blood, for the clumping of the cells may block the flow of the blood in the capillaries (embolism). Consequently, the blood of the donor must be tested for its compatibility with the blood of the recipient.

Precipitins. When an animal receives several injections of an egg albumin solution, a few days elapsing between the successive injections, the serum of this animal acquires the ability to precipitate egg albumin. A drop or two of the serum added to an albumin solution causes a rapid precipitation of the albumin. The agent in the blood causing this is called a *precipitin,* and its action is specific. For example, if the egg albumin was obtained from the egg of a chicken, the serum precipitates the albumin of the chicken egg but not from any other egg.

If a rabbit receives injections of human serum over a period of several days in progressively increasing amounts, the rabbit blood acquires a precipitin that causes a turbidity or a flocculent precipitate in human serum. Such anti-human immune serum causes 100% precipitation in human blood, also 100% in the blood of anthropoid apes, only 92% in the blood of common

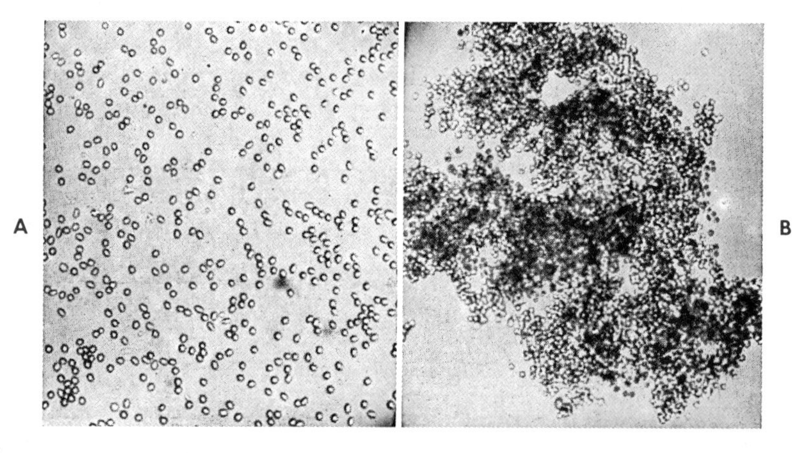

Fig. 12-7
Normal, **A,** and agglutinated, **B,** red blood cells. Agglutination in **B** resulted from transfusion of incompatible blood. (From Watter, C. W.: Hygeia, 1941.)

monkeys, and none in that of lemurs or animals lower in the scale. It may, therefore, be used to differentiate human blood from blood of all common animals.

Lysins. The serum of one species is often destructive to the RBC of another species. For example, if the serum of a dog's blood is mixed with the blood of a rabbit, the RBC of the rabbit become laked (hemolyzed). Normal guinea pig's serum has no hemolytic action on the cells of a rabbit. But if the guinea pig during the course of ten or twelve days receives three or four injections of about 4 ml of defibrinated rabbit's blood, the blood of the guinea pig acquires the power to destroy the rabbit's cells. The agent causing the destruction of a foreign cell is known as a *lysin;* in case the foreign cell is a red blood cell, the lysin is specifically called a *hemolysin.* This may happen in mismatched blood transfusions.

Interferon. A major natural line of defense against viral diseases involves the *interferon system.* Within a few hours after a virus enters the body, a protein compound known as interferon appears in the bloodstream of the infected individual. Interferon is one of the most active biological substances known to man. Although interferons produced by different species of animal vary, i.e., they are species-specific, the interferon produced by a given species is active against many different viruses.

When a virus enters a cell, its chromosome associates with the cell ribosomes (Chapter 2) and causes the production of new viral nucleic acid and protein. This virus multiplication is at the expense of the cell and infected cells may not survive. However, the cell gene responsible for interferon production is stimulated by a mechanism not yet understood and quantities of interferon are excreted by the cell. This interferon reacts with other cells, causing them to produce new RNA and an antiviral protein. Interferon itself apparently does not attack the virus, but the antiviral protein that it induces a cell to produce inhibits the association of the virus chromosome with the cell ribosomes. Consequently, the replication of the virus is inhibited and the spread of the infection is checked.

Interferon has been detected in the serum of humans afflicted with mumps, influenza, chicken pox, yellow fever, and measles. Injections of purified interferon provide passive immunity to virus infection; but the substance is not readily available, since to be effective it must be obtained from human sources for human use. Because of this and because interferon may prove valuable in the treatment of virus-related tumors, a great deal of current research is directed toward the discovery of a substance that will, when administered, safely induce the natural production of interferon.

■ BLOOD GROUPS

Human red blood cells contain at least thirty commonly occurring antigens, certain ones of which are highly antigenic and can cause transfusion reactions if proper precautions are not taken. Two groups, the ABO and the Rh factors, are the most likely to cause difficulties with transfusions; the other antigens are used principally for studying inheritance (Chapter 30) in order to establish parentage.

Agglutination. ■ Agglutination of RBC is brought about by two factors: *agglutinogen* and *agglutinin.* Agglutinogens are antigens. They are associated with the red cells and make it possible for the cells to be agglutinated. Agglutinins are the active agglutinating agents; they are antibodies dissolved in the serum. Some agglutinins are naturally found in the blood.

Agglutinogens are hereditary characteristics and remain unchanged throughout life. A large number of distinct agglutinogens have been discovered; the most important agglutinogens are the A and B, the M and N, and the Rh factor. Their respective presence in the red cells is responsible for the three large groups (or types) of blood: the ABO groups, the MN groups, and the Rh groups.

Table 12-1. Blood types and their agglutinating actions

Blood group	*Its RBC are agglutinated by serum of*	*Its serum agglutinates the RBC of*
A—*b*	B and O	B and AB
B—*a*	A and O	A and AB
AB—*o*	A and B and O	None
O—*ab*	None	A and B and AB

Table 12-2. Determination of blood types

If tested red cells are agglutinated in serum from blood types	*The examined blood belongs to*
Both A and B	AB
A	B
B	A
Neither A nor B	O

ABO blood groups. ■ A red cell may contain either the agglutinogen A or B; it may contain both A and B; or it may contain neither (O). There are, therefore, four groups or types of blood: A, B, AB, and O. In the serum are found two kinds of agglutinins (antibodies); they are known as anti-A (or *a*) and anti-B (or *b*). Agglutinin *a* agglutinates only red cells having agglutinogen A; *b* agglutinates red cells having agglutinogen B. To prevent agglutination of its own cells, if the red cells of a certain sample of blood have agglutinogen A, the serum never has agglutinin *a*. And B is never accompanied by *b*. But if the red cells lack A (or B), the serum has *a* (or *b*). The four blood groups (or types) and their agglutinating actions are given in Table 12-1.

In Table 12-1 it will be noticed that cells from individuals of blood group O, since they lack agglutinogen, are not agglutinated by the serum of any blood group; these persons are called *universal donors.* The serum from individuals of group AB does not cause agglutination of cells of any group; these persons are called *universal recipients.* But no single type of whole blood exists that does not agglutinate red cells of any type and whose cells are not agglutinated by the serum of any type; that is, there is no blood group that renders a person a universal donor and universal recipient at the same time.

However, in transfusion the donor's blood is diluted so largely with the recipient's blood that the agglutinating power of the donor's blood becomes negligible, and, therefore, only the effect of the agglutinin in the serum of the recipient's blood on the agglutinogen of the cells of the donor's blood need to be considered. Hence in transfusion the donor and the recipient should be of the same type or the donor should belong to the O group. In the population of the United States, these blood groups are distributed approximately as follows: O group, 45%; A, 41%; B, 10%; AB, 4%.

To determine the group to which an individual's blood belongs, the behavior of the cells of this individual in serum of type A and of type B are noted. Table 12-2 summarizes the reactions of blood typing.

A and B agglutinogens are synthesized from a common precursor substance (H) which is present in O blood group individuals. The precursor is a glycoprotein of about 300,000 MW containing 80% carbohydrate and 20% amino acids. A and B specificity is determined by the addition of a single, different sugar molecule to the terminal end of the precursor. The precursor and A and B antigens are found in many cell membranes and secretions of the body. They are especially prevalent in the vascular endothelium and in cells of the gastrointestinal tract; they appear in the saliva of many individuals and in secretions from the respiratory tract, lactating mammary gland, and Brunner's glands (Chapters 19 and 20) of the small intestine. There is a statistical association between certain blood groups and increased susceptibility to a number of diseases. ABO incompatibility between mother and fetus may occur and lead to hemolytic disease of the newborn, if maternal agglutinin crosses the placenta. Such incompatibility is relatively rare, much more so than the Rh incompatibility to be discussed next.

Rh factors. ■ In addition to the previously discussed agglutinations, another group of agglutinogens known as the Rh factors* are present in most people. These individuals are said to be Rh positive. If the blood of an Rh-positive person is transfused into an Rh-negative individual, no ill effect is experienced, but there is developed in the blood of the latter an anti-Rh agglutinin. A subsequent transfusion of Rh-positive blood leads to agglutination and hemolysis of the transfused blood. To avoid this, both the donor and the recipient should be either Rh positive or negative. Of course, Rh-negative blood does no harm to an Rh-positive person. There are said to be at least twelve types of Rh factors, with seventy-eight different combinations possible. Of these combinations only six are common (93% of the population). By laboratory test the type to which a specimen of blood belongs can be determined.

The Rh factor is hereditary. If one parent is Rh positive (homozygous†) and one is Rh negative (homozygous), the child will always be Rh positive (heterozygous‡) because the Rh factor is a dominant trait. Suppose the father is Rh positive and the mother is negative and the fetus has inherited the positive factor from the father.

$$\frac{\text{♂} Rh^{+}Rh^{+} + Rh^{-}Rh^{-} \text{♀}}{Rh^{+}Rh^{-}}$$

Some of the Rh agglutinogen finds its way through the placenta§ into the maternal blood; this generates anti-Rh agglutinins. These antibodies, if they pass back into the fetal blood, cause agglutination and hemolysis of the red cells of the fetus. The anemia *(erythroblastosis fetalis)* thus caused frequently proves fatal either before birth (stillborn) or soon after. The first-born child usually shows no ill effects, but the second or third pregnancy may end disastrously. Rh incompatibility of mother and fetus can be treated by the administration of anti-Rh gamma globulin to the pregnant mother. This antibody neutralizes the antigen coming from the fetus and prevents the production of natural anti-Rh antibody by the mother.

M and N factors. ■ The two agglutinogens, M and N, give rise to three blood groups: M, N, and MN. The presence of these agglutinogens is determined by two genes.

Differing from the ABO and Rh blood groups, the agglutinogens M and N associated with the red cells are not accompanied by antibodies (agglutinins) in the serum. Hence in blood transfusions these agglutinogens do not need to be taken into consideration. They are, however, of great value in establishing the parentage of a child when this is in doubt or dispute.

If two babies in a hospital were exchanged and assigned to the wrong parents, the correct parentage can frequently be established by the MN factors as illustrated in Table 12-3.

Table 12-3. Use of M and N factors for establishing parentage

If child is	*The mother*	*The father*	
		Must be	*Cannot be*
M	Must be M or MN	M or MN	N
N	Must be N or MN	N or MN	M
MN	May be M	N or MN	M
	N	M or MN	N
	MN	M, N, or MN	

*So called because the Rh factor was first discovered in the blood of rhesus monkeys.

†Homozygous means that the individual has inherited like genes from each parent, for example, Rh^{+}, Rh^{+} or Rh^{-}, Rh^{-}.

‡Heterozygous means that the individual has inherited unlike genes from his parents, for example, Rh^{+}, Rh^{-}.

§The placenta may be regarded as a membrane that separates the maternal blood from that of the fetus.

READINGS

Brewer, G. J., and Eaton, J. W.: Erythrocyte metabolism: interaction with oxygen transport, Science **171**:1205-1211, 1971.

Clarke, C. A.: The prevention of "rhesus" babies, Sci. Am. **219**(5):46-52, 1968.
Craddock, C. G., Longmire, R., and McMillan, R.: Lymphocytes and the immune response, New Eng. J. Med. **285**(6):324-331; **285**(7):378-384, 1971.
Edelman, G. M.: The structure and function of antibodies, Sci. Am. **223**(2):34-42, 1970.
Harris, M.: Interferon: clinical application of molecular biology, Science **170**:1068-1070, 1971.
Marcus, D. M.: The ABO and Lewis blood-group system, New Eng. J. Med. **280**(18):994-1006, 1969.
Perutz, M.: Hemoglobin: the molecular lung, New Scientist and Sci. J., June 17, 1971, pp. 676-679.
Perutz, M.: Hemoglobin: genetic abnormalities, New Scientist and Sci. J., June 24, 1971, pp. 762-765.
Salzman, E. W.: Cyclic AMP and platelet function, New Eng. J. Med. **286**(7):358-363, 1972.
Solomon, A. K.: The state of water in red cells, Sci. Am. **224**(2):88-96, 1971.
Stormorken, H. S., and Owens, P. A.: Pathophysiology of hemostasis. In Maischer, D. A., Jaffee, E. R., and Luscher, E. F., editors: Disorders of hemostasis: current status, New York, 1971, Grune & Stratton, Inc., pp. 3-29.

13

MECHANICAL FACTORS OF BLOOD CIRCULATION

The chief function of the blood is to aid in maintaining as nearly constant as possible the physical and chemical state of the internal environment of the cells—*homeostasis.* To perform this function the blood must circulate.

The movement of a fluid depends on establishing an *inequality of pressure.* For example, if a circular rubber tube nearly filled with water lies on the table, there is no movement of the water in the tube because the pressure is the same at all points. If pressure is exerted with the hand at some point, the greater pressure forces the water to flow away in both directions until the pressure is equally distributed. On removing the hand the pressure at this point becomes less, and the water now returns. Let us examine such a tube (Fig. 13-1) with a valve at *b* that opens and allows the water to flow from *a* to *b* and to *c* but not in the reverse direction. A valve at *c* allows water to flow from *c* to *a* and not from *a* to *c.* Pressure applied at *a* causes valve *b* to open, but it closes valve *c;* the water, therefore, flows from $a \rightarrow b \rightarrow c \rightarrow a$. By properly timed rhythmic compression and decompression, circulation of the water is established.

To obtain proper circulation of blood in the vascular system, four factors are indispensable:

1. A rhythmically beating heart to generate the necessary pressure
2. A system of blood vessels, without which an active heart would be of no avail
3. An adequate amount of circulating blood
4. A mechanism for ensuring unidirectional flow (valves)

These factors will be discussed in this and the following two chapters.

■ ANATOMY

The circulatory system may be divided into four functionally different parts:

1. The pumping organ—the heart
2. The conducting and distributing vessels—arteries and arterioles
3. The exchanging parts—the capillaries
4. The collecting vessels—venules and veins

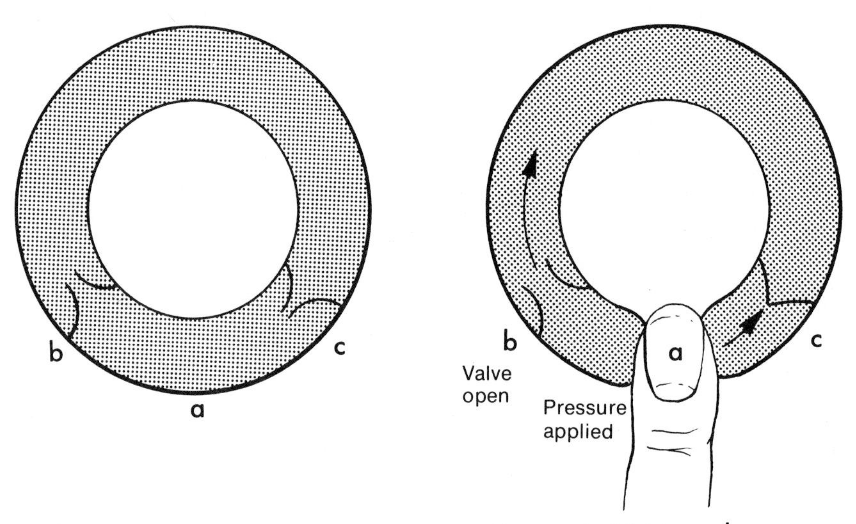

Fig. 13-1
Illustration of circulation of fluid in a closed system; **b** and **c** are the valves; **a,** the point at which pressure is applied.

■ The heart

The heart is a hollow muscular organ; in the human adult male it weighs from 250 to 350 g. In its walls three coats can be distinguished: the endocardium, the myocardium, and the epicardium. The innermost layer, or *endocardium,* is a thin layer of connective tissue fibers and smooth muscle cells covered with a single layer of squamous endothelial cells; this coat is reflected over the valves of the heart. The *myocardium* forms the bulk of the muscular wall. It is covered externally by the *epicardium,* a serous membrane, which is reflected at the upper portion of the heart to form a sac, the pericardium, in which the heart lies. In Fig. 17-2, the pericardium is cut open to show a view of the heart.

The mammalian heart is divided into four cavities: the right atrium, the right ventricle, the left atrium, and the left ventricle (Fig. 13-2). In fact, the mammal may be spoken of as having two hearts, a right and a left heart, each composed of an atrium and a ventricle, but these two hearts are anatomically so intimately united that they are generally regarded as constituting one organ.

The atrial walls are thin and feel flabby to the touch; those of the ventricles are thicker and firmer. This is especially true of the walls of the left ventricle, which may be from three to six times as thick (and therefore as powerful) as those of the right ventricle.

■ Cardiac valves

Between the atria and ventricles there are valves that open into the ventricles: the *bicuspid* or *mitral valve* on the left and the *tricuspid valve* on the right (Fig. 13-2). The bicuspid valve is composed of two triangular flaps of connective tissue; the tricuspid valve has three flaps. These valves open when blood passes through the atria into the ventricles but close during the contraction of the ventricles and thereby prevent regurgitation (backflow) of the blood into the atria. In order that a bulging of the valves into the atria may not nullify their function, there are attached

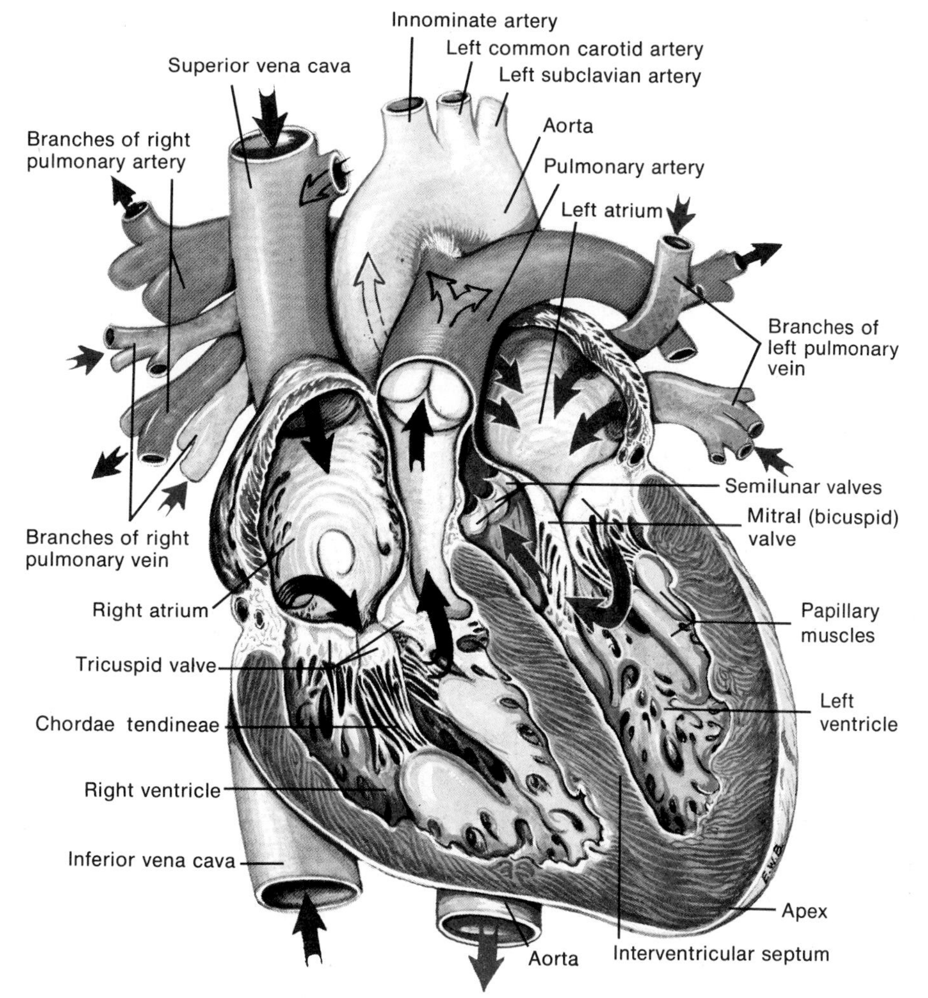

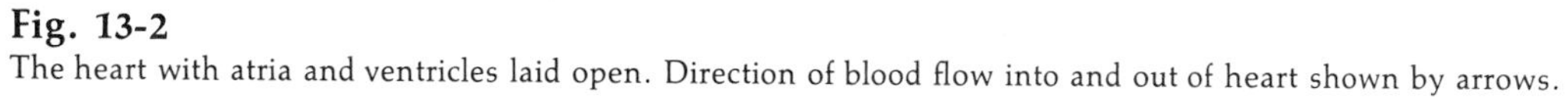

Fig. 13-2
The heart with atria and ventricles laid open. Direction of blood flow into and out of heart shown by arrows.

to the borders of the valves (Fig. 13-2) many slender but strong cords, the *chordae tendineae;* the other ends of the chordae tendineae are attached to fleshy columns, *papillary muscles,* of the ventricular wall; the contraction of these muscles tightens the cords during the contraction of the ventricles. The two other valves of the heart are discussed later.

■ Systemic and pulmonary circulations

Each side of the heart, the right and the left, has its own circulation. The circuit from the left ventricle (Fig. 13-3) through the aorta into the systemic capillaries throughout the whole body and onward to the superior and inferior venae cavae to the right atrium is known as the *greater* or *systemic circulation.* The *lesser,* or *pulmonary,*

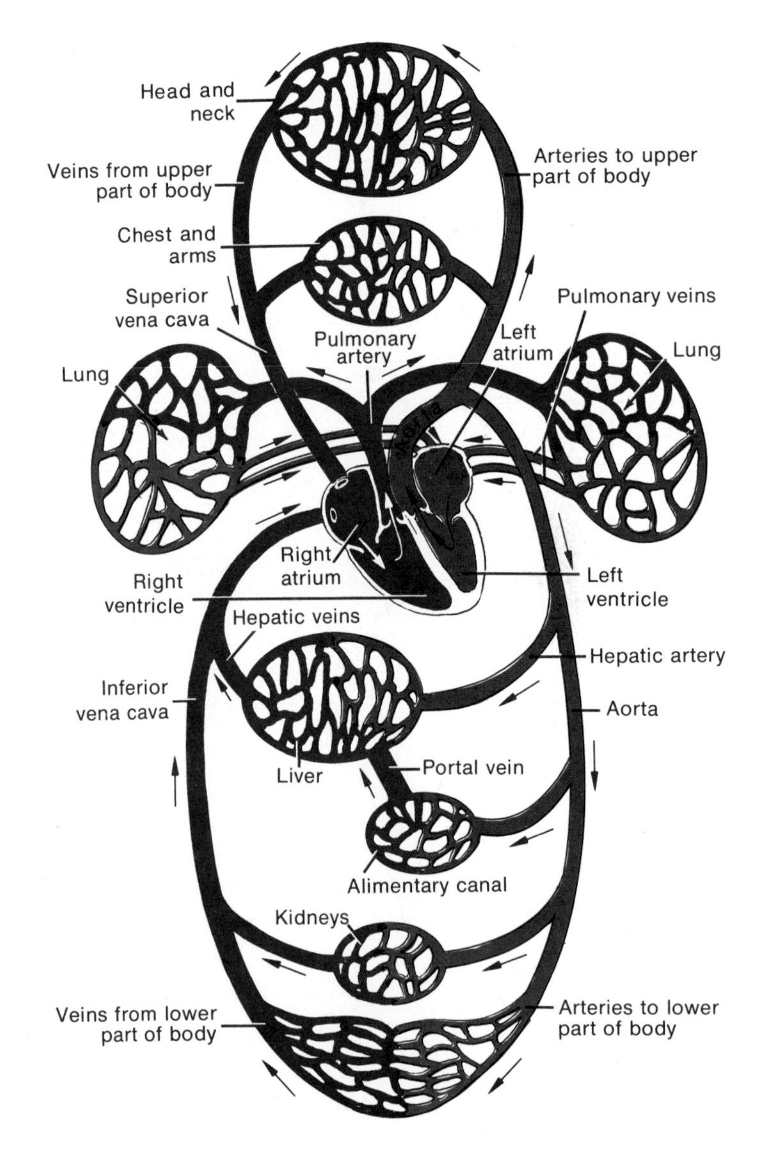

Fig. 13-3
Diagram showing circulation of the blood. Arterial, or oxygenated, blood is shown in red; venous blood is shown in black.

circulation starts in the right ventricle, from which springs the pulmonary artery. The latter gives rise to the pulmonary or lung capillaries; these unite to form the four pulmonary veins that connect with the left atrium. Many of these vessels and their connections with the heart are shown in Fig. 13-2.

In its passage through the lungs the blood acquires oxygen and loses some of its carbon dioxide (Chapter 17); thus oxygenated, the blood becomes arterial, shown as red in Fig. 13-3, and as such it is returned to the left heart. In the systemic capillaries the arterial blood loses some oxygen and acquires carbon dioxide, thereby becoming venous; this is returned to the right heart, to be sent by it to the lungs.

■ Blood vessels

From the ventricles arise tubes, known as arteries, which carry the blood away from the heart; when severed, the blood spurts from the proximal (heart) end. The *aorta* (Fig. 13-3) arises from the left ventricle, and the *pulmonary artery* arises from the right ventricle. At the opening of these arteries the *semilunar* (*semi,* half; *luna,* moon) valves are found, which, by opening up only into the arteries prevent regurgitation of the blood during relaxation of the ventricles (Fig. 13-2). The arteries split into many branches, each smaller than the parent stem; this multiplication of vessels, with progressive decrease in size, continues until vessels about 0.2 mm in diameter are formed—the *arterioles*. The arterioles give rise to microscopic vessels, the capillaries, which exist in countless numbers in the tissues of the body. The capillaries reunite to form larger tubes *(venules)*, and these, in turn, merge to form progressively larger tubes, the veins. Veins return the blood to the atria. From a severed vein the blood flows from the distal end. The course of the blood in the systemic circulation, therefore, is as follows:

Left ventricle → Artery → Arteriole → Capillary → Venule → Vein → Atrium → Right ventricle

Between 15% and 20% of the blood volume is contained within the aorta and arteries; capillaries contain about 5%. The large veins hold most of the circulating volume, about 70% to 75%.

■ Arteries and arterioles

The wall of an artery is considerably thicker, stronger, and more elastic than that of the corresponding vein; this arrangement is necessary because the pressure in an artery is always greater than that in a vein. In the walls of the larger vessels are found small blood vessels *(vasa vasorum)*, by which the vessel wall is nourished. The arterial wall (Fig. 13-4) consists of three coats.

1. In all blood vessels the *endothelium* or inner coat is composed of an elastic membrane upon which is placed a single layer of squamous epithelial cells. Because of its smoothness this coat reduces the resistance encountered by the flowing blood and aids in preventing the coagulation of the blood. It is also nonwettable.
2. The *tunica media,* the thickest coat, is composed of smooth muscle fibers, mostly circularly arranged, and of yellow elastic fibers. This coat constitutes the active part of the artery.
3. The *tunica externa* or outer coat is composed chiefly of white fibrous connective tissue. Being indistensible and tough, it limits the distention of the artery and gives it tensile strength.

Arteries may be classified into *conducting* and *distributing* arteries. The former, of which the aorta and the pulmonary and subclavian arteries are examples, are of larger caliber and have relatively thin walls; in them the tunica media is more liberally supplied with elastic tissue fibers; therefore, they constitute the *elastic type* of arteries. The tunica media of the smaller calibered distributing arteries, although containing some elastic fibers, is more abundantly supplied with smooth muscle fibers. They are the *muscular type* arteries. This difference in construction plays an important part in the functioning of

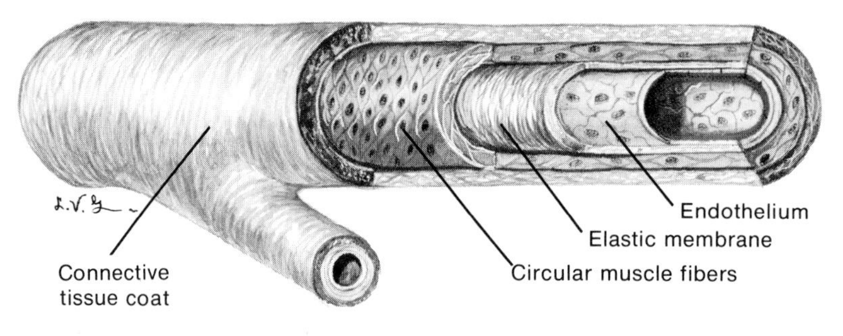

Fig. 13-4
Illustration of coats of a small artery.

these two types of arteries. The walls of the arterioles are composed almost entirely of smooth muscle and an endothelial layer. Arterioles are the major *resistance vessels* of the circulatory system. The smallest of these muscular vessels gives rise to the capillaries.

■ Capillaries

The length of a capillary is from 0.5 to 1 mm; the average diameter is approximately 0.01 mm. The repeated branching of the distributing arteries is accompanied by a progressive thinning of the walls; in the capillary the wall consists only of a very thin layer of flat endothelial cells joined together by intercellular cement. It is important to bear in mind that *blood performs its function of nourishing the tissues during its passage through the capillaries and at no other place.* The function of the other more conspicuous parts (i.e., heart, arteries, and veins) is subservient to that of the capillaries. The study of circulation concerns itself almost entirely with the neural and humoral mechanisms by which the activities of the heart and vessels are regulated in order that an amount of blood sufficient for the nutritional demands of the body may be efficiently supplied to the capillaries.

■ Veins

In veins the muscle and the elastic tissue are less developed, and the white connective (inelastic) tissue is more developed. For this reason veins are less distensible and elastic than are arteries. The venules and small veins constitute the major *capacitance vessels* of the circulatory system. In distinction to the arteries, the veins, especially those of the extremities, are well supplied with valves that prevent backflow of the blood into the capillaries.

■ MECHANICS OF FLUID FLOW

In studying the movement of the blood through the heart and blood vessels we are concerned with: (1) the force (pressure) that propels it, (2) the resistance it encounters, (3) the velocity and nature of the flow, and (4) the volume of the flowing blood.

■ Pressure

The tank in Fig. 13-5, *A*, with its outlet tube, *cs*, is *constantly* filled with water to the line *b*. The tube is of uniform bore and offers the same amount of resistance at all points to a fluid moving through it. When the opening, *x*, is closed, the water in the pressure tubes, *1, 2, 3,* and *4,* rises to the same height as in the tank, *b.* The height of this column of water, *ab,* represents the pressure that the water is exerting on the bottom of the tank. In the same manner the height of the water in the pressure tubes, *4, 3, 2,* and *1,* indicates the pressure in the tube at the points *m, n, r,* and *s,* respectively. The amount

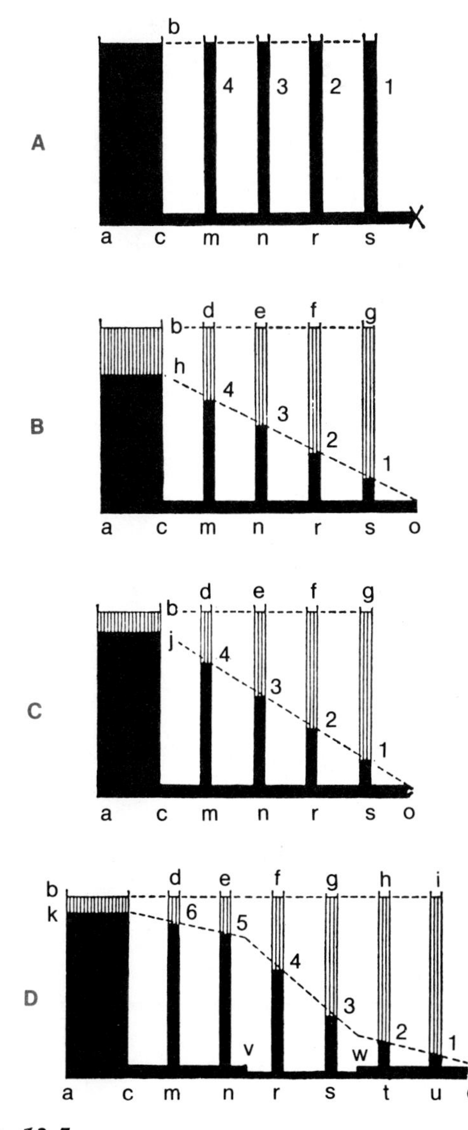

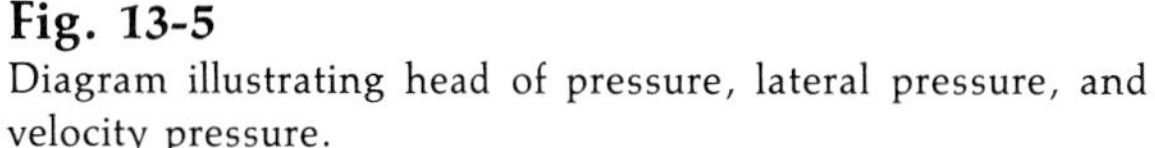

Fig. 13-5
Diagram illustrating head of pressure, lateral pressure, and velocity pressure.

of this pressure can be stated in centimeters of water or millimeters of mercury.

On opening wide the outlet, *o,* the water gushes out, and the levels in the pressure tubes (Fig. 13-5, *B*) fall to *1, 2, 3,* and *4,* respectively. The pressure, *ab,* in the tank is potential energy; it is called the *head of pressure.* Part of this energy is utilized (transformed into heat) in overcoming the resistance that the fluid encounters in its passage from the tank to the outlet. A straight line joining the levels *1, 2, 3,* and *4,* and projected to the tank, meets it at *h.* The pressure, *ah,* is that part of the total energy represented by the head of pressure, *ab,* needed for overcoming resistance; it is the *lateral pressure.*

The kinetic energy of the water as it issues from *o* is what remains of the total energy (head of pressure) after the resistance along the tube, *co,* has been overcome. In *B* this is represented by *hb* and is known as the *velocity pressure.* The line, *ho,* indicates the difference in pressure from one point to another in the tube, *co;* this is known as the *pressure gradient.*

When more resistance is introduced, as by partly closing the opening, *o,* the levels in all the pressure tubes, *d, e, f,* and *g,* in Fig. 13-5, *C,* stand higher, indicating an increase in the lateral pressure, *aj,* but a decrease in the velocity pressure, *jb.*

Due to the uniform distribution of resistance along the tube, the pressure gradient in both *B* and *C* is a straight line, indicative of a uniform fall in pressure from point to point. If we narrow the tube between the points *v* and *w* in Fig. 13-5, *D,* the resistance between these two points is now greater than that to the right or left of this area. The water levels in the pressure tubes *r* and *s* reveal a greater (steeper) pressure gradient between these two points than exists between *m* and *n* or between *t* and *u.* This greater steepness between *v* and *w* indicates a greater fall in lateral pressure and is caused by a greater expenditure of energy as the water moves from *v* to *w* than in flowing from, for example, *m* to *n.* Because of the greater resistance from *v* to *w,* the water level in tube *n* is higher and that in *t* is lower than they would otherwise have been.

The total lateral pressure, *ak*, needed for overcoming resistance is greater and the velocity pressure, *bk*, is less than in *B* or *C*. With a constant head of pressure, the lateral pressure of a moving fluid at any point in its course varies *directly* with the amount of resistance between this point and the outlet and varies *inversely* as the amount of resistance already overcome. The pressure gradient between any two given points is steeper, the greater the resistance between these two points. *Lateral pressure and velocity pressure vary inversely.*

Volume flow. ■ From Fig. 13-5, *D*, it is evident that no matter what variations there may be in the size of the cross section of tube *co,* the amount of fluid passing any one given point of the system in a unit of time must be the same as that passing any other point in that system. For example, if 1 liter of water passes the wide part, *m,* in one minute, the same amount must pass by the narrow part, *r,* in the same length of time; that is, the *volume flow* or *flow rate* is the same at *r* as at *m. The volume flow varies directly as the velocity pressure and inversely as the resistance.*

Velocity of flow. ■ The velocity (the distance a certain particle of fluid travels in a unit of time) varies directly as the pressure. Increasing the height of the water in the tank in Fig. 13-5 would increase the velocity at every point along the tube *co.* From the law of volume flow it follows that the velocity is greater between *v* and *w* (Fig. 13-5, *D*) than at any other point because the

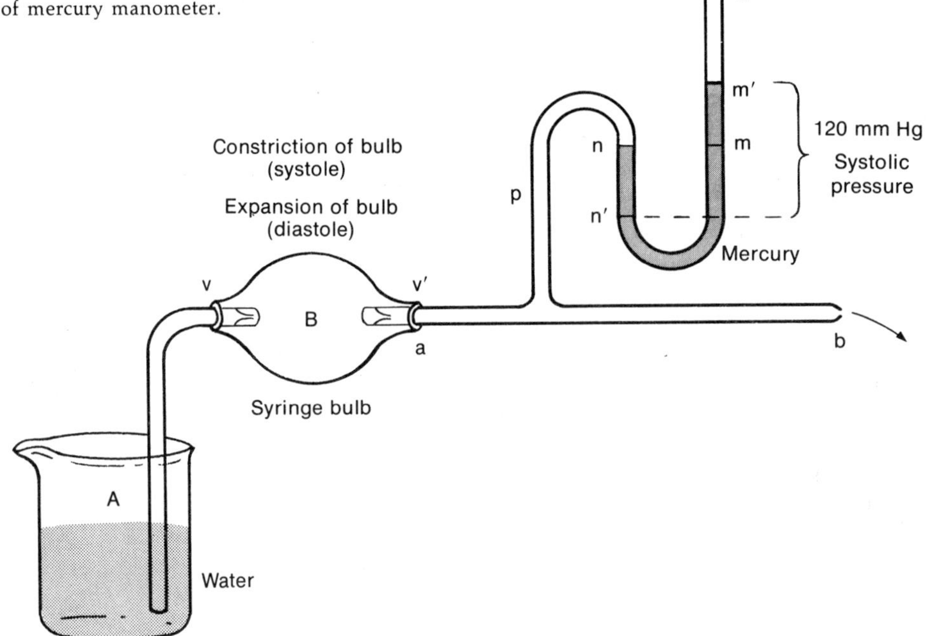

Fig. 13-6
Illustration of apparatus used to demonstrate nature of flow and function of mercury manometer.

pressure gradient is steeper. Hence, the *velocity varies inversely as the cross section of the bed of the stream.*

Nature of the flow: elastic versus rigid tubes. ■ Thus far our discussion has been limited to the flow caused by a constant force. Now examine the flow caused by the intermittent application of a force similar to that created by the heart. Using a syringe bulb (Fig. 13-6, *B*) equipped with valves at *v* and *v′* as a pump, water may be forced *intermittently* into a rigid (glass) tube, *ab*. The tube has a narrow outlet at *b* and has previously been completely filled with water. With each stroke of the pump a certain quantity of water is injected into the tube; this amount may be called the *stroke volume.* Let us call the constriction of the bulb *systole* (sys′-to-le) and its expansion *diastole* (di-as′-to-le). The energy of the hand is transferred to the water and thereby increases the pressure in the tube. Pressure in a tube can be measured by means of a manometer. There are many forms of manometers, but a mercury manometer will be used in this illustration. As shown in Fig. 13-6, this consists of a U tube partly filled with mercury, *mn*. Above and beyond the level of the mercury, *n,* the proximal limb, *p,* is filled with water that is contiguous with that in *ab;* the distal limb, *d,* is open to the air. The increased pressure during systole causes the mercury in the proximal limb to fall to, say, *n′*, and that in the distal limb to rise correspondingly to *m′*. The differences in the levels *m′* and *n′* is found to be, for example, 120 mm; this is the systolic pressure. At the instant the systole ceases, the pressure abruptly drops to zero (the *mn* level of Fig. 13-6). In consequence, the flow which was at maximum velocity during systole comes to an immediate and complete standstill during diastole. This causes the outflow at *b* to be intermittent. It will be noticed that if the energy loss to friction is ignored, all the energy exerted during systole is *directly* expended in moving an amount of water equal to the stroke volume out of the tube at *b* in the same length of time as was consumed in putting it in at *a,* irrespective of the size of the opening at *b.*

Now replace the rigid tube, *ab,* in Fig. 13-6 with an elastic tube. It also has previously been filled with water. As the stroke volume is being forced into the tube, the increased pressure exerted upon the elastic walls causes them to stretch or distend. In so doing, storage space is afforded for a certain amount of the stroke volume, and only a portion of it is forced out at *b* during the systole. The systolic pressure (Fig. 13-6) rose, for example, to 120 mm Hg. When the pump goes into diastole and no more water is forced into the tube, the distended elastic wall by its recoil continues to squeeze upon the water and to force it out at the exit, *b,* in a continual stream. This causes the pressure in the tube to fall slowly during diastole, and when it has fallen to, for example, 80 mm, the next systole begins and once more sends the pressure to 120 mm. Hence, with repeated strokes of the pump, the pressure in the tube fluctuates rhythmically between these two values. To bring about these results, considerable resistance must be present at the opening, *b.* Widening of the opening will alter the ability of the elasticity of the tube to maintain a constant flow.

■ SYSTEMIC BLOOD PRESSURE

■ Cardiac activity

In the cyclical action of the heart, the atria contract first; by this atrial systole, blood is driven through the open atrioventricular valves (bicuspid and tricuspid, Fig. 13-3) into the ventricles. This is followed by the atrial diastole, which allows the blood from the veins to flow into the empty atria. Almost simultaneously with the beginning of the atrial diastole, the ventricular systole takes place. The contraction of the muscles of the ventricle puts the blood under a considerable pressure and, by first closing the atrioventricular valves and then opening the semilunar valves, forces the blood into the arteries. The

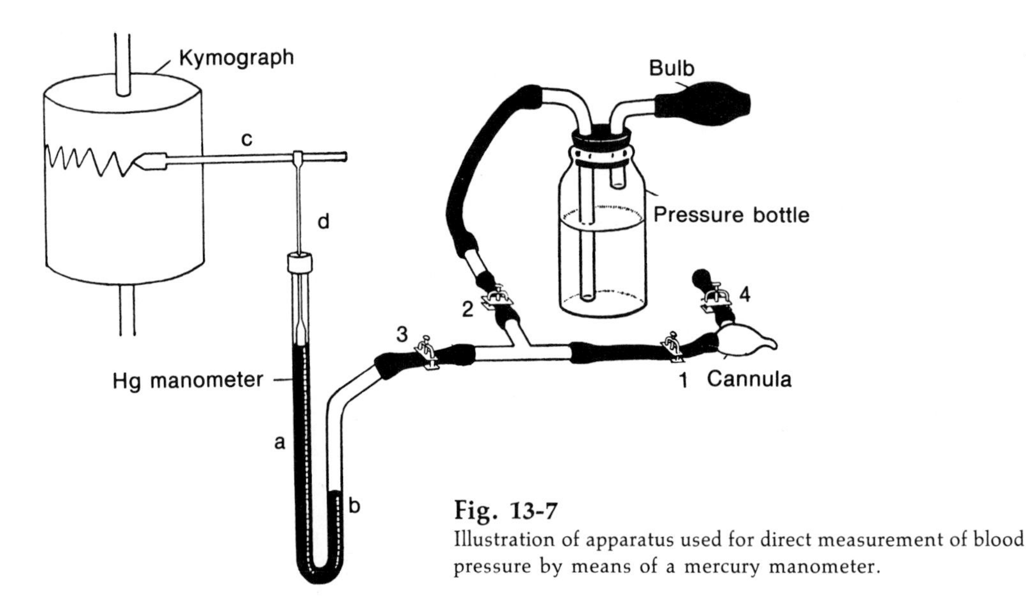

Fig. 13-7
Illustration of apparatus used for direct measurement of blood pressure by means of a mercury manometer.

contraction of the ventricles is followed by diastole, which results in the closing of the semilunar valves in the arteries and the reopening of the atrioventricular valves.

Stroke volume and cardiac output. ■ With each systole a ventricle discharges a certain quantity of blood; this is called the *systolic output* or *stroke volume.* The average amount in a resting adult human is estimated at 60 ml, but the values obtained by various methods differ considerably. The stroke volume of the right ventricle (averaged over any considerable length of time) must be the same as that of the left ventricle, since the circulation is a closed system. The amount of blood leaving the left (or right) ventricle per minute is known as the *cardiac output* or *minute volume.* Here it may be stated that *the major factor that directly influences the cardiac output is the amount of blood flowing into the ventricle during diastole* (diastolic intake) and, therefore, the venous circulation. The determination of the cardiac output is considered in Chapter 14.

Direct measurement of blood pressure. ■ The energy set free during the contraction of the ventricular musculature puts the blood under considerable pressure. The amount of this pressure can be determined in the same manner as the water pressure in the tube in Fig. 13-6.

An apparatus sometimes used for the direct measurement of blood pressure in animals is shown in Fig. 13-7. An artery in an animal, usually the carotid (Fig. 14-20), is exposed and properly prepared by making a V shaped cut into it. An arterial cannula (Fig. 13-7) is inserted into the artery and securely ligated. The whole system is filled, under a pressure estimated to be the same as that of the animal, with an anticoagulating fluid (for example, sodium citrate) from the pressure bottle. In the distal limb, *a,* of the manometer is placed a hard rubber float. Into the float is fixed a stiff vertical wire, *d,* which, in turn, holds in place a horizontal wire, *c.* To the end of this is affixed a stylus that records the rise and fall of the mercury level in *a* on a kymograph.

When the pinchcock, *1,* is opened, the blood in the artery forces itself against the column of sodium citrate and thus exerts its pressure against the mercury, forcing it up in the distal limb, *a*. The difference in the height of the mercury in the two limbs, expressed in millimeters, is the blood pressure in mm Hg.* Obviously, if the excursion of the stylus is measured, this must be multiplied by two to get the correct value because the stylus only records the displacement in one limb of the tube.

The mercury manometer system shown in Fig. 13-7 has been largely supplanted by the use of pressure transducers, even in the student laboratory. Nevertheless, the manometer system, as illustrated, because of its simplicity, provides a useful visual reinforcement of the learning process.

■ Arterioles and peripheral resistance

Arterial blood pressure is determined by the volume of blood that is forced into the aorta (cardiac output) and by the resistance the blood encounters in passing through the vessels.† In the large arteries the resistance is slight, and the fall in blood pressure in these vessels is small.

In coursing through the arterioles (which are narrow, comparatively long, and not very numerous) the blood meets with a large amount of resistance and suffers a greater loss in pressure (from 50 to 60 mm Hg) than in any other part of the vascular system (Fig. 13-8). The resistance offered by the arterioles is spoken of as *peripheral resistance.* An important function of these vessels is to constrict or dilate and thus increase or decrease, respectively, the amount of resistance; they *control the outflow from the arterial system.* This makes them the most important factor in changing the blood pressure. By the same token, *the arterioles largely regulate the blood supply to the capillaries.*

Critical closing pressure. ■ It has been proposed that at the sites of controlled peripheral resistance the vessels should be closed or open. This concept of a critical closing pressure is founded upon the law of Laplace:

$$\Delta P = T/r$$

where ΔP is the pressure in the vessel, T is the tension in the vessel wall, and r is the radius of the vessel. Any increase in smooth muscle activity would increase the tension in the wall and the pressure would have to increase to maintain the original radius. Since the pressure is reasonably uniform, the radius would, in fact, decrease. In turn this would cause the tension to exceed that needed to balance the pressure and the radius would continue to diminish until the vessel was closed. The critical closing pressure would depend upon the contractile capabilities of the muscular elements in the vessel wall. Because it neglects the influence of the wall thickness and because observations have shown that the vessels at sites of controlled peripheral resistance undergo graded variations in diameter, in contrast to the predicted all-or-none behavior, the concept of a critical closing pressure is less important to the study of hemodynamics than once believed. The law of Laplace is, however, of importance in interpreting the different vessel wall thicknesses found in the circulatory system. In the aorta and large arteries the diameter and pressures are relatively large, consequently the tension in the wall must be large, as well. In the smaller vessels the pressure is reduced three- to fourfold, but the diameter may be a hundred- or a thousandfold less; the tension needed to support the reduced pressure may be several thousand times less. Since the tension in the vessel wall depends upon the thickness, in the smaller vessels the thickness diminishes. This accounts for the exceedingly thin walls of the capillaries.

■ Capillary blood pressure

Earlier we stated that the velocity of flow is inversely proportional to the area of the vascular

*Due to the inertia of the mercury, there is a lag during its rise and fall. Hence the pressures are a little too high or too low, depending on the rate of the heartbeat.

†The size of an animal is no index of its blood pressure. The pressure in a small dog is frequently higher than in man.

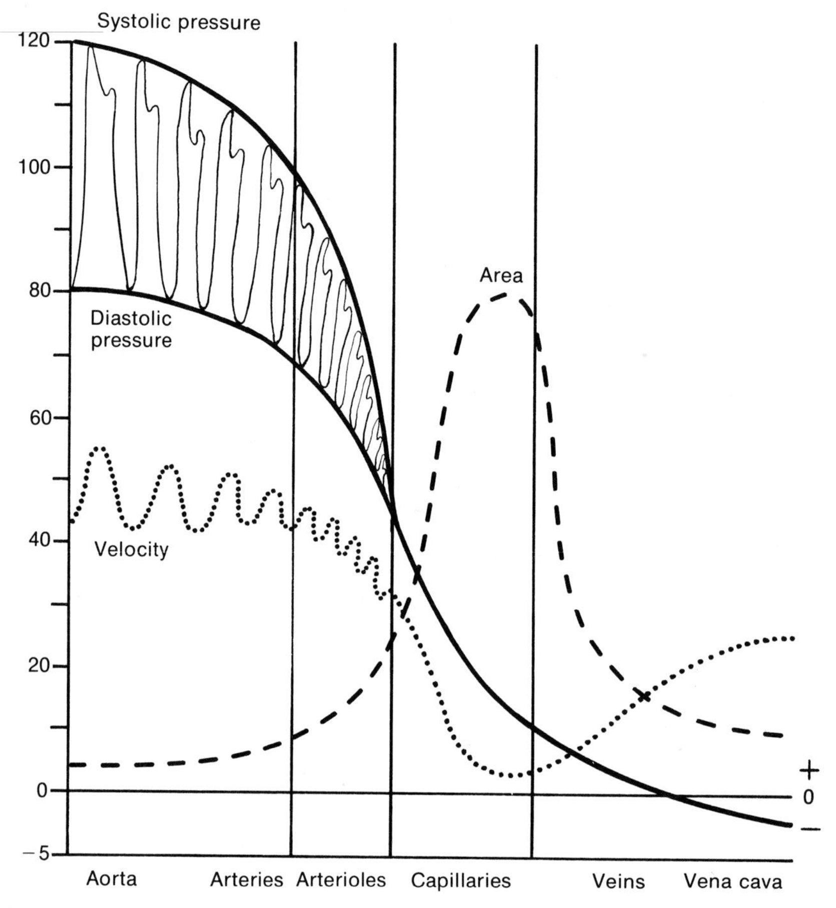

Fig. 13-8
Diagram showing systolic and diastolic pressure in the arterial system, pressure gradient in whole circulatory system, variations in bed (area) of stream and velocity of flow.

bed. The resistance varies directly as the velocity of flow and the exceedingly slow flow in the capillaries (Fig. 13-8) causes the fall in pressure in these vessels to be small (less steep gradient). The short length of the individual capillary also plays a part in this. The systemic capillary pressure varies along the length of the vessel; it may be placed at from 30 to 10 mm Hg.

■ Venous blood pressure

The pressure of the blood in veins is not only much less than in arteries, but here, as in the capillaries, the pressure is steady; that is, it does not undergo rhythmical fluctuations depending on the action of the heart.

In the venous system the pressure gradient falls very slowly, from about 10 mm in venules to about zero in the right atrium (Fig. 13-8). In an arm or hand held at the level of the heart it has been found to be about 6 mm Hg. In the large veins near the heart the blood pressure may be negative; that is, when the cannula of the manometer is inserted into the vein, the mercury is not forced up into the distal limb (Fig.

Table 13-1. Influence of cardiac output and peripheral resistance on blood pressure in areas of the circulatory system

	Arterial pressure	*Capillary pressure*	*Venous pressure*
Cardiac output			
Increased	+	+	+
Decreased	−	−	−
Peripheral resistance			
Increased	+	−	−
Decreased	−	+	+

13-7, *a*) but is aspirated into the proximal limb (Fig. 13-7, *b*). The atmospheric pressure upon the mercury in the distal limb is greater than the pressure that the blood in the large vein (vena cava) exerts upon the mercury in the proximal limb. Opening of such a vein may cause air to be aspirated into the vein and may lead to pulmonary air embolism. Due to the low blood pressure and the great distensibility and collapsibility of the vessels, the venous pressure is readily affected (either aided or retarded) by pressure external to them. This may take place in the abdomen and thorax. We have previously noted that the venules and small veins are the *capacitance* vessels of the circulatory system.

■ Poiseuille's law

Blood pressure is ultimately the result of two factors: cardiac output and peripheral resistance. The first of these is the product of the stroke volume and the heart rate. Peripheral resistance is determined by the constriction and dilation of the arterioles. The influence of these two factors is summarized in Table 13-1.

The manner in which pressure, blood flow, and resistance are related resembles in many ways the relationship of electromotive force, current, and resistance in an electrical circuit (Ohm's law):

$$E = I \times R$$

By substituting the value of the pressure gradient ($\triangle P$) for electromotive force, flow (F) for electrical current, and peripheral resistance (R) for electrical resistance, the equation becomes:

$$\triangle P = F \times R$$

The equation may be rearranged as:

$$F = \triangle P/R \text{ or } R = \triangle P/F$$

Each of these three forms of the equation is useful in hemodynamic studies.

Another important equation, which takes into account various factors that influence resistance, is Poiseuille's law. We can see from Fig. 13-5, *B*, that the resistance to flow increases directly with the length of the vessel, and, therefore, the flow is inversely proportional to the length. Increased viscosity, for example, blood instead of water, likewise increases the resistance and reduces the flow. Flow is directly proportional to the size of the vessel, in fact, to the fourth power of the radius; resistance, on the other hand, is inversely related to the radius. These facts are expressed in Poiseuille's law as:

$$\text{Flow} = \frac{\text{Pressure gradient} \times \pi \text{ Radius}^4}{\text{Length} \times \text{Viscosity}}$$

Poiseuille's law applies to the circulatory system only in a qualitative sense. Because blood is not a homogeneous fluid, the cellular constituents alter its properties, and because the vessels are not rigid tubes, the equation is not quantitatively applicable. However, the velocity and the volume flow in arteries are increased by increased cardiac output and decreased by increased resistance (Fig. 13-8).

■ SYSTEMIC BLOOD FLOW

■ Velocity and volume of flow

According to the law of volume flow, the amount of blood flowing through the combined capillaries in a unit of time is the same as that passing through the aorta (if both are measured over a considerable length of time). Although the diameter of an individual capillary is microscopic, 0.01 mm, because of the countless number of capillaries in the body the area of the total capillary bed is estimated to be from four to eight hundred times that of the aorta.* From the law of velocity it can be deduced that the flow must be correspondingly slower in the capillaries than in the larger vessels. In the aorta this has been found to be about 320 mm/sec. As the blood passes into the branches of the aorta, the bed becomes progressively larger and the velocity decreases correspondingly. When the blood reaches the capillaries with their tremendously large bed, the velocity is from 0.5 to 1 mm/sec. Since the exchange of materials between the blood and the tissues takes place in the capillaries, the value of this greatly reduced velocity can be readily appreciated. As the blood leaves the capillaries and courses through the veins, the bed becomes gradually smaller and the velocity increases correspondingly. But since a vein is generally larger than the corresponding artery,* the blood in a vein never attains the velocity in the artery.

*In the average-sized adult of thirty years of age, the aorta has a diameter of about 2.1 cm.

It is of interest how greatly the volume flow varies from one organ to another; thus, per 100 g of tissue the flow through the stomach wall is about 21 ml/min; for the brain, 54 ml; for the kidney, 400 ml; and for the thyroid gland, 560 ml. In subsequent chapters it will be noted that the quantity of blood passing through an organ is greatly increased or decreased according to the immediate function of that organ.

The enormous size of the capillary bed, as compared with that of the aorta (Fig. 13-8), must cause the volume flow in each capillary (amount of blood discharged in a unit of time) to be exceedingly small. If the length of the capillary is taken as 1 mm and the velocity of flow as 1 mm/

*An artery supplying a definite area of the body and a vein draining this area are said to be corresponding vessels, for example, the renal artery and the renal vein.

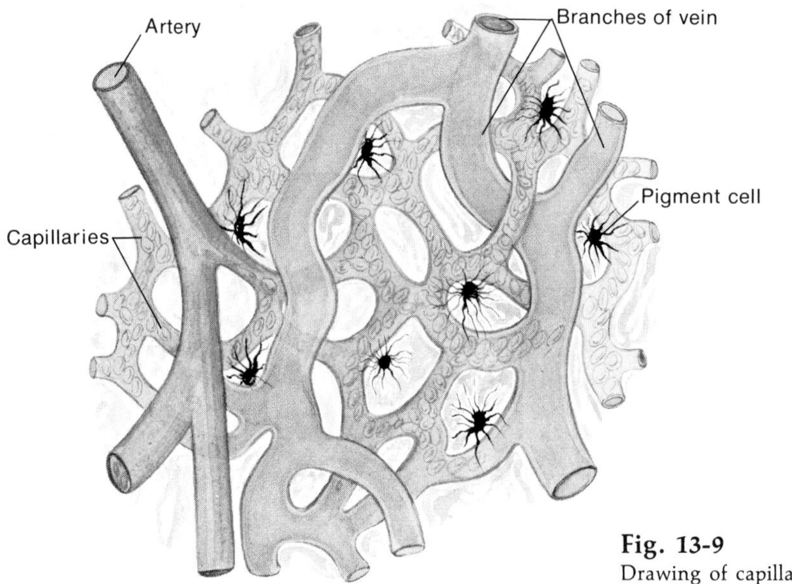

Fig. 13-9
Drawing of capillaries in web of frog's foot.

sec, the quantity of blood passing through the capillary in this length of time is equivalent to the volume of capillary; this can be found by the formula for the volume of a cylinder: $\pi r^2 h$. Assuming that the radius of the capillary is one half of the diameter of a red cell (0.004 mm), the volume becomes 0.00005 mm^3 or 0.00000005 ml. Furthermore, assuming that one drop of blood is 0.067 ml, it would require approximately sixteen days for a drop of blood to pass through a capillary (Fig. 13-9).

■ Advantages of elastic arteries

It has been stated previously that the pulsating flow in arteries is changed to a more nearly uniform flow in the capillaries and veins by the elasticity of the arteries. This offers the following advantages over rigid tubes:

1. It will be recalled that in a rigid system the pressure rises with each systole from zero to maximum; this sudden rise is experienced almost instantly from one end of the system to the other. During diastole the pressure suddenly falls back to zero. Such violent and marked rises in pressure would be destructive to the delicate arterioles and capillaries; the elastic arteries act as buffers for the smaller tubes.

2. The slow, uniform flow in the capillaries is advantageous for the exchange of materials between the blood and the cells.

3. The heart can miss a beat or two without appreciably disturbing the capillary circulation. Several minutes after the heart has ceased to beat, the opening of a vein may cause bleeding, whereas the arteries are often found empty after death.

4. The duration of the discharge of the blood from the ventricle during systole is approximately one-tenth second. Of the cardiac energy expended during this time, only a small part is used for moving the column of blood forward toward the capillaries; the larger part is stored as potential energy in the distended arterial wall. This stored energy is liberated (by the elastic recoil) to keep the blood moving during the remainder of the *cardiac cycle,** which, with a heart rate of 60/min, equals nine-tenths second. In the brief ejection time of one-tenth second it is easier to make room for the systolic output by distending the arterial wall than it is to move the whole column of blood forward.

When the elasticity of the vessels is lessened, as generally occurs in old age (arteriosclerosis), more cardiac energy is needed to move the column of blood toward the capillaries; this puts extra work upon the heart.

The pulse. ■ The ventricular systole generating a pressure of, for example, 120 mm Hg (Fig. 13-10) forces the blood into the aorta, causing the pressure in this tube to become practically the same as that in the ventricle. Being highly elastic, the walls of the aorta distend. During the following diastole the recoil of the aortic wall, having closed the semilunar valves, forces a large quantity of this blood forward into the next por-

*The cardiac cycle is a complete heartbeat, comprising the systole and diastole.

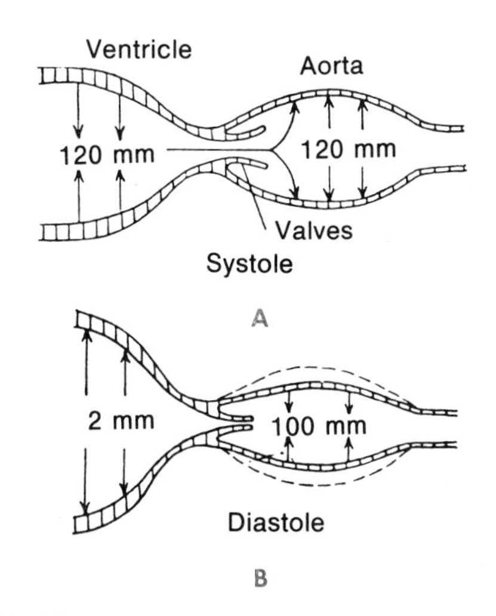

Fig. 13-10
Diagram illustrating origin of pulsatile flow.

tion of the arterial system. This results in a constriction of the previously dilated aorta and in a distention of the more distal part of the arteries. The process of dilation and constriction repeating itself along the whole arterial system constitutes the pulse. It can be felt when a finger is placed over an artery just beneath the skin, for example, the radial pulse at the wrist.

In the smaller tubes (collectively of greater capacity) the pulse wave is gradually extinguished so that in the capillaries the pressure and velocity of flow are steady. The pulse wave must not be confused with the flow of blood. While the velocity of the flow in an artery is 100 to 500 mm/sec, the pulse wave in this length of time travels 6 to 9 m.*

Circulation time. ■ By circulation time is meant the time required for a particle of blood to make a complete circuit through the vascular system. There are extremely short circuits and there are long circuits; hence there is no single circulation time. The pulmonary circulation time in man has been estimated at from four to eight seconds. Injecting sodium dehydrocholate into the vein of the arm evokes a bitter taste in about fifteen seconds; this is the arm → tongue circulation time (partial circulation time).

■ Secondary aids in venous circulation

The main force that drives the blood along the veins and back to the heart is the pressure of the blood behind it in the capillaries; hence the muscular energy of the left ventricle is the driving force. In addition to this, there are at least two other factors that play a part—the respiratory pump which alters the pressure upon the great veins and the rhythmical activity of skeletal muscles.

Venous circulation is aided by the rhythmical contraction of skeletal muscles. When these muscles contract, the veins in and near the muscles are squeezed upon, and because of the valves found in the veins, the blood can be forced only toward the heart. When the muscles relax, the veins open up and the blood from the capillaries flows more rapidly into the almost empty veins. By this milking action the muscles function as auxiliary venous hearts. In addition to aiding the flow of venous blood, this same process accelerates the flow of lymph from the exercised parts of the body (Chapter 16). The improved nourishment of the tissue brought about by these "booster pumps" may be regarded as one of the beneficial effects of rhythmical exercises. For the same reason, passive movement or the massage of a limb may be useful.

The valuable influence of rhythmical muscular action in contrast with that of static or constant contractions readily can be appreciated. Stand erect for two or three minutes and note the prominence of the veins of the feet; also note that they feel hard to the touch. On taking a few steps, a great change in these veins will be seen.

■ ATRIAL PRESSURE AND FLOW

Pressure. ■ It has been stated that the walls of the atria are very thin compared with those of the ventricles; it can be assumed that the pressure generated by them must also be much less; direct observations on animals confirm this. The maximum pressure in the left atrium is 2 to 10 mm Hg, whereas that in the left ventricle is 110 to 150 mm Hg. To overcome the resistance that the blood encounters in passing from the atrium into the ventricle requires little energy. Therefore the primary function of the atrium is not to generate pressure.

Flow. ■ The duration of the atrial systole is much shorter than that of the ventricle (Chapter 14); this subserves the principal function of the atrium. Suppose that the pulmonary veins emptied directly into the left ventricle, the atrium being absent. With each ventricular systole, the

*About eleven seconds is required for the blood to travel from the heart to the arterioles of the foot; the pressure (or pulse) wave traverses this distance in two-tenths to three-tenths second.

flow of blood in the veins would come to a standstill until the ventricle again dilated. The venous flow would be pulsatile and the capillary flow would also be intermittent; this would be highly unfavorable to the performance of the functions of the blood in the capillaries. In an atrium whose systole lasts but one-tenth second, the retardation of the flow in the veins (which is in part the cause of the so-called venous pulse) is almost negligible. The atrium remains open during its long diastole (seven eighths of the entire time of the cardiac cycle), and thereby it serves chiefly as a *reservoir for the blood coming to the heart.* Moreover, the short atrial systole ensures an adequate filling of the ventricle during its diastole; this becomes of great importance when the heart rate is accelerated, as in muscular work.

■ PULMONARY BLOOD PRESSURE AND FLOW

The lesser, or pulmonary, circulation is characterized principally by the fact that the pressure in the pulmonary vessels is much less than that in the systemic vessels. The mean pressure in the pulmonary artery (20 mm Hg) is about one fifth that found in the aorta. This is in accord with the relative size of the musculature of the right and left ventricles. The reason for the much smaller force required to send blood through the pulmonary circuit lies in the smaller amount of resistance encountered there. To clarify this we should recall from our earlier discussion that (1) the output of the right and left hearts must be equal over any reasonable span of time and (2) the pressure is the product of flow and resistance ($\triangle P = FR$). Two factors account for the lesser resistance of the pulmonary circulation: (1) the vessels are shorter and generally of larger caliber than their counterparts in the systemic circulation and (2) the vessel walls are thinner and more distensible; that is, their diameters change more readily in response to pressure changes. Due to this greater distensibility of the pulmonary vessels, a threefold increase in blood flow is required to cause a noticeable rise in pulmonary pressure. It should be borne in mind that the pulmonary arteries carry reduced (venous) blood and the pulmonary veins carry oxygenated (arterial) blood.

We will learn, in Chapter 16, that the mean systemic capillary pressure is about 28 mm Hg. In the pulmonary capillaries this pressure is about 8 mm Hg. Because of the lesser pressures, little if any fluid filtration normally occurs in the pulmonary capillaries, a feature that aids in keeping the alveoli dry. However, if the pulmonary capillary pressure is increased about fourfold, as may occur when in left ventricular failure blood pools in the pulmonary veins, significant filtration occurs and pulmonary edema results.

■ BLOOD PRESSURE IN HUMANS

The blood pressure in a human being is usually determined by the *sphygmomanometer* (sfig″-mo-man-om′-et-er). Of these there are several makes; however, all depend on the same principle. Suppose that while the pulse is being felt at the wrist, the upper arm is encircled by a bandage. The tighter the bandage is pulled, the feebler the pulse becomes; when the external force applied to the arm equals the blood pressure and completely blocks the artery, the pulse wave cannot pass. In the commonly used form of sphygmomanometer, a rubber bag (armlet) is applied to the bare upper arm of the subject (Fig. 13-11). This armlet has two openings; one of these is connected by means of a rubber tube with an inflating bulb for forcing air into it; the other opening is connected with a mercury or aneroid manometer to indicate the amount of pressure in the armlet. The bulb is supplied with a needle-valve, by which the pressure may be reduced slowly.

Indirect determination of blood pressure. ■ There have been two indirect methods described for measuring blood pressure in humans: the palpation method and the auscultatory method.

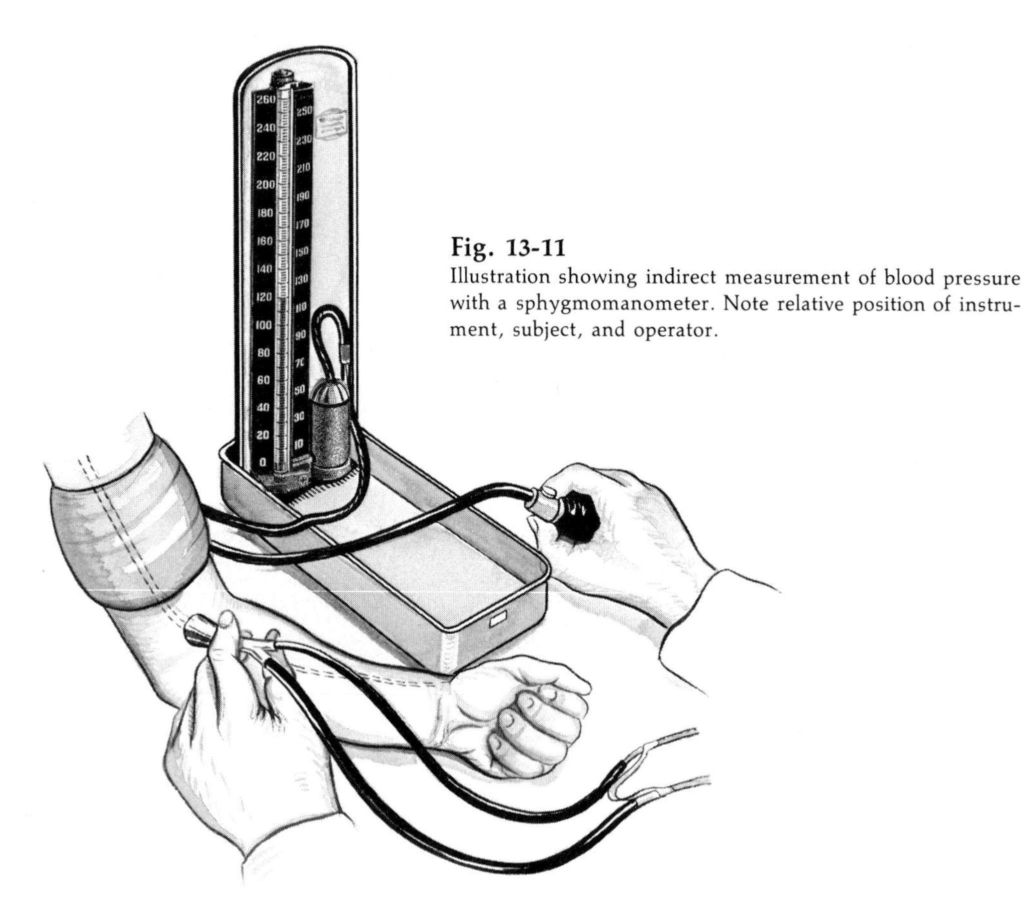

Fig. 13-11
Illustration showing indirect measurement of blood pressure with a sphygmomanometer. Note relative position of instrument, subject, and operator.

The auscultatory method serves the purpose best, since it provides for the determination of both systolic and diastolic pressure. The procedure for the use of this method is as follows: The arm band of a sphygmomanometer is placed around the left arm just above the elbow. The brachial artery is found just below the band. A stethoscope is placed lightly over the artery and air is pumped into the arm band until a distinct sound is heard. When the sound is located, the pressure in the band is increased slowly until the sound disappears. This is the systolic point. The systolic pressure is read from the sphygmomanometer in mm Hg.

There are two diastolic points to be considered. The first point is found as follows: Following the location of the systolic point, the pressure in the arm band is gradually reduced. As this occurs, the sound becomes louder and louder for a time, finally breaking sharply into a soft sound. The break in the tone of the sound is the first diastolic point. Simultaneously with this event, the diastolic pressure is read in mm Hg. The pressure in the band is reduced still further until the sound disappears. This is the second diastolic point, and again the pressure is read in mm Hg. To avoid confusion, the usual practice is to note both diastolic end points and to record the pressure as, for example, 120/75/70 mm Hg, which indicates the pressure at both diastolic points.

The difference between systolic and accepted

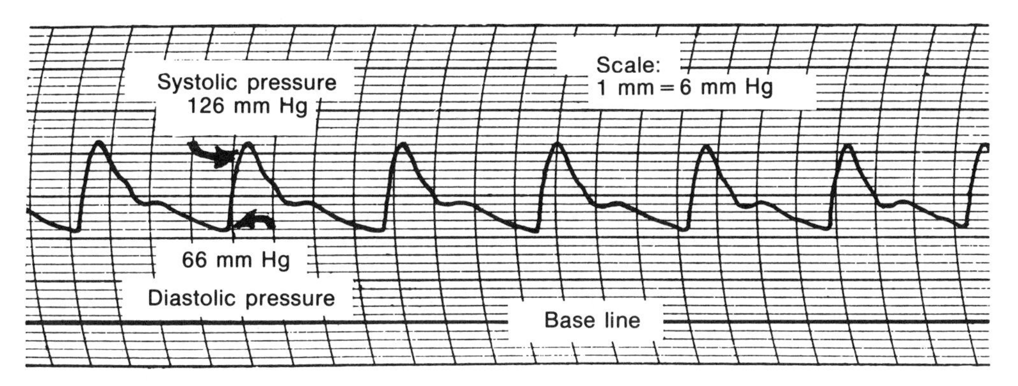

Fig. 13-12
Blood pressure tracing of man taken directly from brachial artery.

diastolic pressure (Fig. 13-12) is called *pulse pressure.* The *mean pressure* is usually stated as the diastolic pressure plus from one third to one half the pulse pressure.

The diastolic pressure defines the load to which the arteries are constantly subjected because of the resistance that the blood encounters in passing from the left ventricle to the right atrium. It is the amount of pressure that must be overcome by the systole of the left ventricle before the semilunar valves open and before any blood can be discharged into the aorta. Whenever the vessels offer greater resistance, the diastolic pressure is increased; *the diastolic pressure may therefore be said to represent the condition or state of the blood vessels,* and for this reason its determination is of great moment. The systolic pressure represents the amount of work the left ventricle does in overcoming the resistance of the vessels. The pulse pressure is caused by the escape of blood from arteries and through the capillaries into the veins between two successive systolic outputs. In a faster beating heart (all other factors remaining constant), less time is given for this escape, and hence the diastolic pressure rises and the pulse pressure decreases.

Basal blood pressure. ■ Since the blood pressure varies from one individual to another and in the same individual under different circumstances, one cannot speak of a normal blood pressure, but it is acceptable to use the term *normal range* of blood pressure. The basal* blood pressure is obtained with the subject in a reclining position, in a comfortably warm room, after resting for at least forty-five minutes, six hours after the last meal, and with his mind at ease (if possible). The objective is to eliminate all factors, physical, emotional, and metabolic, known to increase blood pressure. Basal blood pressure may be regarded as the lowest pressure necessary to maintain a flow of blood sufficient for the needs of the body. Although the basal pressure is constant in any given individual, in different individuals it varies with many factors, among which are age, sex, body weight, and effects of exercise.

Age. In newborn babies the systolic pressure ranges from 60 to 90 mm Hg; at the age of ten years it has risen to 100 mm Hg, and at puberty, to approximately 120 mm Hg. From then on it rises more slowly with advancing years (Table 13-2).

Generally a systolic pressure above 150 mm

*The term *basal* as frequently used in physiology means a condition with the least possible amount of strain or stress.

Table 13-2. Mean blood pressures and standard deviations in a number of apparently healthy persons 20 to 106 years of age*

Age group	*Males*			*Females*		
	N	*Systolic*	*Diastolic*	*N*	*Systolic*	*Diastolic*
20- 24	500	123 ± 13.7	76 ± 9.9	500	116 ± 11.8	72 ± 9.7
25- 29	500	125 ± 12.6	78 ± 9.0	500	117 ± 11.4	74 ± 9.1
30- 34	500	126 ± 13.6	79 ± 9.7	500	120 ± 14.0	75 ± 10.8
35- 39	500	127 ± 14.2	80 ± 10.4	500	124 ± 13.9	78 ± 10.0
40- 44	500	129 ± 15.1	81 ± 9.5	500	127 ± 17.1	80 ± 10.6
45- 49	500	130 ± 16.9	82 ± 10.8	500	131 ± 19.5	82 ± 11.6
50- 54	500	135 ± 19.2	83 ± 11.3	500	137 ± 21.3	84 ± 12.4
55- 59	500	138 ± 18.8	84 ± 11.4	500	139 ± 21.4	84 ± 11.8
60- 64	500	142 ± 21.1	85 ± 12.4	500	144 ± 22.3	85 ± 13.0
65- 69	911	143 ± 26.0	83 ± 9.9	856	154 ± 29.0	85 ± 13.8
70- 74	694	145 ± 26.3	82 ± 15.3	682	159 ± 25.8	85 ± 15.3
75- 79	534	146 ± 21.6	81 ± 12.9	404	158 ± 26.3	84 ± 13.1
80- 84	385	145 ± 25.6	82 ± 9.9	344	157 ± 28.0	83 ± 13.1
85- 89	325	145 ± 24.2	79 ± 14.9	203	154 ± 27.9	82 ± 17.3
90- 94	124	145 ± 23.4	78 ± 12.1	122	150 ± 23.6	79 ± 12.1
95-106	25	146 ± 27.5	78 ± 12.7	28	149 ± 23.5	81 ± 12.5

*The diastolic end point used was the disappearance of the sound. From Lasser and Master: Geriatrics **14**:345, 1959.

Hg or a diastolic pressure above 100 mm Hg is regarded as abnormally high. But no sharp line of demarcation can be drawn between what may be considered normal pressure, hypertension, or hypotension.

Sex. In women both systolic and diastolic pressures are generally lower than in men, up to the age of about forty-five years.

Body weight. Overweight is generally associated with a higher blood pressure; frequently this is reduced when the body weight is reduced by proper dieting.

Effects of exercise. The tendency is to regard work as dynamic activity when evaluating its effect on blood pressure. However, as far as the cardiovascular system is concerned static work, for example, squeezing an object, causes different and more marked changes than running up a flight of stairs.

When a person performs moderate dynamic exercise, there is a sudden sharp rise in systolic pressure, which persists for several minutes after work. This is accompanied by little or no change in diastolic pressure. In contrast to this, when static work is performed there is a sudden sharp rise both in systolic and in diastolic pressure, which disappears suddenly after the conclusion of the work. The explanation for the difference in blood pressure responses to dynamic and static work appears to be that in the former the metabolic costs are quite large and are responsible for the pressure changes, whereas in the latter the metabolic cost is small, the effects being due to a reflex arising from the active muscle.

READINGS

Berne, R. M., and Levy, M. N.: Cardiovascular physiology, ed. 2, St. Louis, 1972, The C. V. Mosby Co., pp. 41-60.

Folkow, B., and Neil, E.: Circulation, London, 1971, Oxford University Press, pp. 1-96.

Rodbard, S.: The burden of the resistance vessels, Circ. Res. **29**(1):1-8, Jan. 1971. (Also published as American Heart Association Monograph No. 33.)

Wood, J. W.: The venous system, Sci. Am. **218**(1):86-96, Jan. 1968.

14

THE HEART

Our study of the circulation has thus far been confined to the mechanical factors that govern the flow of any liquid in a closed system of tubes with a central pump. But the amount of blood needed by any organ in a unit of time varies with the degree of activity of that particular organ. To meet these ever-changing needs efficiently, the work of the heart and the condition of the vessels must be closely regulated. This chapter deals with the activity of the heart and its regulation.

■ ORIGIN AND TRANSMISSION OF THE HEART IMPULSE

■ Origin

The relation between skeletal muscles and the central nervous system is so intimate and the action of these muscles is so thoroughly controlled by, and dependent on, impulses coming from the brain and spinal cord that severing the connection between them causes a complete and permanent cessation of the activity of the muscles—motor paralysis. In this respect the cardiac muscle is very different. The heart is connected with the central nervous system by means of two autonomic nerves, the vagi and the cervical (sympathetic) nerves. In a frog, when these nerves are cut or when the heart is excised, it goes on beating. *Heart action is automatic,* which means that it is not dependent on outside impulses for the initiation of its contraction.

The frog's heart is composed of four parts: a sinus venosus, into which the three systemic veins (venae cavae) empty; a right atrium, which receives the venous blood from the sinus venosus and sends it into the ventricle; and a left atrium, which receives the oxygenated blood from the pulmonary veins and delivers this to the same ventricle. The frog ventricle, unlike that in the mammalian heart, receives both reduced (venous) and oxygenated (arterial) blood; the mixed blood is sent to all parts of the body, including the lungs.

If the sinus venosus is cooled, the whole heart beats more slowly; cooling the ventricle does not have this effect. Warming the sinus venosus accelerates the beat. The impulse causing contraction of the musculature is generated in the sinus venosus. From the sinus the cardiac impulse spreads through the atria, into the ventricle, and finally into the aortic bulb. But on entering each succeeding part, there is a delay of the impulse, so that the atria contract after the sinus and the contraction of the ventricle follows that of the atria. The sinus beats first because its irritability to what is sometimes referred to as the "inner stimulus" is greater than that of any other part of the heart. The systole of the sinus is followed by a relatively long refractory period during which the sinus has lost its irritability; this causes its activity to be rhythmical and not tetanic.

■ Transmission

Sinoatrial and atrioventricular nodes. ■ Two small masses of specialized tissue are found in the mammalian heart. One of these collections, known as the *sinoatrial* (S-A) *node* (Fig. 14-1), lies in the wall of the right atrium and near the mouth of the superior vena cava; this part corresponds embryologically to the sinus venosus of the hearts of lower vertebrates. Another node, situated a short distance from the sinoatrial node and known as the *atrioventricular* (A-V) *node,* gives rise to a bundle of fibers (the *atrioventricular bundle,* or the bundle of His), which passes down into the fibrous ring separating the atrial from the ventricular muscle fibers. After splitting into two branches, it passes along the septum between the two ventricles and ultimately breaks up into many fibers, which make connections with the ventricular muscle fibers.

The mammalian cardiac impulse originates in the sinoatrial node, for this is the first structure of the heart to become electrically negative.

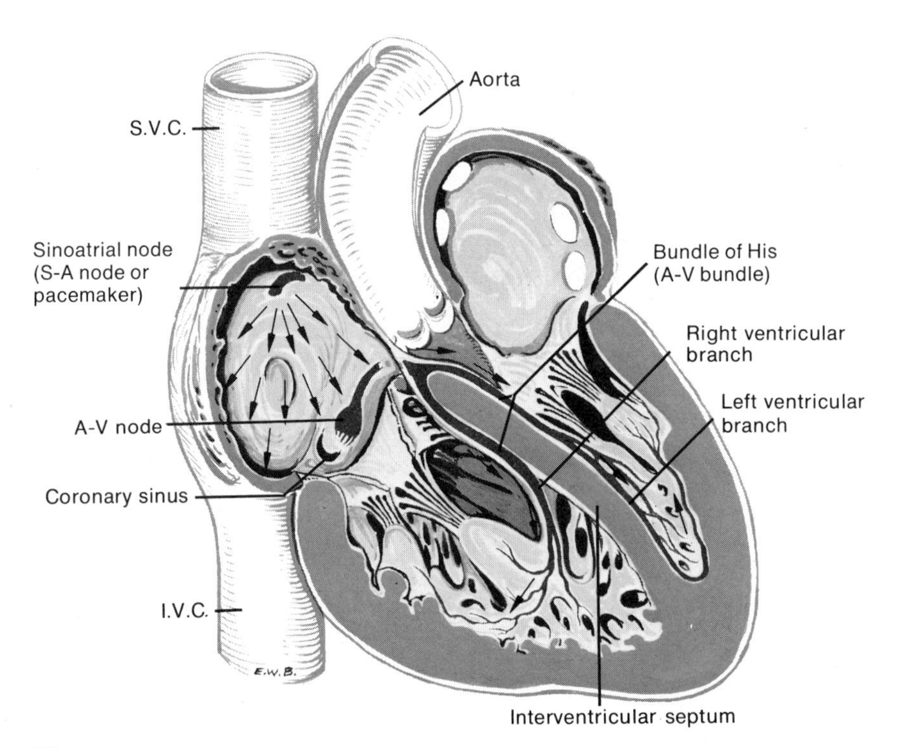

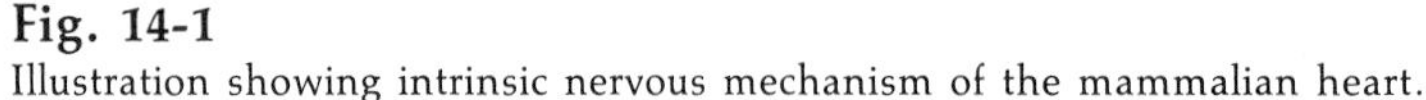
Fig. 14-1
Illustration showing intrinsic nervous mechanism of the mammalian heart.

From here the impulse spreads to and through the muscle fibers of the atria and causes the atrial systole; the muscle fibers convey the impulse to the atrioventricular node. From this node the impulse is carried by the atrioventricular bundle to the ventricles. The conduction rate of the atrioventricular node is slow; this results in a delay of about 0.15 second (the atrioventricular interval), which allows the atrium sufficient time to finish its contraction before the systole of the ventricle begins—an arrangement necessary for properly coordinated heart action. The propagation by the atrioventricular bundle is very fast, so that all the muscle fibers of the ventricles receive the impulse almost simultaneously; the advantage of this is apparent.

Although in the frog's heart the sinus venosus and in the mammalian heart the sinoatrial node must be regarded as the *pacemaker* of the heartbeat, by proper experiment it can be shown that all the various parts of the heart possess a certain degree of automaticity. If the sinus venosus of a frog's heart is cut away from the atria, the sinus continues its usual beat, but the activity of the atria and ventricle ceases for a certain length of time.* After a lapse of time these cavities resume their beat, which is, however, at a slower rate than of the sinus; in this resumption of activity the atria always beat before the ventricle. If now a section is made between the atria and the ventricle, the former continue their beat, and after a period of quiescence the ventricle may resume a very slow rate of beat. The lower part of the ventricle (apex) severed from the rest of the heart never beats spontaneously, although it responds to artificial stimulation. These facts demonstrate the gradual decrease in irritability and automaticity of the different parts of the heart, from the sinus to the apex.

Atrioventricular block. ■ The delay that the impulse encounters in passing from the atria into the ventricle can be exaggerated in the frog's heart by applying pressure (the second Stannius ligature—Fig. 14-2) at the atrioventricular groove. If the pressure is sufficiently great, the impulse cannot pass and the ventricle ceases to beat; after a time it may initiate its own beat, which bears no relation to the atrial beat. If the pressure is less great, only every second, third, or fourth impulse from the atria will pass into the ventricle, and hence the atria makes two, three, or more beats to every ventricular beat. The

*A tight ligature (the first Stannius ligature—Fig. 14-2) placed between the sinus and the atria causes arrest of the heart below the ligature; the sinus continues its beat.

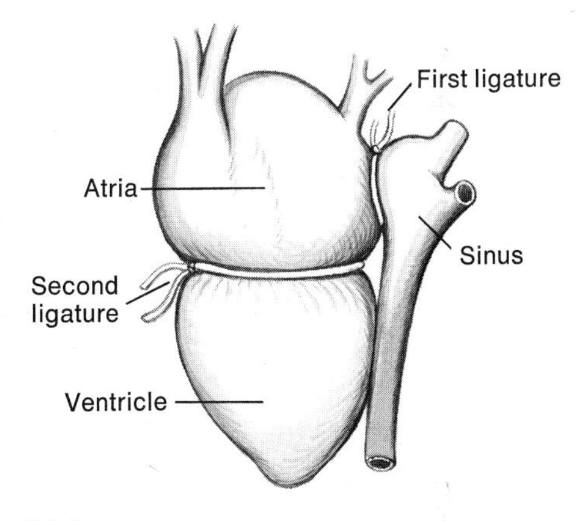

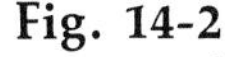
Fig. 14-2
Position of first and second Stannius ligatures in the frog heart.

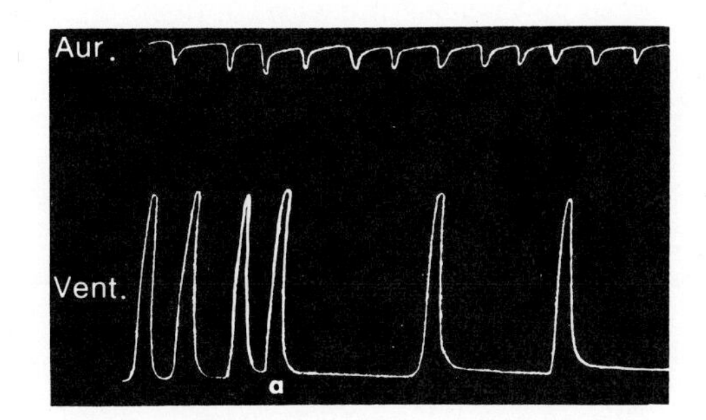

Fig. 14-3
Contraction record showing heart block produced at time **a** by applying clamp to atrioventricular junction.

same experiment can be made with the mammalian heart by pressure applied to the atrioventricular bundle (Fig. 14-3).

Fibrillation. ■ Normally the various muscle fibers of the ventricles contract in an orderly or coordinate manner, so that by their united action the pressure in the cavities is increased and the blood is expelled. When a faradic current is applied to the heart for a few seconds, the fibers no longer act coordinately; some contract before others and relax while the latter are in contraction. This is known as fibrillation, or *delirium cordis*. Because of the lack of harmonious action, in fibrillation no blood is discharged and circulation fails.

Fibrillation may be induced by the lack of oxygen (as by interference with coronary circulation), by mechanical stimulation of the heart (e.g., by rough handling), and by the passage of a strong electric current through the body—electrocution. In man ventricular fibrillation is usually irreversible, and since it causes the coronary circulation through the heart muscle to fail, it is fatal. Since the atrium is not absolutely necessary for the filling of the ventricle, atrial fibrillation may be compatible with life; it may also disappear. Atrial fibrillation may be detected by means of the electrocardiogram (EKG).

■ The electrocardiogram

It will be recalled that the formation and propagation of an impulse (or wave of depolarization) in a nerve or muscle are associated with changes in the membrane potential that can be detected by a suitable recorder. Fig. 14-4 shows the action potential from a single cardiac cell of a frog's ventricle (compare with Fig. 6-6). With an electrocardiograph machine, whole heart electrical

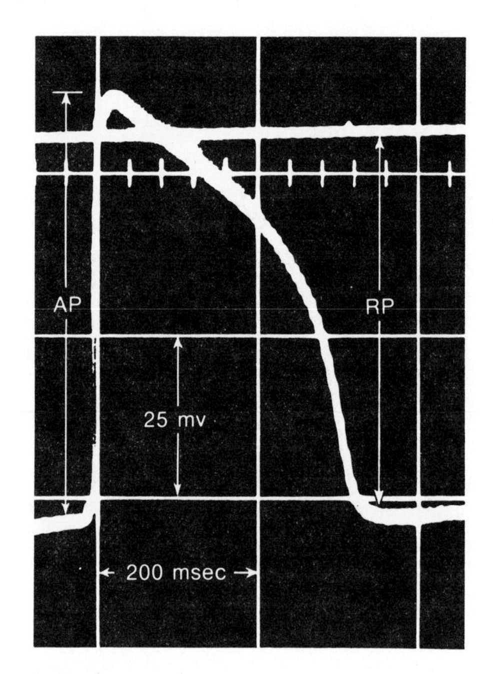

Fig. 14-4
Intracellular recording of action potential from a frog's ventricle. **AP**, Action potential; **RP**, resting potential; voltage and duration calibrations as indicated.

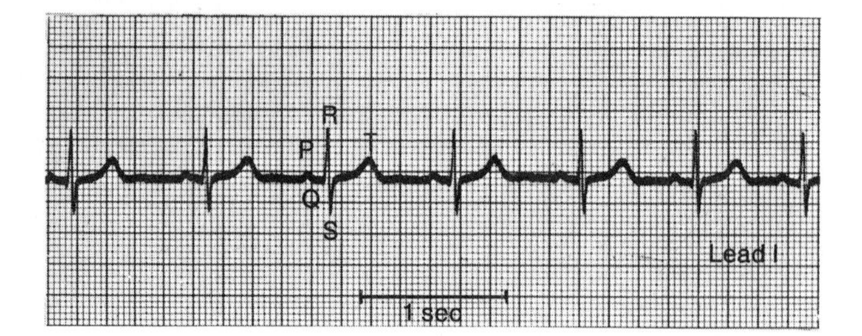

Fig. 14-5
Normal electrocardiogram (EKG), lead I. Chart paper is ruled in mm. Calibration is: horizontal, 25 mm/sec; vertical, 10 mm/mv.

activity can be recorded from the exterior surface of the body; the EKG is a record, taken at a distance, of summed cellular potentials. For this purpose standard limb leads, which are the left and the right arms (lead I), the right arm and the left leg (lead II), and the left arm and the left leg (lead III), make contact with suitable electrodes. From these electrodes the currents are conducted to the electrocardiograph. A record thus obtained is known as an EKG. In order to increase the diagnostic value of the electrocardiogram, numerous other leads are taken from various parts of the body.

From Fig. 14-5, which is a normal electrocardiogram (lead I), it will be noticed that in the heartbeats recorded a number of larger and smaller waves occur. These waves are due to changes in the electrical potential of the heart as the excitation wave, which always precedes the contraction itself, spreads from the sinoatrial node to the ventricular muscle. The impulse is not propagated by all parts of the heart at the same velocity. To determine which particular wave of the electrocardiogram is associated with a certain phase of cardiac activity, an electrocardiogram is obtained at the same time as the record of the atrial and ventricular pressures.

The P wave just precedes the contraction of the atria; it is the wave of depolarization of these structures. The most prominent feature of the EKG, the QRS complex, occurs immediately before the contraction of the ventricles. It signifies the depolarization of the ventricles which continues until the T wave. Repolarization of the ventricles is signaled by the T wave, that of the atria is masked by the QRS complex. The relationship of the extracellularly recorded QRS complex and T wave to the intracellularly recorded action potential can be appreciated by comparing Figs. 14-4 and 14-5.

The distance between the P wave and the QRS complex defines the atrioventricular interval. It is lengthened in partial heart block. Many features of the EKG are of diagnostic value in studying disturbances in the generation and transmission of the cardiac impulse.

Electrical axis of the heart. ■ The depolarization of the heart muscle proceeds from the base toward the apex of the ventricles as previously described. Thus, the base will become electrically negative before the apex and two EKG electrodes properly spaced on the surface of the body record a potential difference. When the electrodes are attached to the right arm (RA) and left arm (LA), the conventional lead I, the RA electrode becomes electronegative with respect to the LA electrode when ventricular depolarization begins. This indicates that the base of the heart is closest to the RA electrode. By convention the two electrodes are connected to the EKG recorder so that it gives an upward deflection when the RA electrode becomes negative. This can be seen in the R wave of Fig. 14-5.

The heart is considered to be in the center of an equilateral triangle, the Einthoven triangle,

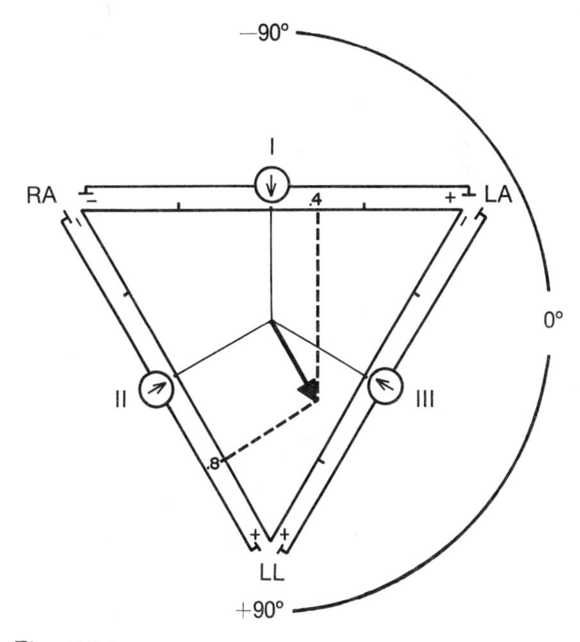

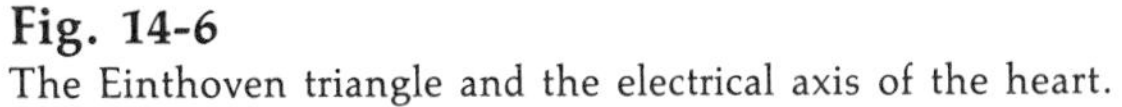

Fig. 14-6
The Einthoven triangle and the electrical axis of the heart.

formed by leads I, II, and III as shown in Fig. 14-6. Lead II also is connected to the recorder so that negativity at the RA electrode with respect to the left leg (LL) electrode produces an upward deflection. Lead III is connected so that LA electrode negativity relative to the LL electrode yields an upward deflection of the recorder.

To determine the electrical axis of the heart, sequential recordings from any two leads are made. From each of these records, the *net* upward deflection (in millivolts) of the QRS complex in a representative cardiac cycle is calculated. These two values are measured off from the midpoints of and plotted on the appropriate arms of the triangle. Perpendiculars to the arms are drawn at these points. An arrow drawn from the center of the triangle to the intersection of the perpendiculars defines the electrical axis of the heart, with the arrow head specifying its direction and the angle from the horizontal its orientation. In normal individuals the mean electrical axis of the heart is about +60°, i.e., downward and to the right as observed from the front (Fig. 14-6). The axis can shift with deep diaphragmatic breathing. It is generally shifted somewhat to the left in short, stocky individuals and to the right in tall, thin persons. Hypertrophy of either side of the heart causes a shift of the axis to that side.

CARDIAC ACTIVITY

Cardiac cycle

The kymographic tracing in Fig. 14-7 is a record of the heartbeat in a frog. Atrial systole causes the record to rise from *A* to *B*. The atrial diastole begins at *B* (in the tracing this is interrupted by the ventricular systole) and lasts till *F*. The contraction of the ventricle begins at *C* and is completed at *E*; here the ventricular diastole begins and continues to *G*. The length of time consumed by each of these four phases in the action of the human heart, beating at the rate of 70/min, is approximately as follows:

Atrial systole, 0.1 sec Atrial diastole, 0.762 sec	} 0.862
Ventricular systole, 0.379 sec Ventricular diastole, 0.483 sec	} 0.862

From these values note the very short time occupied by the atrial systole, this being about one fourth of the time taken by the ventricular systole and less than one seventh of that of the atrial diastole. It will also be noticed that the diastole or resting period of the ventricle is, at the rate of 70 beats per minute, about 25% longer than the systole or active period. This provides for sufficient rest. When the rate of the heart is greatly increased, the diastole is cut short far more than the systole.

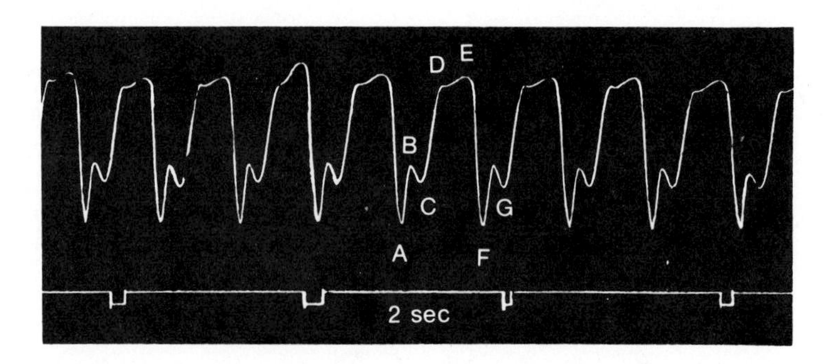

Fig. 14-7
Contraction record showing frog's heartbeat. **AB,** Atrial systole; **BC,** first part of artrial diastole, which is completed at **F; CDE,** ventricular systole; **EF,** first part of ventricular diastole, which is completed at **G.**

On observing the active frog heart, it will be noticed that the ventricle (if bleeding has been avoided) alternately flushes (diastole) and blanches. During its systole the ventricle hardens to the touch and decreases in size. The atrioventricular groove moves downward, but the apex does not move upward; this is caused by the recoil from the discharge of the blood into the aorta.

Rate of heartbeat. ■ The rate of the heartbeat under different conditions and in different persons varies so much that we cannot talk about a normal pulse rate. It ranges in the human adult male from 50 to 100 beats per minute. In the adult female the rate is generally faster. The rate is more susceptible to variations in women and children than it is in men.

Factors influencing heart rate. ■ Among the more common factors causing variations are age, size of body, body position, and physical training.

Age. At birth the heart rate in the human being is generally from 130 to 150/min; at the age of one year, the resting pulse rate is approximately 120; at ten years of age it is 90; in the adult it may be put at about 70.

Size of body. A small animal has a higher pulse rate than does a larger animal; compare the pulse rate of a mouse, about 1,000, with that of an elephant, 28.

Body position. Position of the body causes a variation in most people, the rate being from 5 to 10 beats more in the erect than in the reclining position.

Physical training. In individuals who are well trained physically the resting pulse is frequently found to be no more than from 50 to 60/min. In the normal individual a low heart rate is generally an indication of a strong, efficient heart.

Other factors. The heart rate is increased by such factors as muscular work, high body temperature (fever), eating, warm baths, hot drinks, high altitude, high external temperature, and certain emotions.

■ All-or-none law

Our study of skeletal muscle demonstrated that the extent of the contraction is, within limits, proportional to the strength of the stimulus. In the heart this is quite different. If the spontaneous beating of a frog's heart is halted (by a ligature applied between the sinus and the atria—Fig. 14-2) and is then stimulated at thirty second intervals, the extent of the contraction remains the same for any strength of adequate stimulus; there is no summation. The heart, *as a whole,* follows the all-or-none law (Fig. 14-8).

It will be recalled that this holds true also for the individual fibers of a skeletal muscle, although the muscle as a whole is capable of graded contractions. The reason for this difference between the heart and the skeletal muscle lies in the relation of the muscle fibers to each other. In a skeletal muscle the individual fibers are isolated from each other by a sarcolemma; in the heart the imperfectly cross-striated fibers (Fig. 14-9) are brought into intimate contact with each other by anastomoses. This permits the impulse

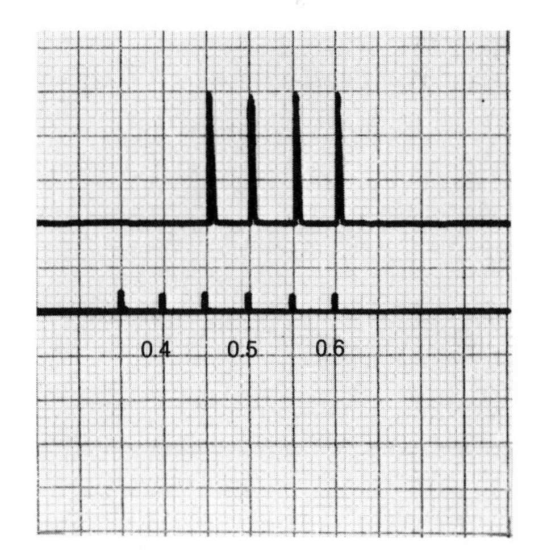

Fig. 14-8
Contraction record (upper trace) illustrating all-or-none phenomenon of the heart. Gradually increasing strengths of stimuli are indicated by numerals.

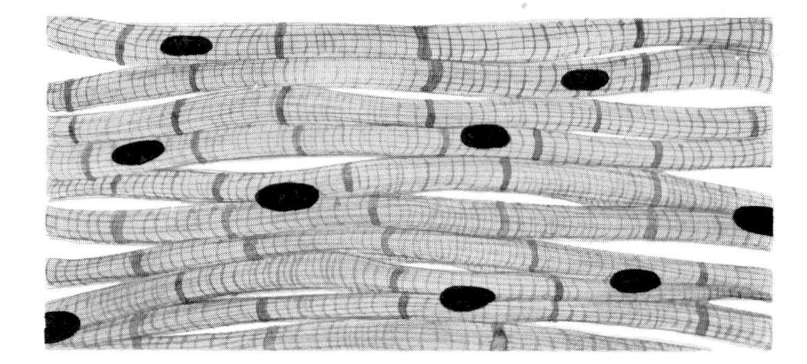

Fig. 14-9
Cardiac muscle.

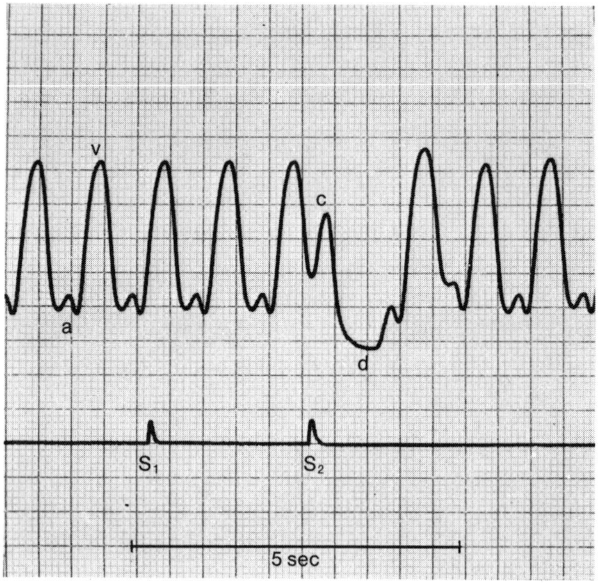

Fig. 14-10
Contraction record illustrating refractoriness in a frog's heart: **a**, atrial contraction; **v**, ventricular contraction; **c**, extrasystole; **d**, compensatory pause. S_1 and S_2 are stimuli. Heart was refractory at S_1.

to spread from one fiber to another throughout the whole atrium (or ventricle), causing them to respond as a single unit. Although *the contraction of the heart is always maximal, this does not mean that the extent of the contractions are always the same. The physiological state of the heart muscle is the determining factor.*

Refractory period. ■ If a beating heart is stimulated with a single electric shock during its contraction, it is not affected. This is illustrated in Fig. 14-10, in which a frog's heart was stimulated at S_1 during the systole of the ventricle; the record shows that the heart continued with its usual rhythm. The irritability of the heart during systole is so greatly reduced that the stimulation is without result. It will be recalled that this period of reduced irritability of a tissue during its activity is known as the *refractory period.* On the

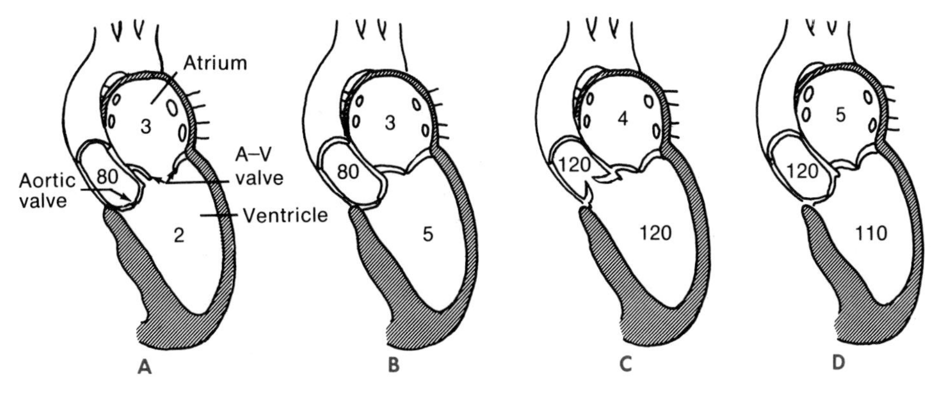

Fig. 14-11
Illustration of action of valves on the left side of the heart. **A,** Ventricular diastole; **B,** early stage of ventricular systole; **C,** final stage of ventricular systole; **D,** early stage of ventricular diastole, which is followed by **A.** Numerals indicate blood pressures.

other hand, when the stimulation was presented during the diastole, at S_2, the heart proved to be irritable, for the stimulation was followed by a hurried, weaker contraction, *c.* This is known as an *extrasystole* (premature or ectopic beat) and is followed by a lengthened pause, *d,* the *compensatory pause.* The human heart sometimes gives these premature beats, which are then followed by a compensatory pause. Because of the long refractory period, the heart, unlike the skeletal muscle, shows no fusion or summation of twitches and cannot be thrown into complete tetanus, which would annul its function.

■ Cardiac valves

In order that the energy expended during the contraction of the ventricles shall cause the blood to circulate efficiently and not to regurgitate, the valves of the heart must operate properly. The opening and the closing of a valve are determined by the differences between the pressures applied to it from opposite sides.

There are no valves between the atria and the veins flowing into them. The reason for this is that the small pressure developed by the atria is not sufficient to force any appreciable amount of blood back into the veins, which are guarded by a sleeve of atrial muscle surrounding the origin of the great veins.

To understand the order in which the various valves of the heart open and close, it is necessary to bear in mind that the pressure in the atrium never rises very high and that arteries are always quite well filled and therefore constantly show a considerable amount of pressure. During the ventricular systole the pressure in this cavity is higher than elsewhere in the vascular system; during its diastole it falls even below that in the atrium.

With the ventricle in a completely dilated and resting state, the pressure in the aorta or in the pulmonary artery is much greater than that in the corresponding ventricle; in consequence, the semilunar valves are closed. The incoming blood from the veins causes the pressure in the atria to be greater than that in the ventricles, and the atrioventricular valves are open (Fig. 14-11, *A*). As the ventricles start to contract, the gradually rising pressure in these cavities causes the changes in the valves shown in *B* and *C.* The gradual reduction in ventricular pressure during the diastole and its effects upon the valves are indicated in *D* and *A.* The action of the valves may be summarized as:

Ventricular systole	1. Closing of the atrioventricular valves, Fig. 14-11, *B*
	2. Opening of the semilunar valves, *C*
Ventricular diastole	3. Closing of the semilunar valves, *D*
	4. Opening of the atrioventricular valves, *A*

Valvular insufficiency. ■ It sometimes happens that at birth one or more flaps of a valve are defective. Furthermore, the endothelium may become pathologically affected by diseases, such as rheumatic fever or syphilis. In healing, the valves may become stiff or distorted by scar tissue, the orifice may be narrowed *(stenosis),* and the valve may be prevented from opening or closing properly *(incompetent).* An incompetent valve may result in regurgitation. The results of regurgitation may be far reaching, as the following illustrates.

Normally the full force of the right ventricular systole is directed to the emptying of the blood into the pulmonary artery; this is indicated in Fig. 14-12, *A,* by the dotted line leading from the right ventricle into the pulmonary artery. But in tricuspid insufficiency, as shown in Fig. 14-12, *B,* while a portion of the blood, *d,* follows the normal pathway, another portion, *b,* regurgitates through the incompetent valve into and partly fills the right atrium. This impedes the flow of venous blood from the venae cavae, *a* and *c,* into the atrium and thereby increases the volume and pressure of the blood in the systemic veins. The damming of the venous flow must sooner or later cause engorgement of the capillaries, reduction in capillary circulation (passive congestion), and interference with tissue nutrition.

Cardiac surgery and catheterization. The diagnosis of congenital or acquired defects can be

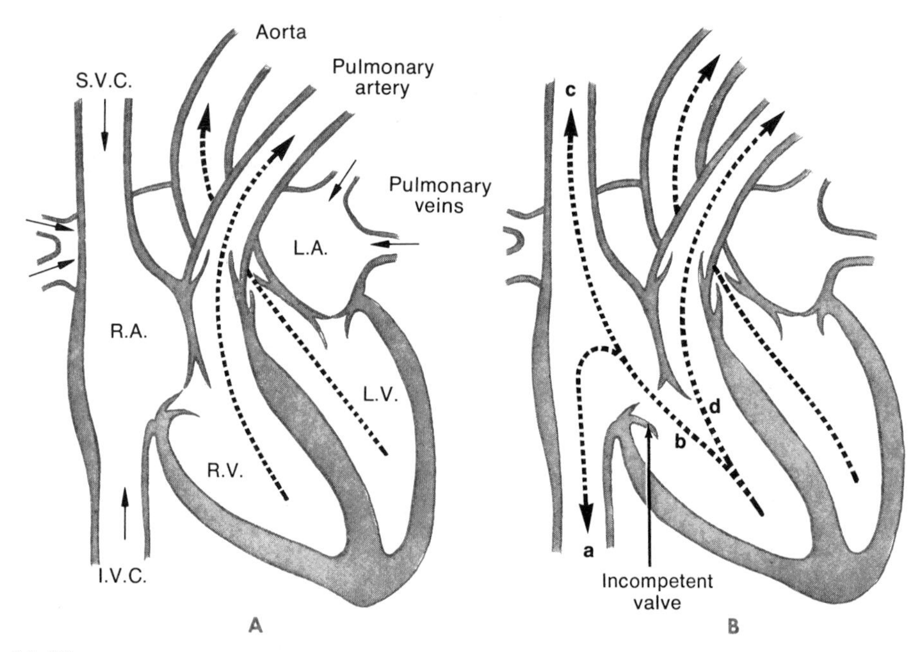

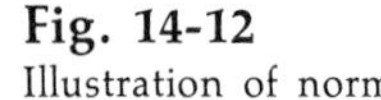
Fig. 14-12
Illustration of normal blood flow during systole, **A,** and tricuspid regurgitation, **B,** due to valvular insufficiency.

precisely accomplished by the catheterization of the heart. The procedure consists of inserting a long tube into a vessel in a limb and gently moving it forward until it enters the heart. With this it is possible to determine pressures in various parts of the heart and to collect blood for the determination of its oxygen content in these regions. An evaluation of these relationships leads to the determination of the nature of the defect.

Modern surgical techniques make it possible to bypass the heart by means of a mechanical heart-lung device, so that the quiescent heart can be surgically repaired.

Heart sounds. ■ The beating of the normal heart is associated with the production of two major sounds that can be distinctly heard through a stethoscope. By noting the apex beat,* it can be determined that one of the sounds is heard during the systole of the ventricle, whereas the other takes place during the diastole. For this reason they are sometimes called the systolic sound and the diastolic sound, but they are more commonly spoken of as the *first* and the *second* sounds, respectively. The first, or systolic, sound is of longer duration and lower pitch than is the second. The pause between the first and the second sound is shorter than that following the second sound. The first sound is heard best if the bell of the stethoscope is placed over the fifth left intercostal space or over the apex of the heart; for the second sound the stethoscope should be placed over the second right costal cartilage (for aortic valve closure) or over the second left cartilage (for closure of the pulmonary valve).

The first, or systolic, sound. The cause of the systolic sound is that during the ventricular systole the atrioventricular valves are suddenly thrown into rapid vibration; these vibrations are communicated to the blood, heart muscle, and thoracic wall and can be heard if a stethoscope is placed on the thoracic wall. When the ventricular musculature contracts, a rumbling noise is produced, which fuses with the sound caused by the vibrations of the closing atrioventricular valves. When the strength of the beat is increased, the first sound is accentuated. If in a beating heart the atrioventricular valves are prevented from closing, the higher pitched part of the first sound disappears, but the lower rumbling part persists.*

The second, or diastolic, sound. The second, or diastolic, sound is caused by the vibration of the closing semilunar valves. Increasing the aortic blood pressure increases the volume of the second sound.†

■ Law of the heart

It will be recalled (Table 6-1) that, within limits, an increase in the load applied to a skeletal muscle increases the force of the contraction. This is attributed to the stretching of the muscle fibers by the larger load. This also is true for cardiac muscle. When a large volume of venous blood arrives at the heart, the increased diastolic filling stretches the fibers of the right ventricle. We speak of the ventricular end-diastolic volume. Up to a certain point, this stretching increases the force of the contraction; thus the ventricular pressure also increases, enabling the ventricle to discharge a greater volume of blood. *The initial length of the cardiac muscle fibers determines the strength of the contraction,* Starling's law of the heart (Fig. 14-13). This change in ventricular output with change in end-diastolic volume is sometimes called the *heterometric autoregulation* of the heart.

Starling's law of the heart is also operative in the opposite condition to that just discussed. For example, in extensive hemorrhage the systolic

*The apex beat is the thump that can be felt in the fourth or fifth intercostal space and about 1 or 2 inches to the left of the sternum.

*A leaking valve is associated with murmurs caused by the regurgitating blood.

†The heart sounds are, in fact, far more complex than described here.

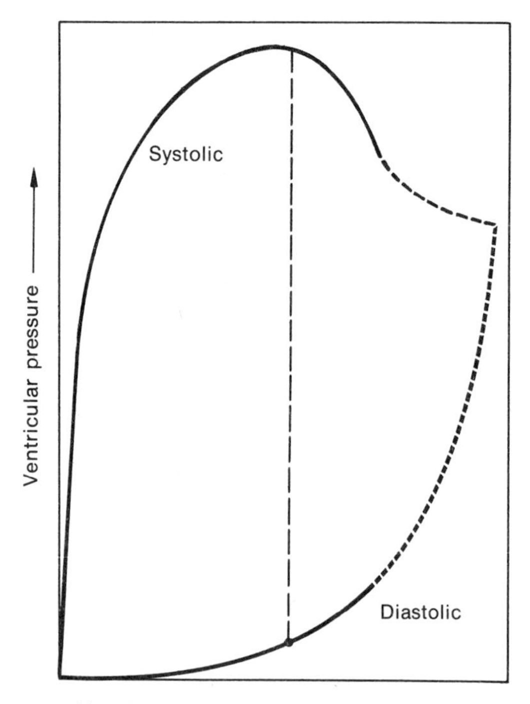

Fig. 14-13
Graph showing relationship of end-diastolic volume to peak systolic ventricular pressure—basis of the law of the heart. Vertical dashed line connects point of optimal end-diastolic pressure (volume) and maximal systolic pressure. Area between two curves represents theoretical maximum amount of pressure-volume work obtainable from the heart. Compare to Fig. 6-17.

and diastolic pressures are very low; the amount of blood delivered to the ventricle is greatly reduced, and, as a consequence, the cardiac fibers are but slightly stretched. This results in a feeble (although rapid) heartbeat and in a systolic ventricular pressure greatly below normal; however, this force is sufficient to overcome the much reduced diastolic pressure in the aorta.

CARDIAC OUTPUT

The term cardiac output *(C.O.)* refers to the volume rate of blood pumped by the heart. As such, it is the product of the stroke volume *(S.V.)* and heart rate *(H.R.)* and is expressed in terms of liters per minute.

$$\text{C.O.} = \text{S.V.} \times \text{H.R.}$$

The resting cardiac output in man is about 5 liters/min. This value may be altered by exercise (Table 15-1), emotional states, or illness. Varying either the stroke volume or the heart rate alters the cardiac output. The stroke volume depends on myocardial contractility and the *venous return,* that is, the flow of blood into the heart. Venous return is a function of the mean systemic pressure that forces blood toward the heart, the resistance of the peripheral vessels, and the pressure in the right atrium that tends to resist the entrance of blood into the heart. How these various determinants of cardiac output are markedly influenced by activity in the autonomic nervous system is discussed in succeeding sections of this text.

Since the cardiac output is vital to the well-being of an individual in that it is the means for providing the body tissues with adequate nutrition, its measurement is often undertaken. Direct measurement of stroke volume or cardiac output in the human subject is impractical because of the surgery involved; therefore, indirect methods are employed. Each of these methods is based on the premise that over a reasonable period of time the quantity of blood entering the heart and the quantity leaving the heart are equal. The indicator dilution technique and the Fick principle are two methods commonly used to determine cardiac output in man.

Indicator dilution technique. ■ In the indicator dilution technique, a precise amount of a dye or radioactive isotope in solution is injected rapidly into a vein. The *average* concentration in the arterial blood during a single, timed passage of the indicator through the circulatory system is determined by appropriate analytical methods. In order that the technique give valid results, the injected material must be harmless, must remain within the circulatory system, and must not produce hemodynamic effects of its own. From the values of the dye concentration and time, the volume rate of flow may be calculated.

To illustrate, let us assume that 10 ml of a dye solution with a concentration of 60 mg/100 ml was injected into a vein. The average concentration of the dye in the arterial blood during a thirty-six second circuit was found to be 0.2 mg/100 ml, that is, 2 mg/liter. The appropriate equation (Chapter 3) would be:

$$C_1 \times V_1 = C_2 \times V_2$$

where C_1 is the average arterial concentration of the dye, V_1 the volume of blood pumped during the single circuit of thirty-six seconds, C_2 the concentration of the injected dye, and V_2 the volume of the dye solution injected. Then one calculates the cardiac output by:

$$V_1 = \frac{60\ \text{mg}/100\ \text{ml} \times 10\ \text{ml}}{2\ \text{mg/liter}}$$

$$= \frac{6\ \text{mg}}{2\ \text{mg/liter}}$$

$$= 3\ \text{liters (in 36 seconds)}$$

$$\text{C.O.} = 5\ \text{liters/min}$$

Fick principle. ■ Measurement of cardiac output by the Fick principle is based on the fact that the amount of a substance taken up (O_2), or eliminated (CO_2), by the body per unit of time is equal to the difference between the arterial and venous levels of the substance times the blood flow.

$$(A_{O_2} - V_{O_2}) \times \text{Volume} = O_2\ \text{consumption/min}$$

To illustrate, let us assume that an individual consumes 200 ml of O_2 per minute, that the arterial blood O_2 concentration is 19 ml/100 ml, and that the venous blood (pulmonary artery) concentration is 15 ml/100 ml. The calculated cardiac output is:

$$\text{C.O.} = \frac{200\ \text{ml/min}}{(0.19\ \text{ml/ml} - 0.15\ \text{ml/ml})}$$

$$= \frac{200\ \text{ml/min}}{0.04\ \text{ml/ml}}$$

$$= 5{,}000\ \text{ml/min}$$

$$= 5\ \text{liters/min}$$

Cardiac reserve. ■ The capacity of the heart to generate sufficient energy for raising the blood pressure above the basal pressure and for expelling a larger quantity of blood is designated as the *cardiac reserve.* In the same manner that all individuals do not possess the same skeletal muscle strength, so the cardiac reserve varies from one individual to another.

By consulting Table 6-1, it will be seen that increasing the load on skeletal muscle beyond a certain point decreases the amount of work accomplished, until finally no mechanical work is possible. Starling's law of the heart also holds good only to a certain limit, which we may call the *physiological capacity of the heart.* When the diastolic filling and the consequent stretching of the muscle fibers exceed this limit, the force exerted by the ventricle becomes less (Fig. 14-13), and, as a result, the whole volume of blood in the ventricle is not discharged into the aorta (or pulmonary artery), that is, the *end systolic volume* is increased. Under these conditions the atrium does not completely empty; this leads to stagnation of the flow in the veins, *venous stasis.*

A marked deficiency in cardiac reserve (in which condition the physiological capacity of the heart is speedily exceeded during even slight muscular exertion) is characterized by breathlessness (dyspnea), fatigue, tendency to fainting, chest pains (angina pectoris), cyanosis, and edema (especially of the feet and ankles). If aggravated, this condition may lead to cardiac failure. Among the many untoward circumstances that reduce cardiac reserve energy is rheumatic fever of childhood.

Cardiac compensation. ■ Its remarkable reserve enables the heart to compensate for extensive impairment in the functioning of its valves. Due to regurgitation in incompetent aortic semilunar valves, only a part of the systolic output reaches the tissues. If the volume of the systolic output were the same as that of a normal individual, the tissues would be ill nourished. But the increase in the release of energy (due to stretching of the ventricular walls by the regurgitated blood) may enable the ventricle to send a larger amount of blood into the aorta. Even though some of this blood regurgitates, enough reaches the tissues to ensure normal nutrition. In consequence, life is possible with a surprisingly large valvular impairment.

However, the victim is constantly drawing on his cardiac reserve, and, as a result, strenuous work is not possible. If the valvular lesion sets in gradually and persists for a long time, the heart muscle may undergo hypertrophy* and thus may build up a greater reserve. If the leakage becomes increasingly greater, the reserve grows correspondingly less, until finally even the generation of basal blood pressure is beyond the physiological capacity of the heart.

In summary: *Every heart has its margin of compensation or reserve above that necessary for the generation of a basal blood pressure.* In a normal individual the reserve of the heart is surprisingly large, but in a weak or so-called "low-toned" heart this margin is so limited that the generation of even a small increase in blood pressure beyond the basal is difficult or impossible.

WORK OF THE HEART

From the stroke volume, the heart rate, the mean blood pressure in the aorta and in the pulmonary artery, and the velocity with which the blood is ejected from the ventricles, the mechanical work done by the heart† can be calculated. In Chapter 3 we made mention of pressure-volume work; in Chapter 1 we distinguished between potential and kinetic energy. The heart develops potential energy in the form of pressure-volume work and kinetic energy by imparting a velocity to the mass of the blood. To calculate the work done by each ventricle at each beat of the heart, the following formula may be used:

$$\text{Work} = PV + \frac{wv^2}{2g}$$

In this equation P is the mean arterial pressure, V the stroke volume, w the weight of the blood ejected, v the velocity of flow from the ventricle, and g the acceleration of gravity (980 cm/sec^2). Pressure-volume work is PV and the kinetic energy is $wv^2/2g$. If we assume a stroke volume of 70 ml, a mean aortic pressure of 100 mm Hg (10 cm), and a specific gravity of mercury of 13.6 g/cm^3, the pressure-volume work of the left ventricle would be:

$$10 \text{ cm Hg} \times 13.6 \text{ g/cm}^3 \text{ Hg} \times 70 \text{ cm}^3 = 9{,}520 \text{ g-cm}$$

If the velocity were 500 mm/sec (50 cm/sec) and the specific gravity of the blood 1.0 g/cm^3, the kinetic energy developed would be:

$$\frac{70 \text{ g} \times 2{,}500 \text{ cm}^2/\text{sec}^2}{1{,}960 \text{ cm/sec}^2} = 89 \text{ g-cm}$$

Because of the lower mean pressure in the pulmonary artery, the work of the right ventricle is one fifth, or less, that of the left ventricle.

$$2 \text{ cm Hg} \times 13.6 \text{ g/cm}^3 \text{ Hg} \times 70 \text{ cm}^3 = 1{,}904 \text{ g-cm}$$

The kinetic energy developed by the right ventricle is of inconsequential magnitude. Under essentially resting conditions, the metabolic energy expended by the whole heart in developing kinetic energy is less than 1% of the total useful work performed.

When the output of the heart is reasonably small, as we have been discussing, the work output may be approximated by the product:

$$\text{Cardiac output} \times \text{Mean arterial pressure}$$

If the cardiac output is 5 liters/min, this gives:

$$5{,}000 \text{ cm}^3/\text{min} \times 10 \text{ cm Hg} \times 13.6 \text{ g/cm}^3 \text{ Hg} = 680{,}000 \text{ g-cm/min}$$

This amount of work equals 6.8 kg-m/min or

*Ventricular hypertrophy may also take place in mitral insufficiency and in high blood pressure.

†Taking the average pulse rate throughout life as 70/min and the output of each ventricle as 5 liters/min, the number of heartbeats during a lifetime of sixty years totals 2.2 billion and the amount of blood pumped by the two ventricles amounts to approximately 335,000 tons.

about 9,800 kg-m/day. Converted to terms of ft-lb per day, the work becomes 71,250; this is equivalent to the work done by a person weighing 150 lb lifting his body to a height of nearly 500 ft.

In the resting individual about 250 ml of oxygen are consumed each minute. Of this quantity the heart uses about 8%, or 20 ml/min. Later, in Chapter 25, we will learn that during the biological oxidation of nutrients 1 ml of oxygen yields about 5 calories of energy. Each calorie is equal to 0.427 kg-m of energy. Twenty milliliters of oxygen/min represents 42.7 kg-m/min of energy (0.427 kg-m/cal × 5 cal/ml × 20 ml/min); the efficiency of the heart would be slightly less than 20%. Two factors principally determine the oxygen consumption of the heart: (1) the developed pressure, the major determinant, and (2) the velocity of the contraction. The efficiency of the heart is altered little by changes in blood pressure, if cardiac output and heart rate are constant, for any increase or decrease in pressure-volume work is matched by a corresponding increase or decrease in oxygen consumption. When the stroke volume is increased, with blood pressure and heart rate unchanged, the efficiency increases, since little extra energy is necessary for handling the increased volume. The improved efficiency of trained athletes is, in part, due to an increase in stroke volume without marked change in heart rate. In contrast, an increase in heart rate, with pressure and cardiac output constant, can produce a decreased efficiency because of the extra oxygen required for the increased velocity of contraction.

■ CARDIAC NUTRITION

Coronary circulation. ■ The cardiac tissue is supplied with blood by coronary arteries, which issue from the aorta, penetrate into the tissue of the heart walls, and break up into a rich network of capillaries around the muscle cells. The blood from these capillaries is collected by the coronary veins and brought to the right atrium. The coronary vessels receive from 3% to 10% of the entire systolic output and are filled with blood during ventricular diastole and are emptied during systole.

To augment the flow according to the needs of the cardiac muscle, the coronary vessels are capable of dilation. For example, during severe muscular work or when the supply of oxygen in the blood is materially reduced (hypoxemia), the flow may be increased fivefold. From this it is evident that, by interfering with the nutrition of the heart, deficient coronary circulation must greatly lessen the efficiency of the heart and the cardiac reserve and may lead to heart failure.

An interruption in coronary circulation (occlusion) results in an insufficient supply of oxygen to the heart muscle. This leads to the accumulation of substances that stimulate the free nerve endings, thereby giving rise to agonizing pain—*angina pectoris.** The pain is frequently referred to the left arm, shoulder, back, and abdominal region. It is far more common in men that it is in women and generally occurs after the age of forty. The cause of the occlusion is nearly always a sclerotic (hardened) condition of the arteries. Other causes are embolism and thrombosis; it will be recalled (Chapter 12) that in these conditions heparin or dicumarol is frequently used with great benefit.

An attack of angina pectoris may occur during strenuous muscular work, anemia, hypoxemia, and excessively low diastolic pressure. A strong outburst of emotion may precipitate an attack. The ingestion of too much food sometimes has this result (often erroneously called acute indigestion). By constricting the arterioles, smoking of tobacco aggravates the angina during muscular effort. In the more severe attack of coronary thrombosis, ventricular fibrillation may set in and cause sudden death.

---

*The pain in angina pectoris is frequently eased by the inhalation of amyl nitrite. This drug, and also nitroglycerin, causes dilation of the coronary arteries.

Energy supply for the heart. ■ In addition to lipids, the preferred fuel for the heart appears to consist of lactic and pyruvic acid, formed chiefly from glucose during skeletal muscle activity. Since these acids are metabolized and yield their energy only when oxygen is available, cardiac metabolism is almost entirely aerobic; hence the heart soon succumbs to lack of O_2. Unlike skeletal muscles, the heart is not able to incur an oxygen debt. Within limits, the mechanical efficiency increases as the stroke volume is increased; in other words, as the heart is called on to do more work, the conditions for the utilization of the liberated energy become more favorable. And because of the dilated condition of the capillaries (brought about by the decrease in O_2 and the increase in CO_2), fuel for the heart muscle is more available.

Functions of inorganic salts. ■ The heart muscle must be supplied for its activity not only with organic materials from which it obtains energy, but also with inorganic salts. A strip of a turtle's ventricle, immersed in 0.7% NaCl solution, continues its activity for a considerable length of time, but finally the beats become feeble and then cease altogether (Fig. 14-14). The addition of a few drops of $CaCl_2$ restores the strength of the contractions. We have discussed the role of calcium ions in excitation-contraction coupling in Chapter 6. By experimentation it has been found that the heart beats best when supplied with the chlorides of sodium, potassium, and calcium. The most favorable concentration of these salts (Ringer's solution) for the mammalian heart is as follows:

NaCl	0.9%
$CaCl_2$	0.024%
KCl	0.042%

This solution must be buffered by the addition of sodium bicarbonate. If the potassium chloride is too concentrated, the heart muscle relaxes and may stop entirely; potassium ions depolarize the muscle and favor relaxation (Fig. 14-15, *A*). On the other hand, calcium chloride increases the tonus of the heart and in excess it causes arrest (Fig. 14-15, *B*). This is an example of the *antagonistic action of cations.* The chief role of sodium

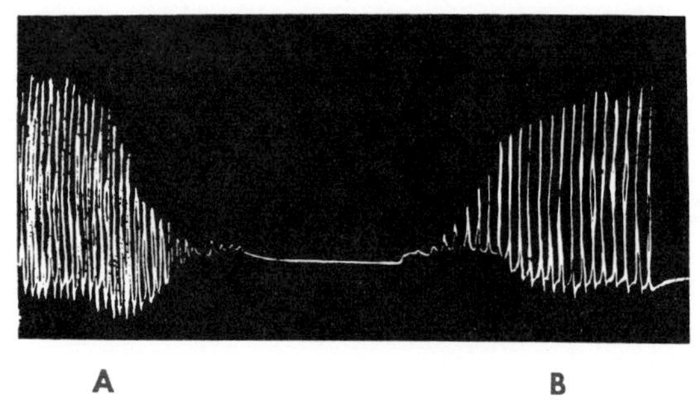

Fig. 14-14
Contraction record of turtle's ventricle in 0.7% NaCl, **A**, and after the addition of $CaCl_2$, **B**.

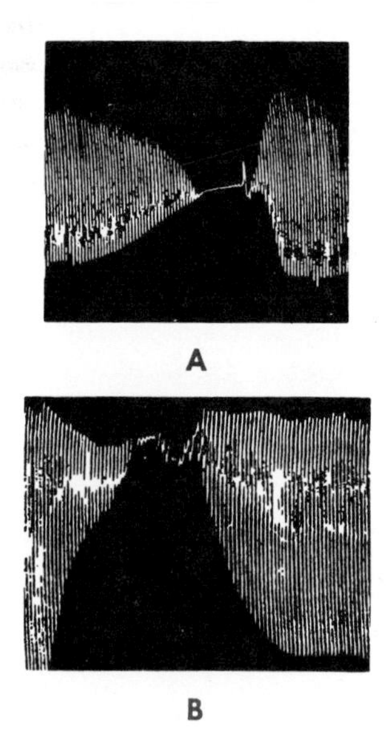

Fig. 14-15
Contraction record of frog's heart perfused with excess salts added to Ringer's solution. **A**, Effect of excess potassium ions, heart is arrested in diastole. **B**, Effect of excess calcium ions, heart is arrested in systole. Restoration of contractions occurs after returning heart to normal Ringer's solution.

chloride is to supply the proper osmotic pressure, although the sodium ion is essential in the generation of action potentials.

To maintain the beat of an excised mammalian heart, the coronary system is perfused with an oxygenated Ringer's solution having the proper osmotic pressure, pH, and temperature. A small amount of glucose is added as an energy source. To supply the necessary oxygen, the pulmonary circulation may be kept intact with the excised heart (a heart-lung preparation); the activity of a rabbit's heart can be maintained for several hours in this way.

Digitalis. ■ We previously noted the effects of a reduced cardiac reserve included increased end-diastolic volume, venous stasis, and edema. These are aspects of congestive heart failure which often can be treated successfully by the administration of digitalis. The term *digitalis* is used to describe a number of naturally occurring sugar-linked steroid compounds (Chapter 23) known as cardioactive, or cardiac, glycosides. Digitalis enables the failing heart to contract more forcefully and efficiently; it does more work without an increased consumption of metabolites or oxygen.

The essential role of calcium ions in the contraction of muscle was described in Chapter 6. In cardiac muscle, sodium and calcium ions appear to compete for the same channels of entry through the sarcolemma. Thus, a slight reduction in external sodium ion concentration or increase in internal concentration, either of which would diminish the electrochemical gradient for sodium, augments the influx of calcium ions and a more forceful contraction ensues. Digitalis exerts its effect by inhibiting the sodium pump and producing an increment in internal sodium ion concentration that facilitates calcium ion influx. Consequently, contractile force is increased and the deleterious effects of congestive heart failure are to a greater or lesser degree alleviated. Because of its action on the sodium pump, digitalis also exerts an effect upon the electrical activity of cardiac cells. Description of this effect is beyond the scope of our inquiry.

■ NEURAL CONTROL OF THE HEART

The function of the heart is to discharge, with adequate force, an amount of blood sufficient for the metabolic needs of the body. The amount discharged in a unit of time is determined by the stroke volume and the heart rate. As the demand for blood by the body varies from moment to

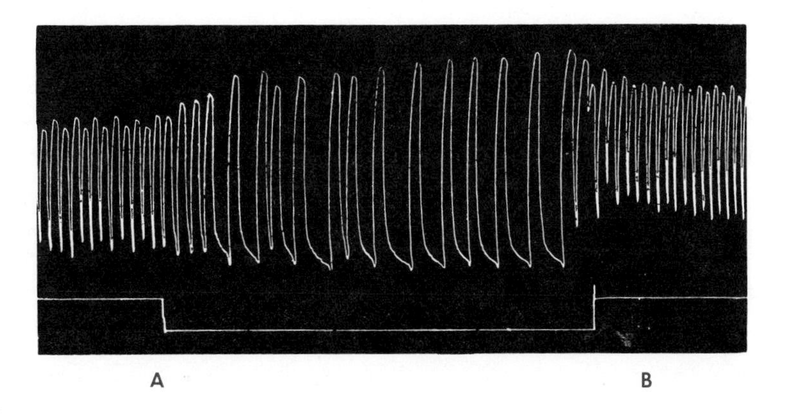

Fig. 14-16
Contraction record of turtle's heart. From **A** to **B** one vagus nerve was stimulated with weak stimuli. Note decrease in rate and increase in force.

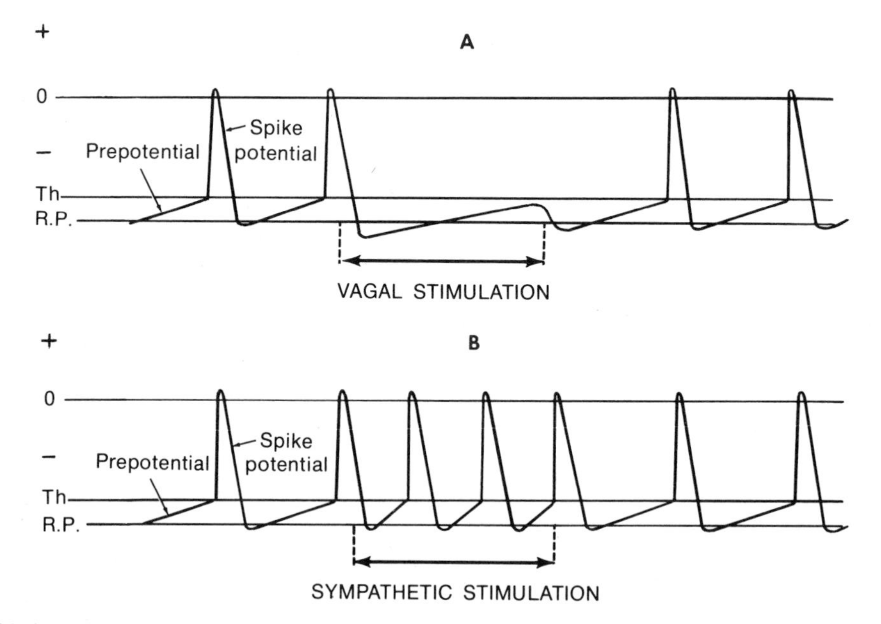

Fig. 14-17
The influence of vagal, **A**, and sympathetic, **B**, stimulation on the intracellularly recorded action potentials of heart pacemaker. In **A** the prepotential rise (depolarization) is slowed due to stimulation of vagus nerves; in **B** the prepotential rate of rise is increased, and the action potentials are more frequent. **Th** is the threshold and **RP** the resting potential.

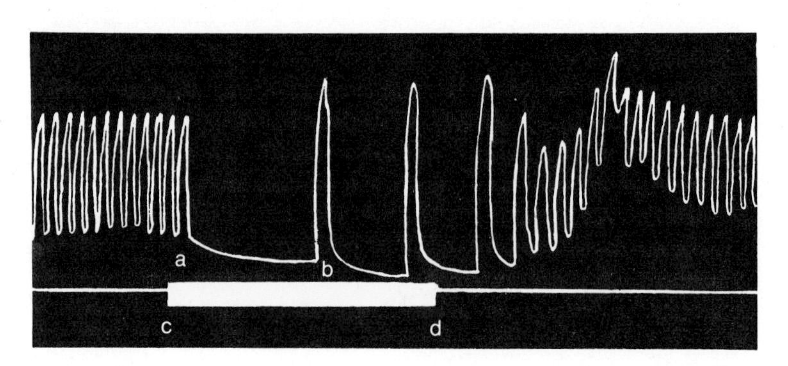

Fig. 14-18
Contraction record of turtle's heart. From **c** to **d** one vagus nerve was stimulated with strong stimuli. Note escape from inhibition at **b** and augmented beats following inhibition.

moment, it is evident that the rate, force, and systolic output must be governed in accordance. It is a general observation that the state of the emotions may profoundly modify cardiac activity. These controls and regulations are both neural and humoral. The neural influences are exercised by the vagi (parasympathetic nerves) and by the cervical sympathetic nerves.

■ The vagi

The vagi originate in the lowest division of the brain, known as the medulla oblongata (Figs. 10-1 and 10-2). These nerves supply both afferent (sensory) and efferent (motor) fibers to the heart. The efferent fibers belong to the autonomic nervous system and are cholinergic (Chapter 11). As preganglionic fibers they end in a small ganglion in the wall of the heart. From there the impulses are carried by very short postganglionic fibers to the sinoatrial and the atrioventricular nodes and the bundle of His (Fig. 14-1). Although the atrial musculature is innervated by this parasympathetic supply, the ventricular musculature is innervated by it only sparsely or not at all.

Cardiac inhibition. ■ In an animal a vagus nerve may be exposed without great difficulty. Under the influence of an anesthetic, the skin in the neck, a little to either side of the larynx, is slit; the structures immediately below the skin are pushed aside, and a carotid artery is brought into view. Alongside this vessel lies a large nerve trunk, the vagus. When the vagus on one side is cut, there is usually little or no result, but when both vagi are severed, there is a marked cardiac acceleration (Fig. 14-22).

Stimulation of a vagus nerve may cause a decrease in the rate, as shown in Fig. 14-16, or it may cause a complete cessation of the heartbeat, as shown in Fig. 14-17. This slowing or stoppage of the heart, cardiac inhibition, is the result of a reduction in the irritability of the sinoatrial node (the pacemaker) and a decrease in the conductivity of the atrioventricular bundle; the vagi have no direct influence on the ventricular musculature. The acetylcholine released by the vagal terminals increases the permeability of the cells to potassium, which brings about an increased internal negativity or hyperpolarization (Fig. 14-17, *A*). Since the rate of sodium ion entry is unchanged, the prepotential rise to threshold is prolonged; thus the action potentials are less frequent and slower. From Fig. 14-18 it will be noticed that notwithstanding the continued stimulation of one vagus (up to *d*), the heart resumes

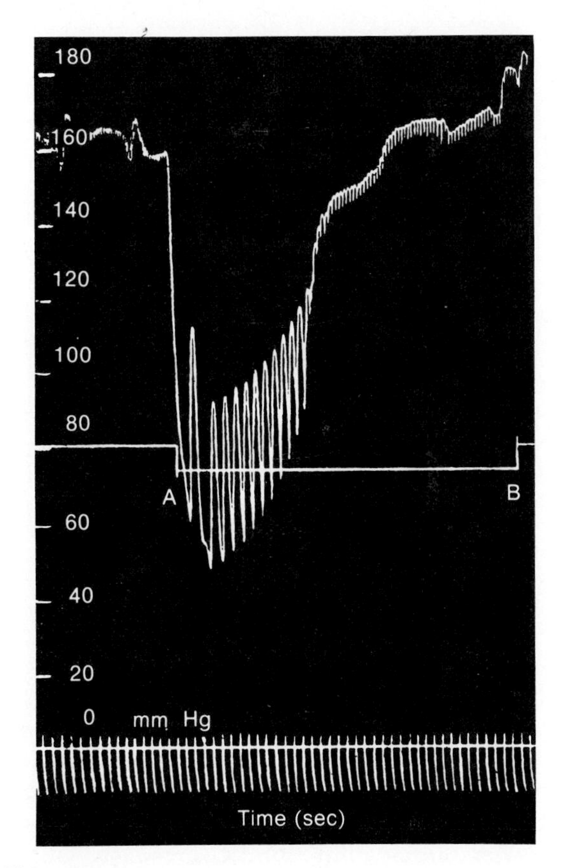

Fig. 14-19
Blood pressure record from dog showing influence of vagal stimulation. From **A** to **B** one vagus was stimulated. Note immediate and severe drop in blood pressure, escape from inhibition, subsequent acceleration of heartbeat, and greater than normal blood pressure.

its beat at *b.* This phenomenon is known as *vagal escape.* The beats immediately following the inhibition are stronger than those prior to vagal stimulation.

Cardiac inhibition and blood pressure. ■ It is evident that a marked decrease in cardiac activity must soon be followed by a fall in the blood pressure. Fig. 14-19 is a carotid artery pressure tracing from a dog; at *A* one vagus nerve was stimulated; as a result the blood pressure fell immediately to less than half its previous value. Here also notice that the escape from inhibition restores circulation to its normal state before the cessation of the stimulation at *B.*

The fall in blood pressure due to cardiac inhibition causes a decrease in the flow of blood, and, as a result, the tissues are not well supplied with O_2 and the CO_2 is not properly removed. This interferes with the functions of the various organs of the body, but the organ influenced most speedily and severely is the central nervous system. Severe cardiac inhibition causes loss of consciousness (fainting, or syncope). When the body temperature is lowered to about 85 F, the consumption of O_2 and the production of CO_2 are decreased to such extent that the heartbeat may be completely stopped for eight minutes without causing any damage to the brain.

■ Cardiac inhibitory center

The inhibitory fibers of the vagi issue from the medulla; the particular area where they originate is known as the *cardioinhibitory center* (Fig. 14-20). It is here that these fibers come in contact with afferent nerve fibers. The center can be influenced by impulses over the afferent nerves from the sensory surfaces of the body and by impulses from the higher brain centers.

Reflex cardiac inhibition. ■ It frequently happens that the stimulation of a sensory nerve in our body is followed by a perceptible slowing of the heart; in fact, the heart may be slowed to such an extent as to cause fainting. No doubt, we can all recall instances of this sort, such as fainting caused by a blow on the abdomen.* The sensory nerve fibers in the wall of the stomach and intestines are mechanically stimulated, and impulses pass up the cord to the medulla; in the cardioinhibitory center the impulses are relayed to the neurons of the vagi. This experiment can be demonstrated in a frog by tapping the exposed intestines with the handle of a scalpel and watching the slowing of the heart. If in this experiment the medulla is destroyed or if the vagus nerves are cut, the action does not take place since the reflex arc has been broken.

Other instances of reflex slowing of the heart are experienced in acute dyspepsia, inflammation of the peritoneum, pain in the middle ear, and so on. It has been supposed that in some instances death brought about by plunging into cold water may be caused in this manner. The inhalation of irritating fumes may stimulate the sensory nerves in the trachea or lungs and thereby reflexly stop the heart. Gentle compression of the eyeball may perceptibly slow the heart (oculocardiac reflex).

Reflexes from the cardiovascular region. ■ The afferent nerve fibers from the cardiovascular region are of special interest; specifically these are (1) the aorta and (2) the carotid sinus (Fig. 14-20).

Aortic reflex. When one vagus nerve is cut and its central end (connected with the brain) is stimulated, the impulse is propagated by the afferent fibers (the *aortic* or *depressor nerve* fibers) of the vagus to the cardioinhibitory center. It is returned to the heart by way of the other, still intact, vagus nerve. Sectioning the other vagus abolishes this reflex. The afferent fibers stimulated here have their endings, known as *pressoreceptors,* in the walls of the arch of the aorta (Fig. 14-20). In the intact animal the pressoreceptors† are mechanically stimulated by the stretching of the aorta whenever the blood pressure becomes

*The solar plexus blow.

†The pressoreceptors are stretch receptors.

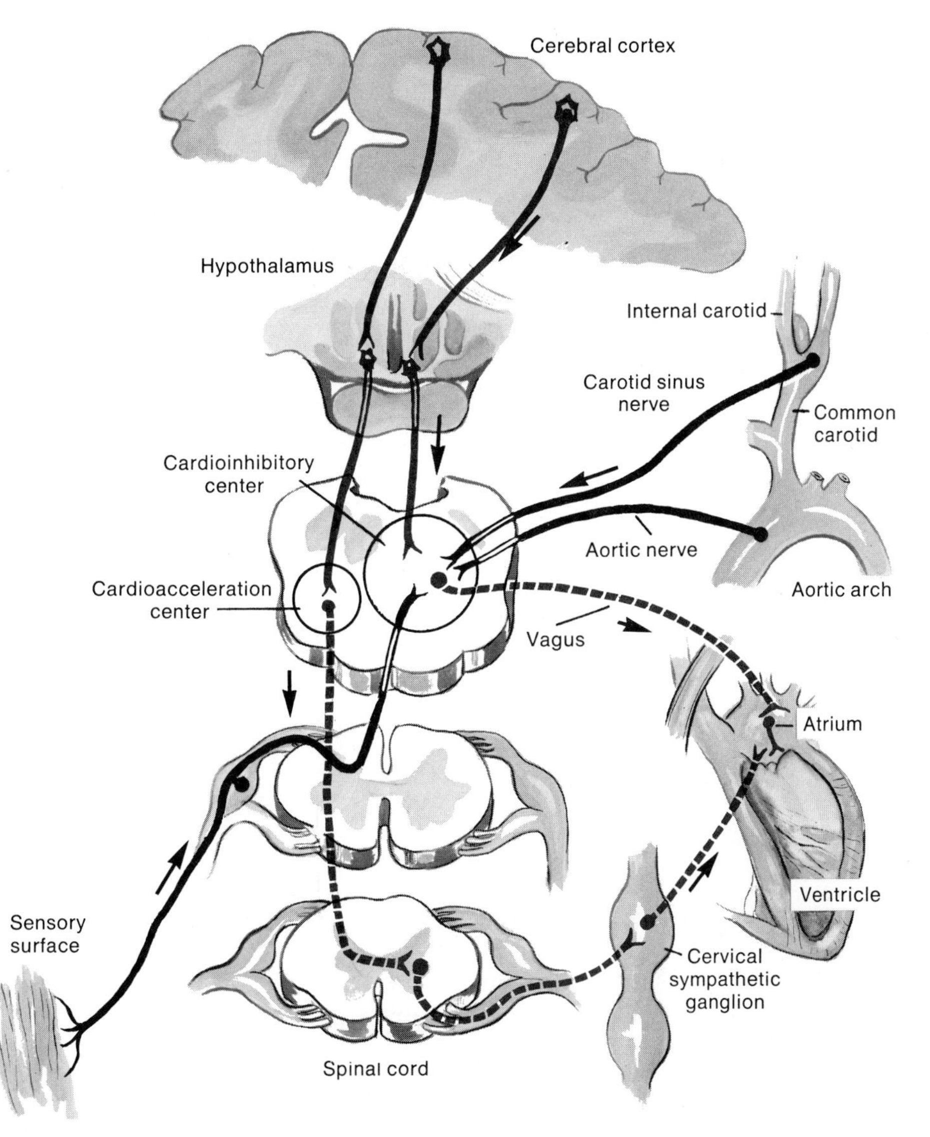

Fig. 14-20
Illustration showing influence of afferent nerves from cardiovascular sensory areas and cerebral centers on cardioinhibitory and cardioacceleratory centers in medulla.

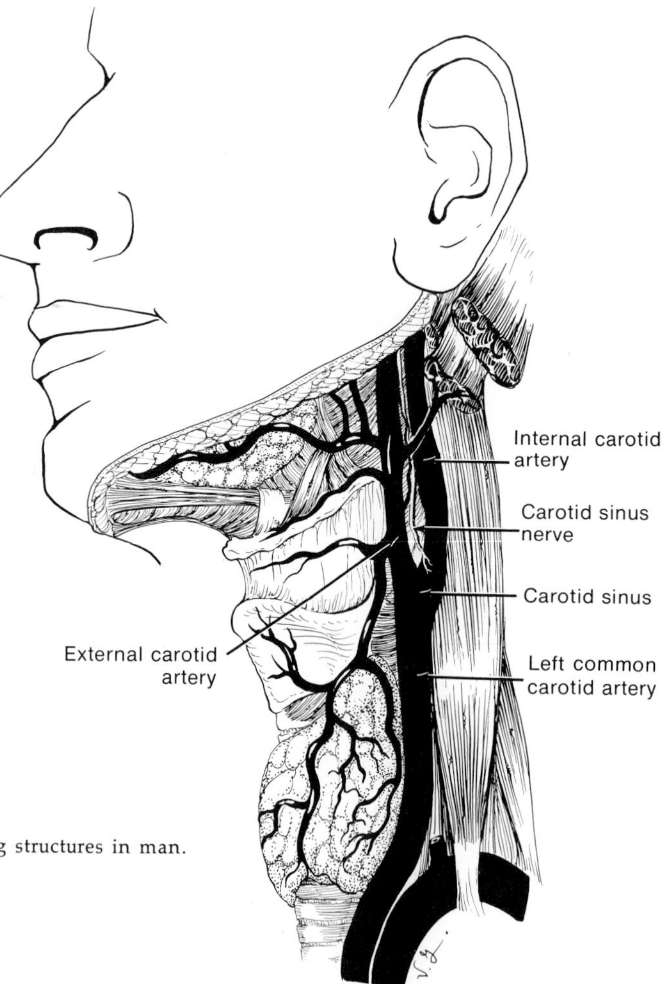

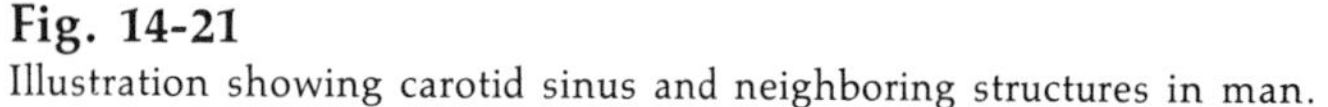
Fig. 14-21
Illustration showing carotid sinus and neighboring structures in man.

excessively high. The result of this stimulation is a reflex slowing of the heart and consequently a fall in blood pressure. This mechanism, therefore, prevents an abnormally high blood pressure and safeguards the heart against excessive strain.

Carotid sinus reflex. From the arch of the aorta three large arteries originate; the innominate, the left common carotid, and the left subclavian artery. The innominate artery gives rise to the right common carotid. The right and left common carotid arteries divide into the internal carotid (which supplies part of the blood to the brain) and the external carotid arteries. Upon the internal carotid artery and near its origin from the common carotid is found a small pocket known as the carotid sinus (Fig. 14-21). The walls of this sinus are supplied with an afferent nerve, the *carotid sinus nerve,* whose endings are pressoreceptor sense organs. The carotid sinus nerve (a branch of the glossopharyngeal or ninth

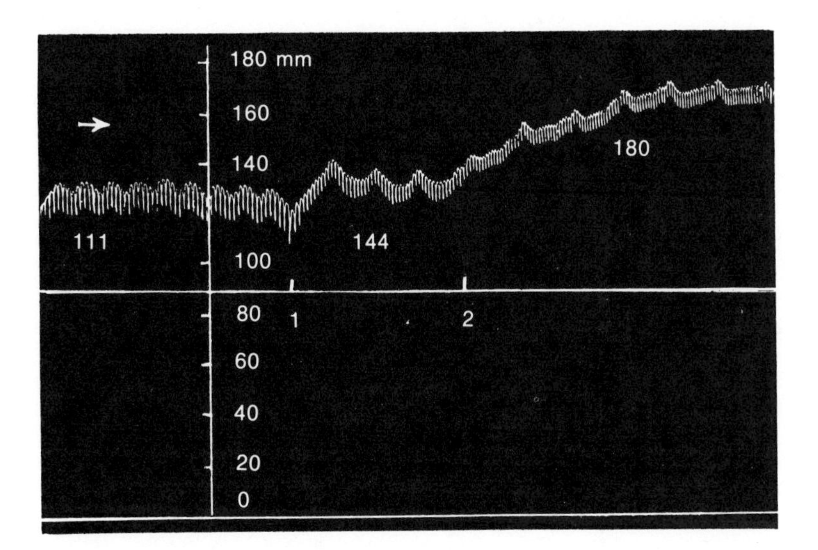

Fig. 14-22
Blood pressure record in dog showing effect of section of both vagi on rate of heartbeat and pressure. **1,** Section of right vagus; **2,** section of left vagus. Numerals on vertical line indicate pressures; numerals below pressure record indicate heart rate.

cranial nerve) carries impulses to the cardioinhibitory center (Fig. 14-20).

When the walls of the carotid sinus are distended by increased blood pressure, the stimulated sinus nerve excites the cardioinhibitory center to greater activity, thus slowing the heart. This result can be obtained also by electrical stimulation of the nerve. In some individuals external pressure applied to the internal carotid artery is also effective.* On the other hand, if the blood pressure in the sinus is decreased by external pressure upon the common carotid (Fig. 14-21), the heart is accelerated because of the lack of impulses transmitted from the sinus to the inhibitory center.†

Carotid, Greek, *Karos* = sleep. Short periods of anesthesia were induced by the ancients by pressure applied to the carotid arteries.

†The pressoreceptors in the walls of the aorta control the general blood pressure throughout the whole body; those of the carotid sinus are concerned with the blood pressure in the brain.

Influence of higher brain centers. ■ By the activity of the cerebral neurons concerned with emotions, impulses are sent to the cardioinhibitory center. Some emotional states accelerate the heart rate, whereas others, especially those that are painful and disagreeable, may cause inhibition even to the point of fainting. The cerebral impulses for these psychosomatic reactions pass through the hypothalamus.

Tonic action of the vagi. ■ As sectioning of both vagus nerves causes cardiac acceleration (Fig. 14-22), it must be concluded that *the cardioinhibitory center has a tonic or constant inhibiting influence on the sinoatrial node;* that is, the rate of the heart is constantly held in check under normal conditions. The vagi may be regarded as "brakes" for the heart. Some states of mind or body cause the center to be more active and the brakes to work more strongly; other conditions take the brakes off. For example, certain gastric disturbances inhibit the inhibitory center, and palpitation of the heart results. The tonicity of the center is maintained chiefly by the aortic and sinus nerves, since cutting these nerves has almost the same result as cutting the vagi.

■ Cervical sympathetic nerves

The adrenergic innervation of the heart arises from the cervical sympathetic ganglia (Fig. 11-2). Both the nodal tissue and musculature of the heart are innervated. Indeed, the ventricular musculature is almost exclusively innervated by sympathetic nerves. Stimulation of cervical sympathetic nerves causes an increase in the rate and in the force of the heartbeat. It is clear from the available evidence that these effects are mediated by cAMP (Chapter 28), but the precise steps involved have not yet been elucidated.

Cardiac acceleration. ■ The *cardioacceleratory* center (Fig. 14-20) is located in the medulla. Afferent impulses arriving at this center from the brain or various areas of the body may increase its activity and produce an increase in heart rate (tachycardia) and force of contraction. For example, anticipation of exercise and pain sensations from the skin accelerate the heart.

The mechanism whereby sympathetic stimulation and subsequent release of norephinephrine increases the heart rate, a positive chronotropic effect, is shown in Fig. 14-17, *B*. By increasing the rate of prepotential depolarization, more spike potentials are generated per unit of time. Maximum sympathetic stimulation can increase the heart rate threefold. It is believed that the sympathetic system plays a lesser role in the tonic regulation of heart rate than does the parasympathetic system.

Inotropic effects. ■ Norepinephrine produces a positive inotropic (Greek, to influence) effect on the heart. The force of the heartbeat may be nearly doubled. Unlike most skeletal muscle, cardiac muscle treated with catecholamines (Chapter 28) exhibits a stronger contraction at a given initial length. Thus, the developed tension curve of Fig. 6-17, *B*, is modified by the presence of inotropic agents. Not only is the force of the contraction enhanced, but its velocity is increased as well. Consequently, cardiac muscle exhibits an array rather than a single load-velocity curve (Fig. 6-18, *A*). This capacity for changing both force and velocity of contraction allows cardiac muscle to adjust its work and power output to accommodate changes in length (end-diastolic volume) and load (aortic pressure) most efficiently.

READINGS

Adolph, E. F.: The heart's pacemaker, Sci. Am. **216**(3):32-37, Mar. 1967.

Berne, R. M., and Levy, M. N.: Cardiovascular physiology, ed. 2, St. Louis, 1972, The C. V. Mosby Co., pp. 61-83 and 139-177.

Brady, A. J.: Active state in cardiac muscle, Physiol. Rev. **48**(3):570-600, 1968.

Braunwald, E.: Control of myocardial oxygen consumption. Physiologic and clinical considerations, Am. J. Cardiol. **27**:416-432, 1971.

Chapman, C. B. and Mitchell, J. H.: The physiology of exercise, Sci. Am. **212**(5):88-96, May 1965.

Epstein, S. E., Levey, G. S., and Skelton, C. L.: Adenyl cyclase and cyclic AMP—biochemical links in the regulation of myocardial contractility, Circulation **43**:437-450, 1971.

Langer, G. A.: The intrinsic control of myocardial contraction—ionic factors, New Eng. J. Med. **285**:1065-1071, 1971.

Robison, G. A., Butcher, R. W., and Sutherland, E. W.: Cyclic AMP, New York, 1971, Academic Press, Inc., pp. 195-204.

Scher, A. M.: The electrocardiogram, Sci. Am. **205**(5):131-141, Nov. 1961.

Singer, D. H., Lazzara, R., and Hoffman, B. F.: Transmembrane potentials of cardiac cells and their ionic basis. In Briller S. A., and Conn, H. L., Jr., editors: The myocardial cell, Philadelphia, 1966, University of Pennsylvania Press, pp. 73-110.

Sonneblick, E. H.: Myocardial ultrastructure in the normal and failing heart, Hosp. Pract. **5**(4):35-43, 1970.

15

Necessity for the regulation of blood vessels
Vasomotor mechanism
Chemical control
Arterioles, precapillary sphincters, and capillaries
Cerebral circulation
Factors influencing circulation
Circulation during muscular work
Hypertension and arteriosclerosis
Hypotension

VASOMOTOR CONTROL

It will be recalled that the circulation of the blood depends not only on the rhythmical activity of the heart but also on a certain state of the blood vessels. Mechanisms must be provided whereby the blood vessels can be regulated so as to adapt to varying conditions in the body.

■ NECESSITY FOR THE REGULATION OF BLOOD VESSELS

Intravascular hemorrhage. ■ Compared with the quantity of blood in the body, the vessels have a very large capacity; all the blood, in fact, can be held and stored away in the blood vessels of the abdomen. If this happens, it may be said that the animal bled to death in its own blood vessels. Technically this is known as intravascular hemorrhage; unless the condition is relieved, death results from anoxia. In the face of this great disproportion between the volume of the blood and the capacity of the vessels, how can there ever be any circulation? The answer is that under normal conditions all the vessels never simultaneously assume their maximum size.

Reserve blood volume; shunting the blood. ■ A working organ needs more nourishment and therefore a larger blood supply than does a resting organ. To meet this situation with the quantity of blood in the body, the blood is shunted from the resting to the working organs. This is possible because the blood vessels are capable of constricting and dilating.

In addition to these minor shiftings of blood from inactive to active organs, the body has, comparable to the cardiac reserve, a reserve blood volume. In the resting state of the body all the blood is not in active circulation; there is no need for it. Some of it is pooled and temporarily stored until there is a demand for it. Capacitance vessels such as the large veins of the abdomen and thorax and the pulmonary veins may serve as reservoirs. The importance of this reserve blood volume will be made clear as we proceed with our subject.

■ VASOMOTOR MECHANISM

■ Vasomotor center

In the medulla oblongata is located, bilaterally, a collection of neurons known as the *vasomotor center.* These neurons send their axons into the cord where they end at various levels in spinal

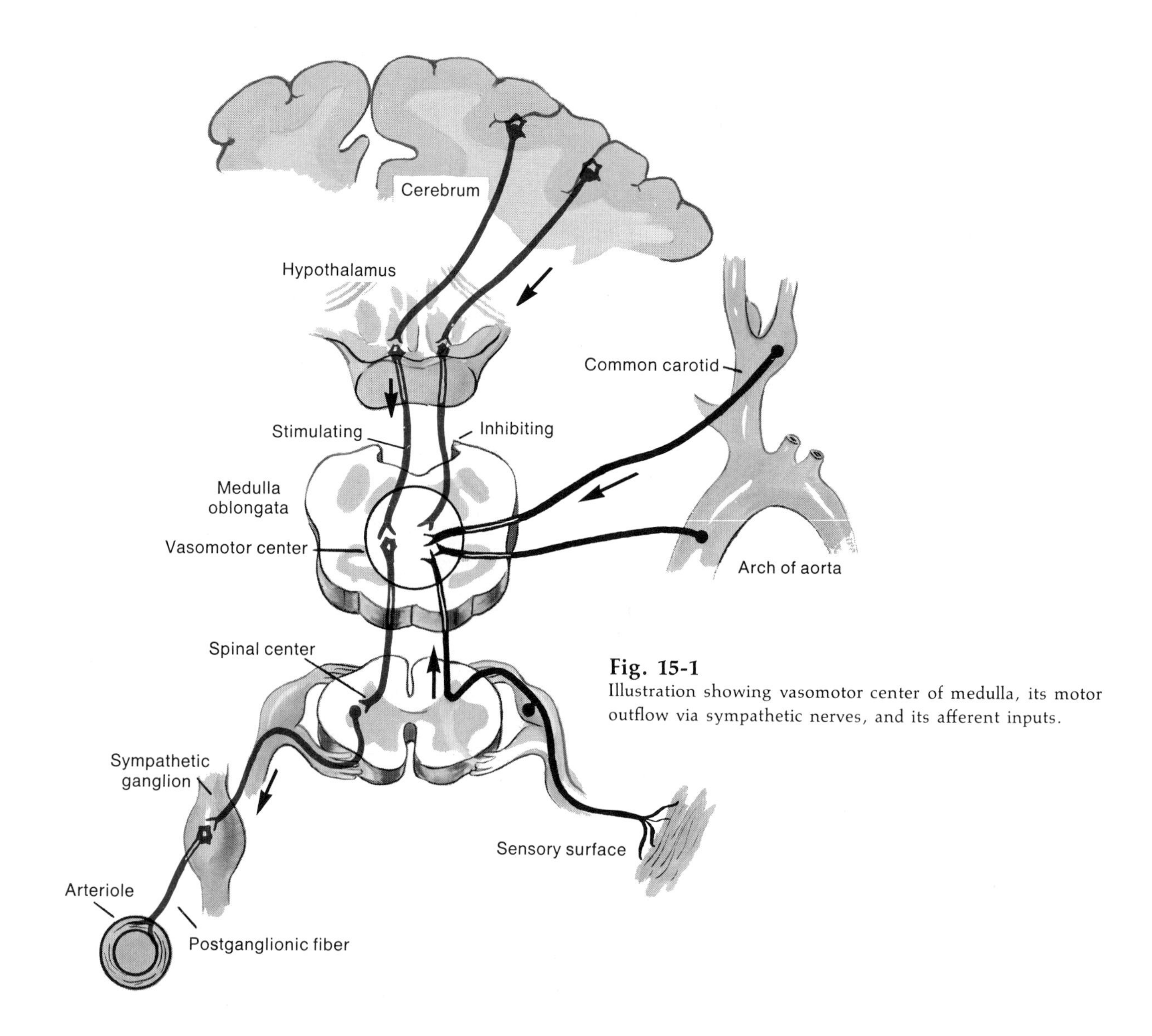

Fig. 15-1
Illustration showing vasomotor center of medulla, its motor outflow via sympathetic nerves, and its afferent inputs.

centers (Fig. 15-1). Here synaptic connections are made with neurons whose axons issue from the cord to end (as preganglionic fibers) in sympathetic ganglia. The postganglionic fibers make connection with the muscles in the walls of blood vessels supplying practically every part of the body. The blood vessels of the skin and of the abdominal organs are particularly well supplied with vasomotor nerves. The preganglionic fibers for the numerous abdominal vessels have their cell bodies in the middle and lower portion of the thoracic region of the spinal cord (Fig. 8-3) and reach the sympathetic ganglia (in the abdomen) by way of the splanchnic nerves (Fig. 11-2). The

Fig. 15-2
Appearance of blood vessels in ears of white rabbit. "To show the contrast between the constricted vessels still connected with the vasoconstrictor center (left ear) and the control dilated vessels that have been disconnected from that center (right ear)."

center in the medulla may be considered to be made up of distinct parts, each of which controls the vessels of a particular area of the body, such as the skin, the intestinal region, and so forth.

■ Vasomotor nerves

The walls of the arteries are supplied with smooth muscle fibers arranged in two layers (Fig. 13-4): (1) the circular muscles in which the individual fibers encircle the lumen; these are more abundant than (2) the longitudinal fibers, which run parallel with the lumen of the artery. It is the circular muscles that are of interest to us. By their contraction the lumen becomes smaller, and therefore the capacity of the arteries is decreased, whereas their relaxation has the opposite effect. These muscles are especially abundant in the smaller distributing arteries and arterioles where the bed of the vascular system is large. Like the heart, they are governed by the central nervous system through autonomic nerves.

Claude Bernard in 1851 was the first to discover these nerves. In a rabbit (but not in man) the vagus and the cervical sympathetic nerves lie as two distinct nerves alongside the carotid artery. On cutting the exposed right cervical sympathetic nerve, it was found that the blood vessels in the right ear appeared much larger than those in the other ear (Fig. 15-2). The vessels in the right ear had dilated, and, as a result of the greater volume of the blood flowing through it, the right ear had a more intense red color than did the left and felt warmer to the touch. The cervical sympathetic nerve therefore exercises some influence on the circular muscle fibers of these blood vessels; this was corroborated by the fact that stimulation of the peripheral end (the end toward the ear) caused the vessels to become smaller. An efferent nerve that governs the vascular muscles is termed a *vasomotor nerve,* and if the stimulation of the vasomotor nerve causes a constriction of the vessels, it is called a *vasoconstrictor nerve.* Nerves having the opposite action were also discovered by Claude Bernard. The submaxillary salivary gland (Fig. 20-2) is supplied with a nerve called the chorda tympani, a branch of the seventh cranial or facial nerve. On stimulating this nerve the arteries of the gland dilate, and the outflow of blood from a severed vein is increased. There are, therefore, *vasodilator* fibers in this nerve.

■ Vasomotor tone

The function of the vasomotor mechanism is not only to shunt the blood from here to there as occasion may demand, but, in conjunction with the regulated cardiac activity (Chapter 14), it also constantly maintains at adequate level the general blood pressure and a sufficient pressure gradient throughout the body. To do so, the vasomotor center constantly discharges impulses to the arteriolar musculature, thereby holding these vessels in a continual, although variable, state of constriction. This activity is spoken of as *vasomotor tone.* This tonic influence has been demonstrated by cutting the spinal cord in the lower cervical region or by cutting the splanchnic nerve; either procedure always results in an extensive fall in blood pressure because of the loss of the tonicity of the vascular muscles. *The blood pressure depends on vasomotor tone; in fact, circulation itself is made possible by means of it.* The tonic action of the vasomotor system is largely due to, and frequently modified by, (1) a constant stream of afferent impulses from the various sensory surfaces of the body, (2) impulses from higher brain centers, and (3) chemical stimulation by compounds (hormones and waste products) found in the blood.

■ Vasoconstrictor nerves and passive vasodilation

It will be recalled (Chapter 13) that by far the larger part of the resistance in the circulatory system is found in the arterioles and that the extent of this resistance determines the amount of blood pressure and blood flow. The amount of resistance depends on the degree of constriction or dilation of the arterioles. Before discussing the control that the vasomotor system exercises over the arterioles, it should be noted that stimulation of a vasoconstrictor nerve to the arterioles of a limited region results in the following:

1. There is an increase in peripheral resistance.
2. A local increase in blood pressure occurs.
3. There is a decreased amount of blood flow through the arterioles of this region and therefore a fall in capillary blood pressure. If the part concerned is a superficial part, such as the skin, a blanching and a lowering of temperature results.
4. The volume of the organ decreases.
5. Since less blood passes through the capillaries into the veins, the venous blood pressure falls. Dilation of the arterioles has, of course, the opposite effects. The arterioles function as controlling elements for the capillaries.

If the part of the body concerned is relatively small, the constriction of its vessels has little or no effect on the general blood pressure. Constriction over a larger area increases general blood pressure (Fig. 15-3), but such constriction is normally associated with a vascular dilation in some other regions of the body. In this manner the pressure is not allowed to rise excessively. On the other hand, arteriolar dilation in a small area increases the volume of blood sent through its capillary bed and into the veins draining this part. However, if the arterioles in a large area dilate simultaneously, the increase in the size of the bed may be so great that a smaller volume of blood circulates; in fact the blood may stagnate in the greatly dilated vessels, and all circulation may come to a standstill.

Reflex activity. ■ An example of reflex constriction or dilation of blood vessels is seen in the paling of the skin in cold weather and flushing in warm weather. When one hand is placed in ice water, the temperature of the other hand is also lowered. Stimulation of the cold receptors (sense organ, Fig. 15-1) in the immersed hand causes impulses to be conveyed to the vasomotor center, which relays them to the blood vessels of both hands, resulting in vasoconstriction.

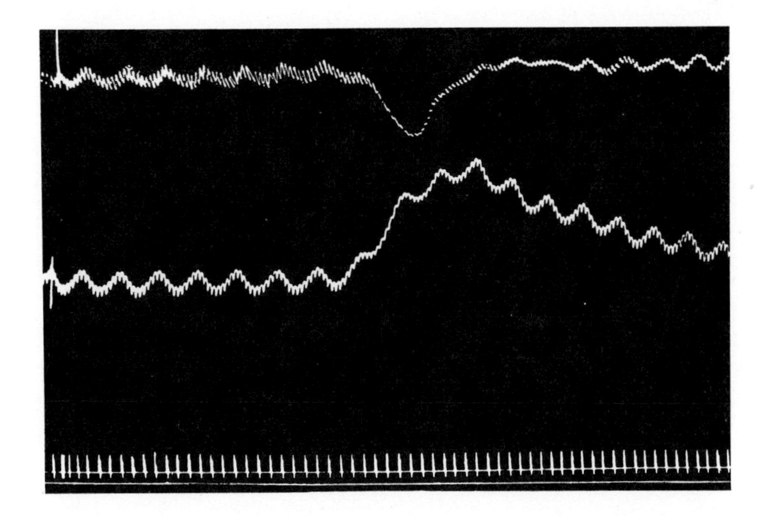

Fig. 15-3
Effect of vasoconstriction produced by stimulation of splanchnic nerve on kidney volume (upper curve) and on carotid pressure (lower curve). (From Wiggers: Physiology in health and disease, Lea & Febiger.)

Stimulation of the bare arms or neck by a cold draft may cause constriction of the vessels of the pharynx and nasal mucosa. On the other hand, stimulation of the heat receptors of the skin causes the superficial vessels to dilate. The marked fall in blood pressure resulting from a widespread dilation may be the cause of the enervation following prolonged exposure of the body to hot baths or intense sunlight.

Sensory nerves coming from the cold receptors just described are spoken of as *pressor nerves,* since by influencing the vasomotor center they increase the blood pressure in the vessels concerned. Sensory nerves from heat receptors of the skin, which, by either inhibiting the constrictor area or stimulating the dilator area in the vasomotor center, cause a dilation of blood vessels and a fall in blood pressure, are called *depressor nerves.* It should be borne in mind that vasomotor nerves are efferent nerves and that pressor and depressor nerves are afferent nerves.

Somatic pressor and depressor nerve fibers. Pressor and depressor nerve fibers are found in almost every sensory nerve. For example, weak stimulation of the central end of the cut sciatic nerve at 1/sec causes dilation of the vessels and a fall in blood pressure; on the other hand, stronger stimulation is followed by a constriction and a rise in pressure. In general, strong or painful stimulation of any afferent nerve results in a higher blood pressure. A startling noise causes vasoconstriction in the fingers.

Cold pressor test. The cold pressor test is frequently employed to detect instability of the vasomotor mechanism and a tendency to hypertension. After the subject has rested for thirty minutes, his hand is immersed in water at 4 C for two minutes. It is believed that a sustained rise of more than 20 mm Hg in systolic pressure and of more than 15 mm Hg in diastolic pressure indicates that the individual is likely to develop high blood pressure in later life.

Aortic, or cardiac, depressor. It will be recalled that in the wall of the aorta are found the pressoreceptors of the afferent vagus nerve, known as the aortic nerve (Fig. 14-20). When the depressor fibers are stimulated by distention of the aortic wall, there is, in addition to the cardiac inhibition studied in Chapter 14, a marked fall in blood pressure due to the dilation of the blood vessels (Fig. 15-4), especially in the abdominal (splanchnic) area and to a lesser extent in the

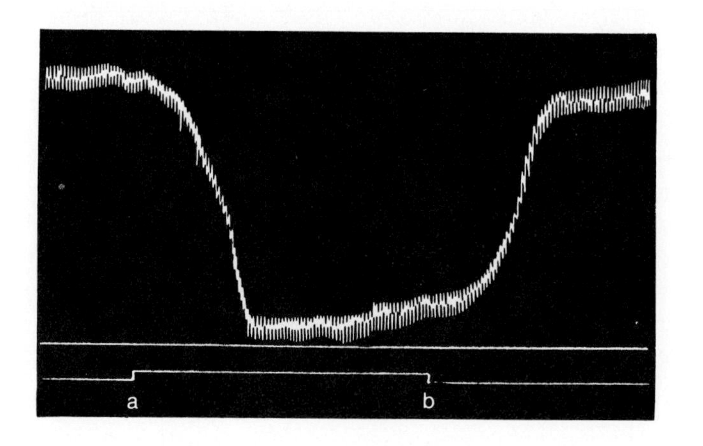

Fig. 15-4
Tracing showing fall in blood pressure caused by stimulation of depressor nerve from **a** to **b.** (From Macleod, after Bayliss.)

skin. That the fall in blood pressure is not merely due to a slowing of the heart can be demonstrated in a dog by first cutting both vagi and then stimulating the central end of one nerve; the usual fall in pressure results although the heart is not affected. By this neural mechanism an excessively high blood pressure caused by too great vasoconstriction is avoided, and thereby the heart is safeguarded against too great a load. It should be noted, however, that in muscular work the blood pressure rises in spite of this depressor mechanism, and the heart rate is much accelerated.

Carotid sinus. Similar to the depressor nerve endings in the aorta, the pressoreceptors in the carotid sinus (Figs. 14-20 and 15-1) are normally stimulated by the distention of the walls of the sinus by the blood pressure. The impulses generated thereby are carried by afferent nerves to the cardioinhibitory center and the vasoconstrictor area of the vasomotor center. The first-mentioned center is stimulated, and the heart rate is reduced; the vasoconstrictor area is inhibited, and the blood vessels are dilated. As a result the blood pressure is reduced. This is shown in Fig. 15-5. In this experiment the common carotid arteries were clamped, preventing blood from flowing into the sinus. The absence of inhibitory impulses to the vasoconstrictor center caused the pressure in the femoral artery to rise to 280 mm Hg. When the clip on one of the carotids was removed at the point *b*, the sinus wall was stretched, and as a result the vasoconstrictor area was inhibited and the blood pressure dropped to 210 mm. Removal at *c* of the clip on the other carotid was followed by a drop of the femoral blood pressure to 180 mm Hg. If, previous to the removal of the clips, the afferent nerves from the sinuses were severed, the removal of the clips at *d* and *e* had no effect on blood pressure.

External pressure applied over the carotid sinus may also lower blood pressure, even to the extent of causing fainting. Persons with hypersensitive carotid sinuses should, for this reason, avoid tight collars and bending the head far backward, as in looking up.

Buffer nerves. From the previous discussion it can be gathered that the aortic depressor and the

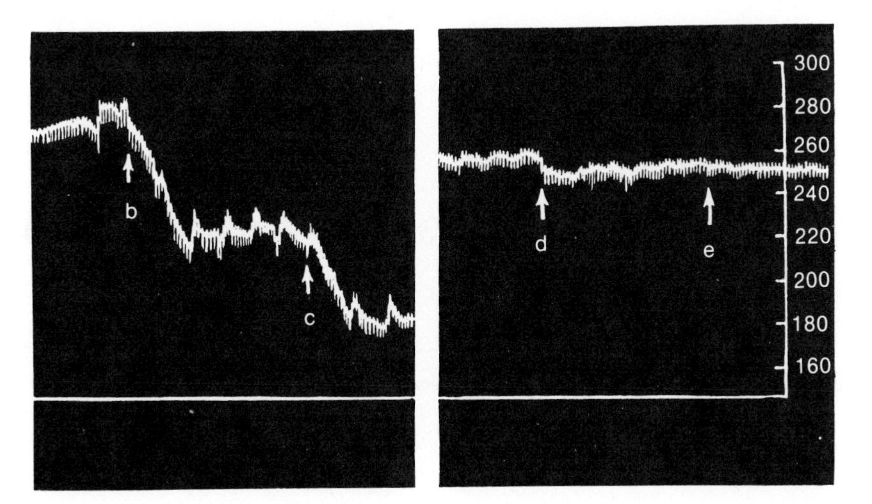

Fig. 15-5
Blood pressure record from dog showing effect of carotid sinus stimulation on blood pressure. Clamps on carotid arteries were removed successively at **b**, **c** and **d**, **e** after nerves from sinuses were severed.

carotid sinus nerves constitute the principal mechanism for regulating blood pressure. Automatically they prevent excessive blood pressure in a twofold manner: by reducing the head of pressure and by decreasing the peripheral resistance. Fittingly, these two nerves have been called the "buffer nerves." Severing these nerves leads to a permanent high blood pressure.

Venomotor nerves. It has been shown that certain veins, for example, those of the portal system, which carry practically all the blood returning from the intestines to the liver, are supplied with nerves whose stimulation results in a diminution of the caliber of these vessels.

Spinal centers and peripheral action. As described in a previous section, the neurons of the vasomotor center in the medulla (Fig. 15-1) send their axons to the preganglionic neurons, which have their cell bodies in the lateral columns of the spinal cord. These second neurons are indirectly connected with the blood vessels. Cutting the spinal cord in the lower cervical region separates all the blood vessels from the medullary center, and their dilation causes a tremendous fall in blood pressure (an aspect of spinal shock). However, if the animal survives, the pressure gradually rises. This must be attributed to the subsidiary spinal centers, which, in the absence of the controlling impulses from the chief medullary center, acquire a certain amount of independent activity, for if now the cord is destroyed, the blood vessels again dilate and the blood pressure falls. Cutting a vasoconstrictor nerve causes the muscles in the vessel wall to relax completely. Eventually, however, the arterioles and capillaries may regain their normal caliber.

Reciprocal vasomotor control. The various parts of the vascular system bear a reciprocal relation to each other. During muscular work the arterioles and capillaries of the active muscles dilate, whereas those of the splanchnic area, if the work is severe, undergo compensatory constriction. The cutaneous vessels also may participate in this constriction, at least until the extra heat produced by the muscular work renders it necessary to increase the cutaneous circulation in order to prevent an excessive rise in body temperature.

We know that the color and temperature of the skin vary with the amount of heat produced in the body by muscular exercise and with the temperature of the environment. The dilated state of the cutaneous arterioles and capillaries during physical work or on a warm day enables a relatively large volume of blood to circulate through the skin, thus affording the body an opportunity to dissipate a large amount of heat. The reverse holds true during inactivity and in cooler weather. Regulating the amount of heat dissipated by the skin and thereby aiding in maintaining the proper body temperature is one of the foremost functions of the vasomotor system.

However, extensive cutaneous dilation may entail a serious fall in blood pressure. To avoid this, the reciprocal relation between the cutaneous and the splanchnic area, discussed previously, affords a simultaneous constriction in the latter area. This mutual action not only maintains a proper blood pressure but also forces the warm blood from the interior of the body to the cooling mechanism, the skin.

In certain instances the collapsing of individuals in hot weather has been attributed to failure of this reciprocal mechanism to operate properly. On the other hand, sudden exposure to cold causes a marked cutaneous constriction with its resultant increase in peripheral resistance. Were this not offset by dilation in the splanchnic area, the high blood pressure might result disastrously in those individuals in whom the vessels have lost, to a large extent, their tensile strength, as in arteriosclerosis.

Influence of higher brain centers. ■ The activity of the vasomotor center is readily modified by emotional states; of this we have such common illustrations as the blushing of shame and the pallor of fear, the first being a dilation and the second being a constriction of the blood vessels of the face. Strong emotions may cause inhibition of the vasoconstrictor center to such a degree that the resulting dilation of the vessels leads to syncope, or fainting. In other acute mental stresses the blood pressure may become abnormally high. Anger has been shown to lower the temperature of the fingers by several degrees.* Because of emotional influences on both the heart and the blood vessels, the first two or three blood pressure readings from an individual not familiar with the procedure may be unreliable.

■ Vasodilators

Vasodilator nerves, in contrast to vasoconstrictors, have a limited distribution; their action is therefore local and is not tonic. Instead of controlling the blood pressure in general and affecting the distribution of blood over large areas and in many organs of the body, they govern the local demand of a limited part or of a single organ. Both sympathetic and parasympathetic vasodilator systems have been described. Each releases acetylcholine at the postganglionic terminals. Dilator nerves are especially supplied to the sweat glands and the glands of the alimentary canal where a large quantity of blood is needed for a relatively short length of time. Although a chief vasodilator area is said to be located in the medulla oblongata, subsidiary dilator areas probably are located in the various parts of the central nervous system, each one governing a limited part of the body in close proximity to it. Thus a vasodilator area for the vessels of the salivary glands is perhaps located in the medulla oblongata and that for the genital organs in the sacral region of the spinal cord (Fig. 11-2). A local vasodilator area can be influenced reflexly and emotionally; it is active only when the part of the body innervated by this particular area is in need of more blood because of increased activity. For instance, stimulation of the gustatory nerves during eating reflexly evokes a dilation of the vessels of the salivary glands; certain emotions have the same effect on the center governing the vessels of the sex organs.

Axon reflex. ■ Mechanical stimulation or application of irritants to the skin evokes a localized vasodilation. This response is commonly attributed to the *axon reflex.* It is presumed that cutaneous sensory nerve fibers possess collateral branches to nearby blood vessels and that intense stimulation of the terminals in the skin causes impulses not only to propagate along the fiber toward the spinal cord, but also to move antidromically down the collateral branches to evoke vasodilation of the vessels.

■ CHEMICAL CONTROL

Many substances are known to increase or decrease blood pressure by influencing the vasomotor mechanism. Some of these are produced in the body.

■ Vasoconstrictors

Carbon dioxide. ■ The tonic activity of the vasoconstrictor area is believed to be largely due to the stimulating action of the CO_2 in the blood. This may account for the great increase in blood pressure during the early stages of asphyxiation. It is a common observation that overventilation of the lungs, as by voluntary deep inspiration and expiration for a period of three or four minutes, brings about a feeling of giddiness. The expulsion of a large amount of CO_2 from the blood by the overventilation deprives the vasoconstrictor area of its proper stimulation by CO_2 and as a consequence vasodilation in the splanchnic area and a fall in blood pressure take place; this lessens the cerebral circulation.

Angiotensin. ■ The most powerful naturally occurring pressor substance in the body is angiotensin (angiotensin II). It is forty times more effective than is epinephrine. Angiotensin is produced by the successive action of the enzyme

*The decreased electrical resistance of the skin caused by emotional disturbances (known as the psychogalvanic reflex) has its basis in the cutaneous vascular changes.

renin (Chapter 29) on a plasma α_2 globulin and of a tissue enzyme especially prevalent in the lungs (Chapter 17).

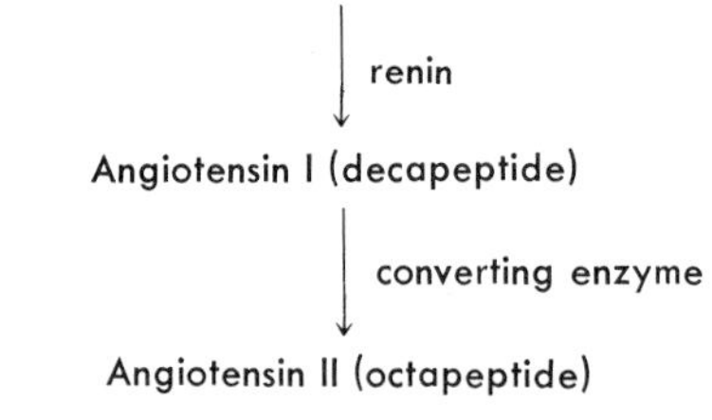

Angiotensin constricts precapillary vessels in the skin, the splanchnic bed, the kidney, and other areas. Cardiac and skeletal muscles receive additional blood flow due to the general rise in blood pressure. The response to angiotensin is augmented by its direct stimulatory effect upon the cells of the vasoconstrictor center of the medulla and upon neurons in the sympathetic ganglia. Angiotensin is also involved in electrolyte balance; it stimulates the adrenal cortex (Chapter 28) to release the sodium-retaining hormone aldosterone. The role of angiotensin in blood pressure regulation is summarized in the diagram below.

Renin has a short half-life in the plasma, about fifteen minutes, before being inactivated in the liver. Angiotensin persists for an even shorter period. Because of this, it is doubted that angiotensin normally has a significant role in the moment-to-moment regulation of blood pressure. It may be important in abnormal, hypertensive states.

Epinephrine. ■ Injected into the blood, the catecholamine (Chapter 28) epinephrine markedly constricts the abdominal and cutaneous arterioles; this results in a precipitous rise in blood pressure (Fig. 15-6). It will be seen that the pressure rises very quickly, but the elevated pressure is maintained for only a brief length of time. Locally applied, epinephrine causes constriction of the arterioles; therefore, it is used in minor operations on the eye, nose, and so on, to stop hemorrhage in small vessels. In contrast, *the coronary and skeletal muscle arterioles are dilated* by epinephrine. The effects of catecholamine release from the adrenal medulla are discussed in Chapter 28.

Vasopressin. ■ A secretion of the neurohypophysis, vasopressin, increases blood pressure. The rise in pressure is not so great as that obtained with epinephrine, but it continues for a longer length of time. The chemistry and action of vasopressin are described in Chapter 28.

Ephedrine. ■ Although much weaker in its action, ephedrine has effects similar to that of epinephrine. It is commonly used as a decongestant for bad colds and hay fever, since it causes an

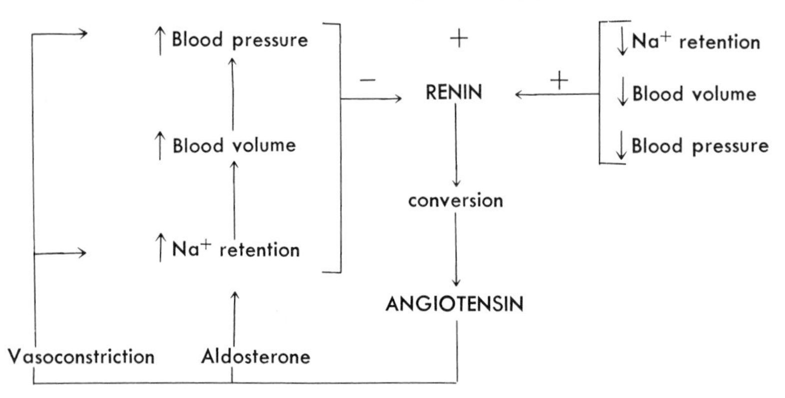

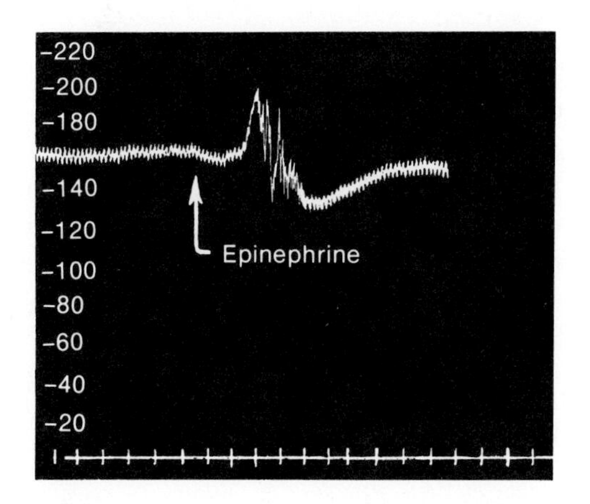

Fig. 15-6
Blood pressure record from dog, showing effect of epinephrine administration on blood pressure.

immediate shrinkage of blood vessels in the nasal mucosa. The effect lasts for two to three hours.

Tobacco. ■ Experiments have shown that smoking increases both the systolic and the diastolic blood pressure. The pulse rate is increased and the temperature in the extremities is decreased. Use of tobacco may be injurious in those diseases that are associated with arteriosclerosis or high blood pressure.

■ Vasodilators

Kinins. ■ The kinins constitute a group of naturally occurring polypeptide compounds that are among the most potent hypotensive agents known. As in the case of angiotensin, they derive from a plasma protein precursor, known as *kininogen*. This precursor can be enzymatically cleaved to form kinins by trypsin, plasmin, certain snake venoms, and the enzyme *kallikrein* that occurs in the plasma and in certain organs (e.g., pancreas and salivary gland) and tissues.

Kallikrein is present in the plasma as an inactive precursor, prekallikrein, which is transformed into the active enzyme by disturbances of the plasma equilibrium (pH and temperature) and by contact with skin or damaged tissues. Factor XII of the clotting mechanism (Chapter 12) can also initiate the prekallikrein to kallikrein transformation. Plasma kallikrein converts plasma kininogen to *bradykinin*, a nonapeptide, the active vasodilator substance. Bradykinin (Greek, *bradys*, slow; *kinein*, to move) is so named because it induces slow contractions of the gut.

Kallikrein in glands and tissues converts plasma kininogen to *kallidin*, a decapeptide. This, in turn, is converted by a plasma enzyme, carboxypeptidase, to bradykinin. The kinins have a very short half-life in the plasma, less than a minute, for they are rapidly inactivated by *kininase*. These reactions are summarized in the diagram below.

Kinins are ten times more potent vasodilators than histamine. They produce a sharp fall in blood pressure, both systolic and diastolic, due to the vasodilation that occurs in skeletal muscle, kidneys, viscera, glands, and coronary, pulmonary, and cerebral arteries. The latter effect causes throbbing headache. Kinins apparently stimulate the terminals of pain fibers and produce intense pain sensations. Plasma levels of

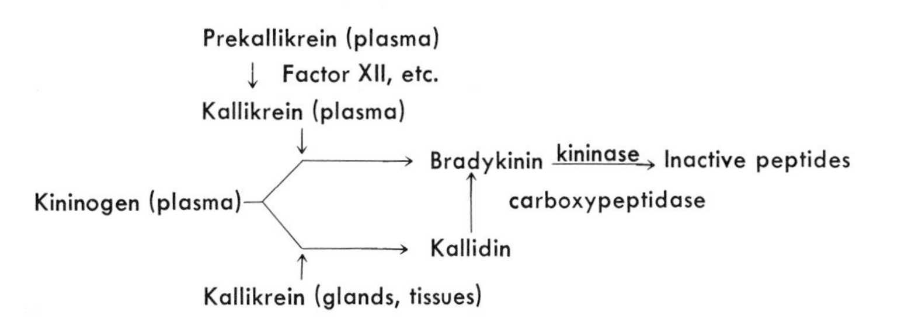

kinins are increased after burns and mechanical trauma and may be a contributing factor in the hypotension associated with traumatic shock; they may increase capillary permeability. The kinins present in wasp and hornet venoms are in large measure responsible for the inflammation and pain associated with their stings.

Acetylcholine. ■ A brief application of acetylcholine (as by perfusion) causes vasodilation. Acetylcholine has a stimulating action on all muscles except the heart and blood vessels.

Histamine. ■ A compound found in many tissues and in blood, histamine, causes a marked dilation of the capillaries and arterioles, resulting in a lowering of blood pressure.

Alcohol. ■ Alcohol dilates the blood vessels, acting as a depressant on the vasomotor center; the cutaneous vessels are especially affected, which accounts for the *feeling* of warmth after consuming alcohol on a cold day.

■ ARTERIOLES, PRECAPILLARY SPHINCTERS, AND CAPILLARIES

In viewing the capillary circulation in a frog's tongue or web (Fig. 13-9), one soon observes that the diameter of a capillary varies from time to time. A capillary may suddenly make its appearance by being opened up (dilated) by oncoming blood; as suddenly it may disappear by being entirely emptied.

It is believed the variations in the diameter of a "true" capillary are passive, in that the dilation of arterioles, arteriovenous shunt capillaries, or precapillary sphincters sends more blood into the capillaries fed by them and in consequence causes the capillaries to dilate. When the supply of blood is lessened by the constriction of the arterioles or sphincters, the elasticity of the endothelial cells reduces the capillary diameter.

Hyperemia, or congestion. ■ Under certain circumstances the capillaries become engorged with blood. This condition is known as *hyperemia,* or congestion. Hyperemia may be classified as active, or arterial, and passive, or venous.

Active, or arterial, hyperemia. Active, or arterial, hyperemia may be produced, for example, by holding one's hand for a few moments in warm water; after withdrawal from the water, the hand appears bright red in color and shows a certain amount of swelling and warmth. Active hyperemia is caused by the dilation of arterioles and capillaries in a *restricted area.* As a result *the amount of blood in and coursing through the capillaries is increased.* Active hyperemia may be functional or inflammatory. The former is due to chemical stimulation by metabolic products normally arising in the body. This favors the exchange of materials between blood and tissues. In inflammatory hyperemia the stimulating agencies are produced by injury to the cells. These vasodilating substances also greatly increase the permeability of the capillary wall. The capillary dilation and the increased permeability of the wall give rise to the four cardinal symptoms of inflammation: color, heat, swelling, and pain.

Passive, or venous, hyperemia. Whenever there is an interference with the passage of blood from the capillaries into the venules, the capillaries become engorged with blood. Raising the venous pressure by an external or internal agency beyond a certain point has this result. It may be quickly produced by tying a string around a finger. A tourniquet or an encircling bandage too tightly applied has the same result. In a severe case the blood may stagnate completely. Soon it loses all its oxygen; the resultant reduced hemoglobin, as seen through the skin, appears blue (cyanosis). The part feels cold. If not relieved in time, the lack of nutrients and oxygen causes destructive cellular changes, which may end in gangrene (necrosis).

■ CEREBRAL CIRCULATION

The brain is more directly dependent than any other organ of the body on a constant supply of oxygen. In man fainting results from more than a momentary stoppage of the oxygen supply to the brain. It is not surprising that the brain is very liberally supplied with blood by the two internal carotid arteries and the two vertebral arteries. There is a question whether the amount of blood in the brain at any given time can be increased by the activity of vasomotors nerves innervating the cerebral vessels. The anatomical relation of these vessels to the neighboring structures makes this supposition very unlikely. The brain vessels are embedded in the incompressible brain tissue that is lodged in a bony cavity, the walls of which are rigid. When a small opening is made in the cranium, the brain matter can be seen to swell out and pulsate with each heartbeat. In fact, this is the cause of the pulsations observed on the head of a young child (in the fontanels). Since normally no opening exists and the brain matter cannot expand, the only way the pulse in the artery can make an impression on the brain matter is to force as much blood out of the cranium as enters through the artery. The shock that the brain receives with each pulse wave is transmitted through its substance to the walls of the veins, causing the flow of blood from these veins to be more or less pulsating. It is not possible to augment the amount of blood in the brain by dilating the cerebral arteries.

Volume flow. ■ More important from a nutritional point of view than the *amount of blood in* a tissue is the *volume flow* through the tissue. It should be recalled that in a relatively small structure outside the cranium (e.g., the salivary glands), this is increased by dilating its arterioles. In contrast, increasing the flow through the brain is accomplished by extracranial means—by increasing the resistance (constriction) of extracranial vessels. This shunts more blood through the cranial vessels, which now offer relatively less resistance. The increase which can be achieved is small; in humans the average rate of flow is about 55 ml/min/100 g of brain tissue. A sudden loss of a considerable quantity of blood may cause unconsciousness. If the loss proceeds more slowly, the vasomotor and cardiac mechanisms may have time to adjust the circulation and maintain systemic blood pressure at a level sufficiently high to assure an adequate flow of blood through the brain.

The blackout of aviators is caused by the drainage of blood from the head when acceleration (with the head foremost) or a sudden change in direction is made. Blood is driven from the head to the large abdominal vessels, and the return of the venous blood to the heart is impeded. To a certain extent these results may be prevented on the part of the aviator by leaning forward as far as possible to bring the head more nearly on a level with the heart and by reducing the capacity of the vessels in the abdomen by powerfully contracting the abdominal muscles or mechanically compressing the abdomen with a special flight suit.

■ FACTORS INFLUENCING CIRCULATION

■ Hemorrhage

Notwithstanding the disproportion between the amount of blood and of the vascular capacity spoken of in the opening paragraph of this chapter, a loss of from 10% to 20% of the total volume of blood has no permanent effect on blood pressure in a dog. In a healthy person a rapid loss of about 30% markedly reduces blood pressure but is not incompatible with life. If the loss of blood extends over a period of twenty-four hours, recovery is possible from a loss of 60%.

Body responses. ■ In combating extensive hemorrhage the body has several negative feedback compensatory mechanisms at its disposal.

Cardiac acceleration and vasoconstriction. Fall in blood pressure due to hemorrhage reduces the activity of the aortic and carotid sinus nerves; thereby the vasoconstrictor area becomes more and the cardioinhibitory area less active. The greater constriction of the arteries and the acceleration of the heart render it possible to maintain at least for a time some circulation, even though feeble. This constriction affects especially the blood vessels supplying organs not immediately necessary for life (chiefly those in the abdomen), thereby allowing most of the blood to be sent to the heart, brain, and lungs. Increased secretion by the adrenal glands of the hormone epinephrine and of vasopressin by the neurohypophysis at this time aids in constricting of blood vessels.

Respiration. The conditions obtaining during hemorrhage bring about increased respiration, which may very favorably affect circulation by increasing venous return during inspiratory effort.

Increase in blood volume. Tissue fluid is drawn from the lymph spaces and the tissues themselves into the blood. In this manner as much as 500 ml of blood volume can be restored in two or three hours. But this does not replace the lost red blood cells.

The spleen. By the contraction of the spleen a large number of red blood cells previously stored in this organ are forced into the blood.

Erythropoiesis. The restoration of the lost red blood cells takes place more slowly since this depends on an increase of the activity of the red bone marrow.

Plasma proteins. Plasma proteins are restored slowly by increased synthesis, chiefly in the liver.

Venomotor nerves. By the activity of the venomotor nerves, the blood in the reservoirs of the large veins is brought into active circulation.

The kidneys. Water and electrolytes are conserved by the reduced renal blood flow and filtration (Chapter 29) and by the action of ADH (vasopressin). Further, the reduced kidney blood flow promotes the release of renin and production of angiotensin. In turn, angiotensin increases the secretion of aldosterone, as discussed earlier. Aldosterone brings about increased retention of sodium and consequently of water as well.

Thirst. The quenching of the great thirst following a large hemorrhage may also be regarded as an indirect aid in increasing blood volume and thereby increasing the blood pressure and flow.

Blood transfusion. ■ No matter how efficient the circulation may be, when there is a serious shortage of red blood cells, transfusion of whole blood is necessary. Failure to supply an adequate amount of oxygen speedily endangers the cells of the nervous system and impairs the selective permeability of the capillary wall.

In transfusion of blood, care must be exercised that the blood of the donor and that of the host are not inimical to each other. In such transfusions about 500 ml of blood may be taken from the donor without any serious results; this volume of blood is restored promptly, and the cells are restored within two months. The minimal safe interval for such large donations is placed at three months for men and four months for women.

Plasma expanders: blood or plasma banks. ■ In certain instances life may be threatened not because of the loss of oxygen carriers but because of the lack of a sufficient amount of fluid to carry on circulation. To prevent death, whole blood is not required; only an inert solution to fill up the vessels is needed. Various solutions such as physiological saline with or without some colloidal material (dextran, a synthetic polysaccharide) can be used for some time. However, blood plasma is superior to all except whole blood. The reason for this is that plasma does not leak out of the vascular system by way of the capillaries quite so readily as do other fluids. When the injection of red blood cells is not needed, plasma is superior to whole blood in that its lower specific gravity and lesser viscosity permit circulation with a smaller expenditure of energy on the part of the heart. In addition to this, plasma or serum can be used without regard to blood groups when properly treated.

■ Body posture; gravity

A moment's reflection will show that the position of the body must influence the distribution of the blood. In the reclining position gravity plays but a negligible part, for none of the

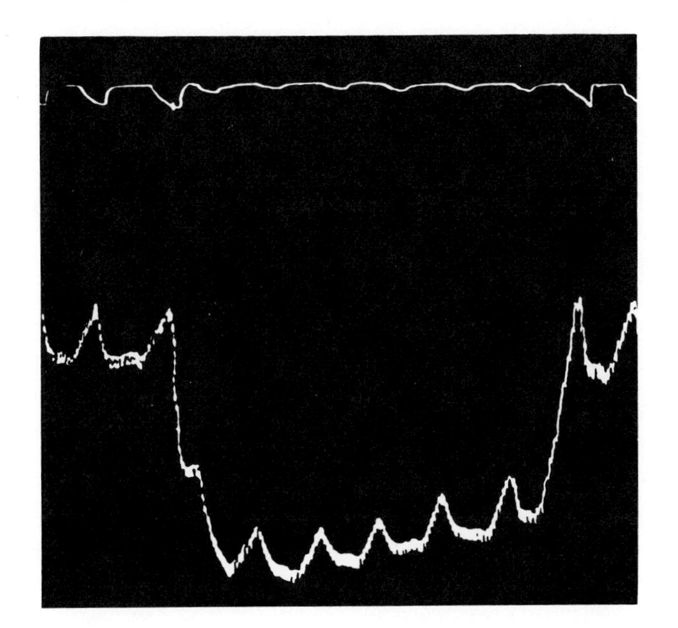

Fig. 15-7
Record showing effect of gravity on blood pressure (lower curve) and on respiration (upper curve) when animal is abruptly tilted from horizontal to vertical feet-down position. (From Wiggers: Physiology in health and disease, Lea & Febiger.)

various parts of the body is either much above or below the level of the heart. In the erect position the blood tends to settle to the lower parts of the body—the abdomen and legs; this may cause a transient, but marked, fall in general blood pressure (Fig. 15-7). The effect of gravity is readily seen by comparing the pale skin and collapsed veins of the hand held above the level of the heart with their appearance when the hand is allowed to hang down.

Were it not for numerous valves in the veins of the legs, there would, in the erect position of the body, rest in these veins a column of blood reaching from the feet up to the level of the heart. This would offer a well-nigh insurmountable obstacle to the return of the venous blood to the heart. Besides this, the hydrostatic pressure would raise the capillary blood pressure in the feet and greatly accelerate the passage of fluid from the capillaries into the tissue spaces and create edema. By means of valves, this long column of blood is divided into several segments, thus protecting the lower capillaries. Even with the aid of the valves, the cardiac output is decreased during the first ten seconds of standing; if no movements are being made, the systolic pressure falls from 5 to 40 mm Hg. Were it not for counteracting mechanisms, fainting would follow. But several agencies are called on to prevent this; among these may be listed the following:

1. Increased vasomotor tone
2. Increased cardiac activity
3. The muscle-pump mechanism
4. The auxiliary action of the respiratory pump
5. The tonicity of the muscles of the abdominal wall

In assuming erect posture, the resultant fall in blood pressure causes, via the aortic and carotid pressoreceptors, an acceleration of heart action or a greater constriction of blood vessels or both. Added to these is the beneficial influence of increased respiratory movements and the massaging of the blood vessels by the rhythmical activity of skeletal muscles, which generally accompanies the assuming of the erect position. In a healthy person, these adjustments assure that circulation goes on without any great effect on the blood pressure, with but little or no increase in the rate of the heartbeat and no significant decrease in cardiac output.* In man the pressure in the carotid artery (supplying the brain) is as great in the erect as in the recumbent position.

A convalescent individual who has spent a long time in bed may faint on being lifted into the erect position. The long period of inactivity has altered the responsiveness of the neural mechanisms. Even in those who may otherwise be regarded as normal persons, this mechanism may act sluggishly, for they may faint when *suddenly lifted* onto their feet. Some persons experience dizziness and a maintained faster heart rate on arising from a reclining position.

In many persons, standing at stiff attention, especially in hot weather, may lower the blood pressure by more than 30 mm Hg and cause fainting. This can be prevented by slight rhythmical activity of the leg muscles—the milking action of skeletal muscles.

A slow pulse in both the reclining and the erect posture and only a slight acceleration on assuming the erect position is frequently used as one test, among many others, of physical fitness.

In this connection, we may call attention to another factor—the proper muscle tone in the abdominal wall. If these muscles have poor tone, the wall is flabby and the size of the abdominal cavity increased; this allows more blood to lodge in this part to the detriment of circulation. A constant slouching posture in standing or sitting weakens these muscles and may cause stagnation of blood in the liver, intestines, and other abdominal organs. It is, therefore, important that the abdominal muscles be strengthened by exercises; this is especially important in persons of sedentary occupations.

*Denervation of the aorta and carotid sinuses may cause gravitational shock in a dog held in a vertical position.

■ Varicose veins

When the blood flowing through a vein meets with great resistance so that stagnation takes place, a varicose vein is likely to result. The superficial leg veins, unsupported by adjoining tissue, are especially susceptible to this disturbance; it is found particularly in those individuals who stand for long periods. In standing, the hydrostatic pressure tends to dilate the veins; the column of blood in the vein is not properly renewed (because of lack of the milking action of rhythmical exercise), and the nutrition of the walls of the veins is impaired; sooner or later the lumen widens, the valves often become incompetent, and the vein, filled with stagnant blood, becomes very tortuous and its wall may degenerate. Neighboring tissue may break down and form ulcers. Obesity favors the development of varicose veins. Varicosities are found more frequently in women than in men. During pregnancy the great pelvic congestion and the enlarged uterus pressing upon the veins may interfere with the return of blood from the legs and cause leg varicosities.

In straining efforts, as in lifting heavy weights, the pressure in the abdomen is increased. A short breath is taken, and the glottis (Chapter 17) is closed; by the contraction of the muscles of expiration, pressure is exerted upon the contents of the thorax and abdomen. The increased abdominal pressure impedes the return of venous blood from the lower parts of the body. Therefore straining efforts should be avoided by persons with inherited weak and thin-walled veins. Straining at stool is frequently responsible for the varicosity of the veins of the rectum, a condition known as hemorrhoids or piles.

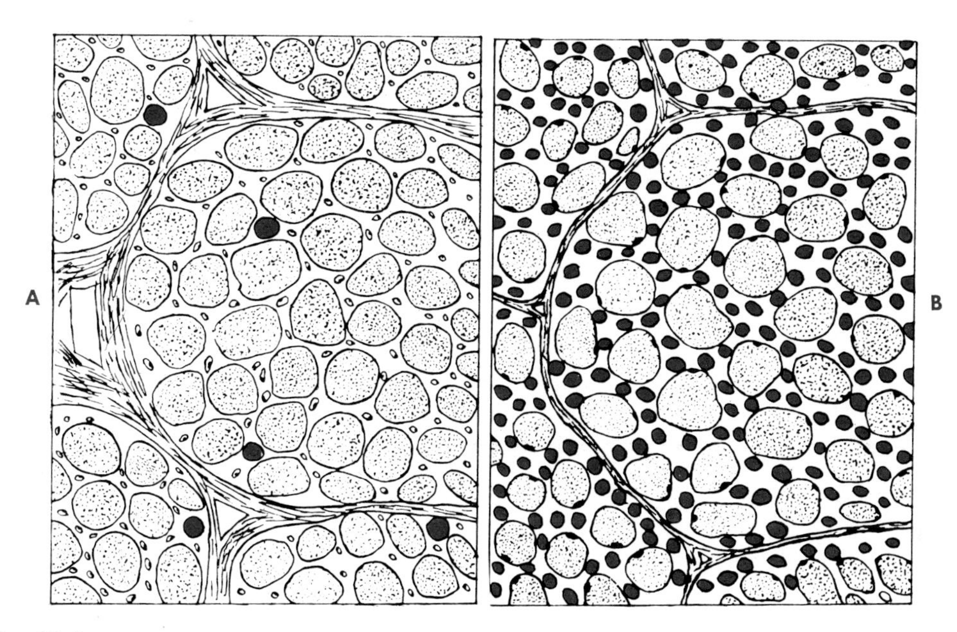

Fig. 15-8
To show the increase in blood flow through active muscle. **A,** The inactive region is of the same area as **B,** the active region. There are five patent capillaries in the inactive area, whereas in the active area there are 190, or thirty-eight times as many. (Courtesy A. J. Nystrom & Co.)

■ CIRCULATION DURING MUSCULAR WORK

The increased demand during muscular activity for nutrients and oxygen and for the removal of waste products and heat can be met only by a great increase in the volume of blood flowing through the active muscles. It has been found that this increase may be more than thirty times the quantity passing through the inactive muscle (Fig. 15-8). We can examine briefly the various methods by which this is brought about.

Heart rate. ■ As a result of the milking action of the rhythmically contracting muscles and the increase in the activity of the respiratory pump, the venous return to the right atrium is greatly increased. The augmented venous return apparently stimulates stretch receptors in the left atrium and cardiac acceleration is reflexly induced. Increased heart rate due to increased venous return is referred to as the *Bainbridge reflex.* It is probable that many other factors (increased temperature and CO_2 content of the blood, augmented secretion of epinephrine, general activation of sympathetic and inhibition of parasympathetic centers) may also be causative agents in the heart acceleration. The degree of increase depends on the severity of the work, as shown in Table 15-1.

A marked difference between the hearts of well-trained and untrained individuals is found. Not only is the increase in rate far greater for the untrained, but the length of time the acceleration continues after work has ceased is much longer. The longer the work is continued, as in athletic endeavors, the longer it takes the heart to regain its original rate (one or more hours), but in this case also, the heart rate of a trained person subsides more speedily than that of the

Table 15-1. Effect of work on pulse rate and systolic and cardiac output*

	Work per minute (kg-m)	*Pulse rate per minute*	*Systolic output (ml)*	*Cardiac output per minute (liters)*
Resting	000	56	57	3.2
Working	720	131	78	9.7
Working	1,260	186	83	15.5

*From Liljestrand.

Table 15-2. Effects of moderate work on blood pressure*

Work	*Systolic pressure (mm Hg)*	*Diastolic pressure (mm Hg)*
Rest	125 ± 5.0	83 ± 5.0
Dynamic work	173 ± 17.0	83 ± 8.5
Static work	161 ± 8.6	109 ± 8.4

*From Tuttle and Horvath: J. Appl. Physiol. **10**:294, 1957.

untrained person. This is another test that has been extensively used for determining the physical fitness of an individual.

Stroke volume and cardiac output. ■ As both the rate and the stroke volume of the heart are increased in exercise, the cardiac output is also augmented (Table 15-1). The increase in stroke volume is much less significant that the acceleration of the heart as a factor in increasing the cardiac output. Since x-ray evidence shows the heart to be smaller in exercise than at rest, any increase in stroke volume is attributable to better ejection of blood. Despite the increased venous return, the venous pressure is relatively constant. It has been recorded that the cardiac output increased from 5 liters in the resting state to 30 liters during severe work—an increase of 600%. As the volume of blood in the average individual is about 5 liters, this entire quantity of blood must have circulated through the left ventricle six times per minute, a notable feat for a pump weighing about 12 oz. *The increase in the cardiac output is the most important factor in the adjustment of the circulation to the increased needs of the body.* Not only is the increase in cardiac output for a specified piece of work greater in the athlete than it is in the nonathlete, but in the former it is obtained with a smaller increase in the rate of heartbeat.

Blood pressure. ■ The effect of moderate work on blood pressure depends on the type of work. Blood pressure reactions to dynamic work (riding a bicycle ergometer 1,250 kg-m/min for one minute) and static work (squeezing a dynamometer at maximum effort for one minute) are shown in Table 15-2.

It will be noted from Table 15-2 that dynamic work caused a marked increase in systolic pressure, whereas the diastolic pressure was unaffected. In the case of static work both systolic and diastolic pressures rose sharply. More strenuous dynamic work also may alter the diastolic pressure.

Due to psychic influence on the circulatory apparatus, systolic blood pressure may increase before work is actually begun. During the early part of dynamic work the pressure rises considerably, but, generally, if the work is not too strenuous, it soon falls to a somewhat lower level; this, no doubt, is to be attributed to a better adjustment of the circulatory-respiratory mechanism to the increased needs of the body. At the cessation of work the pressure may stay slightly above the resting level for a time. For any given piece of work, the blood pressure rises neither as high nor as quickly in the physically trained man as in the untrained; in the former it also returns to the resting level sooner after cessation of work.

Distribution of the blood. ■ That the extra cardiac output may do the largest amount of good, it is preferentially shunted to the organs (skeletal muscles, heart, lungs, and brain) needing it. This is accomplished as follows:

1. During physical work the blood vessels of the abdominal organs are constricted. It is for this reason that digestion may be seriously delayed by physical work and that a full stomach is incompatible with strenuous exercise. Whether the renal arteries are also constricted is a disputed point.

2. The arterioles to the active muscles dilate, thereby ensuring a good supply of blood to the capillaries.

3. There is an opening of previously closed capillaries (Fig. 15-8). The enormous increase in the available capillary bed serves to furnish the larger amount of nutrients and oxygen demanded by the active muscle.

4. Due to the milking action of the rhythmically contracting muscles, the venous blood is more forcibly and speedily removed from the muscle. Thus the capillary blood pressure is kept low so that the arterial blood in the arterioles finds very little resistance to its entrance into the capillaries, and the venous blood is sent hastily to the heart to be returned to the lungs for oxygenation.

5. Soon after the work is begun, there is a marked dilation of cutaneous blood vessels. This aids in the elimination of heat from the body.

6. The volume of the circulating blood is actually slightly decreased due to loss of fluid into the exercising muscle and through perspiration (Chapter 27) and increased ventilation of the lungs. Some compensation for the losses of fluid is achieved by reduced urine formation and fluid movement into the bloodstream from splanchnic regions.

Cardiac nutrition. ■ To enable the heart to accomplish the extra work imposed by exercise, its nutrition is increased. Thus the coronary flow measured in an experimental animal was increased from the usual 140 ml/min to 800 ml during muscular work. This is made possible by the dilation of the coronary vessels that receive vasodilator fibers from the sympathetic nervous system and constrictors from the parasympathetic nervous system.

Although competent authorities believe that the normal adult heart is never injured by physical exercise, there is some doubt as to the validity of this statement in the case of a child's or adolescent's heart. It is well known that animals that work hard generally have a larger heart in relation to body weight than do those that lead a less active life. Whether this is also true for athletes and nonathletes has not been demonstrated conclusively.

■ HYPERTENSION AND ARTERIOSCLEROSIS

Hypertension and arteriosclerosis, popularly called high blood pressure and hardened arteries, respectively, are so common that a few remarks concerning them may not be amiss. Although it is the general opinion that high blood pressure is always and only associated with advanced age, it is common among the younger age groups. Among 5,122 male students at the University of Minnesota, 9% showed, by repeated examinations over a period of three years, a systolic pressure above 140 mm Hg. On the other hand, some elderly people have no higher pressure than do young adults.

An abnormally high blood pressure may be brought on by either a greatly increased cardiac output or by increased peripheral resistance. In hypertension the first factor may be disregarded. The increased resistance is attributed to a narrowing of the blood vessel lumen due to degenerative changes* in the walls of the arteries and arterioles. Stated in this manner, the arteriosclerosis is assumed to be the cause of the hypertension. Another viewpoint regards the sclerotic condition as the result of hypertension. Without

*Some of these changes are known as atherosclerosis (ăth″-e-rō-scle-rō′-sis); others are designated as arteriosclerosis (är-te″-ri-o-scle-rō-sis).

reference as to which is cause and which is effect, let us consider some of the conditions generally said to favor the development of excessive blood pressure.

■ Factors involved in hypertension

Emotional stress. ■ It will be recalled that emotions, expressed or repressed, exert no small influence on blood pressure. That existing hypertension is aggravated by emotional stress is freely admitted. Even in lower animals fear and anger induce marked vasoconstriction. It may be possible that chronic fear (worry) and excessive emotional excitement make this condition permanent. Aside from this, undesirable mental states interfere with the functions of other organs (e.g., digestion and hormone formation) which, in turn, may unfavorably influence the blood vessels or the neural mechanisms governing them.

Obesity. ■ Overweight is very frequently associated with high blood pressure; this is especially true for women after the menopause. Persons of stocky type with short body, broad thick neck, and a tendency to obesity are more prone to exhibit hypertension. But a causal relation between obesity and hypertension is difficult to find.

Hormones. ■ Overactivity of certain glands, such as the adrenal glands and the pituitary, which normally secrete hormones having pressor effects, has also been regarded by some as involved in hypertension.

Renal hypertension. ■ That hypertension is frequently associated with impaired function of the kidneys has been known for years. Constriction of the kidney arteries, either experimentally by clamps or naturally by arteriosclerosis, invokes an increase in blood pressure. However, only about one fifth of all hypertensive subjects appear to suffer from constriction due to degenerative changes. Their increased blood pressure is attributed to the pressor substance angiotensin, which was discussed earlier.

Hereditary factor. ■ Many observers believe that heredity plays a dominant role in hypertension. This conclusion is based on investigations that showed that when both parents were afflicted with hypertension the incidence of high blood pressure in the children was significantly higher than in children whose parents had normal blood pressure.

Food intake. ■ Aside from the ingestion of too much food (which leads to obesity), the nature of the food may have some bearing on hypertension. A diet high in saturated fat, a cholesterol source (Chapter 23), appears to be conducive to the development of atherosclerosis. Feeding an *extra large amount* of this substance to a rabbit or a fowl causes an abnormally large amount of cholesterol to be deposited in the intima (or tunica interna) of the arteries. The infiltration of cholesterol damages the intima and causes it to become excessively swollen. As a result, the lumen of the vessel decreases and the resistance to the blood flow is correspondingly increased; hypertension is the natural result.

Coronary disease nearly always involves coronary occlusion, and in the great majority of these occlusions, sclerosis of the coronary vessels is present. It has been found that the concentration of cholesterol in the blood increases with increasing years; this is also true for the incidence of coronary disease. Moreover, the concentration is considerably higher in people afflicted with coronary disease than in normal people. Cholesterol is commonly found more abundantly in the blood of a man than in that of a woman. This tallies with the fact that coronary disease occurs more often in men than in women. Experimentation has established that the eating of food containing cholesterol (e.g., eggs and saturated fats) does not necessarily cause greater infiltration of cholesterol into the intima. Cholesterol, as described in Chapter 23, is made by the body itself from smaller molecules. The exact role of cholesterol in the development of atherosclerosis has not been well defined.

Other changes. ■ With increasing age, arteries may show degenerative changes. In the large conducting arteries the elastic connective tissue, which predominates, is replaced by inelastic fibers.* Calcium salts are frequently deposited in the walls of the smaller distributing arteries, which thereby become rigid and brittle, and the lumen is decreased. In these degenerative changes (arteriosclerosis) the valuable property of elasticity is gradually lost.

■ Results of hypertension and arteriosclerosis

Although hypertension is not always incompatible with long life, coupled with arteriosclerosis it may prove disastrous for several reasons.

1. The gradually rising diastolic pressure throws extra work upon the heart and causes it to undergo hypertrophy; as long as the heart can maintain a sufficiently high systolic pressure, an adequate general circulation may be achieved. But finally a diastolic pressure is reached at which it is impossible for the heart to deliver an adequate systolic output.

2. A second danger relates to the vessels themselves. A normal blood vessel has a high tensile strength; that is, to rupture a vessel a very great internal force is required. For example, it was found that the carotid artery of a goat could withstand a pressure of 2,250 mm Hg, which is at least fourteen times the usual arterial blood pressure. *Normal blood vessels never rupture.* But when an artery has undergone the changes discussed, the breaking point may be reduced and any sudden increase in blood pressure may cause rupture. If the vessels of the brain are concerned, the resultant destruction of the brain cells and fibers may cause great disturbance (paralysis) in the mental and neuromuscular life of the individual; frequently this proves fatal.

3. An abnormal condition of the intima and the slow flow through the greatly narrowed arterioles favor the formation of thrombi or emboli. The resulting stoppage of circulation and anoxia may be fatal to brain cells, quite similar to that following the rupturing of a vessel. In the heart (coronary thrombosis), if the area affected is quite localized and especially if time is given for establishing collateral circulation around the affected area, the individual may survive.

All the arteries of the body do not exhibit arteriosclerosis simultaneously or to the same extent. The increased resistance offered by a narrowed artery leads to a poor supply of blood to the organ or the part of the organ involved; this, in turn, leads to inadequate nutrition, and the function and structure of the organ suffer. If the artery under consideration supplies the stomach or intestines, impaired nutrition of the whole body may result. Gradual loss of memory or of some other psychic function frequently attends the sclerosis of a cerebral vessel.

The previous discussion affords a reasonable basis for believing that a person is as old as his blood vessels. Physiological age is measured by the capacity of each individual vital organ to perform its specific function.

■ HYPOTENSION

Some individuals have what is generally looked on as a low basal blood pressure. In them a low pressure seems to be normal and compatible with physical endurance and vitality. The minute volume of circulating blood, not blood pressure, per se, determines the amount of nutrition supplied to the tissues and therefore limits the extent of tissue activity. A naturally low basal blood pressure favors longevity.

The term *hypotension* as a pathological condition should be restricted to those instances in which a previously higher blood pressure falls

---

*The cross section of the abdominal aorta has, at the age of twenty years, an area of about 2 cm^2; at fifty years of age this has increased to 4 cm^2, and at seventy-five years of age to 5 cm^2. The distensibility of the thoracic aorta above the undistended level at the age of twenty-two years is 240%; this has dropped to 80% at the age of seventy-five years.

below, let us say, 100 mm Hg systolic and 60 mm diastolic pressure. In consequence, the individual experiences difficulty in maintaining normal activity.

READINGS

Berne, R. M., and Levy, M. N.: Cardiovascular physiology, ed. 2, St. Louis, 1972, The C. V. Mosby Co., pp. 116-138 and 237-253.

Braunwald, E.: The sympathetic nervous system in heart failure, Hosp. Pract. **5**(12):31-39, Dec. 1970.

Chapman, C. B., and Mitchell, J. H.: The physiology of exercise, Sci. Am. **212**(5):88-96, May 1965.

Folkow, B., and Neil, E.: Circulation, London, 1971, Oxford University Press, pp. 285-306 and 560-581.

Kellermeyer, R. W., and Graham, R. C., Jr.: Kinins—possible physiologic and pathologic roles in man, New Eng. J. Med. **279**(14):754-759; **279**(15):802-807; **279**(16):859-866, 1968.

Mellander, S., and Johansson, B.: Control of resistance, exchange and capacitance functions in the peripheral circulation, Pharm. Rev. **20**(3):117-196, 1968.

Spain, D. M.: Atherosclerosis, Sci. Am. **215**(2):48-56, Aug. 1966.

Toole, J. F.: Effects of change of head, limb and body position on cephalic circulation, New Eng. J. Med. **279**(6):307-311, 1968.

Webb-Peploe, M. M., and Shepherd, J. T.: Veins and their control, New Eng. J. Med. **278**:317-322, 1968.

Tissue fluid
Lymph
Edema
Shock

THE CAPILLARIES AND THE LYMPHATICS

■ TISSUE FLUID

In our discussion of blood, it was noted that tissue cells are dependent on the circulatory system for their supply of nutrients and for the removal of their waste products. This exchange between blood capillaries and tissues takes place across intervening spaces that are filled with a fluid generally designated as *tissue* or *interstitial fluid* (Fig. 12-1). The spaces existing between capillaries and tissue cells and between neighboring cells are microscopic. Except in cartilage, bone, epithelium, and the central nervous system, another set of closed-end capillaries, the *lymph capillaries*, occurs in tissues. The contents of these and the larger vessels that they form (Fig. 16-2) are referred to as lymph.

Other fluids found in certain larger spaces are frequently spoken of as tissue fluid. Many of these fluids differ materially from true tissue fluids and serve very special functions. Among these are the fluids found in the pericardial, pleural, and peritoneal cavities, the cerebrospinal fluid in the ventricles and canals of the central nervous system and between its coverings, the humors of the eye, and the endolymph of the inner ear. All these perform special functions that are spoken of in their proper place. There is also the synovial fluid found in the joints; this contains a viscous mucinlike substance that is increased by inflammation.

■ Composition of tissue fluid

In tissue fluid are found essentially the same ingredients as in blood plasma but in different concentrations. Containing practically no red blood cells, tissue fluid has a very low specific gravity (1.015). The number of lymphocytes varies considerably. The amount of protein in tissue fluid found in the limbs is about 0.3%; that in the fluid from the abdominal organs is from 2% to 4% (compare with 6% to 8% in plasma). The salts and glucose have nearly the same concentration as in plasma.

Changes in the amount and composition of tissue fluid occur constantly and are caused by (1) the addition of materials from the blood, (2) the removal of substances from the tissue fluid and their consumption by the cells, (3) the addition of cellular products, e.g., waste products and hormones, and (4) the removal of substances by absorption into the blood.

■ Tissue fluid formation

The passage of fluid from the capillaries through the capillary wall and into the tissue spaces is known as *transudation* and is a process of *ultrafiltration*. Filtration of water and its solutes through the capillary wall is spoken of as *bulk transport*. Additional exchange of water and solutes across the capillary wall is by passive diffusion and is determined by such factors as the permeability, the perfusion pressure, and the available surface area of the capillary wall. In tissue fluid formation there are three factors to be considered: the nature of the membrane serving as a filter, the forces causing the substances to pass through the membrane, and the forces restraining the passage of fluid.

The capillaries. ■ Blood is delivered to the capillaries through muscular walled arterioles of 20 to 50 μ in diameter. Commonly a smaller vessel of 10 to 20 μ in diameter, the *metarteriole*, links the capillary and arterioles. This vessel has a sparse, discontinuous supply of smooth muscle fibers. At the junction of metarterioles (or arterioles in those tissues without metarterioles) and capillaries there is a small cuff of smooth muscle cells known as the *precapillary sphincter*. Metarterioles can bypass the capillaries and link more directly with the venules. The capillaries consist of epithelial cells and are devoid of muscular elements. Capillary density is most marked in glandular structures and muscles, both cardiac and skeletal. It is least in cartilage and subcutaneous tissues.

Selective permeability of the capillary wall. ■ For a membrane to serve as a filter, it must possess a certain degree of permeability and impermeability; filtration is never accomplished with a membrane that does not restrain some things from passing through. A capillary wall is exceedingly thin, being composed of a single layer of flat endothelial cells not more, and sometimes less, than 1 μ ($^1/_{25,000}$ inch) in thickness. Under usual conditions the wall is permeable to water and crystalloids (glucose, salts, urea, amino acids, and lactic acid); indeed, for these the capillary wall is more permeable than any other membrane in the body. Crystalloid permeation of the capillary wall is dependent on both the flow through and the surface area of the capillary. Lipid solutes are limited principally by the blood flow and diffusion distances. The capillaries are not totally impermeable to protein, but the degree of permeability varies in different tissues. In the liver and intestines the permeability is relatively great and the protein concentration of the tissue fluid is half, or more, of that found in plasma; in muscles the concentration is considerably less, and in the CNS it is extremely limited.

The diffusion of lipid-insoluble materials through the capillary wall is considered to take place not through the endothelial cells but through the cement substance found between the cells. Therefore, the greater is the capillary surface, the greater is the opportunity for crystalloid diffusion. The cement is of porous nature, and the porosity varies from one tissue to another. The chief function of the endothelial cells is to maintain the cement substance in its normal state. Proteins in the plasma are evidently essential to maintain the integrity of the capillary wall. If plasma expanders (Chapter 15) are substituted completely for plasma protein, the capillaries become excessively leaky. Even small amounts of protein added to the expander will restore normal permeability. The diffusion of lipid-soluble substances such as O_2 and CO_2 is not limited to the spaces between cells; it occurs across the cells themselves. Consequently, their transfer is not, as with crystalloids, so dependent on the surface area, but is more dependent on a rate of blood flow which provides an optimal concentration gradient for diffusion.

The extent of the filtration surface plays an important part in the rate of filtration. According to August Krogh, the capillaries in the human body, if placed end to end, would form a tube 62,000 miles long. Since a capillary is slightly less than 1 mm in length, the number of capillaries is enormous. This places the cells of the body in close proximity to their base of supply. The surface of the combined capillaries has been placed at 67,000 ft^2 ($1^1/_2$ acres). It has been calculated that each milliliter of blood in the capillaries is exposed to a surface of 7,300 cm^2, or 8 ft^2.

Capillary flow. ■ The capillary system forms an exceedingly capacious set of tubes, with a bed estimated at from 400 to 1,000 times as large as that of the aorta. The usual velocity of flow in the capillaries is about 1 mm/sec, which allows ample time for the exchange of materials between intravascular and tissue fluids. Considering the small amount of blood in the entire vascular system, it is evident that all the capillaries cannot be filled at the same time. Capillary flow is intermittent because of the periodic opening and closing of precapillary sphincters. This phenomenon is known as *vasomotion.* It is a response to the nutritional needs of the tissues nearby to the capillary. As a result, in a resting muscle only about $^1/_{100}$ of the capillaries are open (Fig. 15-8); in man this would mean a muscle capillary blood volume of about 150 ml.

Activity of a muscle or gland is accompanied by dilation of its arterioles and capillaries. Arteriolar dilation or constriction is generally attributed to neural mechanisms. Dilation of metarterioles and relaxation of precapillary sphincters are by and large produced by local factors in the tissues, for example, decreased O_2 tension, increased CO_2 tension, lactic acid, and so forth. In some instances, special vasodilator substances such as acetylcholine and histamine are involved. An increase in temperature of the active tissue may also help.

The augmented inflow of blood under higher pressure from the dilated arterioles results in passive dilation of open capillaries and the opening up of those previously closed. An increase in the volume of blood passing through the capillaries in a given area favors formation of tissue fluid. *Indeed, all the neural and chemical mechanisms for regulating cardiac and vascular activity by which blood pressure, rate of flow, and volume flow are altered exist for but one purpose: to supply the capillaries with an amount of blood sufficient to meet the varying metabolic demands of the tissues.*

The opening up and the dilation of the capillaries not only increase several hundredfold the total filtration surface, but the stretching of the capillaries increases the porosity of the intercellular cement substance and thereby favors filtration. Whether, aside from this mechanical method, metabolites as chemical agents increase the permeability is doubtful.

Factors promoting filtration. ■ Although the magnitude and direction of fluid movement across the capillary wall are determined by the permeability and surface area of the endothelial membranes, probably the chief factor concerned in forcing the fluid part of the blood into the tissue spaces is the blood pressure (hydrostatic pressure) in the capillaries. A second factor that aids filtration is the tissue fluid colloid osmotic pressure, due to the presence of protein. This pressure, of course, varies in different tissues in accord with the concentration of protein. Smaller solutes are in equilibrium between extravascular and intravascular fluids and do not contribute to the difference in osmotic pressures. The greater is the filtration, the greater is the amount of tissue fluid formed. Were it not for some restraining force, the blood pressure would squeeze all the plasma through the capillary wall in a very short time. This would have two disastrous results. (1) It would rob the blood of its water and

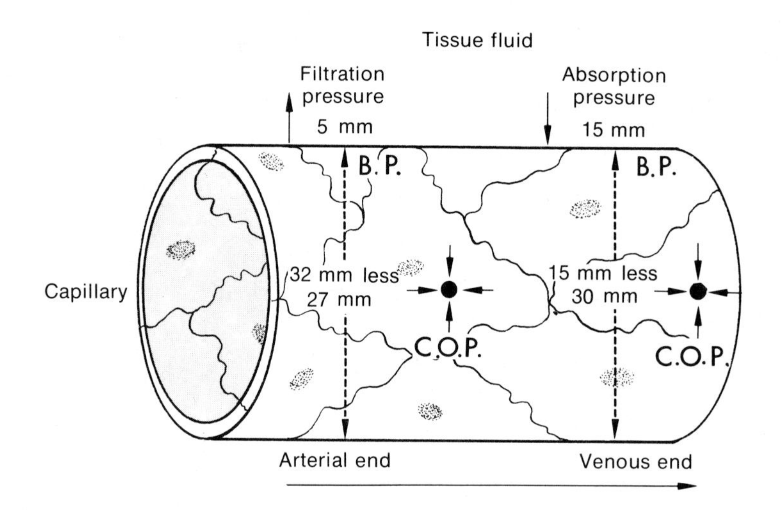

Fig. 16-1
Illustration showing forces (in mm Hg) involved in tissue fluid formation. **B.P.,** Capillary blood pressure; **C.O.P.,** colloidal osmotic pressure of plasma proteins. Tissue hydrostatic and osmotic pressures are not shown.

leave the cells stranded in the vessels, unable to move; all circulation would come to a standstill—*stasis*. (2) The amount of tissue fluid formed would be so great as to cause destructive pressure upon the cells.

Restraining factors. ■ To prevent excessive filtration, two factors oppose the blood pressure (and tissue colloid osmotic pressure): (1) the colloid osmotic pressure (C.O.P.), the so-called *oncotic pressure*, of the plasma proteins, and (2) to a greater or lesser degree, the hydrostatic pressure of the fluid in the tissue spaces. The osmotic pressure of proteins may be thought of as a water-drawing power.

Effective filtration pressure. ■ It is apparent from our discussion of factors promoting and restraining filtration that:

Filtration pressure = [Capillary blood pressure + Tissue colloid osmotic pressure] − [Plasma colloid osmotic pressure + Tissue hydrostatic pressure]

The major factors involved in filtration are illustrated in Fig. 16-1.

The amount of colloidal osmotic pressure varies with the concentration of proteins in the plasma. Whatever increases the blood pressure in the capillaries or decreases the colloidal osmotic pressure also increases the filtration pressure. If these two opposing forces just balance each other, no water exchange between blood and tissue fluid takes place; if the osmotic pressure of the proteins is greater than the capillary blood pressure, tissue fluid is drawn into the capillaries (absorption). As to the quantitative values of these factors, no universal statement can be made since they vary from time to time and from place to place. For the sake of example, the capillary blood pressure may be estimated in the arterial end of the capillary at 32 mm Hg and the colloidal osmotic pressure at 27; this leaves 5 mm Hg as the *effective filtration pressure* (Fig. 16-1).

The tissue hydrostatic pressure under normal conditions is quite low; indeed, it may even be negative in certain tissues at certain times. Its contribution to effective filtration can, in fact, generally be ignored. However, any excessive accumulation of tissue fluids will elevate the tissue pressure and, consequently, increase the restraining force. Because of the relatively smaller contributions to effective filtration attributable to tissue hydrostatic and colloid osmotic pressure under normal conditions, these factors have not been included in Fig. 16-1.

Any increase or decrease in the osmotic pressure of the tissue fluid must affect the cells. For example, if much salt is consumed, it freely enters the blood and tissue fluid. But since the membrane of the cells is relatively impermeable to NaCl, the salt remains in the tissue fluid and increases its osmotic pressure; this draws water from the cells into the tissue fluid.

■ Removal of tissue fluid

In order that the tissue spaces shall not become surcharged with liquid and thereby exert injurious pressure upon the cells, the materials not needed by the cells, whether derived from the blood or from the cells, must be constantly drained away. Two channels are provided for this: the capillaries and the lymph vessels.

Absorption by capillaries. ■ As the blood enters the capillary (Fig. 16-1) the values just quoted for blood pressure *(B.P.)* and colloidal osmotic pressure *(C.O.P.)* yield an effective filtration pressure of 5 mm Hg that drives water and water-soluble solutes (except most proteins) through the capillary wall and into the surrounding tissue spaces. In its passage along the capillary the blood loses some of its fluid by filtration; this reduces the blood pressure and increases slightly the concentration and therefore the colloidal osmotic pressure of the plasma proteins. The capillary blood pressure is further reduced by the friction that the blood encounters in moving toward the venous end of the capillary. The colloidal osmotic pressure *(C.O.P.),* arbitrarily placed at 30 mm Hg in Fig. 16-1, now exceeds the blood pressure *(B.P.)* of 15 mm Hg; this excess pressure gives an absorption pressure and fluid is drawn into the capillary. In this manner a constant circulation of tissue fluid is brought about, and, with some exceptions, excessive accumulation of tissue fluid is prevented. The absorption of fluid at the venous end of the capillary restores the normal oncotic pressure of the plasma proteins. Filtration of water and solutes at the arterial end of the capillary and their reabsorption at the venous end of the capillary is spoken of as the *law of the capillary;* it is also known as *Starling's hypothesis.*

Drainage by lymph vessels. ■ Some of the fluid and all of the protein that escapes from the capillaries are returned to the bloodstream, by means of the lymph vessels or lymphatics. In most tissue spaces there is found a very delicate system of tubes, the lymph capillaries. A lymph capillary is composed of only one layer of flat, broad endothelial cells. The fluid from the tissue spaces finds its way into the lymph capillaries and, by the forces to be discussed presently, is moved onward toward the heart. Lymph capillaries are very permeable (more so than blood capillaries); as a result, foreign materials (e.g., bacteria) can readily enter.

Lymph capillaries anastomose freely with each other and by their union progressively form larger and larger vessels. Although considerably thinner, the walls of the larger lymph vessels resemble those of the veins, being composed of an inner layer of endothelial cells, a muscle layer, and an outer layer of connective tissue. Lymph vessels are abundantly supplied with valves (Fig. 16-3), which gives them a beaded appearance. The lymph vessels from the lower limbs and the organs of the abdomen unite to form the largest lymph vessel in the body, the *thoracic duct* (Figs. 13-3 and 16-2), which, after being joined by the lymphatics of the left arm and left side of head, neck, and thorax, empties its contents into the left subclavian vein. Lipids absorbed from the digestive tract are carried by this route. The lymph from the right side of the head, neck, and thorax and from the right arm is discharged into the right subclavian vein.

With some exceptions, during the inactive con-

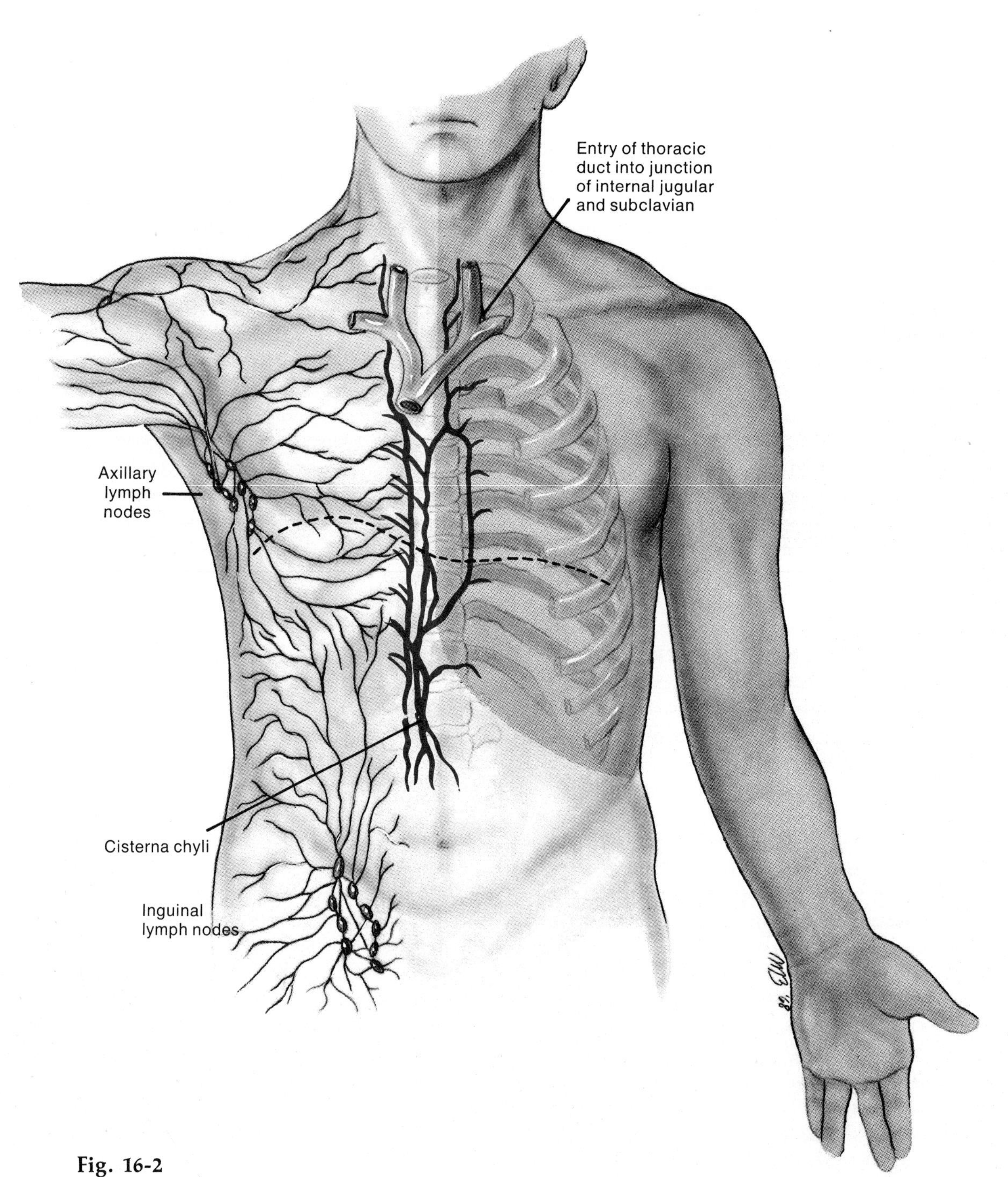

Fig. 16-2

Illustration showing portion of lymphatic system. Cisterna chyli receives lymphatics from small intestine.

dition of an organ, most of the tissue fluid is returned by absorption directly into the blood capillaries; it is only during greater tissue fluid production (as in an active muscle) that any appreciable amount finds its way into the lymph vessels to be carried to the large veins of the thorax. The drainage of lymph from the intestine and liver into the thoracic duct proceeds constantly.

■ LYMPH

■ Flow of lymph

Lymph is a transparent, slightly yellowish liquid of alkaline reaction found in the lymphatic vessels, and since it contains fibrinogen, it will coagulate when shed. Lymph and plasma are nearly identical in composition except that the average protein concentration is about 1.5%. The rate of lymph flow is slow. In the thoracic duct the average velocity is about 4 mm/sec. The total lymph discharged by the duct has been found to be from 1,200 to 2,280 ml/day, nearly equal to the animal's plasma volume. Compared with the volume of blood carried by the aorta (about 5,000 ml/min), the volume rate of flow of lymph is exceedingly small. Lymph flow is increased after ingestion of food. The pressure of the lymph is also low. Since no special organs for the propulsion of lymph exist, the flow is dependent on external forces, among which are the following:

1. Whatever increases the flow of the blood in the veins also aids the lymph flow. Lymph does not flow from a resting limb, but in muscular activity the rhythmically contracting muscles squeeze out the lymph by exerting external pressure upon the lymph vessels, and a considerable flow takes place.
2. Aspiration by the thorax during respiration affects the lymph flow (Chapter 17).
3. Massaging of the body and passive movements of the limbs act on the flow of lymph and venous blood in much the same manner as the rhythmical contraction of muscles. Fatigue after work is lessened by massage, for it removes more speedily the waste products from the tissues. Massaging a bruised part may prevent the accumulation of lymph, which generally takes place and causes the part to swell.

■ Lymph nodes and lymphoid tissue

In the course of the lymph vessels there are inserted at frequent intervals, as can be seen in Fig. 16-2, round or oval masses varying from 1 to 25 mm in diameter. These bodies are known as *lymph nodes* (or glands). In certain areas they are very abundant, as in the neck, axilla, inguinal region (groin), and the mesentery. Lymph vessels are described as afferent and efferent, depending on whether they carry lymph into or out of the lymph node. The afferent and the efferent vessels are supplied with valves (Fig. 16-3).

The capsule of a lymph node is composed of white fibrous tissue, which sends trabeculae (trabek'-u-le) into the medullary portion of the node. The medulla, or interior part, is a network of inelastic, collagenous, reticular fibers. The network bordering on the capsule and trabeculae is more loosely meshed and thus forms open spaces known as the cortical and medullary sinuses. The nodules and the medullary cords

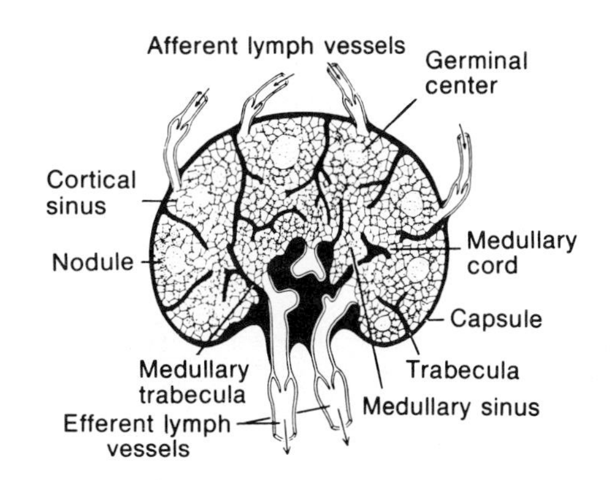

Fig. 16-3
Illustration showing structure of a lymph node. (From Hoskins and Bevelander: Histology, The C. V. Mosby Co.)

shown in Fig. 16-3 are formed by a denser network; in them the sinuses are obscured by closely packed lymphocytes. Among, and attached to, the reticular fibers are certain phagocytic cells known as fixed macrophages (histiocytes); these cells frequently become detached and are found in the sinuses as free macrophages.

Functions. ■ The lymph nodes have four functions: (1) the filtration of foreign particles from the lymph, (2) the formation of lymphocytes, (3) the disposal of bacteria, and (4) the production of antibodies.

Filtration. The lymph brought to a node by the afferent lymph vessels finds its way into the cortical and medullary sinuses and thence into the efferent lymph vessel. It is now en route to the blood by way of the thoracic duct. During its passage through the node, foreign particles are removed by the phagocytic cells. In the macrophages may be seen red blood cells in all stages of disintegration. The nodes of the respiratory tract may become pigmented with particles of carbon, as in the case of coal miners.

Formation of lymphocytes. The germinal center of the nodule is believed to give rise to lymphocytes. These find their way into the efferent lymph vessels. The lymph leaving the node is richer in these cells than that which enters. This function they have in common with the other structures composed of what is known as lymphoid tissue, such as the tonsils and spleen, as also with the more or less scattered patches of this tissue found in the mucous membrane of the intestines and in the respiratory tract.

Disposal of bacteria. From a local infection bacteria frequently enter the lymph vessels. On their arrival at a node they are phagocytosed by the macrophages and thus disposed of. During an infection the macrophages are greatly increased in number. Lymph nodes, therefore, furnish an important barrier against the spread of bacteria throughout the body, and macrophages are the defense mechanism against local infection. The cervical (neck) nodes may become infected through the tonsils, ears, nose, infected pulp of teeth, or pyorrhea. The great multiplication of macrophages in a node causes the node to become enlarged and tender to the touch (adenitis). When the nodes, loaded with bacteria and their products, are unable any longer to cope with the situation and break down, the infection spreads, and septicemia (the presence of pathogenic bacteria in the blood) results. It may be mentioned here that carcinoma (cancer) generally spreads by way of the lymph channels.*

Production of antibodies. In addition to the phagocytic destruction of bacteria, the lymph nodes aid in the defense of the body against bacteria by the formation of antibodies. It will be recalled (Chapter 12) that the introduction of an antigen into the blood calls forth, under the proper conditions, the formation of a specific antibody by which the deleterious effect of the antigen may be neutralized. Gamma globulin of the plasma is an important antibody.†

The following is one of the several views as to the part played by the lymph nodes: When bacteria are present in the lymph stream, the efferent lymph vessel from a node may contain as many as 80,000 lymphocytes per cubic millimeter. Also the amount of antibodies in the efferent vessel may be one hundred times as great as in the afferent vessel. Lymphocytes contain about ten times the amount of antibodies found in the lymph plasma. When the lymphocytes disintegrate in the blood, the antibodies are liberated. This dissolution of the cells seems to be regulated by a chemical compound made in the cortex of the adrenal glands.

■ EDEMA

When the drainage of the tissue fluid cannot keep pace with its formation, the fluid accumu-

*This spreading is technically known as metastasis (metas'-ta-sis).

†Gamma globulins have been satisfactorily used in combating measles and other infectious diseases.

lates in the part of the body affected. This causes the part to be swollen—a condition known as *edema.* From our study it is evident that edema may be caused by a marked increase in capillary blood pressure, by an increase in the permeability of the capillary wall, or by a decrease in the colloidal osmotic pressure of the blood.

Edema and capillary blood pressure. ■ A marked retardation of the venous flow to the right heart (e.g., in an incompetent right atrioventricular valve) causes a rise in venous blood pressure. The capillary pressure is thereby increased and this results in *passive congestion.* When capillary blood pressure exceeds the colloidal osmotic pressure of the blood, filtration at the arterial end of the capillary is unrestrained and edema results.

Decreased circulation through the kidney or a malfunctioning kidney may lead to insufficient excretion of sodium salts. The resulting increased osmotic pressure of the blood retains more water in the blood vessels and the capillary blood pressure may increase sufficiently to induce edema. A salt-free diet is the current treatment for this disorder. Because of the influence of gravity, the capillaries in the lower parts of the body are most affected; as a result, in the erect position the edema is greatest in the feet. In certain instances, increasing the excretion of water from the body by stimulating the kidneys with diuretics is helpful.

Edema and increased permeability. ■ Damage to the capillary wall may increase the permeability sufficiently to enable the plasma proteins to pass into the tissue fluid. The resultant increase in the colloidal osmotic pressure of the tissue fluid opposes the reabsorption of the tissue fluid into the blood and leads to its accumulation in the tissue spaces. The lesion may be local and, if superficial, may result in what is known as a blister or wheal. We are familiar with the wheal resulting from the lash of a whip, the blisters raised on the hands by manual labor, and the great redness and blistering of the skin after exposure to intense light (ultraviolet). In all instances the skin is injured. The redness is, of course, the outward token of a great dilation and hyperemia of the capillaries.

In injury to tissues the capillary circulation undergoes marked changes, as can be observed in the following manner: If the web of a frog's foot is subjected to extreme heat or cold or to irritating substances or allowed to become dry, observation shows an initial increase in the volume flow due to the dilation of the vessels. Gradually, however, the circulation slows down until, finally, it ceases altogether, with the capillaries engorged with red cells—a condition called stasis. The stasis is due to capillary leakage. During stasis more leukocytes (Fig. 16-4) force their way by ameboid movement through the wall of the capillary and gather in the surrounding tissue. It is, therefore, necessary that the *exposed tissues of an animal be kept moist by means of a physiologically balanced solution. In the case of warm-blooded animals, the tissues also must be kept at or near body temperature.*

Injury to the tissues brings on extreme capillary dilation and an increase in permeability due to production of a powerful dilator compound. As to the nature of this substance there is doubt, but it is believed to be a compound called *histamine.* A drop of histamine pricked into the skin causes a redness and blister similar to that observed in mechanical injury to the skin. The rashes and hives in individuals allergic to certain foods are due to substances found in these foods or formed from them in the body after their ingestion. Since epinephrine is a powerful constrictor of the arterioles, it is used in allergy for reducing vasodilation and local edema.

Edema and colloidal osmotic pressure of the blood. ■ Since the effective filtration pressure is equal to the capillary blood pressure minus the colloidal osmotic pressure of the blood proteins, it is apparent that a decrease in blood proteins is associated with increased filtration pressure. When the normal percentage of plasma proteins

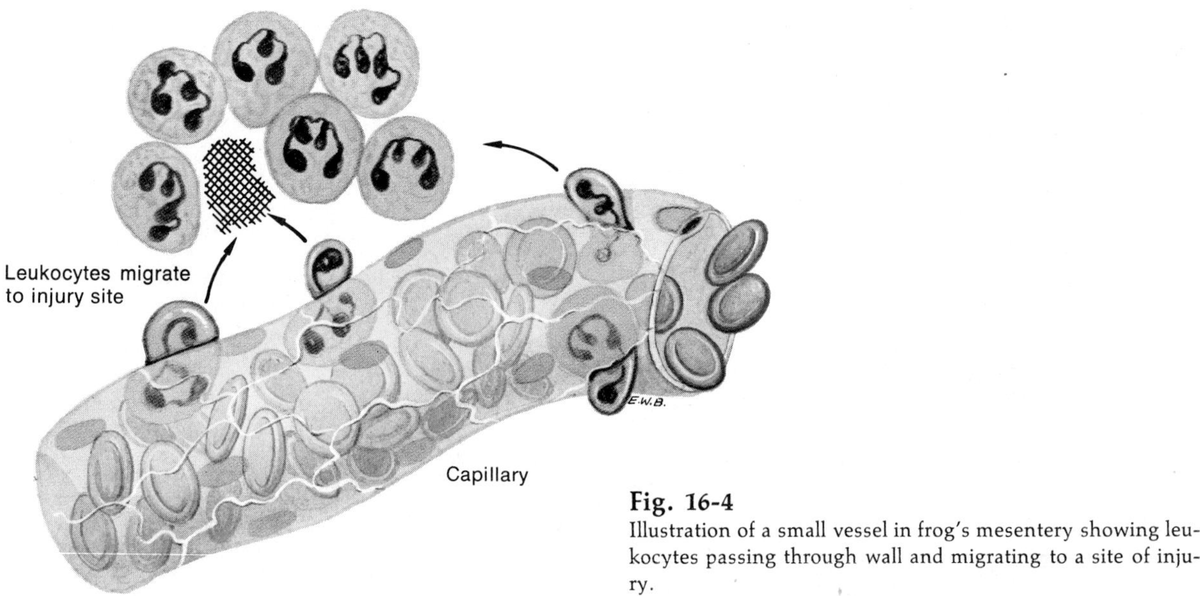

Fig. 16-4
Illustration of a small vessel in frog's mesentery showing leukocytes passing through wall and migrating to a site of injury.

(6% to 8%) falls to about 4%, as in starvation, edema ensues.

Hemorrhage. ■ The reverse of the process just described is seen in an extensive hemorrhage. The volume of the blood and its pressure are considerably decreased, but the osmotic pressure of the proteins in the plasma remains normal. When the osmotic pressure exceeds the capillary blood pressure, fluid is drawn from the tissue spaces into the capillaries. In this manner the body strives to maintain a volume of blood adequate for circulation.

■ SHOCK

So many meanings have been assigned to the term *shock* that it has become almost meaningless. We shall limit this discussion to those conditions in which there is a depression of circulation sufficient to interfere with the proper functioning of the body and, in severe cases, to result in death. Many circulatory disturbances (aside from cardiac failure, incompetence of valves, and so on) are known as shock; these are generally designated in accordance with the manner of their origin, for example, neural, gravitational, hemorrhagic, anaphylactic (allergic), toxic, traumatic, and surgical shock.

Most of us are acquainted with the fact that shock may follow severe mechanical injury (especially great laceration of muscles) and surgical operations, even though very little blood has been lost. A pale, clammy skin is the most common sign; this is frequently accompanied with cyanosis (blueness) of ears and fingers, a feeble but rapid pulse, shallow and rapid breathing, and, to a greater or lesser extent, mental confusion or unconsciousness. All of these are indicative of a great fall in blood pressure. Reduction in pressure could be due to a disturbance (cardiac failure, loss of vasomotor tone, or a reduction in blood volume) in any one of the three necessary elements in circulation. That the heart is not at fault is shown by the very rapid, though less forceful, beat. The arterioles of the cutaneous and splanchnic area may be constricted—a compensatory reaction to the low blood pressure.

It is now generally conceded that the cause of the fall in pressure must be sought in a *decreased volume of blood in the vascular system;* this view is substantiated by the fact that the cavities of the heart and the great veins are poorly filled. We have already intimated that the loss of blood is not due to extravascular hemorrhage. As the concentration of the cells in the blood remaining in the vessels is high, it may be safely concluded that the loss of circulating fluid is due to the escape of plasma from the blood vessels. This increase in hematocrit produces an increase in the specific gravity and viscosity of the blood that adds to the circulatory difficulties.

Several theories have been advanced as to the cause of the increased transudation; it is possible that more than one factor may be operative in various instances. According to the toxemic theory, the injured tissues liberate histamine (or some similarly acting compound), which causes dilation of the capillaries in the injured part and increased filtration of tissue fluid. The sluggish circulation thus produced may result in anoxemia and augment the factors already at work in reducing the blood pressure. According to this view, the factors concerned in normal tissue fluid formation are also the cause of traumatic or surgical shock.

To combat hemorrhagic shock, the lost circulating fluid must be replaced by intravenous injections of human plasma or of human serum albumin dissolved in a proper medium. To treat shock, the patient is placed in such a position as to aid the feeble circulation, especially the cerebral circulation. If the body temperature is below normal, the subject must be kept warm. Heat should be applied with caution for the following reasons: (1) heat causes dilation of the cutaneous vessels and thereby further decreases blood pressure; (2) it may induce sweating and thus still further decrease the blood volume; (3) in shock there is great dearth of oxygen in the tissues; the application of external heat increases the metabolic rate and the consumption of oxygen and thus increases the degree of anoxia.

READINGS

Berne, R. M., and Levy, M. N.: Cardiovascular physiology, ed. 2, St. Louis, 1972, The C. V. Mosby Co., pp. 100-115.

Champion, R. H.: Blood vessels and lymphatics of the skin. In Champion, R. H., Gillman, T., Rook, A. J., and Sims, R. T., editors: An introduction to the biology of the skin, Oxford, 1970, Blackwell Scientific Publications, pp. 164-174.

Folkow, B., and Neil, E.: Circulation, London, 1971, Oxford University Press, pp. 97-131.

Mayerson, H. D.: The lymphatic system, Sci. Am. **208**(6):80-90, Jun. 1963.

Selkurt, E. E.: Status of investigative aspects of hemorrhagic shock in Physiological Basis of Circulatory Shock, Fed. Proc. **29**(6):1832-1835, Nov.-Dec. 1970.

17

Pulmonary ventilation
Chemistry of respiration
Utilization of oxygen by cells
Regulation of pulmonary ventilation
Respiration and acid-base balance
Conditions affecting respiration
Modified forms of respiration

GAS EXCHANGE—RESPIRATION

For the maintenance of life the body must be supplied with a large number of different substances, but *the most urgent need is a continual supply of oxygen.* Muscles obtain their energy directly from various substances; nevertheless, oxygen is required for the utilization of these substances. The body is liberally supplied with reserves of energy-furnishing materials. Since oxygen usually is readily available, there is no need to provide the body with large stores. The aforementioned reserves are of no value unless O_2 is furnished from the environment in a continual, uninterrupted stream. As the ability of the body to defend itself against the deleterious effects of carbon dioxide is exceedingly limited, it is of equal importance that this waste product must be removed almost as fast as it is formed. *The exchange of gases between the organism and its environment is called respiration.*

Unfortunately the word *respiration* is used to designate three very distinct, although closely associated, processes. In their proper sequence these are the following:

1. Moving air into and out of the lungs, better named *pulmonary ventilation.*
2. The reciprocal exchange of gases between the air in the lungs, the blood, and the cells —*pulmonary* and *internal respiration.*
3. *Cellular respiration,* by which the oxygen is utilized in the catabolism of energy-yielding substances for the production of energy.

The production of CO_2 by the body tissues, its transport by the blood, and its elimination by the lungs are counterparts, in reverse order, to the previous sequence of events. Until we speak of the utilization of oxygen by the cells, we shall use the word *respiration* in its original meaning of ventilation. As occasion demands, the term *cellular respiration* will be used.

■ PULMONARY VENTILATION

■ Anatomy

The mechanism by which pulmonary ventilation is brought about involves (1) the respiratory tract and (2) the respiratory muscles. The respiratory tract is, broadly speaking, a tubular structure that includes the nose, pharynx, trachea, bronchi, and the alveoli of the lungs (Figs. 17-1 to 17-4).

Mucosa. ■ Many tubular organs in the body

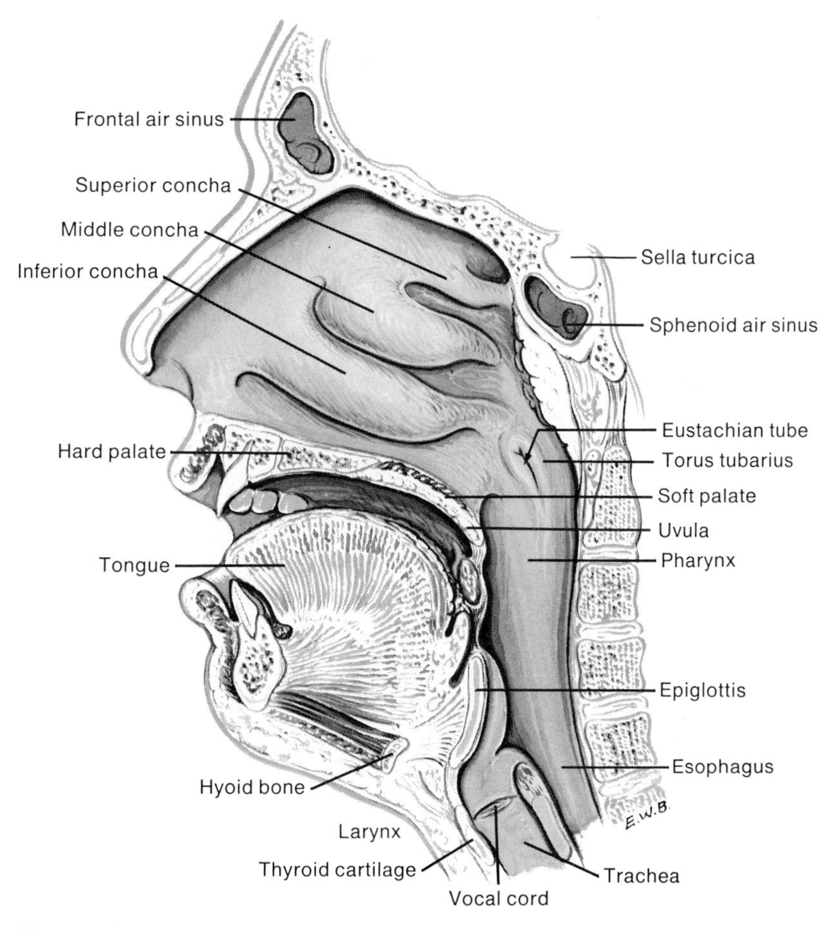

Fig. 17-1
Sagittal section showing upper portion of respiratory tract in man.

have their inner surfaces lined with a membrane known as the mucosa or mucous membrane. The structure of this membrane differs in the various organs, but the part common to all is the epithelial layer bordering on the lumen. This layer may be formed of pavement or columnar cells (Fig. 4-2). In many parts of the respiratory tract the columnar cells are ciliated (Fig. 4-1). Beneath the epithelium is a layer of areolar connective tissue that binds the epithelium to the next layer, the muscularis mucosa, in which are found bundles of smooth muscle fibers. The mucosa is well supplied with blood vessels, and capillaries lie just beneath the epithelial cells. Small glands are frequently found in the mucosa that pour a viscid fluid (mucus) onto the surface, or the columnar epithelial cells themselves secrete this fluid.

Nose. ■ The first function of the nose is to inform us of the presence of noxious gases, if these stimulate the receptors of the olfactory nerves. Certain cranial bones contain cavities or *sinuses* (Fig. 17-1). The frontal sinus is located in the frontal bone behind the eyebrows; the maxillary sinus is found in the maxilla (or upper

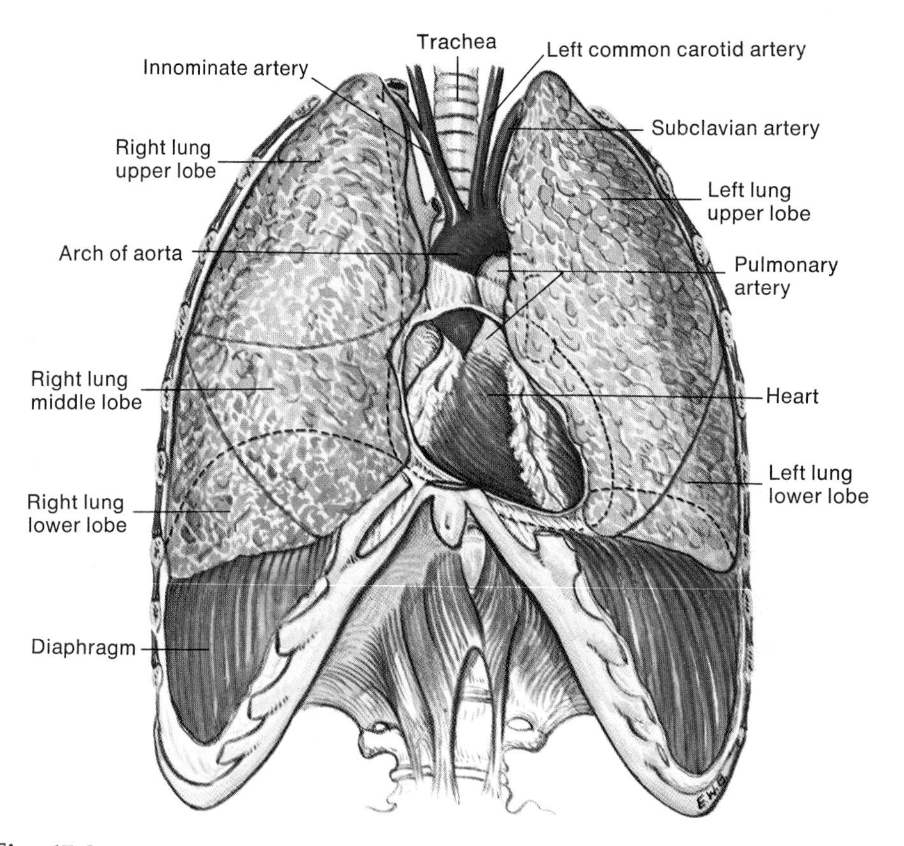

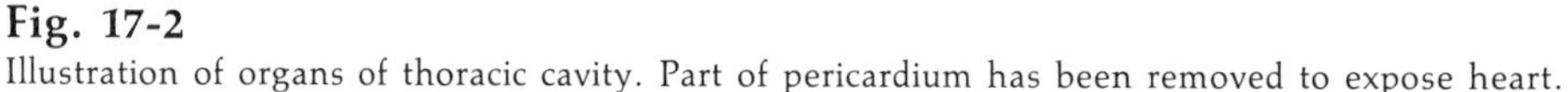

Fig. 17-2
Illustration of organs of thoracic cavity. Part of pericardium has been removed to expose heart.

jawbone). In addition to these there are the ethmoid and the sphenoid sinuses. The sinuses are filled with air, are lined with mucosa, and open into the nasal cavities. Infection of the sinuses (sinusitis) coupled with incomplete drainage (especially true for the maxillary sinuses) may give rise, by the pressure of the confined pus, to severe pain and may eventually involve the eyes and brain. The value of sinuses consists in reducing the weight of the skull and in acting as resonators for the voice. Into the nasal cavities also open the nasolacrimal ducts (Fig. 9-28), which convey the tears from the inner angle of the eyes to the nose. The second function of the nose is, therefore, to drain the secretions of the sinuses and of the lacrimal (tear) glands.

The third function of the nose is to prepare the inhaled air for the lungs; for this the nose is admirably constructed. From the bones forming the side walls of the nasal cavities extend three scroll-like bony processes known as the superior, the middle, and the inferior *conchae* or turbinates (Fig. 17-1). These divide the nasal cavity into three narrow passages known as the superior, middle, and inferior meatuses. The conchae, as all parts of the nasal cavities, are covered with a highly vascular mucous membrane, the epithelial cells of which are ciliated. Passing through

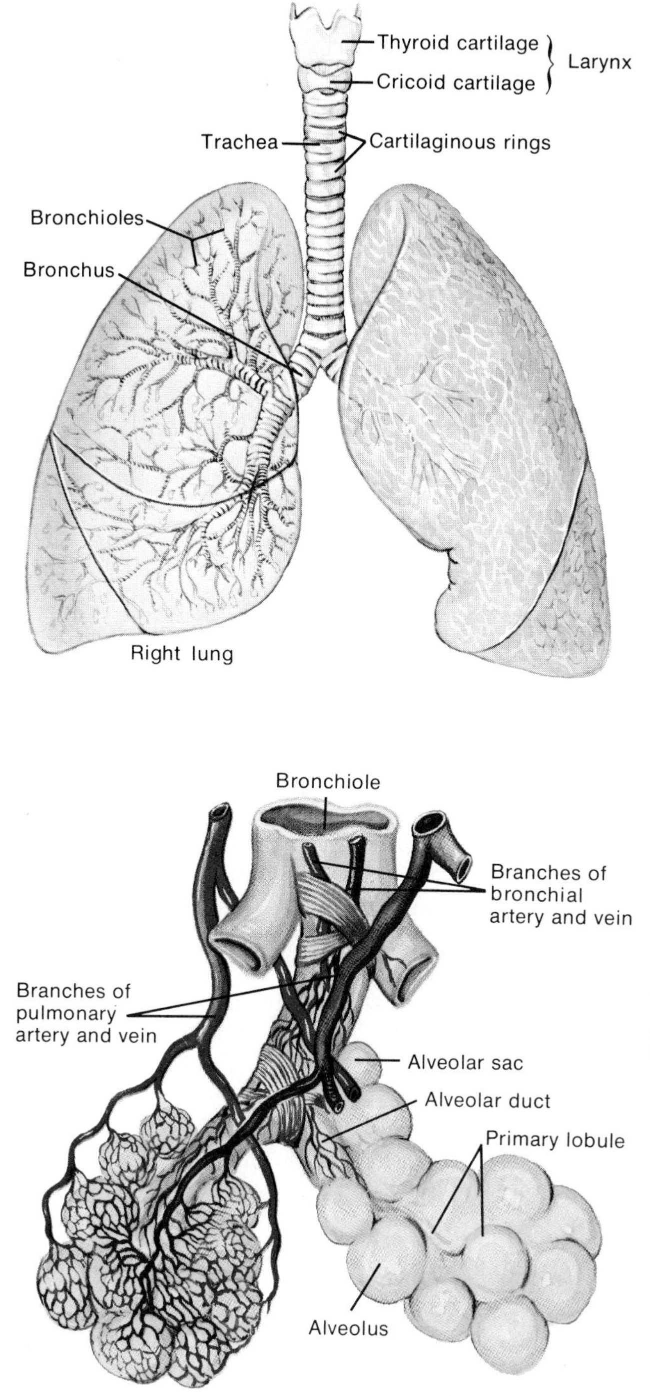

Fig. 17-3
Illustration of respiratory tract with right lung cut away to expose bronchi and bronchioles.

Fig. 17-4
Illustration showing termination of bronchiole in alveolar duct and sacs and relationship of blood supply to these structures.

the narrow meatuses in a turbulent fashion, the inspired air is prepared for the lungs by being warmed, moistened, and filtered. Dust and many bacteria found in the inspired air are precipitated in the mucus that bathes the mucous membrane and, by the action of the cilia of the nasal passage, are moved outward. Some of the dust and bacteria are taken up by the lymph tissue of the respiratory tubes and brought to the lymph nodes.

The inhalation of metal and mineral dust is especially injurious to the respiratory tract because of the sharpness and cutting action of this form of dust and is, therefore, a great predisposing factor in pulmonary diseases. The deposition of sand particles in the lungs is known as silicosis; the blackened condition of the coal miner's lungs is spoken of as anthracosis.

Mouth breathing. Increased resistance offered to the flow of the air through the nasal passages leads to mouth breathing. This resistance may be due to an excessive development of the turbinate bones, the mucous membranes may be swollen, or the bony septum that divides the nasal passage into the two nasal fossae may be deflected to one side. Marked hypertrophy of the adenoid (connective) tissue constitutes the well-known adenoids. In a child, mouth breathing may affect the jaws, the hard palate may become narrow and highly arched, and the lower jaw may recede, causing protruding upper incisors—buckteeth. It may also increase bacterial attack upon the tonsils. Even the chest may not develop properly because of inadequate expansion.

Colds. Closely related to the functions of the nose in respiration are colds. A cold in the head, rhinitis, pharyngitis, and laryngitis are caused by infections. Undue exposure to the elements never starts a cold, but it may in some manner activate the virus or render the body more susceptible to its attack. At present the only assured prophylaxis against colds is the avoidance of contact with infected people. Generally a cold is regarded as annoying, not serious; certainly most people look on it as such. But having a cold may lower the vitality of the body so as to make it more susceptible to other diseases.

Lungs. ■ Fig. 17-2 shows the position of the lungs in the thoracic cavity and their relation to the other organs. The lungs may be regarded as great ramifications of the windpipe or trachea. Posteriorly the mouth gives rise to the pharynx or throat (Figs. 17-1 and 19-1), a funnel-shaped muscular tube about 5 inches in length. The pharynx at its lower end forms two tubes: the esophagus (food tube) for conveying food to the stomach and the trachea (windpipe). The trachea (Figs. 17-2 and 17-3) divides into two large bronchi; these, in turn, split progressively into smaller and smaller branches (bronchioles) until, finally, the smallest branches (alveolar ducts) form little dilated end pockets named alveolar sacs (Fig. 17-4).

Alveoli. ■ The walls of the alveolar sacs are dilated to form alveoli. These are about 0.2 mm in diameter. As the numerous branches of the air tubes become progressively smaller, the walls become simpler and thinner in structure, until there are in the walls of the alveoli only flattened epithelial cells, a network of capillaries, and elastic connective tissue. This exceedingly thin structure is known as the *respiratory epithelium.* A thin film of fluid wets the alveolar walls. About 300 million alveoli are present in the lungs.

Surfactant. In Chapter 3 we discussed the phenomenon of surface tension and described it as the attractive force between atoms and molecules that opposes an increase in surface area. At an air-water interface, for example a bubble, where the surface tension is constant, the pressure required to increase the surface area is inversely proportional to the radius ($\triangle P = k\text{S.T.}/r$).* During inspiration the radius of an alveolus increases and during exhalation it decreases. The decrease

*This is an expression of the law of Laplace, which describes the relationship of pressure, tension in the wall, and radius of any hollow sphere or tube. It applies to tensions created by any recoil forces.

in radius on expiration is due partly to elastic fibers; however, more than one half of the recoil force is attributable to the surface tension of the film of fluid lining the alveolar wall.

Not all alveoli in the lungs are of equal size, and to overcome surface tension forces would require a greater air pressure in smaller alveoli than in larger alveoli. Since the pressure within the lungs is reasonably uniform at any instant, the smaller alveoli would be harder to inflate and the larger alveoli easier to inflate. In fact, because of this, the smaller alveoli would gradually collapse, forcing the air they contain into larger and larger alveoli (Fig. 17-5). Eventually, this would destroy the functional integrity of the lungs and produce pulmonary collapse. However, in the pulmonary secretions there occurs a substance, *surfactant,* which has the capability of reversibly altering the surface tension of the alveolar fluid lining. It serves to stabilize the alveolar surface against collapsing forces. As the pressure in an alveolus is reduced and its radius becomes smaller during expiration, the surfactant molecules become more concentrated and markedly *lower* the surface tension, which facilitates reinflation; on inspiration the molecules spread apart, become less concentrated, and exert much less effect upon the surface tension. Consequently, the surface tension forces are at least fivefold greater in inflated alveoli and facilitate expiration.

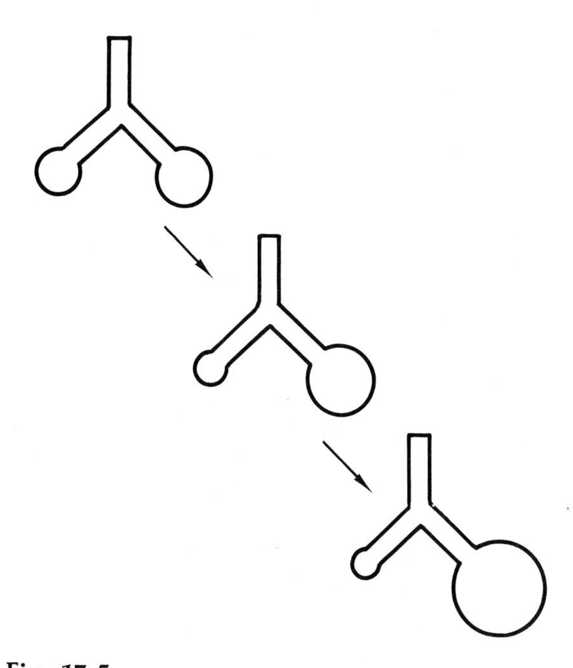

Fig. 17-5
Collapse of smaller alveoli and expansion of larger alveoli due to surface tension.

Surfactant is a term describing a number of phospholipid compounds (Chapter 23) synthesized by the lung's epithelial cells from free fatty acids. The surfactant system also includes carbohydrate and protein. About 50% of the total lipid is dipalmitoyl lecithin. Surfactant is secreted into the alveoli to become a part of the lining. Its turnover is relatively high, due to continuous removal by macrophages. The reduction in surface tension that surfactant achieves not only aids the even distribution of inspired air throughout the alveoli, it also helps to keep the alveoli dry, for a high surface tension favors the movement of liquid from the capillaries into the alveolar spaces. A number of abnormal conditions bring about a deficiency in surfactant, consequently causing uneven distribution of inspired air and decreased compliance of the lungs. Among these conditions are hypoxia, acidosis, pulmonary edema, and hyperoxia. A respiratory distress syndrome observed in some newborn infants, sometimes called hyaline membrane disease, appears to involve a surfactant deficiency.

Blood supply. ■ The blood supply for the lungs is obtained from two sources, the pulmonary and the bronchial arteries. The pulmonary artery arises from the right ventricle (Fig. 13-2) and carries venous blood to the lungs. Branches of the pulmonary artery accompany the bronchi in all of their ramifications (Fig. 17-4) and become progressively smaller. This pulmonary vasculature does not contain typical muscular walled arterioles; all of the vessels are much more distensible than those in the systemic circulation. A dense network of alveolar capillaries with rather large diameters is formed directly from the terminal arteries. It is in the capillaries that the exchange of gases between the blood and air in the alveoli takes place. The average length of each capillary is about 0.1 μ and the total capillary surface area available for gaseous exchange is about 70 m^2. Oxygenated blood is collected into veins and returned to the left atrium.

We have learned (Chapter 13) that a much lower pressure exists in the pulmonary than in the systemic circulation, although the cardiac output of the right and left ventricles is the same, that is, about 5 liters/min. Hence, the total blood volume passes through the pulmonary system each minute. However, the pulmonary vessels only contain, at any instant, some 900 ml or less of blood, of which 150 ml may be in the capillaries. Slightly less than 1 second is required for the transit of a given RBC through an alveolar capillary, and the gaseous exchange is completed in even less time.

The bronchial arteries originating from the thoracic aorta supply oxygenated blood to the lungs for the nourishment of these organs.

Pleura. ■ Each lung is surrounded by a double serous membrane known as the pleura (Fig. 17-7). That part of the membrane applied to the lungs is the *visceral pleura;* that which lines the thoracic cavity is the *parietal pleura.* The space between these two layers is known as the *pleural cavity.* Under normal conditions this cavity is only potential; the two layers are closely applied to each other, being separated only by a very thin layer of a lubricating fluid (lymph) secreted by the pleura. In pleurisy these membranes are inflamed, and, as a result, considerable lymph

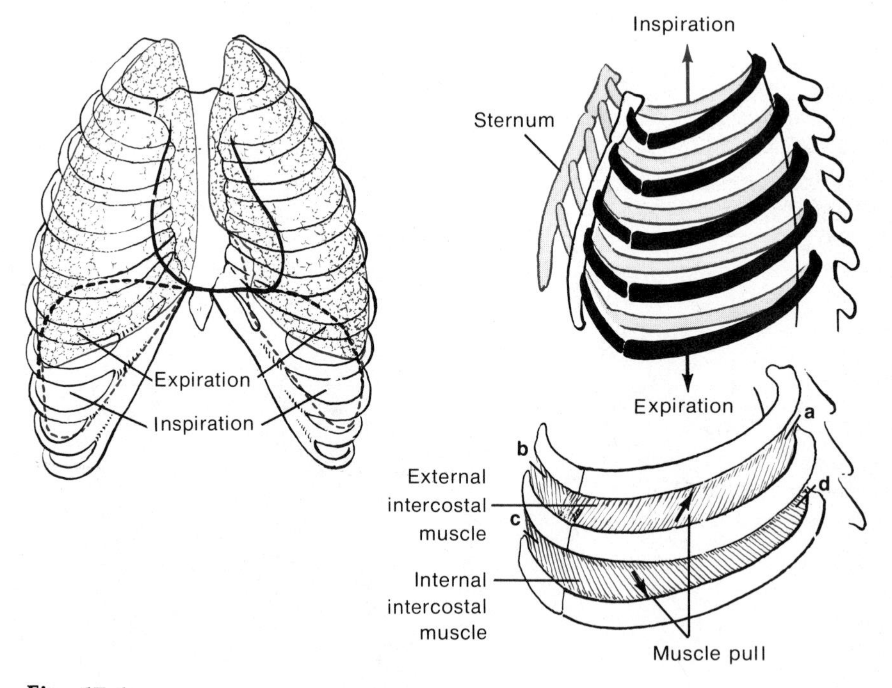

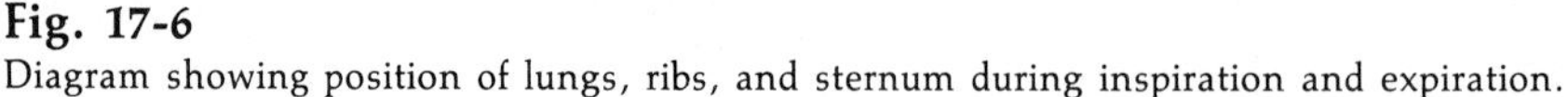
Fig. 17-6
Diagram showing position of lungs, ribs, and sternum during inspiration and expiration.

may collect in the pleural space. The resultant intense pain inhibits the voluntary effort of breathing, and the pressure of the surrounding fluid upon the lungs restricts their movements.

■ Mechanics

Inspiration. ■ The lungs, enclosed in the pleura, are suspended from the trachea in an airtight box, the thoracic cavity. The walls of this cavity (Fig. 17-6), formed by the ribs, sternum, and diaphragm, are movable. By means of the proper muscles the walls can be moved outward and thus enlarge the thoracic cavity; this, together with the flattening and lowering of the diaphragm, constitutes *inspiration.* The return of the chest wall and the diaphragm to their original positions constitutes *expiration* (Fig. 17-7).

Diaphragmatic breathing. The diaphragm (Figs. 17-2 and 17-8) is a dome-shaped muscular sheet between the thorax and the abdomen. It is innervated by the phrenic nerves, which take origin from the third to the fifth cervical nerves (Fig. 17-13). At its periphery it is attached to the chest wall. By the contraction of its radially arranged muscle fibers, the diaphragm is flattened and lowered; this thereby enlarges the longitudinal diameter of the chest and reduces the air pressure in the lungs. As the air pressure in the lungs becomes less than the atmospheric pressure, the air rushes in. The lungs, being elastic bags, distend and follow the expanding thoracic wall. In all their movements the lungs are passive. In its descent, the diaphragm increases the pressure in the abdomen, and, since at this time the tone of the abdominal muscles is relaxed, the wall of the abdomen is pushed outward. *The diaphragm is the chief muscle of inspiration.*

Costal breathing. The chest cavity also is enlarged by the raising of the ribs. Since the ribs slant downward and forward from the vertebral column, the upward rotation of the ribs not only elevates the sternum but also pushes the sternum forward (Fig. 17-6). In consequence, the ribs assume a more nearly horizontal position and thereby increase the anteroposterior diameter of the chest. The raising of the ribs also increases the lateral diameter, for the ribs that form the wall of the thorax vary in size, each rib being larger than the one just above it. The raising of the ribs is brought about by many muscles; of

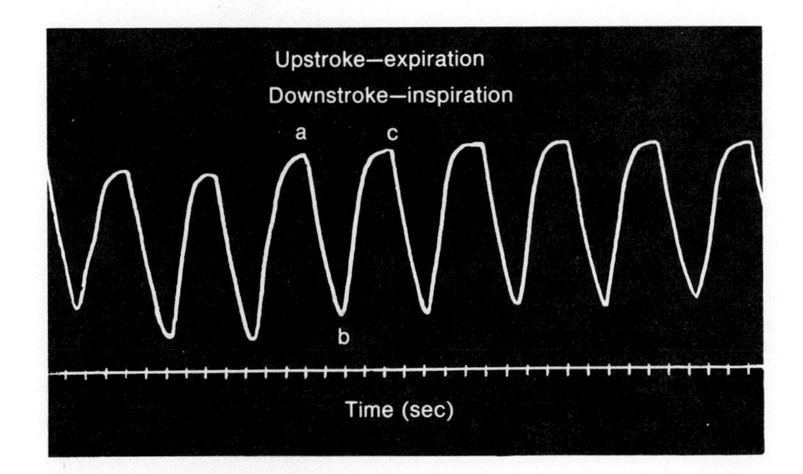

Fig. 17-7
Pneumograph tracing of respiratory movements in man. **ab,** Inspiration; **bc,** expiration.

these, the two most important will be mentioned.

The *scaleni muscles* go from the transverse process of the cervical vertebrae (Fig. 8-1) to the first and second ribs. Their most important function is to hold the first rib in a fixed position so that it will not be pulled down.

The *external intercostal muscles* are attached to the lower edge of one rib (Fig. 17-6, *a*) and to the upper edge of the next lower rib, *b;* they slant downward and forward. Due to this slanting position, the end *a* on the upper rib lies nearer the fixed vertebral column than does the end *b;* by the contraction of the external intercostals, *b* (insertion) approaches *a* (origin), thereby elevating the lower rib.

Expiration. ■ Expiration is either passive or active. During quiet respiration muscular contraction is not needed to bring the chest wall to its resting position, for the recoil of the elastic tissues (the costal cartilages, the abdominal wall, and the lungs) and the weight of these structures are sufficient to bring about expiration. During forced or labored respiration, as occurs in muscular work, expiration is a powerful muscular process. The contraction of the *external* and *internal obliquus, rectus,* and *transversus muscles* of the abdominal wall exerts pressure upon the contents of the abdomen, thereby forcing the diaphragm upward. The *internal intercostal muscles* are attached to the lower edge of one rib (Fig. 17-6, *c*) and the upper edge of the next lower rib, *d.* As the fibers slant downward and backward, their contraction draws the upper rib down to the lower rib. This also pulls the sternum down and in.

Compliance. ■ The measurable distensibility of the chest and lungs is termed the compliance of the pulmonary system. Compliance is expressed in units of volume change per unit of pressure change. Because of the moderate pressures usually involved, the latter are commonly given in cm H_2O; the volume is expressed in liters. The compliance of the pulmonary system, chest structures, and lungs is about 0.1 liter/cm H_2O. That of the lungs alone is about 0.2 liter/cm H_2O. Chest structure compliance is affected by diseases of the muscles, bones, joints, and nerves and by obesity. Lung compliance is altered by abnormalities in the bronchi and alveoli. It is more important than chest compliance because of its closer relationship to the exchange of gases. Surface tension of the alveoli and elasticity of the connective tissues of the airways, lungs, and blood vessels make up the total lung compliance.

Rhythm of respiration. ■ By means of a pneumograph and a tambour the rhythmical movements of the chest wall can be recorded. The reader will find a description of this method in a laboratory manual. From the curve obtained (Fig. 17-7) it is learned that the expiratory phase, from *b* to *c,* is a little longer (usually 1.2 times as long) than the inspiratory phase, from *a* to *b,* and that while a short pause may take place at the end of each expiration (expiratory pause at *c*), there is normally no inspiratory pause at *b.*

Respiratory sounds. The movement of the air through the air passages causes two sounds.

The *bronchial* or *tubular sound* is a blowing sound originating at the glottis and is heard in a healthy person only over the trachea and large bronchi, both during inspiration and expiration. When the lungs are congested, as in pneumonia, this sound is conducted by the solid lungs to other parts of the chest.

The *vesicular murmur* is a much softer, breezy sound and is heard over the whole lung, but only during inspiration. It is said to be due to the distention of the alveoli by air. Hence the absence of this sound indicates that the alveoli are not functioning properly, as in tuberculosis and pneumonia.

Artificial respiration. ■ In many instances, such as apparent drowning, being overcome by carbon monoxide or electric shock, or after an extensive hemorrhage, it may be necessary in order to save the life of the person to institute artificial respiration. Indeed, this should be done whenever respiration has ceased for a short length of time, and it is thought that life may not be extinct. There are several methods of manual artificial respiration.

A method frequently used is the back-pressure arm-lift (Nielsen) method. When artificial respiration is to be maintained for a considerable length of time, mechanical means (e.g., an iron lung) must be resorted to.

A method of artificial respiration commonly used with infants and recommended for use with adults is mouth-to-mouth breathing. This is accomplished by rapidly inspiring a deep breath and then breathing into the mouth of the subject (insufflation).

■ Pressures

In our discussion it may be assumed that the atmospheric pressure is 760 mm Hg. This pressure is exerted on the rigid thoracic wall and is the pressure existing in the lungs (Fig. 17-8).

Intrapulmonary pressure. ■ By alternate expansion and constriction of the chest cavity, the pressure in the lungs is decreased and increased, respectively. The magnitude of the changes in pressure depends on the rapidity and extent of the action of respiratory muscles and on the amount of resistance the air encounters in finding its way into or out of the lungs. Thus, during quiet inspiration, the intrapulmonary pressure falls only 2 or 3 mm Hg, but during labored inspiration the decrease is from 5 to 10 mm Hg, and with the air passages closed this can be re-

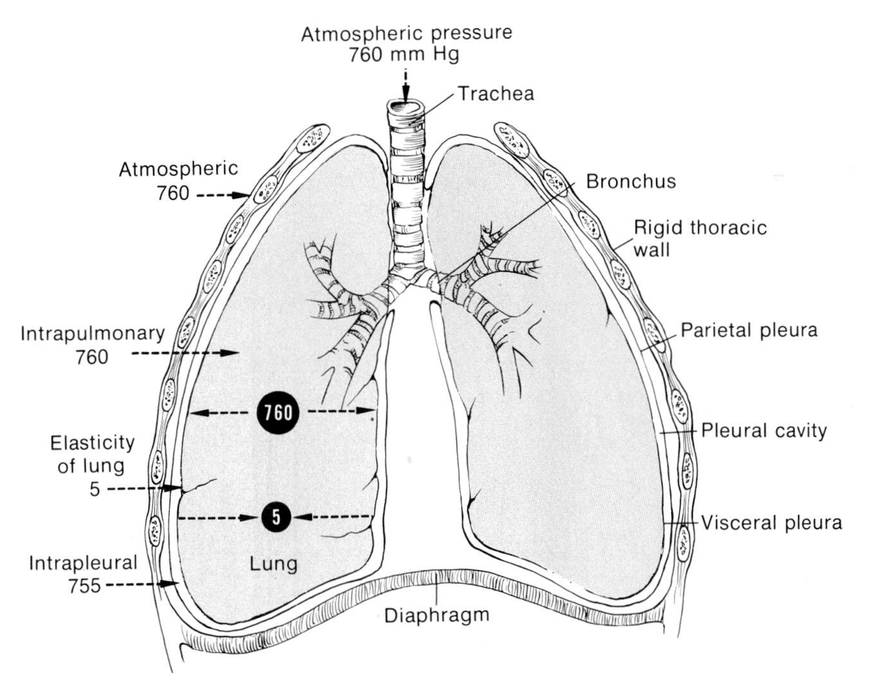

Fig. 17-8
Illustration of thoracic cavity structures showing intrapulmonary and intrapleural pressures with chest wall in resting position.

duced by from 30 to 70 mm Hg. Similarly, during quiet expiration the intrapulmonic pressure is seldom increased by more than 2 or 3 mm Hg, but with the glottis closed the expiratory muscles are able to generate a pressure from 40 to 100 mm Hg in excess of the atmospheric pressure.

Intrapleural pressure. ■ On removal from the body, the lungs, being highly elastic, collapse. The tendency to collapse also exists in the living body. However, the rigid thoracic wall does not follow the retreating lungs, and in consequence a partial vacuum is created in the pleural cavity. The intrapleural pressure is reduced by about 5 mm Hg and has, therefore, a value of 755 mm Hg (Fig. 17-8). The negative intrapleural pressure prevents any further collapse of the lungs; in other words, the collapsing power of the lungs is equal to a force of 5 mm Hg. This may be verified by inserting a cannula, attached to a manometer, into the pleural cavity.

The intrapleural (intrathoracic) pressure also undergoes changes during inspiration and expiration, quite similar to those of the intrapulmonary pressure. It varies in normal respiration from −8 mm Hg at the end of inspiration to −2 mm Hg at the end of expiration.

Pneumothorax. When the chest wall is opened, the inrush of the air causes the pressure in the pleural cavity to become equal to the atmospheric pressure and therefore equal to that in the lungs; as the restraining force has been removed, the lungs collapse due to their elasticity. This condition is called *pneumothorax.* In this condition the act of inspiration causes air to enter through both the natural channel, the respiratory tract, and the artificial opening in the chest wall. The relative amount of resistance to the entrance of air offered by these two passages will determine how well the lungs are ventilated. Pneumothorax is sometimes induced surgically by the injection of nitrogen gas into the pleural cavity to collapse a lung, as in tuberculosis and lobar pneumonia; this ensures complete rest, prevents extension of the disease, and favors healing. Because of the absorptive capacity of the pulmonary circulation, the injected gas is eventually removed and the lung assumes its usual volume.

Influence of respiration on circulation. High intrapulmonary and intrapleural pressures are experienced when a large amount of pressure is needed in the abdomen, as during straining at stool or vomiting. Preliminarily, the person takes a deep breath; by closing the glottis the air is imprisoned in the lungs. This is followed by powerful and steady contraction of the muscles in the wall of the abdomen. The increased abdominal pressure thereby created tends to force the diaphragm up into the thorax and thus to annul the effort. To prevent this, forcible contractions of the expiratory muscles compress the air in the lungs. Forced expiration against a closed glottis is referred to as *Valsalva's maneuver.*

The maneuver creates large intrapulmonary and intrapleural pressures which are applied externally to the arteries in the thorax, and the blood pressure in them is thereby greatly increased for a brief length of time. Pressures as high as 200 mm Hg have been recorded. The high intrapleural pressure exerted on the thin-walled thoracic veins and on the right atrium tends to compress these structures and raise the pressure within them. In consequence, the flow of the venous blood from outside the thorax (low pressure) toward the heart is retarded. As a result the cardiac intake and output are lessened (by as much as 30%), and the arterial blood pressure falls sharply. The stagnation of the blood in the veins causes cyanosis of the face and the reduction in cerebral blood flow may produce unconsciousness in less than 1 minute. The subsequent opening of the glottis suddenly releases the air in the lungs, and the pent up venous blood rushes in larger volume to the heart. The augmented cardiac output raises the arterial blood pressure to a high level for a short length of time. Conditions somewhat similar to this occur in any muscular effort associated with a closed glottis and a fixed thorax, as in childbirth, rowing, or lifting heavy weights.

During inspiratory effort with the air passages closed, the previously described conditions and results are reversed. Because of the decreased intrapleural pressure, the venous blood flows more freely to the heart, and the cardiac output and arterial blood pressure are increased.

Ordinary respiration also affects the blood pressure and blood flow.

Volumes ■

Respirometry. The volume of air that can be taken into or forced out of the lungs can be measured by a respirometer, of which there are many types. Fig. 17-9 shows a typical respirometer. By means of the mouthpiece, *e,* the subject either inhales from or exhales into the tank, *c,* which is suspended over an outer tank filled with water. The amount of air inhaled or exhaled can be calculated from the distance the indicator, *i,* moves when the amount of air displaced per unit of distance is known (calibration factor).

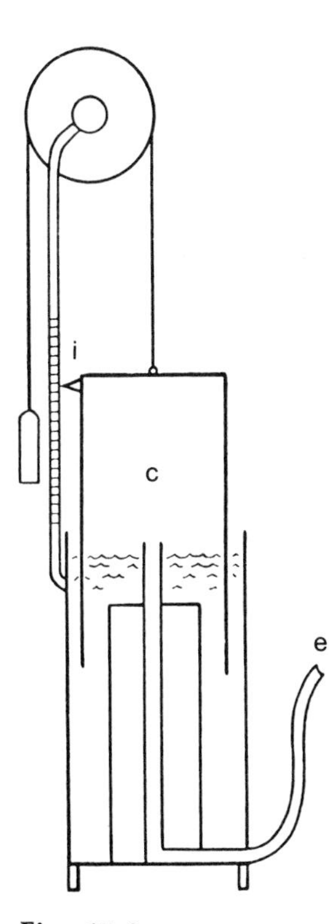

Fig. 17-9
Respirometer.

Vital capacity. Vital capacity is the volume of air expired by the most forceful expiration after maximal inspiration. It represents the total movable air in the lungs; in men this volume is about 4.8 liters. If the vital capacity is stated in terms of body surface area (Fig. 25-4), the average for normal men is 2.64 liters/m^2 of surface area; for women, 2.09.

Residual volume. After the most forcible expiration there remains a quantity of residual air estimated at about 1.2 liters. The residual air is important as it makes possible a continuous exchange of gases between the blood and the air in the lungs during the entire respiratory cycle.*

Total lung capacity. The sum of the vital capacity and the residual volume is, of course, the total lung capacity, about 6 liters. As does the vital capacity, this varies from one person to another.

Tidal volume. How much of the movable air, vital capacity, is actually moved depends chiefly on muscular activity. In the resting state it averages for an adult about 500 ml. This is represented in Fig. 17-10 by the curves in part *A*. During greater activity, *B,* the tidal volume theoretically may be increased by any amount (both by increased inspiration and expiration) up to the limit of the vital capacity.

Inspiratory and expiratory reserve volumes. The volume of air, about 3.3 liters, which can be inhaled by more forceful inspiration after the resting tidal volume has been taken in is called the inspiratory reserve volume. That which can be exhaled, about 1 liter, by more forceful expiration after the tidal volume has been expelled is the expiratory reserve volume. Both inspiratory

*After removal from the body, the collapsed lungs still contain a small amount of air (minimal air).

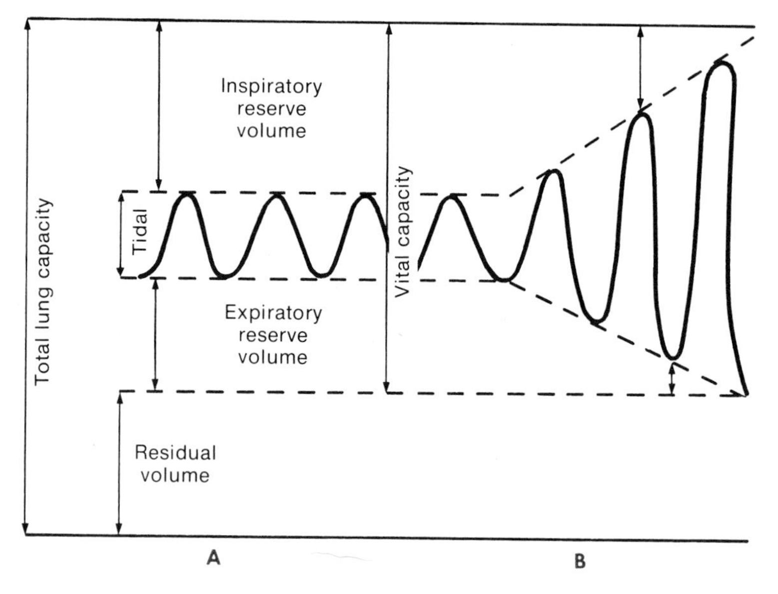

Fig. 17-10
Graph showing relation of various volumes of air inhaled and exhaled during period of rest, **A**, and greater activity, **B.**

and expiratory reserve volume decrease as the tidal volume increases (Fig. 17-10).

Dead space. ■ Of the 500 ml of air taken in by quiet inspiration, about 150 ml remain in the nose, trachea, bronchi, and other parts of the airway where no exchange of gases takes place; this volume is referred to as the *anatomical dead space.* In women, this dead space is smaller than in men. When the nonexchanging space actually is measured, it sometimes is greater than the anatomical dead space as defined above. It is then spoken of as the *physiological* (or *alveolar) dead space.* The difference between the two spaces is a measure of the volume of the alveoli that are not functioning optimally either because of underperfusion by pulmonary capillaries or hyperventilation. There is an important relationship between the physiological dead space and the ventilation-perfusion ratio to be discussed later. Marked increases in dead space occur as a result of hemorrhage, obstruction of the pulmonary circulation, or severe decline in blood pressure. Adequate ventilation cannot be achieved when the dead space exceeds 60%.

Importance of vital capacity. ■ The importance of the vital capacity is evident especially in muscular work and in certain pathological states. Its volume depends on the physical build (especially of the chest) of the person, on the development and tone of the respiratory muscles, and on the resiliency of the lung tissue.

From birth to maturity vital capacity gradually increases, but because of the stiffening of the costal cartilages, it may decrease materially in old age and thereby lessen the ability to carry on vigorous muscular work. That body position is a factor to be considered is shown by the fact that an individual with a vital capacity of 4.8 liters in a standing position experiences a reduction of as much as 680 ml when in a reclining position. In young people physical exercise in which there is a demand for increased pulmonary ventilation tends to increase vital capacity; sedentary occupations conducive to shallow breathing have the opposite effect. Due to the filling of a portion of the lung (or lungs) with fluid, the vital capacity is decreased in pneumonia; because of the destruction of lung tissue, this is also true in pulmonary tuberculosis.

Timed vital capacity. The adequacy of a person's pulmonary function may be tested in many ways. One simple test, the timed vital capacity, measures the volume of air an individual can exhale in a given amount of time. Usually, by maximal effort, an individual can exhale 83% of his vital capacity in 1 second. A much reduced volume is exhaled whenever airway obstructions are present.

■ Rate and volume of breathing

Rate of respiration. ■ The usual rate of respiration in an adult is from 13 to 18/min, but this number is subject to great variation (5 to 22). It is increased by muscular exertion, higher body temperature, and emotional states. In women it is from 2 to 4/min more than in men. The rate changes with age: at birth it varies from 40 to 70/min; at 5 years of age, about 25/min; at 15 years of age, 20/min; at 30 years of age, 16/min.

Respiratory minute volume. The amount of air passing into and out of the lungs in a unit of time, one minute, is determined by the rate and depth of breathing. This amount is the *respiratory minute volume.* In young men it is about 6 liters/min at rest. However, because of the dead space, the *effective* minute volume is reduced; this *alveolar ventilation* is approximately 4.2 liters/min at rest. The following terms are frequently employed in describing the rate and depth of respiration: *eupnea* is quiet breathing; *hyperpnea* is increased breathing (as in muscular exercise); *dyspnea* is very labored breathing associated with great discomfort and distress; *apnea* is the temporary cessation of breathing.

■ Alveolar ventilation

Ventilation-perfusion ratio. ■ Anoxemia, a low oxygen content of the blood, is commonly the result of a poor match between alveolar ventilation and blood flow (perfusion) to the lungs. The normal alveolar ventilation is about 4.2 liters/min and the normal cardiac output is about 5 liters/min. Thus, the ratio of ventilation to perfusion (4.2/5.0) is approximately 0.8. This ratio is the ideal value to assure the most efficient oxygenation of the blood. Although the ideal ratio assumes equal aeration and perfusion in all alveoli of the lungs, this is never the case. Gravitational effects favor hyperventilation of the upper regions and hyperperfusion of the lower regions of the lungs in the erect position. As a result of bronchopulmonary disease certain regions may be either underperfused or hyperventilated. If these deviations are not too severe, the ideal ratio may be maintained by shifts in the ventilation or perfusion of other regions. However, when larger regions are affected and the physiological dead space is increased, no adequate compensation may be possible and the blood leaving the lungs is inadequately oxygenated and, in some instances, CO_2 retention exists.

Work of breathing. ■ Similar to the work of the heart, pulmonary ventilation involves pressure-volume work. Intrapulmonary pressure and volume changes are produced by the respiratory muscles during inspiration and, under conditions of increased effort, also during expiration. Two forces must be overcome by the work of breathing: (1) the elastic properties of the lungs, airways, chest wall, and diaphragm and (2) the resistance to air flow in the airways. The elastic recoil forces include the true elasticity of the tissues and, as we have discussed previously, the surface tension forces of the alveoli. Overcoming these recoil forces requires about 65% of the work output in eupneic breathing. The remainder of the effort, 35%, is required to overcome the resistance to airflow. Elastic recoil forces increase with depth of respiration, whereas resistance to air flow increases with the rate of respiration.

In quiet breathing the total work is about 5 ×

10^4 g-cm/min. About 2.5 ml/min of oxygen (less than 1% of the total body rate of consumption) are consumed by the respiratory muscles. After conversion of both the work and oxygen consumption into comparable units of energy (calories), the efficiency of breathing can be calculated. It is about 10%. As the rate and depth of breathing are increased, the amount of work required increases disproportionately to the minute volume of air moved. Maximal breathing capacity is about 200 liters/min; this is an increase of slightly more than thirtyfold. On the other hand, the work output increases to 2.5×10^7 g-cm/min and the oxygen consumption for respiration to 1,250 ml/min; each of these is five hundred times the resting value. The consumption of oxygen for respiratory work now exceeds 10% of the total oxygen delivered to the blood.

CHEMISTRY OF RESPIRATION

Respired air

Respiration is the exchange of gases between the organism and its environment; in this process the blood acts as the middleman between the tissues and the outer world. We may, therefore, anticipate changes in the air as it passes into and out of the lungs and changes in the blood as it circulates through the lungs and through the tissues.

Changes in respired air. ■ The changes the air undergoes in its passage through the lungs are studied by comparing the inspired with the expired air. The volume and tension (pressure) of the gases in dry inspired and expired air and in alveolar air, under standard conditions of temperature and pressure,* are approximately as given in Table 17-1. The amount is stated in volumes percent and the tension in millimeters of mercury.

At room temperature the expired air is more than 90% saturated with water vapor; the amount of vapor in inspired air is variable but generally less than that found in expired air. Through respiration about 400 ml of water is lost per day; the evaporation of this water entails a loss of heat. Expired air is also ordinarily warmer than inspired air.

*Temperature zero C and 760 mm Hg pressure.

Table 17-1. Volume and tension of gases in dry respired air

	Oxygen		*Carbon dioxide*		*Nitrogen* and rare gas*
	Volume (ml/100 ml)	*Tension (mm Hg)*	*Volume (ml/100 ml)*	*Tension (mm Hg)*	*volume (ml/100 ml)*
Inspired air	21	160	0.04	0.3	79
Expired air	16	120	4.0	32	
Alveolar air	14	100	5.5	40	

*The *nitrogen* in the blood is held in physical solution only and plays no part in the metabolism of the body.

Room ventilation. Crowding a number of people into an unventilated room for any length of time soon causes discomfort, and ventilation is required. The necessity for ventilation is sometimes attributed to the lack of oxygen and the accumulation of carbon dioxide. How unlikely this is the following will show.

At sea level (and zero C) the atmospheric pressure is 760 mm Hg; the content of oxygen is 21% of this or 160 mm partial pressure.

$$760 \text{ mm Hg} \times .21 = 159.6 \text{ mm Hg}$$

On a mountain of 4,000 m elevation, the atmospheric pressure is 450 mm Hg and the partial pressure of the oxygen, 95 mm Hg. This oxygen concentration is equivalent to that of an atmosphere, at sea level, containing 12% to 13% oxygen. Healthy people adapt to living at this altitude.

The average resting adult consumes about 250 ml of oxygen and produces about 200 ml of CO_2

per minute. The daily consumption of oxygen therefore is 360 liters, or approximately 12 ft^3. This amount of oxygen is found in 60 ft^3 of air. Consequently, the oxygen in a room measuring 10 by 10 by 9 ft is sufficient for over five days before its content falls to 12%.

Experiments have shown that the first demand for fresh air has nothing to do with the lungs. Discomfort is caused primarily by those qualities of the air (moisture and temperature) that stimulate the cutaneous nerves and play an important part in the maintenance of the body temperature.

■ Transport of gases

Gases in solution. ■ To understand the uptake and release of O_2 and CO_2 by the blood, it is necessary to discuss briefly the laws governing the absorption of gases by liquids.

Henry's law. Because of their continual motion, the molecules of a gas exposed to the surface of a liquid penetrate and are dissolved in the liquid. When first exposed to the liquid a large number of molecules rapidly enter, but almost immediately a few of the dissolved molecules escape. Gradually, the number entering in a unit of time decreases and the number leaving increases, until these two values are equal. The free gas above the liquid and the dissolved gas are now in equilibrium; the fluid is saturated with the gas under the conditions of the experiment. The *tension* (pressure) of the dissolved gas (i.e., the force with which the molecules tend to leave the liquid) is now equal to that of the gas above the liquid, which drives the free gas molecules into the liquid. Reducing the later pressure upsets the equilibrium, and gas molecules leave the liquid until equilibrium has been reestablished at the new tension.

At a constant temperature, the quantity of a gas dissolved is proportional to the partial pressure of the gas above the liquid (Henry's law). The quantity of dissolved gas also depends on the solubility coefficient (proportionality constant) for that particular gas:

Table 17-2. Volume and tension of gases in the blood

	Oxygen		*Carbon dioxide*	
	ml/100 ml of blood	*Tension (mm Hg)*	*ml/100 ml of blood*	*Tension (mm Hg)*
Arterial blood	20	100	50	40
Venous blood	15	40	54	46
Tissues		30		50

$$\text{Vol. gas} = \text{Vol. liquid} \times \text{Pressure} \times \text{Sol. Coeff.}$$

Increasing the temperature decreases the quantity of gas a liquid can hold. The quantity of a gas dissolved at a certain pressure and temperature depends on the nature both of the gas and of the liquid. For example, 100 ml of distilled water, under standard conditions, takes up 4.9 ml of oxygen; of CO_2 it can hold 171 ml. At 40 C the amount of O_2 drops to 2.3 ml and that of the CO_2 to 53 ml. Since the atmosphere contains 21% oxygen and since the quantity of a gas absorbed is determined by the partial pressure of the gas in question, 100 ml of water at 40 C exposed to air absorbs approximately one fifth of 2.3, or 0.46, ml of oxygen.

Dalton's law of partial pressure. In a mixture of gases, each gas exerts a pressure proportional to its percentage in the total mixture. We speak of the *partial pressure* (P), or tension, of the gas; it is expressed quantitatively in mm Hg. The total pressure of the mixture is, therefore, the sum of the pressures of the individual gases.

Oxygen transport. ■ The approximate volume of gases in 100 ml of blood and their tensions are given in Table 17-2.

The amount of O_2 and CO_2 in the tissues and venous blood varies with the degree of muscular activity; during severe work the blood returning from the muscles may contain as little as 5 ml of O_2 and as much as 65 ml of CO_2 per 100 ml

Table 17-3. Volume and partial pressure of gases in the alveolar air

Alveolar air	*Vol %*	*Partial pressure*
Oxygen	13.6	13.6/100 × 760 ≃ 104 mm Hg
Carbon dioxide	5.3	5.3/100 × 760 ≃ 40 mm Hg
Nitrogen	74.9	74.9/100 × 760 ≃ 569 mm Hg
Water	6.2	6.2/100 × 760 ≃ 47 mm Hg

of blood. The values given in Table 17-2 must be regarded as averages of mixed venous blood.

If the amount of oxygen found in arterial blood (20 vol %) is compared with that held by water exposed to air at body temperature (0.46 vol %), the large amount of the former is impressive. This becomes still more striking when it is borne in mind that blood is exposed not to atmospheric air but to the air found in the alveoli, which, as shown in Table 17-3, contains about 14% oxygen. Exposed to alveolar air, at 40 C, 100 ml of water would absorb about 0.33 ml of oxygen, whereas blood under the same conditions takes on 20 ml of oxygen, about sixty times as much. It is evident that this large amount of oxygen is held in the blood in some other way than in physical solution; it is accomplished by the hemoglobin of the red blood cell.

Oxygen binding capacity. The value of hemoglobin as a carrier and distributor of oxygen lies in the fact that the chemical union of hemoglobin and oxygen depends on the tension of the oxygen dissolved in the plasma surrounding the cell. As the venous blood from the tissues flows through the lung capillaries, the alveolar oxygen enters the blood, and the tension of this gas in the plasma is increased. This enables the hemoglobin to take up oxygen (oxygenation) in loose chemical combination, forming *oxyhemoglobin.* The greater the oxygen tension in the plasma, the more nearly the hemoglobin becomes saturated with oxygen. The capacity of hemoglobin to bind oxygen is great, 1 g being able to combine (at the proper tension) with 1.34 ml of oxygen. Since 100 ml of blood contains 15 g of hemoglobin, this furnishes about 20 vol % of oxygen. In the systemic capillaries this process is reversed.

Oxyhemoglobin dissociation curve. The oxyhemoglobin dissociation curve of whole blood is illustrated in Fig. 17-11. This S-shaped curve is a plot of the oxygen content (the percent saturation is often used instead) on the ordinate and the partial pressure of oxygen (P_{O_2}) on the abscissa. For comparison, the curve depicting the amount of oxygen carried in physical solution in the plasma has been included (the actual volumes in physical solution are exaggerated about threefold in order to make them more evident). The S-shape of the oxyhemoglobin curve has a number of significant consequences:

1. The flatness of the curve at oxygen tensions above 70 mm Hg assures nearly full saturation (20 vol % equals 100% saturation) under circumstances where the oxygen tension in the environment may be reduced, for example, at high altitudes.

2. The steep part of the curve between 10 and 40 mm Hg P_{O_2} allows for the release (dissociation) of rather large amounts of oxygen for small changes in P_{O_2}. This characteristic affords protection for actively metabolizing tissues of the body where the P_{O_2} is in the range of 10 to 40 mm Hg.

3. The curve is shifted by changes in pH, P_{CO_2}, and temperature. Whereas the shift is small at high values, in the range of 50 to 20 mm Hg P_{O_2}, the shift is large. Increased temperature and P_{CO_2} and decreased pH shift the curve to the right, which produces an even greater release of oxygen per unit change in oxygen tension than normal. For example, at P_{O_2} values between 50 and 30 mm Hg blood with a low pH would discharge 7 vol % of O_2, but blood at normal pH would release only 5 vol %. The oxygen tension at which Hb is 50% saturated with O_2 is customarily referred to as the P_{50} value. A low P_{50} means

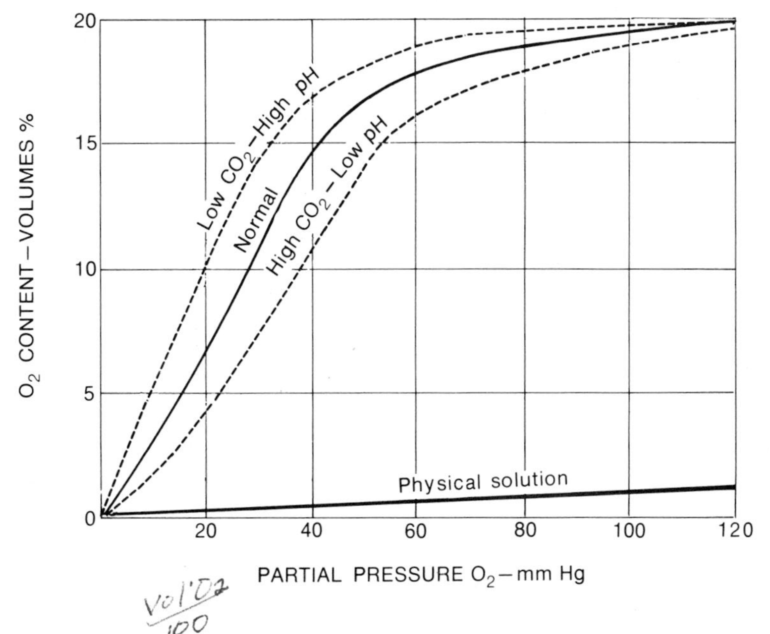

Fig. 17-11
Graph showing O_2 content (vol %) of arterial blood and plasma (physical solution) at various partial pressures of O_2. Note changes in shape of curve for blood with varying CO_2 concentration and pH as well as small amount of O_2 carried in physical solution. Latter is exaggerated for clarity.

a high affinity of Hb for O_2 and vice versa. This response is known as the *Bohr* effect. It is of importance to active tissues of the body where such changes in pH, P_{CO_2}, and temperature are ongoing.

4. By recognizing that a content of 20 vol % represents fully saturated oxyhemoglobin, it becomes evident that under normal circumstances voluntary deep breathing or breathing pure oxygen cannot markedly increase the oxygen-carrying capacity of the blood. Breathing pure oxygen only brings that quantity carried in physical solution to about 2 ml.

Hemoglobin and 2,3 DPG. The chemistry of hemoglobin was discussed in Chapter 12. There it was noted that the red cells contain high levels of 2,3 diphosphoglycerate (2,3 DPG); other cells of the body contain only trace amounts. This compound is produced by the red cell through a *shunt* pathway (Chapter 23) of glycolysis:

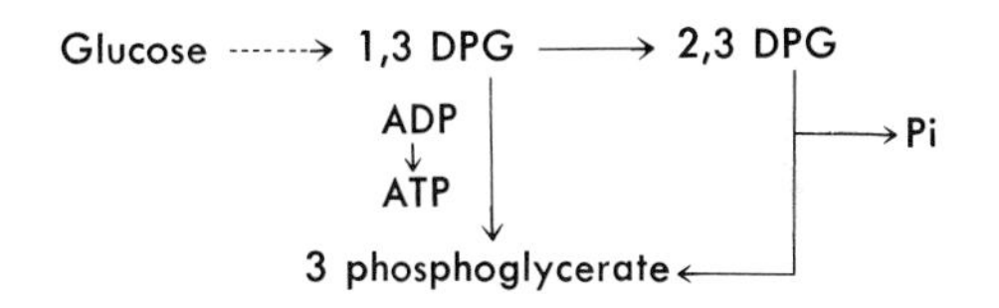

The shunt mechanism is energetically useless to the cell, since there is a decrease in ATP formation. Reduction in ATP levels, however, allows glycolysis to continue because the ATP/ADP ratio is an important determinant of the metabolic rate. Not all of the 1,3 DPG produced during glycolysis is converted to 2,3 DPG, only a very large proportion.

At physiological pH, 2,3 DPG is a highly charged anion and thus capable of binding by electrostatic interaction with positive charges on a protein. It appears to bind reversibly to the β subunits of the Hb molecule:

$$HbDPG + O_2 \rightleftharpoons HbO_2 + DPG$$

The relative affinity of Hb for O_2 is, according to the above equilibrium, determined by the amount of free DPG. In the presence of increased concentrations of free DPG, the oxygen dissociation curve is shifted to the right, i.e., the P_{50} is increased. Such a decreased affinity for O_2, as we have already noted, is important because it facilitates the release of O_2 in the tissue capillaries at relatively high oxygen tensions. Thus, the presence of DPG increases O_2 availability.

A number of observations reveal the physiological role of 2,3 DPG in oxygen transport. Decreased O_2 affinity of the red cells occurs in persons afflicted with hypoxemic diseases, for example, cardiopulmonary disorders. In most instances these individuals exhibit increased amounts of DPG in the cells. Young red cells have a lower O_2 affinity than old cells; they also have greater DPG concentrations. Fetal Hb transports O_2 well but binds DPG less effectively than adult Hb. Consequently, the newborn infant is less able to cope with hypoxia, because it has a lesser ability to shift the dissociation curve to the right and thereby improve the delivery of O_2 to the tissues.

Individuals living at high altitudes have greater amounts of DPG in their red cells than persons living at sea level. When individuals ascend from sea level to high altitude, they improve their oxygenation by hyperventilating, but this response also reduces the CO_2 levels in the blood (alkalosis) and increases the O_2 affinity of the Hb. Any such increase in pH stimulates glycolysis in the red cells and a compensatory increase of DPG content, of about 10%, occurs within 24 hours. A decrease in pH enhances O_2 release by Hb, the Bohr effect that was mentioned previously. Acidosis, however, has a depressant effect on glycolysis and as a consequence there is a fall in DPG levels with time that counterbalances the pH effect. Thus, the enhanced O_2 release evoked by acidosis persists only for a few hours. In contrast, the decreased O_2 affinity that results from elevated DPG content is a sustained effect.

An inverse relationship exists between DPG content and Hb concentration in the red cells. Thus, persons with low Hb levels (anemia) have augmented DPG levels that aid in achieving adequate oxygenation for normal activity. Polycythemic individuals with excess Hb have depressed DPG levels. It appears from the foregoing observations that the metabolism of the red cell is a precisely adjusted mechanism for providing appropriate oxygen delivery.

Oxygen therapy. In anemia that is due to a deficiency of oxygen carrier, oxygen therapy can be of little benefit, for any increase in oxygen transport could come about only from the increase of O_2 carried in physical solution. Under certain conditions, even though the blood is normal, breathing of ordinary air does not bring a sufficient amount of oxygen into the blood *(hypoxemia*)*; that is, when the ventilation-perfusion ratio is disturbed. In these circumstances breathing of air with a higher oxygen content may be of great value; indeed, it may be necessary for maintaining life. This obtains when the available alveolar surface for the absorption of oxygen is greatly reduced (as in pneumonia and pulmonary edema); the higher the concentration of the oxygen in contact with the still-functioning surface, the more oxygen will enter the blood. For this purpose the air in an oxygen tent contains from 40% to 60% oxygen. The body is built on a large margin of safety; one lung may be rendered functionless without the person experiencing anoxemia, except when extra demand for oxygen obtains. Breathing pure oxygen for a considerable length of time produces lesions of the lungs (inflammation, edema, or pneumonia) that diminish lung capacity, lung compliance, and surfactant content. This proves fatal in a few days, due to an impaired oxygen absorption.

---

*The terms *hypoxemia* and *anoxia* are frequently used interchangeably. Anoxia is more commonly used to denote a general want of O_2 throughout the body. Naturally, prolonged hypoxemia leads to anoxia.

Carbon dioxide transport. ■ We noted that at 40 C 100 ml of water exposed to carbon dioxide gas at 760 mm Hg pressure absorbs 53 ml of the gas. The tension of the CO_2 in mixed venous blood is about 46 mm Hg (Table 17-2). Exposed at this pressure, 100 ml of water absorbs 46/760 of 53 or 3 ml. The carbon dioxide content of venous blood averages 54 vol %. If all the CO_2 were in solution, the pH of the blood would be about 5.4. Evidently, by far the larger part of this gas must be held in some other form. To understand this, we must consider the characteristics of a solution of CO_2 and the behavior of the red blood cells.

1. Carbon dioxide passed through water unites with the water to form carbonic acid (H_2CO_3). When a solution of carbonic acid under pressure is exposed to the air, the acid breaks up into CO_2 and H_2O. This is seen in the uncorking of any carbonated beverage. We may express this reversible reaction as:

$$CO_2 + H_2O \rightleftharpoons H_2CO_3$$

2. Being an acid, the H_2CO_3 can form salts; of these, sodium carbonate (Na_2CO_3) and bicarbonate ($NaHCO_3$) are familiar examples.

3. When a solution of sodium or potassium bicarbonate is subjected to a vacuum, the following reaction takes place:

$$2\ NaHCO_3 \rightarrow Na_2CO_3 + H_2O + CO_2$$

One half of the CO_2 is liberated and the other half is more firmly held as sodium carbonate, Na_2CO_3. To drive off all the CO_2 an acid must be added:

$$Na_2CO_3 + 2\ HCl \rightarrow 2\ NaCl + H_2CO_3$$
$$H_2CO_3 \rightarrow H_2O + CO_2$$

4. By subjecting plasma to a vacuum, only a small part of the CO_2 is set free, but the addition of an acid removes all the CO_2; therefore, most of the gas in the plasma must be held as bicarbonate.

5. If instead of plasma, whole blood is subjected to a vacuum, all the CO_2 is liberated without the addition of an acid; from this it must be concluded that the red blood cells behave like an acid.

6. The formation of H_2CO_3 and the spontaneous breakdown of H_2CO_3 into H_2O and CO_2 take place very slowly. These processes are greatly accelerated by an enzyme, *carbonic anhydrase*, found in the red blood cells, but not in the plasma.

7. The amino acid molecules of which a protein molecule is composed act as weak bases (Chapter 18) and are, therefore, able to unite with CO_2; the compound formed is known as a *carbamino* compound. Since hemoglobin is the most abundant protein in the blood, most of the carbamino compounds are carbaminohemoglobin or, more simply, *carbhemoglobin.* The union of CO_2 takes place with the globin part, not with the heme, of the hemoglobin.

8. Hemoglobin has acid properties, although it is even a weaker acid than carbonic acid. Acid hemoglobin may be symbolized as HHb and acid oxyhemoglobin as $HHbO_2$. The hydrogen of HHb and of $HHbO_2$ can be replaced by potassium and thereby yield the potassium salt of hemoglobin, KHb or $KHbO_2$.

The transport of CO_2 in the blood occurs in three forms: about 8% goes in physical solution in the plasma, roughly 27% is carried by RBC in the form of carbamino compounds (principally carbhemoglobin), and the remainder is present in the plasma as bicarbonate.

■ Exchange of gases

Pulmonary exchange. ■ Supplying oxygen to, and removing carbon dioxide from, the cells is of such paramount importance that all possible means are sought to facilitate these processes. We have considered the highly efficient manner in which the blood transports these gases; our attention now turns to the equally efficient physical and chemical mechanisms by which the body acquires oxygen and disposes of carbon dioxide.

In the exchange of gases between the alveolar air and the venous blood in the pulmonary capillaries, O_2 and CO_2, under pressure, pass through the alveolar membranes and capillary walls separating the air and the blood. The following features greatly facilitate this process:

1. The thinness of the respiratory epithelium is essential.

2. Since the cells pass through the lung capillaries single file, every cell is close to the alveolar air.

3. All the blood in the body passes through the combined lung capillaries in one minute under resting conditions; during severe work this does not require more than ten to fifteen seconds. The large number of red blood cells and their large surface area favor the rapid uptake of oxygen. Since the lungs are never empty, because of the residual air, the exchange of gases is a continuous process.

Oxygen exchange. The force driving the gases across the respiratory membrane is the greater tension or pressure of the gas on one side than that on the other. At sea level (760 mm Hg) the partial pressure of O_2 in alveolar air (Fig. 17-12) is 100 mm Hg; the tension of the oxygen in the venous blood in the pulmonary capillaries is 40 mm Hg; the oxygen, therefore, diffuses from the air into the blood. Even though a red cell makes the passage through the capillary in one second, the hemoglobin leaving the pulmonary capillary is almost saturated (96% to 98%) with O_2. The oxygen tension (which is due to the dissolved oxygen, not to that chemically held by the hemoglobin) in the arterial blood leaving the lungs is but little, if any, lower than that of the oxygen in the alveolar air (95 to 100 mm Hg). The pressures of the blood gases vary with the degree of muscular activity.

The difference in content of O_2 between the arterial and mixed venous blood is spoken of as the *arteriovenous difference.* It is usually expressed in terms of volumes percent, as in Table 17-2. In healthy individuals the *a-v* difference at

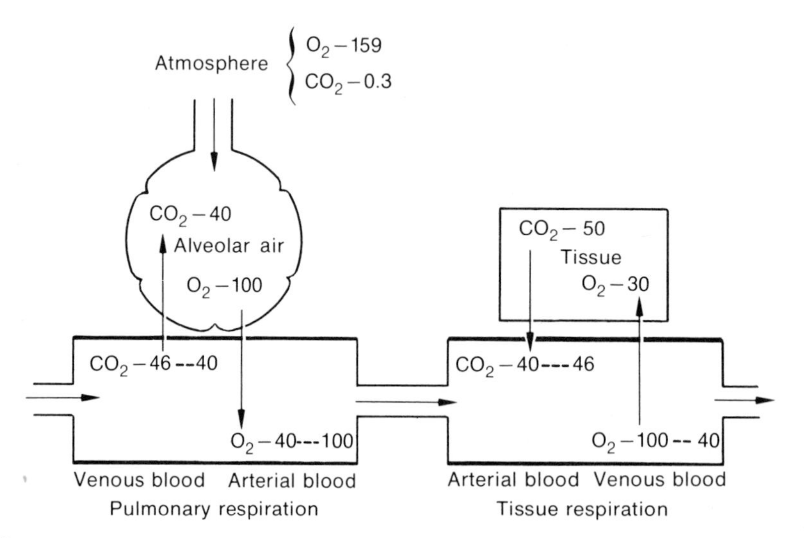

Fig. 17-12
Diagram showing exchange of gases in lungs and tissues. Numerals are partial pressures in millimeters of mercury.

rest is from 4 to 5 vol %. We saw in Chapter 12 how the *a-v* difference is used to calculate cardiac output by the Fick principle.

Carbon dioxide exchange. The venous blood arriving at the lungs contains approximately 54 vol % of CO_2; this exerts a tension of 46 mm Hg (Fig. 17-12). The partial pressure of the CO_2 in the alveolar air being 40 mm Hg, the dissolved gas diffuses from the blood into the alveoli.

The conversion of HHb to $HHbO_2$ in the lungs lessens its ability to hold CO_2; as a result carbhemoglobin liberates its CO_2, which diffuses first into the plasma and then into the alveoli. Plasma HCO_3^- diffuses into the RBC in exchange for Cl^- (the chloride shift), where carbonic anhydrase brings about its conversion to CO_2 and H_2O. The CO_2 diffuses back into the plasma and on into the alveoli.

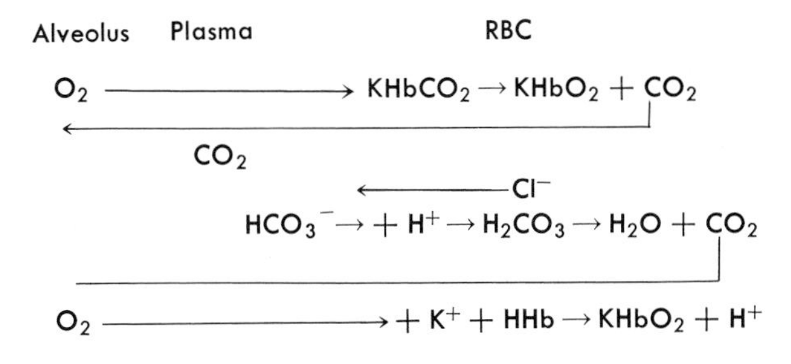

Tissue exchange. ■ In the tissues (Fig. 17-12) the oxygen tension varies with the extent and rapidity of oxygen consumption, let us assume a value of 30 mm Hg. Its greater tension in the arterial blood in the systemic capillaries (100 mm Hg) forces the dissolved oxygen from the plasma into the tissue fluid and then into the cells. This reduction in O_2 tension in the plasma renders it impossible for the oxyhemoglobin to hold all of its O_2; therefore, it is liberated and passes through the plasma and tissue fluid into the cells.

The tension of the carbon dioxide in the tissues is in proportion to the amount of work done. An average value is 50 mm Hg. This being greater than its tension in the arterial blood in the systemic capillaries, causes CO_2 to diffuse into the blood. Venous CO_2 tension averages 46 mm Hg.

The liberation of oxygen in the tissues is materially assisted by the diffusion of carbon dioxide into the blood, since increased CO_2 is associated with a decreased affinity of Hb for O_2 (Fig. 17-11). This is due to the fact that the amount of oxygen that hemoglobin can hold is decreased by the addition of acid (H ion from H_2CO_3) to the blood. The increase in temperature during muscular activity also favors the deoxygenation of oxyhemoglobin. On the other hand, the conversion of oxyhemoglobin into deoxyhemoglobin aids the taking up of carbon dioxide by the hemoglobin in carbamino formation.

Carbon dioxide is transported in three forms. In the exchange of CO_2 between tissues and blood some goes into physical solution in the plasma. A greater amount is involved in the formation of carbhemoglobin. By far the largest amount of CO_2 first enters the RBC, where it is rapidly converted to carbonic acid, which dissociates into H^+ and HCO_3^-. The HCO_3^- diffuses from the RBC, down its concentration gradient, into the plasma in exchange for Cl^- (the chloride shift). Excess H^+ is buffered by the hemoglobin molecule.

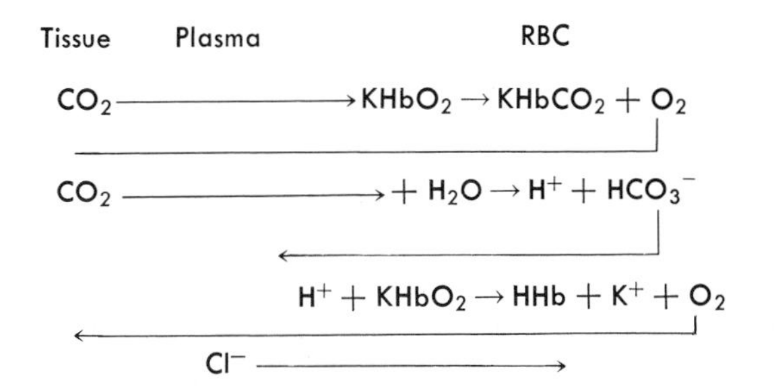

The amount of oxygen released to the tissues is governed by the oxygen tension in the tissues and by the amount of acids added to the blood; both these factors are determined by the degree of activity of the tissues. The extent of the diffusion of carbon dioxide from the tissues into the blood is determined by the tension of this gas in the tissues and by the oxygen content of the blood; these two factors also are in proportion to tissue activity. We have, therefore, *an efficient, physicochemical mechanism by which the exchange of gases is controlled by the needs of the tissues.*

■ UTILIZATION OF OXYGEN BY CELLS

In this discussion only the changes occurring in the oxygen content of arterial, capillary, and venous blood and some factors influencing the cellular use of oxygen will be considered. The changes in the energy-yielding substances resulting from oxidation and the agencies by which oxidations are brought about will be discussed with metabolism in Chapter 23.

Location of the oxidative processes. ■ The oxygen absorbed by the blood in the lungs is used to oxidize the foodstuffs, in which process carbon dioxide is formed. A small amount of oxidation takes place in the lung tissue and in the blood itself. These require only a minor fraction of the total gas consumption by the body. It is in the various other tissues that the significant consumption of oxygen and the production of carbon dioxide occur.

Oxidations and oxygen supply. ■ In a furnace fire the amount of oxidation is directly proportional to the amount of oxygen supplied; under a forced draft a more intense fire is obtained, more heat is liberated, and, other factors remaining constant, more work is performed by an attached engine. In the animal body, however, an increase in the amount of oxygen supplied (provided a minimum amount necessary to carry on the function is available) does not influence the intensity of the oxidation and neither, therefore, the activity of the organ. For example, in an atmosphere of pure oxygen the amount of the oxidations in the animal body is no greater than when ordinary air is breathed. But when a tissue becomes active, the oxygen consumption is immediately increased.

The oxygen reserve of the body amounts to less than 1,400 ml. It consists mainly of that oxygen within the lungs and bound to hemoglobin. The basal metabolic requirement of the body is about 250 ml/min. If the airways are blocked, adequate oxygen *tension* for basal cellular respiration can be maintained for only two to four minutes. If the arterial blood supply to a tissue is occluded, oxygen is available for only ten to fifteen seconds. Some organs are able to function for extended periods under anaerobic conditions, thus the skeletal muscles and the heart may develop a considerable oxygen debt; but other organs, such as the brain and liver, must operate on a "pay-as-you-go" basis.

When the oxygen saturation of arterial blood drops to 85%, a P_{O_2} of about 50 mm Hg, disturbances in the operation of the nervous system (e.g., dizziness, impairment in judgment, headache, EEG abnormalities, and incoordination of muscular movements and vision) and increased heart rate and respiration occur. A 65% saturation (P_{O_2} of 35 mm Hg) may occasion convulsions and collapse. For effect on vision, see Chapter 9.

Coefficient of oxygen utilization. ■ A tissue or organ uses a certain amount of the oxygen furnished it by the arterial blood. If the oxygen concentration in the arterial blood is 20 vol % and that in the venous blood 15 vol %, the arteriovenous oxygen difference is 5. The last factor divided by the arterial oxygen concentration gives the coefficient of utilization; in this instance, $5/20 \times 100$, or 25%. This is said to be the value of the coefficient for the whole resting body; during severe muscular work it may be 70%. The coefficient varies from one organ to another and also with the degree of activity; in an individual muscle during extreme activity it may reach almost 100%.

Table 17-4 shows some approximate oxygen extractions, arteriovenous (AV) differences, and blood flows of various organs and tissues under basal and exercise conditions. It emphasizes how blood flow is restricted in organs with low oxygen extraction in favor of organs with high extraction,

Table 17-4. Oxygen extraction and blood flow in various tissues under basal and exercise conditions

Tissue	*Basal* AVO_2 *difference (vol %)*	*Basal blood flow (ml/min)*	*Exercise blood flow (ml/min)*
Skeletal muscle	8.4	1,200	22,000
Heart	11.4	250	1,000
Brain	6.3	750	750
Splanchnic (liver)	4.1	1,400	300
Kidney	1.3	1,100	250
Skin	1.0	500	600
Other		600	100

when the oxygen supply becomes limited. Changes in cardiac work are closely paralleled by changes in blood flow to the heart. This is due to the high extraction by this organ that obtains even under basal conditions. In skeletal muscle both extraction and flow can increase with exercise. The brain subsists on a fixed flow and extraction. Although the kidney tolerates an appreciable reduction in blood flow and oxygen supply, when the supply falls below one third of normal impaired handling of sodium and urea occur.

Metabolism of the lung. ■ Oxygen utilization by lung tissue amounts to about 1 ml per liter of ventilation. This consumption supports numerous functions among which are the contraction of smooth muscle in the airways, ciliary activity, synthesis of surfactant, and tissue maintenance and repair. The substrates for this metabolic activity are glucose and free fatty acids (Chapter 23). Unlike the brain, which is almost completely dependent upon glucose, lung tissue tolerates a short-term occlusion of its blood supply, since the availability of free fatty acids is not as dependent upon the circulation as is that of glucose.

It is not surprising that the lung is more than a passive exchanger of O_2 and CO_2, for it is the one organ that comes into contact with all of the blood flow and over a broad surface area. The larger alveolar cells synthesize, store, activate, or inactivate a number of important vasoactive substances (Chapter 15). It stores histamine, bradykinin, serotonin, and prostaglandins (Chapter 28). The latter it synthesizes. Angiotensin I is activated by conversion to angiotensin II (Chapter 15) in the lungs.

Macrophages are found in the lungs in large numbers, perhaps as many as 6×10^8 in the human lung. These alveolar phagocytes are the chief defense mechanism of the body against bacteria and particles in the inspired air. Although the mucus and cilia of the respiratory passages trap particles of 2 μ or larger size, the smaller particles and bacteria are dealt with by the alveolar macrophages. They are rich in hydrolase enzymes (Chapter 22) that digest the invaders. They are also rich in immunoglobulins (Chapter 12). However, these macrophages have the highest O_2 consumption of all the different phagocytic cells and their effectiveness is lost under conditions of inadequate O_2 supply. This contrasts with other phagocytes which turn to glycolysis for their energy when anaerobic conditions prevail.

■ REGULATION OF PULMONARY VENTILATION

Innervation of respiratory muscles. ■ The expansion and constriction of the chest are brought about by skeletal muscles. Their action, like that of all skeletal muscle, is initiated by impulses brought to them from the central nervous system. The diaphragm is innervated by the phrenic nerves. The intercostal muscles are innervated by the intercostal nerves originating in the spinal cord below the cervical region. Cutting one phrenic nerve, as indicated at *a,* Fig. 17-13, paralyzes the corresponding side of the diaphragm. When the spinal cord is cut at the seventh cervical vertebra, *b,* the costal respiration ceases, whereas that of the diaphragm goes on. A section of the cord made at the first or second cervical vertebra, *c,* stops diaphragmatic as well as costal breathing. This indicates that the area of the central nervous

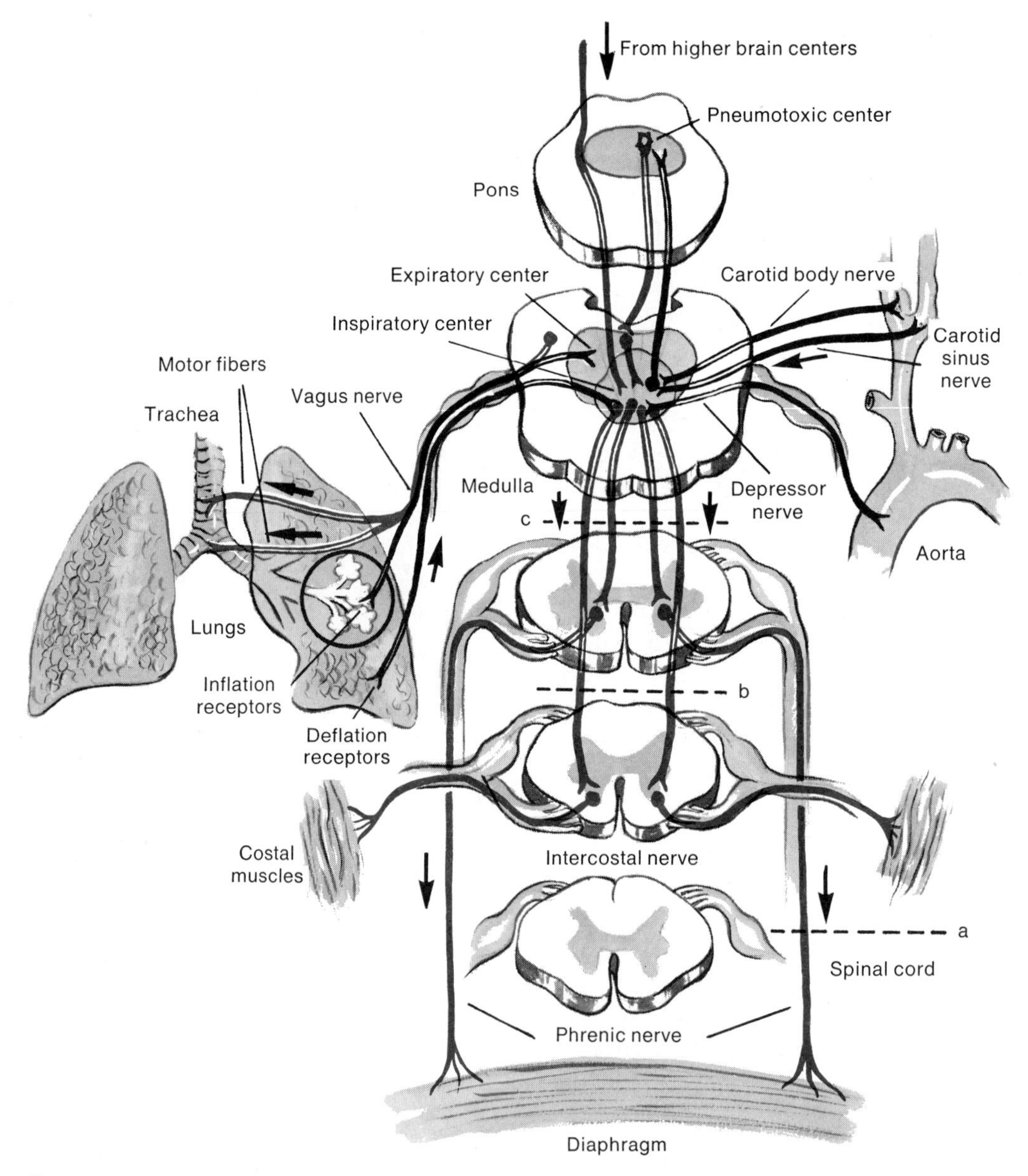

Fig. 17-13
Illustration showing neural mechanisms for the regulation of respiration. Motor fibers are in red. Inspiratory and expiratory centers are not shown bilaterally for purposes of clarity.

system that controls these muscles lies above the last-mentioned transection.

Respiratory center. ■ By making other sections, it has been found that the area from which the nerve impulses for the respiratory muscles proceed lies in the upper part of the medulla oblongata (Figs. 10-1 and 10-2). This area is called the *respiratory center.* On its activity depends all respiration, and for this reason breaking one's neck may be fatal. In the bulbar type of anterior poliomyelitis (infantile paralysis) the efferent fibers are rendered functionless and respiratory movements cease. The respiratory center is bilateral. Each lateral area is composed of an inspiratory and an expiratory center (shown consolidated in Fig. 17-13), which innervate the muscles of inspiration and expiration, respectively, chiefly on the corresponding side of the body. The respiratory centers of the medulla are responsible for the *basic rhythm* of respiration.

■ Neural regulation

It is doubtful whether any other nerve center is as continually subject to such a variety of stimuli as the respiratory center. These stimuli, derived from the environment or from the body itself, may be either of a neural or a chemical nature. Some have an accelerating and others a retarding action on either the inspiratory or expiratory regions.

For continued ventilation of the lungs it is evident that inspiration and expiration should follow each other rhythmically, without either phase being prolonged unduly. When, for example, the activity of the expiratory center is held in abeyance, the inspiratory center discharges a continuous stream of impulses to the inspiratory muscles, and the chest wall remains in a constantly expanded state; this ends all pulmonary ventilation and death follows. Moderation of the basic respiratory rhythm is the function of two distinct neural mechanisms: the vagal (the *Hering-Breuer reflex*) and the pneumotaxic. By these the rate and depth are modified so that pulmonary ventilation can adjust to metabolic demand.

Afferent vagal mechanism. ■ Some of the afferent fibers of the vagus nerve in the lung tissue end in *stretch or inflation receptors* (Fig. 17-13). These receptors are stimulated by the expansion of the lungs during inspiration, and the impulse generated is propagated by the afferent vagal fibers to the expiratory center; this center, in turn, discharges inhibitory impulses to the inspiratory center; inspiration ceases. In quiet breathing the cessation of inspirations is mechanically followed by deflation of the lungs; for this *passive expiration* the expiratory center does not send impulses to the expiratory muscles. When the lung deflation has proceeded to a certain extent, the inspiratory center is released from inhibition and, by its spontaneous activity (i.e., without being stimulated by nerve impulses), discharges impulses for the next inspiration. In this manner normal rhythm, or *eupneic,* breathing is maintained.

When in labored breathing the lungs are greatly deflated, the collapse of the lungs stimulates the *deflation receptors.* The impulses thereby generated stimulate the inspiratory center directly, causing its impulses (normally automatically discharged) to be augmented. This results in a hastened and forced inspiration. The deflation receptors are not active in eupnea.

Experimental stimulation of the *central* end of a cut vagus causes various results. Generally a slow rate of electrical stimulation results in an arrest of respiration in the inspiratory phase; a faster rate of stimulation causes arrest in the expiratory phase (Fig. 17-14). Cutting one vagus nerve has little effect (Fig. 17-15, *A*), but cutting the other vagus is immediately followed by a decrease in the rate and a corresponding increase in the depth of the respiratory movements (Fig. 17-15, *B*). Rhythm and depth of breathing are now being moderated by the pneumotaxic center.

Pneumotaxic mechanism. ■ The second mechanism controlling the inspiratory center is the *pneumotaxic center* in the pons of the hind-

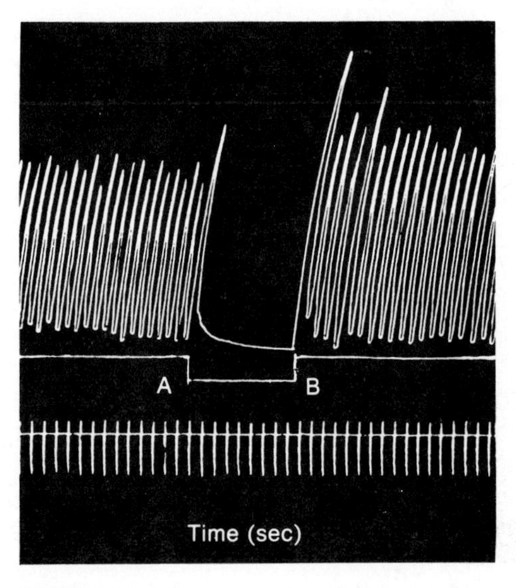

Fig. 17-14
Tracing of respiratory pattern in a dog showing inhibitory effect of vagal stimulation from **A** to **B.** Note slight increase in depth of inspiration on cessation of stimulation.

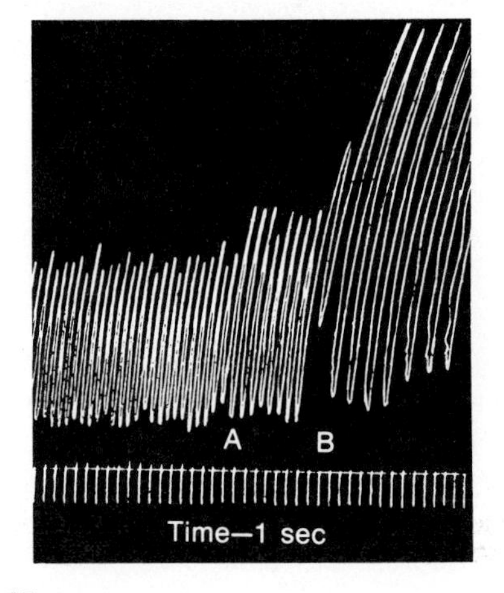

Fig. 17-15
Tracing of respiratory pattern in a dog showing effect of cutting the right vagus nerve at **A,** and a few moments later the left nerve at **B.** Note marked decrease in rate and increase in depth of respiration.

brain (Fig. 10-1). Activity of the inspiratory center sends impulses to the pneumotaxic center; here the impulses are relayed to the expiratory center, which, as already stated, causes inhibition of the inspiratory center (Fig. 17-13). When the activity of the inspiratory center has ceased, impulses no longer flow from this center to the pneumotaxic center, the negative feedback is removed, and the inspiratory center is now free to discharge impulses; another inspiration follows.

The irritability of the pneumotaxic center is of a low order; the center appears to be active only in labored breathing. In quiet breathing the inflation receptors seem to be of greatest importance to rhythmic breathing.

Efferent fibers in the vagi. ■ The vagi also contain efferent or motor fibers that supply the muscles of the bronchi and bronchioles. By stimulation of these fibers smooth muscles become active and the lumen of the air tubes is diminished; as a result, the air in passing into and out of the lung encounters more resistance. In asthma the spasm of the bronchial musculature is responsible for the difficult and labored breathing. Such constriction can be brought about reflexly by stimulating the nasal mucosa. The bronchial muscles receive inhibitory (dilatory) impulses from the sympathetic nervous system.

Other sensory stimuli. ■ In addition to the influences exerted by the vagus nerves and the pneumotaxic center, the rate, depth, and rhythm of the respiratory movements are modified by impulses from sensory surfaces in general (Fig. 17-16) and from higher brain centers and by chemical compounds in the blood.

An individual's respiration is frequently altered by the stimulation of sensory nerves. The deep and audible gasp following a plunge into cold water and the sneeze produced by the inhalation of pepper or by a bright light shining into the eyes are well-known instances. We may mention also the use of "smelling salts" or a dash of cold

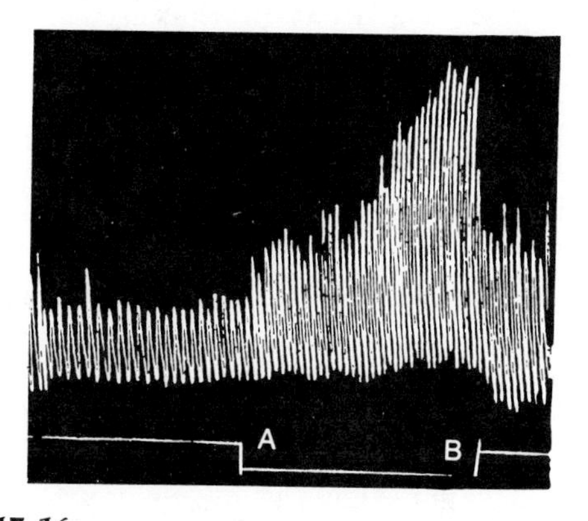

Fig. 17-16
Tracing of respiratory pattern in a dog showing effect of stimulation, **A** to **B**, of sciatic nerve.

water to revive a person who has fainted and the spanking of the newborn infant to induce breathing. In narcotic poisoning the respiratory center may be kept active by stimulating the skin, as by hitting with a wet towel or by applying electrical shocks.

Influence of higher nerve centers. ■ Emotional states may influence respiration. Intense activity of the cerebral cells may have a depressing effect on the respiratory center, thereby causing shallow and slow breathing. This leads to an accumulation of carbon dioxide in the blood, and after a while the stimulating effect of this product is great enough to break through the inhibitory influences coming from the cerebral hemispheres; the result is a deep inspiration followed by a deep, audible expiration—a sigh.

The influence of emotions is apparent in laughing, crying, and other modified forms of respiration. Extreme concentration modifies respiration, for example, when about to shoot a gun or hit a golf ball, the individual takes a deep breath and holds it until after the perfomance, when respiration again becomes normal. Respiration can be voluntarily increased or inhibited, but normally the length of time a breath can be held is limited, generally not exceeding from thirty to fifty seconds.* The capability for voluntary deep breathing not accompanied by physical work is also very limited. Volitional control is limited by chemical stimulation of the respiratory center.

■ Chemical regulation

The respiratory center is stimulated by three chemical conditions of the blood—the tension of the CO_2 and of the O_2 and the concentration of the H ions. The volume of the air respired is governed by the algebraic sum of the influence of these three factors.

Carbon dioxide tension. ■ The most important stimulus for the respiratory center and to which the center is exceedingly sensitive is a slight increase in CO_2 tension of the blood—*hypercapnia.* Evidence suggests that it is not the blood CO_2 that stimulates the center, but the H ions that are formed by the conversion of CO_2 to carbonic acid in the cerebrospinal fluid. In man the addition of 0.22 vol % of CO_2 to the alveolar air, thereby increasing its tension (which normally is 40 mm Hg) by 1.5 mm, doubles the amount of air respired in a unit of time.

If the CO_2 tension in the alveolar air should increase beyond 40 mm Hg, the blood would not be able to rid itself of the usual amount of this gas. This is prevented by increasing the minute volume of respired air, as shown in Table 17-5. As carbon dioxide was added to the inspired air (which normally contains 0.04%), up to 3.07% (about seventy-five times the usual amount) the rate of respiration remained practically unaltered, but there was a greatly increased depth of

*At the so-called *breaking point,* the alveolar CO_2 pressure may have risen from the normal 40 to 50 mm Hg, and the alveolar oxygen pressure may have fallen to 60 mm Hg.

breathing; the percentage and, therefore, the tension of the carbon dioxide in the alveolar air were kept constant.

By voluntary forced breathing for a period of three or four minutes, the CO_2 concentration of the alveolar air may be decreased to as little as 3% and its tension to 15 mm Hg; the oxygen tension may rise to 140 mm Hg. By such breathing the concentration of the CO_2 in the arterial blood can be decreased from the normal 50 vol % to 40 and the pH value increased to 7.7. The influence of this reduction of chemical stimulant on the respiratory center is shown by the following experiment. After normal quiet respiration was registered (Fig. 17-17), overventilation of the lungs was performed for two minutes. It will be seen that this thorough ventilation is followed by a period of apnea.

Table 17-5. Relation of rate and tidal volume of respiration to percent CO_2 in inspired and alveolar air

CO_2 *in inspired air (%)*	*Rate of respiration*	*Tidal volume*	CO_2 *in alveolar air (%)*
0.04	14	673	5.6
2.02	15	864	5.6
3.07	15	1,216	5.5
5.14	19	1,771	6.2

Oxygen content. ■ As it is absolutely necessary that the animal body be supplied with oxygen, it might be surmised that a diminution of O_2 in the blood would quickly stimulate the respiratory center. It is, therefore, surprising to find the irritability of the center toward oxygen deficiency very low. Individuals seem to differ in this respect. It has been found that for most normal subjects the O_2 content of the inspired air had to be decreased from the normal 21% to 18% before any increase in respiration was noticed. Inspired air with only 10% O_2 causes cyanosis, and at 4% consciousness is quickly lost. For some individuals it is possible to lessen the oxygen to such an extent that unconsciousness results without respiration having been much influenced. In fact, there is some question whether the lack

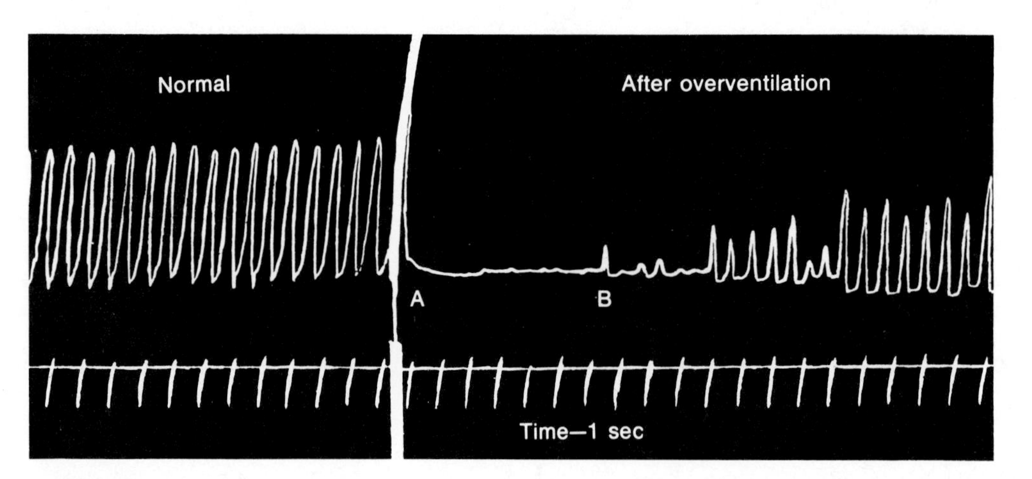

Fig. 17-17
Tracing of respiratory pattern in a dog showing normal, quiet breathing and effects of overventilation. Record was stationary during deep breathing period (heavy, vertical white line). Period of apnea, **A** to **B**, followed overventilation. Subsequently, the pattern gradually returned to normal.

of O_2 directly stimulates the respiratory center. Actually, low P_{O_2} will *depress* the center. In sharp contrast to this is the extreme sensitivity of the center to CO_2; this is well illustrated by the following experiment.

A subject breathed into and out of a rubber bag, the CO_2 being allowed to accumulate; a very severe hyperpnea set in when the CO_2 in the bag reached 5.6%, at which time the O_2 content was still 14.8%. In a second experiment the CO_2 of the respired air was removed by absorption. Now the subject could maintain normal breathing for a much greater length of time, and because there was no increase in CO_2 in the blood, no hyperpnea set in; the subject, however, became cyanotic because of the lack of O_2, which had dropped to 8%. In fact, when the experiment was continued a little longer, the individual passed into *apnea* (no breathing) and, because of the want of oxygen, unconsciousness resulted. If before this took place the person inhaled a little CO_2, the usual rhythm of respiration was reinstated, indicating that the loss of irritability of the respiratory center due to low O_2 could be restored by increasing CO_2 in the blood.

Aortic bodies and carotid bodies. ■ Aortic and carotid bodies, situated close to the pressoreceptors spoken of in Chapter 14, are composed of sensory cells that are chemically stimulated by a deficiency of oxygen or by an excess of carbon dioxide in the blood. For this reason they are spoken of as *chemoreceptors*. The afferent fibers from the aortic bodies run in the vagi; the carotid body nerves supply the carotid bodies (Fig. 17-13). Stimulation of these bodies increases the depth and rate of the respiratory movements. As to the degree of irritability of these bodies to the previously mentioned stimuli and their influence on respiration, opinions are still divided. The facts seem to justify the view that the aortic and the carotid bodies play little, if any, part in regulating respiration in eupnea; it is only in emergencies (e.g., severe hemorrhage) that they assume any importance.

To summarize, under the usual conditions of life the neural mechanism moderating the ventilation of the lungs is activated by the accumulation of carbon dioxide in the blood; the effort to remove the excess of carbon dioxide at the same time charges the blood with oxygen.

■ RESPIRATION AND ACID-BASE BALANCE

As will be recalled (Chapter 12), the hydrogen ion concentration of the blood is expressed by the symbol pH. Its usual value of 7.4 indicates that the negative OH ions slightly predominate over the positive H ions; that is, the blood is slightly alkaline. In the blood the only free acid giving rise to H ions is carbonic acid (H_2CO_3) formed by the union of CO_2 and H_2O. The important alkali from which OH ions are derived is sodium bicarbonate* ($NaHCO_3$). It is, therefore, clear that the pH is determined by the relative amount of each of these present or by the ratio $\frac{NaHCO_3}{H_2CO_3}$. The carbonic acid of the blood is equivalent to $\simeq$0.0015 M solution, and the concentration of the bicarbonate is $\simeq$0.03 M; the ratio of these two substances in the blood is therefore 20:1. This ratio represents the acid-base balance of the blood and is equal to about pH 7.4.

Acidosis. ■ An increase in the carbon dioxide content of the blood, under otherwise constant conditions, decreases the 20:1 ratio and therefore decreases the pH. A tendency toward a decrease in pH, known as *acidosis,* can be brought about in many ways. Some of these are: (1) decreasing the elimination of CO_2 by the lungs (as by voluntarily holding the breath), (2) breathing air rich in CO_2, (3) the inability of the heart adequately to supply O_2 to the tissues, and (4) a decrease in the alkali reserve. In muscular work the lactic acid produced reacts with the $NaHCO_3$ to form sodium lactate and carbonic acid and thus causes

*For the sake of simplicity, the $KHCO_3$ is grouped with the $NaHCO_3$.

acidosis by increasing the acid and decreasing the alkali reserve. This state, however, is transient, for the carbonic acid is eliminated as CO_2 by the lungs. Any acid salts that may rise in this process are eliminated by the kidneys. The kidneys secrete nonvolatile acids in the amount of 40 to 80 mEq/day. About 13,000 mEq/day of volatile acid, CO_2, are removed by the lungs. In whatever manner it may be produced, acidosis results in increased stimulation of the respiratory centers, and the acidity of the urine and the excretion of ammonium salts are increased (Chapter 29). A high degree of acidosis results in coma; a blood pH below 6.9 is customarily fatal.

Alkalosis. ■ By voluntary forced ventilation of the lungs, a larger volume of CO_2 is eliminated from the blood. As a result, the H_2CO_3 concentration is lowered, and the pH of the blood is increased, a condition known as *alkalosis.* The resulting lack of respiratory stimulation induces a cessation of respiration (apnea) until an amount of CO_2 has been formed and retained by the body sufficient to lower the pH to its normal value. Severe alkalosis (pH 7.8) is accompanied by tetany and an increased excretion of alkaline salts by the kidneys. The increased pulmonary ventilation during muscular work does not cause alkalosis because there is a continued increase in the formation of various acids. It may be concluded that, in discharging more or less CO_2 from the blood, *the respiratory mechanism is an important factor in maintaining the proper pH of the body.*

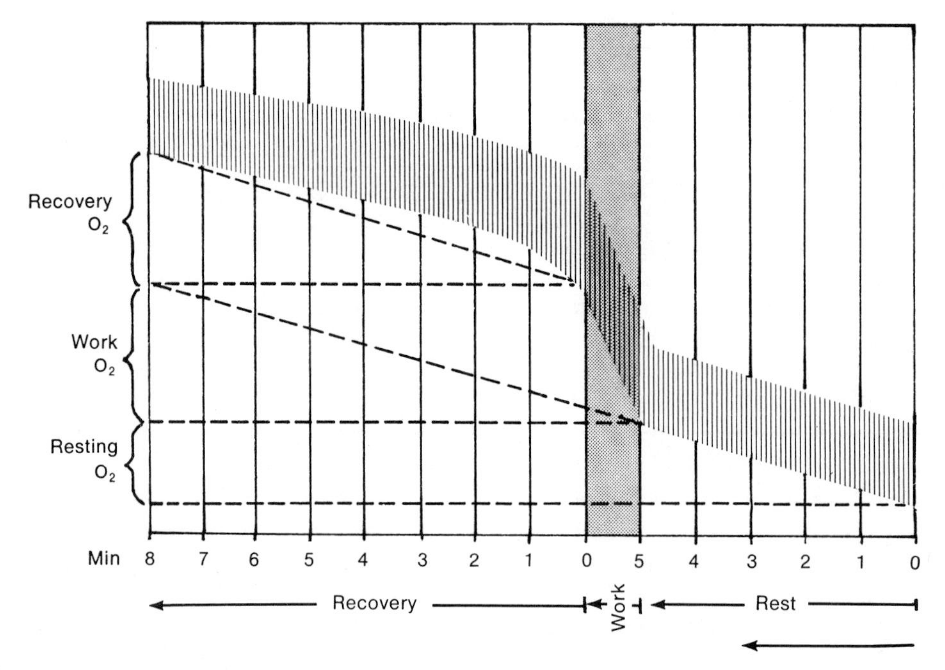

Fig. 17-18
Record of oxygen consumption, slightly schematized, obtained with a recording respirometer. Record reads from right to left and shows total oxygen consumption on left side, consisting of five minutes of rest, one minute of work, and eight minutes of recovery. (From Schottelius, Thomson, and Schottelius: Physiology laboratory manual, ed. 3, The C. V. Mosby Co.)

■ CONDITIONS AFFECTING RESPIRATION

■ Muscular work

Muscle work affects the consumption of O_2 and the production of CO_2 more than any other factor. A frequently used method for investigating these changes is to let the individual breathe from and into a container that is supplied with a measured amount of oxygen; the exhaled carbon dioxide is absorbed by soda lime or a similar chemical. Fig. 17-18 is a schematized drawing of a respirometer record showing the utilization of oxygen during rest, one minute of moderate work, and recovery. The increase in oxygen consumption in response to work is readily apparent.

Walking at the rate of 8 km (5 miles)/hour in one experiment increased the consumption of oxygen from a resting level of 320 ml to 2,500 ml/min and the CO_2 production from 260 ml to 2,300 ml/min. The greatly increased O_2 consumption (during which the arterial blood remains almost saturated with O_2) is made possible by (1) accelerated pulmonary circulation and (2) increased pulmonary ventilation. During the above-mentioned rate of walking, the amount of air respired by the subject increased from about 7 liters/min in the resting condition to almost 50 liters. These factors increase as the severity of the work increases.

An increase in the depth of breathing greatly accelerates the return of venous blood to the left atrium and thereby aids both the systemic and the pulmonary circulation. The trained person obtains the large increase in pulmonary ventilation more by increasing the depth and less by increasing the rate of breathing than is true for the untrained person. All in all, the trained excels the untrained in adjustment of the circulatory-respiratory mechanism to the needs of the active muscles and in performing the work more efficiently.

The cause of the hypernea during muscular work is not fully understood. It seems that the deficiency in O_2, the excess of CO_2, and the change in the blood pH are not altogether responsible. Afferent impulses from the contracting muscles or moving joints (exercise stimuli) or nerve impulses from higher brain centers likely are involved.

Oxygen uptake. ■ The augmented uptake of O_2 from the blood by the active muscles and the greater coefficient of oxygen utilization, which may rise to 0.9, are made possible by several factors:

1. The O_2 tension in a very active muscle may drop to zero, thereby creating a steep O_2 pressure gradient between blood plasma and muscle.
2. The increased concentration of CO_2 and acids in the arterial blood hastens the dissociation of oxyhemoglobin.
3. The release of O_2 is favored by the increase in temperature.

How well an active muscle can be supplied with oxygen depends, among other factors, on oxygen-carrying capacity of the blood, capillary adjustment, and minute volume of the heart.

Oxygen-carrying capacity of blood. The oxygen-carrying capacity of blood is determined chiefly by the relative number of red blood cells and the amount of hemoglobin in the blood, hence the shortness of breath and incapacity for sustained work of an anemic person. During acute vigorous exercise the red cells stored in the more or less stagnant blood pools are forced into the active circulation. After long periods of physical training, there is an actual increase in the number of erythrocytes per unit volume of circulating blood. This chronic increase is attributed to a persistent oxygen lack, similar to that experienced by persons living at high altitudes.

Capillary adjustment. The opening up and dilation of previously closed capillaries in an active muscle are important in decreasing the arteriovenous oxygen difference.

Minute volume of the heart. Both the stroke volume and the heart rate are increased during muscle work, and consequently the cardiac output is increased; the greater these increases, the greater will be the oxygen delivery. So urgent for the active animal is this oxygen delivery that one of the most fundamental physiological responses is the adjustment of the volume of the blood discharged by the heart to the amount of O_2 consumed and CO_2 produced. Because the heart can adjust itself better, the O_2 debt in moderate exercise in a well-trained person is less, and the respiration is restored to the normal resting state sooner than in the untrained individual. This adjustment in a trained individual is made more by an increase in the stroke volume and less by an augmentation in the heart rate than in the untrained person.

Maximum oxygen uptake. ■ The requirement of the body cells for oxygen is met by the combined activities of the cardiovascular and pulmonary systems. Together they determine the limits of oxygen transport which may be achieved. Maximum oxygen uptake is a measure of maximal oxygen transport capacity of a person performing physical work. There is a linear relationship between increasing work load and oxygen uptake until the limit of oxygen transport is reached, but greater work loads may be performed at the cost of an oxygen debt. Only a slight increase in blood lactic acid is observed in moderate exercise. As the maximum oxygen uptake is approached and passed, a marked increase in blood lactic acid occurs.

Maximum oxygen uptake can be twelve times greater than the uptake at rest. This is accomplished by increasing the tidal volume fourfold and the respiratory rate threefold. Concurrently, the cardiac output is increased four times by doubling both the stroke volume and the heart rate. As the extraction of oxygen (average arteriovenous [AV] oxygen difference) increases by three times, the twelvefold increase in maximum oxygen uptake is attained. The limiting factor in maximum oxygen uptake of healthy individuals is the cardiovascular rather than the pulmonary system.

During exercise the AV oxygen difference across the heart, which is normally quite large (Table 17-4), increases only slightly. Cerebral blood and oxygen extraction are virtually unchanged by exercise. In contrast, blood flow to the gastrointestinal tract, liver, and kidneys is markedly diminished, but a wider AV oxygen difference partially compensates for the reduction in blood flow.

A number of factors account for the variation in the maximum oxygen uptake between different persons. Much of the variation can be accounted for by the state of physical condition, body size, age, and sex of the individual. The maximum oxygen uptake of a healthy twenty-year-old man is approximately 45 ml/kg body weight/min. Strenuous physical training can increase this value. Olympic champions in endurance events often achieve values of 75 ml/kg. A decrease in physical activity leads to a prompt fall in the maximum oxygen uptake of an individual. The maximum oxygen uptake of women is about two thirds that of men, and the value for both sexes decreases with age, although less markedly in physically active than in sedentary individuals.

Second wind. ■ The cause of getting one's *second wind* is not well understood. When it occurs, the organs of circulation and respiration and other organs have adjusted themselves better to the increased demands of the active muscles. The heart beats more slowly and more regularly, and breathlessness and the feeling of distress *(dyspnea)* subside.

■ Foreign gases

Foreign gases may be divided into three classes.

Physiologically inert gases. ■ These, in themselves, have no effect on the body. They include hydrogen and nitrogen. They cause ill effects or death only when their amount in the inspired air is so great as to exclude the needed amount of oxygen.

Irritating gases. ■ These cause spasmodic closure of the glottis. They kill, therefore, by strangulation. Among these are ammonia, sulfur dioxide, and chlorine.

Carbon monoxide. ■ Carbon monoxide is injurious because it blocks hemoglobin; the latter has a greater affinity for CO than it does for O_2. Since the affinity of hemoglobin for this gas is about 200 times as great as that for oxygen, the carbon monoxide drives the oxygen from the oxyhemoglobin and forms a rather stable compound known as carboxyhemoglobin. This compound decreases the oxygen-carrying capacity of the blood and, in consequence, the tissues undergo oxygen starvation.

Carbon monoxide is found in mines, in the incomplete combustion of carbon (as in charcoal braziers), in the exhaust gases of gasoline engines, and in the fumes from leaking stoves, furnaces, and flues. In a concentration of 0.03% (3 parts CO in 10,000 of air), CO shows its deleterious effects. Air containing 0.2% may prove fatal if respired for a sufficient length of time, and in a concentration of 2% to 4% death occurs quickly. Frequently the victim of CO poisoning has no warning of its presence, since the gas is odorless.

In small amounts CO is not a protoplasmic poison, and therefore, if it is possible to increase sufficiently the oxygen held in physical solution by the plasma, the victim survives. As deprivation of oxygen for even minutes causes irremediable damage to many of the higher brain centers, artificial respiration should in all cases be instituted without delay. It may be necessary to use transfusion of whole blood. Recovery from carbon monoxide poisoning is frequently associated with temporary or permanent disturbances in the functions of the nervous system, such as blindness and loss of power of speech.

■ Pressure changes

Respiration is influenced by an increased or decreased atmospheric pressure.

Increased atmospheric pressure. ■ That increased oxygen pressure has no effect on the rate of metabolism and that a concentration above 70% causes irritation of the lungs has been discussed previously.

Increased atmospheric pressure is not encountered except by divers and caisson workers. The high pressure of the air in which they work causes a great amount of oxygen and nitrogen to be dissolved in the plasma of the blood. As long as the workmen remain in this high air pressure, no particularly distressing symptoms are experienced. But on suddenly passing from the high pressure to normal pressure, the extra gases dissolved in the plasma can no longer be held in solution and make their appearance as bubbles in the blood.* These bubbles clog the capillaries and lead to anemia of the tissues; this results in many disturbances such as the "bends" and "staggers"; the general paralysis is sometimes called "diver's palsy." By ascending to the surface slowly or passing from the caisson through air locks, the decompression is made very gradually; this allows time for the dissolved gases (chiefly nitrogen) to be discharged by the lungs.†

Decreased atmospheric pressure: acclimatization. ■ It will be recalled that at sea level (760 mm Hg pressure and 21 vol % of O_2) the P_{O_2} of the alveolar air is 100 mm Hg. This is sufficient to cause the arterial blood leaving the lungs to be almost completely saturated (97%) with oxygen. An altitude of 10,500 ft (512 mm Hg barometric pressure and the alveolar P_{O_2} at 67 mm) allows for an 88% O_2 saturation of the arterial blood; this is sufficient for nearly all the usual functions of life. Above this, the O_2 saturation drops rapidly; for example, on Pike's Peak (alti-

*This is similar to the effervescence occurring on removal of the stopper from a bottle of carbonated beverage. Letting the gas flow slowly through a small opening prevents the formation of bubbles.

†Since helium is only one half as soluble and diffuses twice as rapidly as nitrogen, it is frequently used, mixed with the proper amount of oxygen, when work is to be done under high air pressure.

Table 17-6. Atmospheric pressure changes associated with various altitudes

Altitude (ft)	*Pressure (mm Hg)*
0	760
500	746
1,000	733
2,000	707
3,000	681
4,000	656
5,000	632
10,000	523
15,000	429
20,000	349
25,000	282
30,000	226
35,000	179
40,000	141

tude about 14,500 ft and 438 mm pressure) the saturation percentage has dropped to 80. Many people entering in this rarified air experience mountain or altitude sickness: vertigo, nausea, weakness, hyperpnea, incoordination, slowed mental processes, dimmed vision, and increased heart rate and stroke volume. At an altitude of 25,000 ft the O_2 saturation is only 55%, and consciousness is quickly lost. Permanent residence at high altitude, for example at 17,500 ft in Peru, is made possible by acquiring an extremely high count of red blood cells and a great increase in hemoglobin, as well as a constantly increased pulmonary ventilation and minute volume output by the heart. The gradual adaptation to a lower atmospheric pressure is an example of *acclimatization.*

Table 17-6 shows the atmospheric pressures encountered at various altitudes.

In the unacclimated individual, ascent to altitude of 4,000 feet causes impairment of night vision. Susceptibility to hypoxia is increased by cold environments, alcohol, many drugs, and smoking. The CO in tobacco smoke reduces the oxygen carrying capacity of the blood, as described earlier.

In the rapid ascent of unpressurized aircraft to a height of 25,000 feet, or above, the occupants undergo decompression sickness. This condition is an air embolism similar to that found in caisson workers. It is due to the nitrogen dissolved in tissue fluids coming out of solution due to the reduced atmospheric pressure. Should the bubbles of gas form within the central nervous system substance, tingling sensations or even paralysis of a temporary or permanent nature may ensue. Increased susceptibility to decompression sickness develops with increasing age, obesity, reexposure to reduced atmospheric pressure, and previous exposure within forty-eight hours to increased atmospheric pressures, as in scuba diving.

■ MODIFIED FORMS OF RESPIRATION

Periodic breathing. ■ Although there are various kinds of periodic breathing, the most common type is called Cheyne-Stokes respiration. In this there is a waxing and waning. Once initiated, the pattern is probably maintained by the buildup of CO_2 during *apnea* and its reduction during the period of increased respiratory activity. The individual respirations are small at the start but gradually increase in depth to a maximum, subside again, and finally cease for a short time (Fig. 17-19). This type of respiration may occur at high altitudes and is sometimes exhibited by healthy infants and occasionally by healthy adults during sleep. Cheyne-Stokes respiration is sometimes seen in pathologic states and under these conditions is generally caused by damage to the respiratory center.

Cheyne-Stokes respiration is accompanied by variations in blood pressure (Fig. 17-19), which are probably caused by oscillation of the pressoreceptor control systems. There may be an interaction of vasomotor and respiratory centers as well. The arterial blood pressure falls during apnea and rises during the period of breathing.

Coughing. ■ Coughing is generally preceded by an inspiration of greater amplitude than nor-

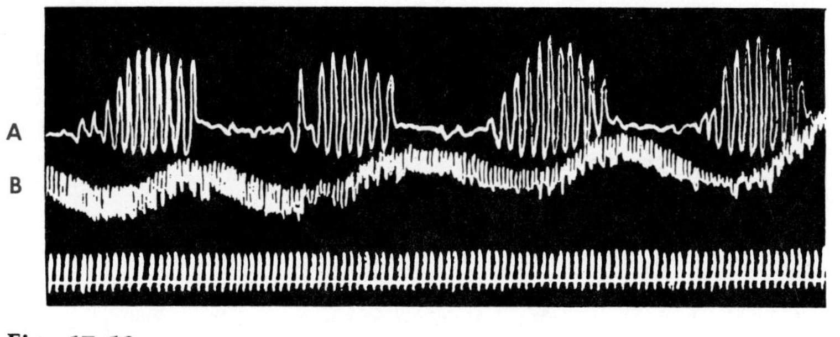

Fig. 17-19
Tracing of Cheyne-Stokes respiratory pattern, **A**, and blood pressure, **B**, in a dog.

mal. Next, the glottis is closed and the muscles of expiration, especially those of the abdomen, contract forcibly, causing a great increase in the intrapulmonic pressure. When a certain amount of pressure has been achieved, the vocal cords part and the imprisoned air escapes with a rush. If movable matter such as mucus or food crumbs is present, it may be dislodged; this is a protective mechanism. Most coughs are the result of the common cold or of throat irritation, but the stimulation may originate in any part of the respiratory tract, in the stomach, heart, skin, and so on. There is also a psychic cough; hearing another person cough is very likely to cause a desire to cough, especially in children.

Swallowing. ■ The act of swallowing is normally associated with an inhibition of respiration due to stimulation of the glossopharyngeal nerves by the food.

Sneezing. ■ Sneezing is a violent expiration, the air being sent through the nose, the contraction of the anterior fauces shutting off the mouth cavity. It can be brought about by such factors as stimulation of the fifth cranial nerve in the nose, or a bright light thrown into the eyes.

Hiccough. ■ Hiccough is caused by a spasmodic contraction of the diaphragm, the sudden inspiration thus produced being cut short by the closure of the glottis. It is generally found in irritation of the stomach and is frequently seen in young children after the overfilling of that organ.

Dyspnea. ■ From our study of the blood, circulation, and respiration, it must be evident that dyspnea can be produced in many ways:

1. Mechanical impediment to expansion of the lungs as in obstruction of the airways (asthma), excess liquid or gas in the abdomen, and postnasal growths
2. Lessened exchange in the alveoli as in pneumonia (inflammation of respiratory surface) and tuberculosis
3. Abnormal conditions of the blood
   a. Accumulation of carbon dioxide
   b. Lack of oxygen (hypoxemia, e.g., at high altitude)
   c. Hemorrhage
   d. Acidosis
   e. Increased temperature
4. Abnormal circulation causing loss of blood pressure
   a. Loss of vasomotor tone
   b. Reduced cardiac output
   c. Insufficiency of the heart valves
   d. Hemorrhage

Speech. ■ The sound produced in the larynx by the movement of air from the lungs is not truly speech, but *phonation.* Speech is a complex process controlled by the cerebrum (Chapter 10) in which vocal sounds are articulated by the tongue, jaw, and lips to form meaningful patterns. Phonation normally occurs upon expiration when the lungs and chest wall acting as a bellows drive air through the vocal folds, or cords (Fig. 17-1). These folds are set into vibration by the air stream and produce the vocal sound at a pitch (frequency) dependent upon the tension (degree of stretch) imposed upon the folds. Pitch increases with tension. The intensity of the sound is a function of the volume rate of flow of the air. In quiet respiration the inner edges of the two folds are about 8 mm apart, about half way between fully open and fully closed. Contraction of the vocalis and cricothyroid muscles increases the tension and length of the vocal folds and decreases the separation of the edges of the folds.

The sound produced in the larynx is modified by the mouth, throat, and nasal cavities, the upper resonators. The lips, jaws, and tongue modify the sound further to determine the quality of the sound and to make intelligible speech. That the larynx and its vocal folds are not the main organs of speech is evident from the phenomenon of esophageal speech. Persons deprived of the laryngeal structures through accident or surgery learn to swallow air and to bring it back in the form of a belch which produces an esophageal sound. The customary actions of the upper resonators form this sound into a hoarse, but intelligible, speech.

READINGS

Comroe, J. H., Jr.: The lung, Sci. Am. **214**(2):56-68, Feb. 1966.

Finch, C. A., and Lenfant, C.: Oxygen transport in man, New Eng. J. Med. **286**(8):407-415, 1972.

Forster, R. E.: Oxygenation of the muscle cell, Circ. Res. **20**:1-115, 1967.

Lenfant, C., and Sullivan, K.: Adaptation to high altitude, New Eng. J. Med. **284**(23):1298-1309, 1971.

Mitchell, J. H., and Blomquist, G.: Maximal oxygen uptake, New Eng. J. Med. **284**(8):1018-1022, 1971.

Morgan, T. E.: Pulmonary surfactant, New Eng. J. Med. **284**(21):1185-1192, 1971.

Slonim, N. B., and Hamilton, L. H.: Respiratory physiology, ed. 2, St. Louis, 1972, The C. V. Mosby Co.

Winter, P. M., and Lowenstein, E.: Acute respiratory failure, Sci. Am. **221**(5):23-29, Nov. 1969.

Food
Chemistry
Carbohydrates
Lipids
Proteins
Vitamins
Inorganic substances

THE FOODSTUFFS

■ FOOD

The animal body performs work, as all mechanisms do, at the expense of energy; the energy thus expended is derived from the potential energy of the foods ingested. In performing its functions protoplasm undergoes wear and tear which must be repaired by the ingestion of food. This gives us a basis for a definition of the term *food.* A food is a substance that:

1. Supplies the body with energy
2. Furnishes the body with material for growth, for repairing the waste the body has sustained, or for the manufacture of cellular products, such as hormones and enzymes
3. Supplies the body in very minute quantities with materials that exercise a tremendous influence on the various organs (the vitamins)
4. In the quantity consumed, has no direct harmful influence on the body.

The foods of civilized man are many and complex, but the real nourishing materials in all the varieties of foods may be divided into five classes of substances.

Carbohydrates	Non-nitrogenous	Organic
Lipids, Proteins, Vitamins	Nitrogenous	
Inorganic substances		

We shall discuss the chemical nature of these substances in this chapter, which will provide a basis for the chapters on digestion and metabolism. The nutritional requirements for all classes of foodstuffs will be considered in Chapter 31. Our foods contain in small quantities other substances that are not true foods since they supply neither energy nor material for repair in the amounts usually consumed; reference is made to the condiments (e.g., mustard and pepper) and to the materials (flavors) already present in the foods that give the foods their distinctive tastes or odors. The taste of a piece of meat, for example, has nothing to do with its real food value; the taste is due largely to the so-called extractives. These extractives can be removed by heating the meat in a large amount of water; most of the food is still found in the tasteless cooked meat, whereas the water contains the tasty materials. Hence meat extracts have very little food value although, like the spices, they are of value in stimulating the flow of saliva and gastric juice.

■ CHEMISTRY

To understand the complicated molecules of food, we will incorporate the review of some basic organic chemistry before we consider the four classes of organic substances.

Hydrocarbons and oxygen derivatives. ■ In an organic compound the atoms in the molecule are grouped around a carbon atom; the carbon atom may be regarded as the nucleus of the molecule. The carbon atom is characterized by its 4 valences, that is, by its ability to unite with 4 univalent atoms (e.g., hydrogen) or with 2 bivalent atoms (e.g., oxygen). A more important characteristic is the linkage of carbon atoms to each other almost without limit.

Hydrocarbons are compounds composed of two elements: carbon and hydrogen. One series of hydrocarbons is known as the methane or paraffin series. The simplest of these is methane or marsh gas; in its molecule the 4 carbon valences are occupied by hydrogen atoms.

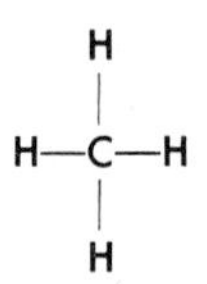

Two or more carbon atoms may be united by single bonds and all the other valences united with hydrogen. The formulas for the next three members of this series are therefore:

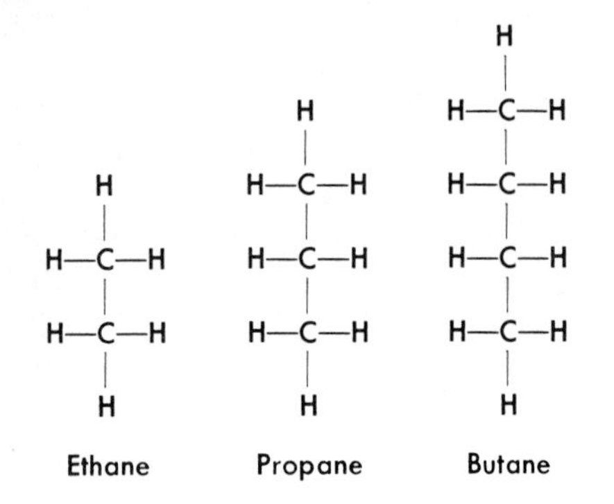

Ethane Propane Butane

We may abbreviate these as follows:

Methane	CH_4
Ethane	$CH_3 \cdot CH_3$
Propane	$CH_3 \cdot CH_2 \cdot CH_3$
Butane	$CH_3 \cdot (CH_2)_2 \cdot CH_3$
Pentane	$CH_3 \cdot (CH_2)_3 \cdot CH_3$
Hexane	$CH_3 \cdot (CH_2)_4 \cdot CH_3$

Alcohols. The introduction of an oxygen atom into a hydrocarbon results in the formation of an alcohol. If the oxidized group is CH_3, a primary alcohol is formed; a secondary alcohol is produced when a CH_2 group is oxidized. Thus, from ethane we obtain ethyl alcohol (always primary).

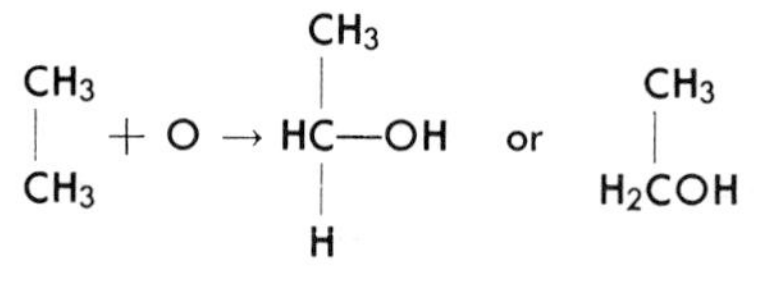

Ethane Ethyl alcohol (primary)

From propane the secondary propyl alcohol is formed when the oxygen is introduced into the CH_2 group.

$$CH_3 \cdot CH_2 \cdot CH_3 + O \rightarrow CH_3 \cdot HCOH \cdot CH_3$$

A primary alcohol contains a primary alcohol group, H_2COH. The secondary alcohol group is HCOH.

A single molecule may contain two or more alcohol groups, thus:

$$\begin{array}{c} CH_2OH \\ | \\ CH_2OH \end{array} \qquad \begin{array}{c} CH_2OH \\ | \\ CHOH \\ | \\ CH_2OH \end{array}$$

Glycol Glycerol

When 6 carbon atoms are present and all the groups are oxidized, a polyhydric alcohol of the following composition is obtained.

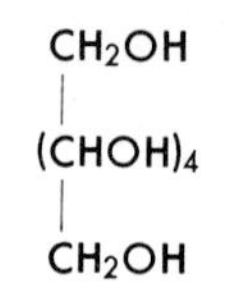

Aldehydes and ketones. When a primary alcohol group, H_2COH, receives another oxygen atom, the alcohol is transformed into an aldehyde and water.

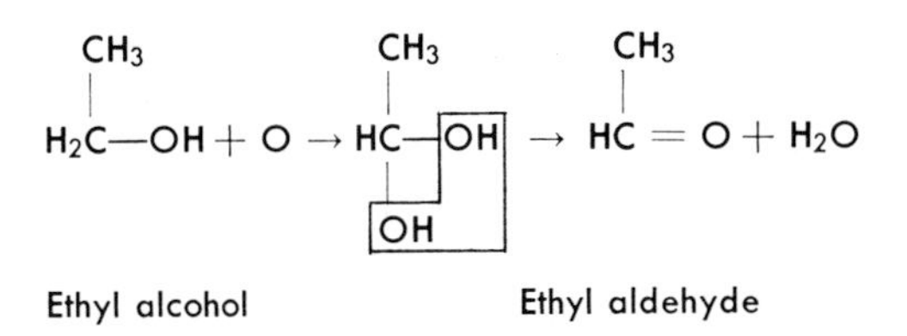

The oxidation of a secondary alcohol group, HCOH, gives rise to a ketone.

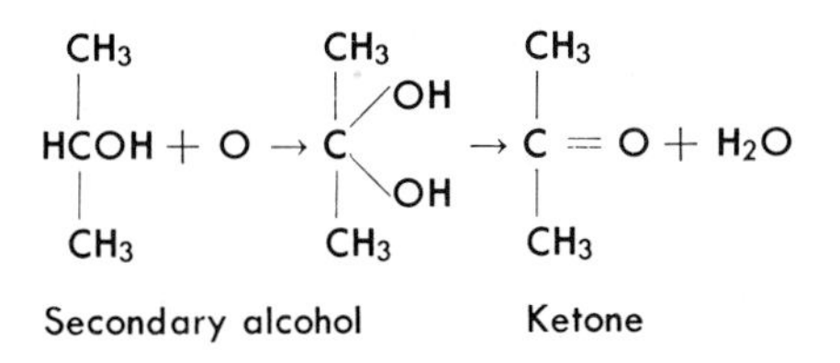

The characteristic group of an aldehyde is the H−C=O group; a ketone has a C=O group.

Hexoses. When one end group, CH_2OH, of a polyhydric alcohol is oxidized, the result is an aldehyde (or aldose); the oxidation of one of the middle groups, CHOH, gives rise to a ketone (or ketose).

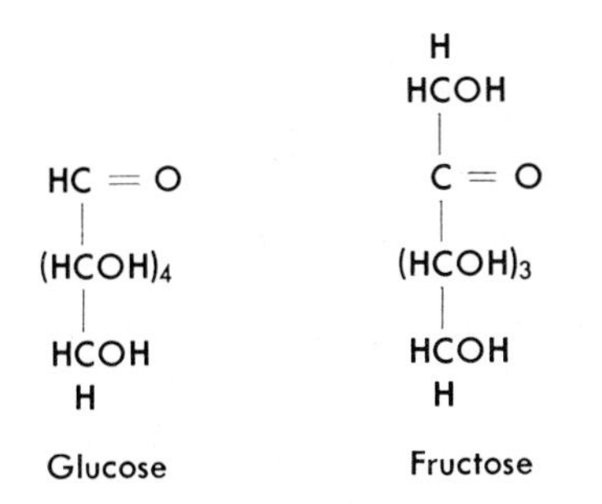

These formulas are for two simple sugars, the first an aldose and the other a ketose. A hexose may be defined as an aldose or a ketose of a 6-carbon polyhydric alcohol. *Trioses* and *pentoses* are the aldoses or ketoses of the 3- and 5-carbon alcohols, respectively.

■ CARBOHYDRATES

Composition. ■ Carbohydrates are chemical compounds composed of carbon, hydrogen, and oxygen. The molecule of the carbohydrates generally spoken of in human physiology contains 6 or a multiple of 6 atoms of carbon, and the ratio of the hydrogen atoms to the oxygen atoms is usually 2:1.

Classification. ■ The carbohydrates may be divided into three classes: monosaccharides, disaccharides, and polysaccharides (*poly*, many).

Monosaccharides. Monosaccharides are sometimes called the simple sugars. They are straight-chain polyhydric alcohols and are named according to the number of carbon atoms in the chain. Three important hexoses of this group are *glucose* (also known as dextrose or grape sugar), *fructose* (also called levulose or fruit sugar), and *galactose.* These sugars are sweet, neutral in reaction, crystallizable, soluble in water, and readily dialyzable. They can undergo alcoholic fermentation. Glucose and fructose are found in many fruits, seeds, and roots of plants and in honey. Glucose is found in the blood (about 0.1%) and is the most essential transportable form of energy in the body.

Two important pentoses (5-carbon chain) are ribose and deoxyribose. Ribose is found in RNA and deoxyribose in DNA (Chapter 2).

Detection of sugars. To detect the presence of hexose and to determine the amount present in a fluid, one of several methods may be employed. The one commonly used is known as Benedict's test. When blue Benedict's solution is heated with sugar, a reddish precipitate is formed which settles to the bottom of the test tube. Sugar at high temperature removes oxygen from the copper salt found in Benedict's solution; therefore it is said the solution has been reduced.

Disaccharides. Disaccharides are double sugars formed by the union of 2 molecules of monosaccharide with the loss of 1 molecule of water.

$$C_6H_{12}O_6 + C_6H_{12}O_6 \rightarrow C_{12}H_{22}O_{11} + H_2O$$

$C_{12}H_{22}O_{11}$ is the general formula for a disaccharide. The following are disaccharides: *sucrose,* composed of glucose and fructose; *maltose,* composed of 2 molecules of glucose; *lactose,* composed of glucose and galactose.

Sucrose is found in certain plants (sugar cane and sugar beet); maltose is formed in grains during the process of malting; lactose (*lac,* milk) is of animal origin, found only in milk.

Like monosaccharides, disaccharides are more or less sweet, crystallizable, soluble, and dialyzable. Except for cane or beet sugar, they reduce Benedict's solution; and, except for lactose, they can undergo alcoholic fermentation by yeast. Lactose or milk sugar is readily changed by certain bacteria into lactic acid.

$$C_{12}H_{22}O_{11} + H_2O \rightarrow 4\ C_3H_6O_3$$

This lactic acid fermentation causes the souring of milk, and the acid is responsible for the subsequent clotting. On being boiled with acids or by the action of proper enzymes, as during digestion, disaccharides are decomposed into monosaccharides.

Polysaccharides. Polysaccharides are anhydride (*an,* without; *hydro,* water) condensation products of monosaccharides; that is, each of several hundred molecules of a monosaccharide loses 1 molecule of water in uniting with each other to form 1 large molecule $(C_6H_{10}O_5)_n$. This is the general formula for a polysaccharide. Among these compounds are starch, dextrins, glycogen, and cellulose. When a polysaccharide is boiled with a strong acid or is acted on by the proper enzyme, the large molecule is changed into monosaccharides; in this change a molecule of H_2O is added to each one of the molecules of the monosaccharides. This chemical action is known as hydrolytic cleavage or hydrolysis.

Although monosaccharides and disaccharides agree quite closely in their various properties, polysaccharides are very different. Moreover, the various polysaccharides differ much from each other; consequently few general statements can be made. They are not crystallizable and are generally devoid of taste; they do not form true solutions and are, therefore, not dialyzable. With the exception of the dextrins, they do not reduce Benedict's solution. They give characteristic color reactions with iodine solution, starch giving a blue, glycogen a dark red, and dextrin (i.e., erythrodextrin) a red color; these color reactions are used as tests for the various polysaccharides.

With the exception of glycogen, the polysaccharides are of vegetable origin. Starches, which are found in all tubers and grains, constitute a large percentage of the dry material of our food. They are a mixture of two types of compounds, amylose and amylopectin. The first of these consists of a long, unbranched chain of maltose units; amylopectin is a branched chain composed of maltose and isomaltose units. The starch granules are surrounded by a very thin capsule of cellulose. Dextrins are produced when starch is heated in the presence of water to 200 C; hence they are found in toast, the crust of bread, and many breakfast foods. Glycogen is found extensively in the liver and has the same relation to the animal economy as starch has to plant; hence glycogen is sometimes called animal starch. It is, like amylopectin, a branched chain polysaccharide and occurs in animal cells as granules.

Cellulose is a peculiar carbohydrate so far as its presence in our food is concerned. Although the various monosaccharides, disaccharides, and polysaccharides thus far discussed are true foods in that they furnish our bodies with energy, cellulose is not a food; this is due to its indigestibility, and therefore its energy is not available. Filter paper is almost pure cellulose. Cellulose also forms the woody part of vegetables and fruits and the outer covering of grains. It promotes intestinal peristalsis (Chapter 19).

■ LIPIDS

Composition. ■ Like carbohydrates, lipids are composed of the elements carbon, hydrogen, and oxygen. Lipids are characterized by the presence of fatty acids or their derivatives and by their solubility in such solvents as acetone, alcohol, ether, and chloroform.

There is, however, a marked difference in the relative number of the atoms of the elements in a lipid molecule compared with that found in carbohydrates. From the formula for a carbohydrate ($C_6H_{12}O_6$), it is evident that the hydrogen and oxygen atoms are in the same proportion as in water. In lipids this is not the case. For example, the composition of a very common fat, stearin, is expressed by the formula $C_{57}H_{110}O_6$. It will be noticed that only 12 of the 110 hydrogen atoms can be united with the oxygen atoms present in the molecule. When a glucose molecule undergoes oxidation, the number of oxygen atoms uniting with the whole molecule is the same as the number of oxygen atoms required to transform the carbon of the glucose into carbon dioxide.

$$C_6H_{12}O_6 + 6\ O_2 \rightarrow 6\ CO_2 + 6\ H_2O$$

On the other hand, when stearin is oxidized, sufficient oxygen must be added not only to form carbon dioxide from the carbon but also to oxidize the hydrogen to water.

$$2\ C_{57}H_{110}O_6 + 163\ O_2 \rightarrow 114\ CO_2 + 110\ H_2O$$

Of the 326 oxygen atoms, 98 are used to oxidize the 196 atoms of hydrogen not already provided with oxygen. This causes the oxidation of fat to give rise to a great deal more heat than is evolved by the burning of a carbohydrate. The potential energy (heat value) of 1 lb of fat is equal to that of $2^1/_4$ lb of sugar. Fats are necessary in the diet to provide the essential unsaturated fatty acids, linoleic, linolenic, and arachidonic, and to provide energy.

Classification. ■ For our purposes, only those lipids that are of physiological importance need be considered. *Fats* are esters of fatty acids with glycerol. *Phospholipids* are substituted fats containing, in addition to fatty acids and glycerol, phosphoric acid, a nitrogen compound, and other components. Phospholipids are essential structural components of mitochondria and neural tissue (Chapters 2 and 5). Compounds that contain a fatty acid, a carbohydrate, and a complex amino alcohol are known as *glycolipids*. *Sterols* are high molecular weight cyclic alcohols; a familiar example is cholesterol. Other important lipids are the *sphingolipids* and the *lipoproteins*.

Chemistry. ■ A molecule of a fat is formed by the union of 1 molecule of glycerol (glycerin) and 3 molecules of fatty acids. There are a large number of fatty acids; acetic acid (in vinegar) is a very simple example. The fats found in our food and in our body are largely formed from three fatty acids; oleic, palmitic, and stearic acid. By the union of these acids with glycerol three common fats are obtained: olein, palmitin, and stearin. Glycerol, it will be recalled, is a polyhydric alcohol.

CH_2OH
|
$CHOH$
|
CH_2OH

We have also learned that the oxidation of a primary alcohol gives rise to an aldehyde. Let us introduce another atom of oxygen in the same group; the aldehyde is now transformed into an organic acid.

Thus:

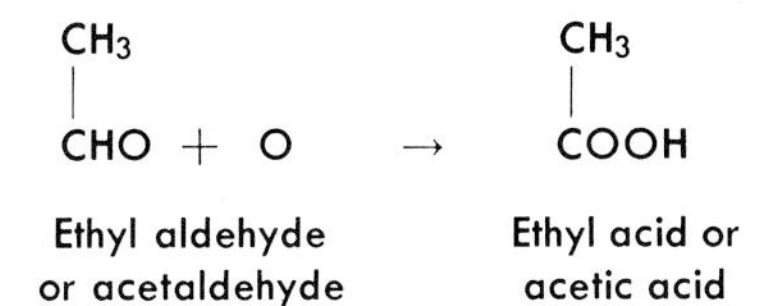

The organic acid is characterized by the *carboxyl group*, COOH, the hydrogen of which is replaceable.

The stepwise oxidation of ethane (a hydrocarbon) into alcohol, aldehyde, and acid is illustrated:

CH_3	CH_3	CH_3	CH_3
|	|	|	|
CH_3	H_2COH	CHO	$COOH$
Hydrocarbon	Alcohol	Aldehyde	Acid

The acids derived from the second to the sixth hydrocarbons of the paraffin series are:

	CH_3	CH_3	CH_3	CH_3
	|	|	|	|
CH_3	CH_2	$(CH_2)_2$	$(CH_2)_3$	$(CH_2)_4$
|	|	|	|	|
$COOH$	$COOH$	$COOH$	$COOH$	$COOH$
Acetic	Propionic	Butyric	Valeric	Caproic

The sixteenth and eighteenth members of the paraffin series and their respective acids are:

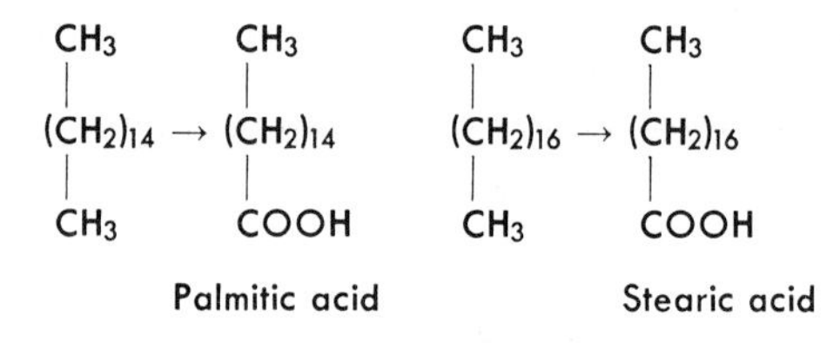

The important chemical characteristic of alcohols is that they act as organic bases, having the ability to unite with acids to form water and compounds corresponding to the salts, quite similar to inorganic bases. For example, sodium hydroxide + hydrochloric acid → sodium chloride + water may be expressed as:

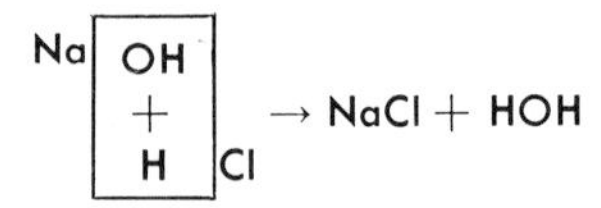

Similarly, ethyl alcohol + hydrochloric acid → ethyl chloride + water:

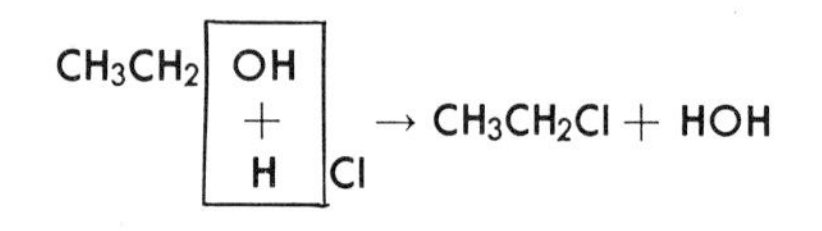

Also, for ethyl alcohol + acetic acid → ethyl acetate + water:

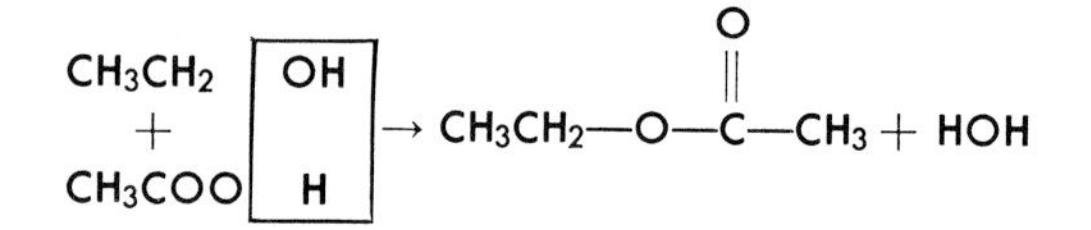

Ethyl acetate is called an ester just as the NaCl is called a salt. Stearin and palmitin are the esters formed by the interaction of the polyhydric alcohol (glycerol) and the fatty acids, stearic and palmitic acids, respectively. As glycerol is a polyhydric alcohol, that is, has three alcohol groups, it can unite with 3 fatty acid radicals, as illustrated.

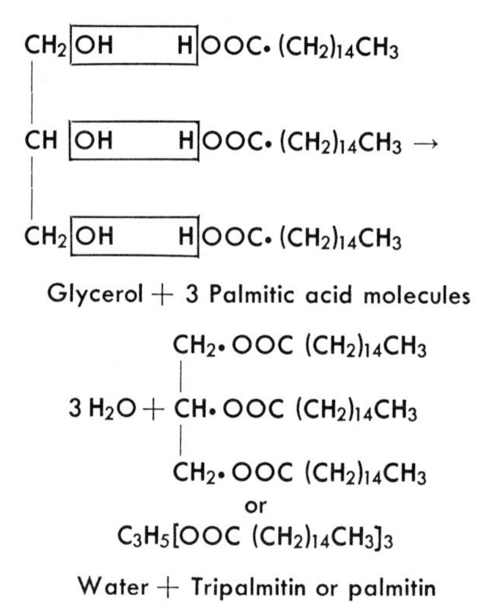

In the same manner tristearin or stearin is formed, having the composition $C_3H_5[OOC(CH_2)_{16}CH_3]_3$. Olein or triolein is the triglyceride of an unsaturated fatty acid called oleic acid. In an unsaturated fatty acid 2 or more of the carbons in the chain are linked by double bonds; for example, the formula of oleic acid is:

$$CH_3(CH_2)_7—HC{=}CH—(CH_2)_7COOH$$

Unsaturated fatty acids are especially found in cottonseed and linseed oils. By treating these fats with hydrogen at a high temperature and pressure and in the presence of a catalyst, the unsaturated fatty acids take up hydrogen at the double bond and thereby become saturated (hydrogenated). Various degrees of hydrogenation result in cooking oils and margarines which are fully utilizable by the body. The fats in the animal body are generally mixed triglycerides; that is, to the glycerol are attached one each of the 3 fatty acid radicals already discussed. Other fatty acids may be present; thus, milk contains all the even-numbered fatty acids, from butyric (4 carbons) to stearic acid (18 carbons).

Soap. A fat may be decomposed into glycerol and fatty acids in many ways, among which the action of certain enzymes, bacteria, and strong alkalies may be mentioned. A fat heated with potassium hydroxide (KOH) or sodium hydroxide (NaOH) is, first of all, decomposed into the fatty acid and glycerol, and then the Na or K ions of the alkali unite with the fatty acid to form soap.

(1) Fat + NaOH + H_2O → Fatty acid + Glycerol + NaOH
(2) NaOH + Fatty acid → Sodium soap + H_2O

This splitting up of the fat into its two component parts is called *saponification.* When fats are decomposed by the action of bacteria, the fatty acids thereby set free are likely to be split to lower fatty acids (i.e., to fatty acids having fewer carbon atoms in the molecule). These lower fatty acids have very penetrating and offensive odors and give the disagreeable taste and smell to rancid fats (e.g., "strong" butter).

Melting point. The three fats, olein, palmitin, and stearin, differ from each other in their melting points; olein melts at about O C, whereas the melting point of palmitin and stearin is variously given from 45 to 60 and from 55 to 70 C, respectively. Most fats we eat (e.g., butter and lard) and also the body fat of an animal are mixtures of the three fats we have been studying. These fats differ from each other in the relative amounts of olein, palmitin, and stearin they contain. The more olein and the less of the other two fats, the lower the melting point of a fat; the melting point of a few common fats is given in Table 18-1.

When a fat has such a low melting point that at ordinary room temperature it is melted, it is called an oil, for example, olive oil, cottonseed oil, and cod liver oil. The human body fat contains from 67% to 80% olein and has a melting point below the body temperature (37 C) and therefore is found in the form of droplets in the cells (Chapter 4).

Solubility. Fats are insoluble in cold or hot water; they dissolve in chloroform, ether, benzol, and hot alcohol. When a drop of olive oil is shaken with a small quantity of water, the two mix as long as the process of shaking continues, but when the shaking ceases, the oil separates from the water and rises to the top. If the oil is agitated with a water solution of soap, a milky fluid is obtained from which the oil separates very slowly. The fat in this condition is broken up into a countless number of microscopic droplets, each surrounded by a film of soap; by this means the fat can be kept in suspension for a long time. Such a mixture is called an *emulsion* and has a milky appearance; in fact, the white color of milk is to a large extent due to the emulsified fat.

Table 18-1. Melting point of fats

Margarine	31 C
Butter	36 C
Lard	44 C
Mutton fat	51 C
Tallow	53 C

Table 18-2. Occurrence of lipids in tissue

Bone marrow	96.0%
Adipose tissue	83.0%
Nerve	22.0%
Egg	12.0%
Milk	4.0%
Liver	2.5%
Blood	0.5%

Occurrence. ■ Lipids occur in many of our foods and in nearly all the tissues of our body, as seen in Table 18-2. Pure fats are odorless and tasteless; the odor of most of the fats with which we are acquainted is generally due to foreign material absorbed by the fat (as in the making of butter).

■ PROTEINS

The third group of organic foodstuffs, proteins, differs from carbohydrates and fats in many respects, especially in regard to composition. The protein molecule contains at least five elements present, by weight, in approximately the proportions given in Table 18-3. Some proteins contain phosphorus; others contain such elements as iron and iodine. Proteins constitute the largest part of the dry material of all cells and are indispensable for life.

Composition. ■ The most complex chemical compounds in the world are proteins. The subject of digestion of proteins and their function and fate in our body is unintelligible unless the student has at least a rudimentary knowledge of the subject. When a protein, for example, egg albumin, is boiled with acid or undergoes digestion or putrefaction, the large protein molecule breaks up into a number of simpler molecules known as amino acids. In somewhat the same manner as a molecule of starch is formed by the union of a large number of monosaccharide molecules, a protein molecule is constructed of a large number of amino acid molecules.

Classification. ■ Of the large number of proteins only those referred to in physiology will be mentioned. Proteins found as such in the animal body are spoken of as native protein; the simple and complex proteins listed herewith are native. Their classification is largely based on their solubility in various media (e.g., water, salt solution, acid or alkali media) and on their being precipitated from solution by various reagents.

Table 18-3. Approximate composition of protein molecules

Carbon	'50% to 55%
Oxygen	21% to 24%
Nitrogen	13% to 17%
Hydrogen	About 7%
Sulfur	0.2% to 7%

Simple proteins. Simple proteins on cleavage yield only amino acids.

Albumins. Egg albumin,* lactalbumin, and serum albumin are of animal origin; legumelin (peas) and leucosin (wheat) are vegetable albumins. They are soluble in pure water, precipitated by saturating their solution with ammonium sulfate but not by magnesium sulfate or sodium chloride. They are coagulated by heat and are then said to be denatured proteins. If denatured as a concentrated solution, a gel is formed; if as a dilute solution, a flocculent precipitate results.

Globulins. Egg globulin, lactoglobulin, serum globulin, fibrinogen, and myogen and myosin of muscle are of animal origin; vegetable globulins include legumin (peas), tuberin (potatoes), and edestin (wheat). Globulins are insoluble in pure water but are soluble in dilute salt solutions; they are coagulated by heat and precipitated by saturation with magnesium sulfate or by half saturation with ammonium sulfate.

Glutelins. Glutelins are found in the seeds of cereal grains. They are insoluble in water and neutral salt solution; they are soluble in dilute alkalies and acids.

Prolamines. Gliadin (wheat), zein (corn), and hordein (barley) are examples of prolamines. These are not soluble in water or absolute alcohol but are soluble in 80% alcohol.

Protamines. Protamines are the simplest of all naturally occurring proteins and possess the least complicated structure. They are soluble in water, dilute acids, and dilute ammonium hydroxide

*White of egg is 88% water, 10% albumin, 1% globulin, and 1% other material.

and are not coagulated by heat. Protamines are strongly basic in reaction and are found in sperm cells.

Histones. Histones occur as part of the nucleoprotein and contain a predominance of basic amino acids. They are soluble in water and dilute acids but are insoluble in dilute ammonium hydroxide.

Albuminoids or scleroproteins. Albuminoids are among the least soluble proteins and are the most difficult to digest. In the body they are found in an insoluble state; they are tough, yet plastic. Being found in all connective tissue, they confer on the organs, and the body as a whole, form, strength, rigidity, and elasticity. Among the albuminoids are collagen, elastin, and keratin.

COLLAGEN. Collagen forms the ground substance of bone and cartilage and is found in white fibrous (inelastic) connective tissue (tendons, aponeuroses, ligaments, dura mater, pericardium, and fascia). By boiling, especially with dilute acid, it is transformed into the well-known *gelatin.* Collagen is very slowly digested; gelatin is more digestible.

ELASTIN. Elastin is found in yellow (elastic) connective tissue in the walls of blood vessels (especially arteries), trachea, and lungs. It is very insoluble and hard to digest.

KERATIN. Keratin is found in the outer layer of the skin and in hair, nails, feathers, hoofs, and so on. It is indigestible.

Compound, complex, or conjugated proteins. These proteins are composed of a simple protein united with some other substance, and, according to the nature of this second component, there are the following groups of conjugated proteins: chromoproteins, nucleoproteins, glycoproteins, phosphoproteins, and lipoproteins.

Chromoproteins. In chromoproteins the simple protein is united with a pigment, for example, hemoglobins, cytochromes, and flavoproteins.

Nucleoproteins. In nucleoproteins the protein is combined with a nucleic acid. The chromatin material of the nuclei of cells and also the substances composing the viruses are mostly nucleoproteins. They constitute the most characteristic material of all living matter. Their molecular weights vary and may run into the millions.

Glycoproteins. Glycoproteins are formed by the union of a carbohydrate with the protein; the most important glycoprotein is mucin. Mucin is found, for example, in saliva and in the secretions of mucous membranes. Its high viscosity and sliminess may be of value as a lubricant and also in furnishing protection to mucous membranes.

Phosphoproteins. Examples of phosphoproteins are vitellin of egg yolk and casein of milk. As the name indicates, these proteins are combined with phosphoric acid. They are soluble in dilute alkalies; hence the addition of acid causes them to be precipitated (as in souring of milk).

Lipoproteins. Lipoproteins are simple proteins combined with fatty substances. Such lipoprotein complexes are found in serum and brain tissue.

Derived proteins. Derived proteins are produced from the previously named native proteins in various ways.

1. Those derived by hydrolysis; for example, by the action of the digestive enzymes, acids, or alkalies
   a. *Acid metaproteins*
   b. *Alkali metaproteins*
   c. *Proteoses or albuminoses*
   d. *Peptones*
   e. *Peptides* (will be spoken of in connection with digestion)
2. *Coagulated proteins* formed by such factors as heat

Characteristics. ■ Since the various proteins differ so largely from each other, it is difficult to make general statements concerning their properties. Very few proteins dialyze through parchment paper; they are, therefore, generally regarded as colloids (Chapter 3).

Proteins give a number of color reactions by means of which their presence is generally detected: the xanthoproteic, Millon's, biuret, and the ninhydrin tests. By these tests one does not discover a protein but substituents of the protein molecule. For example, the xanthoproteic test is given by the benzene ring in the protein molecule; the Millon test is given by a benzene ring to which an OH group is attached, and so on.

Separation of various proteins in solution. ■ Most dietary proteins serve chiefly to supply a mixture of amino acids from which the body tissues can select those they need for the synthesis of their own particular proteins. Many chemical compounds, on being added to a solution of protein, cause the protein to be precipitated from solution. In some instances the protein behaves as an acid and, on reacting with a heavy metal (e.g., lead, aluminum, zinc, iron, mercury, and silver), forms an insoluble salt (e.g., lead proteinate). On this reaction is based the antidotal action of white of egg or milk in cases of poisoning by salts of heavy metals. On the other hand, when a protein reacts with tannic acid (tannin) or picric acid, it behaves as a base. When substances of this nature are applied to the mucosa (e.g., of the mouth) or to the exposed animal tissues, the precipitation of the protein causes a hardening or wrinkling (dryness and puckering of the mouth); such agents are called *astringents.* The precipitated protein forms a barrier to the deeper penetration of the astringent into the tissue and thereby localizes the action of this substance. Their efficacy to precipitate the proteins of the blood enables the astringents to stop the oozing of blood from smaller vessels (hemostatic or styptic action). In some instances, checking the action and growth of bacteria (antiseptics) or destroying them completely (germicides) is the result of the precipitation of the bacterial proteins.

Many proteins can be "salted out" (precipitated) by one or more neutral salts, such as magnesium sulfate, ammonium sulfate, or sodium chloride. Albumin and globulin are coagulated by heat; this may be used as a ready test for protein in urine. These proteins are also precipitated by strong acids; nitric acid is used in Heller's test for albumin.

Since the proteins are so vital to life it behooves us to investigate their components, the amino acids.

Amino acids. ■ From our study of lipids it will be recalled that a molecule of an organic acid contains a group (COOH) known as the carboxyl group. This group gives the compound its acid properties and therefore enables it to unite with basic substances to form organic salts. To illustrate, acetic acid reacting with sodium hydroxide forms sodium acetate (an organic salt) and water.

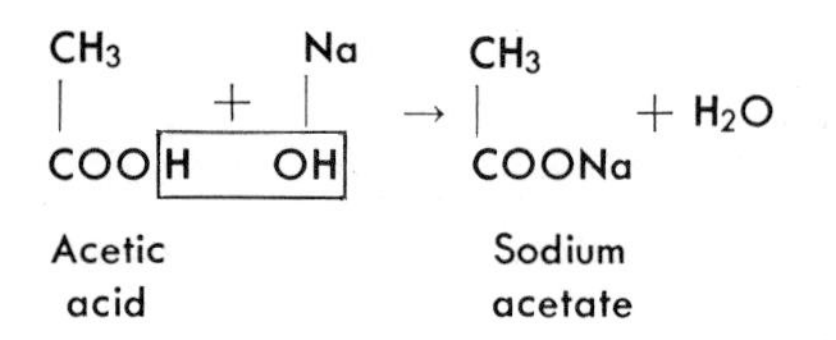

Let us replace one of the hydrogen atoms of the CH_3 group of acetic acid with an NH_2 group:

$$
\begin{array}{c} H_2C\text{—}NH_2 \\ | \\ COOH \end{array}
$$

Amino acetic acid or glycine

The new product is an amino acid. Amino acids may perhaps be called one of the most important groups of compounds in existence. They are characterized by having a carboxyl (COOH) group and an amino (NH_2) group; the former gives them acidic and the latter, basic qualities *(amphoteric).* In solution both the acidic and basic groups are ionized to form dipolar ions or *zwitterions.* For example, glycine in solution gives an electrically neutral ion.

$$
\begin{array}{c} H_2C\text{—}COO^- \\ | \\ NH_3^+ \end{array}
$$

Amphoteric substances will react with both acids or bases to form salts. Since proteins are composed of amino acids, they are amphoteric and thus are able to neutralize both acids and bases. This property of proteins is responsible for their buffering action in blood and other tissues.

Amino acids are soluble in water and are diffusible. The following are the formulas for a few simple members of this group.

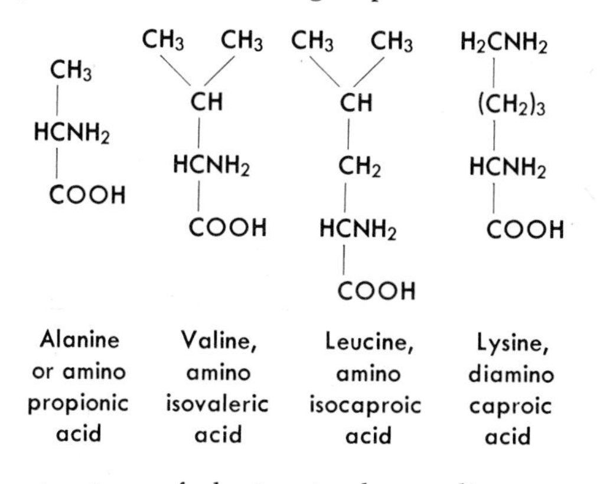

The structure of alanine is shown diagrammatically in Fig. 18-1.

We may add the following:

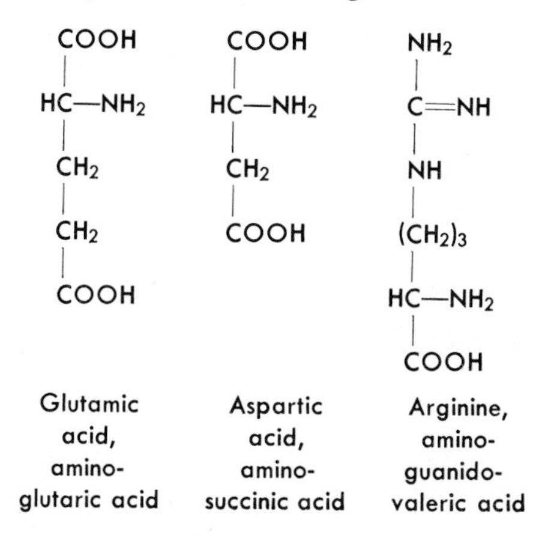

The aforementioned amino acids have their carbons linked in an open chain and are known as aliphatic amino acids. It should be noted that all these amino acids have an amino group attached to the α-carbon, that is, the carbon next to the carboxyl group. They are, therefore, re-

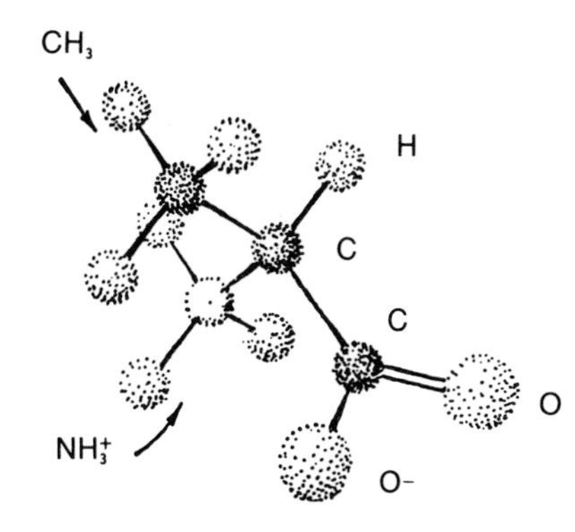

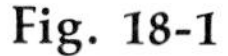

Fig. 18-1

Diagram of structure of alanine. (After Pauling: Sci. Am. **183**:35, 1950.)

ferred to as *α-amino acids.* We spoke earlier of proteins containing basic amino acids, by this is meant there are two or more amino groups, and lysine and arginine are examples. Aspartic and glutamic acids are acidic amino acids since they contain two carboxyl groups.

Some amino acids belong to the cyclic or aromatic compounds. In these compounds the carbon atoms are grouped to form a ring. The parent substance of the aromatic compounds is benzene,* C_6H_6. On oxidation it gives rise to oxybenzene, better known as phenol or carbolic acid. The structural formula for benzene may be written as:

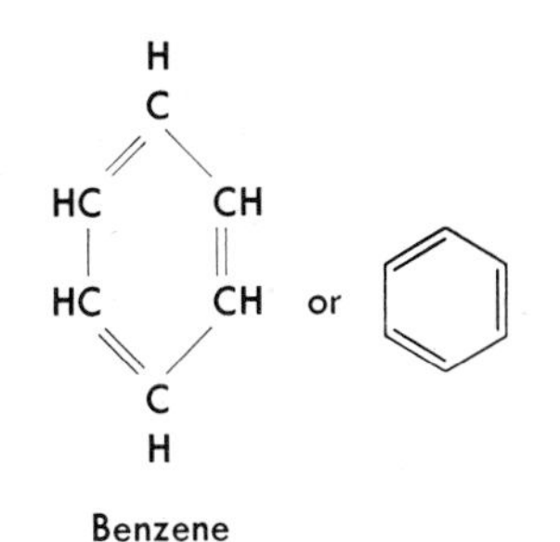

Benzene

Frequently the formulas are abbreviated by omitting the carbon and hydrogen atoms, as shown.

*Not to be confused with benzine, a cleaning fluid.

In the benzene molecule we may substitute methane and obtain methane benzene, toluene.

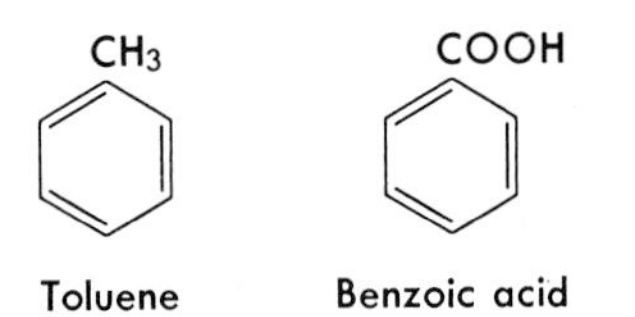

Toluene Benzoic acid

If now the CH_3 group is oxidized, benzoic acid is formed, $C_6H_5 \cdot COOH$. Or suppose that one of the hydrogen atoms of benzene is replaced by alanine (aminopropionic acid); a more complex amino acid, phenylalanine, is obtained.

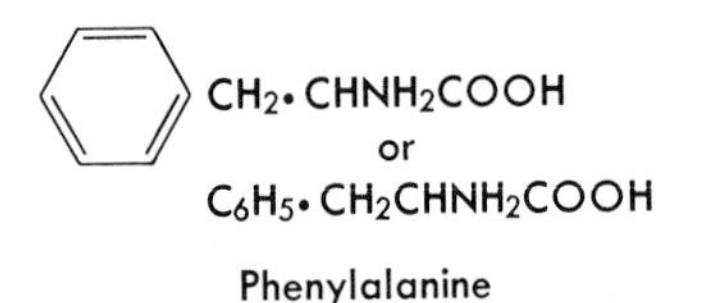

Phenylalanine

Alanine can also unite with oxybenzene or phenol and form para-hydroxyphenylalanine, generally called tyrosine.

The previous discussion will give the reader a fairly good picture of these complex organic substances; for the composition of the other amino acids, he may consult more advanced texts on biochemistry. The following is a list of the remaining more frequently mentioned amino acids: isoleucine, serine, thyroxine (contains iodine), tryptophan, threonine, histidine, cysteine, cystine, and methionine (the last three contain sulfur).

Importance of amino acids. ■ Amino acids are of prime importance because they are the building stones of the protein molecule. Since the molecules of amino acids are both acidic and basic, they can unite with each other to form larger and more complex molecules. To illustrate this, let us take 2 molecules of glycine.

$$\underset{\text{Glycine}}{NH_2CH_2COOH} + \underset{\text{Glycine}}{NH_2CH_2COOH} \rightarrow$$

$$\underset{\text{Glycyl-glycine}}{NH_2CH_2CO{-}NHCH_2COOH} + H_2O$$

It will be noticed that the carboxyl of 1 molecule and the amino group of the other molecule react with each other, resulting in the formation and splitting off of water. This leaves a free valency on each of the glycine molecules; by uniting they form what is called glycyl-glycine. A compound formed by the union of any two amino acids (e.g., glycyl-glycine) is known as a *dipeptide.* It is very important to notice that the dipeptide molecule has a COOH group and an NH_2 group, similar to the original individual amino acids. Consequently, a dipeptide can unite with a third amino acid molecule to form a tripeptide. This union of amino acids with each other is called a *peptide linkage.*

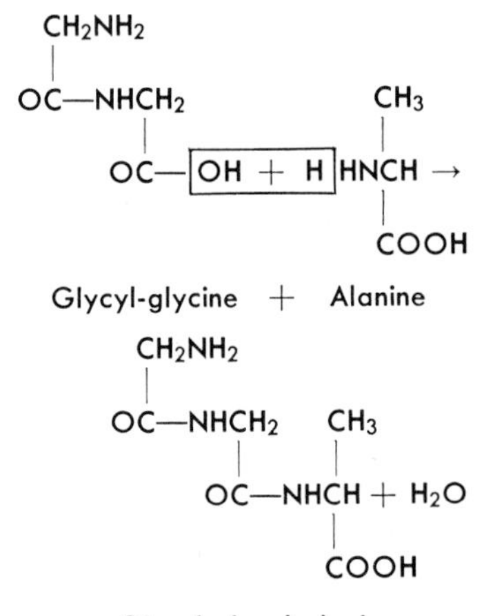

Glycyl-glycyl-alanine

When a still larger number of amino acids unite with each other, *polypeptides* are formed. A number of polypeptides uniting give rise, progressively, to peptones, proteoses, and proteins. In this the molecules become increasingly more complex and the molecular weight larger, ranging from 900 to truly colossal proportions. There are twenty-two kinds of amino acids known to exist in our body. They are similar in that each amino acid has one or more carboxyl and amino groups. They differ from each other in the composition of the remainder of the molecule.

■ VITAMINS

Vitamins do not constitute a class of chemically related compounds, as the proteins or carbohydrates do. Indeed, they have nothing in common chemically except that they are all organic substances. What then is a vitamin? It is a naturally occurring organic constituent of the diet that cannot be manufactured by the body. These compounds, in minute quantities, are indispensable for life and maintenance of normal tissue activities. For convenience, these diverse compounds will be discussed as two groups based on their solubilities, that is, fat-soluble or water-soluble.

■ Fat-soluble vitamins

Vitamin A. ■ Vitamin A occurs in two forms in tissues of animal origin and is available as a precursor from plants in the form of carotenes. Vitamin A, or retinol, is a complex primary alcohol with the empirical formula of $C_{20}H_{29}OH$. It is esterified with a fatty acid for storage in the liver and is redistributed to the various organs in the form of a protein complex. Vitamin A is necessary for maintenance of the chemical integrity and the normal functioning of epithelial tissue throughout the body. The neuroepithelial cells of the retina are the only cells where the metabolic reaction of this vitamin has been demonstrated. Vitamin A, in the aldehyde form, performs a highly specialized role in the visual process. Both the rods and cones of the eye contain light sensitive pigments, which require vitamin A for their formation and functioning. *Rhodopsin,* or *visual purple,* is the pigment contained in the rods. When the eye is exposed to light, rhodopsin is bleached and two substances, retinal and a protein, opsin, are produced. In the dark, rhodopsin is regenerated and the sensitivity of the eye to light is restored (dark adaptation). If there is insufficient retinal present in the retina, rhodopsin cannot be restored and night blindness (the inability to perceive objects in dim light) results. In man, night blindness is one of the first indications of a deficiency of vitamin A. Color vision requires the combination of retinal with different protein opsins to form the pigments found in the cones.

In addition to its role in vision, vitamin A has been implicated in other metabolic reactions. It is apparently involved in carbohydrate and nucleic acid metabolism. It may be concerned in the electron transport system and seems to be involved in the synthesis of mucopolysaccharides, especially chondroitin sulfate. Further experimental evidence is necessary before the mechanisms of vitamin A entry in these reactions can be determined.

Vitamin D. ■ When certain sterols are irradiated with ultraviolet light, a change in their molecular structures produces chemical compounds known as vitamin D. Little is known definitively about the complex physiology of this compound and how it is involved in calcium and phosphorus metabolism. The active form of the vitamin in the animal body is apparently cholecalciferol (D_3), which derives from the irradiation of 7-dehydrocholesterol. Recent evidence indicates that a further enzyme activation mechanism in the liver changes the D_3 molecular structure to another molecule that is considerably more effective. This compound is 25-hydroxycholecalciferol (25-HCC). Vitamin D does increase the absorption of calcium and phosphorus from the intestinal tract, but this does not explain its role in bone formation. The only definite statement that can be made at this time is that calcification of the matrix of the bone occurs when vitamin D is present; when it is absent, calcification does not occur. A current hypothesis suggests that 25-HCC in both the intestine and bone enters the cell nuclei to unmask a specific DNA. Complementary RNA is made which codes for the components of the calcium transport system.

Vitamin E. ■ The tocopherols, isolated first from wheat germ oil, have a pronounced influence on the reproductive system of some laboratory animals. They are also required for the maintenance of function and structure in the muscles and peripheral vascular systems of a variety of laboratory animals. The physiological role of this vitamin in man has not been clearly elucidated, but it may be involved as an antioxidant in the preservation of the lipid configuration of cell membranes.

Vitamin K_1. ■ This naphthoquinone derivative plays an unknown but necessary role in the hepatic synthesis of prothrombin. The role of vitamin K in the electron transport system is still unclear.

■ Water-soluble vitamins

Thiamine (vitamin B_1). ■ This member of the B complex contains both a pyrimidine ① and a thiazole ② nucleus and has the following structure:

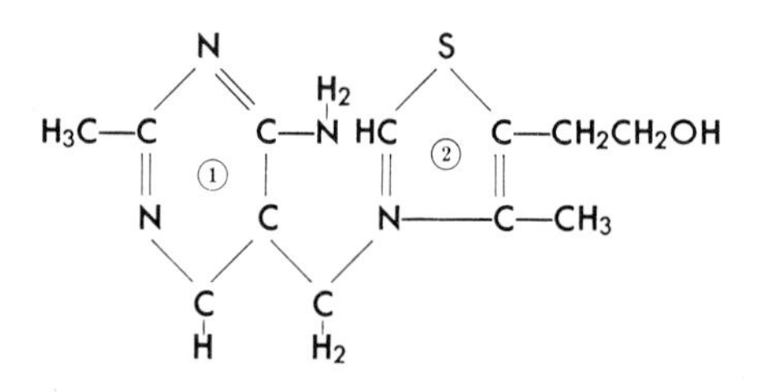

Carbohydrate metabolism in all cells of the body depends on the presence of a coenzyme, thiamine pyrophosphate (TPP), or cocarboxylase, which is essential for the decarboxylation of keto acids, including pyruvic and α-ketoglutaric. Without the presence of TPP, oxidative metabolism of glucose would not be possible, and about 90% of the energy contained in this compound would be lost.

Nicotinic acid (nicotinamide, niacin). ■ The structure of nicotinamide is:

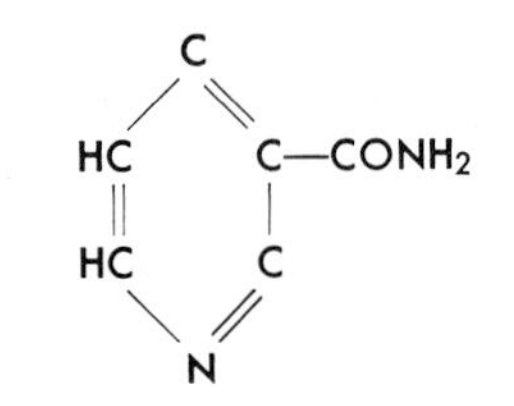

Nicotinic acid, in combination with a nucleotide, forms two important coenzymes which are required in cellular respiration as components of the electron transport system. These two coenzymes, necessary for the activity of the dehydrogenase enzymes, are composed of nicotinamide, ribose, adenine, and phosphoric acid. If 2 molecules of phosphoric acid are present, the compound is nicotinamide adenine dinucleotide (NAD); if 3 molecules of phosphoric acid are present, the compound is known as nicotinamide adenine dinucleotide phosphate (NADP). Previous designations for these two factors were NAD, formerly DPN and coenzyme I and NADP, formerly TPN and coenzyme II. The nicotinamide portion of the coenzyme is responsible for the transfer of hydrogen molecules in the oxidation of substrate by a dehydrogenase. It is possible for the body tissues to transform tryptophan (an amino acid) into nicotinic acid, although this is not an extremely efficient process.

Riboflavin (vitamin B_2). ■ Riboflavin ($C_{17}H_{20}N_4O_6$) is a yellowish compound that also functions in the electron transport system. The two coenzymes formed when riboflavin is phosphorylated are flavin mononucleotide (FMN) and flavin adenine dinucleotide (FAD); these in combination with proteins are involved in a number of oxidation-reduction reactions.

Pyridoxine (vitamin B_6). ■ Pyridoxine and two other closely related compounds, pyridoxal and pyridoxamine, are grouped together as vitamin B_6. The structure of pyridoxine is:

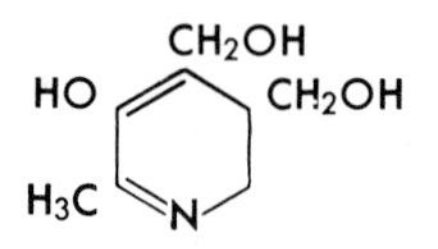

Pyridoxine is essential in the metabolism of fatty acids and in the conversion of protein to fat. Pyridoxal is important in the transport of amino acids and metal ions across cell membranes. Pyridoxal phosphate is a coenzyme involved in the decarboxylation of several amino acids. The formation of pyridoxamine from pyridoxal, and its reversal, is required in transamination reactions.

Pantothenic acid. ■ Pantothenic acid, found in every living cell, functions as a part of coenzyme A. Pantothenic acid is a derivative of a β-alanine and has the following structure:

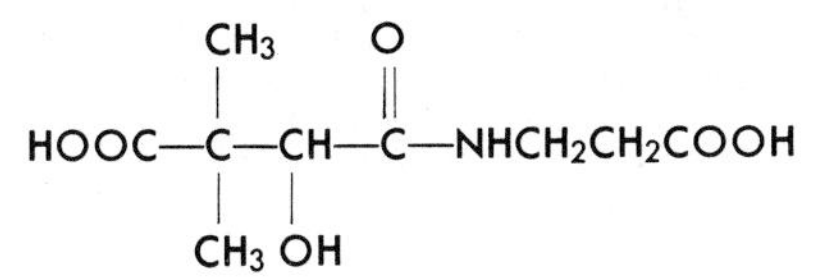

Coenzyme A functions in the transfer of 2-carbon fragments and is, therefore, concerned with carbohydrate, fat, and protein metabolism.

Biotin. ■ As a coenzyme, biotin ($C_{10}H_{16}O_3N_2S$) is necessary for the fixation of CO_2 or carboxylation such as the conversion of pyruvic acid to oxaloacetic acid. Biotin has been implicated in the biosynthesis of arginine and pyrimidines.

Folic acid (pteroylglutamic acid). ■ Folic acid is composed of three major parts: pteridine, *p*-aminobenzoic acid, and glutamic acid. One of its derivatives, tetrahydrofolic acid, serves as a carrier of 1-carbon compounds (such as hydroxymethyl and formyl) in much the same fashion as coenzyme A is involved in 2-carbon transfer. It is involved in the synthesis of purines, pyrimidines, and some amino acids. Folic acid, or folacin, is necessary for the normal formation of red blood cells.

Vitamin B_{12}. ■ The structure of vitamin B_{12} ($C_{63}H_{90}N_{14}O_{14}PCo$) is unique in that it has been found to contain the element cobalt. The cobalt is in the center of a ring structure somewhat similar to that of heme. This vitamin is necessary for erythropoiesis; it also appears to have some growth-stimulating properties. Vitamin B_{12} forms a cobamide coenzyme that participates in methylation reactions.

Inositol. ■ Inositol is a hexahydroxy cyclohexane, which plays an unknown role in lipid metabolism. It is possible that it is an intermediate between carbohydrates and aromatic substances.

Choline. ■ Choline, whose formula is

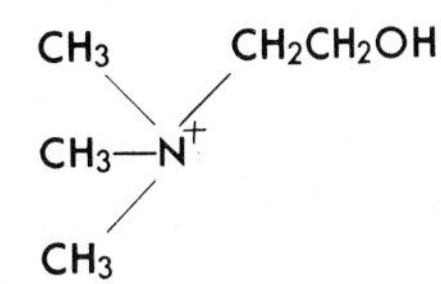

may not be a true vitamin, but it is a dietary requirement. It serves as a source of methyl groups. It aids in the storage and mobilization of fats. Choline is a constituent of some phospholipids and is used in the formation of acetylcholine.

Ascorbic acid (vitamin C). ■ Ascorbic acid is a hexose derivative with the following formula:

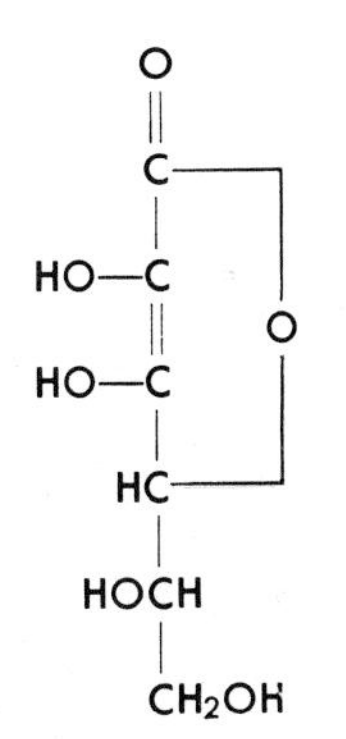

Only primates (including man) and the guinea pig require this substance in the diet. The important function of ascorbic acid is to make possible the construction and maintenance of the intercellular-binding substances that are vital in cartilage, bone, and muscle. Ascorbic acid is necessary for the formation of the amino acid hydroxyproline from proline. Hydroxyproline is present only in collagen and this may be the reason ascorbic acid is required for maintenance of intercellular substances. Although this substance can be readily oxidized and reduced, ascorbic acid has not been revealed as a coenzyme in any biological oxidation system; it appears to be important in the reduction of folic acid and plays a part in the manufacture of adrenocortical hormones.

■ INORGANIC SUBSTANCES

Sodium, potassium, calcium, phosphorus, chlorine, magnesium, molybdenum, manganese, iodine, copper, iron, zinc, cobalt, sulfur, bromine, fluorine, aluminum, arsenic, nickel, and silicon are all inorganic substances which are found in cells and tissues. Not all of these elements have been assigned definite physiological functions and many of them can be toxic if present in too great quantities. Minerals are vital to human nutrition, and disruption of normal mineral metabolism can cause severe physiological reactions. The importance of these elements have been discussed throughout this text, especially in Chapter 24.

One vital and important inorganic substance upon which all life depends is water. Chapter 24 presents a discussion of water metabolism.

READINGS

Borgstrom, G.: Principles of food science, New York, 1964, The Macmillan Company.

Eskin, N. H., Henderson, H. M., and Townsend, R. J.: Biochemistry of foods, New York, 1971, Academic Press, Inc.

Kermode, G. O.: Food additives, Sci. Am. **226**(3):15-21, Mar. 1972.

Scrimshaw, N. S.: Food, Sci. Am. **209**(3):72-80, Sept. 1963.

Stefferud, A., editor: Food, 1959 Yearbook of Agriculture, Washington, D. C., 1959, U. S. Government Printing Office.

General anatomy of the digestive tract
Mouth and pharynx
Deglutition and the esophagus
Stomach
Small intestine
Large intestine

DIGESTIVE TRACT—ANATOMY AND MOVEMENTS

The food that we eat cannot be used directly by the cells of the body but first must be ingested, propelled, digested, and absorbed. The digestive tract is the system in which these physical and chemical changes occur, and we shall discuss them in this chapter and in Chapter 20. First we will be concerned with the anatomy of the digestive tract and the movements of food and water along this tract and later with digestive secretions and absorption of these substances.

■ GENERAL ANATOMY OF THE DIGESTIVE TRACT

The alimentary canal is a coiled tube extending the full length of the trunk. It is divided into five segments: mouth and pharynx, esophagus, stomach, small intestine, and large intestine. Fig. 19-1 illustrates these segments and the closely associated glands and organs that are a part of the digestive tract.

The structure of the wall of the alimentary canal differs in the various regions but is usually composed of several layers. From the outside in, the coats (Fig. 19-10) are arranged in the following order:

1. A *serous* or *fibrous* coat composed of a thin layer of connective tissue
2. The *muscular* coat, composed of two layers of smooth muscle fibers, an outer longitudinal layer and an inner more highly developed circular layer
3. The *submucosa,* made up of loose connective tissue elements containing numerous blood and lymph vessels
4. The *muscularis mucosa,* containing smooth muscle cells
5. The *mucosa,* or mucous membrane, in which the glands peculiar to the alimentary tract are located

Constrictors, or sphincters. ■ At certain points along the alimentary tract the circular muscles are hypertrophied, forming areas of constriction, or sphincters. These areas are found in the upper (hypopharyngeal sphincter) and lower (cardiac sphincter) esophagus, at the antral end of the stomach *(pyloric canal* and *pylorus),* between the

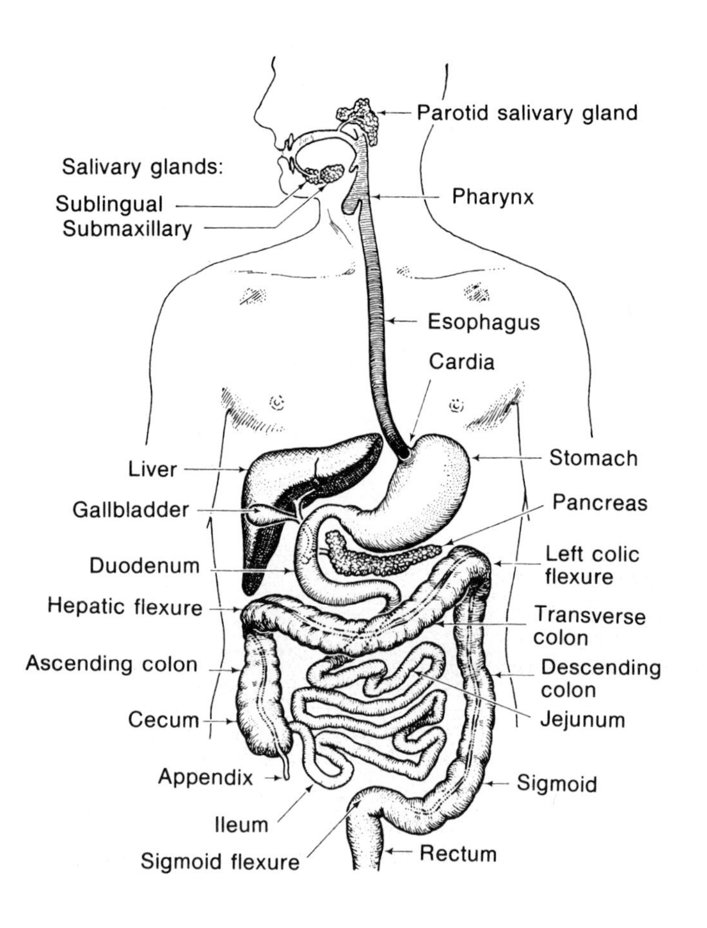

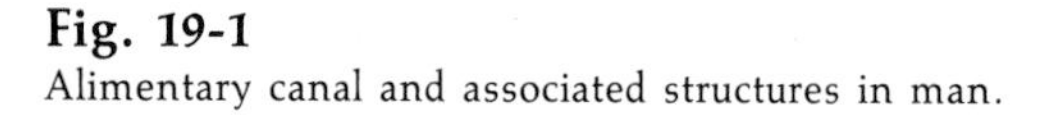
Fig. 19-1
Alimentary canal and associated structures in man.

small and large intestines *(ileocecal sphincter),* and the *internal* and *external anal sphincters.* The latter three sphincters are normally closed and restrict further passage of material while the first three areas assist in movement of the digestive contents.

Innervation of the digestive tract. ■ Except for the striated muscles of the mouth, the upper part of the esophagus, and the external anal sphincter, the musculature of the alimentary canal is composed of smooth muscle fibers. Electrical and mechanical properties of smooth muscle cells were discussed in Chapter 6. All smooth muscle is involuntary and nervous control in the gastrointestinal tract is both by extrinsic autonomic nerves (Fig. 19-2) and by intrinsic neural plexuses (Fig. 19-10). Two main neural networks are located between the longitudinal and circular muscle layers *(Auerbach's plexus)* and between the circular muscle layer and the submucosa *(Meissner's plexus).* Less well-defined networks are also present in the subserosa and mucosa of the gut. These plexuses interact extensively with each other and independently of extrinsic neural control. Many of the local neurogenic reflexes in the gut occur in these plexuses and, since they are interconnected, play an important role in the downward transport of the intestinal content. The intrinsic neural structures are modulated by extrinsic nerves supplied by both divisions of the autonomic nervous system (Chapter 11).

The fibers of the sympathetic nervous system

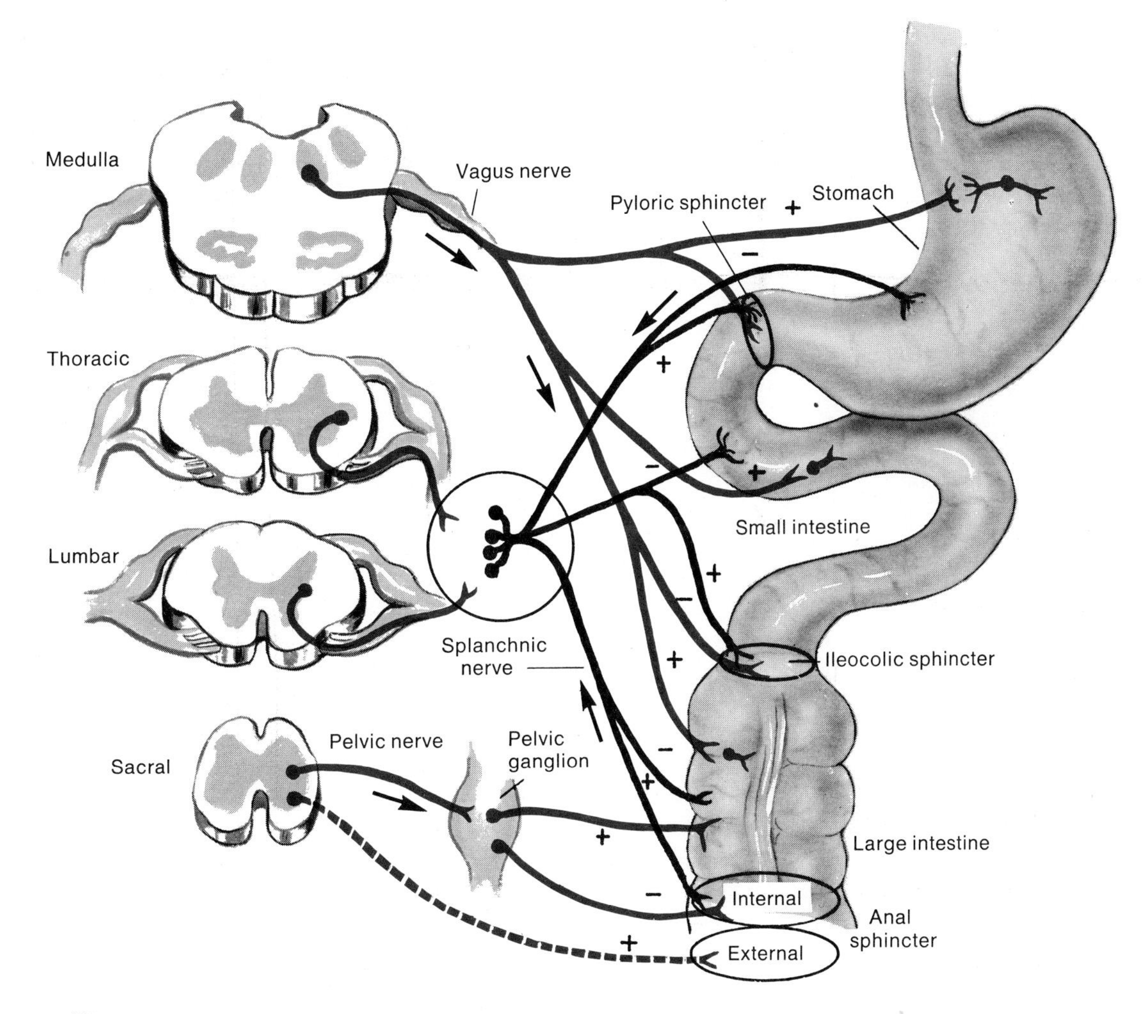

Fig. 19-2
Innervation of alimentary canal. All nerves are efferent. Parasympathetic system is indicated by red lines and sympathetic by black lines. Plus signs indicate excitation; minus signs, inhibition. The external anal sphincter is innervated by a somatic nerve.

reach the alimentary canal via the splanchnic nerves; preganglionic fibers of the parasympathetic are derived from two distinct regions of the central nervous system—the medulla oblongata in the brain and the sacral portions of the spinal cord (Fig. 19-2). Those fibers originating from cells in the medulla are part of the *vagi* and supply the muscles of the stomach, small intestine, and upper half of the large intestine. The vagal fibers make synaptic connections with the intramural plexus. Preganglionic fibers from the sacral cord reach the pelvic ganglia by way of the pelvic nerves; the postganglionic fibers innervate the lower half of the large intestine and the rectum. Although the results of the stimulation of parasympathetic and sympathetic nerves cannot be stated categorically, some generalizations (follow Fig. 19-2) can be made. The parasympathetic system is excitatory (indicated by plus signs) for all the musculature except the sphincters, where it is generally inhibitory (indicated by minus signs). The sympathetic nerves (black lines) are excitatory for the ileocecal sphincter, the internal anal sphincter, and the smooth muscle fibers of the muscularis mucosa throughout the entire gastrointestinal system and increase the number of folds in the tract; the remaining musculature is inhibited. Thus these two systems are antagonistic but as far as the proper propulsion of food *(chyme)* is concerned, they are complementary.

Additional control of gut movement. ■ In addition to the neural control of the gut, enteric hormones may be responsible for integration of gut movement. Gastrin, for instance, may increase tone in the lower esophageal sphincter and may be responsible for the increased motor activity of the lower ileum and colon after eating. Movements of the small bowel are affected by secretin and cholecystokinin. Prostaglandins also affect movement of the gut.

The membrane electrical properties of smooth muscle may constitute a major integrative mechanism. Because of the tight junctions (nexuses), smooth muscle cells spread over great distances act together electrically and *slow waves,* consisting of slow regular uniform fluctuations in membrane potential, are conducted over great distances. Superimposed on these slow waves are action spikes or potentials which can initiate contractions. It would appear that these slow waves which pass over great distances can trigger the action potentials followed by contraction, and thus can integrate the action of whole organs.

Peritoneum. ■ The interior of the abdominal cavity is lined with a serous membrane, the *peritoneum.* This is a double membrane; the outer, or parietal, layer is in contact with the body wall; the inner, or visceral, layer envelops the abdominal organs. A continuation of the peritoneum known as the *mesentery* extends to the small and large intestine from the dorsal body wall. By it these organs are suspended from the body wall and in it are carried the blood vessels, lymph vessels, and nerves for the intestine. The stomach is attached by a special fold, the *lesser omentum,* to the liver. The *greater omentum* hangs like an apron from the greater curvature of the stomach over the intestine to the colon. In this fold fat may accumulate. The organs of the abdomen are also supported by the striated muscles of the abdominal wall.

■ MOUTH AND PHARYNX

Mastication. ■ The most important physical change that food undergoes is that of mastication. This accomplishes two functions: reducing the food particles to a size convenient for swallowing, which also exposes more surface to the action of the digestive enzymes, and proper mixing of the food with saliva. Structures necessary for the proper mastication of food are illustrated in Fig. 17-1. The various movements of the lower jaw that help to accomplish this process and the muscles that produce them are:

1. Raising—masseter, temporal, and internal pterygoids
2. Lowering—digastric, mylohyoid
3. Projection—both external pterygoids when they act at the same time
4. Retraction—posterior fibers of temporal and the geniohyoid
5. Lateral movements—external pterygoids acting alternately

The muscular movements described are voluntary, and the motor nerves innervating these muscles are from a somatic source: the inferior maxillary division of the fifth cranial, or trigeminal, nerves (Fig. 10-2).

The *chewing reflex,* which is initiated by the tactile stimulus of food in the mouth, is responsible for much of the mastication process. Although chewing can be a voluntary act, the presence of food in the mouth evokes an inhibition of the muscles that normally keep the jaw closed.

As the jaw lowers, the strength of stimulus is decreased and the muscles contract, closing the jaw.

Teeth. ■ The movements of the lower jaw provide the muscle power for the teeth, which are extremely important in the mastication process. The force exerted by the molars of man can be from 29 to 90 kg; by the incisors, from 11 to 25 kg.

Man and certain animals have two dentitions: the temporary, composed of two incisors, one canine, and two molars for each lateral half of each jaw, and the permanent, composed of two incisors, one canine, two premolars (bicuspids), and three molars for each lateral half of each jaw (Fig. 19-3). The temporary teeth erupt from about the sixth or eighth month (lower central incisors) to the twenty-fourth month (second molars); these are gradually replaced by the permanent teeth starting about the fifth or sixth year and continuing until young adulthood.

A tooth (Fig. 19-4) consists of a *crown* (that part projecting above the gum), a *root* (which fits

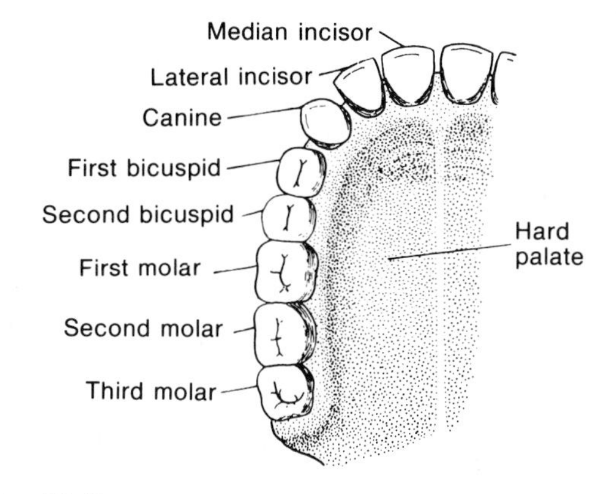

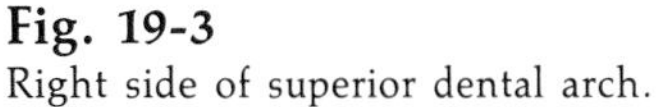

Fig. 19-3
Right side of superior dental arch.

Fig. 19-4
Longitudinal section of human molar tooth.

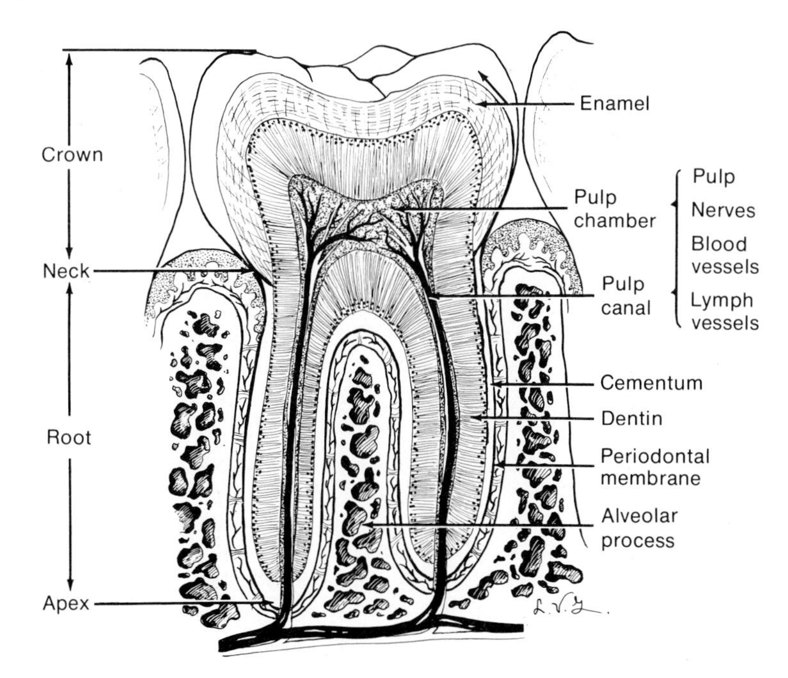

into an alveolar socket in the jawbone), and a *neck* (which is the junction between the crown and the root and lies just below the gum line). The inner structure of the tooth is composed of *dentin,* which surrounds the *pulp canal* and resembles bone chemically and structurally but is harder and contains tubules radiating from the pulp cavity. The *cementum* surrounding the root outside the dentin is also a bonelike structure. It is penetrated by bundles of coarse fibers from the *periodontal membrane.* This membrane acts as a bond between the jawbone and the cementum. At the apex of the tooth there is a foramen through which the blood vessels and nerves enter the pulp cavity. The pulp cavity also contains connective tissue and *odontoblasts,* which send fine processes into the dentinal tubules. *Enamel* covers the crown of the tooth and is a white, translucent, very hard material. Cementum, dentin, and enamel are formed of an organic base, or matrix, composed of proteins in which are deposited calcium phosphate, calcium carbonate, and magnesium phosphate in the form of apatite crystals similar to those found in bone. It has been determined with the aid of radioactive tracer substances that this structure is in a state of flux, with the minerals, calcium and phosphorus, being continuously removed and replaced.

Dental caries. In caries the insoluble calcium salts of the hard structures of the teeth are transformed into soluble salts and are washed away, so that a cavity is formed. The problem of dental caries is not a simple one, and several factors are involved in their production. Some of these are bacterial decomposition of food with the production of lactic acid, especially in crevices between teeth and at the margin of the tooth and gum; action of proteolytic bacteria on the protein matrix of the tooth; and the structure of the teeth, as in maldevelopment or dietary deficiencies (especially vitamin D). The presence of fluorine in drinking water (0.5 to 1 part per million) has proved to be beneficial in preventing caries. This element may act either by inhibiting the acid production of bacteria or by making the tooth enamel more resistant to acid. The importance of good dental hygiene in preventing caries is obvious; the use of a toothbrush and dental floss is commendable; periodic examination and prophylactic treatment are of prime importance.

■ DEGLUTITION AND THE ESOPHAGUS

From the food in the mouth a *bolus,* a soft rounded mass larger than a pill, is separated into the pharynx for swallowing. This deglutition may be initiated voluntarily but is completed by the *swallowing* reflex. This reflex can be elicited by stimulation of a number of areas in the mouth and pharynx. The swallowing center in the medulla evokes the complete act of swallowing through nuclei and motor neurons that control the complex musculature involved. The sequential relaxation and contraction of muscles opens a passage ahead of the bolus and closes the esophagus behind it; this creates a pressure gradient and moves the contents.

Deglutition from the mouth to the stomach may be divided into three stages. During the first stage the food is placed on the back of the tongue (Fig. 17-1), and the tip of the tongue is elevated and pressed against the hard palate. When the tongue is drawn back, this elevation travels toward the root of the tongue and propels the food to the back of the mouth and into the pharynx. The pharynx is a funnel-shaped tube between the mouth and the esophagus and trachea. The posterior nares (openings from the nasal cavities) and the Eustachian tube (Figs. 9-29 and 17-1) open into the pharynx, the common pathway for food that must go into the esophagus and air passing through the larynx into the trachea.

During the second, or pharyngeal, stage of swallowing, the palate and the contracted palatopharyngeal muscles close off the nose from the mouth. As the bolus reaches this area, respiration is briefly inhibited; the larynx raises abruptly to meet the tongue, and the *glottis* is closed, cutting off the airway. These movements can be observed and felt by following the Adam's apple as it bobs up and down with swallowing. The *epiglottis* is moved into the path of the bolus by the upward movement of the larynx and is tilted backward over the glottis; however, it does not act as a lid. Pressures of from 4 to 10 mm Hg are generated by movements of the tongue and larynx as the bolus moves over and around the epiglottis through the hypopharyngeal sphincter, and into the upper esophagus. Following the passage of the bolus, the larynx descends, the glottis and upper respiratory passages open, and respiration resumes. This entire stage occurs within one second.

The third stage of swallowing involves the esophagus, a muscular tube between the pharynx and stomach. The upper one fourth to one third of the esophagus has a layer of longitudinal striated muscle fibers surrounding a circular layer of fiber bundles. Between these layers and the squamous epithelial cells lining the esophagus are thick submucosal and muscularis mucosa layers. These create a number of folds in the inner lining, which obliterate the lumen. The bolus smooths out these folds when it is swallowed. Smooth muscle fibers arranged in circular and longitudinal layers constitute the musculature of the lower third of the esophagus. The middle third is a transition region between striated and smooth muscle fibers. There is no anatomical muscular structure at the lower end of the esophagus separating it from the stomach, although a slight thickening may occur about 3 to 4 cm above the cardia. However, the lower end of the esophagus does act as a physiological sphincter. The striated muscles of the esophagus are innervated by portions of the glossopharyngeal and vagus nerves, which contain efferent motor fibers and afferent sensory fibers. Autonomic fibers of the vagus supply the smooth muscle regions of the esophagus.

Pressures at the pharyngeal side of the pharyngoesophageal junction rise to about 100 mm Hg. As the bolus reaches the sphincter it relaxes, and the bolus passes through; the sphincter immediately closes, and a peristaltic wave begins at the upper end of the esophagus. This peristaltic wave pushes the solid bolus ahead of it. If the pharyngeal phase of swallowing is rapidly repeated, the esophagus remains relaxed and the peristaltic wave occurs only after the final swallow. It has not been determined whether the higher center controlling the esophageal phase of swallowing is identical with that controlling the pharyngeal phase, but there is a definite sequential activation of the esophageal muscles by efferent nerves. A *secondary peristalsis* can occur in the esophagus without any movement of the mouth or pharynx, and it is initiated whenever there is an abnormal distention of the wall by a large piece of food remaining in the canal after a swallow. The propulsive force of the esophageal peristalsis is weak, and gravity provides a valuable assist in movement of the bolus through the esophagus.

The gastroesophageal sphincter is relaxed when swallowing is normal. Thus the barrier between the esophagus and stomach is reduced when swallowing begins; the pressure remains low until the peristaltic wave of the esophagus is complete. After the food arrives in the stomach, the pressure increases in the sphincter for a time and then gradually returns to a resting value. The time taken for the bolus to traverse the esophageal canal depends on its consistency; water takes about one second, whereas a large firm bolus may take nine seconds.

■ STOMACH

Structure and innervation. ■ The stomach is the dilatable portion of the otherwise narrow digestive tract, whose shape and position vary considerably in normal individuals. It is divided functionally into two parts: a storage and secretory area, the body, composed of the *fundus* and the *corpus;* and the *antrum,* a more heavily mus-

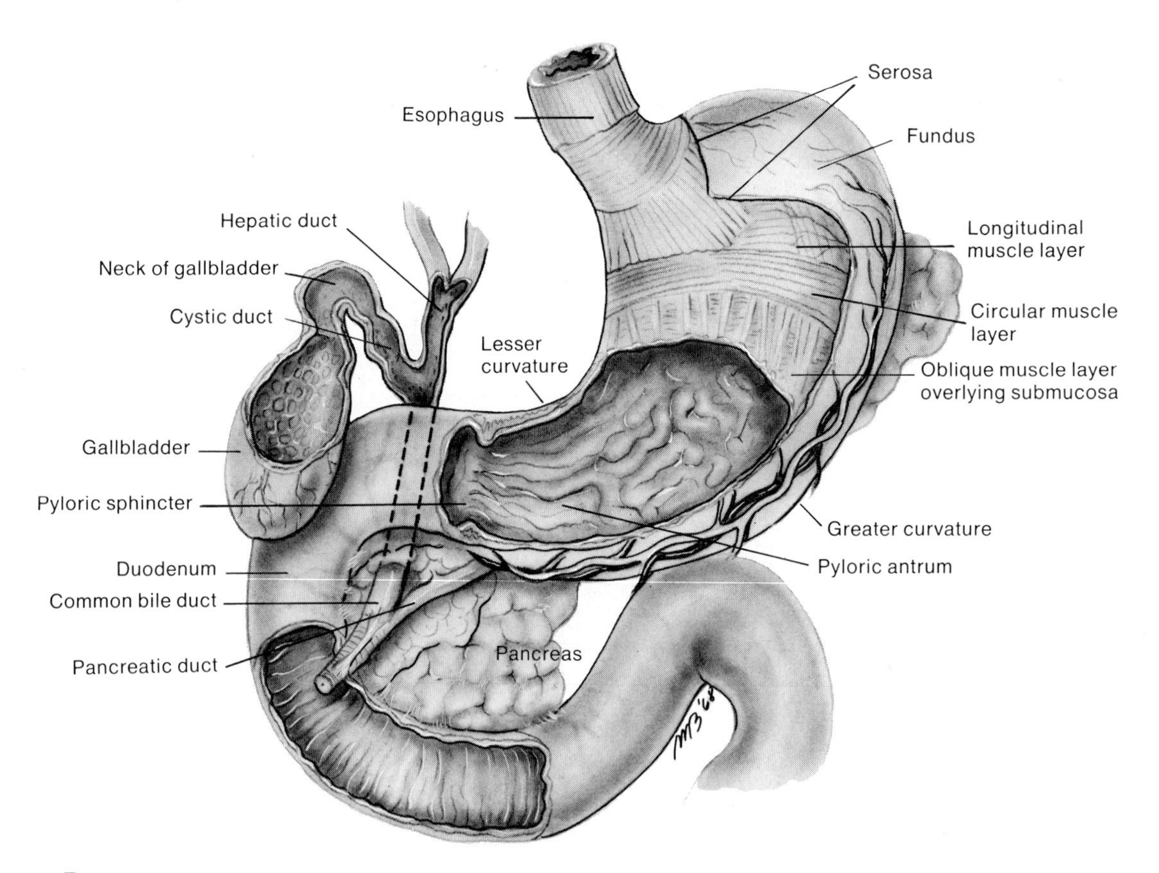

Fig. 19-5
Illustration of structure of stomach and upper small intestine. Relationship of gallbladder, pancreas, and ducts is shown.

cularized region. The area of the opening of the esophagus into the stomach is the *cardia,* and the opening into the duodenum of the intestine is the *pylorus,* which is closed by the *pyloric sphincter.* These are illustrated in Fig. 19-5. The longitudinal and circular muscles of the stomach wall are continuous sheets and therefore function as a unit. The mucosa of the stomach is divided into two parts—that present in its body and that present in the pyloric region. The mucosal layer in the body is covered with simple columnar epithelial cells, and its surface contains many depressions, or pits. Several glands empty into each pit (Fig. 19-6). *Chief* cells and *parietal,* or *oxyntic,* cells line the lumen of the glands. The mucosa of the pyloric region also is covered with epithelial cells; the glands in this region are composed of cells resembling chief cells; parietal cells are few in number or completely absent. Only mucus-secreting glands are found in the cardia.

The splanchnics (Figs. 11-2 and 19-2) and vagus provide the extrinsic sympathetic and parasympathetic innervation. Continuous intramuscular plexuses of the stomach regulate the response of the muscle to excitation. Stimulation of the sympathetics reduces peristaltic strength,

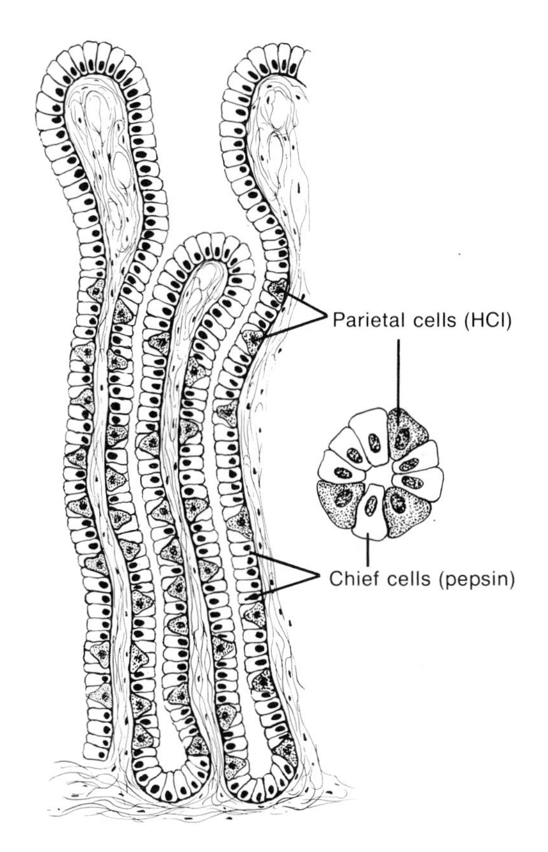

Fig. 19-6
Diagram of a gland from pit of stomach. The lumen of the gland is lined with chief cells and parietal cells. Chief cells contain and secrete pepsinogen (precursor of pepsin) and are most numerous; parietal, or oxyntic, cells secrete acid (HCl).

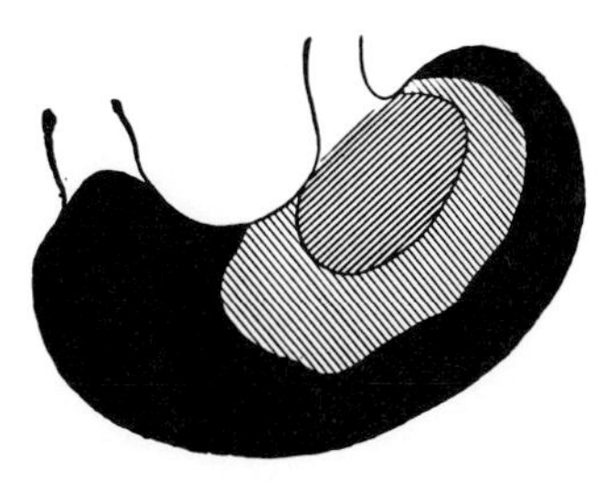

Fig. 19-7
Section of frozen stomach (rat) some time after feeding. The dark area was consumed first, and the material closest to the esophagus was consumed last. (From Howell: Textbook of physiology, W. B. Saunders Co.)

whereas stimulation of the vagus enhances it; the basic frequency of the wave is not changed. The stomach is an automatic organ and sympathectomy (surgical or other interruption of the sympathetic nerve supply) usually produces little change in gastric activity. Vagotomy produces difficulties, since the strength of the peristaltic wave is reduced and gastric retention may occur. Stimulation of many parts of the central nervous system affects gastric motility, but the effect is unpredictable.

Filling the stomach. ■ One of the important functions of the stomach is *storing of food;* in the human adult the capacity is approximately 1 liter. Because of the tonus of its smooth muscle fibers, the walls of the empty stomach are practically in contact with each other; the cavity is almost obliterated. As food enters the stomach, the cavity is enlarged without any appreciable change in the intragastric pressure. This is due not only to the plasticity of the smooth muscle but also to the shape of the stomach, which allows it to fill without changing pressure. The pressure in the abdomen outside of the stomach is not increased by eating, since the muscular wall accommodates itself to the volume of the gastric contents.

The first food swallowed is placed against the greater curvature in the body of the stomach, with successive layers lying progressively nearer the lesser curvature. The last food eaten lies near the upper end in the vicinity of the cardia (Fig. 19-7). Fluids tend to flow along the lesser curvature, although large volumes do flow around the food mass over the entire area of the stomach.

Gastric movements when full. ■ A few minutes after food enters the stomach, gentle peristaltic waves begin, which occur at a rate of 3/min in man. A pacemaker in the cardia initiates the waves of activity, which continue on to the pylorus. These waves are usually shallow throughout the stomach for about an hour after the ingestion of food. As many as two or three waves may be present at any one time. Since the musculature is thin in the body of the stomach, the contractions are weak; in the antrum the contractions are stronger. These gentle movements serve to mix the food with the secretions of the stomach; the resulting mixture is called *chyme.* As digestion proceeds, the peristaltic waves deepen as they reach the antrum, which is the site of vigorous churning of the chyme. Because of the viscosity of the gastric contents, pressure increases ahead of the wave. If the intragastric pressure is high enough, it will overcome the pressure barrier in the pylorus, and a small amount of the contents will pass into the duodenum. As the peristaltic wave continues, the terminal portion of the antrum, which includes the pyloric sphincter, contracts, forcing gastric chyme backward into the antrum and preventing chyme from returning from the duodenum. The sphincter then relaxes and is again open. About 1% of the stomach contents passes through the sphincter with each deep peristaltic wave.

Rate of gastric emptying. ■ The activity of the gastric musculature is influenced by the chemical and physical properties of the chyme in the duodenum. In this manner the *rate of gastric discharge is adapted to the secretory and motor capacity of the intestine.* Filling of the duodenum restrains gastric motility; this allows the duodenum time to propel its contents further along before the next gastric discharge. The physical state of the chyme, the size of its particles, its volume, its osmotic pressure, and its acidity also can influence the rate of gastric emptying. All these factors elicit the *enterogastric reflex,* which is transmitted to the stomach by the intramuscular plexuses, the vagus, and the sympathetics. Foods containing fat slow gastric motility (Fig. 19-8) and delay gastric evacuation by reducing the amplitude of the peristaltic waves. Presence of fat in the duodenum exerts an inhibitory influence on gastric motility by a humoral agent, *enterogastrone,* manufactured in the intestinal wall and carried by the bloodstream to the stomach wall. Thus two types of "feedback" mechanisms exist by which the duodenum can control gastric motility and emptying: the enterogastric reflex, which acts rapidly, and humoral control, which is somewhat slower. Actually, a combination of these may occur when any of the above factors affect gastric emptying.

Hunger. ■ Hunger sensations have two origins: local (gastric) and systemic. These unpleasant and sometimes painful sensations that are referred to the region of the stomach are due to the rhythmic contractions of the gastric musculature—hunger contractions. In an investigation of this matter, the individual under observation swallowed one end of a rubber tube; this end was provided with a small, soft rubber balloon; the other end was attached to a water manometer. Gastric contractions compressed the air in the rubber balloon; this, communicated to the manometer, caused the water in the distal limb

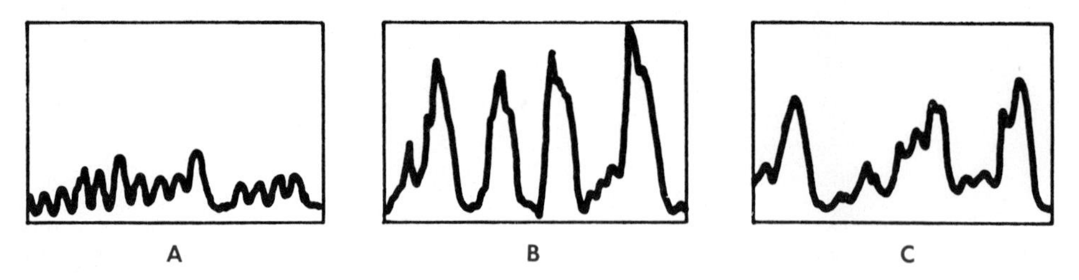

Fig. 19-8
Tracings illustrating gastric motility in man with gastric fistula. **A,** Curves obtained after fat-containing breakfast. **B,** Same as **A** but with subject in situation of anxiety and tension. **C,** Same as **A** but with fat omitted from food. (From Wolff and Wolff: Human gastric function, Oxford University Press.)

to move when the sensation of hunger was felt. At this time the individual pressed an electric key and, by means of a signal magnet, caused a heavy straight line to be written below the manometer curve on the kymograph (Fig. 19-9). The periods of hunger pangs, it will be noticed, coincide with the hunger contractions. These contractions (each lasting for about thirty seconds) occur in periods of from thirty to forty-five minutes; they are followed by periods of repose lasting from one-half to two and one-half hours.

Partaking of food stops the contractions for the time being. They are also inhibited by fatigue. Intellectual work seems not to affect them, and they continue during sleep. Drinking ice-cold water inhibits them; strenuous exercise also has this effect. Tobacco smoking stops them, and this inhibition may continue for from five to fifteen minutes after the cessation of smoking.

It has been found that in a cat, gastric peristalsis and evacuation ceased immediately when the animal became enraged or exhibited signs of fear or anxiety. In a dog even a strange surrounding may stop all gastric activity for two or three hours. In man, anxiety and nervousness increase, whereas other emotions decrease, stomach motility (Fig. 19-8).

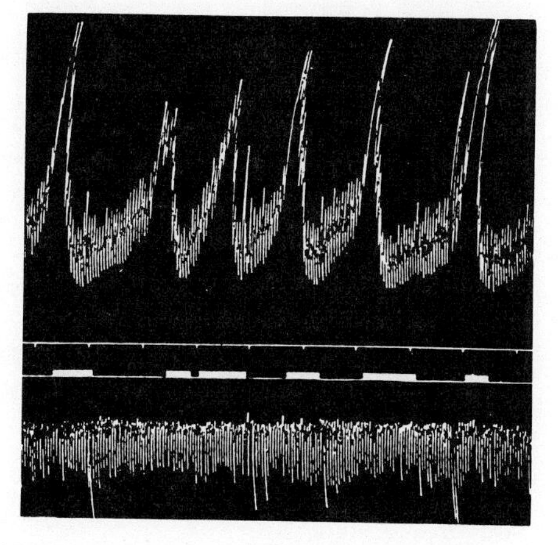

Fig. 19-9
Tracings of intragastric pressure and respiration in man. Top record is intragastric pressure (large oscillations due to contraction of stomach, and small oscillations due to respiration); second record is time in ten-minute intervals; third record is report of hunger pangs; fourth record is respiration registered by means of a pneumograph about the abdomen. (From Cannon: Bodily changes in pain, hunger, fear, and rage, D. Appleton-Century Co.)

Appetite. ■ Although generally thought of as closely related, hunger differs from appetite. The latter term is more closely associated with the desire for and the ability to enjoy food. As such, it is to a large extent acquired, being based on previous eating experiences. In contrast, hunger is an inherited physiological phenomenon. Appetite and hunger may exist independently of each other. Many individuals never experience hunger; scheduled times of eating and a good appetite forestall any real hunger. It is certainly true that conditions which decrease the tonicity of the gastric musculature also inhibit the appetite. This view is supported by the fact that deficiency of vitamin B_1 reduces the motor activity of the alimentary canal and is associated with lack of appetite, anorexia.

Vomiting. ■ This reflex act relieves the upper gastrointestinal tract of its contents and occurs either because the contents are irritating or the organs themselves are more irritated than normal. Vomiting begins with a feeling of nausea accompanied by excess salivation. A strong contraction in the upper intestine and pylorus empties the contents into the fundus of the stomach. A deep inspiration followed by closure of the glottis and relaxation of the esophagus occurs. Immediately the voluntary muscles of the abdomen contract in a jerky manner, and the stomach is compressed, causing evacuation through the esophagus. Coordination of the vomiting reflex is centered in the medulla.

■ SMALL INTESTINE

The small intestine with its digestive glands is the most important area of the digestive tract, since most of digestion and absorption occurs here.

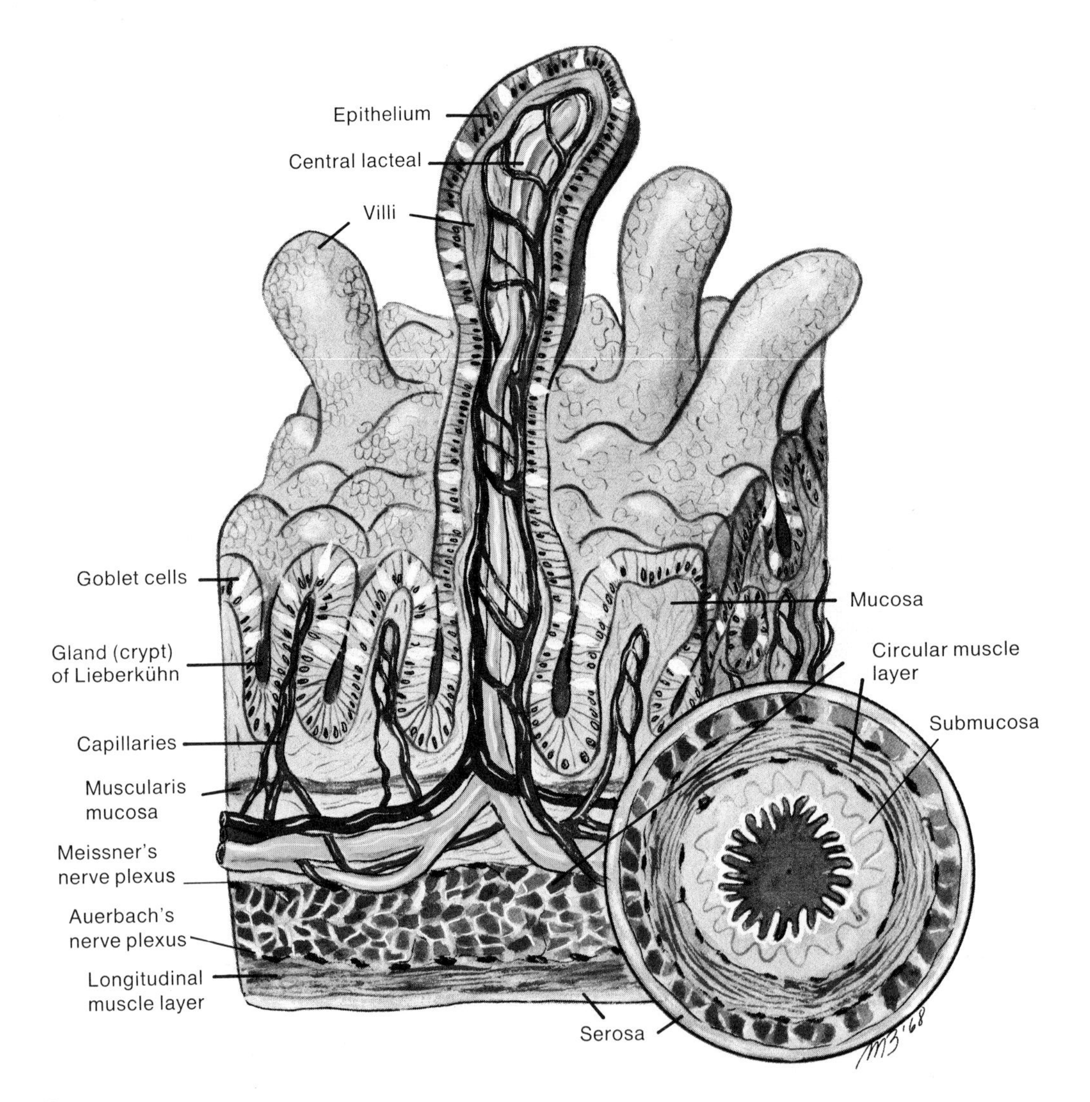

Fig. 19-10
A section of small intestine showing the intestinal villi. Microscopic sections of the columnar epithelial cells of the villi show these to have numerous projections, the brush border. Each villus contains blood and lymph vessels important for absorption in the small intestine. Cross section of intestine appears at lower right.

Structure and innervation. ■ The small intestine is composed of three segments: the *duodenum* (about 20 cm long), the *jejunum,* and the *ileum* (2 to 3 m long). Ducts from the liver, gallbladder, and the pancreas (a long, narrow, thin gland lying back of and below the stomach) (Figs. 19-1 and 19-5) open into the duodenum, generally through a common orifice about 7.5 cm from the pylorus. The gallbladder provides a reservoir for the storage of bile, which is continually secreted by the liver. Evacuation of this organ is controlled primarily by the presence of food in the duodenum. Fats, hydrochloric acid, etc., promote the production of a hormone cholecytokinin, which stimulates the contractions necessary to empty the gallbladder. The circular muscle layer of the intestine is somewhat thickened and is probably responsible for most of the intestinal movements. In order to accomplish its functional role of absorption the small intestine requires a vast epithelial surface. The mucosa of the upper two segments of the small intestine has a large number of circular folds, and, in addition, the surface is studded with small projections about 1 mm in height. These are the *intestinal villi* (Fig. 19-10). The surface of these villi is covered with columnar epithelial cells, and a further increase in the surface area is produced by the *brush border* (composed of microvilli) of these cells. At the base of the villi are located the *crypts of Lieberkühn,* and below these in the submucosa are the *glands of Brunner.* These glands and specialized cells perform secretory activities, which will be considered in Chapter 20.

Automaticity and rhythmicity are two characteristics of the muscles of the small intestine. A basic electrical rhythm originates near the entrance of the bile duct and moves down the duodenum at a velocity of 19 to 20 cm/sec and at a rate of 17 to 18/min. Muscular contractions occur at intervals of some multiple of 3.4 seconds and do not travel very far along the intestine. The activity of the small intestine is regulated by

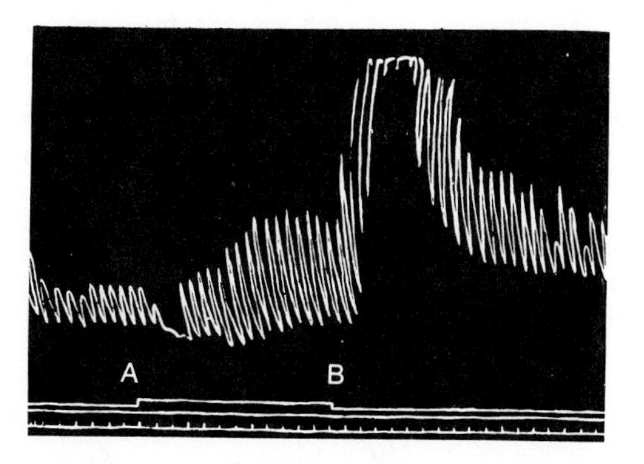

Fig. 19-11
Tracing showing effect of stimulation, **A** to **B**, of right vagus on intestinal contraction. (Starling.)

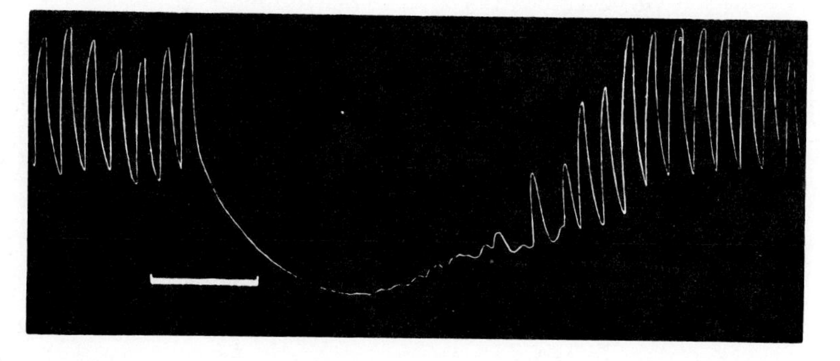

Fig. 19-12
Tracing showing effect of stimulation, during duration of heavy white line, of splanchnic nerves on intestinal contraction. (Starling.)

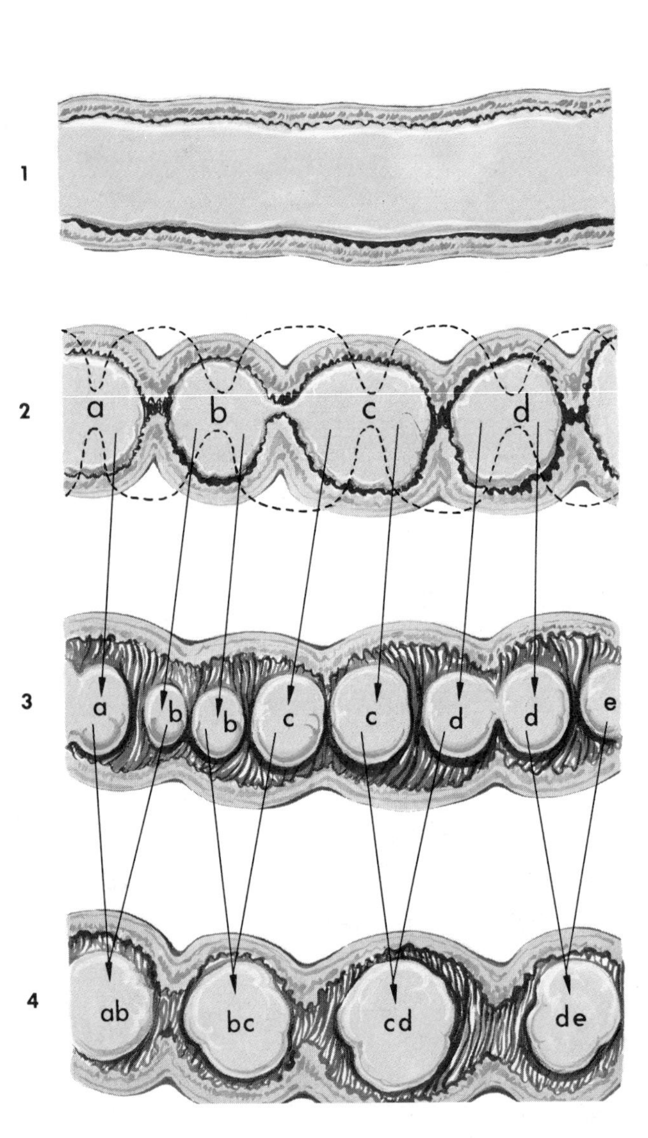

Fig. 19-13
Diagram showing process of rhythmic segmentation in a segment of the intestine. The sequence of events proceeds from **1** to **4**.

the vagi (Fig. 19-11) and the splanchnics (Fig. 19-12). Action of these nerves is reversed in the control of the ileocecal sphincter (Fig. 19-2). The normal stimulus for the intrinsic neural mechanism is the presence of food in the canal.

Movements. ■ In the intestine two distinct types of movements are observed: *rhythmic segmentation* and *peristalsis.*

Rhythmic segmentation is best described by following the accompanying diagram (Fig. 19-13). A quiescent loop of the intestine, *1,* suddenly divides itself by the contraction of its circular muscle fibers into a number of segments, *2.* After a few seconds each of the segments divides into two parts, *3;* next, one of the halves, *a,* formed from a previous segment, unites with a neighboring half, *b,* formed from another segment, and thus forms a new segment, *4,* of the same size as the original, *2.* The segmentation rhythm in the duodenum of man is 11/min; the rate slows somewhat in the ileum. This type of activity is ideal for mixing purposes, and it also aids the circulation of blood and lymph.

Peristaltic movements are responsible for propelling chyme slowly down the digestive tract. Luminal content moves down the tract about 1 cm/min and chyme reaches the ileocecal valve about three hours after the onset of gastric emptying. When the small intestine is locally stimulated (Fig. 19-14), *c,* the circular muscle 2 to 3 cm above this point contracts, *b,* and relaxation occurs a few centimeters below, *d;* this is the law of the intestine. If these waves move only a short distance at a rate of 1 to 2 cm/sec, the chyme will proceed slowly down the small intestine. In some animals and in man, under abnormal conditions, a compound wave of contraction and relaxation can move downward as a *peristaltic rush,* which pushes the chyme more rapidly through a long section of the intestinal tract. Peristaltic waves characteristically travel in an aboral direction, and they are superimposed on the segmental movements already discussed.

A third type of movement, which is indepen-

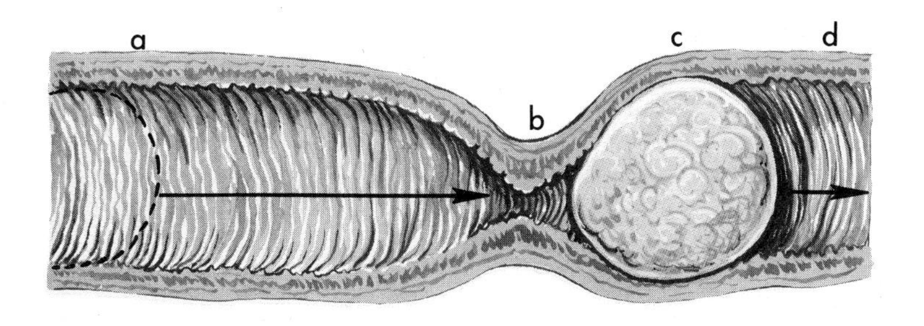

Fig. 19-14
Diagram showing peristalsis. The presence of chyme at **c** provides a local stimulation, causing the circular muscle at **b** to contract and at **d** to relax and thus propelling the material down the intestine.

dent of the contractions of the muscular walls of the gut, occurs in the small intestine; it is the movement of the *mucosa* and *villi.* This is responsible for the numerous folds and the changing pattern of these folds seen in the small intestine. Contraction of the muscularis mucosae, located just beneath the mucosa, causes these movements, and the presence of chyme is a sufficient stimulus for the movement of the villi and the folding of the mucosa. The villi move either by swaying back and forth or by shortening and lengthening. These contractions of the villi can be independent of activity of the muscularis mucosa, since even an individual villus may contract in some areas of the intestine.

Ileocecal sphincter. ■ Separating the ileum and the colon is the ileocecal sphincter, which is normally closed. It serves two functions: (1) to delay passage of chyme from the small intestine and (2) to prevent regurgitation of material from the colon into the small intestine. Two reflexes that serve as examples of the influence of activity in one part of the digestive tract on that of another part are the *gastroileal reflex* and the *ileogastric reflex.* The first reflex is so named because shortly after a meal the ileocecal valve relaxes each time a peristaltic wave passes along the last few centimeters of the terminal ileum. If, however, the ileum is distended, gastric motility is inhibited, and the reflex has been appropriately designated.

■ LARGE INTESTINE

Anatomy and innervation. ■ The large intestine consists of the cecum, the vermiform appendix, the ascending, transverse, descending, and pelvic colons, and the rectum (including the anal canal) and terminates at the anus. The longitudinal muscular coat is incomplete in the proximal colon and is largely collected into three flat bands, the *taeniae coli.* Since these bands are shorter than the intestine, the walls of the proximal colon have a saccular appearance. The mucosa of the large intestine possesses no villi, but it is quite thick, and the crypts of Lieberkuhn contain many goblet cells. Columnar epithelial cells line most of the surface except for the last 2 cm of the anal canal, which is lined with stratified squamous epithelium. The longitudinal coat becomes complete in the lower colon and rectum; its contractions elevate and shorten the anal canal.

Both the myenteric and submucosal plexuses are present in the large intestine. Vagi supply parasympathetic innervation to the proximal colon, and parasympathetic fibers for the remainder of the colon come from the sacral segments of the spinal cord (Fig. 19-2). Sympathetic fibers are also present in this part of the gut. Stimulation of the parasympathetics increases the activity of the large intestine (excepting the internal anal sphincter), whereas stimulation of the sympathetics is inhibitory.

Movements. ■ The cecum and ascending colon have a type of segmenting movement that thoroughly mixes the material received from the small intestine; in the transverse colon the circular muscles contract, forming *haustra*, and exert kneading movement on the material. Propulsive movements, consisting of waves of contractions moving from one segment to the next, push the material along the large intestine.

Mass movements, which may occur three to four times a day, are initiated by entrance of food into the stomach, the *gastrocolic reflex*. These movements empty the proximal colon into the distal colon rather rapidly, and on some occasions fecal matter is moved into the rectum.

Defecation. ■ Distention of the rectum by the entrance of fecal material is the normal stimulus for defecation. The sensory impulses are carried to the lumbrosacral portion of the spinal cord, and the return via motor nerves induces increased activity in the musculature of the colon and rectum; pressure rises within the rectum; the internal and external sphincters relax, and this activity can serve to empty the distal colon as high as the splenic flexure. The coordination of the defecation reflex occurs in the medulla; the center lies near the vomiting center and is probably associated in some manner with the respiratory centers and influenced by even higher levels of the central nervous system. In a normal adult the defecation reflex may be assisted or suppressed volitionally. By lowering the diaphragm with a full inspiration and closing the glottis, contraction of the chest and abdominal muscles increases intrathoracic and intra-abdominal pressure; this aids in expulsion of the feces. Voluntary inhibition is expressed by strong contraction of the striated muscles of the pelvic diaphragm and external anal sphincter. The rectum then relaxes, and although feces remain, the stimulus for defecation is removed. Continued voluntary inhibition of the defecation reflex will lead to *constipation* and should be avoided when possible. Since fecal material consists, in part, of material not digested (chiefly the cellulose of plants), it is logical that diet has an influence on the quantity of feces produced.

READINGS

Brooks, F. P.: Control of gastrointestinal function, New York, 1970, The Macmillan Company, pp. 148-240.

Christensen, J.: The controls of gastrointestinal movements: some old and new views, New Eng. J. Med. **285**:85-98, 1971.

Davenport, H. W.: Physiology of the digestive tract, ed. 2, Chicago, 1966, Year Book Medical Publishers, Inc., pp. 11-78.

Texter, E. C., Jr., Chou, C.-C., Laureta, H. C., and Vantrappen, G. R.: Physiology of the gastrointestinal tract, St. Louis, 1968, The C. V. Mosby Co., pp. 73-138.

20

DIGESTIVE TRACT—SECRETIONS AND ABSORPTION

The various processes taking place in the alimentary canal change the composition of the ingested food to render it suitable for cell utilization. Secretions of the digestive tract and accessory organs contain the necessary enzymes for this purpose. These enzymes split larger molecules into smaller ones by *hydrolysis,* or *hydrolytic cleavage.* This can be simply illustrated by the splitting of a disaccharide molecule into two monosaccharides.

$$C_{12}H_{22}O_{11} + H_2O \rightarrow C_6H_{12}O_6 + C_6H_{12}O_6$$

The digestive tract also furnishes the means for transferring material from the lumen of the canal into the bloodstream—*absorption.*

Secretory glands. ■ Secretion, which involves both active and passive transport of material (Chapter 3), is brought about by specialized epithelial cells. These cells may be organized into complex or simple organs, referred to as *secretory glands* (Fig. 20-1). These glands can be thought of as a tube closed at one end and, in the case of *external secretions (exocrine glands),* opening into a lumen at the other end. This lumen connects either directly or by means of a special duct to a free surface such as the skin or the mouth. A dense network of capillaries surrounds the gland.

In general, two distinct processes take place in a gland (Fig. 20-1): the gland cell may transfer certain materials such as H_2O and NaCl from the blood to the lumen of the gland (Fig. 20-1, *a*), and it may synthesize a new compound from materials supplied by the blood (Fig. 20-1, *b*). All the digestive enzymes are formed in the digestive glands by the latter process. That secretion by the digestive glands is an *active* phenomenon and not merely a passive filtration is supported by the following facts:

1. Secretions contain substances manufactured by the gland cells.
2. The salts in the secretions have a concentration different from those in the blood, which requires an expenditure of *energy* by the cell.
3. The consumption of oxygen and production

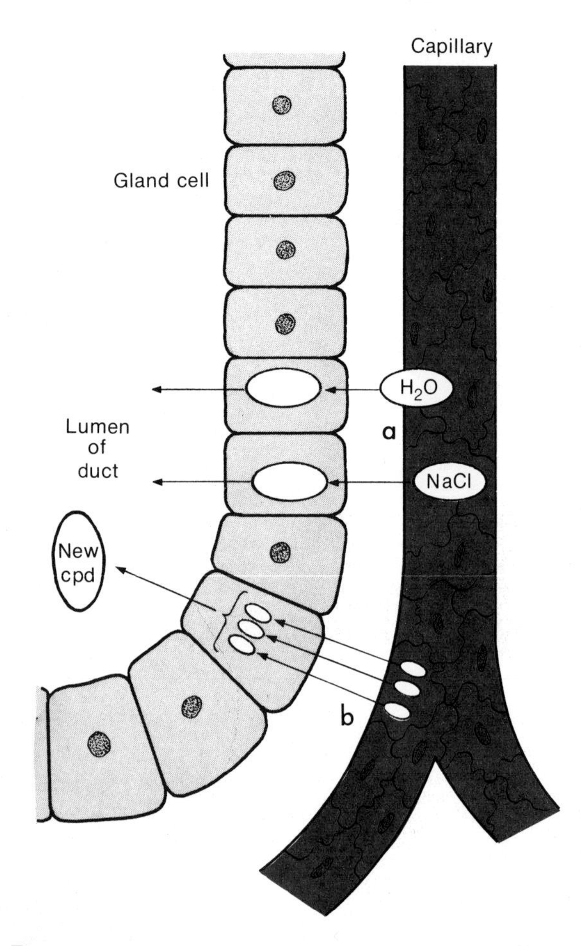

Fig. 20-1
Diagram showing processes occurring in a gland. Material can be transferred across gland cells from the blood to the lumen, for example, water and sodium chloride, **a.** Materials also can be removed from blood and synthesized into a new compound in the gland cell and transferred into the lumen of the gland, **b.**

of CO_2 are increased during activity of the gland, which indicates an increase in catabolism to supply the necessary energy.

4. Activity of the gland cells is accompanied by changes in electrical potentials.
5. Histological changes occur in certain gland cells during different phases of their activity.

We shall now consider the various parts of the digestive tract and associated structures as to their specific secretions.

■ MOUTH AND SALIVARY SECRETION

Nature of the secretions. ■ Saliva is secreted in man primarily by three pairs of large glands: the submaxillary, the sublingual, and the parotid (Fig. 20-2). Some small buccal glands and mucous glands are located in the mouth and pharynx, but these contribute only very limited quantities of material. An adult secretes 1 to 2 liters of saliva per day; the rate ranges from 0.5 ml/min to as high as 4 ml/min during maximal stimulation. The submaxillary gland produces about two thirds of this volume, the parotid one fourth, with the remainder deriving from the sublingual and small glands of the mouth. The histology of these glands and the composition of the saliva secreted vary; the parotids are serous glands, for their acini contain only one type of cell, which secretes a saliva devoid of mucin. The other two glands contain both serous- and mucin-secreting cells, with the submaxillary containing more serous cells and the sublingual more of the mucin-secreting cells. Serous secretion is thin and composed primarily of water and electrolytes, whereas mucinous secretion is clear and viscous. Mucin, a mixture of mucopolysaccharides and glycoproteins, is responsible for the lubricating action of saliva.

Functions of the secretions. ■ Saliva moistens and holds particles of food together, thus aiding in mastication, formation of a bolus, and deglutition. It enables us to taste dry foods by dissolving them. By moistening the oral cavity, saliva facilitates speech. During dehydration, secretion of saliva is decreased, making the mouth dry and contributing to the sensation of thirst. Salivary secretion has some bacteriostatic properties and performs an important role in oral hygiene by continuously rinsing the buccal cavity. The presence of the α-amylase enzyme, *ptyalin,* secreted by the parotid gland, is responsible for the

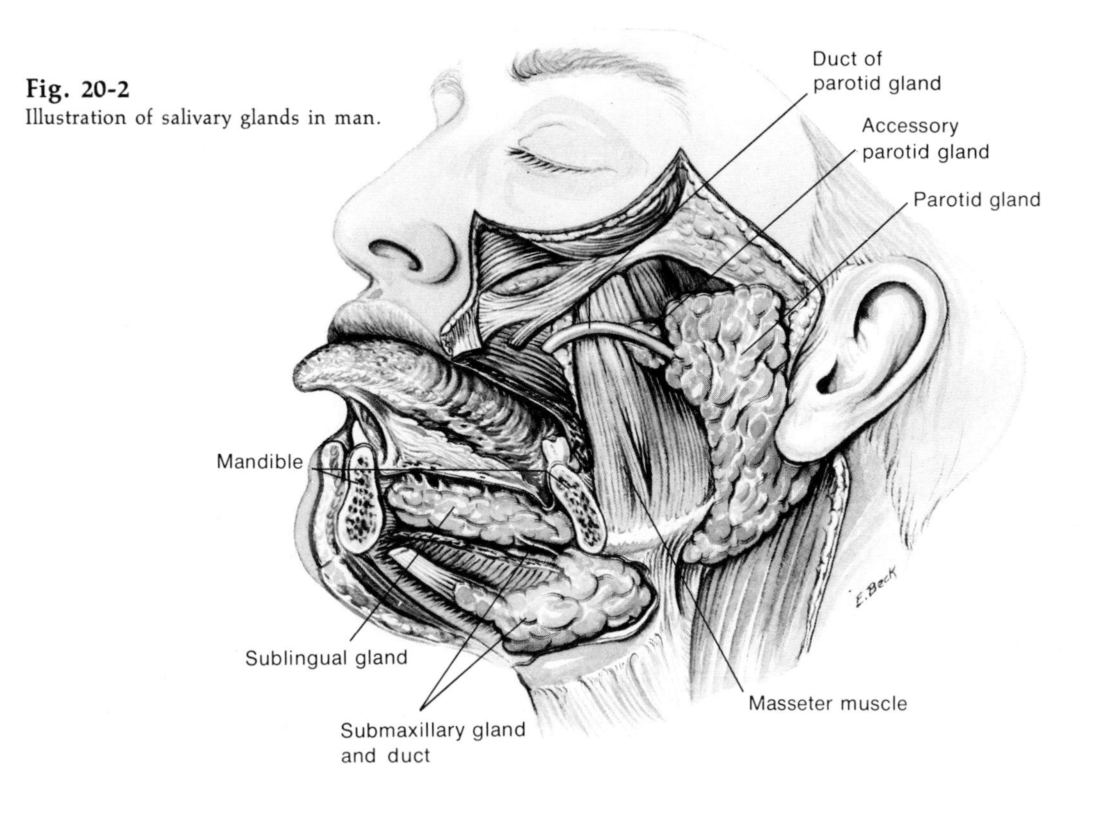

Fig. 20-2
Illustration of salivary glands in man.

principal digestive function of the saliva. If the cells of the parotid are examined after the gland has been at rest for some time, numerous granules are present; immediately after secretion the number of granules is decreased. These are the *zymogen* granules and are thought to be the locus of the storage of enzymes about to be secreted. Salivary amylase is a hydrolytic enzyme, which splits the α-1,4-glucosidic linkage of starch, producing smaller fragments. This enzyme has a pH optimum of 6.9 and is stable between pH 4 and 11. It requires the presence of anions for activity, with Cl^- being the most effective. The enzyme will continue to digest starch in the stomach until its inactivation by the gastric juice, which takes some time to penetrate the food mass. Human saliva also may contain insignificant quantities of other enzymes, but these contribute little to the digestive functions.

Composition. ■ Saliva is variable in composition, depending on the gland from which it originates, on the rate at which it is secreted, and to some extent on the stimulus that evokes the secretion. Water represents 97% to 99.5% of the saliva; the total osmotic pressure is always hypotonic in man; at maximum secretory rate it is about two thirds the plasma value. The pH of mixed saliva varies from 6.3 to 6.8, but it rapidly becomes more alkaline in the mouth as the result of a loss of dissolved bicarbonate. These bicarbonates can act as buffers, and the pH of the mouth remains quite stable regardless of the material ingested. The main electrolytes in saliva are Na^+, K^+, Cl^-, and HCO_3^-, with all but the K^+ at lower concentrations than that occurring in the plasma. It has been postulated that the acinar cells secrete a primary juice containing K^+, Cl^-, HCO_3^-, and H_2O, with the first two ions coming from the bloodstream and the third from both the blood and cellular metabolism. The primary secretion contains little Na^+ and is probably isosmotic with plasma; however, it is modified by exchanges with the bloodstream as it passes through the tubules to the excretory ducts. Urea may passively diffuse into the tubules; iodide is actively secreted into the tubules and is present in the saliva at a concentration many times that found in plasma. In general, K^+ is reabsorbed from the tubular fluid, and Na^+ enters it; since the K^+ is reabsorbed faster than the Na^+ is delivered, the tubular fluid becomes hypotonic. The tubular walls are apparently relatively impermeable to water, since the saliva remains hypotonic.

Organic substances found in saliva, other than previously mentioned enzymes, are *kallikrein* and the specific *blood group substances* (Chapter 12). Kallikrein has been demonstrated to be the agent responsible for the fall of blood pressure when saliva is injected intravenously into an animal; it acts enzymatically on a plasma protein to produce a polypeptide known as *bradykinin.* Vasodilation in the salivary gland during secretion may be due to kallikrein release. The specific soluble blood group substances have the same characteristics as the agglutinogen on the erythrocyte.

Neural control of salivary secretion. ■ All salivary secretions are in response to nerve impulses; there is little humoral control. The salivary glands are innervated by two sources. Centers governing secretion are located in the medulla and receive afferent impulses from the mouth, pharynx, and olfactory areas; they supply efferent innervation by both sympathetic and parasympathetic outflow. The parasympathetic fibers are carried by way of the chorda tympani (a branch of the seventh cranial nerve) to the submaxillary and sublingual glands and by way of the glossopharyngeal (ninth cranial nerve) to the parotid gland. All three glands receive fibers from the cervical sympathetic trunk. Stimulation of the parasympathetic efferent fibers causes vasodilation and an increase in volume of secretion in the glands; stimulation of the sympathetics results in vasoconstriction and some flow of saliva; both sources of stimulation result in a reduction in number and size of zymogen granules in tubular epithelium.

The nature and intensity of the stimuli acting through the receptor-center-efferent system determines the rate of flow of the secretion and its composition. During sleep the glands are in a truly basal state and there is very little flow. Flow is reduced by dehydration, severe mental effort, fear, or anxiety. Chewing stimulates the flow of saliva, and it increases as the size of the bolus and pressure required to chew it increase. Sour substances are potent stimulators of salivary secretion. Smells also stimulate to some extent the flow of saliva. A conditioned reflex can be established in an experimental animal in which sound becomes a sufficient stimulus for salivation (Chapter 10). Whether this same type of conditioned reflex has great importance in man has not been sufficiently established. The sight of food or the sound of cooking may only make one aware of the saliva always present in the mouth but may not actually increase the rate of flow.

■ GASTRIC SECRETION, DIGESTION, AND ABSORPTION

Secretions. ■ The surface epithelial cells of the stomach secrete both an alkaline juice (composed of water and ions) and an organic substance, *mucus.* The latter material is a gel-like substance that entraps the alkaline juice and adheres closely to the walls of the stomach, forming a layer 1 to 1.5 mm thick. This material provides lubrication and serves as both a chemical and mechanical protector of the gastric mucosa. It is present at all times and is increased with mucosal stimulation. Glands found in the cardiac region secrete mucus and very little else. The glands in the pyloric region also produce a great deal of mucus, which provides lubrication for the back and forth

During the interdigestive periods bile is stored in the gallbladder (volume 50 ml), where bile salts and pigments are concentrated twenty times by the rapid absorption of water and electrolytes. Contraction of the gallbladder occurs about thirty minutes after the ingestion of a meal and produces a flow of bile into the duodenum. Control of the gallbladder is by both neural (through the vagus) and humoral means. The hormone *cholecystokinin* is released from the duodenal mucosa in response to the presence of chyme. Fat, egg yolk, and meat are the most effective stimuli for the release of cholecystokinin.

Intestinal secretions. ■ *Brunner's glands,* located in the first part of the duodenum, secrete a viscous material rich in bicarbonate, which serves to protect the mucosa in the same manner as the gastric mucin in the stomach. The intestinal juice varies in volume and composition throughout the divisions of the intestine. It is slightly alkaline and is isotonic with blood. In addition to the usual electrolytes, it contains mucin, amylase, and enterokinase. The most effective stimulus for secretion of intestinal juice is local mechanical or chemical stimulation of the intestinal mucosa. Clear experimental evidence has not been obtained for neural regulation, but the vagus seems to have some role.

Brush border of the villus. ■ Terminal digestion by hydrolytic enzymes (Chapter 22) is one of the functions of a specialized substructure—the *brush border*—found in the apical portion of the epithelial cell. This border consists of finger-like projections, the *microvilli,* which form a boundary between the lumen of the intestine and the interior of the cell. The membranes of these microvilli have been found to contain the following enzymes: alkaline phosphatase, ATPase, cholesterol ester hydrolase, retinyl ester hydrolase, β-glucosidase, sucrase, maltase, isomaltase, trehalase, lactase, and leucylglycine hydrolase. The second important function of this organelle is that of absorption and active transport of materials from the lumen of the gut into the cell.

■ Digestion and absorption in the small intestine

We shall now consider the digestion and absorption of the major foodstuffs, carbohydrates, proteins, and fats, as well as water and electrolytes in the small intestine.

Carbohydrates. ■ The major dietary carbohydrates are plant starch and sucrose. Both salivary and pancreatic amylase attack dietary starch, and the final end products of their activities are maltose, isomaltose (which contains the branching points), and glucose. Disaccharidases (lactase, isomaltases, maltase, trehalase, and invertase) are found in the brush border of the mucosal cells, and they complete the hydrolysis of the disaccharides to monosaccharides. The proximal part of the small intestine appears to be the major site of absorption of sugars, although sucrose can be digested and absorbed in the jejunum and ileum. Glucose is the most abundant monosaccharide resulting from digestive hydrolysis; it is absorbed by an active transport process (against a concentration gradient). All sugars that have a structure containing the requirements shown below are actively transported.

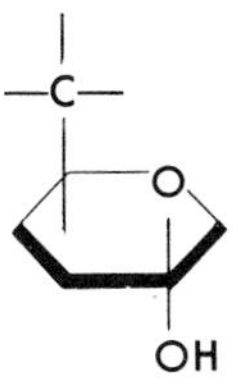

Glucose and galactose are transported by the same carriers; the exact nature and mechanisms are not known, but phosphorylation and insulin are not involved, and Na and K ions are required.

Fructose is not absorbed by diffusion alone, since some of it is converted in the mucosal cell to glucose and lactic acid. All other sugars appear to be absorbed by passive diffusion.

Proteins. ■ The proteinases in pancreatic juice are secreted as the inactive zymogens *trypsinogen, chymotrypsinogen,* and *procarboxypeptidases.* Enterokinase, liberated by the intestinal mucosa, catalyzes the conversion of trypsinogen to *trypsin*, which, in turn, activates the other two proenzymes to *chymotrypsin* and *carboxypeptidase*. Trypsin and chymotrypsin both can hydrolyze peptide bonds located within a protein chain and are reasonably specific. Trypsin attacks those linkages adjacent to lysine or arginine, and its products are amino acids and various polypeptides. Chymotrypsin preferentially splits peptide bonds involving the aromatic amino acids but will attack those of leucine, histidine, asparagine, and methionine. Again, amino acids and simpler polypeptides result. The carboxypeptidases attack peptide chains at the end, liberating the amino acid with a free carboxyl group. Dipeptidases located in the brush border complete the hydrolysis of protein to free amino acids.

Pancreatic juice also contains *elastase* and two nucleolytic enzymes, *ribonuclease* and *deoxyribonuclease*. These enzymes assist in the digestion of ingested collagen and nucleoproteins, that is, RNA and DNA.

By the time ingested protein reaches the ileum it is 60% to 70% digested and absorbed. Amino acids do not accumulate in the intestinal contents; they are absorbed as rapidly as they are liberated. An active process supported by the oxidative metabolism of the mucosal cells is involved in the transport of the amino acids. At least three and possibly five transport systems have been identified by *in vitro* studies: (1) A single transport system exists for the neutral amino acids. Both sodium ions and pyridoxal phosphate (vitamin B_6) seem to be involved in the transport of neutral amino acids. (2) A separate system operates at a slower rate for the basic amino acids, including arginine, lysine, ornithine, and cystine. (3) Proline, hydroxyproline, sarcosine, dimethylglycine, and betaine are transported by a third system. Possibly a fourth system for the specific transport of glycine may be present, and some evidence has been obtained for an independent system for valine, leucine, and isoleucine. The L forms are usually transported preferentially. There seems to be no special transport system involved for glutamic and aspartic acids, but they are metabolized to some extent in the mucosal cell.

Lipids. ■ Dietary fat composed primarily of triglycerides arrives in the duodenum undigested. Bile salts, cholesterol, and lysolecithins assist in emulsifying the neutral fats, and pancreatic lipase then hydrolyzes the triglycerides. The triglycerides are hydrolyzed at the 1 and 3 positions, forming monoglycerides and free fatty acids. A stable emulsion or micelle is formed from the free fatty acids, monoglycerides, bile salts, cholesterol, and Na ions, and this penetrates the brush border of the mucosal cells. The exact mechanism is not known, but the material probably goes through the lipid phase of the membrane by simple diffusion. In the endoplasmic reticulum of the mucosal cell the free fatty acids and monoglycerides are resynthesized into triglycerides. At the same time, phospholipids and proteins are synthesized in the cell. The protein, phospholipid, and cholesterol combine to produce a specific lipoprotein that coats the triglyceride, forming *chylomicrons* 0.1 to 3.5 $m\mu$ in diameter. The latter are primarily removed by the lymph system, since they are too large to penetrate the capillary endothelium. Absorption of fat usually occurs in the jejunum, with the ileum completing the process if necessary.

Cholesterol, present in almost all diets and in the intestine, is mixed with fat and bile salts to participate in the micellar formation discussed above and then absorbed. Cholesterol synthesis in the liver is increased with lowered amounts of dietary cholesterol. The small intestine also plays a role in cholesterol synthesis, which is regulated by the amount of bile salts in the intestine—when bile salts are diverted, the amount of cholesterol synthesized in the intestinal cell increases.

Other organic materials. ■ Dietary nucleic acids are digested by pancreatic nucleases and are completely absorbed in the small intestine.

The fat-soluble vitamins (A, D, E, and K) are probably absorbed by diffusion through the proximal parts of the small intestine. Their absorption seems to be related to fat absorption, and bile salts are necessary for their transfer.

The water-soluble vitamins (C and B vitamins, except B_{12}) appear to be absorbed by passive diffusion in the proximal area of the intestine. Vitamin B_{12} is complexed with the intrinsic factor produced by the gastric mucosa and is actively transported at a specific absorption site in the ileum.

Bile salts, as previously indicated, are absorbed by an active process in the distal small intestine and enter the enterohepatic circulation.

Absorption of water and electrolytes. ■ The absorptive capacity of the small intestine is in excess of normal needs, since 50% or more of it can be removed without harmful effects. It has been estimated that the total absorptive area is 4,500 m^2. The absorption of water by the small intestine is a vital function. It is absorbed at an average rate of 200 to 400 ml/hr. From 5 to 10 liters of water derived from food, drink, and secretions enter into the gastrointestinal tract every day, but only 0.5 liter enters the colon. Water is chiefly absorbed through the upper part of the small intestine. Water can and does move in both directions across the intestinal wall, and when chyme enters the duodenum, it is quickly brought to isotonicity either by the addition or removal of water. The contents of the lumen are maintained in osmotic balance throughout the small intestine; therefore, water leaves only with particle absorption.

Sodium also moves in both directions across the intestinal epithelium, but there is also an active transport of this material from the lumen to the interstitial fluid, and this is responsible for the net absorption of sodium. Potassium moves across the intestinal epithelium; when sodium concentration decreases in the lumen, potassium concentration increases. There is, however, a net absorption of potassium, probably as a passive consequence of water movement.

Chloride appears to follow the movements of sodium, since electrical neutrality must be maintained; however, in some areas of the small intestine chloride movement seems to occur independent of the movement of sodium. Chloride concentrations are reduced to very low values in the jejunum and ileum. Passive absorption of bromide, iodide, thiocyanate, and nitrate occurs at a rate equal to that of chloride.

Bicarbonate is secreted and absorbed by different areas in response to changes in intestinal content.

Calcium is absorbed along the entire length of the intestine, and a metabolite of vitamin D is required for transport of this divalent ion. The amount of calcium absorbed depends on body needs, and parathyroid hormone may be one of the regulating factors controlling the absorption of the dietary calcium (Chapter 24). Magnesium and phosphate ions are absorbed passively along the small intestine.

Iron is absorbed in the duodenum and upper jejunum, and two steps are apparently necessary. Iron moves rapidly into the mucosal cells, where it can exist in two forms—either diffusely distributed throughout the cytoplasm, divalent iron, or in combination with ferritin, trivalent iron. Iron is released from the mucosal cell into the blood when there is a need for it (Chapters 12 and 24); otherwise it remains in the cell and because of the short life-span of the mucosal cell (three to four days) is lost in the feces. A high concentration of iron in the cell does not completely depress absorption, and continuous ingestion of excess iron may have serious consequences.

■ SECRETION, ABSORPTION, AND EXCRETION IN THE LARGE INTESTINE

The large intestine in man contains no villi, and the mucosa consists of crypts; its free surface is covered with epithelial cells. A high density of mucus-containing goblet cells exists; large intestine secretion is rich in mucus. The secretion is alkaline and approximately isotonic; it is rich in bicarbonate and potassium. Secretion occurs

Table 20-1. Secretions, digestion, and absorptive locations of the digestive tract—summary

Anatomical division	*Secretion*	*Main components and enzymes*	*Enzyme substrate*	*Product of enzyme*	*Products absorbed*
Mouth	Mucus				
	Saliva	Ptyalin (α-amylase)	Carbohydrate, α-1,4-glucosidic linkage of starch	Maltose, isomaltose, dextrins	None
Stomach	Mucus				
	Gastric juice	HCl			Fat-soluble materials to a limited extent
		Intrinsic factor			
		Lipase	Tributyrin	Fatty acids	
		Pepsin	Proteins	Polypeptides	
Small intestine	Mucus				
Duodenum	Pancreatic juice	HCO_3			
		Trypsin	Proteins	Polypeptides, amino acids	
		Chymotrypsin	Proteins, polypeptides	Smaller polypeptides, amino acids	
		Carboxypeptidase	COO^- end of peptides	Amino acids	
		RNase	RNA	Nucleotides	
		DNase	DNA	Nucleotides	
		Amylase	Starch	Maltose, isomaltose, glucose	
		Lipase	Fats	Fatty acids, glycerol, monoglycerides	
	Bile	Bile salts, pigments, cholesterol			Glucose and other monosaccharides H_2O Most water-soluble vitamins
	Intestinal juice	Enterokinase Disaccharidases, peptidases, nucleosidases, etc., which may be present as intracellular enzymes in brush border	Trypsinogen Disaccharides Peptides Nucleotides	Trypsin Monosaccharides Amino acids Bases, sugars	Amino acids Fats
Jejunum	Mucus Intestinal juice				Fat-soluble vitamins H_2O, Na^+, Ca^{++} Mg^{++}, iron
Ileum	Mucus Intestinal juice				Fats, amino acids Vitamin B_{12} Bile salts H_2O
Large intestine	Mucus K^+, HCO_3^-				H_2O Na^+

independent of extrinsic innervation. The colon absorbs approximately 300 to 400 ml of water per day as well as sodium and chloride.

Fecal material consists of water and solid materials made up of mucus, undigested food residue, and microorganisms.

■ SUMMARY

Table 20-1 provides a summary of the secretions, digestion, and absorptive locations of the digestive tract.

READINGS

Brooks, F. P.: Control of gastrointestinal function, New York, 1970, The Macmillan Company, pp. 11-147.

Davenport, H. W.: Why the stomach does not digest itself, Sci. Am. **226**(1):87-93, Jan. 1972.

Eichholz, A.: Fractions of the brush border, Fed. Proc. **28**:30-34, 1969.

Gardner, J. D., Brown, M. S., and Laster, L.: The columnar epithelial cell of the small intestine: digestion and transport, New Eng. J. Med. **283**:1196-1222, 1264-1271, 1317-1324, 1970.

Kimberg, D. V.: Effect of vitamin D and steroid hormones on the active transport of calcium by the intestine, New Eng. J. Med. **280**:1396-1405, 1969.

Mackenzie, I. L., and Donaldson, R. M., Jr.: Vitamin B_{12} absorption and the intestinal cell surface, Fed. Proc. **28**:41-45, 1969.

Porter, K. R.: Independence of fat absorption and pinocytosis, Fed. Proc. **28**:35-40, 1969.

Sleisenger, M. H.: Malabsorption syndrome, New Eng. J. Med. **81**:1111-1117, 1969.

21

Anatomy
Function

THE LIVER

■ ANATOMY

The liver is the largest organ in the adult human body, weighing approximately 1.5 kg. It lies, as shown in Fig. 19-1, about level with and to the right of the stomach. This reddish brown organ is covered with a capsule composed of connective tissue. On the underside—the visceral surface—there is a deep transverse fissure—the porta—where the supporting connective tissue enters into the substance of the liver. This connective tissue branches extensively, and the portal vein, the hepatic artery, the bile duct, and one or more lymphatic vessels accompany these branches, providing pathways for the movement of materials to and from the liver. Blood from the intestines is brought to the liver by the portal vein, and blood from the arterial circulation is brought by the hepatic artery; it is carried away by hepatic veins draining into the vena cava.

The human liver is a mass of *parenchymal* cells arranged into plates one cell thick, as shown in Fig. 21-1. These plates form walls around the continuous spaces, or lacunae, that pervade the liver. Suspended within the lacunae are the specialized capillaries of the liver known as *sinusoids*. The size of the sinusoids is a function of the amount of blood within them. They differ from other capillaries in having a greater permeability for macromolecules, especially protein. Reticuloendothelial cells—Kupffer cells—line the walls of the sinusoids and by the increase in their size apparently assist in the passage of large molecules through the sinusoidal walls. There is a space between these lining cells and the parenchymal cell—the space of Disse—which also assists in the transfer of materials and waste products from the liver cell to the blood. Under normal conditions of blood pressure, there is a radial distribution of sinusoids and liver plates around the smallest divisions of the hepatic veins, the central veins; this arrangement constitutes what is referred to as a liver lobule. Blood flow through the liver can be traced as follows: The portal vein from the intestine divides into very fine branches, which discharge portal blood into the sinusoids. Each branch of the portal vein is accompanied by anastomosing branches of the hepatic artery, which provide capillaries for the parenchymal cells and bile ducts; this blood eventually flows into the sinusoids. The content

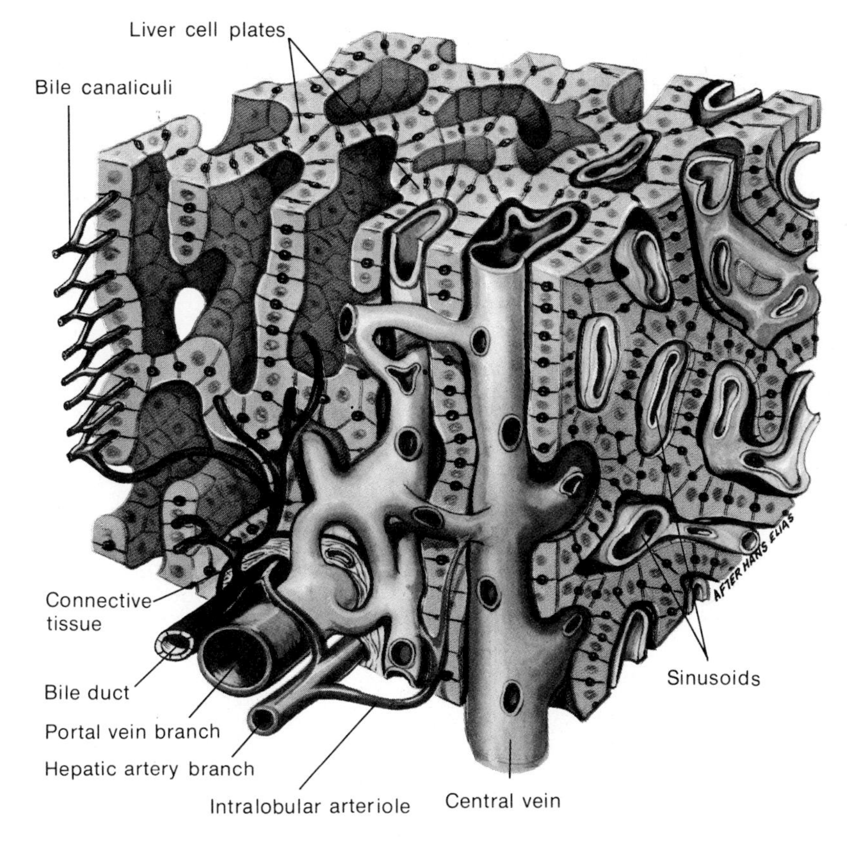

Fig. 21-1
Illustration of a section of the liver. The parenchymal cells are arranged in plates one cell thick, which form continuous spaces, or lacunae, where specialized capillaries of the liver—sinusoids—are located. Bile canaliculi are formed by grooves in the contact surfaces of two liver cells. (After Hans Elias.)

of the sinusoids enters the central vein, is carried to the hepatic vein, and finally reaches the lower vena cava. A network of lymphatic vessels exists in the portal canals; it drains the space of Disse. Blood flow in liver tissue is controlled by specfic mechanisms; thus sphincters on the portal venules can block flow into the sinusoids; Kupffer cells can enlarge and thereby control and steer the flow in the network of sinusoids; finally, sphincters of the central veins and arteries control exit and entrance of blood at these points.

The *bile canaliculi* are formed by grooves in the contact surfaces of two liver cells that fit together to provide a cylindrical lumen. Bile canaliculi do not have a wall of their own until they drain into the small ductules that combine to make up the larger bile ducts.

The liver cell is well equipped to undertake and carry out its many diversified metabolic and excretory tasks. It has either one or two nuclei (Fig. 2-3) with prominent nucleoli. It contains numerous mitochondria and shows a well-defined endoplasmic reticulum. The cytoplasm, which contains numerous inclusions, especially glycogen, has a large number of enzymes present in it. Lysosomes are also present which contain large amounts of hydrolytic enzymes. These cells are polygonal in shape and their surfaces may be exposed to the sinusoidal space (space of Disse), to the lumen of bile canaliculi, or to another parenchymal cell. Microvilli are present on those surfaces in contact with the space of Disse to increase the surface area for transport of materials.

■ FUNCTION

Liver is the principal organ for biochemical homeostasis. The specialized functions of the liver cell include the elaboration of bile—the external secretion of the liver; the conversion of the variety of nutrients transported from the gut into the limited number of substances utilized by the body; regulation of the blood level of a multitude of substances including glucose, amino acids, vitamins, and plasma proteins; and conversion by several different mechanisms of a number of toxic and foreign substances to compounds which can be readily excreted by the body. In addition to these unique functions the enzymes necessary for deriving energy from glucose and lipid metabolism are present in the liver cell. For convenience of study liver function will be discussed under the following major aspects: secretion and excretion, metabolic regulation, blood clotting factors, storage, and detoxification.

■ Secretion and excretion—the production of bile salts and bile pigments

Bile salts. ■ The formation of bile acids and their secretion as bile salts are important func-

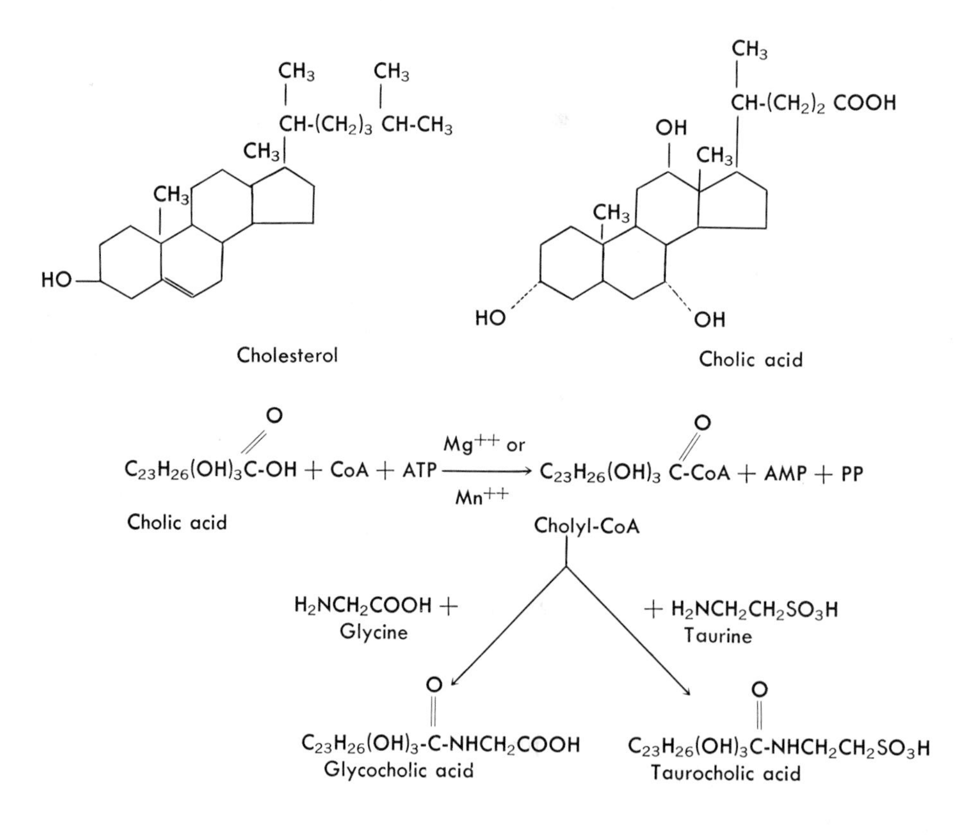

tions of the liver. Approximately 80% of all cholesterol is metabolized by transformation into bile acids. The detailed mechanisms for this transformation are beyond the scope of this text, and we will present only the formulas of cholesterol and the bile acid—cholic acid—to illustrate their relationship.

In man, these acids are conjugated with the amino acids glycine and taurine and secreted as the salts of *glycocholic acid* and *taurocholic acid.* The linkage is a two-step, energy-consuming, enzymatic process, which occurs in liver cells as shown on p. 402. Bile salts are necessary for the absorption of lipids and fat-soluble vitamins from the intestinal tract (Chapter 20).

Bilirubin. ■ Breakdown of hemoglobin by the reticuloendothelial system located in the spleen, liver, and bone marrow results in the production of bilirubin. The process involves the scission of the methene bridge (CH) connecting the pyrrole rings of hemoglobin (Chapter 12) to give a biliverdin-iron-protein complex; the iron and globin are removed, and the biliverdin is reduced to bilirubin. The bilirubin is subsequently transported in the bloodstream to the liver, where it is combined with a diglucuronide (a carbohydrate metabolite). Heme, which appears in the plasma, complexes with a globulin, *hemopexin,* and is transported to the liver where the hepatocytes convert it to bilirubin. This soluble compound then enters the bile canaliculi to provide the yellow pigment characteristic of bile. In the intestine, bilirubin is further reduced to mesobilirubinogen and is excreted in the feces as stercobilinogen. The yellow tint in the skin and conjunctiva characteristically found in *jaundice* is due to the accumulation of bile pigments in the plasma. This accumulation can result from three primary causes: excess destruction of red blood cells (hemolytic jaundice), failure of liver cells to conjugate bilirubin with diglucuronide and to secrete it (infectious hepatitis and cirrhosis), and obstruction of the biliary passages so that bile cannot flow into the intestine (obstructive jaundice).

■ Metabolism

The intermediary metabolism of the three main classes of foodstuffs occurs in liver cells, just as it does in many other cells of the body. Therefore, we will leave detailed consideration of intermediary metabolism to another chapter (Chapter 23) and discuss here only those metabolic functions that are performed solely or primarily by liver cells. It should be reemphasized that the end products of carbohydrate and protein digestion, monosaccharides and amino acids, pass by way of the portal veins to the liver before entering the rest of the systemic circulation, so that this organ can serve as a prime controlling point for these substances. The liver plays an important part in all lipid metabolism.

Carbohydrates. ■ One of the main metabolic functions of the liver is the maintenance of a normal blood glucose concentration. Glucose is always present in the blood to the extent of about 70 to 100 mg/100 ml of blood.* This quantity is known as the *normal fasting level* of blood sugar. As this supply is drawn upon for the expenditure of energy, it is replenished by the intake of food. Since most persons eat only two or three meals per day, the sugar concentration may rise during the height of absorption to values of 130 to 150 mg/100 of ml blood,† causing a temporary condition of *hyperglycemia.* Such a concentration of sugar would greatly increase the specific gravity of the blood and raise its osmotic pressure. The former would require more work by the heart; the latter would draw an excessive amount of water from the tissues. To prevent

---

*The concentration of glucose differs from one individual to another. It also varies slightly in a given individual during changing conditions. To avoid confusion, the values quoted in the text should be regarded merely as fairly normal ranges or averages.

†It may be of interest to note at this concentration the total amount (5 g) of sugar in the blood equals about 1 teaspoonful. This is equivalent to 20 Cal, or 61,720 ft-lb, of energy. It would enable a person weighing 150 lb to climb a stairway 400 ft high.

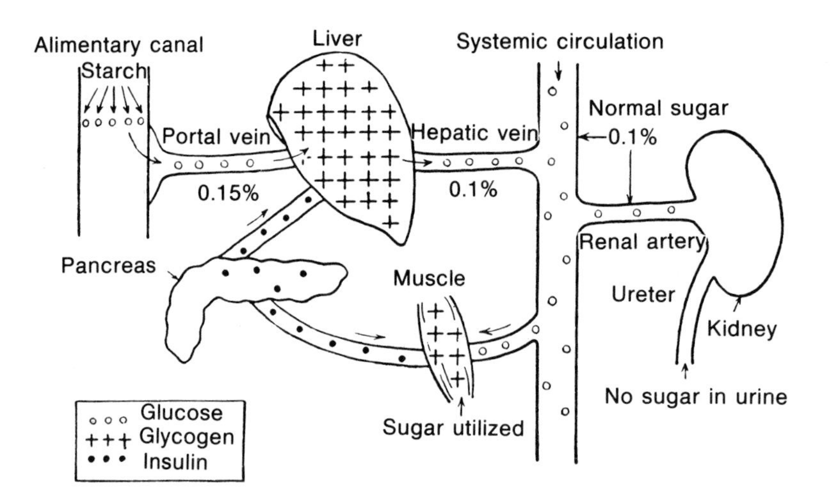

Fig. 21-2
Diagram showing glucose utilization in the normal individual. Excess glucose present in the bloodstream is stored in the liver in the form of glycogen. Additional glucose is stored in muscle in the form of glycogen.

these untoward results, the excess sugar (beyond 0.1%) is withdrawn from the blood by the liver and stored as glycogen. This process is known as *glycogenesis.* The liver has the highest glycogen content of all the tissues in the body, as much as 10% of its weight; thus it is a significant storehouse for energy. Fig. 21-2 illustrates in diagrammatic form how the liver handles the excess glucose that comes to it from the alimentary canal through the portal vein. The liver can, in fact, convert many of the different end products of digestion (amino acids, fatty acids) into glycogen by the process known as *glyconeogenesis.*

If the blood sugar level attains a concentration of approximately 170 mg/100 ml, the renal threshold is reached and the kidneys excrete glucose; this condition is termed *glucosuria.** When the tissues are utilizing glucose at a rapid rate or when the organism is fasting, the blood sugar level drops. Stored glycogen in the liver is reconverted into glucose and released into the blood. This process is known as *glycogenolysis.* The liver makes galactose available for tissue utilization by converting it to glucose in a series of enzymatic reactions.

In the process of glycogenesis, glucose-6-phosphate is converted to glucose-1-phosphate. This compound, in turn, combines with uridine triphosphate to form uridine diphosphate glucose. The glucose is then transferred to a glycogen molecule, thereby increasing its chain length by one unit (Chapter 23). All these steps are catalyzed by specific enzymes. In the process of glycogenolysis, the formation of glucose from glucose-6-phosphate is accomplished only by the cells of the intestine, the liver, and the kidney. Glucose so produced is released into the interstitial fluid and into the blood. This provides a source of energy that can be utilized by all body tissues.

The glycogen content of the liver depends not only on the quantity of food, but also on the type; there is a higher content with a diet rich in carbohydrates. Furthermore, liver glycogen content is under endocrine control, which ap-

*The term *glycosuria* refers to the excretion of any reducing sugar. However, when the urine contains large quantities of glucose, it is appropriate to speak of the condition as *glucosuria.*

pears to regulate certain metabolic reaction rates in the liver. The liver may act as a buffer between other tissues and the influence of hormones on them. Epinephrine and glucagon both cause a degradation of glycogen (glycogenolysis) and consequently an increase in blood sugar. This hormonal effect is mediated by the "second messenger," cyclic AMP, whose production is accelerated in the liver cell by glucagon and epinephrine. The increase in cyclic AMP converts an inactive form of an enzyme to an active form which results in the degradation of glycogen. Steroid hormones tend to increase liver glycogen by the promotion of gluconeogenesis. Excess thyroid hormone results in a depletion of liver glycogen.

Proteins. ■ Protein metabolism by the liver is essential to the maintenance of life, and both catabolic and anabolic reactions are involved. Some of these reactions can be carried out by other cells in the body, but the rates attained in the liver cell are much greater than in the other cells. In the liver cell the amino acids can be handled in several ways. They may be synthesized into structural, enzymatic, or circulating (plasma) proteins, or they can be metabolized to provide energy through conversion to carbohydrate intermediates. Their metabolism also provides the nitrogen and carbon skeletons needed for production of other compounds in the liver. With the exception of immune globulins, all the plasma proteins are produced by hepatic cells. As much as 50 g of plasma proteins can be synthesized by the liver in twenty-four hours. One of the most important proteins synthesized by the liver is albumin and it, along with other proteins, is produced in the ribosomes and then transported to the extracellular space. Hemopexin, a β-globulin, is produced in the liver; it has a molecular weight of 70,000 and contains 20% carbohydrate material. It is, therefore, a glycoprotein. Nitrogen transfer between amino acids is one of the essential functions of liver cells; it is possible for the liver to synthesize the nonessential amino acids by these transamination reactions between amino and keto acids. Amino acids provide the nitrogen and carbon skeletons for the synthesis of creatine, purines, and pyrimidines that occurs in liver cells.

Ammonia, formed by liver deamination of amino acids prior to their use for energy production and continually produced in the gut by bacteria and released into the bloodstream, is toxic. The production of urea by the liver provides the mechanism for the removal of this ammonia from the body. Urea is the major nitrogenous excretion product in man; that is, it is the main end product of protein catabolism, and on an average protein diet it accounts for 80% to 90% of the nitrogen excreted in the urine. All the enzymes required by the *ornithine cycle*—the mechanism responsible for urea formation—are present in liver, and various experimental studies have proved that the liver is the primary site of such formation. A detailed discussion of the biochemical steps involved in urea formation does not fall within the scope of this text; hence, we will present only a generalized description of the ornithine cycle to aid the understanding of this process. Initially, ammonia, carbon dioxide, and ATP combine in the presence of the enzyme carbamyl phosphate synthetase to form an active compound, carbamyl phosphate. This compound reacts with the amino acid ornithine to form a second amino acid, citrulline. Next, arginine is produced from the citrulline by a two-step process involving aspartic acid and ATP. Finally, the arginine is hydrolyzed with the assistance of the enzyme arginase to yield urea and ornithine. The latter is once more available for combination with carbamyl phosphate—thus a cycle is established. This cycle can be diagrammed as shown on p. 406. The enzymes necessary for operation of the ornithine cycle, with the notable exception of arginase, also exist in other cells in the body, but only liver cells contain this enzyme, which is essential for urea formation.

The liver plays a paramount role in the regula-

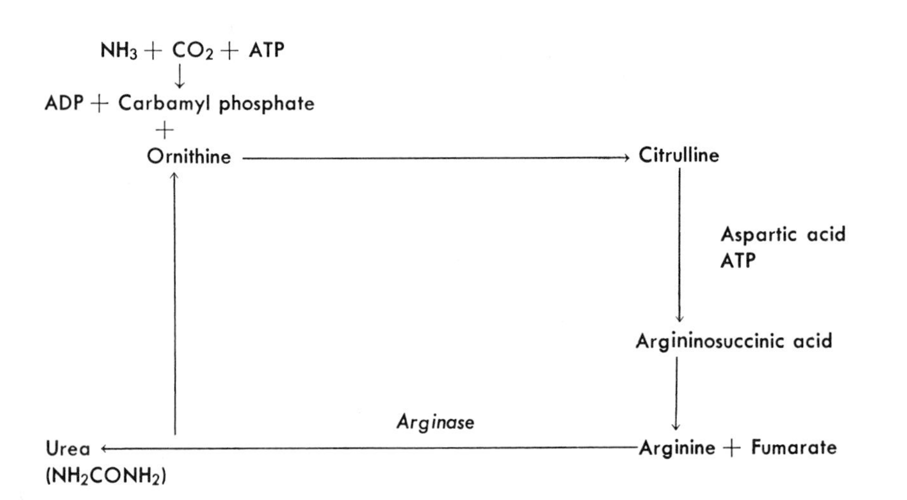

tion of the amino acid content (both qualitative and quantitative) of the body. It serves as a regulatory mechanism between dietary intake and the amino acid needs of the body for protein synthesis. The entire output of free amino acids by the gut (with the exception of valine, leucine, and isoleucine) after a meal are absorbed by the liver cell. The liver cell then maintains an almost constant supply of essential amino acids to the rest of the cells for synthetic purposes.

Lipid metabolism. ■ The liver plays a prominent role in lipid metabolism. It is important in fatty acid synthesis and degradation; it plays a dominant role in cholesterol and phospholipid metabolism. Since other tissues in the body also have a functional role in lipid metabolism, we shall postpone discussion of the biochemistry involved to Chapter 23. Liver extracts the dietary lipids, in the form of chylomicrons, from the blood and, in a fed animal, is responsible for redistributing this material to other tissues. In a fasted or starved animal, the liver oxidizes about 45% of the available dietary lipid for energy needs. A predominate, if not exclusive, role is played by the liver in production of ketone bodies.

Phospholipids are synthesized by the liver from the dietary fatty acids, and this organ is considered to be the major contributor to the circulating phospholipids. The biosynthesis of cholesterol can occur in many tissues, but regulatory effects on the blood level of this substance are exerted chiefly by the liver. They are accomplished by a decrease in synthesis and by oxidation of cholesterol in the liver. The esterification of cholesterol and the maintenance of a specific ratio between the free and esterified forms in the blood are likewise metabolic functions of the liver. A portion of the cholesterol removed from the blood by the liver subsequently appears in the bile.

Although it is no longer believed that the liver is the only organ essential to lipid metabolism, it undoubtedly serves an important regulatory role in conjunction with the other tissues in the overall metabolism of these materials.

■ Blood coagulation

In Chapter 12, the mechanism for the coagulation of blood was discussed. Two of the important proteins essential for this process are fibrinogen and prothrombin, both of which are exclusively synthesized in the liver. Factors V and VII (proaccelerin and proconvertin), IX, and X are also synthesized by the liver. Vitamin K is vital to the synthesis by the liver of many of these factors essential in blood clotting. The liver also plays a role in the removal of these compounds since the Kupffer cells are one site for their destruction. The liver is responsible for maintaining optimal circulating levels of these protein precursors necessary for blood coagulation.

■ Storage

Vitamins. ■ Hepatic concentrations of vitamins A, riboflavin, pyridoxal, B_{12}, folic acid, and pantothenic acid are particularly high; this either may be due to specific reactions occurring in the liver, or it may be simply the result of storage. For example, liver tissue converts carotene to vitamin A, and, in many animals, the liver is the main site of storage of this vitamin. Good liver function is important because bile is necessary for the absorption of vitamin D from the intestine; it is stored in the liver. The levels of both riboflavin and pyridoxal are somewhat higher than in other tissues, and both of these vitamins play a necessary role in the intermediary metabolism of carbohydrates, lipids, and proteins. Vitamin B_{12}, which has a role in erythropoiesis, is stored in the liver. Hepatic storage of vitamin K_1, an important factor in blood coagulation, is of some importance.

Minerals. ■ The storage of iron by the liver, in the form of *ferritin,* is important in the management of this essential element. The adult male liver contains approximately 700 mg of iron. Excess iron accumulates in the liver as *hemosiderin,* a normal constituent of most tissues. Hemosiderin granules are much larger than ferritin molecules and have a higher iron content. Since excess iron cannot be excreted, continued unnecessary intake of iron may lead to hemosiderin accumulations in the liver of sufficient amounts to damage this organ.

Blood. ■ The portal system of the liver may act as a blood reservoir, since a decreased flow through the portal vein would provide a relatively large volume of blood to other parts of the body.

■ Detoxification

Since most of the materials absorbed from the intestine are carried by the portal vein to the liver, it can exercise a selective action on those materials. Several specific reactions which take place predominantly or exclusively in the liver have been called detoxification. This includes not only materials which are foreign to the body but also many normal intermediary metabolites which if allowed to accumulate are toxic. While the same types of processes may occur with both the foreign substances as with the metabolites, different enzymes are usually involved. The microsomes of the liver seem to be particularly important in these reaction processes.

The following are characteristic types of reactions which are important pathways for detoxification.

1. *Oxidations.* Hydrocarbons, very inert substances, can be attacked by liver enzymes and detoxified. The first step involves hydroxylation which takes place in the smooth endoplasmic reticulum by the following general scheme:

$$RH + O_2 \xrightarrow[\text{Reducing substance to furnish 2H}]{} ROH + H_2O$$

   A number of unrelated compounds follow this pathway, such as those which contain aromatic rings as well as long straight chain molecules. Many drugs used in modern medicine are detoxified by this process. In general hydroxylation renders the material less lipophilic, which promotes excretion.
2. *Methylation.* The liver cell can transfer the CH_3 group from the methionine to toxic compounds, thus making them more water soluble and suitable for excretion.
3. *Conjugation.* Compounds are conjugated with glucuronic acid and sulfates which can then be removed by the kidney and excreted in the urine. Bilirubin, as mentioned, is

excreted as a glucuronide. Various phenolic compounds are detoxified in this fashion.

4. *Specialized.* Mercapturic acid formation for the elimination of aromatic hydrocarbons takes place by a complicated enzymatic process involving an initial reaction with glutathione. Hippuric acid is formed by a reaction of benzoic acid with glycine and is then excreted.

READINGS

Bittar, E. E., and Bittar, N., editors: The biological basis of medicine, New York, 1969, Academic Press, Inc., pp. 145-315.

Elias, H., and Sherrick, J.: The morphology of the liver, New York, 1969, Academic Press, Inc.

Gartner, L. M., and Arias, I. M.: Formation, transport, metabolism and excretion of bilirubin, New Eng. J. Med. **280**:1339-1345, 1969.

Netter, F., and Oppenheimer, E.: Liver, biliary tract and pancreas, Ciba Collection of Medical Illustrations, Part III, 1957.

Shoemaker, W. C., and Ehoyn, D. H.: Liver: functional interaction within the intact animal, Ann. Rev. Physiol. **31**:227-268, 1969.

22

INTRODUCTION TO METABOLISM AND THE ENZYMES

■ METABOLISM

Thus far we have studied the nature of the various foodstuffs, their preparation, if necessary, for use by the body, and their distribution by the circulating blood and by lymph to all tissues and organs. We have also become acquainted with the more obvious activities of the body, such as the contraction of muscles, the heartbeat, and the secretion of saliva and gastric juice, and with the less conspicuous passage of an impulse along a nerve. These activities are the results of energy transformations occurring within the cells. It is these energy changes, together with the material changes inseparably bound up with them, that form the subject matter of metabolism.

Some constituents of the cells are obtained from the environment via the alimentary and the respiratory tracts; others are manufactured by the body itself from environment-furnished materials. Cell constituents vary from one type of cell to another, but, in general, the following may be enumerated: proteins, amino acids, nucleic acids, carbohydrates, lipids, minerals, vitamins, enzymes, and hormones; to this must be added water and oxygen. Since in each of these groups the number of discrete substances range from a few to hundreds, tracing the history of each in its course throughout the body is a large undertaking.

Some cell constituents are concerned either with directly supplying energy for cellular activity or with controlling, to a certain degree, the occurrence, extent, and duration of the energy liberation. The releasing of energy and all it entails is spoken of as *catabolism,* or *dissimilation.* Other cellular components serve as material for the growth and maintenance of the protoplasm. This is a constructive process—*anabolism,* or *assimilation.*

■ METHODS OF INVESTIGATION

To ascertain the particular part played by a given substance in anabolism and catabolism, various methods are used. Only some of the more important ones will be discussed.

Isotopes. ■ In a study of the metabolic transformation of a particular chemical substance, the primary objective is to observe the fate of that substance *in vivo* and under conditions that cause as little disturbance to the organism as possible. The use of the isotopic tracer technique has proved to be one of the most fruitful methods; in this technique one or more atoms of the metabolite under consideration is labeled by means of an isotope. Isotopes, as we recall, may be defined as atoms that have the same atomic number but different atomic weights. For example, carbon (atomic number, 6) is found predominantly in nature with a mass of 12 (^{12}C), but it also exists with a mass of 13 (^{13}C) and a mass of 14 (^{14}C). The first two are referred to as stable isotopes; that is, their nuclei do not undergo spontaneous decomposition. The third, ^{14}C, is referred to as a radioactive isotope, since its nucleus decomposes spontaneously with the emission of radiation. If a compound containing an increased amount of either ^{13}C or ^{14}C is synthesized, it is possible to follow the progressive metabolism of that compound when administered to animals or man. By employing a mass spectrometer, it becomes possible to identify the ^{13}C, and the ^{14}C may be determined by a Geiger-Müller tube, which detects the beta radiation emitted by the isotope.

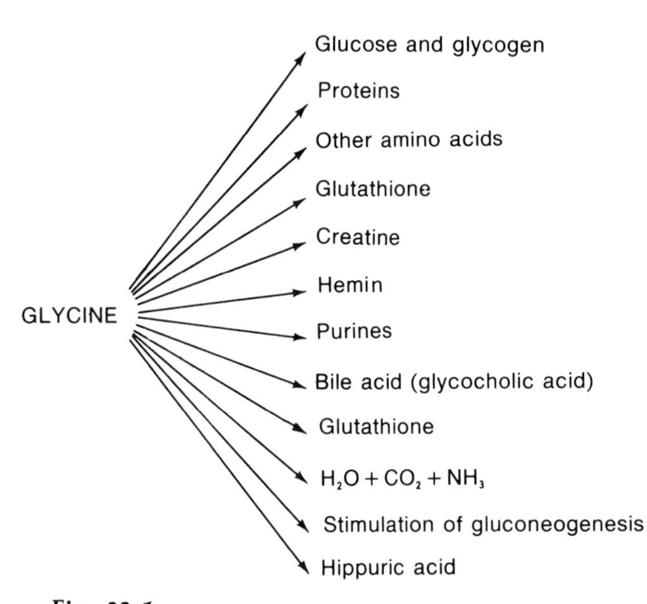

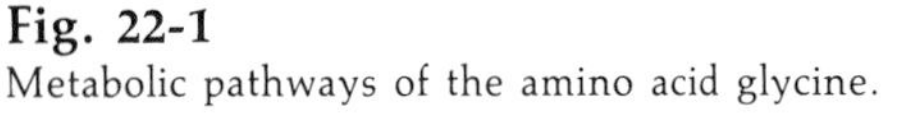

Fig. 22-1
Metabolic pathways of the amino acid glycine.

For example, through appropriately labeling atoms in the amino acid glycine ($H_2{}^{15}N^{13}CH_2{}^{14}COOH$), it has been possible to determine the many pathways of synthesis and degradation which this compound may take (Fig. 22-1).

Isotopes have been extremely useful in elucidating the various metabolic pathways that will be considered in a later chapter. Despite the importance of this technique, it would be incorrect to suppose that it is the only experimental approach to the study of intermediary metabolism in intact organisms.

Growth studies. ■ A great deal of information has been gained by dietary studies with immature or growing organisms. By selective omission of various components from the diet of growing organisms, knowledge has been obtained about those that are essential for growth. Subsequent studies may bring forth other compounds that can partially or fully replace this dietary essential. The ability of substances to substitute for an essential nutrient may reveal metabolic relationships between the compounds tested for their growth-promoting capacity. This type of investigation has been useful in the study of the metabolism of microorganisms and mammalian cell cultures.

Perfusion. ■ An isolated organ can be perfused with blood (or a specially chosen solution) to which has been added the chemical substance whose changes in its passage through the organ are to be investigated. Proper analysis of the blood issuing from the organ (with reference to the added substance) may yield an answer.

Tissue slices. ■ A method much used at present is that of the tissue slice. From the fresh organ a slice, about 0.3 mm thick, is cut. Immersed in a solution of definite composition (medium), the tissue slice is placed in a Warburg apparatus, by which the amount of oxygen consumed and of CO_2 produced by the tissue can be measured. To the medium surrounding the tissue slice may be added the specific food material under investigation or any enzyme, vitamin, or hormone whose influence on the tissue is to be explored.

Homogenates and differential centrifugation. ■ By the technique of homogenization, tissue organization is destroyed and the individual cells are ruptured so that the solid components of the cell (such as nuclei, mitochondria, and microsomes) become suspended in a solution containing the soluble components of the cell. Differential centrifugation separates these components, and thereby the various enzymes associated with them can be studied readily. Such studies have led to further purification and isolation of the individual enzymes. One difficulty inherent in this method is that, since the delicate morphological relationships have been lost, there is always some degree of uncertainty in interpreting the results.

All the methods previously discussed and many others (e.g., histochemistry, autoradiography, and biochemical genetics) have been utilized in the elucidation of the metabolic role of a given chemical compound in a particular organism.

■ SOME CHARACTERISTICS OF FOOD UTILIZATION BY PROTOPLASM

Before discussing the metabolic changes each of the three organic foodstuffs undergo after entering the cells, it may be well to consider briefly some basic facts underlying these changes, irrespective of the nature of the foodstuffs.

Dynamic equilibrium of the cell. ■ Notwithstanding the apparent permanency of the adult body throughout the years, by the use of stable or radioactive isotopes (Chapter 3) it has been demonstrated that extensive changes are constantly taking place in the composition of the body tissues. By following the course of the marked elements (C, H, N, or S) throughout the body, it was learned that the structural elements of the cell (which, we may say, form the machinery of the cell) are not the fixed and stable things they were once considered to be. Rather, they are in a constant state of flux in which there is a fairly rapid exchange of certain substances (e.g., a particular amino acid of a protein molecule) between the cellular structures and the plasma. The protoplasmic constituents are constantly undergoing a regulated tearing down and resynthesis. As an example of this: one half of the nitrogen found in the proteins of the liver is renewed every seven days. Glycine constitutes one of the amino acids found in liver proteins; 10% of the glycine is replaced daily. Even the constituents of the most rigid and enduring structures, bones, are in a state of flux. To quote Donald Van Slyke, "Every protein molecule in the body is itself alive in the sense that it is continually changing and renewing its structure."

Machine-fuel structure. ■ The exceedingly complex and highly organized content of the cell, protoplasm, is generally regarded as constituting the machine which in some manner releases and utilizes the potential energy of the food. The mechanical changes in the muscle cell due to the proteins actin and myosin favor this view. What relation exists between the unstable, ever-changing protoplasm and the dead and quite inert food molecule? Perhaps it is best to speak of these two ingredients as a machine-fuel structure. In functioning, the machine-fuel undergoes a limited and carefully controlled disintegration, resulting in the release of energy. Reconstruction takes place perhaps simultaneously with, or soon after, the disintegration.

Basic energy requirements for each cell. ■ The protoplasmic machine expends energy not only in adjusting the body or parts of the body to environmental changes (both external and internal), but energy must also be expended to maintain its own structure and thereby ensure its existence. The highly labile materials and combination of materials entering into the makeup of protoplasm are constantly disintegrating; evidence of this was presented in a previous section. Its reconstruction from the more stable food requires the expenditure of energy. It takes life to keep alive.

Storage of energy. ■ To carry on life, energy is needed every moment, day and night. Yet we supply ourselves with food only two or three times per day. It is apparent, therefore, that part of the food not immediately needed on its arrival in our blood must be stored for future use. In our study we have learned that for carbohydrates a rather limited storage capacity exists in the liver and in muscles. Protein is not stored to any great extent. But for fats the storage space in adipose tissue is large. It can readily be seen that, in certain unusual conditions of the body or the environment, the storage of energy-yielding substances may be essential.

Catabolism and energy. ■ The chemical disintegration or catabolism of the three organic foodstuffs (carbohydrates, lipids, and proteins) by which potential energy is released is sometimes loosely spoken of as oxidation. Food, it is popularly said, must be burned, or oxidized, in the body. Let us note some similarities and some sharp differences between the combustion or burning of food outside of the body and its catabolism in the cells.

1. For either the complete combustion in air or its catabolism in the body, oxygen is required. Taking the all-important fuel food, glucose, as an example, this may be expressed in its most abbreviated form.

$$C_6H_{12}O_6 + 6\ O_2 \rightarrow 6\ CO_2 + 6\ H_2O + \text{Energy}$$

Potential energy   Waste products

2. In both instances, the same final waste products (CO_2 and H_2O) are formed.

3. Essentially the same amount of energy is set free.

4. The combustion of food in air is a *direct union* of O_2 with the food; no auxiliary agents are needed. It is an *uncontrolled chemical reaction,* the velocity and extent of which are limited by the supply of oxygen and the removal of the products. The burning process is irreversible. Were the so-called oxidations in an animal's body to take place in this manner, it would be the animal's pyre.

5. Glucose (and this is true for foodstuffs in general) is a relatively stable substance. Protected from such lower organisms as molds and bacteria, it can be kept in air indefinitely. To cause it to unite with atmospheric oxygen, it must be heated to several hundred degrees (kindling temperature). In the body, catabolism of glucose takes place at body temperature and without the use of strong chemical agents.

6. The chemical elements of food are said not to unite *directly* with oxygen. For example, CO_2 (an abundant waste product originating from all three organic foodstuffs) is not formed by the oxygen in the cells combining with the carbon of, for example, glucose.

7. Catabolism of food materials is normally exquisitely timed as to occurrence and carefully controlled as to extent by the activity of the tissue needing energy. *When the need has been met, the catabolism for that particular purpose stops automatically.*

Anabolism and energy. ■ The constant renewal of cellular constituents requires the utilization of energy. Synthesis of protein, deoxyribonucleic acid, and ribonucleic acid has a high-energy demand, which is met primarily by the enzymatically controlled degradation of carbohydrates and lipids. Adenosine triphosphate (ATP) is apparently the most important source of energy for these synthetic processes.

Metabolism and vegetative functions. ■ Increased cellular activity, as during muscular work or exposure to cold, is dependent on an increase in the disintegration of food and is associated with a greater formation of waste products and an increase in heat production. This demands a greater supply of properly digested and absorbed food and removal and disposal of a greater quantity of waste products and of heat. In consequence, environmental conditions remaining constant, the activity of the vegetative organs is directly proportional to the extent of metabolism.

■ ENZYMES

Cellular metabolism would not occur without the presence of enzymes; therefore the nature and activity of these substances will be considered. Nearly all the chemical changes occurring during catabolism and anabolism are the results of enzyme activity. *Enzymes* are proteins formed by, and are important constituents of, the protoplasm of the cells.

It is not only in the animal body that enzymes play such a prominent part, for many of the common phenomena occurring around us are also enzymatic processes—fermentations. Attention should be called to a few common examples of fermentation—the production of alcohol from sugar, the production of vinegar, the decay of vegetable matter, the souring of milk, and the ripening of cheese. In these instances the enzymes are formed by the protoplasm of lower organisms, such as yeast, molds, and bacteria.

■ DEFINITION OF AN ENZYME

An enzyme has the unique property that during its action it is not consumed. At the end of the reaction there is almost as much enzyme as at the start; hence it contributes nothing to the end products; these are derived solely from the materials—*substrate*—on which the enzyme acts. Neither does the enzyme furnish any of the energy that may be liberated during the reaction. In these respects enzymes are like certain agents in chemistry, the catalysts. A catalyst is defined as a substance that will accelerate a reaction without itself being altered during the reaction. Therefore enzymes might be defined as *complex organic catalysts produced by living cells* but capable of acting independently of the cells that synthesized them. Enzymes enable the cell to permit complex reactions to occur under conditions where little or no response would be expected. The efficiency with which an enzyme can accelerate a reaction can be illustrated by its *turnover number* (the number of moles of substrate acted on by 1 mole of enzyme per unit time). Turnover number can range from 10,000 to 1 million, that is, 1 mole of enzyme can convert from 10,000 to 1 million moles of substrate per minute.

Because of their colloidal nature, enzymes undergo "aging" and destruction; therefore they must be constantly replenished by further synthesis in the body. They differ from inorganic catalysts in many other respects. Unlike the latter compounds, each of which may catalyze many kinds of reactions, most individual enzymes are rather specific in that they act only on certain types of substances

■ ENZYME PROPERTIES

Chemical nature of enzymes. ■ Many enzymes have been obtained in a highly purified form, and, indeed, many have been isolated as crystals. Since all of these crystalline enzymes have been proved to be protein, it is generally accepted that all enzymes are protein in nature. Therefore it would be expected that enzymes would be heat labile, that is, inactivated or destroyed by excessive heat, and this is true. Like other proteins, enzymes are precipitated and thus lose their activity when heated or when treated with excess amounts of alcohol, salts of heavy metals, or concentrated inorganic acids. Some enzymes contain only protein, some incorporate within their structure a metallic ion, whereas others have a separate compound closely associated with the protein. In the latter case the protein portion is referred to as an *apoenzyme;* it is a heat-labile nondialyzable molecule. The associated compound is referred to as a *coenzyme* or *prosthetic group;* this portion is heat stable and dialyzable. Both portions are necessary for enzyme activity.

Specificity. ■ One of the most striking and unique properties of an enzyme is its specificity; that is, it will react with only certain groups of compounds or in some cases with only a single compound. Therefore the enzyme molecule must have a structural configuration such that it will combine with only one particular substance or only with those substances containing a given chemical linkage. To gain some appreciation of what is meant by configuration, the *molecular configuration* of two common hexose sugars, glucose and fructose, is illustrated. These have the same chemical formula, $C_6H_{12}O_6$, but when the structural formulas are compared, it is readily apparent that the spatial arrangement of the atoms is different.

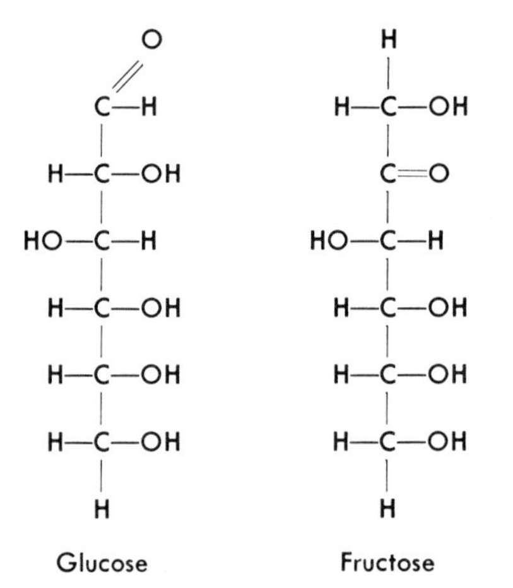

An enzyme molecule, of course, has a much more complicated arrangement than do these hexose molecules, since it is a very large protein composed of many amino acids. Nevertheless, this complicated molecule must be geometrically fitted to a specific substrate in order to allow intimate contact and activity. Since the substrate molecule may be small compared with the size of the catalyst molecule, only a small portion of the enzyme may be in contact with the substrate. This has resulted in the concept of an *active site* for enzymatic activity and is usually thought of as a three dimensional constellation of amino acids. Two theories have been proposed to explain enzyme specificity: (1) the *lock-key* hypothesis, based on a rigid enzyme molecule into which the substrate fits, and (2) the *induced-fit* theory (Fig. 22-2), based on the attachment of the substrate to the active site, which causes a change in the geometry of an enzyme. The latter theory is attractive because the protein molecule is flexible enough to allow conformational changes and because it allows some explanation for the influence of hormones on enzymatic activity.

The degree of specificity of enzymes is variable—in some cases it is absolute, and in others it is relative. For example, urease will act only on urea, whereas lipase will catalyze the rupture of a specific chemical link in all lipids but will not affect that linkage if present in carbohydrates or proteins.

■ TERMINOLOGY AND CLASSIFICATION OF ENZYMES

As the number of known enzymes increased, the problem of classification became more complex. Quite naturally a formal classification system has been proposed to standardize the nomenclature. However, those enzymes that have been recognized for a long time and whose names have been so firmly established in the literature, such as pepsin and trypsin, also retain their trivial names. These will continue to be found in the literature. Commonly the suffix "ase" is appended to the name of the substrate (the substance acted on by the enzyme) in naming enzymes, for example, urease and lipase. A useful and broad classification of enzymes will be outlined in this discussion. In general there are six large major groups, which of necessity have been subdivided further. For our purposes, only

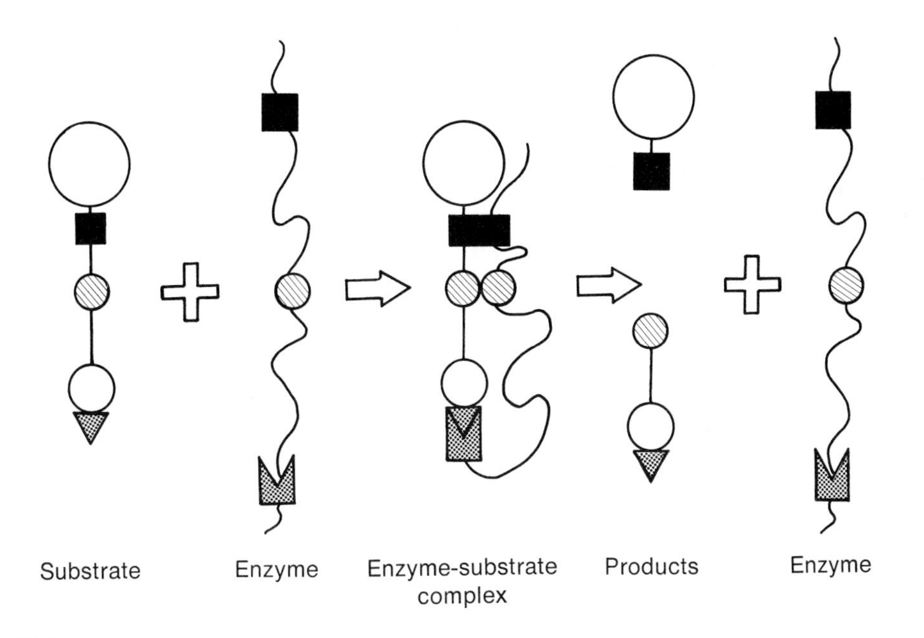

Fig. 22-2
The induced fit hypothesis of enzyme specificity. The active site of the large protein molecule undergoes the necessary conformational changes to align with specific groups on the substrate molecule. A complex (ES) is formed and the enzyme has altered its shape. The enzyme changes the substrate in such a fashion to liberate the products and then resumes its former configuration. Enzyme inhibition occurs when another substance interferes with the proper alignment of the active site of the enzyme to the substrate. Hormones may affect enzyme activity in this fashion. They may accelerate substrate breakdown by unmasking active site of the enzyme and making it available to attach to the substrate.

the six main groups will be considered, and an example of a typical enzyme for each will be offered.

Hydrolases. ■ Hydrolases bring about the addition of water to the substrate and thereby initiate the splitting of the molecule, a hydrolytic cleavage. Acetylcholinesterase, which catalyzes the cleavage of acetylcholine to choline and acetic acid, may be used as an example. This reaction is important in transmission of nerve impulses. The reaction is shown on p. 416. Many of the enzymes that catalyze the breakdown and utilization of carbohydrates, lipids, and proteins belong to this category.

Lyases. ■ Enzymes of the lyase group split or cleave the substrate without another reactant being present, for example, the action of carbonic anhydrase on carbonic acid.

$$H_2CO_3 \rightleftharpoons H_2O + CO_2$$

Carbonic anhydrase is found abundantly in red blood cells, renal tubule cells, and parietal cells of the gastric mucosa.

Transferases. ■ The enzymes that effect the transfer of a radical (a group of atoms) from one compound to another belong to the transferase group. Creatine phosphoryl transferase, an important enzyme in muscle, will be used as an example (p. 416). This enzyme transfers the "energy-rich" phosphate group from creatine phosphate to adenosine diphosphate to form adenosine triphosphate. All of these are rather complicated molecules, but familiarity with their formulas will aid in understanding other areas of study.

Oxidoreductases. ■ Oxidoreductases form a large group of enzymes concerned with biologi-

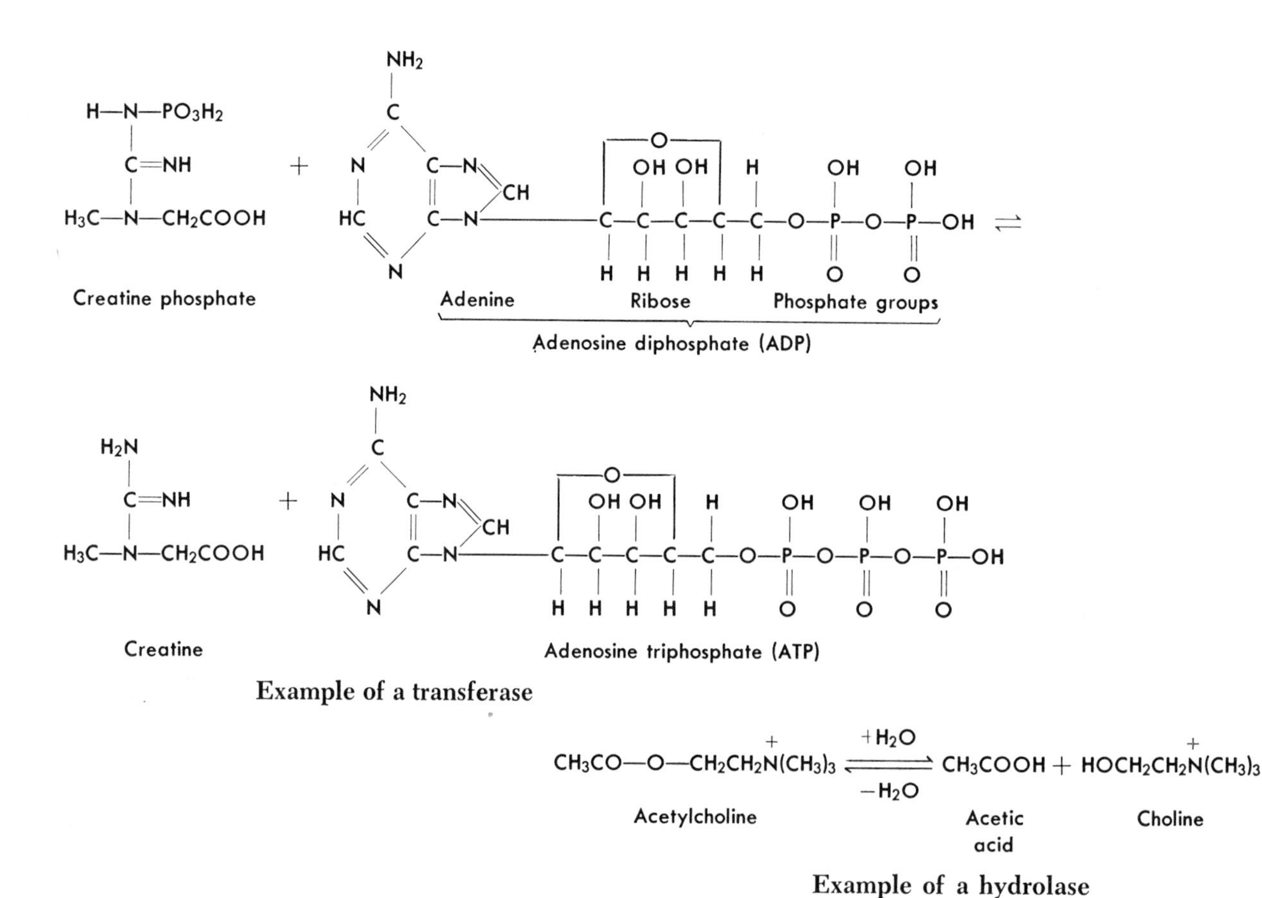

Example of a transferase

Example of a hydrolase

cal oxidations and reductions—the removal or addition of hydrogen atoms. An important example is lactic dehydrogenase.

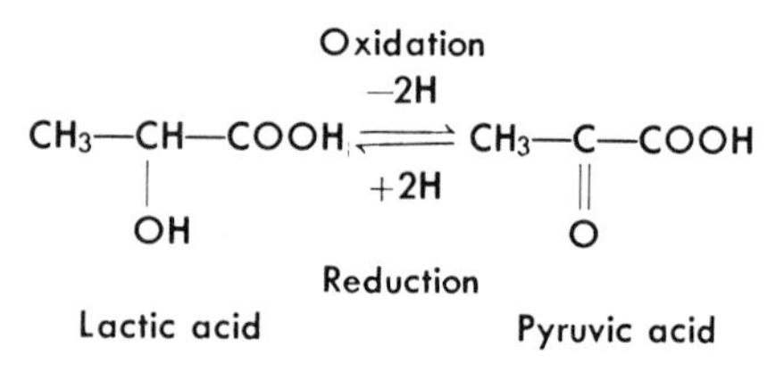

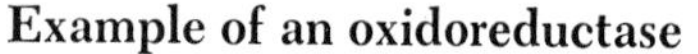
Example of an oxidoreductase

Coenzymes are typically required for these oxidoreductase enzymes, and, in fact, it is the prosthetic group that accepts or donates the hydrogen atoms. Nicotinamide adenine dinucleotide (NAD) and NAD phosphate (NADP) are examples of these coenzymes.

Isomerases. ■ Intramolecular rearrangements are catalyzed by isomerases. For example, in the anaerobic breakdown of carbohydrate to lactic acid, triose phosphate isomerase converts 3-phosphoglyceraldehyde to dihydroxyacetone phosphate.

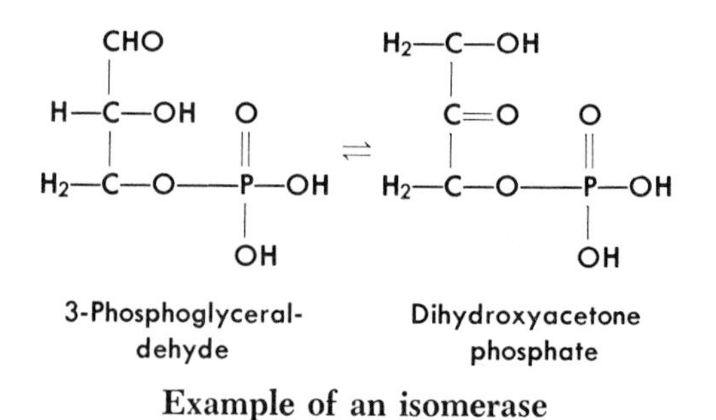

Example of an isomerase

Synthetases. ■ Enzymes that catalyze the joining together of two separate molecules are known as synthetases. This reaction is coupled with the breakdown of an "energy-rich" bond, such as the terminal phosphate bond of ATP. In Chapter 2 reference was made to the synthesis of protein by the joining together of "activated" amino acids. Another excellent example of a synthetase is acetyl-coenzyme A (CoA) synthetase.

$$\text{Adenosine triphosphate (ATP)} + \text{Acetate} + \text{CoA} \rightleftharpoons \text{Adenosine monophosphate (AMP)} + \text{Acetyl-CoA} + \text{Pyrophosphate (PP)}$$

Acetyl-CoA is an important compound formed during the utilization of carbohydrate, lipids, and proteins in body metabolism.

• • •

It should be noted that in the examples given the reactions are reversible; that is, they may proceed in either direction, and in theory all enzyme reactions are reversible. But in many instances the point of equilibrium is so far in one direction as to make it impossible in practice for the reverse reaction to occur.

In the intact cell the reaction is usually not reversible, since the end product of one reaction is the substrate for the following reaction. Indeed, most enzymes do not function alone but in a multienzyme apparatus, which is responsible for the many steps necessary for both the degradation and synthesis of essential materials for the cell.

■ MECHANISM OF ENZYME ACTION

Enzyme-substrate complex. ■ A theory of enzyme action, which has been generally accepted as the best explanation of the sequence of events that results in the change of substrate to products, was postulated by Henri and later mathematically formulated by Michaelis and Menten. According to this theory an enzyme combines with its substrate to form an unstable intermediate complex. This complex rapidly breaks down to yield the reaction products plus the original enzyme. This may be represented in simple form as follows:

$$\text{E(enzyme)} + \text{S(substrate)} \underset{k_2}{\overset{k_1}{\rightleftharpoons}} \text{ES(complex)} \overset{k_3}{\longrightarrow} \text{E} + \text{P(products)}$$

The Michaelis constant, K_m, which combines the three velocity constants above, is equal to $k_2 + k_3/k_1$. This constant is of value, since for every enzyme-substrate system, K_m has a characteristic value and can be used to determine the rate-limiting step in a metabolic series of consecutive reactions.

Since the enzyme-substrate complex is a very transitory compound, some difficulty in actually proving its existence was incurred. However, Chance and others have been able to identify its existence by means of a specialized spectrophotometer. It is undoubtedly the formation of this complex that enables enzymatic reactions to occur, since the energy of activation is lowered. For any chemical reaction to occur, molecules must achieve a specific energy level; that is, they must attain a certain level of activation to overcome an energy barrier. This can be achieved by increasing the temperature; however, this is not possible within a cell; therefore the complex permits the reaction to occur by lowering the energy level.

■ Factors influencing the rate of enzyme action

Enzyme and substrate concentration. ■ It has been observed with purified enzymes in reaction mixtures that increasing the concentration of the enzyme produces a proportional increase in the rate of the reaction, providing other conditions remain the same and that a constant but excess amount of substrate is present. That is, if the amount of the enzyme is doubled, the reaction proceeds twice as fast. A similar situation occurs when the concentration of the substrate is progressively increased and when all other conditions are kept constant. Now, however, the reaction rate does not increase indefinitely, but only until a well-defined maximum is reached. At this point the enzyme-substrate complex concentration is optimal, the enzyme is saturated, and further addition of substrate evokes no additional increase in reaction rate.

pH. ■ Enzyme reactions proceed at their fastest rate at an optimum pH and are considerably slowed or even stopped at higher or lower pH values. The majority of intracellular enzymes have a pH optimum close to neutrality (pH 7.0). Many enzymes, however, have optimums close to either end of the pH range; for example, pepsin (about pH 2.0) and alkaline phosphatase (about pH 9.0 to 10.0).

How pH influences the rate of enzymatic reactions is not completely understood. It may affect the substrate if the latter is an electrolyte; certainly pH affects the ionic groups on the protein molecule, and it may even denature the enzyme. Within the cell, pH may have an important role in controlling the rate of the various enzyme reactons, but insufficient knowledge is available about pH values in different portions of the cell. These three factors are illustrated diagrammatically in Fig. 22-3.

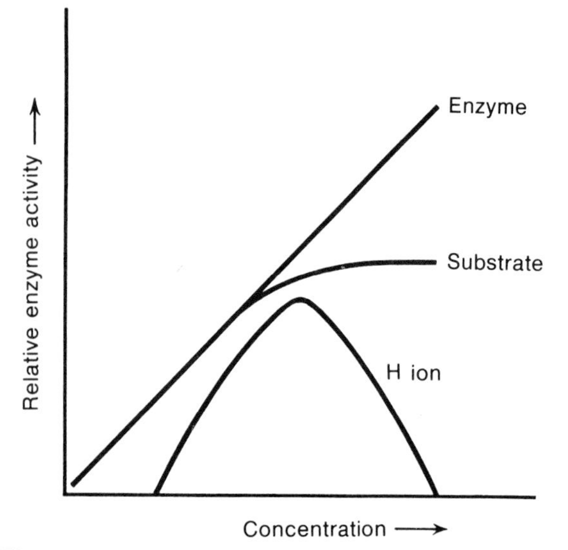

Fig. 22-3
Influence of enzyme, substrate, and hydrogen ion concentration on the rate of enzyme reaction. Rate will continue to increase with an increased concentration of enzyme, provided unlimited substrate is available. Increasing the substrate concentration with a constant amount of enzyme will increase the rate of reaction until all the enzyme present is saturated. As indicated, there is an optimum hydrogen ion concentration (pH) for maximum rate of activity.

Temperature. ■ The rate of chemical reactions is usually increased two to three times for each 10 C rise in temperature. The change in rate per unit temperature change is known as the temperature coefficient of the reaction rate and is designated as the Q_{10}. For enzymes the Q_{10} ranges from about 1.1 to 3. The maximum temperature range for enzyme activity is quite narrow (from 10 to 50 C), and at temperatures above 60 C they are denatured or destroyed. In Chapter 18 the effects of temperature on some proteins were discussed.

End products. ■ The end products of the reaction have a very definite influence on the rate of the reaction. If they are allowed to accumulate, these end products will inhibit the activity of the enzymes. This inhibition is an example of *negative feedback*. One reason for this retarding effect on the rate of the enzyme reaction is that some of these end products may be structurally similar to the substrate and therefore will combine with the enzyme. It also has been postulated that these products, as well as other agents, may combine with the enzyme at a site other than the active site and cause a conformational change in the enzyme protein. This has been termed an *allosteric effect*. This effect may be inhibitory or stimulatory. The enzyme phosphofructokinase, which is utilized in carbohydrate metabolism, is subject to allosteric control—ATP is necessary for its activity, but in higher concentrations ATP will inhibit enzymatic activity.

Hormones and other biochemical substances. ■ Recently it has been demonstrated that hormones, amino acids, and other substances can alter enzymatic activity. As mentioned earlier, hormone action on the rate of enzyme activity may be due to altering the conformation of the protein. This may provide the mechanism for understanding the effect of steroid hormones on the rate of protein synthesis. The influence of hormones may be mediated in many instances by a "second messenger" located within the target cell. The production or removal of this substance is influenced by the hormone in question. Cyclic AMP has been shown to be a "second messenger." For example, epinephrine and glucagon (Chapter 28) accelerate the production of cAMP in the liver cell; an increased level of cAMP promotes the conversion of the phosphorylase enzyme to its active form, which augments the breakdown of glycogen. Cyclic AMP plays an important part in the activation of several other protein kinases found in many cells.

Light and other physical factors. ■ Light and other physical factors may inhibit or accelerate the rate of enzyme activities; however, they are usually not of great importance physiologically.

■ Activation, acceleration, inhibition, and induction of enzymes

Activators and accelerators. ■ For protective purposes, many enzymes are present in cells and secretions in an inactive precursor form, that is as *proenzymes.* Trypsinogen, pepsinogen, and chymotrypsinogen are the proenzymes of trypsin, pepsin, and chymotrypsin, respectively. Some proenzymes can be converted into the active enzyme autocatalytically. Conversion of pepsinogen to pepsin is an example of such self-activation. Others require the intervention of a separate enzyme (in the case of trypsinogen it is enterokinase) for activation.

Coenzymes are another important activator of certain enzymes and are necessary to their function. These heat-stable, dialyzable organic compounds are usually separable from the enzymes that they activate. In all cases the apoenzyme determines the substrate specificity of the enzyme, whereas the coenzyme determines the chemical nature of the enzymatic reaction. The latter moiety is frequently the donor or acceptor of the part of a molecule added to or removed from the substrate. Many of the known coenzymes are chemically closely related to or incorporate vitamins into their structure. Specifically, NAD and NADP contain nicotinamide, whereas cocarboxylase contains thiamine.

Investigation reveals that many enzymes need metal ions for their activity. Such metals as zinc, magnesium, iron, and cobalt are essential parts of the active centers of such enzymes. For example, carbonic anhydrase requires zinc; it is inactive when this metal is removed from the enzyme molecule.

Many inorganic ions such as Na^+, K^+, and Ca^{++} either accelerate or inhibit certain enzyme reactions, although their special mechanism of action has not been thoroughly elucidated.

Inhibitors. ■ As already indicated, many agents in addition to end products inhibit the rate of enzyme reactions. Such enzyme inhibitors may be classed as *noncompetitive* or *competitive.* Noncompetitive inhibitors act directly on or combine with the enzyme. These may be specific for either the coenzymes, as are azides and fluorides, or for the protein portion, as are heavy metals such as mercury. Competitive inhibitors are substances that compete with the substrate for the available enzyme because of a similar chemical structure. Although these substances resemble the substrate closely enough to actually combine with the enzyme, the reaction ceases at this point without formation of end products. In consequence, the enzyme is not free to act with its usable substrate. The enzyme that catalyzes the conversion of succinic acid to fumaric acid—an important step in the oxidative metabolism of glucose—is competitively inhibited by malonic acid.

There are many nonspecific substances not properly included in these two classifications but which, nevertheless, influence the rate of enzyme reactions. Among these are anesthetics, antiseptics, and some therapeutic drugs.

Enzyme induction. ■ This phenomenon was first recognized and studied with microorganisms, that is, yeast and bacteria. If a strain of microorganisms grown in a medium containing glucose was transferred to one containing another sugar, for instance, galactose, after a time lag, the ability to oxidize the first sugar diminished as the oxidation of galactose increased. This occurred with a nondividing cell, and it could only mean that the cell had adapted its biochemical machinery; further experiments proved that indeed there was a new protein synthesized. Therefore the cell had been induced to form an enzyme by the presence of a substrate. This phenomenon is not limited to microorganisms; it has been demonstrated in mammalian cells as well. A tenfold increase in activity of rat liver tryptophan pyrrolase has been demonstrated following the injection of the amino acid tryptophan. Other examples have also been shown in mammalian cells. In some cases the inducer has been a hormone rather than a specific substrate, which opens new avenues for investigation of hormonal regulation of metabolism.

■ ENZYMES AND THE CELL

The multiplicity and variety of chemical reactions taking place within a cell during metabolism are, in nearly all instances, dependent on enzymes for initiation or acceleration. Within the living, intact cell, individual enzymes do not act independently but are linked together in multienzyme systems so that the end products of one reaction are utilized as the substrate for the next reaction. This linking, or coupling, of reactions is clearly influenced by the morphological location of the enzymes and substrates within the cell. The mitochondria are the sites of intracellular energy production and transduction, since the enzymes of the citric acid cycle and fatty acid oxidation cycle appear to be localized on the outer membrane, whereas the inner membrane contains the enzymes concerned with oxidative phosphorylation (the generation of oxidative energy and its conversion to chemical energy). The endoplasmic reticulum and its associated ribosomes contain the enzymes necessary for protein synthesis. The Golgi apparatus appears to be a storehouse for proteins produced by the ribosomes. The cytoplasm contains numerous enzymes, among which are the glycolytic enzymes. The nucleus, which contains most of the cellular DNA, must contain enzymes necessary for the production of this material.

Interrelationships must exist between all these parts of the cell, and there is an ordered coordination of functions. A cell should not be thought of as a sack of enzymes, even though there are large numbers of them present. Common pathways do exist for carbohydrate, protein, and lipid metabolism, which reduce the number of enzymes. Since the various organs have specialized functions to perform, it is reasonable to predict that the nature of the intracellular enzymes will vary and that therefore the characteristic or individuality of any given cell will be determined by its particular complement of enzymes. The organized activity of enzymes in the protoplasm may be the most important characteristic of those features designated as life.

READINGS

Bernhard, S.: The structure and function of enzymes, Menlo Park, Calif., 1968, W. A. Benjamin, Inc.

Koshland, D. E., Jr., and Nett, K. E.: The catalytic and regulatory properties of enzymes, Ann. Rev. Biochem. **37**:359-410, 1968.

Lehninger, A. L.: Biochemistry, New York, 1970, Worth Publishers, Inc., pp. 147-187.

Phillips, D. C.: The three dimensional structure of an enzyme molecule, Sci. Am. **215**(5):78-90, Nov. 1966.

Carbohydrate metabolism
Lipid metabolism
Protein and nucleoprotein metabolism
Interrelations of carbohydrate, lipid, and protein metabolism

METABOLISM—FOODSTUFFS IN THE CELL

In this chapter we will discuss the major metabolic pathways taken by the carbohydrates, lipids, and proteins after they reach the cell. It should be pointed out that not all these pathways are present in every cell in the body to an equal extent; many cells have become highly specialized in their metabolic functions, for example, adipose cells and lipid metabolism. Interrelationships of the various metabolic pathways will be examined. The advantages accruing from the use of common pathways for the various compounds is obvious—it would decrease the total number of enzymes needed by each cell. Cells must use these materials to obtain energy for their functional roles, to provide the raw material for synthesizing new compounds necessary to maintenance, and to provide a storage of material to be used at a later time for either energy or synthesis.

It should be recognized that the anabolic and catabolic pathways between a precursor and its end product are usually not identical. Several benefits accrue to the cell from this apparently wasteful and illogical system of independent pathways. It is in many cases energetically impossible for biosynthesis to take place by a direct reversal of the degradative pathway. Separate pathways can be located in different parts of the cell and thus allow catabolism and anabolism to occur independently and simultaneously. Regulation of anabolic and catabolic sequences can be and usually are independent.

We will not attempt to present the detailed chemistry involved in these metabolic transformations, since it is not necessary for our purposes. Students who are interested should consult one of the many excellent biochemistry textbooks.

Before considering these pathways we should briefly examine the most important intermediary compound in the transfer of energy, adenosine triphosphate (ATP), whose formula has been given in Chapter 22. The last two phosphate radicals of this molecule are connected to the remainder of the molecule by energy-rich bonds. During hydrolysis approximately 7,000 to 8,000 cal of energy are liberated, and, conversely, the same amount of energy is necessary for the formation of these bonds. ATP is coupled with the many reactions occurring in the cell that liberate or require energy for their completion. Thus in the following sections, when energy is required for the reaction, ATP is converted to ADP in a coupled reaction; when energy is released by a reaction, the cell may utilize it to reform ATP and thereby maintain its supply of this material. Other energy-rich compounds do occur in many cells and are important in specific reactions; however, none fulfills the diverse roles of ATP.

■ CARBOHYDRATE METABOLISM

That which characterizes carbohydrates in our body and sets them apart from proteins and lipids is their great mobility and their availability as a source of energy. About 60% of ingested protein and 10% of ingested fats are transformed by the body into carbohydrates. As indicated in Chapter 21, a supply of glucose is always present in the blood and is readily available to all cells for utilization. In addition, cells can store this material to a greater or lesser extent in the form of glycogen granules. Tissues, such as muscle and glands, which perform a great deal of work, have a rather large supply of glycogen available for energy.

Interconversion of monosaccharides. ■ Before any of the monosaccharides can be utilized by the cell for energy or synthesis of glycogen, they must be converted to glucose-6-phosphate or fructose-6-phosphate. Entry of the monosaccharides into a cell is by a carrier system whose capacity is greatly increased by the presence of the hormone insulin (Chapter 28). Immediately on entry into the cell, glucose, fructose, and galactose are phosphorylated by means of a specific hexokinase; for example, glucose in the presence of *glucokinase* and ATP is converted to glucose-6-phosphate. Galactose is converted to galactose-1-phosphate, which is then converted to glucose-1-phosphate ⟶ glucose-6-phosphate.* Fructose is converted to fructose-6-phosphate.

$$\text{Glucose} \xrightarrow[\text{Glucokinase}]{\text{ATP}} \text{Glucose-6-phosphate}$$

$$\text{Glucose-6-phosphate} \xrightleftharpoons{\text{Phosphoglucomutase}} \text{Glucose-1-phosphate}$$

Glycogen. ■ All cells in the body are capable of storing carbohydrates in the form of glycogen to some extent, but liver and muscle cells can store up to 8% and 1%, respectively, of their weight in this material. The formation of glycogen from glucose (glycogenesis) occurs by the steps shown above and below. The transformation of the glucose-1-phosphate to glycogen requires uridine triphosphate (UTP) to form an active nucleotide, uridine-diphosphate-glucose (UDP-glucose). These activated molecules unite in the presence of the enzyme UDP-glycogen-transglucosylase (glycogen synthetase) and a branching enzyme to form glycogen:

$$\text{Glucose-1-phosphate} \xrightleftharpoons{\text{UTP}} \text{UDP-glucose}$$

$$\text{UDP-glucose} \xrightarrow[\text{Branching enzyme}]{\text{Glycogen synthetase}} \text{Glycogen}$$

Glycogen has an average molecular weight of 5 million and is in the form of solid granules in most cells, and therefore it can be stored without disturbance of the osmotic equilibrium of the cell. When the need for glucose arises, the glycogen is broken down one unit at a time by *phosphorylase a,* with the resulting formation of glucose-1-phosphate. Energy is not required for this conversion. Phosphorylase is widely distributed in the cells and exists in two forms—*b,* which is inactive, and *a.* Conversion of the *b* to *a* form is necessary before glycogenolysis can occur, and it provides a mechanism through which hormones may influence the breakdown of glycogen. This hormonal control has been amply veri-

*Arrows will be employed hereafter in the text to show reactions and their direction.

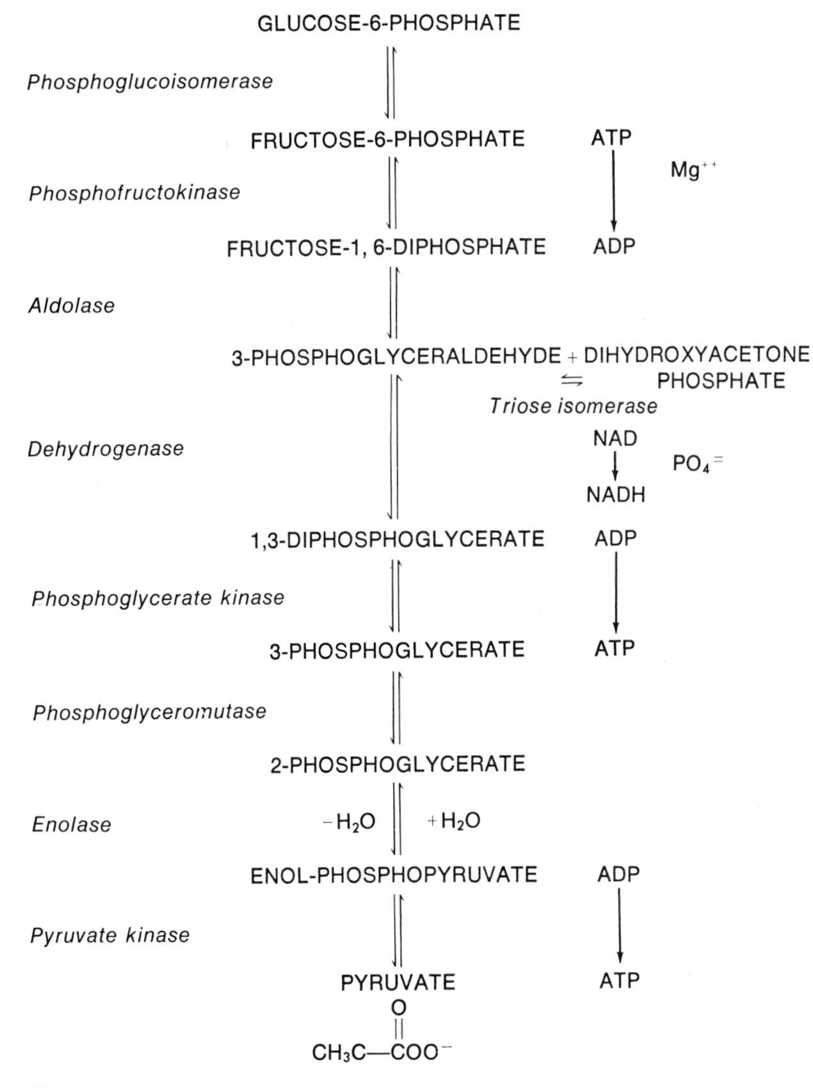

Fig. 23-1
Sequence of reactions occurring in glycolysis.

fied to occur by way of cyclic AMP, which is the activator for an enzyme phosphorylase kinase, which in turn activates phosphoryl kinase, the enzyme necessary for the conversion of phosphorylase *b* to *a*. Glucose-1-phosphate is converted to glucose-6-phosphate by phosphoglucomutase. The conversion of this compound back to glucose can occur only in the liver.

Anaerobic breakdown of glucose-6-phosphate —glycolysis. ■ By a number of enzymatically catalyzed steps, glucose-6-phosphate is converted from a 6-carbon molecule into two molecules of a 3-carbon compound; if no oxygen is present, the end product is lactic acid. In Fig. 23-1 the sequence of reactions illustrates glycolysis as it occurs in muscle and other tissues, primarily in the cytoplasm of the cell. It should be noted that by reversing this series of steps it is possible to form glycogen. In this scheme, only some of the necessary enzymes and cofactors are indicated.*

The last step in this anaerobic sequence is the conversion of pyruvate to lactate; it is catalyzed by the enzyme lactic dehydrogenase, which requires NADH obtained from the fourth reaction of the sequence.

$$CH_3\overset{\overset{O}{\|}}{C}—COO^- + NADH \rightleftharpoons CH_3\overset{\overset{OH}{|}}{C}H—COO^- + NAD$$

When there is an accumulation of lactate, as in prolonged muscle contraction, this material may enter the bloodstream and be carried to other cells, where it is utilized, or to the liver, where it is converted back to glycogen. The lactic acid cycle is illustrated in Fig. 23-2. In the biosynthesis of glycogen most of the steps are a reversal of those employed in glycolysis. Two steps, however, cannot proceed by simple reversal because of the large energy changes required. Alternate pathways are utilized which are favorable in the direction of synthesis. Pyruvate is phosphorylated to enol phosphopyruvate by a circuitous route which first involves its conversion to oxaloacetate in the mitochondria. The oxaloacetate is then reduced to malate (NADH is oxidized) and it diffuses out of the mitochondria. The malate is reoxidized to oxaloacetate which is acted on by phosphoenolpyruvate carboxykinase to yield enol phosphopyruvate. The phosphate donor is guanosine triphosphate (GTP).

*NAD and NADP are synonymous with DPN and TPN, which in the past have been called coenzymes I and II. In our text, NAD and NADP represent the oxidized forms; NADH and NADPH represent the reduced forms.

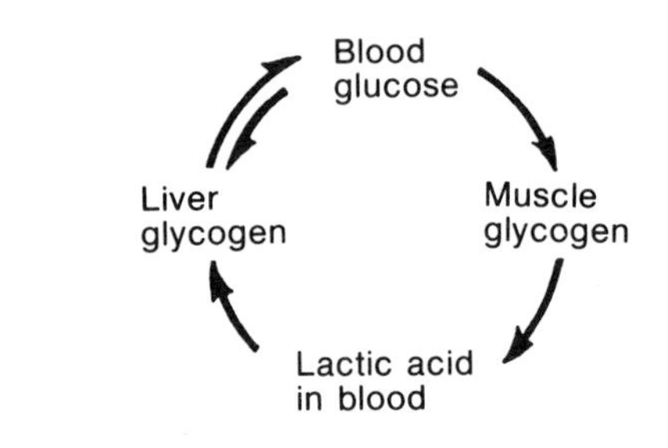

Fig. 23-2
Lactic acid cycle.

The second reaction in carbohydrate synthesis which does not utilize the glycolytic sequence is the conversion of fructose-1,6-diphosphate to fructose-6-phosphate. Instead an irreversible hydrolysis catalyzed by fructose diphosphatase occurs as follows:

$$\text{Fructose-1,6-diphosphate} + H_2O \rightarrow \text{Fructose-6-phosphate} + H_2O + Pi$$

In some cells glucose-6-phosphate is not used as a precursor of glycogen formation but is dephosphorylated to form free glucose. This also cannot occur by a reversal of the glucokinase reaction and again a hydrolytic reaction brought about by glucose-6-phosphatase is used.

The overall reaction from pyruvate to glucose involves the consumption of 6 energy-rich phosphate bonds and the utilization of two molecules of NADH.

In muscle work of moderate intensity the O_2 supply may lag somewhat at first and some lactic acid may be formed, but by increased cardiac activity and pulmonary ventilation, the intake of O_2 soon becomes sufficient to furnish all the energy needed during the recovery period. This condition, which is attained by the pulmonary intake of oxygen keeping pace with the increased muscle activity, is known as the steady state.

In prolonged hard work it is not possible for the circulatory-respiratory machinery to furnish sufficient O_2, and lactic acid accumulates. Since the acid must be oxidized, this constitutes an *oxygen debt.* The labored breathing following the cessation of work is the consequence of the need for paying the debt. The greater the rapidity with which lactic acid is produced, the shorter is the period during which the work can be maintained. To illustrate: A 100-yard dash can be made in about ten seconds and requires about 6 liters of O_2, but the maximum intake of O_2 by respiration does not exceed 4 liters/min. If our muscles were dependent on a full supply of O_2 while work is in progress, running up even one flight of stairs would be impossible.

During the glycolytic process there is a net production of three molecules of ATP from ADP. It is a somewhat inefficient process, yielding only a small percentage of the available energy in the form of ATP. It should also be noted that the glycolytic cycle will proceed more rapidly under anaerobic conditions; the presence of oxygen inhibits glycolysis and promotes the resynthesis of glycogen. This has been termed the *Pasteur effect.*

Although the pathway just described constitutes the major pathway, another scheme is necessary to explain the utilization of glucose when anaerobic glycolysis is blocked. This second pathway is an aerobic one and has been called the pentose-phosphate shunt, or direct oxidative pathway. In addition to an alternate route for glucose, this shunt provides the NADPH necessary for fatty acid synthesis and the intermediate pentose esters necessary for nucleic acid synthesis. Briefly the steps for this pathway are shown below.

Glucose-6-phosphate + NADP ⇌ 6-Phosphogluconic acid + NADPH

6-Phosphogluconic acid + NADP ⇌ Ribulose-5-phosphate + CO_2 + NADPH

Ribulose-5-phosphate + ATP ⇌ Ribulose-1,5-diphosphate + ADP

2 Ribulose-1,5-diphosphate ⇌ Ribose-5-phosphate + Xylulose-5-phosphate

Ribose-5-phosphate + Xylulose-5-phosphate ⇌ Sedoheptulose-7-phosphate + 3-Phosphoglyceraldehyde

Thus we have two 5-carbon sugars, ribose and xylulose, uniting to form a 7-carbon sugar, sedoheptulose, and a 3-carbon sugar phosphoglyceraldehyde, which can be utilized in glycolysis as originally described. It is also possible for these latter two sugars to combine to yield fructose-6-phosphate and erythrose-4-phosphate. This last compound may combine with xylulose-5-phosphate to yield fructose-6-phosphate and 3-phosphoglyceraldehyde.

Another shunt in the glycolytic sequence is present in the red blood cell (Chapter 12).

1,3-Diphosphoglycerate ⟶ 2,3-Diphosphoglycerate ⟶ 3-Phosphoglycerate + Pi

The first reaction is catalyzed by diphosphoglycerate mutase, the second by 2,3 diphosphoglycerate phosphatase.

Aerobic breakdown—the terminal oxidative pathway. ■ Although we shall discuss the *tricarboxylic acid cycle* (Krebs' cycle, citric acid cycle) with the catabolism of carbohydrates, it should be recognized that this pathway is a common one for the final degradation of the 2- and 3-carbon end products of lipid and protein catabolism. It also performs an anabolic role by providing compounds necessary for the biosynthesis of amino acids, nucleic acids, long-chain fatty acids, and other essential cellular components.

The next step beyond glycolysis in the catabolic sequence is the formation of acetyl-coenzyme A (acetyl-CoA) from pyruvate. The transformation occurs by means of the multienzyme pyruvate dehydrogenase, which is associated with the mitochondria of the cells. This multienzyme includes NAD, thiamine pyrophosphate, and lipoic acid; the overall result is given in the following reaction.

$$CH_3\overset{\overset{\displaystyle O}{\|}}{C}—COO^- + CoA—SH + NAD \rightleftharpoons$$

$$CH_3\overset{\overset{\displaystyle O}{\|}}{C}—S—CoA + NADH + CO_2$$

The fate of the reduced coenzyme will be considered later when the electron transport system is discussed. We shall now consider some of the more essential steps involved in the citric acid cycle. Acetyl-CoA enters the cycle by combining with oxaloacetate to form citric acid; the cycle proceeds as shown in Fig. 23-3.

All of these reactions are localized in the mitochondria of the cells. Condensation of acetyl-CoA and oxaloacetate is catalyzed by the citrate condensing enzyme; the reversible hydration-dehydration of citrate to isocitrate is catalyzed by the enzyme aconitase. Two forms of isocitrate dehydrogenase, one requiring NAD and the other NADP, catalyze the conversion of isocitrate to α-ketoglutarate, with the reduction of NAD or NADP and the production of CO_2. The conversion of an α-ketoglutarate to succinyl-CoA is catalyzed by the α-ketoglutarate dehydrogenase complex of enzymes, which is similar to the complex responsible for the pyruvate to acetyl-CoA conversion. Succinyl-CoA, unlike acetyl-CoA, does not undergo a wide variety of metabolic reactions, and its main route is to continue on in the cycle. Conversion of succinyl-CoA to succinate provides the energy necessary to form an energy-rich bond, and the reaction is catalyzed by the enzyme succinic thiokinase. Guanosine diphosphate is the phosphate acceptor, and the triphosphate formed is readily converted to ATP. Dehydrogenation of succinate to fumarate is catalyzed by succinic dehydrogenase, which requires a flavoprotein (FAD) as the hydrogen acceptor. Fumarase catalyzes the conversion of fumarate to malate. The final step in the cycle is the removal of two hydrogens from malate, which yields oxaloacetate; this is catalyzed by malic dehydrogenase, which requires NAD as a hydrogen acceptor.

It is clear that this cycle can continue to function only if there is provision for the reoxidation of the reduced coenzymes produced during its operation. A sequence of carriers called the *electron transport system,* located in the mitochon-

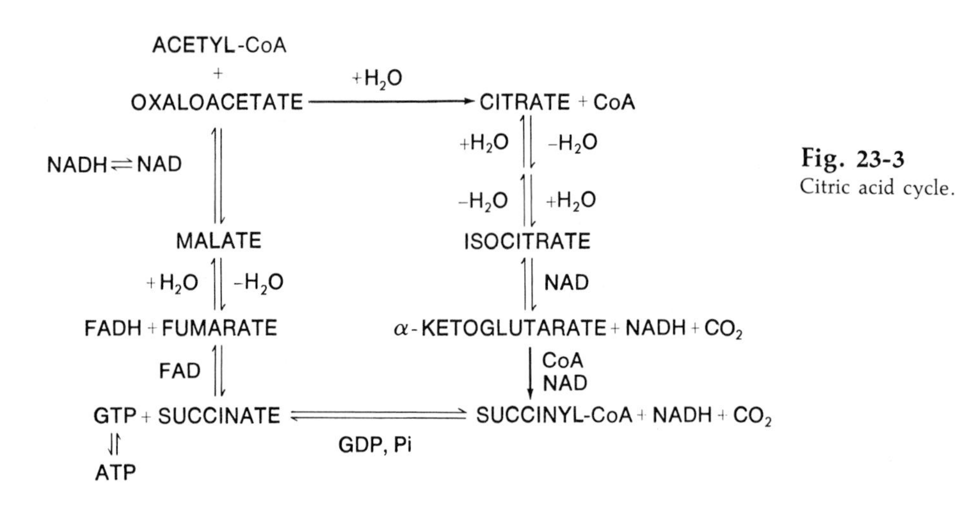

Fig. 23-3
Citric acid cycle.

dria of the cell, provides such a mechanism. This system provides the means for transferring hydrogen and electrons to oxygen (oxidation), with the energy released being utilized to form the energy-rich bond of ATP (oxidative phosphorylation). A great deal of doubt still exists concerning the mechanism by which the hydrogens and/or electrons are transferred along this chain and at what points the available energy is utilized for the production of ATP. The following sequence appears to fit much of the experimental data. Specific enzymes catalyze each step in the reaction sequence.

$$\text{NAD or NADP} \longrightarrow \text{FAD} \longrightarrow \text{Cytochromes b} \longrightarrow \text{c} \longrightarrow \text{a} \longrightarrow \text{a}_3 \longrightarrow O_2$$

Each carrier is reoxygenated as it passes the hydrogen atoms or electrons along to the next member until the final step is reached when oxygen accepts them to produce the terminal product, H_2O. The cytochromes are heme proteins in which the iron atom can change from the ferrous (Fe^{++}) and ferric (Fe^{+++}) states, with reduction and oxidation, respectively. In the cytochromes, unlike hemoglobin, the fifth and sixth positions of the iron are occupied by amino acid residues from the protein portion of the molecule and are unavailable to bind with oxygen, etc. There are at least three sites along this chain where ATP is produced; the first is at the NAD-FAD junction, the second is at the cytochrome b to c junction, and the third is at the cytochrome a to a_3 junction. Therefore, for each 2 electrons that pass along this chain the following reaction occurs:

$$3\ \text{ADP} + 3\ \text{Pi} \rightarrow 3\ \text{ATP}$$

It is now possible to prepare a set of equations for the complete oxidation of glucose to CO_2 and water by means of the glycolytic pathway and the tricarboxylic acid cycle.

Glycolysis

$$C_6H_{12}O_6 + 2\ \text{ADP} + 2\ \text{Pi} + 2\ \text{NAD} \rightarrow 2\ \text{Pyruvate} + 2\ \text{ATP} + 2\ \text{NADH} + 2\ H^+ + 2\ H_2O$$

The NADH which results is located in the cytoplasm and cannot penetrate the mitochondrial membrane directly. Electrons derived from it, however, are transported in by means of a shuttle system which enters the transport sequence at the FAD level. The net result is the oxidation of the NADH with the production of only 2 molecules of ATP.

$$2\ \text{NADH} + 4\ \text{ADP} + 4\ \text{Pi} + 2\ H^+ + O_2 \rightarrow 4\ \text{ATP} + 6\ H_2O + 2\ \text{NAD}$$

Tricarboxylic acid cycle

$$2\ \text{Pyruvate} + 30\ \text{ADP} + 5O_2 + 36\ \text{Pi} \rightarrow 30\ \text{ATP} + 6\ CO_2 + 34\ H_2O$$

When two pyruvates are converted to two molecules of acetyl-CoA, there is a reduction of two NAD; since these are inside the mitochondria the passage of these through the electron transport chain would result in six ATP. For every passage of two acetyl-CoA through the citric acid cycle, six NADH were produced, and their reoxidation would result in eighteen ATP; two FADH and their oxidation would yield four ATP. Finally, two ATP would be directly produced at the succinyl-CoA to succinate step.

The overall equation for the complete oxidation of glucose gives:

$$\text{Glucose} + 6\ O_2 + 36\ \text{ADP} + 36\ \text{Pi} \rightarrow 6\ CO_2 + 36\ \text{ATP} + 42\ H_2O$$

LIPID METABOLISM

The lipids include three classes of substances of great importance to our body: (1) simple fats, or trigylcerides, (2) phospholipids, and (3) sterols. Triglycerides are a valuable energy source and can be stored in adipose tissue as important energy reserves. Adipose tissue also provides insulation when located subcutaneously; it serves as a protective cushion for certain organs such as the eyeball; when properly distributed, adipose tissue gives a pleasing contour to the body.

Phospholipids are essential for life and are found in all cells primarily as part of the membranes. Phospholipids, along with cholesterol and certain proteins, form the structural basis for the cell membrane and contribute to its selective permeability. Cholesterol is the most important sterol and this lipid is also essential for life. A number of physiologically important agents are derived chemically from cholesterol, such as the bile salts, adrenal cortical hormones, sex hormones, and vitamin D. Body lipids can be obtained not only from dietary fat, but from carbohydrates and proteins as well. Only the major pathways concerning the degradation and synthesis of the various lipids will be discussed.

Oxidation of fatty acids. ■ The glycerol portion of the triglyceride molecule is converted to phosphoglyceraldehyde and can enter the chain of reactions followed by the three carbon derivatives of carbohydrate metabolism. It is completely oxidized in the citric acid cycle.

The essential steps in the *beta oxidation* of fatty acids are shown in Fig. 23-4. The first step (1) is the formation of an *active* fatty acid by combination with coenzyme A. This is actually a three-stage process in which the fatty acid is first esterified with extra mitochondrial CoA, transferred to *carnitine*, and then to intramitochondrial CoA. These steps are necessary since fatty acids have only a limited ability to cross the mitochondrial membranes. Three different enzymes can catalyze this formation of acyl-CoA, depending on the length of the fatty acid chain. One of these enzymes utilizes GTP instead of ATP as an energy source. In some cells, succinyl-CoA supplies the CoA group, and this is catalyzed by a thiophorase enzyme. The second step (2) is catalyzed by an acyl dehydrogenase, with FAD required as the hydrogen acceptor. Addition of water across the double bond occurs in step (3). The second dehydrogenation (4) requires NAD as the hydrogen acceptor. In the final step (5) acetyl-CoA is split off, with the formation of an acyl-CoA two carbons shorter than the initial fatty acid. This

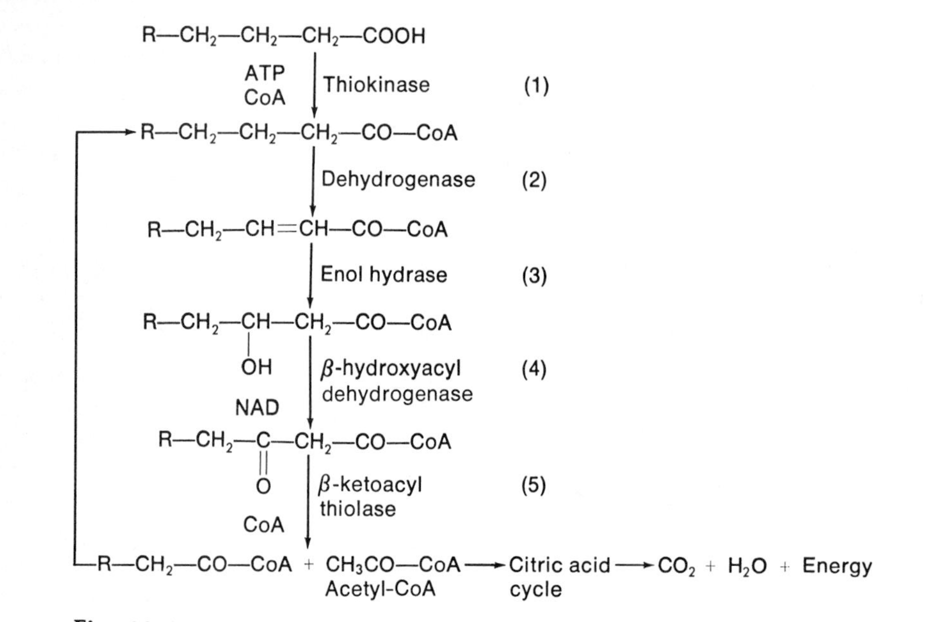

Fig. 23-4
Beta oxidation of fatty acids.

acyl-CoA can be recycled, starting at the second step, until the entire fatty acid has been converted to acetyl-CoA. Molecules of acetyl-CoA can enter the citric acid cycle for breakdown to CO_2 and H_2O with formation of ATP. The enormous number of ATP obtained from the complete oxidation of an 18-carbon fatty acid illustrates the importance of lipids in energy metabolism. The oxidation of the eight NADH and eight FADH formed during the fatty acid cycle would yield forty ATP. In addition, the nine acetyl-CoA formed would enter the citric acid cycle, and, since each turn of that cycle yields twelve ATP, this would give 108 ATP. A total of 148 ATP are produced. This example also demonstrates the important contribution of the citric acid cycle to fatty acid metabolism. Fatty acid oxidation occurs primarily in the mitochondria of the cells of a variety of tissues, including liver, kidney, and cardiac muscle.

Ketone bodies (acetoacetic acid, acetone, and β-hydroxybutyric acid) are present normally in blood in small amounts. Since the latter two are derived from acetoacetic acid, our consideration will be limited to its origin and fate. The last acyl derivative formed in the sequential oxidation of an even-chain fatty acid would be acetoacetyl-CoA, which can follow any of three pathways. It can be cleaved to produce two acetyl-CoA molecules; it can combine with an acetyl-CoA to form an intermediate compound necessary for cholesterol synthesis, or it can be converted to acetoacetic acid. When acetoacetic acid is formed in the liver, it enters the blood and, under normal conditions, is activated and utilized by muscle and other body tissues. Only in abnormal conditions, such as starvation and diabetes mellitus, is more acetoacetic acid produced than can be utilized, a condition known as ketonemia.

When an odd numbered–chain fatty acid is oxidized, the final end product of cleavage would be one molecule of acetyl-CoA and one of propionyl-CoA. This latter substance can be converted to succinyl-CoA by two separate enzymes, one which requires biotin for its activity and the other, vitamin B_{12}. The succinyl-CoA can then enter the citric acid cycle. Propionyl-CoA also can be converted enzymatically to pyruvate or acetyl-CoA in the cell and enter the citric acid cycle.

Biosynthesis of fatty acids. ■ Two pathways exist for the synthesis of fatty acids. One is essentially just the reverse of the pathway for fatty acid oxidation, except NADPH is the necessary coenzyme in place of NADH. This system catalyzes the elongation of preformed fatty acids through condensation with acetyl-CoA and occurs only in the mitochondria. The second pathway is found primarily in the soluble portion of the cell, although some evidence exists for its presence in the mitochondria. Palmitic acid formed from acetyl-CoA is the principal product of this second pathway. The first step in synthesis is the formation of a 3-carbon intermediate, malonyl-CoA. This can be formed in three ways: malonic acid can react with CoA in the presence of ATP; malonic acid and succinyl-CoA in the presence of a transferase can form malonyl-CoA and succinate; or CO_2 can be fixed to acetyl-CoA by a biotin-containing enzyme, acetyl-carboxylase, as follows:

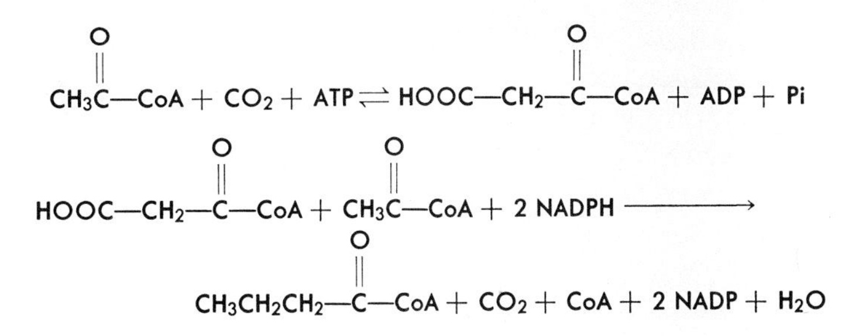

$$CH_3\overset{O}{\overset{\|}{C}}\text{—CoA} + CO_2 + ATP \rightleftharpoons HOOC\text{—}CH_2\text{—}\overset{O}{\overset{\|}{C}}\text{—CoA} + ADP + Pi$$

$$HOOC\text{—}CH_2\text{—}\overset{O}{\overset{\|}{C}}\text{—CoA} + CH_3\overset{O}{\overset{\|}{C}}\text{—CoA} + 2\ NADPH \longrightarrow$$

$$CH_3CH_2CH_2\text{—}\overset{O}{\overset{\|}{C}}\text{—CoA} + CO_2 + CoA + 2\ NADP + H_2O$$

Malonyl-CoA and another molecule of acetyl-CoA combine in the presence of NADPH to form a 4-carbon chain as shown on p. 429. Repetition of the above steps would yield, successively, 6-, 8-, 10-, 12-, 14-, and finally the 16-carbon palmitic acid. The NADPH utilized in this pathway is provided by the pentose-phosphate shunt. Synthetic pathways exist in mammals for most of the fatty acids except *linoleic* and *linolenic* acid, and therefore these are referred to as *essential fatty acids* (Chapter 31). *Arachadonic acid* is formed from linoleic acid and is an important precursor of prostaglandins (Chapter 28).

Synthesis of triglycerides and phospholipids. ■ Fatty acids can be converted to triglycerides or phospholipids by combination with glycerol. First, glycerophosphate is produced either by a reaction between glycerol and ATP or by the reduction of dihydroxyacetone phosphate. The latter reaction occurs in adipose tissue and intestinal cells, since they lack the glycerokinase required for the former.

$$\text{Dihydroxyacetone phosphate} + \text{NADH} \rightleftharpoons \text{Glycerophosphate} + \text{NAD}$$

Glycerophosphate then reacts with two molecules of activated fatty acids to yield phosphatidic acid. Phosphatidic acid is used to form either a triglyceride or any one of the phospholipids, *phosphatidyl serine, phosphatidyl choline* (lecithin), or *phosphatidyl ethanolamine.* Sphingolipids (sphingomyelins, gangliosides, and cerebrosides) are complex and, together with the phosphatides, are the principal lipids found in the brain and the myelin sheath of nerve. The synthetic pathways have not been firmly established for these lipids. Sphingomyelin is composed of equal parts of sphingosine (a long-chain aminoalcohol), choline, phosphate, and fatty acid (R), and its structure is shown below. Cerebrosides contain sphingosine, fatty acids, and galactose. The structures of the gangliosides are even more complex in that they contain sphingosine, fatty acids, one or more sugars, and neuraminic acid (a 9-carbon amino acid).

Biosynthesis of cholesterol and other steroids. ■ Cholesterol (Chapter 21) is the key intermediate in the biosynthesis of the steroids of physiological importance. These include the bile acids, the adrenocortical hormones, the androgens, and the estrogens. Therefore, elucidation of the pathway for the synthesis of cholesterol was of great importance in steroid chemistry. Many of the steps in this pathway have not been fully established; consequently we will not attempt to present the intermediates involved. All of the twenty-seven carbon atoms in cholesterol are derived from acetate (CH_3COO^-). Acetyl-CoA, ATP, NADPH, Mg^{++}, glutathione, and a variety of enzymes are involved in the formation of this molecule. Cholesterol is essential for life and can be synthesized by the animal if it is absent from the diet. It is formed by the cyclization of *squalene,* an open-chain hydrocarbon.

The steroids (Chapters 28 and 30) derived from cholesterol are formed by various modifications of the basic four-ring compound illustrated below. This basic cyclopentoperhydrophenanthrene skeleton:

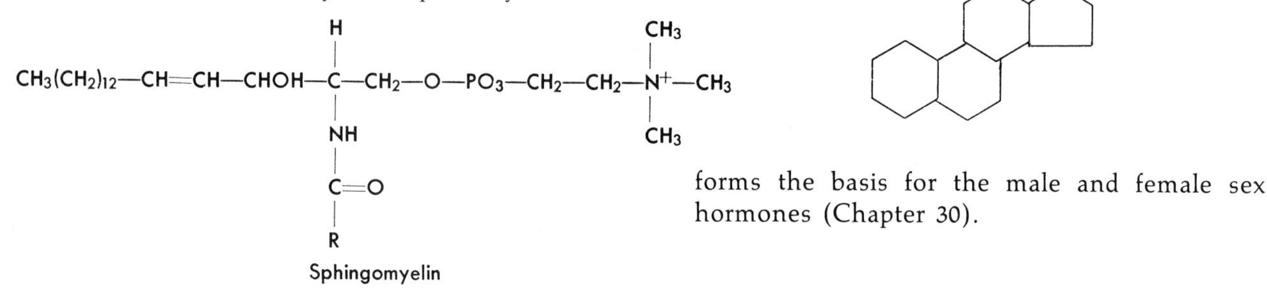

Sphingomyelin

forms the basis for the male and female sex hormones (Chapter 30).

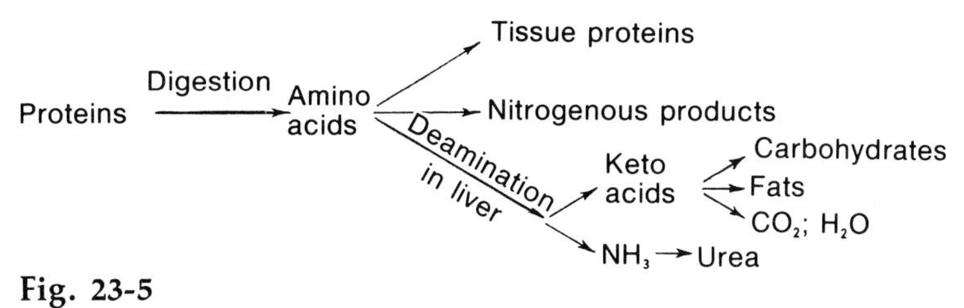

Fig. 23-5
Pathways of protein metabolism.

PROTEIN AND NUCLEOPROTEIN METABOLISM

Protein and nucleoprotein metabolism is quite different from that of carbohydrates and lipids. Proteins stand at the very center of life and form the basic material of all protoplasm. They are utilized in the construction of specialized cellular products such as enzymes, and, in addition, along with nonnitrogenous compounds they can supply energy (Fig. 23-5). Each particular protein of an organ is unique in its chemical composition. Since a protein molecule is composed of a large number of amino acids, proteins can differ in the total number of amino acids, in the number of each kind of amino acid, and in the manner in which the amino acids are arranged. Table 23-1 illustrates how some proteins differ in the relative amounts of the nine essential amino acids. It should be emphasized that not only are the proteins of each tissue structurally distinct from each other but also that there is a species distinction and, indeed, an individual distinction. This diversity in structure can be achieved in part by the arrangement of the amino acid units. Let us imagine a protein molecule of only four different amino acid units, each of which is used but once. These four molecules could be arranged in twenty-four ways. For example:

Glycine-alanine-leucine-lysine
Glycine-leucine-alanine-lysine
Alanine-glycine-leucine-lysine, etc.

In each case a different compound with its peculiar chemical and physical properties would be obtained. If only one single molecule of each of the twenty-two known amino acids combined with each other to form a protein molecule, the number of possible combinations amounts to 39 $\times$ 10^{25} (i.e., 39 followed by 25 zeros). Suppose that the number of the amino acid molecules in

Table 23-1. Percentage of the essential amino acids in some common proteins*

	Lactoglobulin	*Ovoalbumin*	*Casein*	*Gelatin*	*Zein*	*Gliadin*
Leucine	15.7	9.6	9.2	3.5	15.4	6.5
Lysine	11.4	6.1	8.2	4.6	0	1.1
Tryptophan	2.6	1.4	1.7	0	0.1	0.6
Valine	5.5	6.9	7.1	2.7	3.3	2.6
Histidine	1.6	2.4	3.0	0.7	1.3	2.1
Threonine	5.1	4.1	4.5	2.2	3.4	2.1
Methionine	3.2	5.0	3.0	1.1	1.5	1.7
Isoleucine	6.1	8.0	6.1	1.7	6.9	5.0
Phenylalanine	3.8	7.4	5.5	2.3	7.0	5.7

*From Greenberg: Amino acids and protein, Charles C Thomas, Publisher.

the protein remains twenty-two but that two or more of the twenty-two kinds are replaced by increasing the number of the others. The number of possible combinations is increased enormously. When 600 or more amino acid units are present, as they often are, the number of possible combinations staggers the imagination. Further specificity of proteins is achieved by the manner in which these long chains of amino acids are organized (as in helical structures, bridges, or connections between the chains).

Determination of this specificity and diversity of protein molecules is a function of two other types of complex molecules—the nucleic acids. Deoxyribonucleic acids (DNA) and ribonucleic acids (RNA) are found in all cells in combination with basic proteins. RNA are composed of a series of nucleotides and, when completely hydrolyzed, yield phosphoric acid, ribose (a pentose sugar), and nitrogenous bases, primarily adenine and guanine (purines) and uracil and thymine (pyrimidines). The ribose of each nucleotide is linked to the next pentose by means of the phosphate group (Fig. 2-6), and helical molecules with molecular weights as large as 300,000 are formed. Many different molecules are found in each cell, which vary only in the sequence of their nucleotides, much as different proteins vary in the sequence of their amino acids. RNA are found primarily in the cytoplasm of the cell, with minute amounts located in the nucleolus. RNA play a very important role in protein synthesis and, indeed, are undoubtedly the templates for determining the amino acid sequence in proteins.

Deoxyribonucleic acids (DNA) also are composed of a series of nucleotides, but on complete hydrolysis yield somewhat different components than do RNA. The nitrogenous bases are adenine, guanine, cytosine, and thymine; the sugar is deoxyribose; and a phosphate group again serves as the link between the nucleotides (Fig. 2-6). DNA molecules are somewhat larger than RNA molcules and, indeed, may have a molecular weight over 1 million. The nucleus of the cell contains nearly all the DNA. A great deal of evidence has accumulated which indicates that DNA is the primary component of genes—the carriers of heredity. It has been shown that the somatic cells of an organism contain a constant amount of DNA, and this amount doubles just prior to cell division. The reproductive cells, on the other hand, contain only one half the amount of DNA of somatic cells. The double helix of the DNA molecule provides the necessary scheme for exact duplication of the molecule, which is a necessary property if the gene is to be a perfect model of itself.

Emphasis in this section will be primarily on the synthetic machinery in the cell, which provides the amino acids, purines, and pyrimidines to form these complex molecules.

■ Amino acid metabolism

The ability to synthesize amino acids varies in different organisms, and lack of this ability means these compounds must be supplied in the diet; that is, they are essential. An *essential amino acid* is one that must be included in the diet to maintain growth or nitrogen balance or both. For the adult human, the essential amino acids are leucine, lysine, tryptophan, valine, threonine, methionine, isoleucine, and phenylalanine. Histidine and arginine may be necessary in the diet of children for adequate growth. The remainder of the amino acids are synthesized as required. Amino acids are not stored to any large degree in the cells, and formation of new protein depends on having all the necessary amino acids present simultaneously. It is necessary, therefore, for all the essential amino acids to be fed simultaneously. Some of the proteins in a cell can be degraded to supply amino acids for synthesis or even energy, but this seldom involves the structural proteins or those present in the nucleus.

There are several general reactions that are basic to amino acid metabolism. We shall not attempt to discuss the reactions of individual

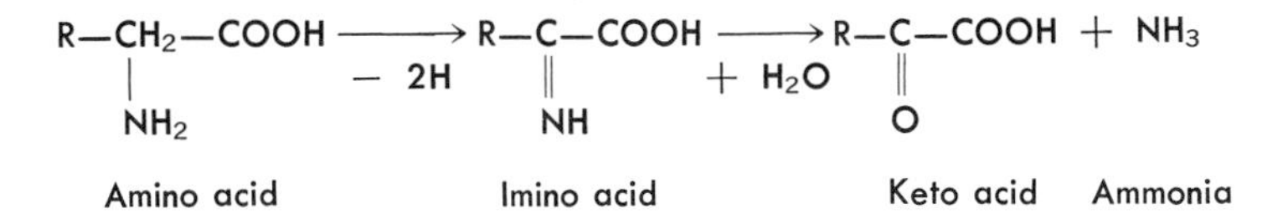

amino acids, since it is beyond the scope of this text.

Transamination. ■ Transfer of the amino group of an amino acid to a keto acid, with the resulting formation of a new amino acid and keto acid, is known as *transamination* and can be illustrated as follows:

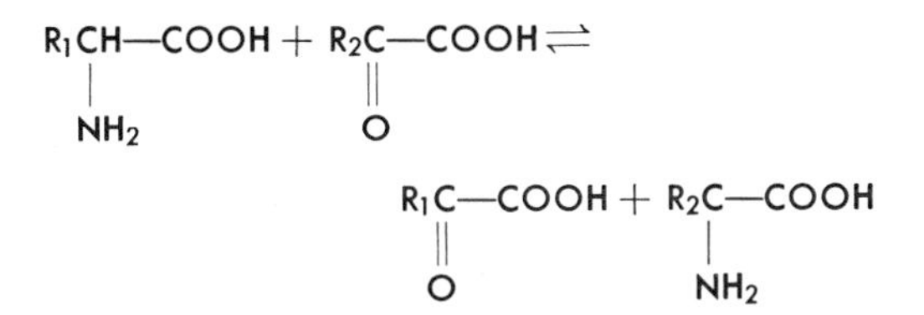

This reaction is catalyzed by specific *transaminases,* which require pyridoxal phosphate as a coenzyme and represents the most important pathway in the synthesis and degradation of amino acids. It provides an important link in joining protein, carbohydrate, and lipid metabolism. A few specific examples that illustrate the interconversion of these compounds may be useful:

Alanine + α-keto glutaric acid ⇌ Glutamic acid + Pyruvic acid

Glutamic acid + Oxaloacetic acid ⇌ Aspartic acid + α-keto glutaric acid

Since all cells can fix ammonia by forming glutamic acid from α-keto glutarate, glutamic acid is of fundamental importance in transamination reactions.

Decarboxylations. ■ Tissues contain various decarboxylase enzymes that attack amino acids at the carboxyl end rather than the amino group, to yield an amine and carbon dioxide. Several physiologically important compounds are formed by these reactions; for instance, the decarboxylation of glutamic acid yields γ-aminobutyric acid (GABA), of histidine yields histamine, of 5-hydroxytryptophan yields 5-hydroxytryptamine (serotonin).

Deamination. ■ Oxidative deamination catalyzed by specific *amino acid oxidases* is the most common pathway of amine removal and leads to the formation of the corresponding α-keto acid and ammonia. These enzymes are present in liver and kidney and catalyze the type reaction shown above.

As a result of either transamination or deamination, a keto acid is formed from the original amino acid, and this product may enter either the carbohydrate or fatty acid cycles for completion of its degradation. Amino acids have been historically classified as being either glycogenic (capable of forming glycogen), ketogenic (forming ketone bodies), or both, when fed to an animal. Examples are given in Table 23-2. The urea cycle discussed in Chapter 21 provides the mechanism for the excretion of the ammonia formed in deamination.

Nonprotein substances synthesized from amino acids. ■ Several physiologically important compounds besides proteins and nucleoproteins are formed from amino acids. We shall mention only a few of these as examples.

Creatine, creatine phosphate, and creatinine. *Creatine* is synthesized from the amino acids

Table 23-2. Glycogenic and ketogenic amino acids

Glycogenic	*Ketogenic*	*Glycogenic and ketogenic*
Alanine Glutamic acid Glycine Methionine Serine	Leucine	Isoleucine Tyrosine Phenylalanine

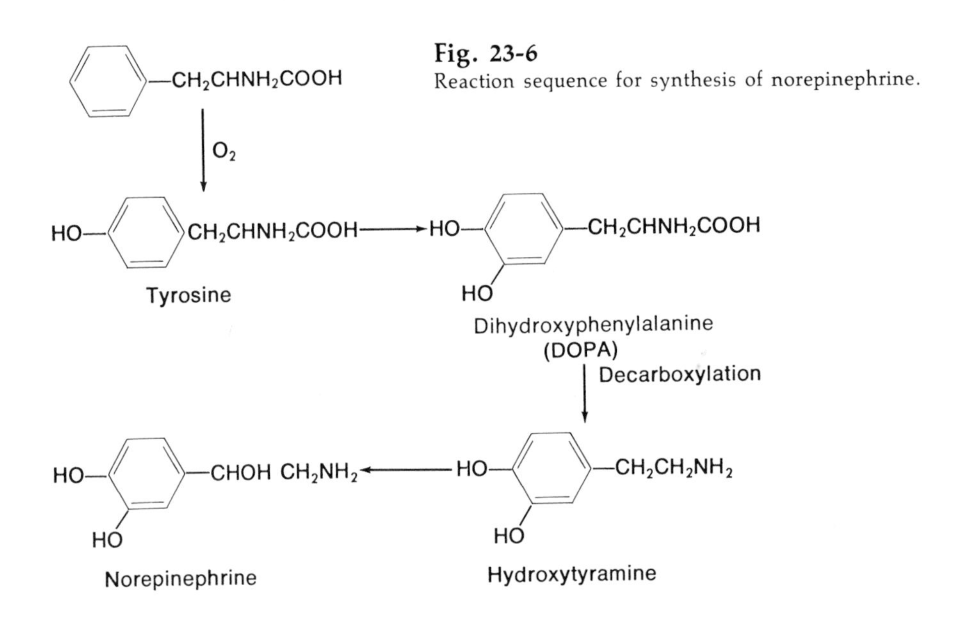

Fig. 23-6
Reaction sequence for synthesis of norepinephrine.

glycine, arginine, and methionine; it may be considered as an end product of these amino acids, since neither the methyl group nor the nitrogen is utilizable by the mammalian cell. This substance occurs in relatively large amounts in muscle tissue as creatine phosphate, which contains an energy-rich bond. *Creatine phosphate* serves as a reservoir of energy readily transferable to ADP in the reformation of ATP hydrolyzed during muscular contraction (Chapter 6). *Creatinine,* formed from creatine, is found in the urine and should be considered as one of the normal end products of nitrogen metabolism.

Epinephrine and norepinephrine. These two hormones are synthesized in the adrenal medulla and have an important role in the sympathetic nervous system activity (Chapter 11). Phenylalanine, tyrosine, and methionine are the amino acids necessary for the production of these compounds according to the scheme in Fig. 23-6.

Norepinephrine is converted to epinephrine by a transmethylation reaction. Methionine is employed as a methyl donor in this and in a variety of other reactions. It does so by first reacting with ATP to form 5-adenosylmethionine, and this energy-rich compound converts the norepinephrine to epinephrine, when a specific transferase enzyme is present.

The porphyrins. Many compounds necessary for life contain porphyrins, and most organisms have the capacity to synthesize them. Protoporphyrin IX, which serves as a nucleus for the heme in hemoglobin and cytochrome enzymes, is synthesized from glycine and acetate. Isotopic labeling studies have indicated that the acetate has been metabolized through the tricarboxylic acid cycle and actually participates as succinyl-CoA. The reactions shown in Fig. 23-7 indicate the overall results of this important synthetic mechanism. Several steps are required for the generation of the protoporphyrin IX molecule (Fig. 23-8), but not all of them have been completely elucidated. It requires condensation of four molecules of porphobilinogen in a head-to-tail fashion to produce a ring structure. Addition of an iron atom completes the formation of heme (Chapter 12).

■ Nucleic acid metabolism

Purine biosynthesis and degradation. ■ The nine-membered purine ring system is synthesized from glycine, formate, carbon dioxide, aspartic

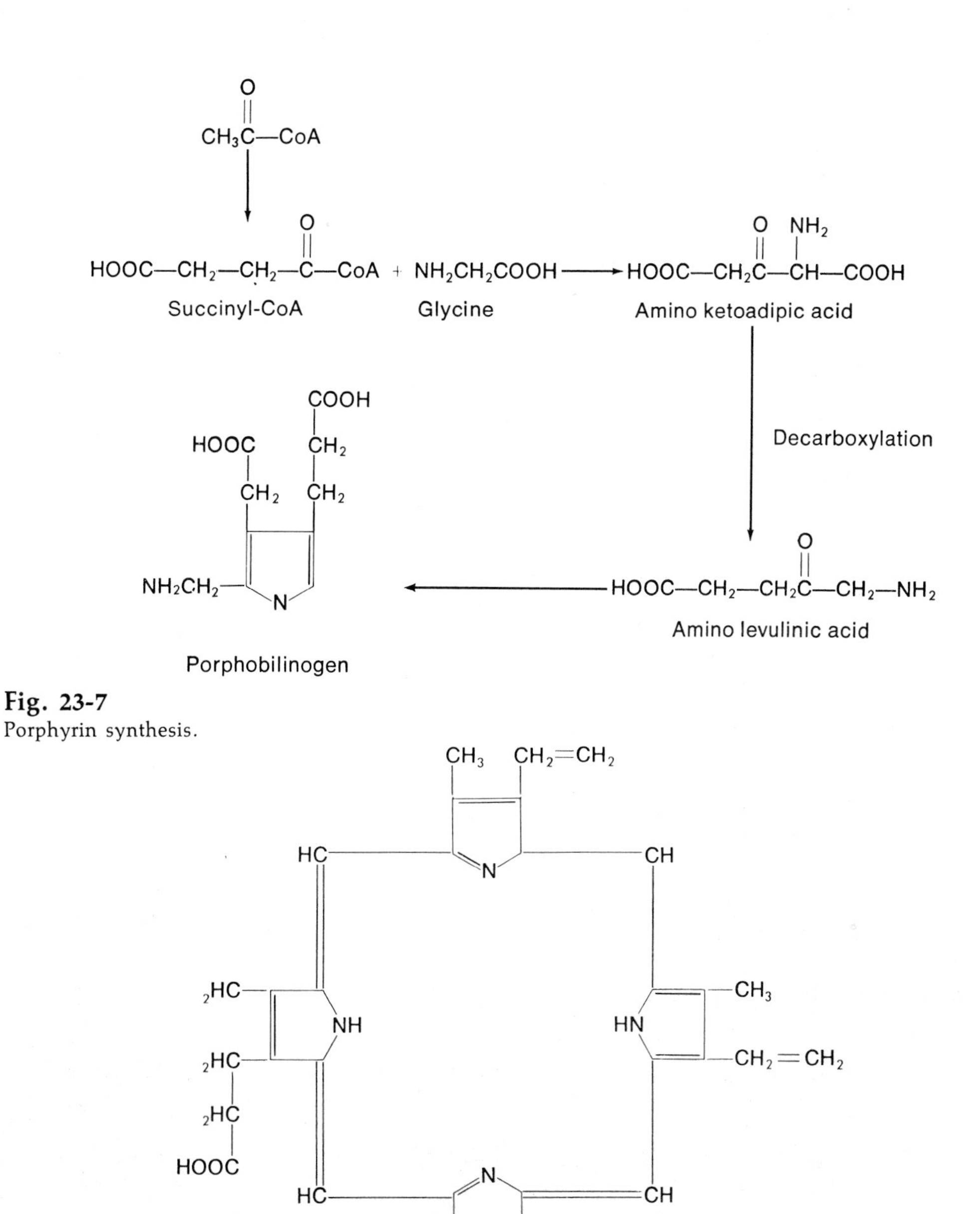

Fig. 23-7
Porphyrin synthesis.

Fig. 23-8
Condensation of 4 molecules of porphobilinogen to produce protoporphyrin IX. Addition of iron to this structure results in formation of heme (Chapter 12).

acid, and glutamine in a series of complex enzymatic reactions. Since these compounds are necessary to form RNA and DNA, we shall present the major pathways in word form and the interested student may consult a biochemistry text for the detailed chemistry involved. Purines are synthesized as nucleotides; therefore the initial step is the formation of the "active" sugar 5-phosphoribosyl-1-pyrophosphate (PRPP) from ribose-5-phosphate and ATP. Several compounds provide the structural components of the imidazole ring; glycine, amine groups from glutamine, and formyl from formyl folic acid. ATP is the energy source for these combinations and ring closure to form the following:

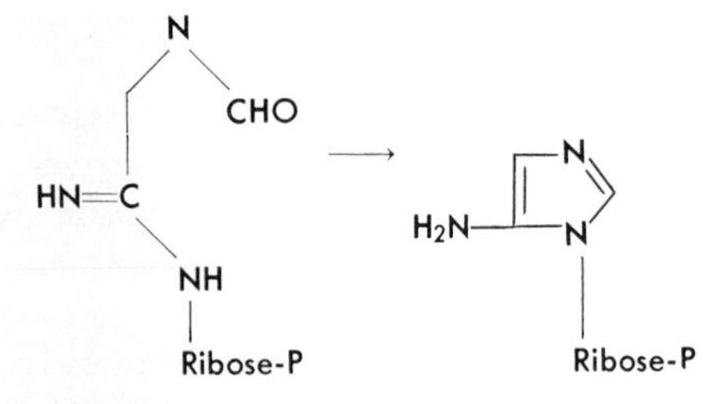

Carbon dioxide and another formyl group furnish the additional carbon atoms, and aspartic acid donates the nitrogen atom to complete the purine ring system, with *inosinic acid* as the product.

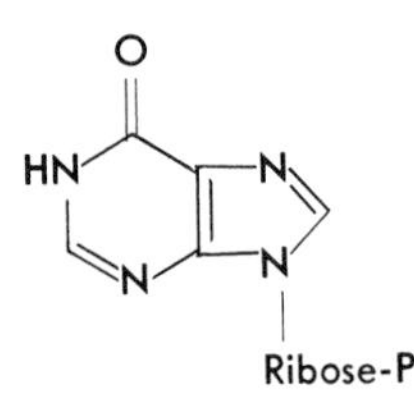

Inosinic acid is then converted to adenylic or guanylic acid by further enzymatic reactions.

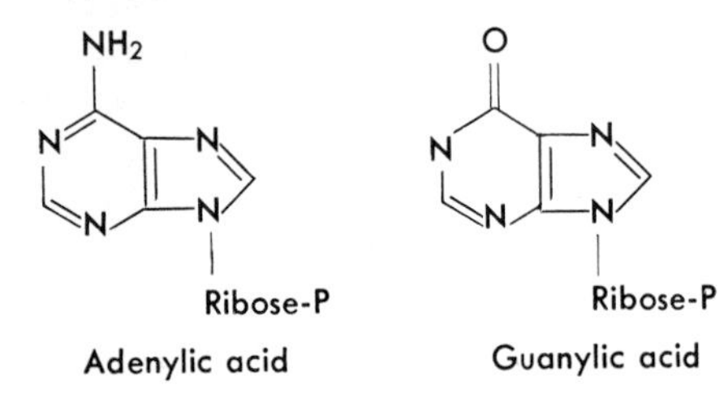

Adenylic acid

Guanylic acid

The corresponding deoxyribonucleotides may be formed using a similar pathway, or the ribose may be reduced to deoxyribose in the final stages.

Catabolism of these compounds in man and other primates leads to uric acid as the principal excretory product. The glycosidic bond is ruptured, and the compounds are converted to xanthine and then to uric acid.

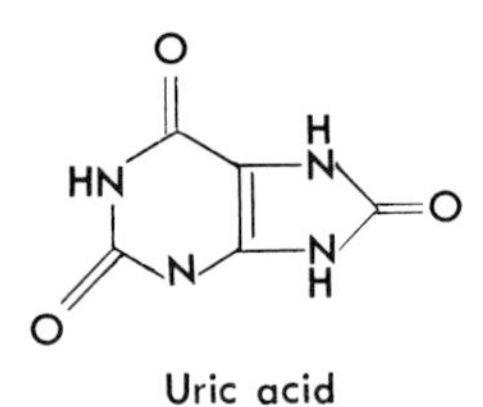

Uric acid

Pyrimidine biosynthesis and catabolism. ■ Unlike the purines, the complete ring structure of pyrimidines is formed prior to the attachment of the sugar molecule. The initial step is the formation of carbamyl phosphate as discussed in connection with urea formation in Chapter 21. Aspartic acid and carbamyl phosphate combine to produce *orotic acid.*

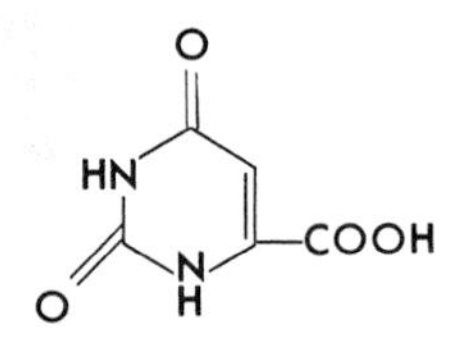

The sugar moiety is now attached by coupling orotic acid with PRPP, yielding orotidine-5-phosphate. Decarboxylation converts this compound to uridylic acid (UMP).

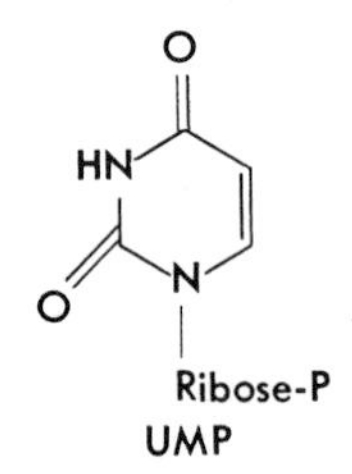

UMP

Cytidine nucleotides (CMP) are formed by amination of UMP.

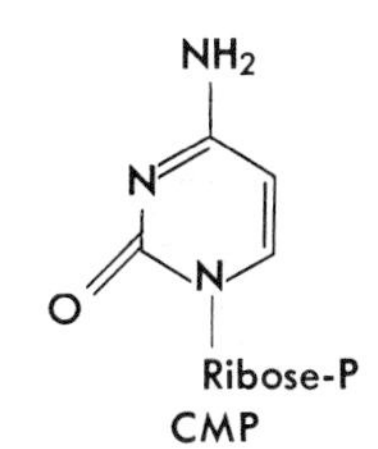

Conversion of the ribose to deoxyribose and methylation of CMP produces thymidilic acid (TMP).

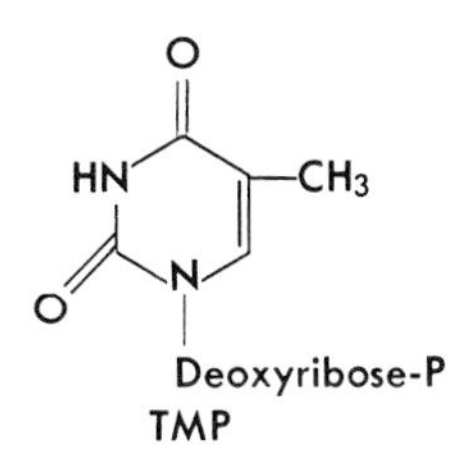

Catabolism of the pyrimidines involves both reduction and hydrolysis. The end products are β-alanine (from uracil and cytosine) and β-aminoisobutyric acid (from thymine).

Biosynthesis of DNA and RNA. ■ Enzymes, termed polymerases, that catalyze the synthesis of DNA and RNA have been isolated from bacteria. Although the function of the DNA polymerases has not been well defined, its seems probable that enzymes possessing the same characteristics must be present in all cells capable of catalyzing the precise duplication of DNA.

The DNA polymerase isolated from *Escherichia coli* catalyzes the formation of the DNA polymer under the following conditions: (1) presence of all four deoxyribonucleosides as triphosphates (ATP, GTP, CTP, and TTP) and (2) the presence of a diavalent metal ion, usually magnesium. The overall reaction is reversible, and the presence of a primer DNA is necessary. Primer DNA apparently serves as a template for the formation of new DNA. A similar polymerase enzyme has been isolated from mammalian tissues.

RNA has been synthesized by several enzymes that differ in their substrate and primer requirements. One, isolated from particulate fractions of liver cells, has many characteristics similar to DNA polymerase and has been named RNA polymerase. This enzyme catalyzes the formation of a polymeric material when all four ribonucleoside triphosphates (ATP, GTP, UTP, CTP) are present and requires the presence of DNA as a primer. Evidence suggests that DNA serves as a template for the RNA synthesis and that the base sequence (of the DNA primer) is transcribed into the corresponding sequence in RNA. Several other enzymes of this nature have been isolated in which RNA may serve as a primer. The ultimate role of DNA and RNA is, of course, precise protein synthesis.

Biosynthesis of protein. ■ Biosynthesis of protein was discussed in Chapter 2. Here we shall present a generalized picture of protein synthesis. The following participants must be present: (1) an adequate amount of the twenty amino acids, the *amino acid pool;* (2) an enzyme for each of the twenty amino acids and sufficient ATP capable of synthesizing active amino acids; (3) a pool of transfer RNA (tRNA) to specify the amino acids; (4) messenger RNA (mRNA), which has transcribed the nucleotide sequence of a particular DNA; (5) a pool of active ribosomes; (6) two transfer enzymes concerned with polypeptide formation; and (7) a pool of GTP, a source of reducing groups, K and Mg ions. A series of sequential reaction steps occur:

1. Amino acids are activated by reactions with ATP, and then amino acyl-tRNA molecules are formed. This step is very similar to the formation of active acetate.

2. A complex is formed between ribosomes and mRNA. A single ribosome attaches to one of the terminal ends and travels along the mRNA away from that end, while peptides are formed on the complex.

3. Active amino acyl-tRNA is bound to this mRNA-ribosome complex. The energy required for moving the mRNA relative to the ribosome and positioning the amino acyl-tRNA is obtained by hydrolysis of GTP. When the next active

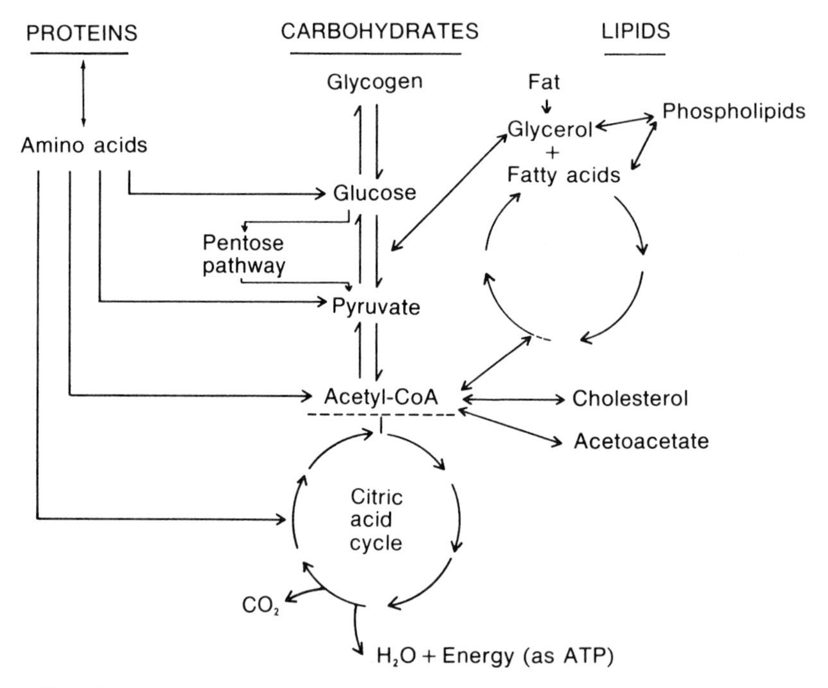

Fig. 23-9
Some of the pathways common to protein, carbohydrate, and lipid metabolism.

amino acid is in position, a peptide bond is formed that requires a transfer enzyme activated by K ion.

4. As soon as a complete peptide chain is formed, it is released from the complex. As yet, little is known concerning the signal necessary for this release.

■ INTERRELATIONS OF CARBOHYDRATE, LIPID, AND PROTEIN METABOLISM

The simplified diagram presented in Fig. 23-9 will serve to remind us of the various points at which the products of the three major foodstuffs enter into common metabolic pathways. Although the same enzyme systems are not used, these pathways indicate the synthetic as well as the degradative interrelationships.

READINGS

Changeux, J. P.: The control of biochemical reactions, Sci. Am. **212**(4):36-45, Apr. 1965.

Dickerson, R. E.: The structure and history of an ancient protein, Sci. Am. **226**(4):58-72, Apr. 1972.

Dietschy, J. M., and Wilson, J. D.: Regulation of cholesterol metabolism, New Eng. J. Med. **282**:1128-1138, 1179-1184, 1241-1249, 1970.

Larner, J.: Intermediary metabolism and its regulation, Englewood Cliffs, N. J., 1971, Prentice-Hall, Inc.

Lehninger, A.: Biochemistry, New York, 1970, Worth Publishers, Inc., chaps. 13-20 and 22-25.

Levine, R., and Haft, D.: Carbohydrate homeostasis, New Eng. J. Med. **283**:175-183, 237-246, 1970.

Masaro, E. J.: Physiological chemistry of lipids, Philadelphia, 1968, W. B. Saunders Co.

Orten, J., and Neuhaus, O. W.: Biochemistry, ed. 8, St. Louis, 1970, The C. V. Mosby Co., chaps. 8-14.

24

Water metabolism
Mineral metabolism
Bones
Teeth

WATER AND MINERAL METABOLISM

■ WATER METABOLISM

On a number of occasions it has been stated that living material, protoplasm, is an intimate mixture of crystalloids and colloids in which water forms the solvent for the first and the medium for the dispersion or suspension of the second. From this it is evident that water plays a very important part in the existence and activity of the living being. This view is substantiated by the fact that, although a fasting animal may survive a loss of practically all its fat and of half its proteins, a loss of one fifth of its water content is fatal.

■ Body water

The amount of water in the body of any given individual is reasonably constant. Its concentration varies from one tissue to another, being least in the dentin of teeth (10%) and greatest in the gray matter of the brain (85%). The younger and the more active the protoplasm, the greater is the amount of water it contains. The human embryo at 6 weeks contains 97% water. Of the approximately 49 liters of water in the body of a man weighing 70 kg, 14 liters are *extracellular* (3.5 liters in the plasma and 10.5 liters in the tissue fluids); the balance of about 35 liters is found within the cells—*intracellular* (Fig. 24-1). Although the extracellular fluid (ECF) volume is less than half that of the intracellular fluid (ICF) volume, it is of great significance, since all exchanges between the tissues and the environment must occur through this compartment.

Measurement of body water. ■ The measurement of body fluid compartments, plasma volume, interstitial (tissue) fluid volume, and intracellular fluid volume (Fig. 24-1) is undertaken for a variety of clinical and research reasons. Whereas the interstitial fluid volume and intracellular fluid volume cannot be determined directly, the volumes of the plasma, the entire extracellular compartment, and the total body water can be estimated with some degree of accuracy. The difference between the total body water and the extracellular water yields a value for the intracellular water; plasma volume is sub-

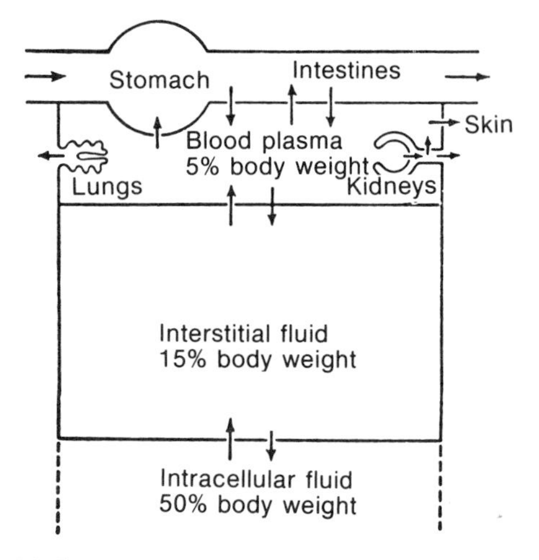

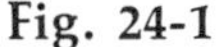
Fig. 24-1
Diagram showing distribution of body water. (From Gamble: Extracellular fluid, Harvard University Press.)

tracted from the extracellular volume to calculate the interstitial fluid volume (TBW − ECF = intracellular; ECF − PV = interstitial). All measurements are based on the dilution principle and each depends on the choice of an indicator (dye, radioactive substance, drug) that satisfies the criteria established in Chapter 3. Many otherwise suitable substances are excreted from the body at readily determinable rates. Even in these circumstances, the substance can be and often is used with proper correction applied for the losses of the administered indicator. When we recall our basic formula, $C_1 \times V_1 = C_2 \times V_2$, it is evident that $C_1 \times V_1$ is the quantity administered and that:

$$\frac{\text{Quantity administered} - \text{Quantity lost}}{\text{Concentration in compartment}} = \text{Volume distribution}$$

Total body water may be measured by using the drug antipyrine or by using heavy water (D_2O), among other compounds. Antipyrine is excreted and metabolized, and D_2O is slowly bound up in the tissues; nevertheless, appropriately corrected estimates indicate that the total body water is 50% to 60% of the body weight. The extracellular space may be determined by examining the distribution of a substance such as inulin, sucrose, thiocyanate, or radioactive sulfate. Such measurements indicate that this fluid compartment constitutes about 20% of the body weight. Plasma volume is estimated from the distribution of compounds like Evans blue dye (T-1824) or radioiodinated serum albumin (RISA). Their mixing in the plasma is relatively rapid, and the determination can be undertaken after a reasonably short wait, perhaps as little as ten minutes. The values obtained indicate that the plasma volume is between 45 and 50 ml/kg of body weight (4.5% to 5.0%). From such information and a knowledge of the hematocrit it is possible to calculate the total blood volume (blood volume = plasma volume/[1 − hematocrit]).

Functions of water. ■ No other chemical compound has so many distinct and vital functions as water. This is largely due to its great solvent power, to the fact that it is chemically a neutral substance, and that ionization of most materials takes place more freely in water than in any other medium. These functions will be briefly summarized.

Solvent. Water dissolves or holds in suspension the other materials in protoplasm.

Medium. Water furnishes a medium for digestion, absorption, metabolism, secretion, and excretion. All these processes, chemical and physical, can take place only in a water medium.

Moistens surfaces. Water moistens the surfaces of the lungs for gas diffusion.

Temperature regulation. Water plays a dominant part in equalizing the temperature throughout the body and in maintaining it at a fairly constant value (Chapter 27). In this function three physical properties of water are concerned: (1) *thermal conductivity,* (2) *specific heat,** and (3)

*Specific heat is the amount of heat in calories required to raise the temperature of 1 g of the substance 1 C. For water this is 1; for iron, 0.11; for silver, 0.057.

*high latent heat of vaporization.** Its great thermal conductivity† enables the circulating blood to take heat very rapidly from the active parts of the body (e.g., muscles and liver); due to its high specific heat, a large amount of heat can be absorbed by the blood with only a small rise in its temperature. The warm blood arriving at the less active regions (connective tissues and bones) parts with some of this heat. Due to its high latent heat of vaporization, the evaporation of sweat through the skin occasions the loss of considerable heat.

Cushion. Cerebrospinal fluid serves as a cushion for the brain and spinal cord.

Transportation. Water furnishes a vehicle for the transportation of nutrients, waste, hormones, gases, and so on.

Hydrolysis. Water takes part in hydrolytic cleavages, as during digestion.

Lubricant. Water serves as a lubricant for moving surfaces, such as joints (synovial fluid), the heart, and intestine.

Sense organs. Water plays an indispensable part in sense organs. Taste and smell are the result of stimulation by chemical compounds in solution. Sound is conducted through the inner ear by a liquid, which is chiefly water. The function of the semicircular canals as sense organs of equilibrium depends on the presence of water in these canals. The transparency of the media of the eye to light is maintained by water.

■ Regulation of water balance

Intake and outgo of water. ■ Water is taken into the body by direct ingestion; it is derived from foods that contain various quantities of water, for example, green vegetables, 90% to 97%; meat, 50% to 75%; bread, 35% to 38%; it is formed by the oxidation of organic foodstuffs in the body (Fig. 24-2). From their formulas it can be calculated that 100 g of the foodstuffs produce the following amounts of water: proteins, 40 g; fats, 105 g; carbohydrates, 55 g. Ordinarily, about 2,000 ml of water are taken into the body per day.*

Water leaves the body to the extent of about 2 liters/day by four channels: kidneys, lungs, skin, and alimentary canal (Fig. 24-2). How much water is excreted depends on many external and internal conditions. Whatever increases the elimination by one organ decreases correspondingly the output by the other organs. In diarrhea the elimination by the alimentary canal is largely increased at the expense of the amount voided by the kidneys. Again, the warmer and drier the air, the more water is lost by the skin and lungs and the less by the other channels. The large amount of water taken into the alimentary canal with food and drink and that poured into it by the secretion of the various digestive juices are almost completely absorbed into the blood.

Water balance. ■ The regulation of the intake and output of water is directly related, respectively, to the deficit and the excess of the normal water level in the body. The normal volume and osmolarity of the ECF and ICF compartments are indicated by the solid lines in Fig. 24-3; it should be noted that these volumes are not, however, drawn to scale. When the water retained above this level (Fig. 24-3, *A*) amounts to more than 8% of the body weight, the tissues become waterlogged *(water intoxication),* which may bring on convulsions and coma. The body controls its water content by loss through the skin and kidneys. How efficiently this is accomplished is at-

*The latent heat of evaporation is the amount of heat required to evaporate 1 g of liquid into vapor at the same temperature. For water (at 100 C) this is 537 g-cal; for alcohol, 208; for ether, 91.

†To the hand, water at 50 F *feels* much colder than air at the same temperature; the heat of the hand is conducted to water at a more rapid rate than it is to air.

*The kangaroo rat (a desert rodent that drinks no water) is able to maintain the proper water balance by what we may call a metabolic water supply. A camel's hump, largely composed of fat, is a reservoir of energy and water.

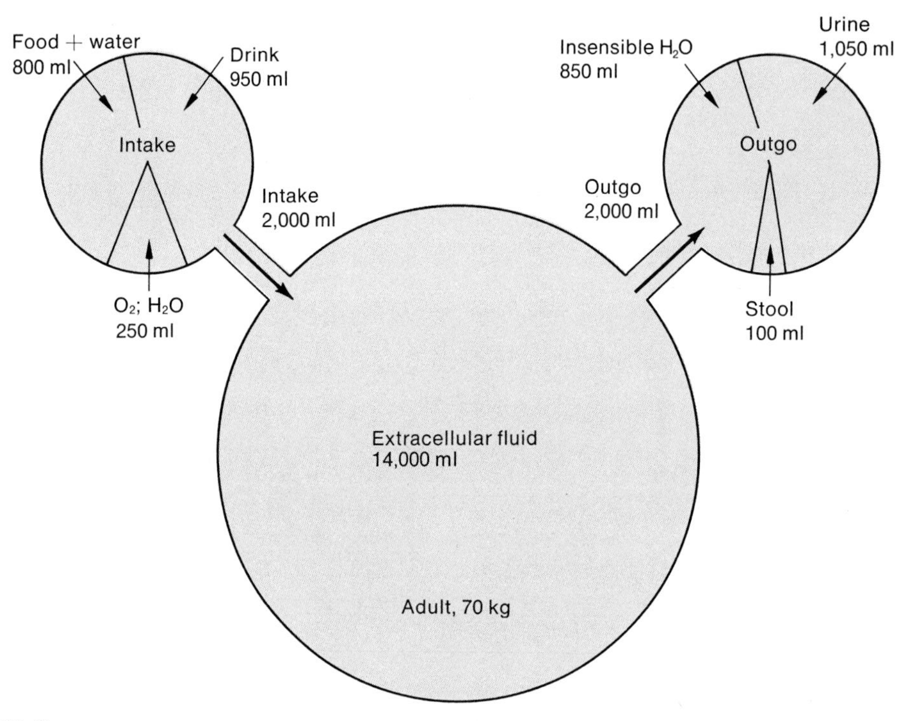

Fig. 24-2
Diagram showing daily intake and outgo of water in a 70 kg adult. Extracellular fluid exchanges with rest of body water.

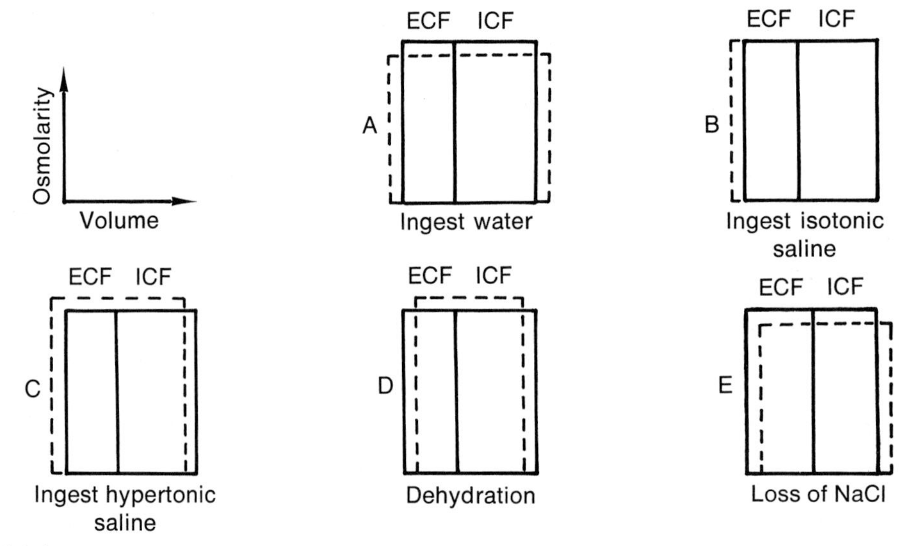

Fig. 24-3
Effects of various conditions on ECF and ICF volume and osmolarity.

tested by the fact that the consumption, in the course of six hours, of as much as 5.5 liters of water (a quantity more than the entire volume of the blood) causes only a temporary dilution of the blood, as shown by the hematocrit. We may note three possible results of this large intake of water.

1. The absorbed water reduces the osmotic pressure of the blood. This favors the passage of water from the blood into the tissue spaces (especially in the skin) and into the cells.

2. The extra blood volume finds room by distension of previously closed, or partly closed, capillaries and by storage in the sinusoids of the liver, spleen, and other organs.

3. However, the ability of the body to store water is rather limited, and while the reservoirs are being filled, some water is excreted by the kidneys.

The ingestion of isotonic saline (Fig. 24-3, *B*) increases only the volume of the ECF, since the isotonic solution causes neither exit nor entry of water into the cellular compartment. On the other hand, the ingestion of hypertonic saline (Fig. 24-3, *C*) diminishes the ICF volume, as water is withdrawn from the cells due to the hypertonicity of the ECF.

A lowering of the water level, either because of too great loss or too limited intake, increases the osmotic pressure of the blood. As a result, water is drawn from the tissue spaces and from the tissues themselves into the blood. Both the ECF and ICF compartments have decreased volumes and increased osmolarities (Fig. 24-3, *D*). A loss of water equal to 6% to 10% of the body weight constitutes serious *dehydration.* Depletion of sodium, through diet or excessive sweating, for example, leads to hypotonicity of the ECF compartment and a shift of water from latter into the ICF compartment, which thereby becomes diluted (Fig. 24-3, *E*).

Drinking and thirst. ■ There are two categories of situations in which drinking occurs. Primary drinking is a response to an absolute or relative (i.e., hypertonicity) lack of water in one, or both, of the major body fluid compartments. Secondary drinking occurs despite no apparent internal need. Dryness of the mouth (e.g., from smoking or mouth breathing), consistency of the diet, activity (thermogenic stimulation of intake), and climatic conditions induce secondary drinking. Certain areas of the limbic (hippocampus and amygdala) system of the brain (Chapter 10) that connect with the lateral hypothalamic nuclei (Fig. 10-3) appear to exert both stimulatory and inhibitory influences on secondary drinking. On a day-to-day basis, under reasonably stable environmental conditions, we seldom experience true thirst but consume fluids in anticipation of need.

Primary drinking and true thirst are essentially emergency mechanisms for a response to an actual need for fluid. Receptors which detect the need and initiate thirst and drinking are located in both of the major fluid compartments, the intracellular and the extracellular. Diminution in the ICF volume through water deprivation or potassium depletion of the cells is detected by *osmoreceptors* in the preoptic and supraoptic regions of the hypothalamus. The role of these receptors in the release of ADH is discussed in Chapter 28, the action of ADH in Chapter 29. Extracellular receptors are located in the capacitance vessels near the heart and in the atria. These detectors are stretch receptors that respond to volume changes (hypovolemia) brought about by water or sodium deprivation, hemorrhage, vomiting, severe exercise, or diarrhea. They are reflexly connected to the centers for the secretion of ADH and, in some way, to the renin-angiotensin-aldosterone mechanism for volume regulation. The extracellular receptors are important because all exchanges between the cells and the environment must occur through the ECF and, furthermore, the circulating volume of the blood must be protected.

Body water balance is often disturbed by disease. Diabetes insipidus (Chapter 28) in which there has been damage to the hypothalamic-hypophyseal area of the brain produces polyuria and polydipsia. Many renal diseases are associated with excess fluid loss and a resultant thirst. Hypercalcemia due to an increased absorption of calcium (vitamin D toxicity), or of bone decalcification, induces water loss and thirst. Chronic dehydration can result from diabetes mellitus (Chapter 28) wherein glucosuria and metabolic acidosis produce water and electrolyte loss from both fluid compartments. In hyperthyroidism there is an increased fluid loss due to the metabolic rate increase and, consequently, thirst develops.

■ MINERAL METABOLISM

Inorganic materials in the body constitute about 5% of the body weight. Among these are the chlorides, sulfates, carbonates, and phosphates of sodium, potassium, calcium, and magnesium; also very small quantities of iodine, copper, iron, and what are sometimes referred to as trace metals.

Minerals must be looked on as forming an integral part of the protoplasm, as truly as fats, carbohydrates, and proteins. Physiological activity is impossible without them. Activity of many of the enzymes depends on the presence of minerals (Chapter 22). Frequently the inorganic material is united chemically with the organic material, as, for example, iron in the hemoglobin of the red blood cells and iodine in the thyroxin of the thyroid gland. On a diet entirely devoid of salts, digestion is seriously impaired and soon the animal refuses to eat; weakness and finally paralysis ensue, ending in death. The addition of NaCl alleviates these disturbances for a long time, but eventually other salts must also be added in order to maintain life.

The physiologically correct osmotic pressure of the extracellular and intracellular fluids depends on the proper ratio of the mineral and of the water content of these fluids. Excreting, as conditions may demand, more or less of the one or the other of these two constitutents is the part played by the kidneys in maintaining this ratio. In Chapter 28 we shall learn that hormones also influence the mineral content of the blood.

Sodium. ■ Many of the functions of sodium salts have been discussed in previous pages. Some of these functions may be recalled:

1. Being the chief inorganic constituent of extracellular fluids, they contribute preponderantly to the osmotic pressure of these fluids and therefore are an important factor in maintaining the proper balance and distribution of water in the body.

2. Sodium bicarbonate forms the chief alkali reserve.

3. They aid in maintaining the irritability of muscles, nerves, and the heart.

Among the results of sodium deficiency is interference with the heat-regulating power of the body (Chapter 27) which may, therefore, cause a rise in body temperature. It also augments the excretion of water; this leads to a decrease in the volume of the blood and thereby to an increase in the concentrations of its cells and proteins; the harmful effects of this condition are apparent. A decrease in sodium intake has proved in some persons of aid in reducing hypertension. When the bodily stores of sodium are running low, the elimination of salt is sharply curtailed.*

Potassium. ■ In the body, sodium and potassium cannot replace one another to any great extent. Their distribution within the body is noteworthy: potassium predominates in the cells, and sodium is chiefly found in the extracellular fluids and secretions of the body.† The specific action of potassium in cardiac activity, in which it antagonizes the action of calcium, will no doubt be recalled (Chapter 14). For the transmission of nerve impulses and for muscle contractions, potassium is essential. It is necessary for growth; on a diet containing less than 15 mg of potassium per day young male rats fail to grow.

*The necessity of sodium salts for the body economy is seen in the traveling of herbiverous animals at great risk to very distant salt licks.

†Of the approximate 175 g of potassium in the body, only 3 g are found outside the cells.

The concentration of both sodium and potassium ions in the body is regulated by the adrenal cortex (Chapter 28).

Iron. ■ The upper duodenum is the site of the active absorption of ferrous (Fe^{++}) iron into the plasma. The iron (Fe^{+++}) combines with a beta globulin in the plasma to form *transferrin.* Transferrin carries the iron to the various tissues of the body for utilization in various enzymes and heme; the excess iron is stored chiefly in the liver and, to a lesser extent, the spleen and red bone marrow. In the cells iron combines with another protein *apoferritin* to form *ferritin.* A single molecule of ferritin may contain as many as 4,300 atoms of iron. When the amount of iron in the plasma is decreased, ferritin releases the iron back into the plasma. The enzyme xanthine oxidase may be involved in this procedure. The iron then can be used for synthesis of hemoglobin (Chapter 12), enzymes (Chapter 22), etc. The increased formation of red blood cells causes the withdrawal of ferritin from storage.

The amount of iron absorbed by the body is determined by the amount of unbound transferrin in the blood. When transferrin is saturated, less iron is absorbed from the digestive tract. If the transferrin is saturated, the iron which enters the epithelial cells of the intestine is chelated to form ferritin and is temporarily stored there. If it is not needed before the lifetime of the epithelial cell (Chapter 20) has expired, the iron is lost by way of the sloughing of these cells and excretion of them in fecal matter.

Magnesium. ■ This element is a necessity for life. Deprived of magnesium, animals become highly irritable and are thrown into convulsions by the slightest disturbance; in the majority of cases this proves fatal. This is in agreement with the fact that increased magnesium has an anesthetic effect on many animals, which is counteracted by calcium. Magnesium forms a small part of bone tissue.

Iodine. ■ Iodine plays an important part in the physiology of the thyroid gland. We shall deal with this in Chapter 28.

Calcium. ■ The importance of calcium hardly can be overrated. Although 99% of the body calcium is found in the skeleton and teeth, the remaining calcium is vitally needed by every cell.

The absorption of calcium (Chapter 20) is enhanced by acidity of the intestinal contents. Absorption from the lumen into the epithelial cells of the gut probably occurs by facilitated diffusion (Chapter 3) across the brush border of the mucosal side of the cell. Because of the negativity of the interior of the cell, the transfer of calcium ions into the interstitial fluids at the serosal side of the cell requires an active transport mechanism—a calcium pump. Although the mechanism of its action is unclear, the role of vitamin D in the absorption of calcium must be significant, since only a minimal transfer occurs in its absence. The amount of calcium in the blood is 10 mg/100 ml plasma. Its absolute value is determined not only by the dietary intake but also by the activity of thyrocalcitonin, parathormone (Chapter 29), and vitamin D.

The functions of calcium have been stressed in preceding chapters; they can be recapitulated: Calcium is indispensable for hemostasis (Chapter 12). Platelets have a five to six times higher concentration of this ion than is found in other soft tissues. It is tightly bound to the lipids of platelets and their secretion of ADP and aggregation is dependent upon the calcium concentration in body fluids. Furthermore, it is essential in the clotting mechanism as a cofactor for the enzymatic reactions. It also is influential in many metabolic reactions involved in the production of energy-rich compounds. For example, it affects glycolysis (Chapter 23) because it is required for the conversion of phosphorylase *a* to *b.* The intensity of muscular contraction is determined by the concentration of free calcium ions in muscle cells (Chapters 6 and 14); the duration of contraction depends upon the rate of removal of calcium ions by the sarcoplasmic reticulum.

Most of the calcium within cells is not free but is bound by chelation with proteins, lactate, citrate, ATP, ADP, and other compounds. Calcium is bound to lipoproteins in the cell membrane (Chapter 2) and plays a key role in the maintenance of membrane structure. It is important in promoting cell-to-cell adhesiveness; wound healing and embryonic development require it. The deformability of cells is regulated by the presence of calcium ions; their outer surfaces are stiffened and hardened by calcium. Such tightening may well enhance the permeability.

It is involved in synaptic transmission and membrane excitation. In the intact body, if the calcium in the plasma falls to 5 mg/100 ml, the irritability of the neuromuscular system is increased to such an extent that twitchings and *tetany* of the skeletal muscles result. An increased blood calcium level is associated with decreased tonicity and excitability of the neuromuscular system, including both smooth and striated muscles.

Phosphorus. ■ Phosphorus is present in all cells of the body, but, like calcium, the greater part of the 700 g is found in the skeleton. The amount in the plasma in adults is generally about 4 mg/100 ml plasma; in children it may be 5 or 6 mg.

Phosphorus is a constituent of many important organic compounds; reference has been made to adenosinetriphosphate as a reservoir of energy for muscle work, to phospholipids in the absorption and metabolism of fats and in the permeability of the plasma membrane, and to phosphoproteins (e.g., casein), lipoproteins, and nucleoproteins.

Trace elements. ■ Notwithstanding the minute quantities of copper, cobalt, zinc, manganese, and other elements in the body, they are indispensable for life. They are frequently referred to as *trace elements.* Some of them are essential for the activity of certain enzymes. We may mention magnesium in carboxylase, zinc in carbonic anhydrase, and iron in catalase and peroxidase. Copper, functioning in the manufacture of hemoglobin, is stored in the liver, especially in young animals. It is found in many oxidases. Copper is stored and transported in the plasma by means of a blue copper protein *ceruloplasmin.* A deficiency of this protein, referred to as Wilson's disease, results in diffusion of copper into the tissues, and it may accumulate to a high level in liver and brain. Deficiency in cobalt leads to anemia; this may find its explanation in that cobalt is a constituent of vitamin B_{12}. Lack of manganese disturbs the reproductive function in both the male and female; a deficiency causes a nursing animal to lose all interest in her offspring. It is also important in enzyme reactions.

■ BONES

Composition. ■ Bone is a living tissue unique to vertebrates. Unlike the soft structures thus far studied, bones contain a comparatively small amount of water (about 25%) but are rich in mineral matter. Similar to other connective tissues, bones are composed of numerous, widely scattered cells and much intercellular material or matrix (Fig. 24-4). The cells are enclosed in small spaces (lacunae) within the bony tissue and are connected with each other and the bloodstream by fine canaliculi. The matrix is composed of both organic and inorganic substances; in a dry bone these constitute from 30% to 40% and from 60% to 70%, respectively. The organic material, consisting chiefly of collagen (a protein), is laid down in the form of fibers; in this framework the inorganic material is deposited. This latter material is composed of highly insoluble calcium salts in the form of a hydroxyapatite, $3\ Ca_3(PO_4)_2 \cdot Ca(OH)_2$. Other mineral salts are present in the intercrystalline material. The hydroxyapatite crystals are relatively stable but can exchange ions at their surface. These salts give *hardness* and *rigidity* to bones. The organic material may be freed from these insoluble salts by treating bone with hydrochloric acid, which dissolves the mineral constituents; the remaining organic part of the matrix is translucent and flexible; it gives *toughness* to bone.

Bones have all the appearance of firm, solid

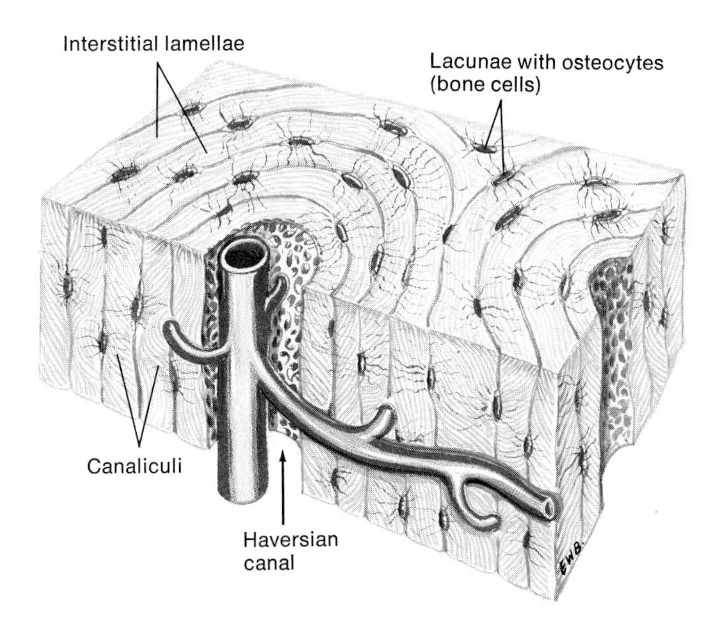

Fig. 24-4
Illustration showing both cross and longitudinal section of compact bone.

structures, and, indeed, nearly all of the mineral is relatively inert and nondiffusible. However, they are living organs. Similar to the soft tissues, there is a constant interchange between the chemical ingredients of the surface bone tissue and those of the fluids that bathe it. *The calcium and phosphorus of bone are in a state of dynamic equilibrium with the calcium and phosphorus of the blood.*

Bone exists in essentially two forms: compact bone and trabecular, or spongy, bone. The former makes up the shafts of long bones, the latter is found in the neck and head of long bones such as the femur (Fig. 8-8) and in the vertebrae of the spinal column (Fig. 8-1). Compact bone is arranged in a tubular fashion, but trabecular bone is arranged as fine plates. About four fifths of the weight of the skeleton consists of compact bone, the remainder is spongy bone. The mechanical properties of bone are comparable to those of cast iron, although it is only one fourth the density and much more flexible.

Functions. ■ Bones perform a number of functions.

1. They furnish protection to the softer tissues, as is well illustrated by the cranium, which almost entirely encloses the brain.
2. They support various organs, for example, the vertebral column to which is affixed, by means of the mesentery, the intestinal canal.
3. They act as levers for the performance of movements executed by the muscles, as is seen in the bones of the extremities.
4. Bones are a storehouse for indispensable calcium and phosphorus. About 98% of body calcium and 80% of the phosphorus are contained in bone. Calcium and phosphorus are continuously excreted from the body. The calcium excreted in the urine amounts to about 150 mg/day. Naturally the amount of these elements in the blood is reduced during starvation. To meet the urgent demands for these elements by the soft tissues, calcium is removed from bones. This loss (which may be as much as one third of their en-

tire mineral content) renders the bones soft and porous; subsequent ingestion of calcium may restore them to normal. By the reciprocal shifting of this element between the bones and blood, the concentration of blood calcium is kept fairly constant.

The withdrawal of calcium from and its incorporation into bone are attributed to the specialized functions of two types of bone cells, *osteoclasts* and *osteoblasts*, respectively (Fig. 24-4). Withdrawal of calcium may occur when extra demands are made on it, as in pregnancy and lactation. This, however, does not apply to the calcium of teeth; the dentin and enamel, when fully formed, are not affected by calcium deficiency in the diet or blood.

5. Red blood cells are formed in the bone marrow.

6. Nearly one fourth of the body's supply of sodium is adsorbed onto the surface of bone.

Formation. ■ Most bones (those of the thorax and limbs) are laid down in previously formed cartilage. Of the many highly complex and still imperfectly understood histological and chemical changes taking place during the transition from cartilage to bone, we can discuss briefly only that remarkable phenomenon, the heavy depositing of insoluble calcium salts.

The explanation generally advanced for the deposition of calcium in growing bone is the *phosphatase theory.* Part of the phosphorus in the blood is held in combination with organic materials to form, for example, hexose phosphates and glycerophosphates; these are esters of phosphoric acid. Osteoblasts contain a high concentration of alkaline phosphatase that hydrolyzes these esters, releasing the phosphoric acid and giving rise to an excess of phosphate ions. A complex mucopolysaccharide in the connective tissue probably attracts calcium to provide a surplus of this ion. When the concentration of both calcium and phosphate ions is increased beyond the saturation point, colloidal calcium phosphate precipitates. Conversion of this material to the hydroxyapatite molecule involves a number of steps, the exact nature of which is unknown. The osteoblasts apparently also control the development of the organic matrix, which is being ossified. The phosphatase enzyme appears at the time when the first evidence of ossification is noticeable. Teeth of young animals also contain large amounts of the enzyme.

Bone behaves piezo-electrically, that is, when deformed, measurable voltages are developed between two opposite surfaces. This is a characteristic of many crystalline structures, for example, those used in phonograph cartridges. When subjected to deformation by compression or bending, bone becomes negatively charged on its convex surface. This semi-conductor–like behavior is ascribable to the collagen component. The practical consequence of the phenomenon lies in the fact that new bone deposition occurs at the cathodal (negative) surface where it will strengthen the structure most effectively. It is probable that the electrical current causes an increase in the pH which favors the action of alkaline phosphatase and osteogenesis.

Repair. ■ The breaking of a bone releases blood into the space between the broken ends. This blood clots and thereby produces a sort of framework for the repair process. Osteoclasts appear in this framework and begin laying down collagen, which upon mineralization turns into new bone that bridges the broken ends. Providing the broken ends have been properly aligned, the bone when repaired is as good as it was previously.

Dietary factors in bone formation. ■ The formation and maintenance of bony structures involve many diverse factors. Their number necessitates limiting our remarks to those of greatest concern.

Calcium and phosphorus. The diet must contain a sufficient amount of calcium and phosphorus for the building of bones. Diets, it is stated, are more frequently deficient in calcium than in any other element, although many foods (e.g., eggs, fish, and dairy products) are rich in it, as we shall learn in Chapter 31. A lack of these elements in the diet of an infant and young child or, more particularly, the inability of the body to make use of them, leads to the lack of proper calcification—a condition called *rickets.* In consequence, the bones lack rigidity and are soft and pliable; because of gravitational and postural stresses placed upon them, they bend and become misshapen. The disease exists in all degrees of severity. The most noticeable defects are bowed or knock-kneed legs and a very narrow pigeon chest; *kyphosis* (humpback) or *scoliosis* (lateral curvature of the spine) may result. The wrists and ankles may be swollen. In some instances the head, due to delayed closing of the fontanels, may be large and flat, with the forehead protruding on either side. The pelvis is distorted by bearing the weight of the body and becomes narrow and contracted. The eruption of the teeth is delayed and their structure is defective. Some observers claim that more important than the absolute amount of calcium and of phosphorus in the diet is the proper ratio of these two elements. For a growing child this should be as 1:1 or as 2:1.

Although a minimum of calcium and of phosphorus must be obtained, it seems that the mere presence of a sufficient amount of these elements in the food does not assure sound bones; severe rickets may develop in children fed on an abundance of cow's milk, which is rich in both calcium and phosphorus. In rickets the amount of calcium ions in the blood is generally (but not always) normal, whereas that of phosphorus has fallen from the normal 6 mg to as little as 1 or 2 mg per 100 ml of plasma. *In rickets it is not so much the lack of calcium as a want of blood phosphorus—hypophosphatemia.* The normal concentration ratio of the blood calcium and phosphorus is maintained by vitamin D.

Vitamin D. Cod-liver oil is a remedial agent for rickets. The active principle of this oil has been isolated; it is known as vitamin D. Being an almost perfect preventive as well as a cure, it is quite appropriately spoken of as the *antirachitic vitamin.*

How vitamin D brings about its beneficial results is not clear. Generally this is attributed to increasing the retention of calcium and phosphorus ions in the blood. Whether this is due to their decreased elimination or to their increased absorption is not agreed on. In some manner the vitamin establishes the ratio between the amount of calcium and phosphorus in the blood most suitable for the calcification of bones. Recently it has been suggested that vitamin D serves as a coenzyme for phosphatase, not only in bone construction but also in other functions in which phosphatase influences calcium and phosphorus metabolism (e.g., in intestine and kidneys).

Irradiation by light. Rickets has been called the disease of sunless areas and of the winter months. This observation led to the study of the influence of light on bone development. Young rats placed on a diet lacking in vitamin D and kept in complete darkness develop rickets. Other rats receiving the same diet and also kept in darkness, with the exception of a fifteen- or twenty-minute period each day during which they are exposed to sunlight, show perfect growth of bone tissue. Rats kept in complete darkness but receiving a supply of vitamin D are also free from rickets.

The relation between the antirachitic power of sunlight and vitamin D is shown by the fact that if certain foods devoid of antirachitic power (such as yeast, olive oil, and lettuce) are exposed to light, especially ultraviolet light, they acquire the property of curing or preventing rickets.

Vitamins D_2 and D_3. The material found in foods of plant origin, which by irradiation acquire the antirachitic property, is a sterol known as *ergosterol,* closely allied to cholesterol. Its activated form is known as D_2 or *calciferol,* but this is not the vitamin found in cod-liver oil. The an-

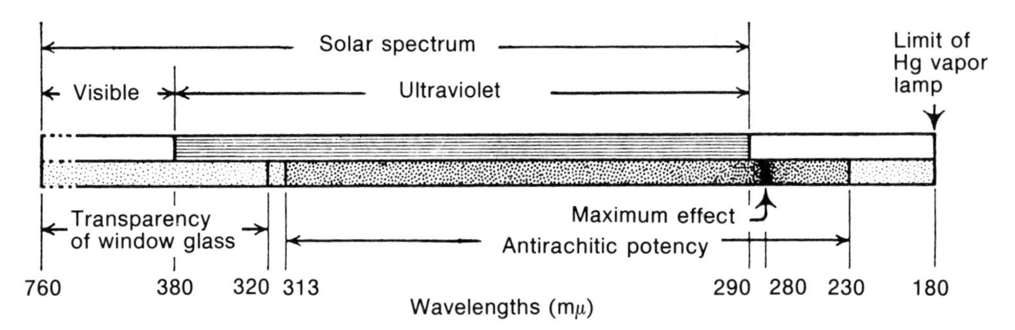

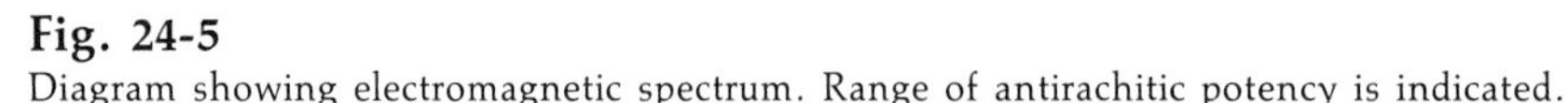

Fig. 24-5
Diagram showing electromagnetic spectrum. Range of antirachitic potency is indicated.

tirachitic action of cow's milk when cows are fed irradiated yeast is due to D_2.

The antirachitic effect produced by irradiation of the skin is caused by the transformation of a sterol known as 7-dehydrocholesterol into D_3. This sterol is found abundantly in the skin. On being absorbed by the blood, vitamin D_3 is carried throughout the body and may be stored in the liver. It is identical with the D vitamin in cod-liver oil and is more potent than D_2.

Ultraviolet light. The visible sunlight (Fig. 24-5) extends from the longest wavelength (red) of 760 millimicrons* to the shortest (violet) of 380 mμ. Beyond 380 mμ the invisible solar spectrum extends to the wavelength of 290 mμ; this portion is spoken of as the ultraviolet region. The wavelengths able to confer antirachitic power on erstwhile inert foods are found between 313 and 265 mμ; the maximum effect is at 280 mμ. It will, therefore, be seen that only about one half of the effective rays are found in the energy derived from the sun and that the most beneficial rays (280 mμ) fall outside of the solar spectrum. Since the wavelengths of the carbon arc lamp extend to 220 mμ and those of the mercury vapor lamp to 180 mμ, these lights possess marked antirachitic powers.

The short wavelengths of light are largely shut off from the body by dark and heavy clothing; they are also absorbed to a great extent by dust, smoke, and vapor suspended in the air. These atmospheric pollutants are more abundant, especially in densely populated districts, during the short winter days when the ultraviolet rays reaching the earth are markedly less than in summer. The shortest wavelength able to pass through ordinary window glass is 320 mμ (Fig. 24-5). Hence, a growing child, unless plentifully supplied with vitamin D, is in danger of rickets during the winter months.

Activity and bone. ■ The size, internal structure, and strength of a bone are also determined by the stress or pressure to which it is subjected; that is, *structural needs determine bone growth.* Physical exercise is a valuable factor for forming strong bones. This is also well illustrated in the jawbone, which, after the loss of the teeth and the consequent partial loss of function, decreases in size by as much as 50%. In a child the pressure exerted in chewing by the deciduous teeth stimulates the growth in size and strength of the jawbones and thereby influences the amount of space available for the permanent teeth. The place-preserving function of the deciduous molars is of great importance.

Changes with age and inactivity. ■ The composition of bone changes with the degree of activity and age of an individual. Weightlessness,

*A millimicron (mμ) = 0.001 micron (μ) = 0.000001 mm.

as experienced by astronauts, brings about an increased excretion of calcium, which suggests that bone resorption occurs even in the short span of space flight and that longer flights may lead to severe bone loss. Immobilization of limbs in plaster casts, denervation of muscles, tenotomy (freeing one tendon attachment), and spinal cord injuries also produce the condition of *osteoporosis* in the bones of the afflicted extremities. Experimental administration of sex hormones (Chapter 30) delays or prevents the osteoporetic process by counteracting the action of the parathyroid hormone (Chapter 28).

Osteoporosis which affects the whole skeleton develops in everyone with advancing age. Only about 5% of the compact bone is lost, but between 40% and 50% of the trabecular bone disappears between the ages of 50 and 80. The spongy bone is a more readily available source of calcium than compact bone. The total loss of skeletal weight amounts to about 15% between youth and old age. Changes in bone associated with aging develop more rapidly in the female than in the male, and such changes increase the risk of fracture. In the postmenopausal state, the loss of bone by women probably is due to the reduction in circulating sex hormones and therefore a relative increase in the effectiveness of the parathyroid hormone's reabsorptive influence. Beyond the age of 65, both sexes experience increased bone loss. The best evidence indicates that this may be due to a progressive malabsorption of calcium, as well as to declining activity and possibly poor dietary habits.

Hormones. ■ The part played by certain hormones in the growth and development of the skeletal structures will be discussed further in Chapter 28.

■ TEETH

Lack of vitamin D during the period of tooth development causes delayed dentition and poor calcification of the dentin. Malposition of the teeth may occur because of the poorly calcified jawbones. There is no sound evidence that either calcium or vitamin D can prevent caries in fully formed teeth. It is possible that vitamins A and C are also essential for normal tooth development. In scurvy the rarefaction of the alveolar bone* causes teeth to loosen.

---

*An alveolus is a socket in either jawbone into which the root of a tooth is fixed.

READINGS

Andersson, B.: Thirst—and brain control of water balance, Am. Sci. **59**:408-415, Jul.-Aug. 1971.

Crichton, R. R.: Ferritin: structure, synthesis and function, New Eng. J. Med. **284**:1413-1421, 1971.

Erlander, S. R.: The structure of water, Sci. J. **54**(5):60-65, Nov. 1969.

Fitzsimons, J. T.: Thirst, Physiol. Rev. **52**:468-561, 1972.

Frieden, E.: The biochemistry of copper, Sci. Am. **218**(5):103-114, May 1968.

Manery, J. F.: Calcium and membranes. In Comar, C. L., and Bronner, F., editors: Mineral metabolism III, New York, 1969, Academic Press, Inc., pp. 405-452.

Calorimetry
Basal metabolic rate
Factors influencing basal metabolism
Factors influencing total metabolism
Energy balance

GENERAL ENERGY METABOLISM

Life is dynamic. Maintenance of life demands a constant flow of energy from the environment through the organism. When this flow ceases and the small amount of reserve energy in the body (also obtained from the environment) has been expended, life ends. The intake of energy is, as we have noted on many occasions, in the form of chemical potential energy of the organic foodstuffs. We have seen how by various chemical processes, catabolism, the food energy is released in a form suitable for the functions of the organs. The released energy is finally returned to the environment in two forms, as heat and as mechanical work. The mechanical work may be external (as in walking, lifting, and talking) or internal. Internal work may be mechanical, as in the work of the heart and alimentary canal, or it may be chemical, as in growth and the formation of organic compounds; this finally leaves the body as heat. In the study of energy changes in the body, it is necessary, first of all, to determine the energy value of the various foods consumed —the study of calorimetry.

■ CALORIMETRY

Unit of energy. ■ The energy unit usually used in physiology is the kilocalorie or Calorie, written with a capital C to distinguish it from the gram calorie. The Calorie is the amount of heat required to raise the temperature of 1 kg of water from 15 to 16 C. The relation of the Calorie to other common units of energy measurement is as follows:

1 Cal = 426.85 kg-m
1 Cal = 3,087.4 ft-lb
1 Cal = 1,000 cal

The heat that the foodstuffs generate when they are oxidized is measured by means of a calorimeter. A metal vessel, Fig. 25-1, *A,* is surrounded by a layer of water, *B,* which in turn is surrounded by a poor conductor of heat, *E.* The food, dried and weighed, is placed in the inner vessel and is ignited by an electric spark (the wires shown at *D*). By means of a tube, oxygen is supplied; the gaseous products are removed by means of tube *C,* which is coiled

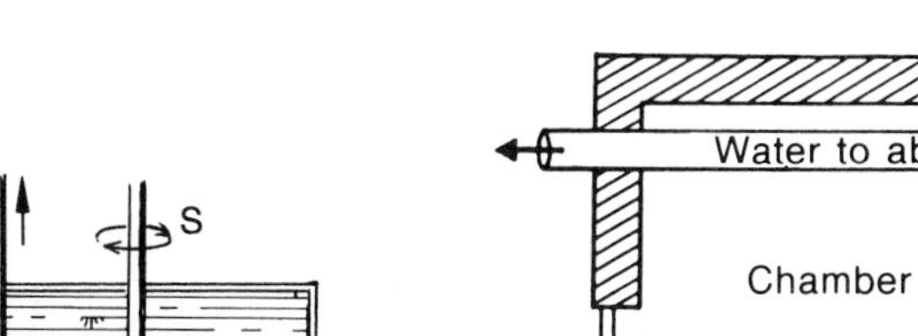

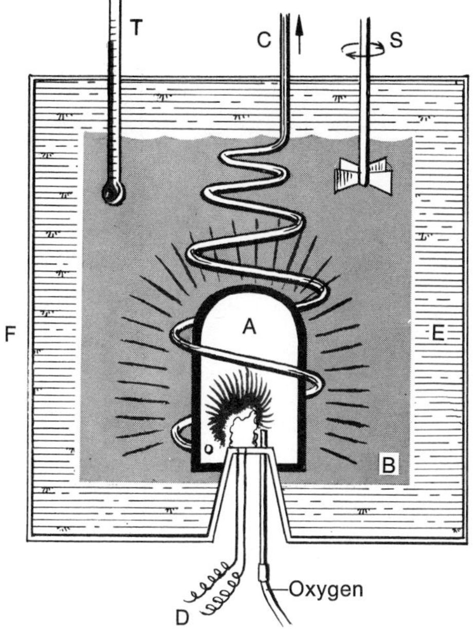

Fig. 25-1
Diagram of bomb calorimeter. **A**, Chamber in which food is placed; **B**, chamber containing water; **C**, outlet tube for gases; **D**, electric wires for igniting food; **E**, outer chamber containing some poor conductor of heat; **F**, outer wall of calorimeter; **S**, stirrer; **T**, thermometer.

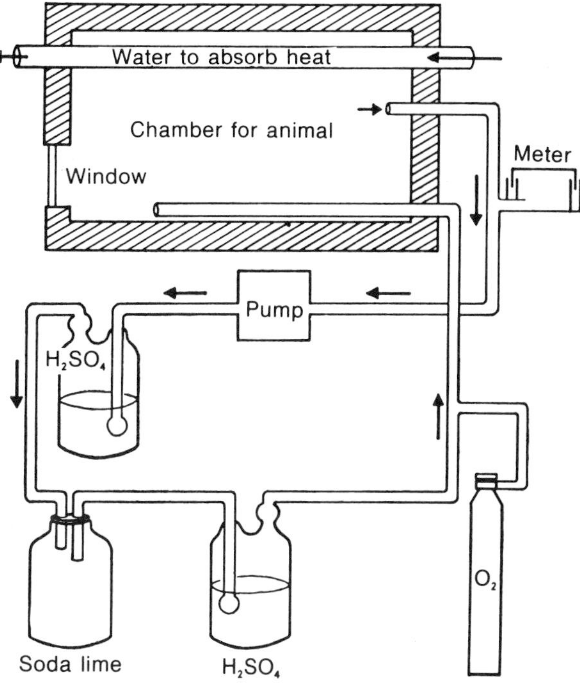

Fig. 25-2
Diagram of Atwater-Benedict calorimeter. (Macleod.)

through the water. The heat generated by the burning of the food is absorbed by the water. By observing the temperature of the water before and after the experiment and by knowing the quantity of water in the calorimeter and the quantity of material oxidized, the number of Calories produced by 1 g of the substance can readily be calculated. For 1 g of the foodstuffs the following average values have been found.

Carbohydrates—4.1 Cal
Fats—9.3 Cal
Proteins—5.6 Cal, physical heat value
Proteins—4.1 Cal, physiological heat value

Fats and carbohydrates are completely oxidized in the body, and, therefore, the amount of heat just given is available to the animal body as well as in the calorimeter. In the body, some of the total energy of foodstuffs is stored in energy-rich compounds, rather than being liberated as heat. It will be recalled that proteins give rise to nitrogenous waste products—urea, uric acid, and creatinine. All of these substances can be oxidized further and heat thereby liberated; 1 g of urea sets free 2.5 Cal. The energy of proteins is, therefore, not perfectly utilized in the body; the amount available is generally placed at about 4 Cal; this is called the *physiological heat value of proteins.* The caloric value and composition of some food items in terms of portions commonly used are given in Table 25-1. It should be noted that the values for the heat of combustion of all foodstuffs differ in any one class. For example, the protein of milk yields 4.27 Cal/g, whereas that of gelatin yields 3.90. In Table 25-1 these differences have been taken into account and the average values have not been used.

Table 25-1. Values of some food items in terms of portions commonly used*

Food	Calories	Proteins (g)	Fat (g)	Carbohydrates (g)
Cream, 20%, 1 oz	61	0.9	6.0	1.2
Butter, 1 pat	50	0.0	5.7	0.0
Milk, Whole, 8 oz	166	8.5	9.5	12.0
Skim, 8 oz	87	8.6	0.2	12.5
Bread, 1 slice, white, enriched	63	2.0	0.7	11.9
Breakfast cereal†, 1 oz	104	3.1	0.6	22.0
Egg, Boiled	77	6.1	5.5	0.3
Fried	110	6.1	9.2	0.3
Fat, Lard, 1 tbsp	126	0.0	14.0	0.0
Vegetable, 1 tbsp	110	0.0	12.5	0.0
Bacon, fried, 1 strip	48	1.8	4.4	0.2
Beef, Rib roast, 1 slice	96	7.2	7.2	0.0
Steak, broiled, 1 serving	104	8.1	7.7	0.0
Hamburger on bun	332	17.1	21.9	15.4
Pork, Chop, 1 medium, fried	233	16.1	18.2	0.0
Loin, roast, 1 slice	100	6.9	7.8	0.0
Liver, calf, 1 slice	74	8.1	3.6	1.7
Lamb, Chop, rib, 1 fried	128	7.9	10.5	0.0
Leg, roasted, 1 slice	82	7.2	5.7	0.0
Halibut, 1 serving	205	21.0	12.2	0.0
Scallops, fried, 6 medium pieces	427	23.8	28.4	19.3
Chicken, Broiler, fried, 1/4 bird	232	24.4	13.6	3.1
Roasted, 1 slice	79	11.3	3.4	0.0
Potatoes, White, 1 medium, baked	98	2.4	0.1	22.5
Sweet, 1 medium, baked	183	2.6	1.1	41.3
Apple, 1 medium, raw	76	0.4	6.5	19.7
Banana, 1 medium	132	1.8	0.3	34.5
Cantaloupe, 1/2 small	30	0.9	0.3	6.9
Grapefruit, 1/2 medium	72	0.9	0.4	25.3
Orange, 1 medium	68	1.4	0.3	16.8
Juice, fresh, 6 oz	81	1.5	0.4	20.3
Peach or pear, in syrup, 2 halves	68	0.4	0.1	18.2
Strawberries, 10 large	37	0.8	0.5	6.0
Asparagus, 2/3 cup	20	2.4	0.2	3.6
Beans, green, 1 cup	27	1.8	0.2	5.9
Broccoli, 2/3 cup	29	3.3	0.2	5.5
Carrots, 1/2 cup	23	0.5	0.4	4.8
Lettuce, head, 1 small leaf	2	0.1	0.0	0.3
Tomato, Juice, 1/2 cup	21	1.0	0.2	4.3
Raw, 1 medium	30	1.5	0.5	6.0
Soup, Bean, 3/4 cup	195	6.1	11.2	18.6
Consomme, 1 serving	34	6.7	0.8	1.8
Onion, 1 serving	64	5.2	3.2	3.7
Tomato, creamed, 3/4 cup	196	6.6	10.8	19.5
Fudge, chocolate, 1 oz	118	0.5	3.1	23.7
Jellies, assorted, 1 tbsp	50	0.0	0.0	13.0
Sugar, 1 tsp	16	0.0	0.0	4.2
Coca Cola, 6 oz	78	0.0	0.0	20.4

*Most values are from Bowes and Church: Food value of portions commonly used, ed. 8, Philadelphia, 1956, J. B. Lippincott Co.

†From Breakfast source book, Chicago, Ill., 1959, Cereal Institute, Inc.

Direct calorimetry. ■ The heat generated by an animal or a human being can also be determined by a calorimeter. Some of these calorimeters not only give information as to the amount of heat set free but are so constructed that the amount of oxygen consumed and the carbon dioxide and water eliminated by the lungs and skin are also measured. Such an apparatus, called a respiration calorimeter, is sketched in Fig. 25-2. The animal whose metabolism and heat production are to be studied occupies the calorimeter chamber. The walls of the outer chamber are electrically heated to exactly the same temperature as that of the inner chamber. Along the inside of the wall runs a system of tubes through which constantly circulates a stream of water. By taking the temperature of the water as it enters and again as it leaves the chamber and by measuring the amount of water passing through the chamber, the amount of heat generated by the subject can be determined. Except for one inlet and one outlet tube, the chamber is hermetically sealed after the subject is placed inside. Food and other materials are passed into and out of the chamber by means of a double window. By means of an inlet tube, fresh air is forced into the chamber, and air is removed through an outlet tube. The outlet tube is connected with a vessel containing sulfuric acid for the absorption of water and connected with a vessel containing soda lime for the absorption of carbon dioxide. The air, deprived of

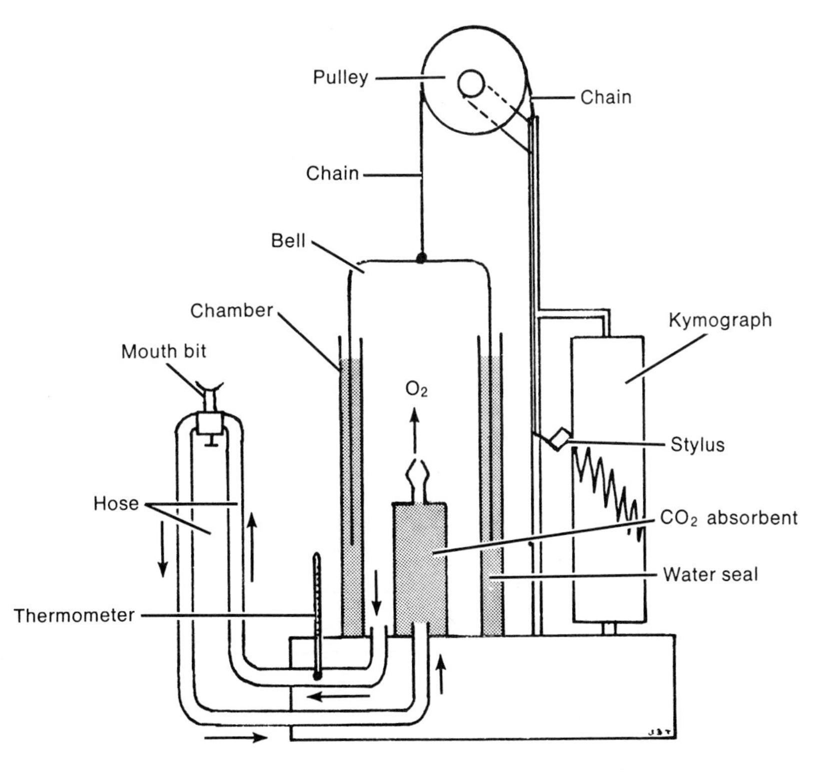

Fig. 25-3
Diagram of recording spirometer. (From Schottelius, Thomson, Schottelius: Physiology laboratory manual, The C. V. Mosby Co., 3rd ed., 1973.)

its water and carbon dioxide, is mixed with fresh oxygen and returned to the chamber. The amount of oxygen passing into the chamber during a unit of time is known. By weighing the soda lime and the sulfuric acid, the amount of carbon dioxide and water eliminated can be calculated. As the amount of nitrogen in the food and in the excreta (urine and feces) is also measured, a nitrogen balance and, therefore, a protein balance, can be determined. In this manner a complete metabolism balance sheet can be drawn up, showing the gain or loss of proteins, fats, and carbohydrates in the body.

Indirect calorimetry. ■ Since all the energy spent by the body is derived ultimately from the oxidation of organic foodstuffs, a simple, accurate, and more commonly used method of determining the total energy expenditure consists of measuring the amount of O_2 consumed over a short period of time—indirect calorimetry (Fig. 25-3). The subject breathes for, for example, four to six minutes from a spirometer filled with oxygen. The expired air, after being passed over soda lime to absorb CO_2 and H_2O, is returned to the spirometer. The decrease in the volume of gas in the spirometer is the measure of the volume of O_2 used by the subject.

For its combustion, 1 g of carbohydrate requires 812 ml of oxygen and liberates 4.1 Cal; hence, the heat generated when 1 liter of oxygen is used equals 5 Cal. For fats 1 liter of oxygen yields 4.7 Cal and for proteins, about 4.5 Cal. If all three foodstuffs (in the proportion usually found in the average diet) are oxidized, 1 liter of oxygen generates about 4.825 Cal. A person taking 1.25 liters of oxygen (reduced to standard conditions) from the spirometer during the course of four minutes generates 4.825 × 1.25 or about 6 Cal of heat. Assuming this rate obtains during the course of a day, the total energy expended is 2,160 Cal.

If we desire to know how much of the total energy was supplied by the catabolism of proteins, we may proceed as follows:* By analysis we find the amount of nitrogen excreted by the kidneys. One gram of urine nitrogen represents the oxidation of 6.25 g of protein, and 1 g of protein supplies the body with 4.1 Cal of heat. One gram of protein requires for its oxidation 910 ml of oxygen. The amount of oxygen utilized to catabolize the protein may be subtracted from the total amount of oxygen consumed; the remainder will be that used in the fat and carbohydrate combustion.

■ BASAL METABOLIC RATE

As the extent of metabolism increases with the amount of energy expended by the body in muscular work and in heat loss, it is evident that the minimum requirement will be found when the body is in a resting condition and surrounded by an atmosphere calling for a minimum heat production. After one or two hours of complete rest in bed in a comfortably warm room and ten to twelve hours after the last meal, with the subject kept as free as possible from all environmental distractions, the average amount of heat produced by an adult is somewhere between 1,500 and 1,800 Cal/day. This amount of metabolism is known as the *basal metabolism* or basal metabolic rate (BMR). It is the minimum expenditure of energy compatible with life†; that is, it represents the amount of energy expended for the internal needs of the body, such as the heartbeat, respiratory movements, activity of the alimentary canal, tone of skeletal muscles, and so forth.

By far the larger part of our energy is lost in the form of heat. Since a greater amount of sur-

---

*The following is a useful index:

1 g nitrogen	=	2+ g urea
1 g urea	=	3 g protein
1 g nitrogen	=	6.25 g protein
1 g protein	=	5 g muscle tissue

†It should be stated that during sleep the metabolism is 10% below the basal metabolic rate as this is generally determined; this is due to the decrease in muscle tone.

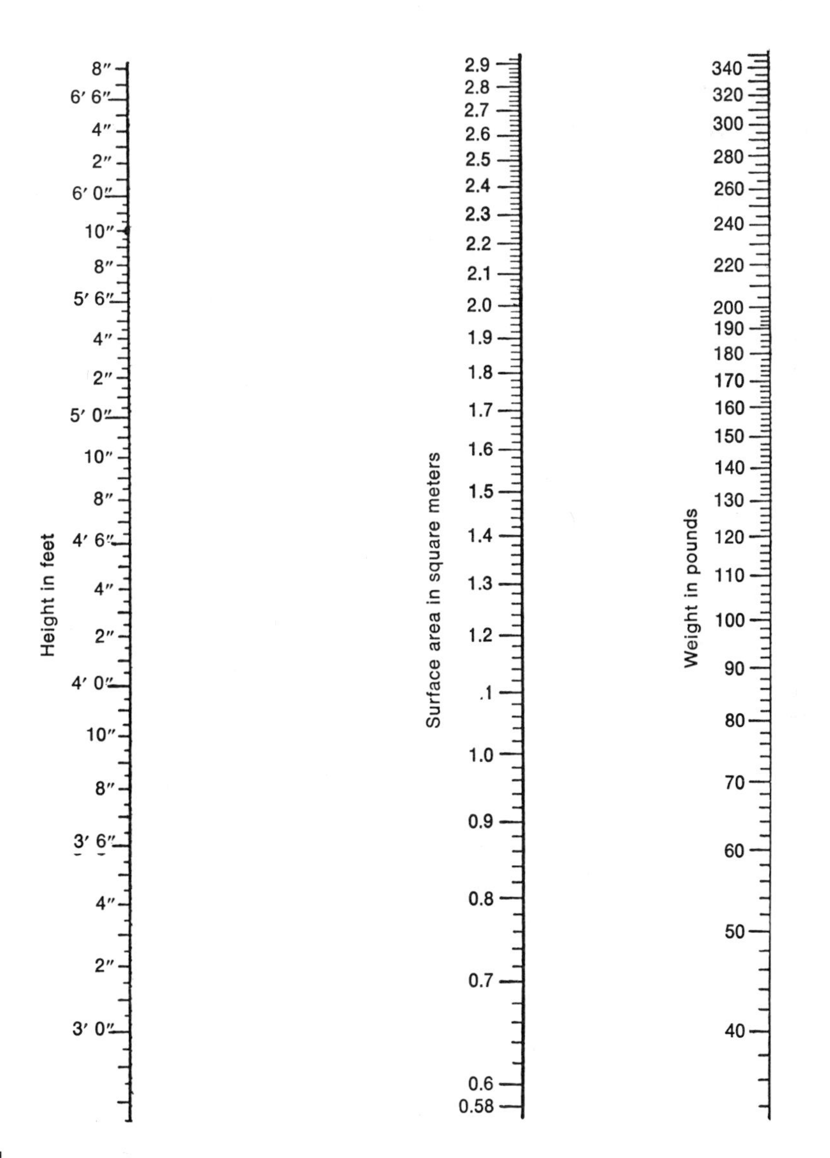

Fig. 25-4
Nomogram for determining surface area of man in square meters from weight in pounds and height in feet and inches. (From Schottelius, Thomson, Schottelius: Physiology laboratory manual, ed. 3, The C. V. Mosby Co.)

face area exposed to the environment means an increase in the dissipation of heat, the heat production varies with the extent of the surface area. For this reason the basal metabolic rate is usually stated in terms of body surface area (Fig. 25-4). For the average adult man it may be stated at about 37 Cal/m^2 per hour; for the adult female, at 35 Cal; for 1-year-old infant, at 53 Cal.

Stated in terms of body weight, the average basal metabolism (or basal heat production) in a man between the ages of 20 and 50 is about 1 Cal/kg of body weight per hour; hence a person weighing 70 kg produces 1,680 Cal daily. Variations of $\pm 10\%$ are within normal limits.

■ FACTORS INFLUENCING BASAL METABOLISM

Age. ■ Age markedly affects basal metabolic rate (Fig. 25-5). From the values given in Table 25-2, it is noted that the BMR is highest in the young child. Moreover, because of growth, some of the energy taken in by the child will not be expended but will be stored. The BMR decreases steadily with increase in age.

Sex and body weight. ■ Women generally have a BMR from 5% to 10% lower than that of men. When stated in terms of body surface, there is very little difference between the BMR of an overweight or underweight person as compared with one of average weight. Living for a considerable time on a diet low in calories reduces, whereas overfeeding increases, the BMR. Within limits the body adjusts the rate of burning its food to the amount available.

Internal secretions. ■ The central nervous system and certain of the endocrine glands are *energy-releasing organs.* Thyroxin, the main secretion of the thyroid gland (Chapter 28), has a significant effect on basal energy needs. The BMR may be depressed as much as 30% in cases of *hypothyroidism* and elevated as much as 50% to 75% in *hyperthyroidism.* One secretion of the adrenal glands, *epinephrine* (Chapter 28), produces an intense stimulation of metabolism. This is usually of short duration. To a lesser extent the gonads have an influence on basal metabolism.

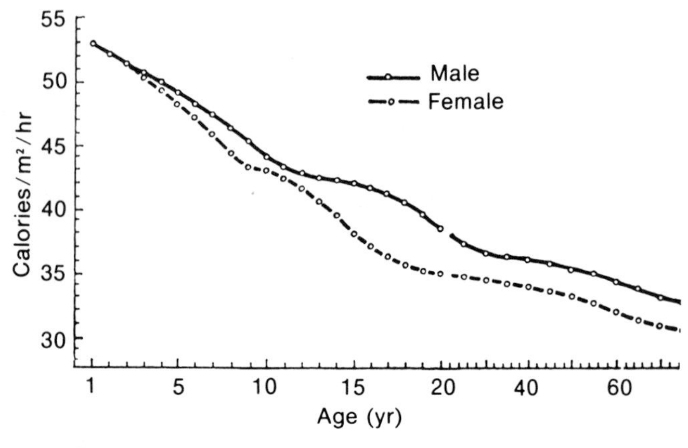

Fig. 25-5
Relation of age to basal metabolic rate. (After Fleisch: Helvet. Med. Acta **18**:23, 1951.)

Table 25-2. Average values for basal metabolic rate*

	Calories per square meter surface area	
Age in years	*Male*	*Female*
1	53.0	53.0
2	52.4	52.4
5	49.3	48.4
10	44.0	42.5
15	41.8	37.9
20	38.6	35.3
25	37.5	35.2
30	36.8	35.1
35	36.5	35.0
40	36.3	34.9
50	35.8	33.9
60	34.9	32.7
70	33.8	31.7
80	33.0	30.9

*From Fleisch: Helvet. Med. Acta. **18**:23, 1951; cited in Ann. Rev. Physiol. **16**:125, 1954.

Body temperature. ■ An elevation in body temperature of 1 F can cause a 7% to 15% increase in the basal metabolic rate.

■ FACTORS INFLUENCING TOTAL METABOLISM

Muscular work. ■ During the production of the energy for muscular activity there is concomitantly a great liberation of heat, which cannot be utilized for the performance of mechanical work. For this reason and because the muscles constitute more than 45% of the body, it can be readily understood that no other factor so powerfully influences the amount of energy liberated in the body as the activity of the skeletal muscles. This is well illustrated in Table 25-3.

To the 2,400 Cal/day produced by an adult not steadily engaged in muscular work, there must be added per hour of work the following:

For light work	50 Cal
For moderate work	50 to 100 Cal
For hard work	100 to 200 Cal
For very hard work	200 Cal

The ability of the body to utilize fats in muscular work is seen in the behavior of ketone bodies. A diet very low in carbohydrates and high in fats causes an increased production of ketone bodies (e.g., acetoacetic acid); if continued for several days, ketosis may set in. Muscular work at this time causes a diminution of ketones in the blood, since they can be used by the muscle tissue.

The respiratory quotient (CO_2/O_2) is at first increased during heavy work, evidence of a greater catabolism of carbohydrates. Later in the work the quotient falls, indicating that now fats are metabolized. Carbohydrates are more readily mobilized. We have seen that carbohydrates in uniting with 1 liter of oxygen yield 5 Cal of energy; fats yield 4.7 Cal. In marathon races the greatest exhaustion is shown by those contestants whose blood sugar level is the lowest. Injecting or eating sugar prior to the work has been shown to delay the onset of fatigue. Although all three foodstuffs can be used in muscular work, it appears that the carbohydrates are used by preference; they are used directly and with greater economy, perhaps by as much as 10%. As can be seen in Fig. 25-6, the oxygen consumption in a unit of time increases progressively with the rate of work.

Table 25-3. Effect of activity on energy loss*

Man in bed for 24 hours	1,840 Cal
In bed 8 hours; sedentary occupation for 16 hours	2,168 Cal
In bed 8 hours; in chair 14 hours; walking 2 hours	2,488 Cal
Active outdoor life, as that of farmer	3,500 Cal

*From Lusk.

External temperature. ■ Next to muscular work no factor has so much influence on metabolism as the temperature of the environment. This can be gathered from Table 25-4, which gives the Calories produced by a dog at various external temperatures.

The increased heat production at lower temperatures originates almost entirely from increased muscle tone and involuntary shivering.

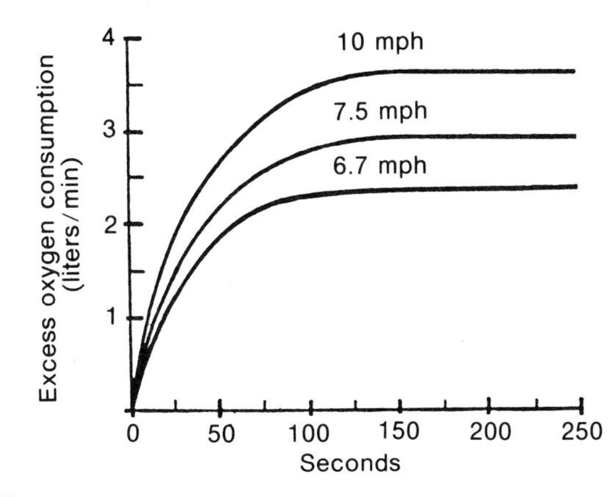

Fig. 25-6
Graph showing relationship between speed of running and oxygen consumption, in excess of resting level. (After Hill.)

Table 25-4. Effect of environmental temperature on heat loss*

External temperature (C)	*Calories per kilogram*
30	56.2
20	55.9
15	63.0
7.6	86.4

*From Rubner.

Table 25-5. Effect of ingestion of food on metabolism

Day	*Diet*	*Heat liberated*
1	Fasting	718 Cal
2	400 g protein	1,046 Cal
3	Fasting	746 Cal
4	400 g protein	1,105 Cal

The greater loss of heat at the lower temperatures must be restored by an increased food consumption. That the cutaneous stimulation by cold sharpens the appetite is a familiar fact. Exposed to a temperature of 52 F, lower animals eat more food than at a temperature of 88 F, and by instinctive food selection the increase is confined largely to carbohydrates. United States ground troops voluntarily increased their daily consumption of food from 3,100 Cal when the external temperature was 92 F to 4,900 Cal on being subjected to arctic cold of −30 F. The percentage of protein foods selected by them was almost constant, irrespective of the environmental temperature. The response to cold stimulation is mediated by the endocrine glands.

Influence of food. ■ The ingestion of food causes a marked increase in metabolism. As shown in Table 25-5, a person fasted one day and consumed 400 g of protein the next. On the eating days he liberated about 300 Cal of heat over and above that produced during the fasting days. This extra heat, associated with the consumption of food, is known as the *specific dynamic action* (SDA) of food. Since on all four days the person was sheltered against cold and was perfectly at rest, what caused the production of this heat? Many and sharply differing explanations have been offered; space will allow us to make only one suggestion.

The extra heat liberated is not due to the work of the alimentary canal, for feeding agar (indigestible) or purgatives does not have this effect. Furthermore, the intravenous injection of amino acids is followed by SDA; however, this does not occur if the liver is removed. Eighty percent of the SDA originates in the abdominal organs. These and other facts favor the view that SDA is associated with the deamination of amino acids and the formation of urea. The specific dynamic action is not the same for all three organic foodstuffs; that for proteins is generally stated as 30%, for carbohydrates, 6%, and for fats, 4% of the Calories found in these foodstuffs. Of the energy in a mixed diet, about 10% is liberated in the SDA. The extra heat that the mere presence of proteins in the blood calls forth cannot be utilized by the organs of the body (muscles and glands) for the performance of their work. But in a cold environment, the extra heat is used to warm the body and thereby saves other fuel material.

■ ENERGY BALANCE

Expenditure of energy. ■ Our body expends energy in four distinct ways: (1) in maintaining life, as shown by the BMR, (2) in the SDA, (3) in muscular activity, and (4) in maintaining the body temperature when the external temperature falls below a certain point. In this latter case, the SDA may aid. A diet furnishing more energy than is needed to cover the losses just discussed leads to the storing up of potential energy in the body, generally in the form of fat.

The majority of normal adult people, without giving the matter any thought, are in energy equilibrium, as is evident from their constant body weight. Over a sufficiently long period of time they automatically eat enough to cover the energy output and no more.

Hunger. ■ The need for energy-furnishing food, coupled with the need for oxygen, is the primordial driving force in all animal life; indeed, without this urge (hunger) life is impossible. But, more than a sensation, *hunger is an energy-needing state of the body,* which leads to the acquisition of food. As a sensation, hunger ceases soon after the partaking of a small amount of food; nevertheless we continue to eat.

Satiety. ■ The feeling of satiety is an important cog in the energy-regulating machinery. Normally this feeling leads to the cessation of eating. It is possible that in extremely lean people an abnormal sensitivity to satiety causes them to stop eating before the demand for energy is met. In contrast, in other people the lack of this sensation or the ability to ignore it may lead to obesity. As to the cause of the feeling of satiety, not much is known. The sensation of fullness in the abdomen may contribute to it. Recently a change has been observed in olfactory (smell) acuity associated with the eating of food; the usual normal sense of smell is noticeably dulled at the end of a meal. The same is true for the sense of taste for sweets. Keen gustatory and olfactory sensations are conducive to appetite; their waning may be a factor in creating satiety.

The hypothalamus. ■ The hypothalamus (Fig. 10-3) plays an important part in determining the food intake. The destruction of a certain area of the hypothalamus causes a complete cessation of eating, and death results from starvation; this area may be termed the *feeding center.* An experimental lesion of another area leads to a marked increase in the food intake; this may be looked on as the intake inhibitor, or a *satiety center.* In man, some hypothalamic lesions (as by a tumor) lead to a loss of the sense of satiety; in consequence, an excessive amount of fat is deposited in the adipose tissue. Instances are recorded of people having a body weight of from 400 to 800 lb.

How these two centers are informed of the energy state of the body is not known. The hypothalamus governs many of the mechanisms and processes employed in the production, loss, and conservation of body heat. Destruction of certain parts causes decreased metabolism and great lethargy. It is conceivable that by nerve impulses from muscles, skin, and other sensory surfaces and by the condition of the blood (especially its sugar content) the hypothalamus is kept informed of the energy state of the body. By efferent impulses or perhaps by hormones it controls the special organs concerned with the taking in of energy or with its expenditure.

READINGS

Blaxter, K. L.: Methods of measuring the energy metabolism of animals and the interpretation of results obtained, Fed. Proc. **30**:1436-1443, 1971.

Carlson, L. D., and Hsieh, A. C. L.: Control of energy exchange, New York, 1970, The Macmillan Company, chaps. 1-3.

Consolazia, C. F., and Johnson, H. L.: Measurement of energy cost in humans, Fed. Proc. **30**:1444-1453, 1971.

Hartsook, E. W., and Herschberger, T. V.: Interactions of major nutrients in whole animal energy metabolism, Fed. Proc. **30**:1466-1473, 1971.

Anatomy
Functions of the skin
Cutaneous burns

THE SKIN

■ ANATOMY

Skin is a specialized organ containing both living and nonliving components. It has remarkable properties and many functions. The skin is composed of many different tissues, including blood vessels, connective tissue, fat, glands, sense organs, and nerves. Three distinct tissue layers are recognizable: the epidermis, the dermis, and the subcutaneous fat.

The epidermis. ■ As shown in Fig. 26-1, the outermost layer of skin, known as the stratum corneum, consists of thin, stratified epithelial cells. It evolves from a dense population of actively dividing epithelial cells in the Malpighian layer. The oldest of these dividing cells migrate or are displaced toward the skin surface, where they differentiate into flat layers. The epidermis, being devoid of blood vessels, is dependent for its nutrition on the vessels found in the dermis. These vessels end as capillary loops in the papillae. The more superficial cells of the epidermis, being far removed from the nutrient supply, gradually degenerate and die; their proteins are transformed into a special type known as *keratin* by an enzymatic process.

Keratin, the most insoluble of all proteins, is tough and extremely resistant to heat, cold, changes in pH, and enzymatic digestion; it gives the stratum corneum a hornlike consistency. However, the outermost dead cells are gradually worn, rubbed, or scraped away but are continuously replenished from below. Keratin is also found in the appendages of the skin, that is, in the nails and hair; in lower animals it is found in horns, hoofs, and claws.

Epidermal thickness is remarkably constant over the body. Except on the palms and the soles, where the stratum corneum may be six to ten times thicker, the usual thickness is 60 to 100 μ. By intermittent pressure the horny layer of the epidermis may be caused to hypertrophy and produce a swelling not only outward but also downward into the next layer of the skin, the dermis, giving rise to a callus or corn. The squamous epithelium of the oral cavity, esophagus, and so on is not covered with keratin.

The dermis. ■ The true skin—corium, or dermis—lies below the Malpighian layer; it supports and binds the epidermis to conform to the underlying bone and muscle; it is formed of

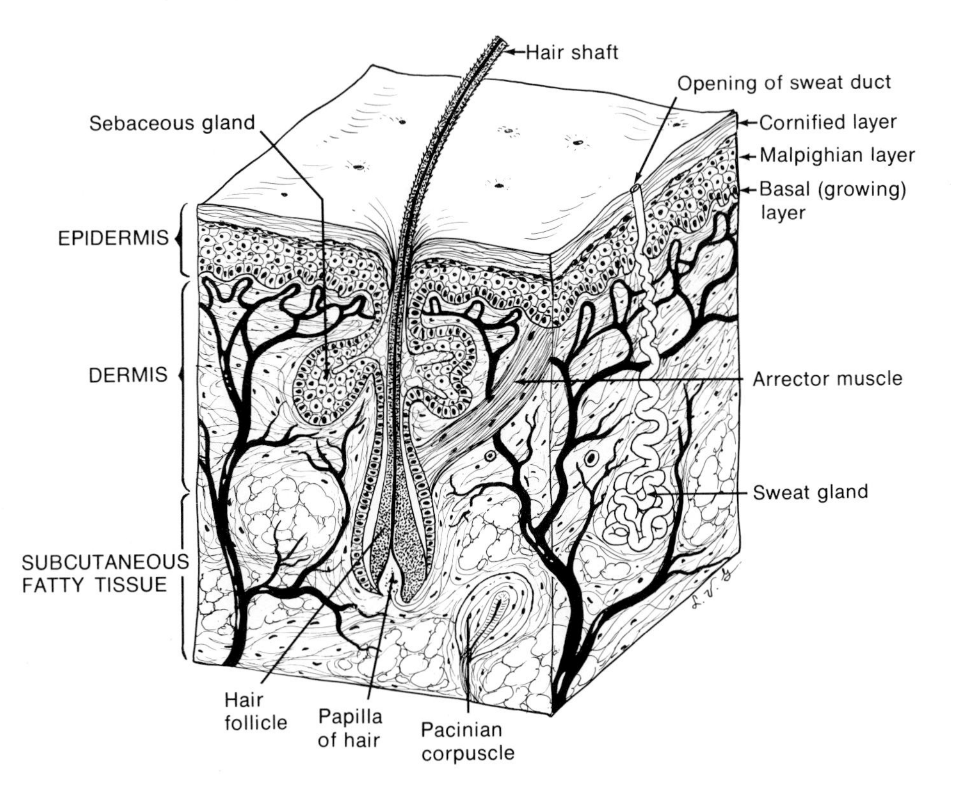

Fig. 26-1
Illustration of a section of human skin showing the several layers and many of the other structures appearing in skin.

loose connective tissue, chiefly collagen, whose fibers provide tensile strength equal to that of thin steel wire. This skin layer produces leather when chemically processed. Thickness of the dermis varies in different body regions. The order of decreasing thickness is back, thigh, abdomen, forehead, wrist, scalp, and palm. Small elevations, the papillae, originate in the dermis and penetrate the epidermis. Many papillae are furnished with modified nerve endings or tactile corpuscles; these are most abundant in the palm of the hand and in the fingers. In a wart the papillae are much enlarged, and the epidermis overlying the papilla is thickened.

The subcutaneous fatty tissue is the deepest skin layer. Its boundary with the dermis is poorly demarcated, and the thickness of this layer is variable. The fat layer serves as a cushion for the overlying skin elements and, because of the flexible linkage to them, permits considerable lateral displacement of the skin in many regions of the body.

Although the epidermis is not supplied with blood vessels, they are diffusely distributed throughout the dermis and to a lesser extent in the subcutaneous fat. The skin has a blood supply far in excess of its metabolic need. Small bundles of smooth muscle fibers are also found in the skin. Most of these are attached to the hair follicles (arrector muscles) and by their contraction elevate the hairs (goose bumps).

Hair is an outgrowth (and a subsequent keratinization) of the dermis. The amount of pigment (known as *melanin*) in the hair shaft varies from albinos to blondes and brunettes. Graying is due to failure of the hair follicle to deposit the usual amount of pigment. Contrary to what occurs in some animals, in man it is doubtful whether graying is caused by any dietary deficiency. Three factors are known to influence baldness: heredity, aging, and stimulation by the male sex hormone.

Sebaceous glands. ■ In the dermis are located sebaceous (oil) and sudoriferous (sweat) glands. The oil glands, lying in the deeper parts of the skin, are present in all areas of the body except the soles and palms. The meibomian glands of the eyelids are also classed with the sebaceous glands. In nearly all instances the ducts open into the follicles or depressions from which the hairs develop (Fig. 26-1). The sebaceous secretion, known as *sebum*, contains fats, proteins, water, salts, and remnants of epithelial cells. Sebum is produced abundantly on the forehead, less so on the trunk, and to a limited extent on the extremities. By oiling the skin and the hair, the sebum prevents these structures from drying unduly and breaking, thus preserving the physical integrity of the skin. It also helps prevent too great imbibition or loss of water by the skin. No secretory nerves governing the activity of the sebaceous glands are known, but sebaceous activity is markedly stimulated by certain sex hormones. We will see that the nutrition of the skin and hair depends on the proper activity of the thyroid gland (Chapter 28).

The sebaceous glands are very active during adolescence. If the secretion cannot be properly discharged, it lodges in the ducts, a "whitehead." The outer portion of this material may be blackened by oxidation and then constitutes a "blackhead." Their presence is due neither to faulty hygiene nor to lack of cleanliness.

An acute inflammation of a hair follicle and its sebaceous gland, brought about by the entrance of staphylococci through the follicle, gives rise to a boil. A carbuncle may be said to be a number of boils closely set together and involving the subcutaneous tissue to a greater or lesser degree. Such eruptions generally indicate a lowered vitality; a diet rich in carbohydrates seems to predispose of boils. Boils and abscesses are effectively treated with antibiotics.

Sudoriferous, or sweat, glands. ■ The secreting part of these glands lies deep in the true skin and consists of a tube coiled up into the form of a little ball; the discharging duct passes through the outer layers of the skin in the form of a corkscrew. These glands are found in almost every part of the skin but are most abundant in the palms and soles. The glands of the ear, which secrete the ear wax, *cerumen*, are regarded as modified sweat glands.

Sweat is composed of water (99%), salts (chiefly sodium chloride), and traces of urea. The amount of solids is very small, in consequence of which the specific gravity is low (1.004). In abnormal conditions the sweat may contain many compounds usually not present, such as bile pigments, albumin, sugar, and blood. The sweat sometimes has an acid and sometimes an alkaline reaction; its odor may be very marked and differs in the various regions of the body.

The amount of water lost through the skin may be placed at 0.5 to 2 liters/day.* During strenuous work this may be increased to 5 liters. Two methods exist for losing water by the skin: (1) by the sweat glands and (2) by osmosis—*insensible perspiration.* The skin is not completely impervious to water; that is, water can pass through it by osmosis without the aid of glands. It has been reported that a boy destitute of sweat glands evaporated, at an external temperature of 25 C, about the same amount of water as a normal subject. With his body (not his head) enclosed in a bag through which air at 48 C was

*In the Tropics the amount of urine voided is reduced to a minimum; the quantity of water lost by the skin may amount to 10 to 12 liters/day.

circulated, he lost only 33 g of water per hour; a normal person, under identical conditions, lost 336 g by osmosis and perspiration.

Regulation of perspiration. The work of the sweat glands is governed by secretory nerves of the autonomic (sympathetic) nervous system. Stimulation of the sciatic nerve may cause drops of sweat to appear on a cat's foot. It is generally believed that a sweat center exists in the brain (perhaps in the hypothalamus); it is stimulated when the temperature of the blood going to the brain is increased by 0.2 to 0.5 C or, reflexly, when the skin temperature exceeds 94 F (in a nude subject). The rise in the temperature of the blood is largely determined by the external temperature and amount of muscular activity. The sweat center is also influenced by nausea, asphyxiation, and emotional states, as seen in the "cold sweat" of fear.

■ FUNCTIONS OF THE SKIN

The functions of the skin are many; some are of vital importance.

Protection. ■ The epidermis, being composed of hard, resistant cells, forms the body's first line of defense against mechanical and chemical injuries and bacterial invasion. Human skin very effectively retards the diffusion of gases, water, and chemicals. Most irritants and poisons, with the exception of "poison ivy" and mustard gas, are unable to penetrate the skin. Were the skin as permeable as the lung membranes, exposure to dry air would rapidly produce fatal dehydration. The unbroken skin is almost germproof.* How truly the skin protects us in this last respect is seen in the frequent infections taking place when the skin is injured so as to expose the underlying tissues. Recently it has been demonstrated that the skin not only serves as a mechanical barrier but that it also possesses immunizing powers. Dry skin is highly resistant to the flow of electrical current and offers some degree of protection.

The skin absorbs, in varying degrees, electromagnetic radiation such as soft x-rays, ultraviolet light, visible light, and infrared wavelengths. The damaging effects of ultraviolet light are familiar to those who have suffered severe sunburn. In the deeper layer of the epidermis there is found a variable quantity of melanin, giving color to the skin. The amount of the pigment is increased as tan and freckles under the influence of intense sunlight and is thought to be of value in protecting the individual against the actinic (ultraviolet) rays of the hot sun; blonde people reflect about 45% of the sunlight, as against 35% by darker people. Proteins and pigments absorb radiation maximally in the 280 $m\mu$ region (Fig. 24-4); thus ultraviolet light is almost completely absorbed within the first 100μ below the surface of the skin. Visible light penetrates deeper than ultraviolet; infrared wavelengths penetrate the deepest of all.

Absence of all pigment in the skin, hair, and iris of the eye is known as albinism. This is a hereditary trait.

Contrary to popular opinion, it is very doubtful that exposure to sunlight has any beneficial influence on growth and metabolism, provided a sufficient amount of vitamin D is included in the diet; in fact, an excess may be deleterious. Psychological effects of moderate sunlight, as compared with darkness, may influence bodily functions; this, however, is mediated by means of the eyes and not the skin. It is generally conceded that prolonged exposure to intense sunlight may cause loss of elasticity of the skin, early aging, and possibly even cutaneous cancer.

Excretion. ■ The skin is frequently said to have a minor excretory function. From 4 to 8 g of CO_2 are excreted per day by this organ; compared with the 700 to 800 g exhaled by the lungs, we can hardly call the skin an important excretory organ for this waste product. The same is

---

*The germ of tularemia (rabbit fever and deer fly fever) is said to penetrate healthy human skin.

true for the small quantities of other wastes, such as urea and the various salts.

Water reservoir. ■ Although sweating is regulated by the autonomic nervous system, certain observations suggest that it may be affected by local conditions on the skin surface. A dry air environment increases sweat production. Heavy sweating from exercise or exposure to heat is reduced markedly when the skin is covered with water, either by immersion or by accumulated sweat. When the skin is exposed to air with more than 90% relative humidity, water moves into it at a rate of 10 to 50 ml/m^2 of surface per hour. The skin seems to be totally impermeable to salts; only pure water enters the stratum corneum when the latter is exposed to salt solutions. The role of the skin in water metabolism also was discussed in Chapter 24.

Body temperature. ■ An important function of the skin is the role that it plays in the regulation of body temperature. Warm-blooded animals must maintain a balance between heat production and heat elimination to sustain critical enzymatic reactions. The body is able to store a large amount of heat because of its thermal capacity; thermoregulatory mechanisms minimize heat storage. Capillary loops from the network of blood vessels in the dermis extend nearly to the surface, and sweat glands are surrounded by clusters of capillaries. The quantity of heat dissipated or conserved is precisely regulated by controlling the blood flow and blood content in the skin vessels. Perspiration accelerates heat loss by evaporation. The role of the skin in body temperature regulation is also discussed in Chapter 27.

Sensory organs. ■ The sensory organs for heat, cold, touch, and pain (Chapter 7) are found in the dermis. These remarkably sensitive transducers—a transducer converts one form of energy into another, in this instance into electrical signals—convey vital information about the immediate environment to the central nervous system. The sensory functions of the skin play an obviously vital part in our daily lives.

■ CUTANEOUS BURNS

Although strictly speaking not within the province of physiology, the high incidence of cutaneous injury, both by flame and by scalding, may be sufficient reason for including a few pertinent remarks. The fatal results following severe burning of a large area of the body have been attributed to many causes, but in the great majority of cases death results from shock.

An extensive burn kills and denudes the skin so that capillary permeability is increased, and tissue fluids escape. The exuded fluids contain large amounts of protein, with consequent decrease in colloidal osmotic pressure of the plasma. Severe burns may result in such a fluid volume loss that hemoconcentration results. Replacement of fluid volume is usually accomplished with plasma rather than whole blood, which would further augment the hemoconcentration and impede the circulation through increase in viscosity.

Burns are classified as (1) first-degree burns, in which there is reddening of the skin, (2) second-degree burns, which give rise to blisters, and (3) third-degree burns, in which deeper tissues are destroyed.

READINGS

Champion, R. H.: Sweat glands. In Champion, R. H., Gillman, T., Rook, A. J., and Sims, R. T., editors: An introduction to the biology of the skin, Oxford, 1970, Blackwell Scientific Publications, pp. 175-183.

Daniels, F., van der Leun, J. C., and Johnson, B. E.: Sunburn, Sci. Am. **219**(1):38-46, Jul. 1968.

Ebling, F. J. G.: Sebaceous glands. In Champion, R. H., Gilman, T., Rook, A. J., and Sims, R. T., editors: An introduction to the biology of the skin, Oxford, 1970, Blackwell Scientific Publications, pp. 184-196.

Edelberg, R.: Electrical properties of skin. In Elden, H.R., editor: Biophysical properties of the skin, New York, 1971, John Wiley & Sons, Inc., pp. 513-550.

Fraser, R. D. B.: Keratins, Sci. Am. **221**(2):87-96, Aug. 1969.

Rook, R.: Hair. In Champion, R. H., Gillman, T., Rook, A. J., and Sims, R. T., editors: An introduction to the biology of the skin, Oxford, 1970, Blackwell Scientific Publications, pp. 164-174.

Tregear, R. T.: Physical functions of skin, New York, 1966, Academic Press, Inc.

27

Body temperature
Regulation of body temperature (thermotaxis)
Disturbances in thermoregulation

REGULATION OF BODY TEMPERATURE

It is common knowledge that the temperature of the human body is almost constant. Yet many factors tend to upset this constancy. In an active muscle there is, in addition to the activity characteristic for this tissue, a liberation of heat; in fact, about 80% of the energy expended during the work is transformed into heat (Chapter 6). Again, the body is generally warmer than the environment, and hence there is a continual loss of heat. On the other hand, on certain occasions the temperature of the environment is higher than that of the body, and under these conditions the body tends to gain in heat. But, notwithstanding all this, the body temperature varies little; evidently the body is equipped with a heat-regulating mechanism.

It is customary to classify animals as warm-blooded (homoiothermic) and cold-blooded (poikilothermic). The body of warm-blooded animals (mammals and birds) is not only warm but has practically the same temperature day in, day out, in winter, and in summer. Cold-blooded animals, on the other hand, are cold when they are subjected to external cold, but when placed in a warm environment, their temperature rises. Hence it is better to classify animals as those with a constant body temperature and those with a varying body temperature. Hibernating mammals during their waking state are homoiothermic, but during their winter sleep the body temperature falls.

■ BODY TEMPERATURE

The body temperature of man, generally stated at about 37 C (98 F), varies in different parts of the body. The mouth temperature is a little lower than the rectal temperature; the liver is the warmest and the skin, the coldest part. Body heat is produced primarily in the skeletal muscles and liver during the resting state. During exercise and in fever the skeletal muscles are the most important generators of heat. Temperature also varies slightly with changing conditions of the body and of the environment; among these are muscular work, sleep, diurnal variation, age, external temperature, and cold and warm baths.

Muscular work. ■ As we have seen, muscular work influences the heat production in our bodies very pronouncedly, yet ordinarily it has but a slight effect on the temperature of the body, the increase being from 0.1 to 1 F. Soon after the cessation of work, this rise subsides. When, however, large groups of muscles are simultaneously in a state of severe and rather prolonged contractions, as in tetany and during a rapid series of epileptic attacks, the body temperature may rise considerably; in marathon races, temperatures of 104 F have been recorded. Since the rise in body temperature increases the rate of respiration and heartbeat and aids in the liberation of oxygen from the oxyhemoglobin, it may be of benefit in severe work.

Sleep. ■ Because of muscular inactivity, sleep results in a slight fall in body temperature.

Diurnal variation. ■ In the early morning hours, the body temperature is at its lowest; it is highest in late afternoon or early evening. These variations amount to 1 to 1.5 F and are reversed on changing the daily routine of living.

Age. ■ A newborn child has a slightly higher temperature than an adult. It is, however, more important to remember that the body temperature of a young child is more variable due to the lack of control exercised by the nervous system over the blood vessels and other organs operative in temperature regulation; the young of some warm-blooded animals must be regarded as poikilothermic.

External temperature. ■ The ordinary changes in atmospheric temperature, such as experienced in winter and summer, have very little effect on body temperature.

Cold and warm baths. ■ Cold and warm baths have a far greater influence than air at the same temperature,* but since the ordinary exposure to these baths is of short duration, they have little effect on the normal body temperature. Of 250 children swimming for forty-five minutes in water at 73 F, only 30 experienced no change in body temperature; in some of the others the temperature dropped as low as 95 F. When the body temperature is abnormally high (fever), a cold bath reduces it.

*The heat conductivity of water is twenty-five times as great as that of air.

■ REGULATION OF BODY TEMPERATURE (THERMOTAXIS)

In order to maintain an optimum environment for the cells of a warm-blooded animal, it is necessary to keep the body temperature fairly constant, notwithstanding the external and internal conditions that tend to raise or lower it. When the human body temperature falls to about 24 C (77 F) and when this temperature is maintained for several hours, death occurs. On the other hand, a temperature higher than 44 or 45 C (111 to 113 F) maintained for more than a brief length of time is also fatal. Normally the human body generates enough heat to raise the body temperature 3 F per hour, if all the heat were retained.

Increasing or decreasing of the body temperature is brought about in two ways:

1. Regulating the loss of heat *(thermolysis),* that is, *physical heat regulation*
2. Regulating the production of heat *(thermogenesis),* that is, *chemical heat regulation*

At normal environmental temperatures fine adjustment of the body temperature is attained by variation of the blood supply at the body surface. This is mediated through the autonomic nervous system (ANS) (Chapter 11). Coarse adjustment is achieved through somatic motor nerve outflow, which causes shivering, or through ANS outflow, which causes sweating.

Physical control of heat loss. ■ Heat is lost from the body by three channels: the skin, lungs, and excretions. Of these, the most important is the skin, the relative amount lost by this organ being placed at about 85%, although this depends on external and internal conditions.* Heat is lost from the skin by the following:

*The amount of heat given off by an average-sized resting man is about equivalent to that of a 60-watt electric bulb.

1. By *radiation* when the temperature of the surrounding objects (such as walls and furniture) is lower than that of the body, the heat passing from the body by invisible infrared rays having a wavelength of more than 760 mμ (Fig. 24-5)
2. By *convection* when the air, having a lower temperature than the body, is warmed by air currents from the body
3. By *vaporization* of water from the skin, because the evaporation of any fluid absorbs heat from the surrounding objects and air
4. By *conduction,* as in coming in direct contact with a cold object; generally the least important of the methods of heat loss

The external factors that determine the amount of heat lost are the temperature and humidity of the air, the velocity of air currents, and the temperature of the surrounding objects. The higher the external temperature, the less heat is lost by radiation; at 95 F this practically comes to a standstill, and nearly all heat is lost by vaporization. In this case the loss of heat by the evaporation of water from the skin becomes of vital importance. Although only 100 Cal are required to raise the temperature of 1 liter of water from 0 to 100 C, to evaporate this quantity of water, having a temperature of 33 C, 580 Cal must be applied. The rate of evaporation varies, under otherwise equal conditions, inversely as the relative humidity (RH) of the air. By relative humidity is meant the ratio of the actual amount of moisture in the air to the greatest amount it can hold at a specified temperature. When at a certain temperature the air is saturated with moisture and, therefore, can take up no more, the relative humidity is 100%; all evaporation ceases. If the temperature is raised, the relative humidity is lowered; for example, when air at 20 F has a relative humidity of 50%, warming it to 70 F reduces the relative humidity to 10%. In addition to influencing the amount and rapidity of evaporation from the skin, the relative humidity also affects the loss of heat by conduction; the greater the amount of moisture in the air, the greater is its heat conductivity.

The regulation of the loss of body heat is, to a certain extent, voluntary; for by artificially sheltering the body (e.g., clothing and housing), by warming the food and drink taken into the body, and by warming the air with which the body comes in contact, the body is protected against a too great loss. Without such voluntary regulation, human life in very cold climates would be impossible.

Fig. 27-1 is a wind chill chart that shows the equivalent still air temperature for different wind velocities. For example, at a thermometer reading of 0 F and a wind velocity of 10 m.p.h., an exposed individual is subjected to a cooling environment equivalent to −22 F. The zigzag vertical divisions in the table separate zones of increasing danger to exposed flesh. For equivalent temperatures on the left side of the chart there is little danger to a properly clothed person. In contrast, at equivalent temperatures on the far right of the chart there is extreme danger from freezing of exposed flesh.

The involuntary regulation of body heat is brought about by the vasomotor mechanism and by perspiration. The cutaneous vasodilation on a warm day and constriction in cooler weather are familiar to all. By the former, warm blood from the interior is sent to the surface to be cooled

Table 27-1. Heat production in a short-haired dog at various external temperatures

Air temperature (C)	*Calories per kilogram*	
35	68.5	Rise in body temperature
30	56.2	Physical regulation
25	54.2	
20	55.9	
15	63.0	Chemical regulation
7.6	86.4	

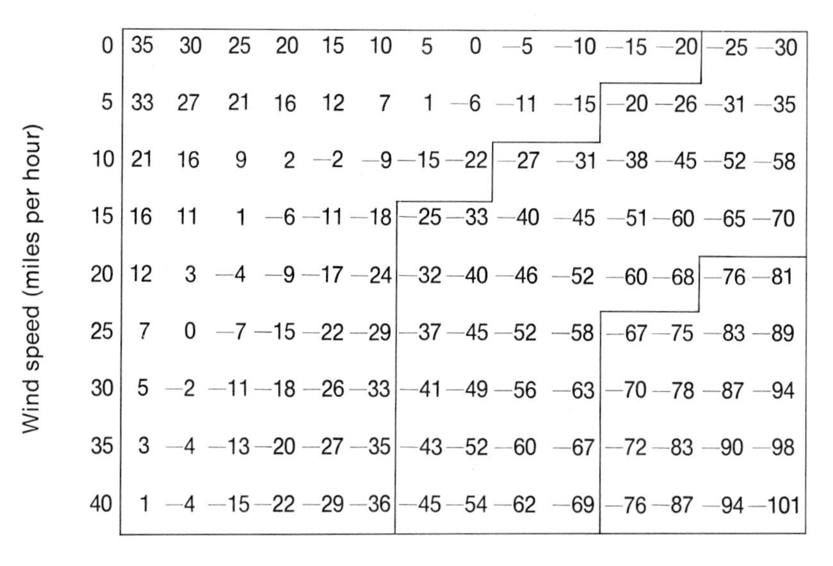

Equivalent temperatures (°F)

Wind speed (miles per hour)														
0	35	30	25	20	15	10	5	0	−5	−10	−15	−20	−25	−30
5	33	27	21	16	12	7	1	−6	−11	−15	−20	−26	−31	−35
10	21	16	9	2	−2	−9	−15	−22	−27	−31	−38	−45	−52	−58
15	16	11	1	−6	−11	−18	−25	−33	−40	−45	−51	−60	−65	−70
20	12	3	−4	−9	−17	−24	−32	−40	−46	−52	−60	−68	−76	−81
25	7	0	−7	−15	−22	−29	−37	−45	−52	−58	−67	−75	−83	−89
30	5	−2	−11	−18	−26	−33	−41	−49	−56	−63	−70	−78	−87	−94
35	3	−4	−13	−20	−27	−35	−43	−52	−60	−67	−72	−83	−90	−98
40	1	−4	−15	−22	−29	−36	−45	−54	−62	−69	−76	−87	−94	−101

Fig. 27-1
Wind chill chart.

and as much as 12% of the entire cardiac output may pass through the skin.

In a resting, thinly clad person, the vasomotor mechanism is solely operative over a range of external temperatures from 25 to 29 C (77 to 85 F). Below 25 C the body temperature drops, and to maintain it heat production must be increased (Table 27-1). At 29 C sweating (and therefore cooling by evaporation) begins. As the temperature rises above 29 C, the vasomotor mechanism becomes progressively less and the sweating mechanism more important. At an external temperature slightly above 35 C, the body cannot rid itself of heat and heat prostration occurs. The importance of sweat evaporation is well illustrated by the following: A person devoid of sweat glands staying for thirty minutes in a room having a temperature of 109 F experienced a rise in body temperature to 102 F; in a normal person the body temperature remained constant.

The range of physical regulation in man extends from about 25 to 35 C (77 to 95 F), depending largely on the amount of fat, the nature of the clothing, and the nature of the environment. Within this range the loss of heat is the least and the oxidations are at a minimum; it is the *physiologically economical range.* The minimum temperature suitable for resting men is stated at about 65 F (18 C), but a more comfortable temperature is nearer 75 F. For physical activity a range from 40 to 65 F is generally preferred.

In young children the mechanism for controlling the loss of heat is poorly developed, and hence their body temperature is more likely to undergo variations. Although in an adult muscular work affects the body temperature little, a child may develop a fever by a fit of crying. A pathological condition that causes a very slight fever in an adult produces a much more pronounced rise in the temperature of a child. In elderly people, not only may the rate of heat production be on the wane but also the temperature-regulating machinery may be somewhat impaired.

Chemical control of heat production. ■ Practically all the heat in the human body is derived from the oxidation of foodstuffs. The amount taken in with warm food and drink is almost negligible. The amount of heat produced varies with circumstances, which were discussed in Chapter 25. Oxidations and, therefore, heat production take place in the tissues themselves. Every tissue contributes to this, but since the skeletal muscles constitute about one half of the active structures of the body, they furnish the largest amount. Glands are of minor importance. Increasing heat production is, therefore, achieved principally by increasing the activity of muscles.

When a cold-blooded animal, for example, a frog, is subjected to a cold environment, the amount of metabolism gradually decreases. The heat production is correspondingly reduced; this results in a fall of body temperature. The poikilotherm follows the rule that chemical action varies directly as the temperature of the reacting agents.

In warm-blooded animals matters are more complicated, for in them both the amount of heat produced and that lost can, within limits, be regulated. In Table 27-1 are indicated the Calories of heat produced by a short-haired dog at various external temperatures. A fall in external temperature from 30 to 25 C did not increase heat production; the animal was able to maintain its body temperature by merely decreasing the amount of heat lost, that is, by physical regulation. However, when the air temperature was lowered to 15 and still further to 7.6 C, the physical regulators were unable to keep the body temperature normal; at this point it was necessary to step up heat production. It will be noticed that at an air temperature of 35 C the animal was not able to lose a sufficient amount of heat to prevent a rise in body temperature.

The onset of chemical regulation is ushered in by an increase in muscle tension, shivering, chattering of teeth, and goose pimples. Shivering (which someone has called involuntary exercise) usually begins when the skin temperature has dropped to approximately 19 C (66 F) and may increase oxidations by as much as 400%.

Heat-regulating center. ■ How is the balance between heat production and heat loss maintained? There are many neural centers in the spinal cord and the lower portions of the brain concerned either with the production or with the loss of heat. These several centers govern such activities as vasoconstriction, vasodilation, sweating, muscle tone, and shivering. To maintain a nearly constant body temperature, it is necessary to integrate these various activities properly. This is the function of the hypothalamus. When the hypothalamus is separated (by a section) from the lower part of the brain and the spinal cord (in which are found the subcenters just mentioned), the animal is unable to regulate its body temperature.

The thermostatic control exercised by the hypothalamus is twofold.

1. A center located in the anterior part controls the loss of heat and thereby prevents overheating of the body. On its destruction, experimental or pathological, the animal or person behaves normally in a cold environment, but on exposure to heat the usual methods of losing heat are inoperative and the body temperature rises. On the other hand, heating the anterior center sets the mechanisms for thermolysis into action—the animal pants, and sweating of the pads of the feet and vasodilation occur; this results in a fall of body temperature of several degrees.

2. The posterior center governs heat production and thereby prevents chilling of the body. After its destruction, exposure of the animal to cold does not increase metabolism or heart rate; heat production lags, and the body temperature falls.

The hypothalamic heat-regulating centers are influenced two ways—reflexly from the skin and by the temperature of the blood flowing through them. The most important factor initiating heat loss through vasodilation, sweating, and panting is the elevated temperature of the blood perfusing the hypothalamus. Recent evidence indicates that thermosensitive neurons of the hypothalamus are influenced also by the sodium-to-calcium ratio of the perfusate. Excess sodium causes the body temperature to rise, and excess calcium has the opposite effect. How the blood temperature and the sodium-to-calcium ratio are related is unknown. The primary mechanism initiating heat conservation through vasoconstriction, heat production, and shivering is not a settled point. It is probable that both surface and deep temperatures are important stimuli that bring about the reflex responses required.

Hormonal control. ■ At least two endocrine glands participate in the regulation of either heat production or heat loss, the thyroid and the adrenal medulla (Chapter 28). Removal or hypofunction of the thyroid lowers the body temperature; this is in agreement with the control exercised by the thyroid over basal metabolism and basal heat production. It is claimed that thyroid secretion is increased by cold; this is a slow process but of long duration and may be responsible, in part, for the greater heat production in winter over that in summer. The part played by the thyroid depends on the interrelationship between it, the hypothalamus, and the anterior pituitary.

Epinephrine increases the basal metabolic rate and therefore augments heat production. Since it also induces great constriction of the cutaneous blood vessels, it is of value in conservation of body heat. Its activity, however, is of short duration. Whether in man epinephrine or the thyroid hormone increases heat production in muscles in the absence of stimulation is a question.

■ DISTURBANCES IN THERMOREGULATION

Heat cramp. ■ When the humidity is zero, surprisingly high temperatures can be borne with impunity for a considerable length of time. However, the profuse sweating occurring in such an environment may dehydrate the body to an extent imperiling circulation. If this is prevented by copious drinking of water, the loss of NaCl in the sweat may induce heat cramp. In this condition the body temperature is not elevated; the main disturbance lies in exceedingly painful and, perhaps, spasmodic contractions of muscles, especially of those that had been employed in work. The addition of 0.2% NaCl to the drinking water prevents the occurrence of heat cramp.

Heatstroke or sunstroke. ■ A high degree of external temperature coupled with high humidity renders it difficult for the body to lose heat by either radiation or vaporization; sunstroke may then occur. Sunstroke is a failure of the heat-regulating mechanism. The outstanding features are cessation of sweating (dry, hot skin) and a very sharp rise in body temperature. Pulse rate and blood pressure are above normal. The person is generally unconscious, with reflexes greatly in abeyance or lacking altogether; he may be delirious or in convulsions. The body temperature may rise to 110 F. At this temperature the brain cells are quickly affected and irreparably destroyed, unless the temperature is speedily reduced by ice packs and cold baths.

Heat exhaustion or heat prostration. ■ In this condition the body temperature may be normal or even a trifle below normal, and, in sharp contrast to heatstroke, the skin is cool and moist (clammy). It is, therefore, primarily not a question of temperature regulation. In heat exhaustion the cardiovascular mechanism is at fault, as is indicated by a low blood pressure and a very rapid, weak, and soft pulse. Elderly people and those in enfeebled health are more prone to suffer heat exhaustion. Generally there are premonitory signs. Complete rest (with protection against further loss of heat, if needed) is generally sufficient for recovery.

Fever. ■ Many pathological disturbances are accompanied by a rise in body temperature, a condition known as fever. This rise in temperature may be due to a disturbance in the regulation of heat production (more heat being formed than can be lost by the normal channels), or the regulation of the loss of heat is upset. Most investigators believe that, although in fever there may be a slight increase in the production of heat, this is too small to account for a rise of several degrees in body temperature. For example, it has been found that the greatest increase in heat production during fever was only 20% above normal. We have learned that during muscular work there may be an increase of 100% to 300% in heat production with only a very slight rise in body temperature; it is, therefore, more likely that fever is generally not due to a disturbance in heat production. There is a disturbance in heat dissipation due to the marked constriction of the cutaneous blood vessels. The decreased flow of blood to the skin chills the individual, who, notwithstanding a rise in his temperature (fever), feels cold and shivers violently. There is an absence of sweating.

The cause of fever almost always is attributable to a disturbance in the regulation of the body temperature by the hypothalamus. The "set point" around which the body temperature is usually maintained is elevated, say from the usual 98 F to 102 F, during the onset of fever. When this new thermal level is reached throughout the body, the center functions as before in creating a balance between the production and loss of body heat. The actual mechanism used to raise body temperature depends upon the environmental temperature. In a neutral environment vasoconstriction occurs, whereas in a hot environment the elevation of body temperature can be achieved simply by cessation of sweating.

In heat stroke and in brain lesions the febrile response appears to result from a malfunction of the hypothalamic thermoregulatory centers. Many fevers, however, are due to the release of an endogenous pyrogen from circulating leukocytes, the monocytes (Chapter 12) being the major source. The endogenous pyrogen that is synthesized in the leukocytes is a protein of 10,000 to 20,000 molecular weight and contains small amounts of carbohydrate and lipid. Its synthesis is stimulated by the presence of infectious agents such as microbes, as well as by noninfectious antigen-antibody reactions and inflammatory responses such as sickle cell disease, gout, and myocardial infarction. The set point of the hypothalamus is raised by the presence of endogenous pyrogen in the circulation. Certain prostaglandins are known to induce fever, and it has been suggested that they may be the molecular transmitters for the effect of endogenous pyrogens in the hypothalamus. Aspirin (salicylate) lowers the set point of the hypothalamic thermoregulatory center, and its antipyretic action may be related to suppression of the pyrogenic prostaglandins. On recovery of the individual, the thermoregulatory center acquires its former, normal setting.

Nothing is known presently about the ultimate disposition of endogenous pyrogen. It may be removed by the reticuloendothelial system, metabolized to inactive products, or excreted intact. It is likewise difficult to specify what purpose fever serves. Raised body temperature most certainly does not destroy infectious agents directly. Fever must confer some advantage to the host's defense mechanism.

READINGS

Atkins, E., and Bodel, P.: Fever, New Eng. J. Med. **286**:27-34, 1972.

Benziger, T. H.: The human thermostat, Sci. Am. **204**(1):134-147, Jan. 1961.

Carlson, L. D., and Hsieh, A. C. L.: Control of energy exchange, New York, 1970, The Macmillan Company, pp. 42-94.

Hardy, J. D., Stowijk, J. A. J., and Gagge, A. P.: Man. In Whittow, C. C., editor: Comparative physiology of thermoregulation, vol. 2, New York, 1971, Academic Press, Inc., pp. 327-380.

Meyers, R. D., and Veale, W. L.: Body temperature; possible ionic mechanism in the hypothalamus controlling the set point, Science **170**:95-97, 1970.

HORMONES OR INTERNAL SECRETIONS

■ INTRODUCTION

The importance of hormones has been stressed on many occasions. In this chapter we shall hopefully, without further reference to the gastrointestinal hormones, present this subject in a more unified manner, although well aware that the subject is often complex and not fully understood by the most competent investigators.

Internal versus external secretion. ■ Similar to the glands of external secretion, the ductless glands are composed of epithelial cells. In some of these glands the epithelial cells are arranged in the shape of a sac or vesicle; the whole structure is supported by connective tissue and surrounded by a network of capillaries (Fig. 28-1, *A*). In other glands the epithelial cells are in small solid masses without any vesicles, as shown in Fig. 28-1, *B*. The secretion manufactured or stored by the epithelial cells is poured directly into the blood or lymph; for this reason these glands are called glands of internal secretion or *endocrine glands.*

Chemistry of hormones. ■ Hormones do not form any special chemical group of compounds. Some are steroids (e.g., hormones of adrenal cortex), others are polypeptides (e.g., those of the posterior pituitary), and some are closely related to amino acids (e.g., epinephrine of the adrenal medulla). They are generally specific in origin, and they are also specific in their activity or function.

Functions of hormones. ■ Hormones are chemical agencies that severally perform specific functions. Their activities may be classified under the headings metabolic, morphogenic, mental, or personality.

Metabolic. In that the hormone either stimulates or retards the rate of chemical changes befalling the various ingredients in the cell (e.g., insulin influence on carbohydrate metabolism), this, naturally, implies a stimulation or retardation of the function of the organ concerned.

At the cellular or molecular level hormones may exert their metabolic effect in a variety of ways, none of which is fully understood and most of which are subject to continuing specula-

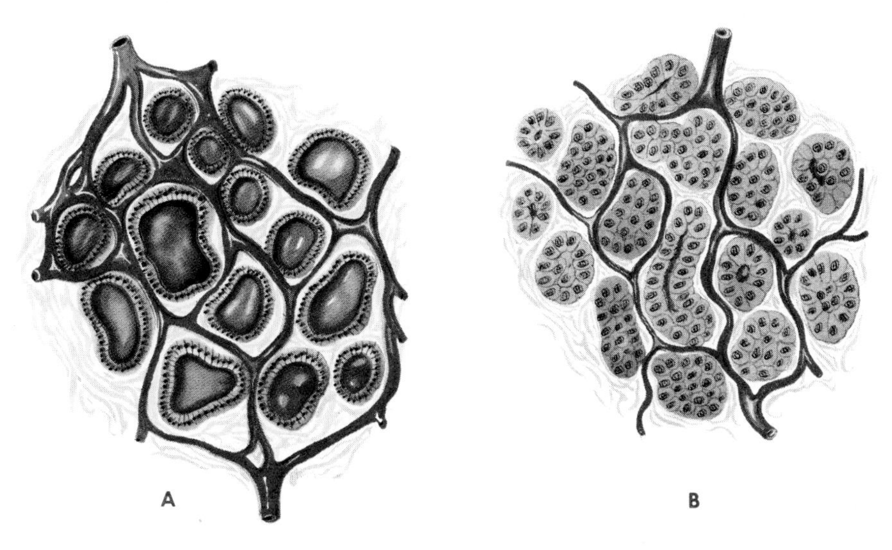

Fig. 28-1
Illustration of two types of endocrine glands. **A,** Gland is composed of irregular sacs (follicles) surrounded by connective tissue and blood vessels; this type includes the thyroid and ovaries. **B,** Gland in which epithelial cells are in clumps surrounded by dense capillary networks; this type includes the islets of Langerhans.

tion and investigation. If we accept the premise that hormones accelerate or retard rate-limiting enzyme reactions, then it is possible that a hormone may (1) act directly on the apoenzyme or its precursor; (2) control the availability of substrate by setting its rate of transfer into the cell or its rate of release from a bound intracellular storage form; (3) perform as, or regulate the availability of, coenzyme or cofactor; and (4) determine the rate of the reaction by adjusting the rate of removal of reaction end products which might bring about inhibition due to their accumulation (Chapter 22).

Morphogenic. Hormones influence the growth and development of the body as a whole or a part (e.g., the thyroid, pituitary, and gonads).

Mental. This aspect is clearly demonstrated in hypothyroidism, which results in mental sluggishness, torpor, and somnolence. The interrelationship of the central nervous system and endocrine glands is obvious.

Personality. Insofar as hormones influence the manner in which an individual responds to daily situations (stimuli), some hormones may be said to take a part in shaping an individual's personality.

Neural versus hormonal control. ■ The two systems of organ intercommunication, the neural and the hormonal, differ in two or three important respects. The action mediated by the nervous system is generally localized, specific, and speedily executed; that following the production of a hormone is vastly slower, frequently more widespread throughout the body, and, because of the continued presence of the hormone, of a more lasting nature. To illustrate: the result of the stimulation of a certain vasoconstrictor nerve may affect the blood vessels in a very circumscribed area and may disappear the moment the stimulation ceases. But in a vasoconstriction brought about by the hormone vasopressin (ADH), there occurs a relatively long latent period in the formation of the hormone and its absorption into, and distribution by, the blood; its wide and indiscriminate distribution causes a multitude of vessels to constrict and to remain constricted until the hormone is either removed or destroyed. The hormonal regulation is, therefore, more primitive and confined to such activities as may or must be continued for a considerable length of time, as in the flow of pancreatic juice brought about by secretin, and in the growth of the skeleton. The degree of interchangeability of hormones from one animal to another attests to the phylogenetic age of the elaborating organs.

Endocrine balance. ■ The number of hormones known at present is large. Some of the endocrine glands are anatomically composed of two distinct structures, each supplying its own hormone or hormones, for example, the adrenal glands, gonads, and pituitary. In not a few instances the hormones secreted by one gland stimulate or inhibit the formation of hormones by other glands. This is well illustrated by the thyroid-stimulating hormone of the anterior pituitary, which stimulates production and secretion of the thyroid hormone and is itself inhibited by *feedback* of thyroid hormone. Hardly any part of the body is exempt from their influence. In their activities the various hormones are intricately interrelated; for this there is evidence in the part played by the hormones of the pancreas, pituitary, thyroid, and adrenal cortex in carbohydrate metabolism. Either hyposecretion or hypersecretion of a hormone may cause physical and mental disturbances. The aforementioned facts, coupled with the extreme potency of the hormones, render it necessary that the amount of each individual hormone be carefully regulated and that this amount be in proper proportion to that of the other hormones. The mechanism of this regulation is not fully understood. In a normal animal the secretions are formed only as they are needed. Increasing the sugar content of the blood evokes an increase in the amount of insulin. The amount of calcium in the blood determines the activity of the parathyroid. The external temperature may affect the production of thyroxin. The pituitary gland governs the activity of a number of endocrine glands and is itself regulated by these other glands by way of the central nervous system. Except perhaps in one or two instances (e.g., the adrenal medulla), the central nervous system exercises no direct control over these glands but an indirect control through the trophic hormones of the pituitary. In fact, most of them have no secretory nerve supply; for this reason they can be transplanted to other parts of the body.

Potency. ■ The potency of hormones borders on the unbelievable; for example, the presence of one of the pituitary hormones can be detected by its activity when one part of the hormone is dissolved in 20 billion parts of solvent. In this respect, hormones, enzymes, and vitamins resemble each other. Several hormones have been obtained in crystalline form and some have been made synthetically.

Methods of investigation. ■ The functions of an endocrine gland are studied by many methods. A suspected or known hormone-producing gland in an animal can be extirpated and the resulting deficiencies observed. In the case of the pituitary gland, selective lesioning or stimulation of the brain tissue immediately above the gland (the hypothalamus) can be undertaken. Furthermore, the pituitary itself can be transplanted to another site, for example, to the anterior chamber of the eye. Some endocrine glands can be blocked effectively by the administration of chemical agents, for example, the pancreas by alloxan and the thyroid by thiouracil. Crude extracts or purified preparations of an endocrine gland product can be administered to either normal animals or animals previously deprived of the gland in order to study the effects of hyperfunctioning of the gland or to assay the potency of the hormone. Such studies in the intact animal sometimes employ as indices of hormonal activity measurements of respiratory gases, determinations of the chemical composition of body fluids (blood, urine, sweat, or tears), and balance studies—intake vs. excretion—of specific nutrients. Valuable information about the action of hormones is derived from *in vitro* studies of tissue homogenates, slices, and cultures. Microscopic examination of target organ cells often reveals histological changes of importance. Finally, the role of a hormone can be investigated in man by noting the disturbances produced following hypofunctioning or hyperfunctioning of the endocrine gland.

Importance of hormones. ■ Few other disturbances in the body are followed by such grave results as those caused by the malfunctioning of the endocrine organs. Not only physical life, but in some instances, mental and emotional life also may be disastrously affected. Fortunately for afflicted persons the hormones are, as a rule, interchangeable from one animal to another. Thus, the severe effects of deficiency of thyroid secretion in man can, to a large extent, be abolished by the administration of desiccated thyroid gland or its extract taken from some animal (sheep or cow). Such treatment, known as replacement therapy, is not a cure, merely a palliative; *no gland is ever stimulated by its own hormone.* The storage of hormones in the body is of minor importance.

■ PITUITARY GLAND

The pituitary gland, or hypophysis cerebri (Figs. 10-2 and 10-8), is a small body (weight, 0.5 g) at the base of the brain, which is lodged in a depression of the sphenoid bone (sella turcica, Fig. 17-1) just back of the optic chiasma. It is connected with the brain by means of a short stalk, the infundibulum. The pituitary (Fig. 28-2) is composed of three parts: an anterior lobe (the *adenohypophysis*), a posterior lobe (the *neurohypophysis*) and the *pars intermedia.* Extirpation of this gland is not necessarily fatal, but the animal requires extreme care and protection. When the pituitary gland is removed *(hypophysectomy)* by design or accident, there is a marked slowing of growth, adrenal atrophy, secondary hypothyroidism, and failure of gonadal function.

■ Anterior lobe or adenohypophysis

The adenohypophysis originates embryologically as an outpocketing of the nonneural primitive foregut. Although it arises from a different precursor than the central nervous system, it does not function in splendid isolation from the latter, for it is intimately related to the hypothalamus by means of an important vascular connection, the *hypophyseal portal system,* which begins as a capillary plexus in the hypothalamus and ends in the sinusoids of the adenohypophysis. The direction of flow in these vessels is from the hypothalamus to the adenohypophysis. Two major classes of cells occur in the anterior lobe, acidophils and basophils. These can be further

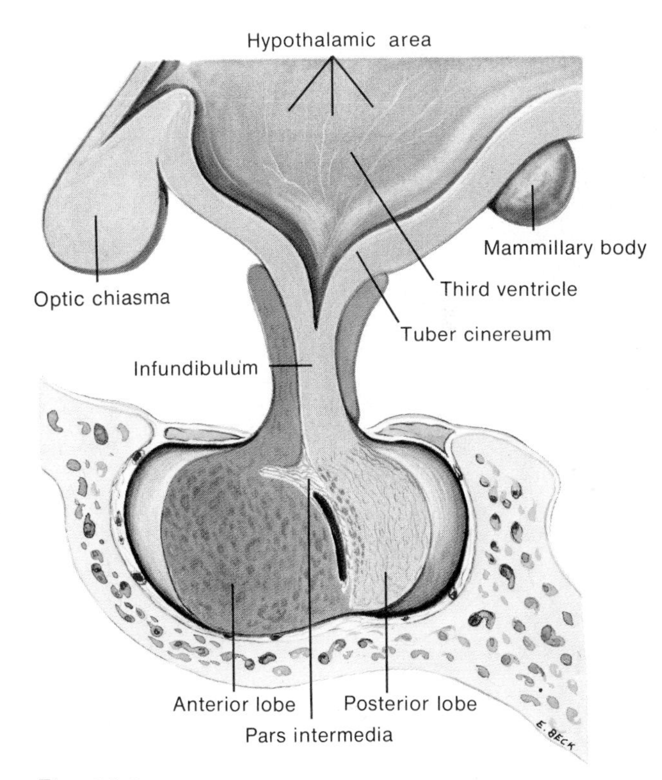

Fig. 28-2
Illustration showing the three lobes of the pituitary and the relationship of the gland to the hypothalamus.

subdivided on the basis of selective staining, and it is thought that individual cell types specialize in production of a single hormone.

The six recognized hormones secreted by the adenohypophysis, with one exception, are *trophic hormones;* they have as their target organ another endocrine gland. Three of these are spoken of as gonadotrophins—*follicle-stimulating hormone* (FSH), *luteinizing hormone* (LH) in the female or *interstitial cell-stimulating hormone* (ICSH) in the male, and *prolactin.* The two other trophic hormones are *adrenocorticotrophic hormone* (ACTH) and *thyroid-stimulating hormone* (TSH). The sixth hormone, *somatotrophic hormone* (STH) or *growth hormone* (GH), is not strictly speaking a trophic hormone since it does not require another endocrine gland as a target but produces its widespread effects by direct action.

Transection of the pituitary stalk may not seriously or permanently interrupt adenohypophyseal function provided no barrier to revascularization by the portal vessels is interposed. This observation constitutes one piece of evidence for the belief that neurosecretory cells in the hypothalamus secrete into the portal system several chemical messengers which, after passing down into the sinusoids of the adenohypophysis, stimulate the various hormone-producing cells of this gland to discharge their products. The chemical nature of these neurohumoral-*releasing factors* is not fully known, but

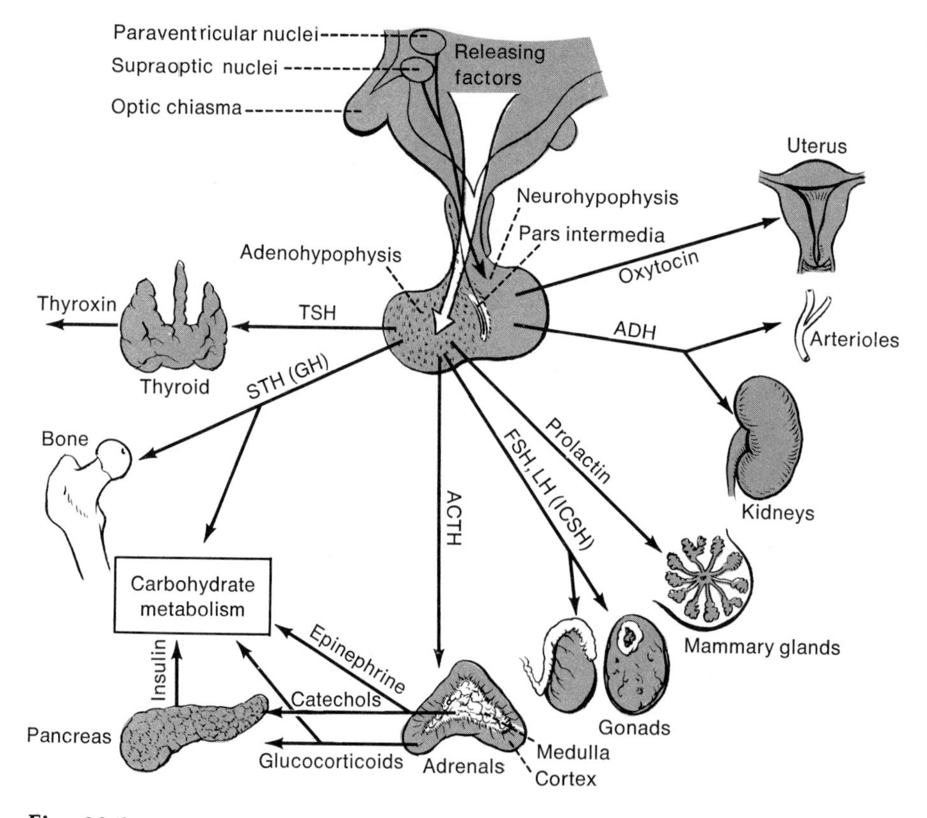

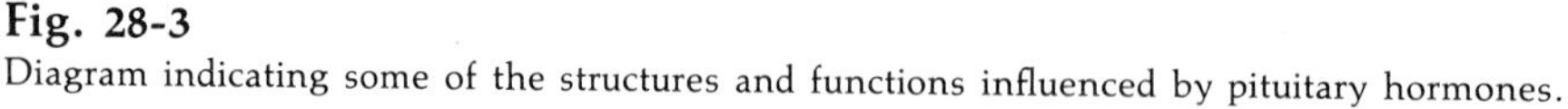

Fig. 28-3
Diagram indicating some of the structures and functions influenced by pituitary hormones.

they seem to be small polypeptides. One such substance has been obtained in a highly purified form and is referred to as *corticotrophin-releasing factor* or CRF. There is probably a specific releasing factor for each hormone of the anterior lobe, with the exception of prolactin. The cells that produce the latter hormone seem to be inhibited by a hypothalamic secretion. Many of the facts discussed so far are summarized in Fig. 28-3. Secretion of trophic hormones is generally regulated by a negative feedback mechanism; that is, the response or output acts as a signal to reinforce (positive feedback) or retard (negative feedback) the original stimulus or input. Thus, the level of glucocorticoids in the plasma determines the further output of ACTH.

Somatotrophic (STH) or growth hormone. ■ The term *growth** is difficult to define. It is an exceedingly complex and highly integrated process, as is evidenced by the proper balance maintained during the period of growth between the various bodily organs and between the various tissues constituting the individual organs. In this process many factors, both external and internal, are involved; among these are nutrition, heredity, and hormones.

Growth hormone is a protein, a globulin, produced by certain acidophilic cells of the adenohypophysis. The molecular weight of bovine STH is about 45,000 and that of primates is about 28,000. Although the hormone obtained from primate pituitary glands is effective in other species as well, that derived from cattle, sheep, and pigs is not effective in primates.

Growth in the human body occurs during two fairly distinct periods: prepuberty and puberty. Most rapid growth occurs during the first year after birth; it then gradually slows down until the onset of puberty. Generally a marked increase takes place during adolescence (the period between puberty and maturity). In the main, growth of the body is characterized by an increase in the length of the long bones and an increase in the amount of soft tissue (e.g., muscle, viscera, and skin). The former is due, in part, to increased deposition of minerals (calcium and phosphorus); the latter is due to the retention of protein.

Functions of growth hormone. The effects of growth hormone are particularly striking in bone, muscle, kidney, adipose tissue, and liver. It appears to facilitate the transfer of amino acids from the blood into muscle cells, which would account for its marked influence on nitrogen retention in the intact animal. Injection of STH into a young rat results in the storing of proteins (as shown by a high positive nitrogen balance), a diminution of urinary nitrogen output, and a decrease in the amino acid content of the blood. In the hypophysectomized animal, the kidneys show functional defects, which are corrected by administration of growth hormone, along with a small amount of thyroid hormone. Free fatty acids are liberated from adipose tissue by treatment with STH. The effects of insulin in promoting fatty acid synthesis from glucose are antagonized by growth hormone. Animals treated with STH tend to develop fatty livers. Hypophysectomy causes a reduction in the size of liver cells and their RNA content; administration of STH restores the cells and stimulates cell division.

Gluconeogenesis, insulin insensitivity with decreased peripheral utilization of glucose, hyperglycemia, and an increased total oxygen consumption with a depressed respiratory quotient (indicative of fat metabolism) are characteristic effects of excess growth hormone. These responses are similar to the effects of starvation and are *diabetogenic*. Indeed, diabetes mellitus can be produced by administration of pituitary extracts, which probably contain ACTH as well as STH, providing insulin is not given simultaneously.

*Growth consists of an increase in the size of a cell or an increase in the number of cells. Differentiation implies the transformation of cells having the same characteristics into two or more classes of cells, which differ in structure and function.

Growth of the skeleton. The middle portion or shaft of a long bone is known as the diaphysis and the ends as epiphyses. In the young animal or child the epiphysis is separated from the diaphysis by a disc of cartilage. In growth both the cartilage and the bone tissue increase. Toward the end of the growth period, the growth of the cartilage becomes progressively less and that of the bone tissue continues. Finally, the cartilage disc disappears and the diaphysis and the epiphysis join; this is known as the *closure of the epiphysis* and marks the end of bone growth (in length). The injection of growth hormone in the immature animal is followed by an increase in the inorganic phosphorus and of phosphatase (an indispensable enzyme in growth of both the cartilage and bone tissue) and it stimulates mitotic activity at the epiphyses. STH increases the length of the bone and keeps the epiphysis open, making still further growth possible. Excessive formation of growth hormone may therefore lead to gigantism. For example, a boy of 13 years of age, having a hyperactive pituitary gland, measured 7 feet and 1 inch and weighed 278 pounds. In hypoactivity of the pituitary, the epiphyses are prematurely closed (disappearance of the cartilage disc) and growth is stunted—dwarfism. In some forms of hypopituitarism, the dwarf is mentally retarded as well.

In discussing growth, it should be mentioned that the hormones secreted by the gonads (both male and female) have two distinct and somewhat contradictory effects on body growth. By stimulating the growth of bone tissue and inhibiting the growth of cartilage, they set a limit to the length of long bones—closure of the epiphyses. Dwarfism may therefore result from two powerful gonadal influences, as well as from a reduction in the production of somatotrophin. On the other hand, gonadal hormones favor the increase in thickness of already formed bones and thereby augment their tensile strength. They also stimulate the growth of skeletal muscles.

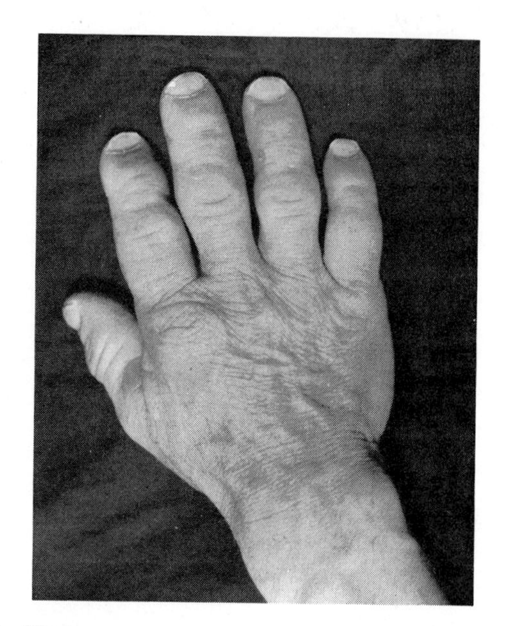

Fig. 28-4
Hand showing characteristics of acromegalic condition.

In the adult in whom ossification of the bones is completed, hyperactivity of the pituitary causes the bones to become misshapen. The face acquires a coarse appearance, with beetling brows, a protruding lower jaw, and widely spaced teeth. Due to the excessive growth of fibrous tissue, the eyelids, nose, tongue, and lips become thick and swollen. The hands and feet are much enlarged (Fig. 28-4). This condition is known as *acromegaly.**

FSH, LH (ICSH), and prolactin. ■ The gonadotrophic hormones stimulate the gonads or sex glands—the testes and ovaries. These glands undergo atrophy when the hypophysis is destroyed or removed. Although the effects of these hormones will be discussed again in Chapter 30, a brief outline of their actions is warranted at this time. FSH functions to maintain the spermato-

*Acromegaly—ak″-ro-meg′-al-e (*akros*, extremity; *megale*, large). The bulldog is an acromegalic, the Great Dane is a giant, and the Pekinese is a dwarf, fixed by heredity.

genic epithelium in the male and promotes the early growth of ovarian follicles in the female. The final maturation of ovarian follicles (with estrogen production) and ovulation (with progesterone production) is under the influence of LH. In males, LH (renamed ICSH for its function) stimulates the interstitial cells (Leydig cells) of the testes and induces the secretion of the male hormone testosterone. Prolactin is known as *luteotrophic* hormone (LTH) in some species in which it maintains the corpus luteum of pregnancy (Chapter 30). Until recently, it was an unsettled question whether prolactin and STH were separate hormones in humans. A variety of biological, immunological, and chemical studies now have established prolactin as a separate entity responsible for lactogenesis. Indeed, a second human lactogenic hormone—*human placental lactogen* (HPL)—also has been identified recently. Prolactin is synthesized in the adenohypophysis primarily during pregnancy and the postparturition period. Normally, prolactin synthesis is under tonic inhibition by a hypothalamic prolactin inhibitory factor (PIF). The latter is probably a small polypeptide similar to those known to be releasing factors for other pituitary hormones. A variety of chemical, cerebral, visual, tactile, and oral stimuli (suckling) inhibit the production of PIF and thereby promote prolactin synthesis and lactogenesis. Whether there is a separate prolactin releasing factor is undetermined. Gonadotrophic hormones are glucose-containing proteins of 25,000 to 30,000 molecular weight. The LH and FSH are secreted by basophilic cells; prolactin (LTH) is secreted by acidophilic cells. A negative feedback between the gonads and the adenohypophysis operates to stabilize the production of gonadotrophins.

Thyrotrophic hormone (TSH). ■ The marked influence of the pituitary on the development and activity of the thyroid is seen in the degeneration of this gland when the pituitary is extirpated and in its hypertrophy resulting from injection of pituitary extract. TSH controls the rate of iodine uptake by the thyroid gland and the synthesis and release of thyroxin. The feedback regulation appears to be a direct effect of thyroxin on the pituitary gland and is not relayed via the hypothalamus, as is the case with regulation of some other trophic hormones. TSH is apparently a glycoprotein of about 28,000 molecular weight, although it may, when purified, turn out to be a much smaller molecule. It is secreted by basophilic cells.

Adrenocorticotrophic hormone (ACTH). ■ The pituitary hormone controlling the glucocorticoid output of the adrenal cortex is ACTH. This hormone is now known to be a single-chain polypeptide containing thirty-nine amino acids, of which only twenty-three constitute the active core, which has been synthesized. It is normally secreted by basophilic cells and in accord with a diurnal rhythm; that is, the plasma level is highest early in the morning and lowest in the evening.

Glucocorticoid concentration in the blood produces a feedback regulation of pituitary ACTH secretion. The ACTH molecule is structurally similar to a skin pigment-regulating hormone (melanophore-stimulating hormone, MSH) found in fish, amphibia, reptiles, and man. After adrenalectomy in man, a hyperpigmentation occurs, which is probably due to overproduction of ACTH and possibly MSH. Hypopituitarism is characteristically accompanied by skin pallor. In mammals, MSH is found in all parts of the pituitary; in species possessing melanophores it occurs in the intermediate lobe, the *pars intermedia.* Moreover, in the latter species a compound issuing from the pineal gland, *melatonin,* antagonizes the action of MSH; that is, it aggregates melanophore granules and lightens the skin in response to light stimuli. By and large, mammalian MSH may be a vestigial hormone.

■ Posterior lobe or neurohypophysis

The posterior lobe, an outgrowth of the primitive brain, is considered to secrete three active principles—*vasopressin, antidiuretic hormone* (ADH), and *oxytocin* (Fig. 28-3). However vasopressin and ADH are *identical chemical substances,* and their treatment as separate principles is a pedagogical convenience for describing different functions.

Both vasopressin (ADH) and oxytocin are octapeptides of about 1,100 molecular weight. Their amino acid sequence is known and commercial synthesis has been undertaken. These hormones apparently are naturally synthesized in the hypothalamus, vasopressin in the *supraoptic nuclei* and oxytocin in the *paraventricular nuclei,* and migrate into the neurohypophysis along the nonmyelinated nerve fibers of the *hypothalamohypophyseal tract.* They are stored in the neurohypophysis until released into the blood by neural impulses carried in the same fibers. In the blood they are transported in loose association with plasma proteins. They are destroyed (inactivated) primarily in the kidney and liver by enzymatic reactions.

Vasopressor activity. ■ The hormone vasopressin in pharmacological amounts causes generalized vasoconstriction, even of the coronary and pulmonary vessels. Indeed, it has a stimulating action on almost all smooth muscle. Because of its constrictor effect on coronary vessels, its therapeutic use is avoided in persons afflicted with or suspected of having coronary disease. However, the only known physiological circumstance when vasopressin output is large enough to evoke the pressor response is hemorrhage. Under this circumstance, it may conceivably assist in maintaining an effective circulating blood volume.

Antidiuretic activity. ■ The antidiuretic effect is the only physiologically significant action of vasopressin (ADH). Removal of the neurohypophysis (or lesions of the neural tracts from the hypothalamic nuclei) results in the excretion of a large amount of urine—up to 20 liters/day *(polyuria)*—and in the intake of a large quantity of water *(polydipsia).* This condition is known as *diabetes insipidus.* Unless the animal has access to a sufficient supply of water, death may result from dehydration.* Diabetes insipidus is relieved by the administration of ADH. The increased urinary excretion that normally follows soon after drinking a large amount of water is delayed considerably by ADH. The amount of hormone normally secreted is in proportion to the need for it. When the body is dehydrated, more hormone is secreted and, consequently, less water is excreted; in the hydrated state of the body, less hormone is produced—an automatic regulatory process.

The normal stimulus for release of ADH appears to be the osmolar concentration of the plasma. Osmoreceptors in the hypothalamus, possibly the supraoptic and paraventricular nuclei themselves (Fig. 10-3), detect changes in the plasma osmolality as small as 2%. Increased osmotic pressure of the plasma stimulates secretion of ADH; decreased pressure inhibits secretion. Loss of blood volume by hemorrhage results in hemodilution which would not, of course, be a stimulus for ADH secretion. Nevertheless, significant amounts of vasopressin (ADH) are secreted in response to this threat to vascular volume homeostasis. It must be assumed that the pressure receptors (baroreceptors) in the vascular tree also are capable of stimulating production and release of the hormone via nerve impulses to the hypothalamus. ADH exerts its water-conserving effect primarily by increasing the permeability of the distal convoluted tubule and collecting duct cells of the kidney (Fig. 29-3). This water returns to the body, and hypertonic urine is excreted. Pain, trauma, acetylcholine, and nicotine are among the varied stimuli that can increase ADH secretion; alcohol inhibits ADH secretion.

Oxytocic activity. ■ Oxytocin has a pronounced stimulating effect on the smooth mus-

*There are cases on record in which an animal lost one sixth of its body weight within eight hours after removal of the posterior pituitary body.

culature of the uterus. This may assist in sperm transport and is of probable importance in parturition. It is, therefore, sometimes employed by physicians to augment the contractions of the uterine muscles during childbirth, although there is some evidence that parturition (the expulsion of the fetus from the uterus) is not prevented by lack of hormone. More certain is its function in "let down" of milk from the breasts. Whereas prolactin initiates milk formation, it is a neural stimulus (suckling) that elicits the secretion of oxytocin and consequent contraction of the mammary myoepithelial cells with consequent ejection of milk.

■ ADRENAL OR SUPRARENAL GLANDS

The adrenals (*ad*, near; *ren*, kidney) are two small yellowish masses of tissue lying above or near the kidneys (Fig. 29-2). Each adrenal gland weighs about 5 g. Their importance is indicated by their extreme vascularity. Histologically the gland is composed of two distinct parts: a cortical (outer) layer and a medullary (inner) portion (Fig. 28-5). These two parts differ in function and development. The medullary portion is derived from the embryonic tissue, which also gives rise to the sympathetic nervous system. The cortex originates from tissue from which the reproductive glands (gonads) are developed.

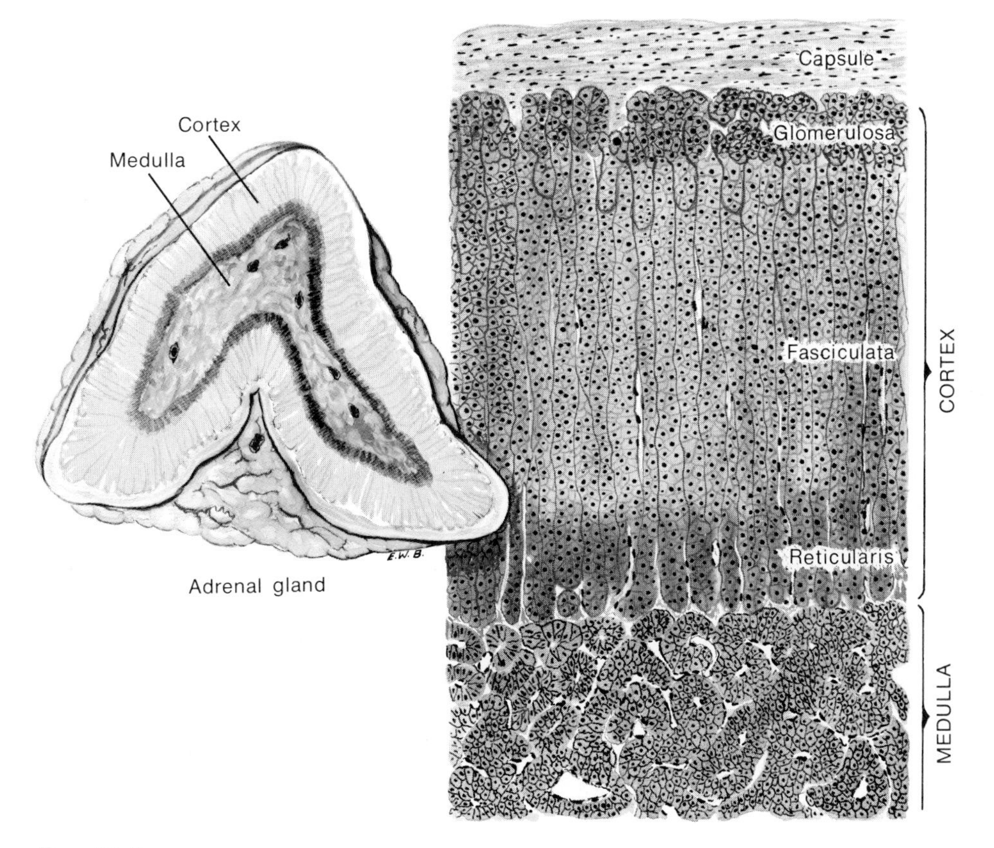

Fig. 28-5
Illustration showing structure of adrenal gland.

■ Adrenal medulla

Removal of the medullary portion of the adrenal gland is not fatal; neither does the administration of its active principle, known as *epinephrine* (*epi,* upon; *nephros,* kidney) or Adrenalin (the proprietary name for epinephrine), save the life of an animal from which the entire glands have been removed. The medulla may be considered as a specialized sympathetic ganglia innervated by the usual long preganglionic, cholinergic, autonomic nerve fiber. On stimulation, the neural signal is converted to a blood-borne message by the secretion of hormone. Medullary cells contain granules, which are the storage form of the endocrine products.

Catecholamines ■ Chemical analysis and synthesis have shown that what had been considered a single hormonal substance was in reality a mixture of two *catecholamines;* these are known as *epinephrine* and *norepinephrine.*

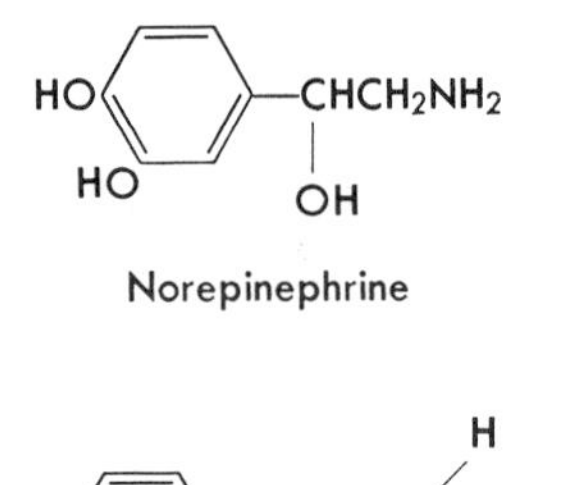

Norepinephrine

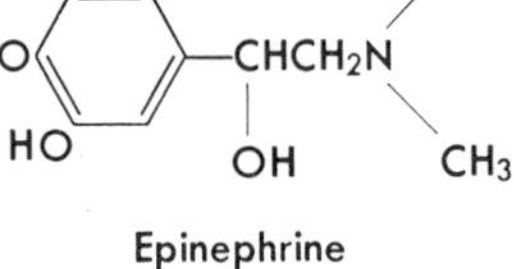

Epinephrine

These catechols are amino acid derivatives. They are synthesized in the cell from tyrosine by way of dihydroxyphenylalanine and its decarboxylation product dopamine (Chapter 23). The hormones are rapidly inactivated by the enzyme *O*-methyl transferase and possibly monoamine oxidase. The ratio of norepinephrine to epinephrine production varies among individuals and is probably determined hereditarily. Stimuli for release of catechols from the medulla include pain, cold, hypotension (hemorrhage), hypoglycemia, and hypoxia.

Epinephrine. Assay of this catechol is currently done by very sensitive fluorimetric methods (catechols produce a fluorescence in alkaline solution), but an interesting and important older technique was the observation of the vasoconstriction produced in the denervated rabbit ear (Fig. 15-1) by the injection of epinephrine. As little as 10^{-11} g/ml can be detected in this way. Injection of epinephrine hormone causes many changes throughout the body, some of which have been noted in our previous study. We may briefly restate some of these.

Cardiovascular effects. The rate and force of the heartbeat are increased, especially if the inhibitory nerves (vagi) are previously cut. The ventricles empty themselves more completely and the output is largely increased.

Epinephrine causes powerful contraction of many arterioles, especially in the skin, mucous membranes, and kidneys, but dilation of others, such as those of the coronary system, the skeletal muscles, and the lungs. As the constrictions are more pronounced than the dilations, the general blood pressure throughout the body rises (Fig. 15-6); it is the systolic pressure that increases; the diastolic is generally unaffected. Its rather rapid subsidence is due to the destruction of the hormone by enzymes.

Smooth muscles. Practically all the smooth muscles of the body are affected by epinephrine, some being stimulated to contract and others inhibited. Those stimulated are the muscles of the spleen, the radial muscle of the iris (dilators of the pupil), the pyloric, ileocolic, internal anal, and bladder sphincters, and the pilomotor muscles of the hair follicles. The muscles of the stomach, intestine, bronchioles, and urinary bladder are inhibited.

Central nervous system. A marked excitatory effect on the central nervous system accompanies injection or endogenous release of epinephrine. Awareness to the environment is increased and reflex responses are facilitated. It is thought that these effects occur because of the hormone's action in a small part of the brainstem known as the reticular formation.

Skeletal muscles. Another effect of the injection of adrenal medullary extract is the increased power it bestows on the skeletal muscles. When an extract of the adrenal glands is injected into a frog, the contractions of the subsequently excised gastrocnemius muscle are greater and continue longer before fatigue sets in than in an untreated animal. Although under usual conditions the irritability of a fatigued muscle is restored in from fifteen minutes to two hours, if a small dose of adrenal medullary extract is given, the normal threshold may be restored in from three to five minutes. Since the effect has been observed in excised muscles, adrenal medullary extract must act, at least in part, directly on muscle tissue. Epinephrine liberation serves to mobilize the reserves of the body to meet situations of stress—the so-called "fight or flight" reaction. The student can usually recall the effects of epinephrine on an organ or tissue by deciding whether a response would serve a useful or adaptive purpose in "fight or flight" situations. Unusual feats of speed, strength, and endurance in emergency situations have been attributed to epinephrine release.

Metabolism. Oxygen consumption is increased up to 30% and the heat production may be raised as much as 17% above the basal level by injection of epinephrine. Epinephrine, by increasing glycogenolysis in the liver, raises the sugar content of the blood (hyperglycemia). In muscles it also accelerates the breakdown of glycogen. However, this does not contribute to the blood sugar, since muscle lacks the necessary *phosphatase* enzyme to produce glucose from glucose-6-phosphate, a product of glycogenolysis. Thus, skeletal muscle glycogenolysis produces lactic acid, which enters the blood and is either used by the heart as a fuel or carried to the liver for resynthesis into glycogen. The mechanism by which epinephrine increases glycogenolysis in liver and skeletal muscle involves an activation of the enzyme *phosphorylase.* Another important effect of epinephrine is the release of free fatty acids from adipose tissue.

Medicinal uses. Because of its pronounced action on blood vessels, epinephrine is often used locally to stop small hemorrhages, as in nosebleed and in minor operations, but not in pulmonary bleeding. For the same reason it has been employed to counteract the histamine-like compound that causes the great vascular dilation in hives. By constricting the blood vessels, it shrinks the mucous membranes and thereby clears the nasal passages which are blocked by a cold.* It is also used in conjunction with Novocain or similar drugs to produce local anesthesia; by its vasoconstrictor action, epinephrine lessens the removal of the anesthetic from the injected area and thus prolongs and intensifies the anesthesia. Because of its inhibiting influence on the bronchial muscles, epinephrine is used in asthma, a condition in which the excessive contraction of these muscles by constricting the bronchi makes breathing difficult. Also it has proved to be of great value as a heart stimulant in emergencies.

Norepinephrine. This hormone is also known as *arterenol.* Epinephrine and norepinephrine are probably secreted by the adrenal glands in varying proportion, depending on needs. Because epinephrine has a dilator as well as constrictor effect, norepinephrine, which has only a constrictor effect, causes a greater rise in blood pressure than epinephrine. It is at least 50% more powerful as a vasoconstrictor than epinephrine and raises both systolic and diastolic pressure. The main function of norepinephrine is the normal control of circulation, whereas epinephrine produces a variety of metabolic effects. Norepinephrine does play a role in the release of free

*The chemically related ephedrine and Benzedrine also have this effect.

fatty acids from adipose tissue; it has no other effect on metabolic functions.

■ Sympatheticoadrenal system

From our study of the innervation of the heart, blood vessels, alimentary canal, and liver, it will be evident that all the previously mentioned results produced by epinephrine and norepinephrine are similar to those obtained by the stimulation of the sympathetic nerves supplying the organs enumerated, except the sweat glands (Chapter 26). According to the neurohumoral theory of intercellular transmission, the stimulation of a sympathetic nerve produces or releases at the ending of the nerve fiber a compound that transmits the impulse to the next protoplasmic structure. This compound has been called sympathin. The result of sympathetic stimulation is the release of a mixture of epinephrine and norephinephrine. The latter is predominant in sympathetic nerves and the former in the adrenal glands. For these reasons the sympathetic system and the adrenal glands frequently are referred to as the sympatheticoadrenal system.

■ Adrenal cortex

The cortical portions of the adrenal glands are necessary for life. Their removal in animals or their hypofunction in man (Addison's disease) is always fatal unless replacement therapy is instituted.

The hormones secreted by the cortex (Fig. 28-5) of the adrenal glands, in the zona fasciculata and zona reticularis, are steroids synthesized by the glands from a precursor, esterified cholesterol. Unlike many other endocrine glands (e.g., pituitary, thyroid, and adrenal medulla), the cortex does not store large quantities of hormone. Its synthesis is largely dependent on the availability of NADPH from the extremely active pentose shunt pathway of carbohydrate metabolism (Chapter 23) found in the glands. Many adrenal steroids have been discovered; how many of these can qualify as hormones is questioned; several have been obtained in crystalline form. Most are doubtless intermediates in the synthesis of the true hormones but are capable of exerting hormonal effects when secreted into the blood. Indeed, a number of *inborn errors of metabolism* are recognized wherein such intermediates are produced and secreted in quantity due to a genetic defect affecting the enzymes of the synthesizing mechanism. Hypersecretion of these glandular products results in masculinization in the female and precocious puberty in the male, for they are *androgenic* (Chapter 30) in their effects.

Steroid hormone actions. ■ The steroid hormones produced by the adrenal cortex and by the gonads (Chapter 30) have three general effects at the target tissue level: stimulation of cell growth, mediation of immature cell differentiation, and induction of specific protein in differentiated cells. Evidence indicates that these effects are brought about by regulation of protein synthesis. Androgens and estrogens stimulate the synthesis of messenger RNA and ribosomal RNA (Chapter 2) in gonadal tissue. Glucocorticoids promote messenger RNA synthesis in fully differentiated cells and thus exert a "fine control" of cellular metabolism; for example, hydrocortisone increases the content and activity of several enzymes involved in amino acid metabolism in the liver. The effect of aldosterone on sodium transport involves the synthesis of new protein. Specific steroids stimulate production of messenger RNAs which, at the translation stage, prescribe the formation of enzymes that are essential in porphyrin and heme synthesis by erythropoietic cells. It is clear that steroids regulate the synthesis, activity, and possibly degradation of tissue enzymes and structural proteins through their effects upon the synthesis of nuclear RNA.

Glucocorticoids and mineralocorticoids. ■ It simplifies matters to speak of the major hormones of the adrenal cortices as *glucocorticoids* and *mineralocorticoids,* although, in fact, there is some overlap in their functions. *Cortisol* is the major glucocorticoid in man and dog (corticosterone in the rat), and *aldosterone* is the principal

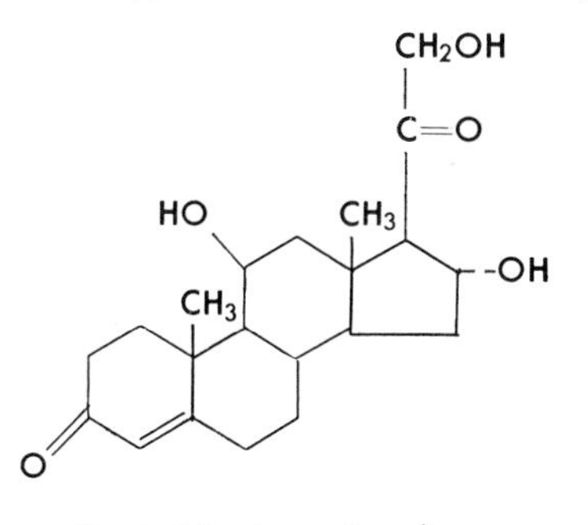

Cortisol (hydrocortisone)

mineralocorticoid. As soon as it is synthesized, cortisol is released into the blood and transported to the tissues in combination with a globulin protein called *transcortin.*

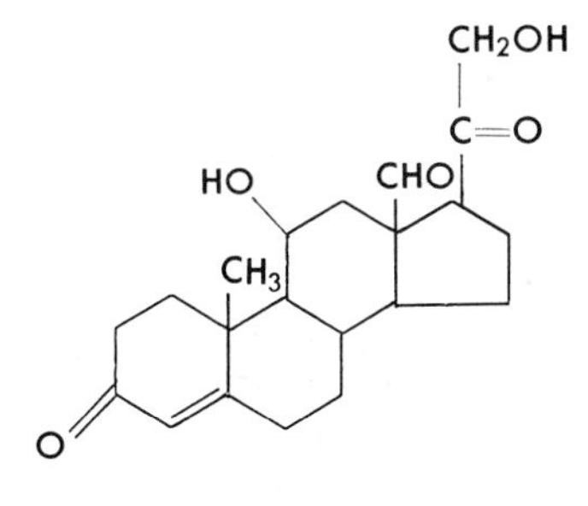

Aldosterone

Corticoid functions. ■ The corticoids are chiefly concerned with maintaining homeostasis by mobilizing the various resources of the body: (1) organic metabolism and (2) the water-mineral balance. To these we may add the influence they exert on (3) the sex organs and (4) white blood cells. Their effects at the cellular level, which were discussed earlier, are not fully understood; however, cortisone appears to stabilize the lysosomal membrane. This may be the basis of its anti-inflammatory action.

Metabolic effects. Experimental removal of the adrenal glands or hypofunction of their cortices in man results in a lowered basal metabolic rate. This is accompanied by a decrease in glycogen in muscles and in liver and a lowering of the glucose content of the blood. As can be readily understood, this curtails the production of energy and thereby leads to (1) extreme muscular weakness and great susceptibility to fatigue, infections, anesthetics, and anoxia, (2) lowering of body temperature, (3) an enfeebled heart that cannot maintain a normal blood pressure, that is, hypotension, (4) appetite failure and food absorption below par because of increased irritability of the intestines, and (5) loss of body weight; inability to retain NaCl and H_2O. Administration of cortisone, which is converted to cortisol in the body, alleviates these symptoms.

Administration of excess cortisol or the hyperfunction of the adrenal cortices in man (Cushing's disease) causes (1) hyperglycemia, (2) marked deposition of glycogen in the liver, (3) increased protein catabolism (negative nitrogen balance) and muscle weakness, (4) characteristic obesity with redistribution of adipose tissue to the abdomen, face, and upper back, (5) bone dissolution and weakening, (6) hypertension, (7) poor wound healing, and (8) possible retention of body water and NaCl.

Water and mineral balance. Deficiency in mineralocorticoids (aldosterone) is associated with an *increased excretion of sodium salts by the kidneys.* In consequence, the concentration of these salts in the blood decreases, and the osmotic pressure of the blood (for which these salts are largely responsible) is lowered. In our study of the kidneys we shall learn that a hypotonic condition of the blood increases the excretion of water and, therefore, causes a decrease in blood volume (hemoconcentration). The change in blood volume is, of necessity, shared by the intercellular (tissue) fluid and eventually also by the fluid within the cells themselves—causing a negative water balance (or dehydration).

In addition to this, in adrenalectomy or in Addison's disease the ability of the kidneys to excrete potassium salts is diminished; this still further upsets the normal Na:K ratio. When we recall the part played by these salts in the beating of the heart and the maintenance of membrane potentials, we can appreciate the fatal results of loss of adrenal cortical functions.

In an adrenalectomized animal, hunger for sodium is greatly increased; it prefers a 3% NaCl solution to tap water. When given an opportunity to satisfy this sodium hunger, the animal survives indefinitely. By properly increasing the sodium and curtailing the potassium (and administering natural or synthetic hormone), the life of a patient afflicted with Addison's disease is lengthened greatly.

An increase in the production of aldosterone has the opposite result, a retention of sodium in the body, which usually, but not always, causes an increase in the volume of the body fluids. This leads to a higher blood pressure, to edema, and to the loss of potassium due to its greater excretion by the kidneys.

We may summarize: By influencing the function of the kidneys, the mineralocorticoids (and to a lesser degree the glucocorticoids) maintain the necessary balance between the various electrolytes (especially Na and K ions) and aid in preserving the proper amount of water in the body.

Secretion of sex hormones. We have already mentioned certain inborn errors of metabolism that result in androgen production by the adrenal cortex. This subject can be reopened to greater advantage in our study of the development and functioning of the sex organs (Chapter 30).

Destruction of leukocytes. In experimental animals the injection of either cortisone or ACTH causes the rapid destruction of eosinophils, lymphocytes, and lymphoid tissue, whereas it increases the number of platelets, RBC, and neutrophils.

Regulation of adrenal cortical activity. The activity of the adrenal cortex is largely controlled by the adrenocorticotrophic hormone (ACTH), for removal of the anterior pituitary causes a marked atrophy of the cortex of the adrenal glands and a severe reduction in the output of glucocorticoid hormones. Injection of ACTH prevents this. An increased manufacture of corticoids coincides with depletion of ascorbic acid and cholesterol normally present in the gland. As in other instances, the relationship between the anterior pituitary and the target organ (here the adrenal glands) is reciprocal. Increase in adrenocortical hormones in the blood lessens the production of ACTH by the pituitary body until the need for it by the adrenal glands calls for a greater supply. ACTH is destroyed by digestive enzymes and is rapidly inactivated by both the liver and kidneys. Its half-life in the blood is about five minutes.

The mineralocorticoid aldosterone apparently does not have any important pituitary interrelationship in its secretory regulation. Although the details of its regulation are not well defined and there is no consensus among investigators, two mechanisms may be involved. The first mechanism may utilize volume receptors in the vascular tree, which initiate the release of *adrenoglomerulotrophin* from the brain—possibly the pineal gland—and, therefore, bring about increased aldosterone secretion. The second mechanism may operate through certain specialized cells in the kidney that release *renin* into the blood. Renin, in turn, produces *angiotensin;* the latter stimulates the secretion of aldosterone by the glomerulosa cells (Fig. 28-5) of the adrenal cortex. Finally, a definite, but small, effect of ACTH on aldosterone output has been observed.

Stresses and alarm reactions. ■ Stresses are exaggerated changes occurring either in the external environment or within the body itself. These changes threaten, if not properly met, the safety of body and mind. They therefore call for greater activity than usual on the part of the organs concerned with maintaining the *status quo* of the body. Among the more common stresses are extreme heat or cold, toxins and infections, trauma and surgical operations, shock, fever, anoxia, chronic underfeeding, strenuous muscular work, and emotional disturbances. In the adjustment of these stresses, the adrenal hormones (of both the medulla and the cortex) play a prominent part.

Ascorbic acid (vitamin C) is found in fairly high concentration in the adrenal cortex. Although the exact connection between this vitamin and adrenal activity is still obscure, a large number of observations leaves no doubt as to a close relationship. Animals afflicted with scurvy, a vitamin C deficiency disease, show a hypertrophic condition of the adrenal glands and a greatly diminished ascorbic acid content thereof. Hypertrophy of the gland and a lessening of its ascorbic acid content accompany and follow severe muscular work, exposure to intense cold, chronic underfeeding, and ether anesthesia; the administration of ascorbic acid or of the adrenal hormones prevents these results.

The adrenal hormones are protective and defensive agents, which, by their stimulating influence on metabolism, make possible a greater and speedier liberation of energy needed by the various mechanisms in making additional efforts for adjustment. In extreme or prolonged conditions of stress these hormones or the reserve material (ascorbic acid and cholesterol) for their reconstruction becomes exhausted; when this occurs, proper adjustment fails and death may follow. This explains why animals deprived of their adrenal glands or human beings suffering from Addison's disease are unable to withstand stresses or undue hardships. The previously discussed reactions to stresses have been designated as *alarm reactions*. It has been discovered that the alarm reactions fail to materialize in the absence of the pituitary body. Hence, stress influences the adrenal cortex by way of the pituitary.

Adrenal gland hormones and function are summarized in Table 28-1.

Table 28-1. Hormones of the adrenal gland and their function

Anatomical division	*Hormone*	*Function*
Medulla	Epinephrine	Affects skeletal muscles Cardiovascular effects Metabolism of carbohydrates and fats
	Norepinephrine	Vasoconstrictor
Cortex	Glucocorticoids	Essential to fat, protein, and carbohydrate metabolism Increases liver gluconeogenesis Resistance to stress
	Mineralocorticoids	Proper kidney function Regulation of fluid and electrolyte balance
	Sex hormones	Influence sexual characteristics

THYROID GLAND

The thyroid gland, weighing about 25 g (Fig. 28-6), is composed of two lobes, one placed on each side of the upper part of the trachea just below the larynx, or voice box, and a connecting part, the isthmus. The gland consists of a large number of small closed vesicles, the walls of which are formed, as shown in Fig. 28-7, of a single layer of columnar or cuboidal epithelium. The vesicles are held together by areolar tissue and are surrounded by a rich network of capillaries. The lumen of the vesicle is filled, to a greater or lesser extent, with a colloidal material made by the epithelial cells.

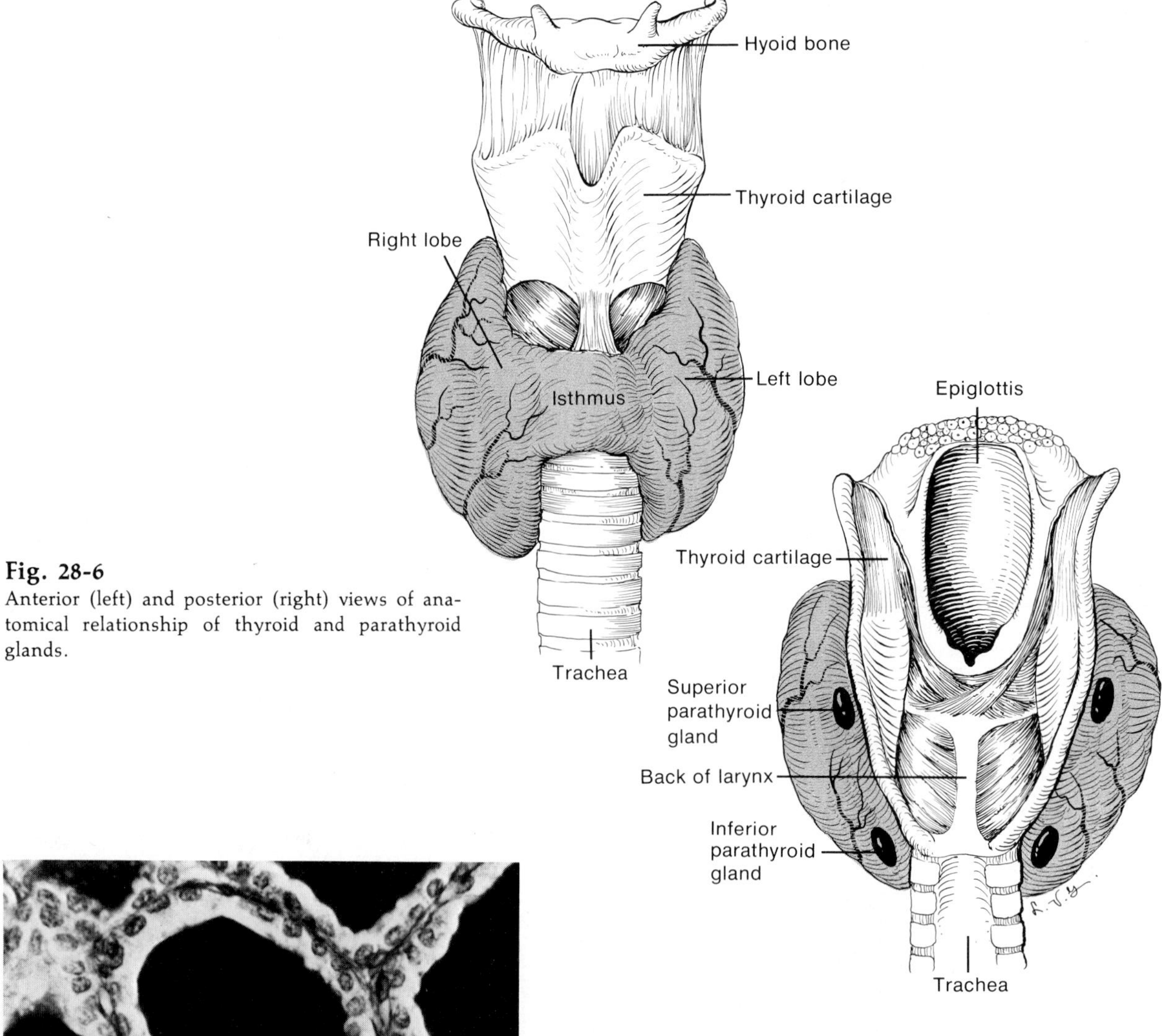

Fig. 28-6
Anterior (left) and posterior (right) views of anatomical relationship of thyroid and parathyroid glands.

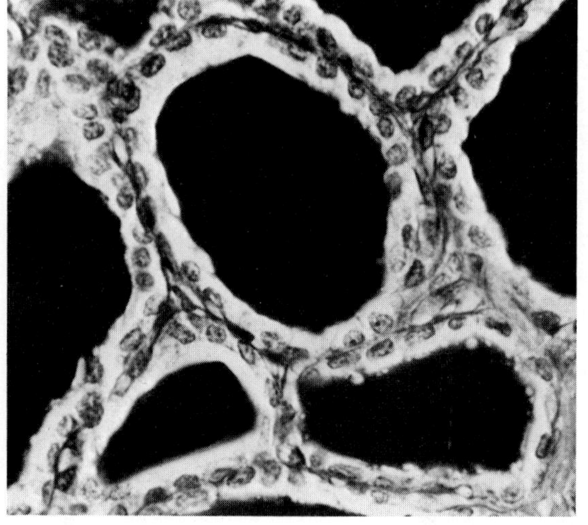

Fig. 28-7
Photograph showing follicles from an active thyroid gland.

of elevation in serum calcium. Its significance in calcium homeostasis in the adult animal has not been clearly defined. Thyrocalcitonin may be useful in control of hypercalcemia resulting from increased bone resorption.

■ PARATHYROID GLANDS

Closely connected with, and sometimes embedded in, the dorsal surface of the thyroid are four small glandular bodies, from 2 to 4 mm long in man, known as the parathyroids (Fig. 28-6).

Function. ■ The function of the parathyroid glands is to maintain the proper ratio of calcium and phosphorus in the blood. Removal of these glands (which is fatal in a few days) causes the Ca ion in the blood to fall from the normal 10 to 7 mg/100 ml or lower (about one half of the plasma calcium is ionized; the remainder is bound to protein); at this time the phosphorus may increase by as much as 10%. The loss of Ca ion seems to be due to its increased excretion by the kidneys. In a growing animal or in a child an insufficient supply of the hormone *(parathormone)* interferes with the development of bones; the teeth are poorly constructed and more susceptible to caries. Animals showing this marked deficiency of blood calcium have a great craving for calcium and, when given an opportunity, eat four times as much calcium salts as normal animals do, and they avoid phosphorus.

Parathormone is a large polypeptide of about 9,500 molecular weight. Its influence is primarily exerted on three structures—bone, gastrointestinal tract, and kidneys. The hormone stimulates the osteoclasts directly, thereby releasing calcium from the bone. Absorption of calcium from the gastrointestinal tract is partly by active transport and partly passive. In conjunction with vitamin D, parathormone increases the active absorption. Renal excretion of Ca ion is determined by the hormone's influence on the threshold. When the hormone level is low, the threshold is also low and calcium is lost from the body; administration of parathormone raises the excretory threshold. Renal reabsorption of phosphate is inhibited by the hormone.

Hyperparathyroidism. ■ An excess of parathormone results in extensive decalcification and may lead to bony deformities and fractures. Calcification of soft tissues, especially the kidneys, may occur. If calcium salts precipitate in the kidneys and ureters, kidney stones result and renal insufficiency develops.

Hypoparathyroidism. ■ A prominent characteristic of hyposecretion by the parathyroids in both man and animals is a greatly *increased excitability of the nervous system;* this shows itself in muscular twitchings and spasms *(tetany).* It will be recalled from our study of calcium metabolism (Chapter 24) that a diminution of this element in muscles, nerves, or central nervous system has a similar effect. The mechanism responsible for these disturbances is unknown, but some evidence suggests calcium plays an important role in determining the permeability of membranes to sodium and potassium ions. The processes of depolarization and repolarization would be modified by alterations in the calcium ion concentration. Seeing that calcium metabolism is closely governed by the parathyroid glands, it is not surprising to learn that the administration of the hormone relieves all the aforementioned disturbances; the administration of $CaCl_2$ or calcium gluconate produces the same result. Maintaining the normal excitability of the neuromuscular machinery is a function of the parathyroid glands. The activity of these glands does not appear to be under nervous or hormonal control. Instead, the secretion of parathormone is regulated by the level of ionized calcium in the plasma perfusing the gland.

In a normal animal or person, a decrease in the Ca ion content of the blood, either because of extra demand (as in pregnancy and lactation) or because of a deficient intake of Ca ion or vitamin D, stimulates the parathyroid to a greater production of the hormone. When this occurs (or when the hormone is injected), the kidneys are stimulated to a greater excretion of phosphates due to decreased reabsorption. This leads to a lower concentration of phosphorus in the blood and, in consequence, calcium migrates from bone to blood, a condition spoken of as the *resorption of bone.* Resorption assures a continued supply of Ca ion to the fetus (especially during the last two months of pregnancy) and to the breast-fed infant, even though the maternal diet may be somewhat deficient in calcium. However, the withdrawal of Ca ion from the bones of the mother softens them and renders them susceptible to bending. Calcium gluconate or lactate, given orally, alleviates the symptoms of parathyroid insufficiency.

■ PANCREAS

■ Insulin

In the body economy, insulin is the primary regulator of storage and conservation of the three major foodstuffs (Chapter 18). Even STH requires the presence of insulin for optimal anabolic action. Its biological potency is very high. As few as ten molecules per cell may elicit a characteristic effect.

The regulation of the blood sugar level (Chapter 21) is made possible by a precise balance of hormones. Insulin affects one or more processes involved in the metabolism of glucose. Its overall effect is one of lowering the blood sugar. When the pancreas fails to produce a sufficient quantity of insulin, a pathological condition known as *diabetes mellitus* results, in which hyperglycemia and glucosuria are prominent.

Insulin is produced in the islets of Langerhans' *beta* cells (Fig. 28-11), which constitute about 75% of the total islet tissue. It consists of two parallel peptide chains of twenty-one and thirty amino acids, respectively. There are slight differences in the amino acid sequence of insulin from different species, which may lead to allergic responses when these are administered to man.

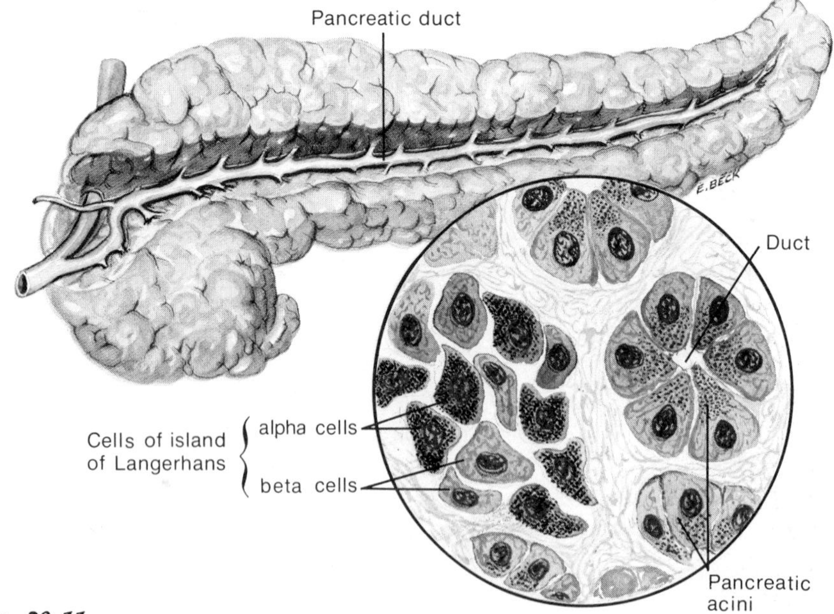

Fig. 28-11
Illustration of pancreas with inset showing cellular structure of gland in greater detail.

For example, sheep and cattle insulins sometimes evoke immune responses, but pig insulin is usually well tolerated.

Cellular effects of insulin. ■ Transfer of free glucose from extracellular fluids into cells, at least skeletal muscle cells, is facilitated by insulin. However, it has no effect on glucose metabolism of brain cells and red blood cells. On entry the glucose is phosphorylated to glucose-6-phosphate and metabolized. A barrier to glucose entry into cells, which is maintained by expenditure of energy in the form of ATP, seems to exist. Factors that reduce energy-rich compounds in tissues (e.g., exercise and anoxia) also permit glucose entry. Insulin apparently interferes with the barrier even in the absence of the aforementioned factors. Insulin and thyroxin produce mitochondrial swelling; this is associated with increased metabolic activity of the cell.

In adipose tissue cells insulin promotes fatty acid synthesis by blocking cAMP and providing glucose for conversion into acetyl-CoA and glycerol (Chapter 23). Insulin exerts an influence on protein metabolism that is not associated with glucose transport; the incorporation of amino acids into cell protein is stimulated by insulin, but the full effect is observed only when growth hormone is present.

Regulation of insulin secretion. ■ Insulin is normally secreted from the beta cells in response to a rise in blood sugar. Glucose is the single most important stimulus to secretion; although certain metabolic intermediates, drugs, and hormones will initiate insulin release, they do so more effectively in the presence of glucose. Products of protein metabolism, especially arginine and leucine, and of fat metabolism evoke insulin secretion. In man a protein as well as carbohydrate meal can activate the beta cells. Some gastrointestinal hormones invoke insulin secretion in a kind of anticipatory manifestation of blood sugar regulation. Isomers of glucose which can be metabolized (e.g., mannose) by the beta cells stimulate insulin, but those which cannot (e.g., galactose) are ineffective. This observation implies that the secretory process is linked to a reaction in the pathway of glucose metabolism. Low but definite concentrations of calcium and potassium ions are essential for insulin secretion to proceed. STH and glucocorticoids enhance insulin secretion, as does obesity, but a gradual development of insulin resistance occurs.

The beta cell contains a large number of insulin storage granules which move toward the cell membrane when the cell is stimulated. The granules are secreted by fusing with the membrane in a sort of reverse pinocytosis called emiocytosis. About 0.4 mg of insulin/hr constitutes the basal secretion rate. Upon adequate stimulation, insulin is released in two phases: a rapid phase of expulsion that lasts only ten to fifteen minutes and peaks within six minutes, and a slower phase that continues for over an hour. The second phase probably represents the secretion of newly synthesized insulin, since agents which block protein synthesis also block the slow phase of insulin secretion. In the body, insulin has a half-life of only eight to ten minutes. Its lack is detectable as an increase in blood sugar and free fatty acids within a few minutes.

The stimulus for insulin release appears to involve the enzyme adenylcyclase and cAMP. Glucose may provide the ATP that is the immediate precursor of cAMP (Chapter 2). One can at this time only speculate that emiocytosis is a response to calcium ion shifts in the membrane brought about by cAMP. Epinephrine specifically inhibits insulin secretion by inhibiting cAMP production in the beta cells. This is in contrast to its effect on other tissues, for example, the heart, where this catecholamine raises cAMP levels (Chapter 14).

An enzyme or groups of enzymes in liver and kidney tissue, *insulinase*, inactivates insulin. The primary treatment of diabetes mellitus involves the administration of insulin. However, when a subject has a reserve of functioning, although sluggish, beta cell tissue, certain chemical compounds (sulfonylureas) that stimulate secretion can be given.

Effects of insulin lack. ■ Continued insulin deficiency, as in uncontrolled *diabetes mellitus,* leads to coma and death through serious alterations in carbohydrate, lipid, protein, electrolyte, and water metabolism. These changes are reflected in the respiratory, cardiovascular, renal, excretory, and central nervous systems.

The decreased peripheral utilization of glucose due to insulin deficiency produces (1) an increased blood sugar, (2) the mobilization and release into the circulation of free fatty acids from the adipose tissues, and (3) a decreased protein synthesis and increased proteolysis due to the unopposed action of corticoids and growth hormone. Glycogenolysis and gluconeogenesis (from the additional amino acids available in the blood) by the liver adds to the already high blood sugar levels. Since the liver can metabolize the increased lipid presented to it in the blood only to a certain upper limit, there is an accumulation of intermediate products, ketone bodies, which produce *metabolic acidosis.* Respiratory rate and depth are increased in an attempt to compensate for the acidosis. The renal threshold for glucose is exceeded and glucose is excreted. Ketones and urea are also excreted in added quantities as a result of lipid and protein catabolism; these and glucose obligate additional water excretion and thus *polyuria* occurs. Loss of body water entails hemoconcentration, with consequent circulatory deficits that bring about tissue anoxia. This anoxia increases anaerobic metabolism and the production of lactic acid, which heightens the acidosis. The "vicious circle" continues, the central nervous system is depressed, and, eventually, death ensues.

■ Glucagon

The remainder of the islet tissue, *alpha* cells, produces the hormone *glucagon,* whose effect on carbohydrate metabolism is opposite to that of insulin. It is a polypeptide containing twenty-nine amino acids and is secreted in response to hypoglycemia. By its activation of liver phosphorylase enzyme (but not skeletal muscle phosphorylase), it produces glycogenolysis and hyperglycemia. It does not have any of the adrenergic effects of epinephrine.

■ OTHER ENDOCRINE SOURCES

■ Prostaglandins

The prostaglandins constitute a family of more than a dozen fatty acid compounds exerting a wide diversity of hormonal effects. They are produced in the body in very small quantities and are rapidly broken down by catabolic enzymes. These characteristics have made their isolation and study difficult and time consuming.

Composition. ■ Prostaglandins are twenty-carbon carboxylic acid compounds which are synthesized in the body from polyunsaturated fatty acids (Chapter 18). An important precursor is arachidonic acid, which is a constituent of the phospholipids in cell membranes. The enzymes required for the synthesis of prostaglandins from free fatty acids occur in a variety of tissues. During the synthesis from the precursor, a five-member, cyclopentane ring is formed and three oxygen atoms are added to the molecule, the site depending upon the particular prostaglandin being produced. Prostaglandins are identified by letters of the alphabet and subscript numerals and Greek letters; e.g., PGA_1, PGB_2, PGE_1, PGF_2-alpha. The various compounds can be interconverted by change in the primary structure.

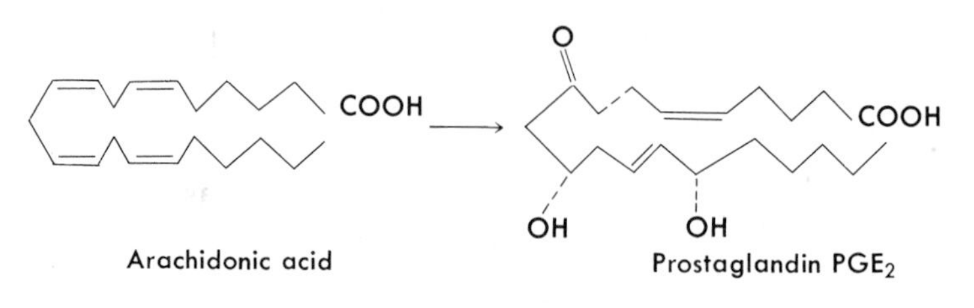

Sources. ■ Prostaglandins have been isolated from most mammalian tissues: intestine, liver, kidney, pancreas, heart, lung, thymus, brain, and both male and female reproductive system tissues. The richest source is human seminal fluid (Chapter 30), but the name derives from an early assumption that they were produced in the prostate gland instead of the seminal vesicles. Only a few tenths of a milligram of prostaglandins are synthesized in the human body each day. Because of the origin of arachidonic acid in membrane phospholipids, it is assumed that the cell membrane is the prime site for prostaglandin synthesis and that the prostaglandins have an important role in regulating membrane functions. They are released in response to a variety of neural and chemical stimuli. Storage of prostaglandins by cells is not evident; therefore, they are most probably produced as required. The half-life of prostaglandins appears to be less than one minute in duration. Such an ephemeral existence may be a consequence of the extreme biological potency of the compounds.

Actions. ■ The effects of individual prostaglandins are very specific and often opposite actions are produced by closely related compounds. In general the effects of prostaglandins are upon smooth muscle contraction, secretion, blood flow, and metabolism. Prostaglandins PGE_2 and PGF_2 are powerful stimulators of uterine contraction and may play an important role in parturition. They are involved in fertilization of the ovum (Chapter 30) and inhibit progesterone secretion by the corpus luteum.

Inhibition of gastric secretion is attributed to prostaglandins PGE_1 and PGE_2, which are compounds normally produced by the stomach. Prostaglandin PGE_1 improves air flow to the lungs by relaxing smooth muscles in the airways. It also produces increased urine flow and excretion of sodium ions, since it opposes the increased water permeability of the kidney tubules that is evoked by ADH (Chapter 29). Prostaglandin PGF_2-alpha increases, whereas PGE_2 lowers, blood pressure. The effects of prostaglandins on heart rate and force are variable. In experimental animals, it has been shown that PGE_2 inhibits the release of norepinephrine in response to neural stimulation; thus prostaglandins may be important negative feedback mechanisms for regulation of transmission in the autonomic nervous system.

Among the metabolic effects of PGE_1 is its inhibition of the lipolysis that normally is produced by catecholamines, vasopressin, glucagon, etc. This depression of lipolysis is accomplished by prostaglandin antagonism of adenylcyclase, the target of catecholamines and other stimulatory "first messengers" (Chapter 2). When adenylcyclase is blocked, the immediate activator of triglyceride lipase, cAMP, is not produced. It has been suggested that prostaglandins are general modulators of adenylcyclase in many tissues; as such they would be inhibitors of cAMP formation. The efficacy of aspirin in reducing fever and inflammation may be due to its blocking of one, or more, prostaglandins. Recently, PGE_2 has been shown to suppress arthritic inflammation in experimental animals.

■ Kidney

It is well established that anoxia, due either to reduced atmospheric oxygen (hypoxic anoxia) or to hemorrhage (anemic anoxia), produces a marked increase in the number of circulating red blood cells. When plasma from a rabbit made anemic by controlled reduction in blood volume is injected into a normal rabbit, there is very soon a measurable increase in the concentration of red blood cells in the recipient animal. Such evidence has led to the concept of a humoral regulation of red blood cell formation by the bone marrow. The circulating hormone responsible is a glycoprotein known as *erythropoietin.* Removal of both kidneys prior to subjecting an experimental animal to anoxia reduces the erythropoietin titer of the blood by 80% to 90%. Thus, although it is produced in limited amount by other unknown tissues, the major source of the enzyme is the kidney. Renal disease is sometimes associated with *polycythemia.* The production of *renin* by the kidney has been discussed elsewhere.

Thymus

The thymus (popularly known in animals as the neck sweetbread) is a glandular structure situated between the upper part of the sternum and the pericardium. It is a temporary organ, reaching its greatest development at the age of 14 to 16 years, after which it gradually atrophies because of activity of the sex glands. Because of its position in the thorax, enlargement in infants may interfere with respiration and circulation. Various functions have been assigned to this structure; aside from aiding in the formation of white blood cells and functioning in immune reactions, little is positively known. Removal of the thymus gland in fetal or postnatal life impairs the production of antibodies throughout the remainder of life.

The suspected endocrine function of the thymus is based largely on evidence that the structure contributes in some unknown way to *myasthenia gravis,* a neuromuscular disorder. Extirpation of the gland in patients often produces a temporary remission of the disease.

A reciprocal relationship between the thymus and the sex glands is demonstrated in that castration retards and sexual activity (mating and pregnancy) hastens the involution (retrogressive changes) of the thymus.

Pineal body

The pineal body is a cone-shaped structure lying between the anterior corpora quadrigemina on the dorsal aspect of the brain (Fig. 10-8). Before puberty it undergoes degenerative changes, and in the adult it consists principally of fibrous tissue. The two postulated endocrine products of the pineal body, adrenoglomerulotrophin and melatonin, have already been discussed.

INTERRELATIONSHIP OF HORMONES

Our study of the pituitary body and its target glands has given an idea of the involved interrelation between the various endocrine organs. Further evidence of this will be encountered in the functions of the reproductive organs (Chapter 30).

We have seen repeatedly that hormones are exceedingly powerful agents and that, in some instances, the field of their activity covers practically the whole body. The hypoactivity or hyperactivity of many of the endocrine glands seriously affects the body and may even result in death. What determines the amount of each hormone formed? In some instances the need of the body for a particular hormone will automatically regulate its production. It may be recalled that an increase of sugar in the blood stimulates the production of insulin. When this need has been met, the production is slowed down or ceases altogether until a fresh demand is made. Again, we have seen that a target gland that is stimulated by a pituitary hormone will, by its own hormone production, check the activity of the pituitary. By this reciprocal interaction, a state of balance between the various hormones is maintained.

READINGS

Butcher, R. W.: Role of cyclic AMP in hormone actions, New Eng. J. Med. **279**:1378-1383, 1968.

Catt, K. J.: An ABC of endocrinology, Boston, 1971, Little, Brown and Company.

Gillie, R. B.: Endemic goiter, Sci. Am. **224**(6):93-101, Jun. 1971.

Goldsmith, R. S.: Hyperparathyroidism, New Eng. J. Med. **281**:367-374, 1969.

Levine, R.: Mechanisms of insulin secretion, New Eng. J. Med. **283**:522-526, 1970.

Li, C. H.: The ACTH molecule, Sci. Am. **209**(1):46-53, Jul. 1963.

O'Malley, B. W.: Mechanisms of action of steroid hormones, New Eng. J. Med. **284**:370-377, 1971.

Pike, J. E.: Prostaglandins, Sci. Am. **225**(5):84-91, Nov. 1971.

Rasmussen, H., and Pechet, M. M.: Calcitonin, Sci. Am. **223**(4):42-50, Oct. 1970.

Sherwood, L. M.: Human prolactin, New Eng. J. Med. **284**:774-777, 1971.

Role of the kidney in homeostasis
Urinary system
Renal function
Micturition

RENAL PHYSIOLOGY

The purpose of the vegetative organs is to maintain within the rather narrow limits of normality the osmotic pressure, the pH, the concentration of the crystalloids and colloids, the quantity of metabolites, the temperature, and the total volume of the internal medium and its freedom from nocuous substances. Any but the slightest variation in these characteristics is followed by harmful results. The part played in this by the respiratory, digestive, and circulatory systems and by the cellular elements of the blood, the skin, and the organs where energy-yielding materials are stored has been dealt with in previous chapters.

The kidneys are of exceeding importance in this homeostasis. The fitness of the tissue fluid to serve as an internal environment for the cells of the body depends on the kidneys in many distinct ways. In addition to the conservation of useful substances and excretion of wastes, the kidneys act to preserve the constancy of the extracellular fluid in composition, volume, and pH. They also have endocrine functions (Chapter 28). Release of red blood cells from bone marrow (erythropoiesis) is stimulated by the hormone *erythropoietin* secreted by the kidneys, and the release of *aldosterone* from the adrenal cortex is influenced by the kidney production of *renin.* In this chapter we will, first, examine the role of the kidneys in homeostasis and, second, look at the renal mechanisms by which this regulation is accomplished.

■ ROLE OF THE KIDNEY IN HOMEOSTASIS

■ Excretion of wastes and excess materials

Useless or harmful materials in the plasma are to a greater or lesser extent removed, or cleared from the plasma, by the kidneys. Some foreign proteins entering the blood and certain excess catabolic products, such as urea, uric acid (as urates), creatinine, and various salts such as nitrates, sulfates, and phosphates (Table 29-1) are eliminated. Benzoic acid, produced in the body or ingested with certain preserved foods, unites in the liver with glycine (amino acetic acid) to form hippuric acid; as such it is excreted by the kidneys. This is an example of the detoxification function of the kidney.

When the concentration of a substance nor-

Table 29-1. Chemical composition of twenty-four hour urine collection

Water	1.2 liter
Urea	30.0 g
Uric acid (as salts)	0.5 g
Creatinine	1.0 g
Hippuric acid (as salts)	1.0 g
Ammonia (as salts)	0.7 g
Sodium chloride	15.0 g
Other salts	10.0 g
Other organic material	3.0 g

mally present in the plasma rises above a certain level, the *renal threshold,* the excess is excreted in the urine. For example, when glucose is present to an extent not above 0.17% (170 mg/100 ml), the kidneys conserve it, but any excess above this level is excreted.

■ Conservation of useful substances

In the process of clearing the plasma by filtration of wastes and excess materials, certain useful substances are likewise cleared. Among these are proteins, amino acids, vitamins, sugars, intermediary metabolites of the citric acid cycle (Fig. 23-3), and ions such as sodium, potassium, calcium, magnesium, chloride, bicarbonate, etc. When these substances are not in excess of the bodily needs, they are conserved through the mechanism of reabsorption, either by active or by passive transport. Moreover, water is passively reabsorbed by osmotic diffusion. Urea and urates are partially reabsorbed and, in circumstances where the plasma levels may become too low, even nitrates, sulfates, and phosphates are reabsorbed.

■ Maintenance of osmotic pressure and fluid volume

The kidneys aid in defending the osmotic pressure and the volume of the internal medium by conserving or eliminating water or salt. In this manner the tissue fluid is maintained as a balanced salt solution for the protection of the cells. This renal function is also a factor in maintaining a constant body weight.

Osmolarity. ■ When the osmolarity of the extracellular fluid and plasma, which depends primarily on the water content, exceeds the normal value of 300 milliosmols/liter (Table 3-2), water is conserved; when the concentration is less, water *diuresis* ensues. This type of control is mediated by the effect of plasma osmolarity on hypothalamic receptors, *osmoreceptors,* which, acting by way of the supraoptic nuclei and the neurohypophyseal tract to the posterior pituitary (Fig. 28-3), determine the output of *antidiuretic hormone* (ADH) from the latter. ADH output is stimulated by increases in extracellular fluid osmolarity and inhibited by decreases. It has been estimated that an increase of only 1% in the effective osmotic pressure, the difference between extracellular and intracellular pressure, will evoke such a response. Through its action on the water reabsorptive capacity of the kidney, ADH markedly influences the urine output. The origin and function of ADH have been discussed in Chapter 28.

Volume. ■ The volume of the extracellular fluid, which depends primarily on its sodium content (Fig. 24-3), also exerts a regulatory effect on urine formation. Maintenance of the extracellular volume is primarily attributable to renal regulation of sodium excretion, since this ion represents nearly 90% of the cation content of extracellular fluid. Obviously, the extracellular volume also depends upon water, but the latter follows sodium wherever a barrier to its diffusion does not intervene. An increase in extracellular volume results in augmented sodium and water excretion by the kidneys; a decrease in extracellular volume causes retention of sodium and water.

The presence of *aldosterone,* a mineralocorticoid (Chapter 28), in the plasma enhances sodium retention by the kidneys and, consequently, expands the volume of the extracellular fluid. The plasma level of aldosterone is controlled, according to a current and commendable hypothesis, by the kidneys. Specialized kidney cells forming the juxtaglomerular complex (Fig. 29-3) are presumably sensitive to sodium ion concentration and, when this is low, secrete into the blood an enzyme known as *renin.* This enzyme catalyzes the production of *angiotensin* (Chapter 15), a small polypeptide, from *angiotensinogen,* a protein synthesized by the liver and always present in the blood. Angiotensin, in turn, stimulates the release of aldosterone from the adrenal cortex. Regulation of plasma aldosterone levels is by a feedback system; when aldosterone causes sodium retention, the secretion of renin and the production of angiotensin are reduced. The output of aldosterone in normal individuals is decreased during periods of high sodium intake and increased during periods of dietary sodium restriction. The role of mineralocorticoids in health and disease was described in Chapter 28. Sodium retention is not solely due to aldosterone activity, however, for a great deal of evidence suggests that reabsorption of sodium is also regulated by renal hemodynamics and possibly by another, as yet, unidentified hormone.

The role of water in maintenance of the extracellular volume depends, in part, upon ADH and its effect upon water reabsorption. The hypothalamic cells mentioned in connection with regulation of the extracellular osmolarity are also stimulated by baroreceptors located in several parts of the vascular tree, but principally in the left atrium. These baroreceptors are stimulated by an increased atrial pressure due to expansion of the extracellular fluid volume. Under these circumstances, their effect upon the ADH-secreting cells is one of inhibition. This diminishes water reabsorption and contributes to the desired reduction in extracellular volume. Conversely, a diminished extracellular volume and removal of inhibition results in an increased ADH secretion and conservation of water for volume restoration.

Cyclic AMP. ■ ADH increases the levels of cAMP in kidney cells. It has been shown that this increase occurs through the action of the hormone upon adenylcyclase. The mechanism by which cAMP may initiate an increased permeability of kidney cells to water is unknown, but currently it is the subject of intensive investigation. Although the action of aldosterone upon sodium retention is believed to involve cAMP, such a relationship has not been confirmed. The mineralocorticoid effect is mediated by induction of RNA and protein synthesis, which correlates with the observed latent period and time to peak effect. In the rat, the response to injected aldosterone is characterized by a one hour latent period and maximum sodium retention at three hours. In contrast, the action of ADH exhibits a rapid onset and peak.

Parathyroid hormone also appears to produce an increase in kidney cell cAMP. Although there is a relationship between the increased cAMP and increased phosphate excretion, the mechanism underlying this relationship is unknown.

■ Quantity and concentration of the urine

The quantity of urine excreted varies considerably because of marked variations in the amount of water taken into the body and in that eliminated by the lungs and the skin. Depending on conditions, from 0.3 to 15 liters of urine may be excreted each day by a normal individual. The average amount is 1 to 1.5 liters/day. The amount of water lost by the lungs is fairly constant; that excreted by the skin varies largely with the temperature and humidity of the air and with the amount of heat generated by muscular activity.

The urine excreted during twenty-four hours, on an ordinary diet, has approximately the composition shown in Table 29-1. The total amount of solids excreted per day is about 60 to 70 g; this varies chiefly with the amount of sodium chloride and protein consumed, the latter determining the amount of urea excreted. As both the solids and the water content of the urine are subject to variations, the specific gravity shows marked fluctuations, even during normal conditions. The usual variation is from 1.015 to 1.020. After much sweating the specific gravity may be above (1.035) the normal limits, and after much water ingestion it may be below (1.001) the normal limits.

Specific gravity is an expression of the concentration of the urine. A more useful expression is the ratio of the urine osmolarity to the plasma osmolarity, Uosm/Posm. These values can be determined by measuring the respective freezing point depressions (Chapter 3) of urine and plasma. The maximum Uosm:Posm ratio attainable is approximately 4, i.e., 1,200 milliosmols:300 milliosmols.

■ Acid-base balance

The production of acids is a constant phenomenon in the body. Oxidation of proteins gives rise to such fixed, nonvolatile acids as sulfuric and, in the case of nucleoproteins and phospholipids, also to phosphoric acid. A volatile acid, CO_2, is universally formed in cells and, in the presence of water, is converted by carbonic anhydrase to carbonic acid. Muscular work gives rise to lactic acid. The dissociation of these acids increases the hydrogen ion concentration of the blood; since this is inimical to life, various mechanisms are provided for regulating the acid-base balance. About one half of the metabolically produced acids can be neutralized by base available in the diet; the remainder has to be neutralized by buffer systems in the body.

Bicarbonate–carbonic acid buffer. ■ The bicarbonate–carbonic acid buffer system is the most important and, as noted in Chapters 12 and 17, plays a key role in the regulation of body pH. Although the ratio (20:1) of the concentration of the salt ($NaHCO_3$) to the acid (H_2CO_3), existing at pH 7.4, is not the best for effective chemical buffering, the importance of this buffer pair is based on the effectiveness of the respiratory and renal systems in stabilizing their concentrations. Carbonic acid concentration (P_{CO_2}) is regulated by the respiratory system, and bicarbonate ion concentration by the kidneys. In man the normal values for these are: bicarbonate, 26 to 28 mM/liter, and carbonic acid, 1.3 to 1.4 mM/liter.

The neutralization of fixed acid in the blood by this buffer pair may be illustrated by the reaction of sulfuric acid and sodium (or potassium) bicarbonate, which gives rise to acid sodium sulfate and carbonic acid.

$$NaHCO_3 + H_2SO_4 \rightarrow NaHSO_4 + H_2CO_3$$

In this reaction the strong sulfuric acid is exchanged for the very feebly ionizing carbonic acid. The acid salt is eliminated by the kidneys. Arriving at the lungs, the carbonic acid is rapidly converted by carbonic anhydrase in the red blood cells to CO_2 and H_2O, and the gas is expelled. The part played by the lungs is of great importance, for the quantity of volatile acid expelled by the lungs (in the form of CO_2 and H_2O) is many times greater than that of any other acid eliminated from the body. By excreting carbonic acid in this manner, the valuable bases* of the body (sodium and potassium) are not lessened, as occurs when the kidneys excrete the acid salts of sulfates and phosphates. However, the kidneys do operate to conserve base by *acidification of the urine* and by *ammonia synthesis.*

Acidification of urine. ■ Neutral salts of weak acids are converted to acid salts or even to free acids, with consequent increase in urine acidity. Consider the reaction:

$$Na_2HPO_4 + HHCO_3 \rightarrow NaH_2PO_4 + NaHCO_3$$

Here the neutral, disodium salt of phosphoric acid in urine undergoes an exchange of Na ion for H ion and is excreted as an acid, dihydrogen salt, whereas the sodium bicarbonate is reabsorbed into the blood, saving one equivalent of valuable base. The excretion of acid in a free form (e.g., H_3PO_4) is severely limited by the fact

*Base is used here in a classical sense to indicate cations.

that the kidneys cannot produce urine with a pH lower than 4.4. Nevertheless, by excreting urine of this maximal acidity, the kidneys manage to salvage large amounts of base. The titratable acidity of the urine is a measure of this conserved base.

Ammonia synthesis. ■ Strong acids, H_2SO_4 and HCl, are finally excreted as ammonium salts. In the case of H_2SO_4 it arrives at the kidneys as the acid salt, having been buffered previously by bicarbonate in the blood. The reaction in the kidneys is:

$$NaHSO_4 + H^+ + 2\ NH_3 \rightarrow (NH_4)_2SO_4 + Na^+$$

Hydrochloric acid is buffered in the blood by bicarbonate to form the neutral salt NaCl. In the kidney, the chloride anion forms ammonium chloride:

$$NaCl + H^+ + NH_3 \rightarrow NH_4Cl + Na^+$$

The H ion in these reactions is formed in kidney cells from ionization of carbonic acid and is actively exchanged for the Na ion in the urine. Ammonia, produced by the kidney cells from amino acids, principally glutamine, diffuses into the urine and forms the ammonium salt that is excreted; the reabsorbed sodium combines with the bicarbonate anion and returns to the plasma. The excretion of NH_4^+ salt reduces the amount of sodium that would otherwise be required to accompany the acid radical into the urine and thereby safeguards the total available base of the body. When the ability to manufacture ammonia is decreased, as in certain kidney diseases, the acids in the blood are not adequately neutralized or removed and acidosis results.

Alkalinization of urine. ■ In certain circumstances in which the acid-base balance of the body is disturbed so that the plasma bicarbonate is elevated or the bicarbonate-to-carbonic acid ratio is increased, as by hyperventilation, the kidneys excrete an alkaline urine. This is always associated with loss of fixed base. Since the maximum alkalinity of the urine is around pH 8, the base is accompanied by bicarbonate ion. To replenish the basic materials lost in the effort to maintain the acid-base balance is one of the functions of the ingested organic acids and the salts of these acids. In the intestines the organic acids (e.g., citric acid of citrus fruits) unite with alkalies to form salts (sodium citrate in our example). Sodium lactate (from sour milk) and sodium malate (from apples) may also be mentioned as examples of organic salts. By means of sodium acetate, we may illustrate how the organic salts replenish the alkali reserve. The composition of this salt is $CH_3COO \cdot Na$. In the body the acid radical, that is, the CH_3COO^- group, enters the citric acid cycle (Chapter 23) to form CO_2 and H_2O, which give rise to carbonic acid (H_2CO_3). This acid, uniting with the sodium of the original organic salt, forms sodium bicarbonate ($NaHCO_3$). Acid-forming foods are rich in sulfur and phosphorus; they include meat of all kinds, eggs, wheat, oatmeal, and peanuts. Nearly all fruits and vegetables are base-forming and tend to reduce the acidity of the urine.

Urine pH. ■ Twenty-four hour urine is usually acid in reaction, the pH ranging from 4.8 to 7.4 (average, about 6). The acidity is due chiefly to monosodium phosphate (NaH_2PO_4). During starvation the relatively increased protein metabolism increases the urinary acidity. In severe diabetes mellitus the urine is acid because of the large amount of organic acids (ketoacids) formed from fat. Production of acid may increase ten times or more and severely tax the capacity for renal buffering and ammonia production.

Following sleep or after a meal the urine generally becomes less acid and even may be alkaline—the *alkaline tide.* This would appear to be due to the increased elimination of CO_2 on arising and to the accumulation of base set free by the secretion of acid gastric juice.

■ URINARY SYSTEM

The kidneys. ■ The kidneys are bean-shaped organs about 4 inches in length and together

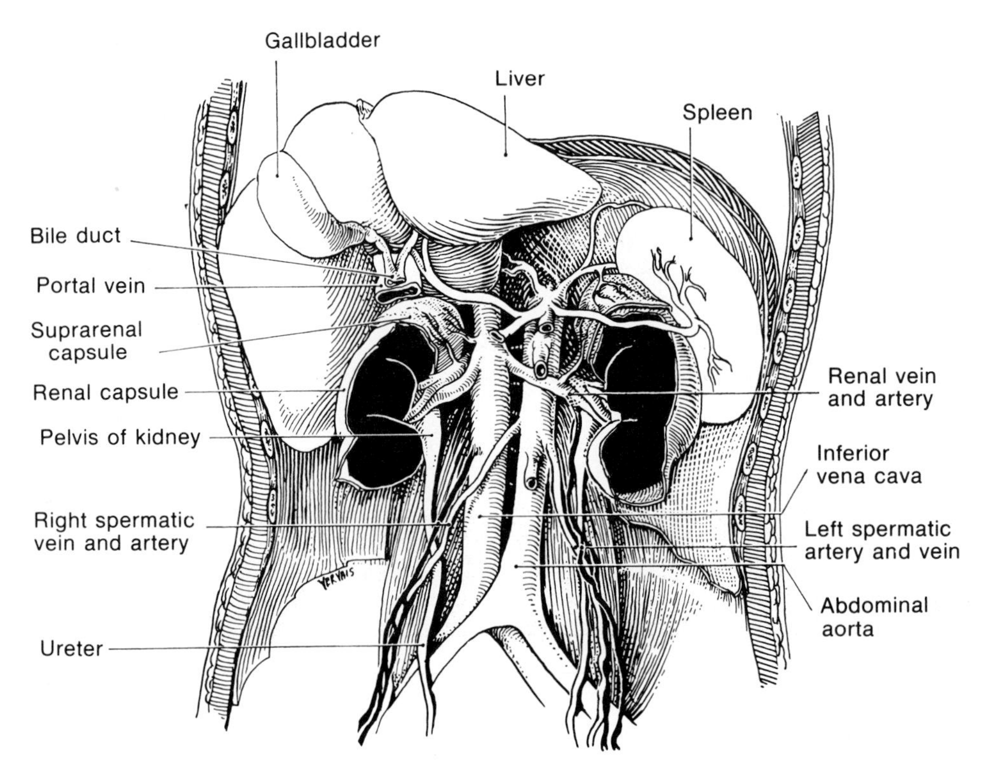

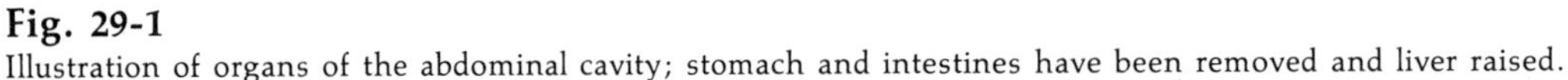
Fig. 29-1
Illustration of organs of the abdominal cavity; stomach and intestines have been removed and liver raised.

weigh about 300 g. They are situated in the lumbar region, one on each side of the vertebral column (Fig. 29-1). Each kidney is composed of thousands of minute tubes, known as the *uriniferous tubules* or *nephrons,* which originate in the cortex of the organ. These constitute the physiological units of the kidney. Their number in the two kidneys is estimated at 2 million and their total length at approximately 75 miles. Nephrons empty into collecting ducts, which course downward through the medulla of the kidney into the renal pelvis (Fig. 29-2). From each kidney springs a tube, the *ureter* (Fig. 29-1), which carries the urine to the urinary bladder; from the bladder arises another tube, the urethra, by which the urine is voided (Figs. 30-1 and 30-3). At its origin the ureter is much dilated and thereby forms the pelvis of the kidney. The fleshy mass of the kidney is divided into what are termed the *pyramids* (Fig. 29-2), the pointed ends of which dip into the pelvis. Each of these pyramids is a collection of nephrons and collecting ducts.

The nephron. ■ The nephron is a highly complex tubule of about 50 μ in diameter except in the loop of Henle, where the diameter is about 20 μ. It begins as a blind tube in the outer portion or cortex of the kidney (Fig. 29-2). The blind end is modified in a special manner so as to give rise to what is usually called the *Malpighian corpuscle,* or *glomerulus* (Fig. 29-3). The glomerulus in man is about 100 μ in diameter and just visible to the unaided eye when engorged with blood. This corpuscle is formed by the invagination of

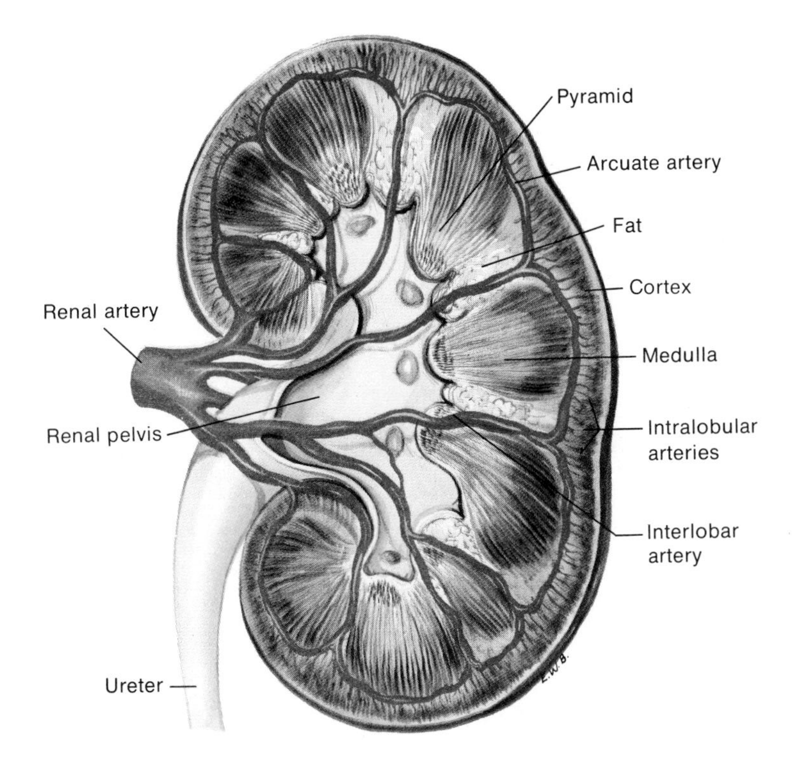

Fig. 29-2
Section through the kidney showing arrangement of lobes, lobules, and blood supply.

the blind end of the tubule (much like a pocket formed when a lead pencil is pushed down upon the closed end of the finger of a glove); the little sac thus formed is called *Bowman's capsule.* From this sac issues the tubule, which, as shown in Fig. 29-3, describes a very devious course; the first part is known as the *proximal convoluted tubule,* the second as the *loop of Henle,* and the last as the *distal convoluted tubule.* After many changes in its course and structure, it finally joins the *collecting tubule,* which empties, as stated before, into the pelvis of the kidney. In the capsule and loop of Henle the walls are composed of a single layer of broad, thin epithelial cells (Fig. 29-3); the cells forming the convoluted tubules are thicker and more cuboidal.

Some 20% of the glomeruli in the human kidney lie deep in the cortex adjacent to the medulla. These are termed *juxtamedullary glomeruli.* They differ from cortical glomeruli in that they have very long loops of Henle, which project deep into the medullary pyramids. The number of juxtamedullary nephrons correlates with the renal capacity to concentrate urine.

Blood supply. ■ The blood supply to the kidneys is derived from the abdominal aorta by way of the renal artery (Fig. 29-1). The renal artery subdivides into interlobar arteries and eventually into intralobular arteries (Fig. 29-2). From the intralobular artery a small branch arises as an *afferent* arteriole (vas afferens), which, as shown in Fig. 29-3, enters Bowman's capsule and there breaks up into a large number of capillaries; to this tuft of capillaries, which is actually the *glo-*

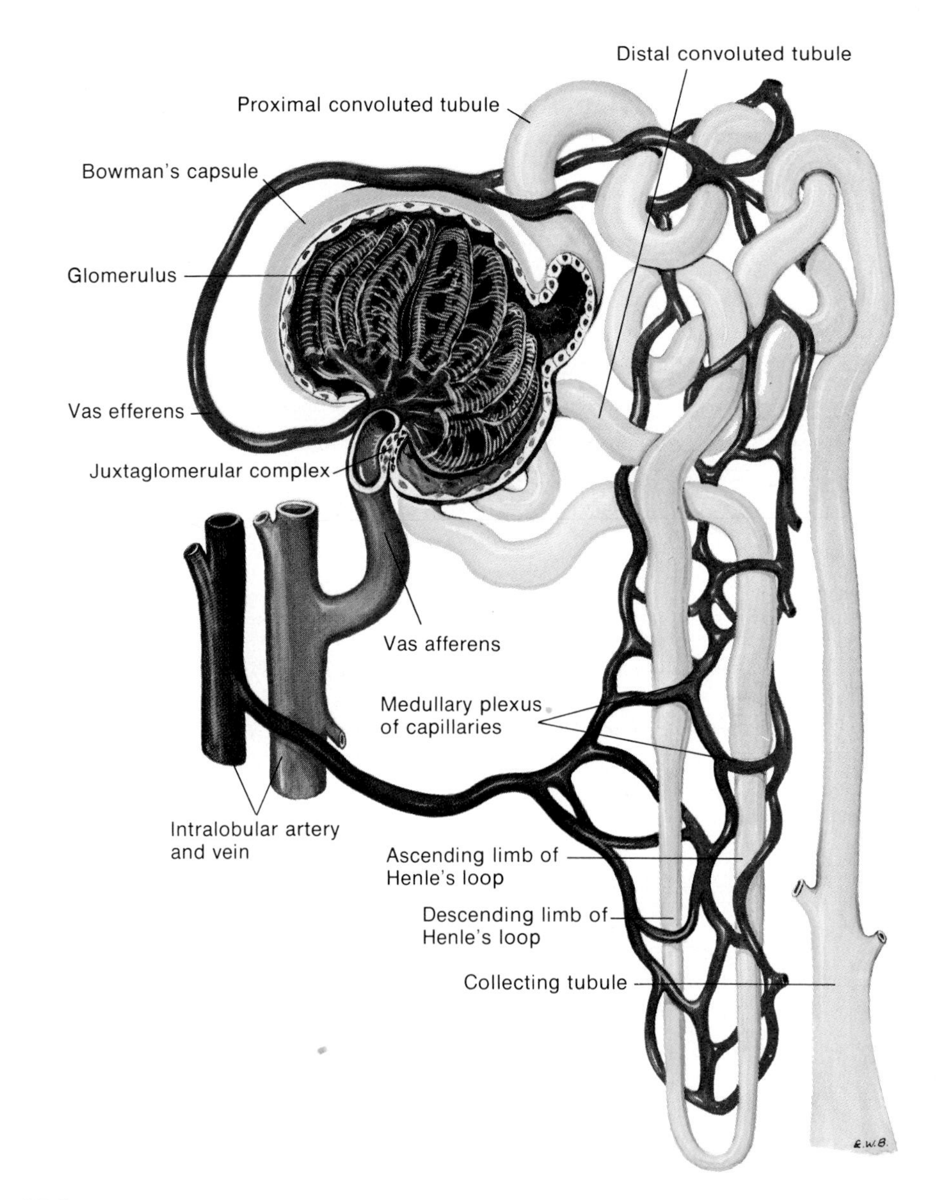

Fig. 29-3
Illustration of nephron and its associated blood supply. Wall of Bowman's capsule has been cut away to reveal detail of glomerulus.

merulus, the inner wall of Bowman's capsule is closely applied. By uniting with each other, the capillaries give rise to an *efferent* arteriole (vas efferens) that leaves the capsule.

As the afferent arteriole approaches the Malpighian body, an increase in the number of cells in the wall is seen to form a thickened structure on one side of the glomerulus. The distal convoluted tubule, as it returns toward the glomerulus, also possesses a distinct modification of its epithelial cells, which face the glomerulus. Together, these altered cellular structures are known as the *juxtaglomerular complex* or *apparatus.* It is believed that this complex is the site of *renin* production and release.

The vas efferens, it will be noticed from the illustrations, is considerably narrower than the vas afferens. Soon after leaving a cortical glomerulus, the vas efferens breaks up into a second group of capillaries, the *peritubular capillaries,* that surround the convoluted tubules; these capillaries discharge their blood into a small intralobar vein, which is drained eventually by the renal vein. The efferent arterioles from juxtamedullary glomeruli break up into peritubular capillaries, as do those from cortical glomeruli, but, in addition, they form long hairpin loops of thin-walled vessels, the *vasa recta,* which dip down alongside the loops of Henle into the medullary regions of the papilla.

The blood supply to the kidney is very abundant, 1,100 to 1,200 ml/min or about one fourth the total output by the heart. Most of the blood flow nourishes the cortex of the kidney, since it has been calculated that 10% or less reaches the medulla. In most animals all glomeruli are constantly perfused when the blood pressure is adequate; the rabbit is an exception.

The abundant lymphatic system of the kidney is very important; it eventually drains into the thoracic duct. Protein concentration in renal lymph averages 2.9 g/100 ml, whereas the average in thoracic duct lymph is 1.5 g/100 ml. Furthermore, renal lymph has a higher concentration of sodium, chloride, and urea than found in plasma or thoracic lymph. These observations are related to and offer support for the countercurrent hypothesis, which will be discussed later.

■ RENAL FUNCTION

In their excretory function the kidneys exercise a fine selective discrimination. The urine-to-plasma ratio in Table 29-2 is an index of the concentrating and reabsorptive capacity of the kidneys. The production of urine begins with the *filtration* through the glomerular capillaries of a fluid that resembles plasma. This glomerular filtrate passes down the tubules, and its volume is reduced and its composition altered by processes of *tubular reabsorption* and *tubular secretion.* The former accounts for the removal of water and solutes, the latter for the addition or exchange of solutes.

■ Glomerular filtration

Glomerular membrane. ■ The Malpighian body is well adapted for filtration in three respects. First, the tufts of capillaries forming the glomeruli (Fig. 29-3) are very numerous; in consequence, the surface for filtration in the kidneys is very large (estimated at 1.5 m^2). Second, it will be recalled that the vas efferens taking the blood from the glomerulus is smaller than the vas af-

Table 29-2. Composition of plasma filtrate and urine

Substance	*Plasma filtrate concentrations*	*Urine concentrations*	*Urine: filtrate ratio*
Protein g/100 ml	8.0	—	—
Glucose g/100 ml	0.10	0	0
Urea g/100 ml	0.03	2.0	66
Creatinine g/100 ml	0.001	0.19	152
Uric acid g/100 ml	0.004	0.05	12
Sodium mEq/L	150	135	0.9
Potassium mEq/L	5	60	12
Chloride mEq/L	110	143	1.3
Bicarbonate mEq/L	30	15	0.5

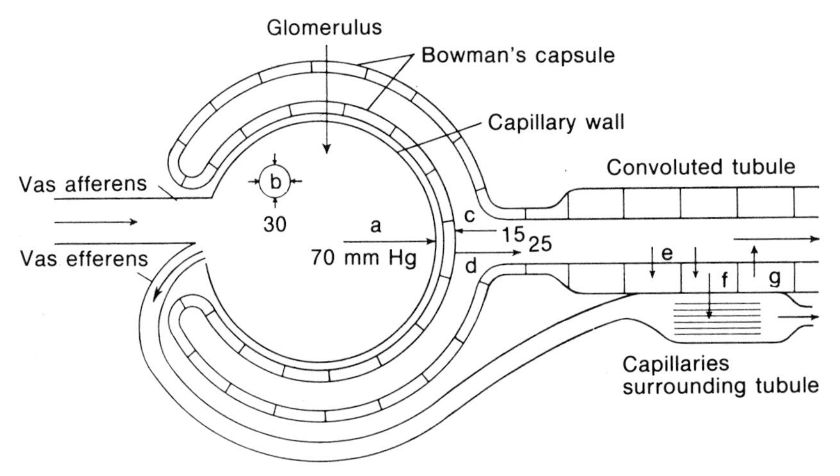

Fig. 29-4
Diagram illustrating factors involved in filtration, secretion, and reabsorption in nephron. **a,** Capillary pressure in glomerulus; **b,** the osmotic pressure of blood proteins; **c,** the hydrostatic pressure of urine in the tubule; **d,** the effective filtration pressure. Also indicated are tubular secretion, **g,** and reabsorption, **e** and **f.**

ferens and breaks up into a second set of capillaries around the convoluted tubules. This introduces considerable resistance to the outflow of the blood from the glomerulus and therefore creates a higher blood pressure in the glomerular capillaries; this is estimated at from 60 to 70 mm Hg, as compared with 25 mm pressure in other capillaries throughout the body (Fig. 29-4). The third factor of great importance in filtration is the very thin membrane (0.1 μ in thickness) separating the blood from the cavity of the capsule. As a result of these combined factors, the flow of fluid through the capillary wall of the glomerulus is estimated to be more than a hundred times that through the walls of muscle capillaries.

Ultrafiltration. ■ The function of the glomerulus can be quite satisfactorily explained by the physical laws of ultrafiltration; the force concerned is the capillary blood pressure. The glomerulus behaves as an entirely passive membrane. Insofar as it exercises some selectivity in regard to materials to be or not to be filtered, this seems to depend on the size of the pores of the membrane (sieve theory). For example, gelatin (molecular weight, 35,000) and egg albumin (34,500) are readily filtered; serum albumin (67,500) is filtered in insignificant amount, less than 3 mg/100 ml of filtrate. Even this small amount is generally reabsorbed in the proximal tubules.

The glomerulus is normally permeable to all other materials (e.g., water, glucose, salts, and nitrogenous waste products) of the plasma except the colloidal blood proteins. Therefore, nearly all the ingredients of the urine leave the blood by ultrafiltration in the glomerulus, and the glomerular fluid formed has the same composition as the plasma, except that it usually contains no proteins. This has been substantiated for lower animals by direct examination. The filtrate has been obtained, by a micropipette (about 10 μ in diameter), from Bowman's capsules of frog and rat kidneys. Except for the absence of proteins, the ultrafiltrate was found to have about the same osmotic pressure, urea, sodium chloride, and glucose concentration and electrical conductivity as protein-free plasma.

In about 4% or 5% of healthy individuals, a very small amount (0.2%) of the albumin of the blood may find its way into the urine. After severe muscular exertion or after a cold bath this condition is quite general in most subjects. In nephritis (Bright's disease), inflammation and degeneration increase the permeability of the glomerulus and allow the albumin of the blood to escape into the filtrate and into the urine. This condition also prevails in many other diseases such as scarlet fever, pneumonia, sinusitis, streptococci throat infection, and decaying teeth. A poorly balanced diet and exposure to inclement weather are likewise said to be predisposing factors. Albuminuria depletes the proteins of the blood, and, as we learned in our study of lymph formation, the resultant decrease in the colloidal osmotic pressure of the plasma leads to edema (Chapter 16).

Recalling our discussion of lymph formation, it can be readily understood that the effective filtration pressure in the glomerulus (Fig. 29-4, *d*) is capillary blood pressure, *a*, minus the colloidal osmotic pressure of the blood proteins, *b*, and the hydrostatic pressure of the urine in the tubule, *c*. The capillary blood pressure in the glomerulus, although fluctuating, may be placed at 70 mm Hg. From the values given in Fig. 29-4, the effective filtration pressure is approximately 25 mm Hg.

$$a - (b + c) = d$$
$$70 - (30 + 15) = 25$$

When the arterial pressure falls to about 50 mm Hg, all renal secretion ceases. The foregoing readily explains why a physiological salt solution injected after a hemorrhage leaves the body by way of the kidneys. The injected salt solution lessens the concentration of the blood proteins and, therefore, their colloidal osmotic pressure; this increases the effective filtration pressure, provided the blood pressure returns to its normal value.

Glomerular filtration rate. ■ The glomerular filtration rate in man has not been measured directly, as it has in the frog and the rat by micropuncture; however, indirect methods have been employed that utilize the concept of clearance (Chapter 3).

Clearance. If a substance uniformly distributed in the plasma is freely filtered through the glomeruli but neither reabsorbed nor secreted by the tubules, it will appear in the urine at a concentration different than that in the plasma due to the reabsorption of water by the tubules. The apparent volume of plasma filtered or cleared per unit of time—the *glomerular filtration rate* (GFR) —may be measured if we know the plasma concentration, the urine concentration, and the volume of urine excreted per unit of time. GFR is calculated by:

$$GFR = U \times \dot{V}/P$$

in which GFR is the clearance or the apparent minute volume of plasma from which the unknown substance was completely removed by both kidneys during the period of collection of $\dot{V}$, the volume of urine. U and P are the concentrations of the substance in the urine and plasma, respectively.

Inulin clearance. A suitable test substance for measuring GFR must meet other criteria in addition to being freely filtrable and neither reabsorbed nor secreted by the tubules. For example, it must not be metabolized in the body, stored in the kidney, bound by nonfiltrable proteins, toxic, or have any other properties that affect the filtration rate. Furthermore, the substance should be reasonably easy to measure with accuracy in the plasma and urine. Such a substance is *inulin*, a polymer of fructose, with a molecular weight of 5,200, which is obtained from plants. We may calculate a typical GFR from the following hypothetical data.

$$U_{in} = 30 \text{ mg/ml}$$
$$\dot{V} = 1 \text{ ml/min}$$
$$P_{in} = 0.25 \text{ mg/ml}$$
$$GFR = \frac{U_{in} \times \dot{V}}{P_{in}} = \frac{30 \times 1}{0.25} = 120 \text{ ml/min}$$

The volume of filtrate formed by the glomeruli of man averages 125 ml/min; this adds up to the enormous value of 180 liters/day. We have noted that the urine volume per day is between 1 and 1.5 liters; therefore, better than 99% of the filtrate must normally be reabsorbed. The total filtration, 180 liters, is about sixty times the total plasma volume and about four times the total body water volume. To make this amount of filtration possible, the volume of blood flowing through the kidneys must be very large.

Renal blood flow. ■ Renal blood flow also can be estimated by means of the clearance concept. Any substance that meets the requirements for calculating GFR can be used, providing previous direct measurements of the concentration of the substance in renal arterial and venous blood have established that at low plasma concentrations the substance is essentially completely extracted from the blood while in transit through the kidneys. *Para*-aminohippuric acid (PAH) has been found to be 90% removed from arterial blood in a single circulation through the kidney by a combination of filtration and active tubular secretion. Since clearance is by definition a minute volume (in this case, of plasma), we can equate the clearance of PAH to the effective renal plasma flow (ERPF):

$$ERPF = \frac{U_{PAH} \times \dot{V}}{P_{PAH}}$$

in which U_{PAH} is the urine concentration, $\dot{V}$ is the urine minute volume, and P_{PAH} is the PAH concentration in circulating plasma. A typical ERPF can be calculated from the following:

$$U_{PAH} = 12 \text{ mg/ml}$$

$$\dot{V} = 1 \text{ ml/min}$$

$$P_{PAH} = 0.02 \text{ mg/ml}$$

$$ERPF = 12 \times 1/0.02$$

$$= 600 \text{ ml/min}$$

The ERPF is, of course, the volume of plasma cleared and does not represent the total plasma flow, since the extraction of PAH is only 0.9. The actual renal plasma flow (RPF) in our example would be:

$$RPF = ERPF/0.9 = \frac{600}{0.9} = 667 \text{ ml/min}$$

If we know RPF and the hematocrit (e.g., 45%), we can readily determine the renal blood flow (RBF).

$$RBF = RPF/(1 - Ht)$$

$$= 667/(1 - 0.45)$$

$$= 1{,}212 \text{ ml/min}$$

Filtration fraction. ■ The ratio of GFR to RPF is referred to as the *filtration fraction.* Its usual value is approximately 0.2. The filtration fraction is sometimes a useful index of kidney malfunction or cardiovascular disease; various diseases cause an increase in the glomerular permeability, and in some the glomeruli are destroyed. A rise in filtration fraction is observed in early congestive heart disease.

Regulation of RBF and GFR. ■ Both extrinsic, neural and humoral, and intrinsic mechanisms are operative to regulate the renal blood flow and glomerular filtration rate.

Extrinsic. The kidneys are richly supplied by sympathetic vasoconstrictor fibers. This extrinsic regulation of renal blood flow probably only functions in emergency, or stress, states. The goal of the regulation is to stabilize the effective filtration pressure and GFR; this is accomplished by an appropriate constriction of the afferent and efferent arterioles of the nephrons. Reduction in the flow due to vasoconstriction associated with a relatively constant GFR brings about an increase in the filtration fraction. Among the agents that reduce renal blood flow are exercise, upright posture, severe hypoxia, and acute hemorrhagic shock. In severe stress both renal blood flow and GFR may be reduced by the degree of vasoconstriction achieved. In effect, this provides a significant augmentation of the amount of blood available for the general circulation, for the flow may be reduced from 1,200 ml/min to as little as 200 ml/min.

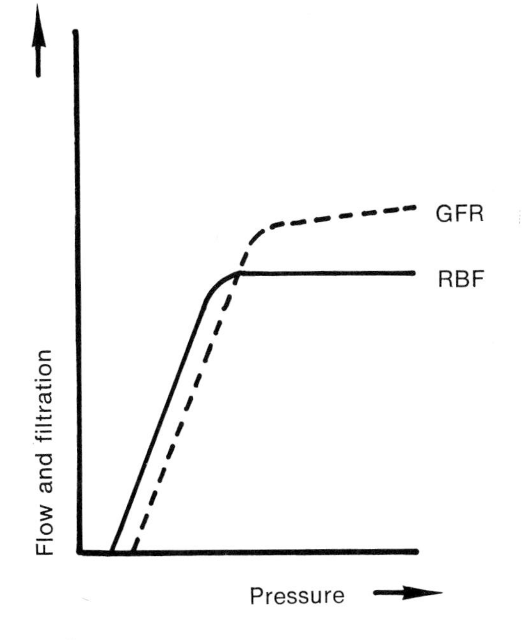

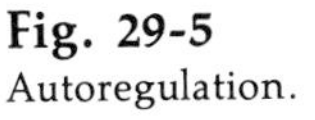
Fig. 29-5
Autoregulation.

Catecholamines in the blood produce a relatively greater afferent arteriolar than efferent arteriolar constriction. The GFR is thereby reduced. Pyrogens, substances that induce hyperemia, produce vasodilation and renal hyperemia without change in the GFR. In consequence, the filtration fraction declines.

Autoregulation. When the systemic blood pressure is suddenly raised in an animal whose kidneys have been denervated, there is a momentary increase in renal blood flow. However, the original flow is quickly reestablished. This intrinsic regulation is known as *autoregulation.* A relative constancy of renal blood flow and GFR at arterial pressures of 80 to 200 mm Hg is maintained (Fig. 29-5). Since flow is determined by pressure and resistance:

$$F = P/R$$

the relative constancy of the flow must be due to arteriolar constriction. Whereas the GFR increases slightly with increasing pressure, there is a pronounced increase in the volume of urine excreted. This occurs because the reabsorptive capacity (tubular maximum) of the kidneys for certain solutes in the filtrate is exceeded. These substances, in turn, reduce the quantity of water reabsorbed and the urine volume is increased. This is sometimes referred to as the *overflow phenomenon.* At pressures below 80 mm Hg, both the renal blood flow and GFR are linearly related to the pressure. Below 55 to 60 mm Hg arterial pressure, GFR ceases, as does urine formation.

No fully adequate explanation of autoregulation is available. The phenomenon is the subject of intensive investigation. Its importance lies in the belief that in terms of overall function of the kidney, extrinsic control is probably superimposed upon intrinsic control only in emergency situations. Autoregulation is attributed to myogenic (smooth muscle) reaction to changes in perfusion pressure and to juxtaglomerular receptors that detect changes in hydrostatic pressure or solute (sodium) concentration in tubular fluid. The juxtaglomerular theories involve the release of renin and the formation of angiotensin; the latter produces vasoconstriction of the renal arterioles and regulation of blood flow and GFR.

■ Tubular reabsorption and secretion

The large amount of filtrate formed by the glomeruli is very dilute as compared to the urine; the 180 liters of daily filtrate are reduced by passive diffusion of water out of the tubules and collecting ducts in response to an osmotic gradient to approximately 1 to 1.5 liters, the usual amount of urine voided per day. Some solutes move into or out of the tubules in accord with their chemical (e.g., urea) or electrical (e.g., chloride) gradients. However, many solutes are transferred into or out of the tubules by energy-requiring active transport mechanisms (Chapter 3). The magnitude of this active transport is shown by the fact that the kidneys, which constitute only

0.4% of the body weight, account for 8% of the resting oxygen consumption of the body.

Tubular load. ■ We have already mentioned that glucose only appears in the urine when its concentration in the plasma exceeds 170 mg/100 ml. Because the normal plasma concentration of glucose is about 90 mg/100 ml, we can conclude that approximately 112.5 mg of glucose, the *tubular load,* are filtered through the glomeruli each minute.

$$90 \text{ mg/100 ml} = 0.9 \text{ mg/ml}$$
$$0.9 \text{ mg/ml} \times 125 \text{ ml/min} = 112.5 \text{ mg/min}$$

If, as is the case, no glucose appears in the urine, all 112.5 mg/min must undergo reabsorption. In much the same manner we can estimate the secretion of PAH into the tubules. Previously we gave an example in which P_{PAH} was 0.02 mg/ml, U_{PAH} was 12 mg/ml, and $\dot{V}$ was 1 ml/min. The quantity of PAH filtered per minute must be 2.5 mg/min (125 ml/min × 0.02 mg/ml), but the excretion in the urine is 12 mg/min (12 mg/ml × 1 ml/min); we conclude that 9.5 mg of PAH are secreted into the tubular fluid each minute.

Tubular transport. ■ In calculations of renal function, the *net tubular transport* (reabsorption or secretion) is indicated by the symbol *T.* We may express the tubular activity in the form of equations for reabsorption and secretion of any substance *x* as follows:

$$\text{Reabsorption } T_X = \text{Filtered} - \text{Excreted} = GFR \times P_X - U_X \times \dot{V}$$

$$\text{Secretion } T_X = \text{Excreted} - \text{Filtered} = U_X \times \dot{V} - GFR \times P_X$$

Tubular maximum. A common observation in regard to transport mechanisms is that they can be saturated; that is, at a certain critical concentration of the substance being transferred, the system becomes loaded to its maximum capacity. The effect of substrate saturation on enzyme activity is illustrated in Fig. 22-4. When this occurs in the kidney as a result of high plasma concentrations of a solute normally reabsorbed, any filtered load in excess of the *tubular maximum* (Tm) appears in the urine. It should be noted, however, that the renal threshold and the Tm are not identical; solute begins to appear in the urine prior to the complete saturation of tubular capacity. For example, in man the renal threshold for glucose is about 1.7 mg/ml, which corresponds to a filtered load of approximately 212 mg/min (1.7 mg/ml × 125 ml/min). The tubular maximum for glucose, Tm_g, is 340 mg/min, which would represent a plasma concentration of 2.7 mg/ml (340 mg/min ÷ 125 ml/min). Glucose reabsorption, which takes place in the proximal tubule and is unaffected by insulin, is an active process requiring the expenditure of energy—a characteristic of all tubular mechanisms exhibiting a Tm. Some other substances that show a definite Tm for reabsorption are amino acids, vitamin C, uric acid, phosphate, sulfate, and certain organic anions of the citric acid cycle.

Materials that undergo active tubular secretion also manifest a Tm. Among these are choline, histamine, PAH, creatinine, and penicillin.

A summary of tubular reabsorption and secretion is offered in Fig. 29-6.

■ Urine formation

The formation by the kidney of 1 to 1.5 liters of urine from 180 liters of plasma filtrate results from the interaction of water reabsorption and mechanisms for handling osmotically active solutes, principally NaCl and urea. The magnitude and effectiveness of reabsorption in man are shown in Table 29-3 for typical daily filtered

Table 29-3. Daily filtration and excretion of selected osmotically active solutes in man

	Filtered mM/day	*Excreted mM/day*	*Percent reabsorbed*
Sodium	23,950	105	99.6
Potassium	690	50	92.8
Bicarbonate	5,100	2	99.9
Chloride	19,850	105	99.5

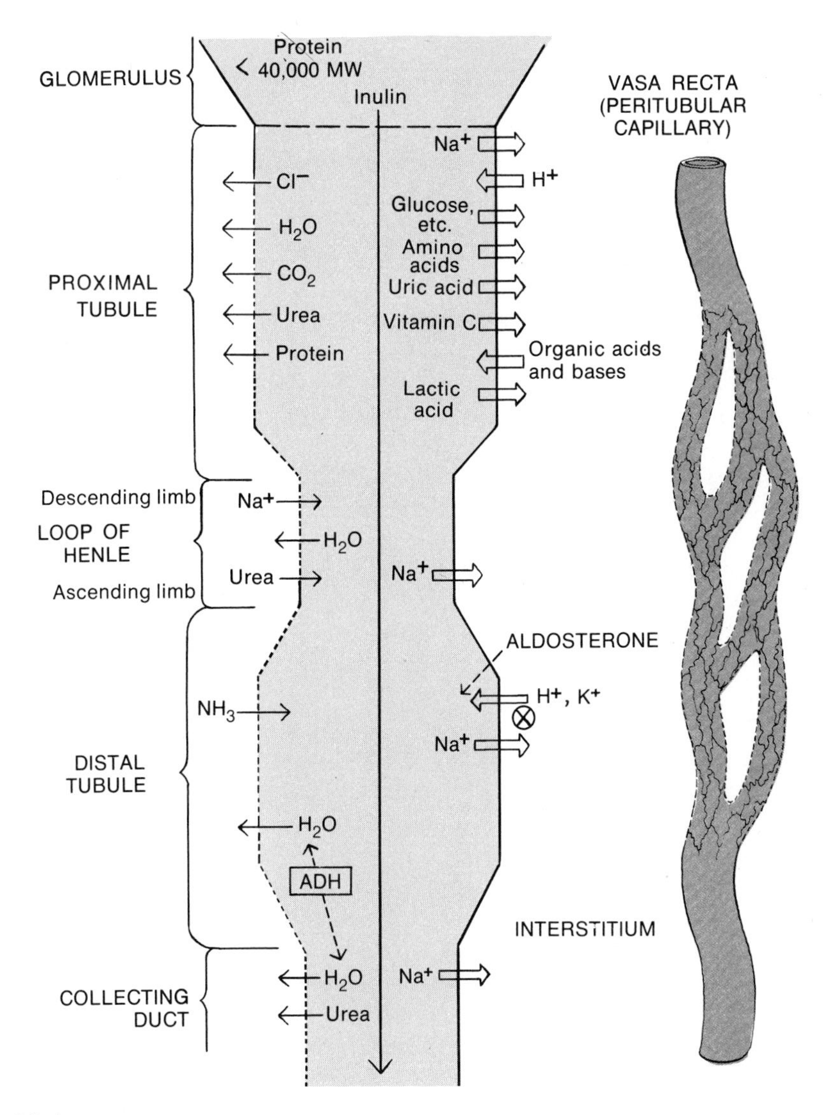

Fig. 29-6
Diagram summarizing tubular reabsorption and secretion. Transport mechanisms indicated by ⇨, diffusion by ⟶, exchange pumping by ⊗.

loads and excretion of certain osmotically active solutes.

Proximal tubule. ■ The glomerular filtrate (Fig. 29-7) enters the proximal convoluted tubule with a solute concentration of 300 milliosmols. Sodium ions are actively transported (against an electrical gradient, since the lumen of the tubule has a potential of −20 mv relative to the interstitial fluid) into the interstitium. Chloride and water diffuse out of the tubule in response to the electrical and osmotic gradients established by the active transport of Na ions. Potassium ions are actively reabsorbed in the proximal tubules. Throughout the length of the proximal tubule the filtrate remains isosmotic with plasma, whereas its volume is reduced by 80%. The water and solute reabsorbed from the proximal tubule enter the cortical peritubular capillaries due to colloidal osmotic pressure and are rapidly carried away by this extensive vascular system. The remaining 20% of the original filtrate proceeds into the loop of Henle, which dips deeply into the medullary pyramids.

Loop of Henle. ■ Although it is impractical to

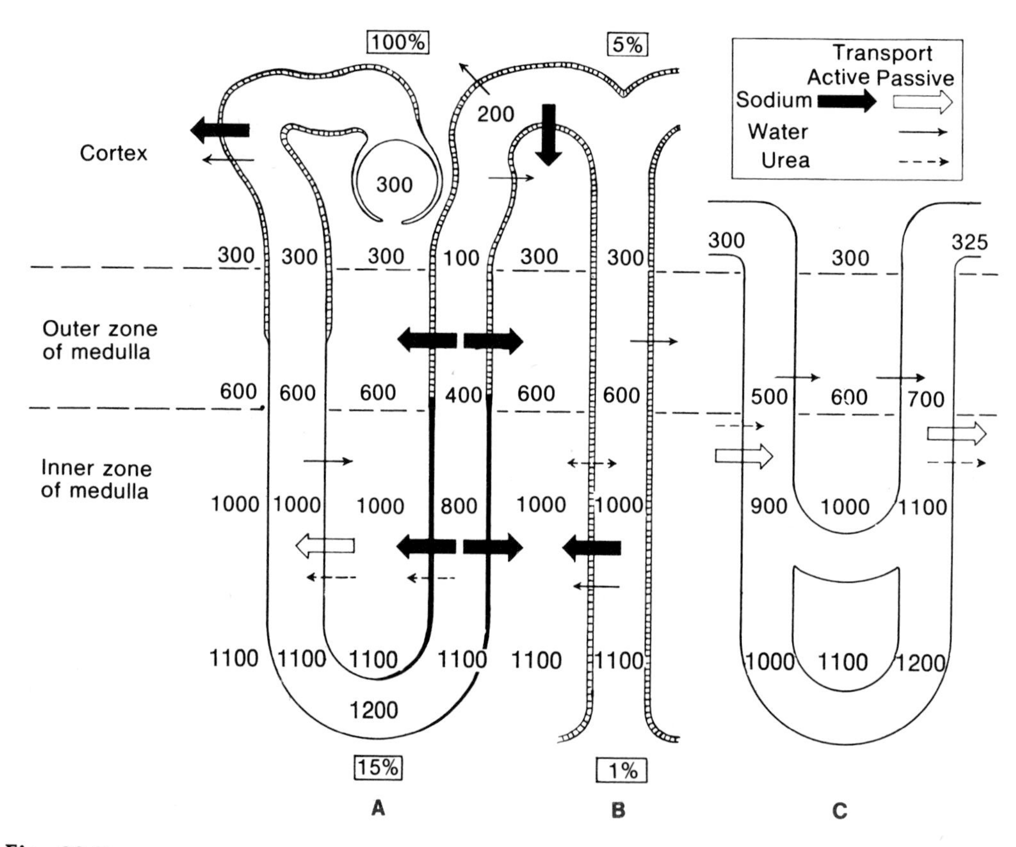

Fig. 29-7
Diagram of countercurrent mechanism. Numbers are hypothetical osmolality values indicating relative tonicity. Percentages indicate relative remaining volumes at different levels. **A,** Nephron; **B,** collecting duct; **C,** vasa recta. Interstitium is area outside of these structures. (Modified from Gottschalk and Mylle: Am. J. Physiol. **196**:935, 1959.)

undertake a thorough exposition of the subject, with the aid of Figs. 29-7 and 29-9, we can sketch in the outlines of the *countercurrent hypothesis* of urine concentration. The anatomical basis for the hypothesis is the loop of Henle system and the associated vasa recta of juxtamedullary glomeruli. Each of these form hairpin loops so that the flow of fluids is in opposite (countercurrent) directions in the two limbs, descending and ascending.

In the descending limb of the loop of Henle the solute concentration increases progressively as the fluid approaches the tip, because Na ions are being actively extruded from the ascending limb into the interstitium. The resultant increase in solute concentration of the interstitial fluid produces an osmotic gradient in the medulla and facilitates the passive diffusion of Na ions into the descending limb and of water out into the interstitium. Such a process is termed *countercurrent multiplication of concentration.* The cells of the ascending limb, although able to transport Na ions, are impermeable to water, and the filtrate becomes hypotonic by the time it reaches the distal convoluted tubule.

Distal tubule and collecting duct. ■ When the circulating amount of ADH is low, as in water diuresis, the hypotonicity of the fluid issuing from the ascending limb is maintained in the distal tubule and the collecting ducts because these structures are impermeable to water. Indeed, the hypotonicity is even increased by the continued extrusion of ions from the distal tubule. The urine that results is dilute and its volume is relatively large. On the other hand, when the titer of ADH is high, the epithelial cells of the distal tubule and collecting duct are quite permeable to water. The tubular fluid now becomes isotonic in the distal tubule and progressively hypertonic in the collecting duct due to the diffusion of water into the interstitial fluid (Fig. 29-7). The final urine is nearly as concentrated as the fluid in the tips of the loop of Henle and its volume is much reduced.

Urea appears to have only a passive role in the excretion of ions and water by most mammalian species. Its concentration in urine and the interstitial fluid of the medulla is essentially the same. The movements of the solute are adequately accounted for by passive diffusion down concentration gradients established by the prior movements of salt and water. About 70% of the filtered load enters the collecting ducts; about 13% appears in the urine. Chloride passively follows sodium whenever and wherever the latter is actively transported.

Acid-base regulation. ■ Hydrogen ions are actively secreted into the tubular fluid by the cells of the proximal and distal tubules and the collecting ducts. These hydrogen ions derive from carbonic acid produced in the cells by the action of the enzyme carbonic anhydrase. They are exchanged for sodium ions in the tubular fluid, as was discussed earlier. The quantity of hydrogen ions secreted depends upon the pH of the extracellular fluids and the quantity of buffer in the filtrate. Secretion cannot continue when the urinary pH reaches a value of about 4.4. If the buffer in the filtrate, especially phosphate, is limited, then excess hydrogen ions can be excreted after combination with ammonia ions produced and secreted into the tubular fluid by the collecting duct cells. Chloride is the most prevalent anion in the filtrate. The secretion of hydrogen ion would produce a strong acid, HCl, which would rapidly lower the pH of the filtrate and inhibit further secretion. The simultaneous secretion of ammonia makes possible the formation of ammonium chloride, a neutral salt, and the continued secretion of excess hydrogen ions.

When the pH of the extracellular fluids is too high (alkalosis), a greater quantity of bicarbonate buffer is filtered. In this situation, potassium ions are secreted by the cells of the collecting ducts in exchange for sodium ions. The same response occurs if the extracellular fluids contained too great a concentration of potassium ions.

These renal mechanisms for compensation of

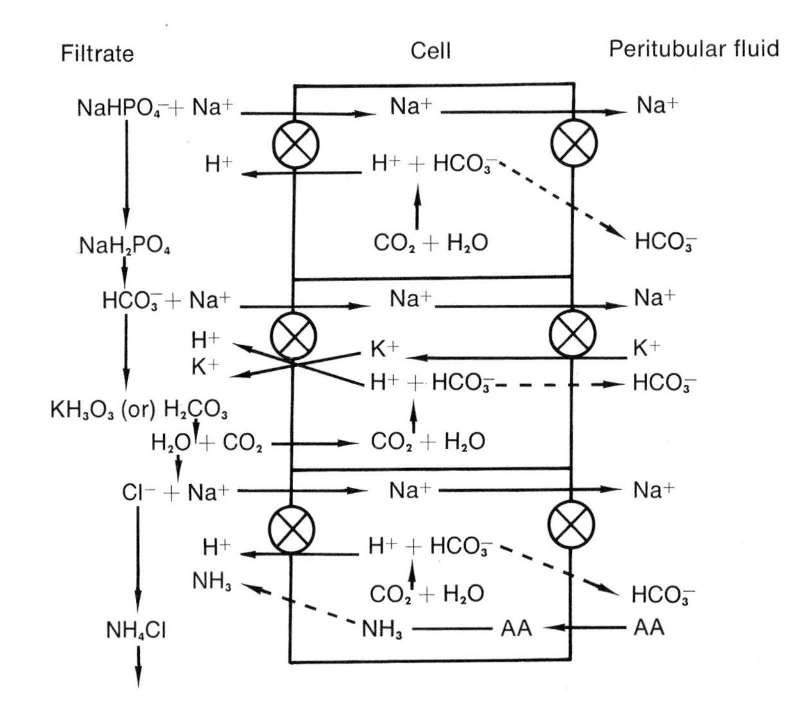

Fig. 29-8
Regulation of acid-base balance by tubule cells. Exchange transport is indicated by ⊗ .

acidosis and alkalosis are schematized in Fig. 29-8.

Vasa recta. ■ Once the medullary osmotic gradient has been established by countercurrent multiplication of concentration, additional sodium and water, which enter the interstitium during the progressive reduction in tubular volume, must be carried away by the blood. Since isotonic fluid goes into the nephron and hypotonic fluid emerges from the ascending limb, it follows that proportionately more sodium than water must be reabsorbed. Therefore, the blood leaving the medulla must be slightly hypertonic (Fig. 29-7). Of more consequence, however, is that the capillaries should not remove excessive amounts of solute and thus disrupt the medullary osmotic gradient and the continued operation of the countercurrent multiplication system, which is dependent, in the final analysis, on the not unlimited ability of the tubule cells to transport sodium. High rates of blood flow would permit a rapid dissipation, by diffusion, of the medullary osmotic gradient. Fortunately, the medullary blood flow is low and constitutes only a small fraction of the total renal blood flow. Moreover, the hairpin loops of the vasa recta provide a *countercurrent exchange,* which further reduces the rate of dissipation of osmotically active substances. Consider Fig. 29-9, *A,* in which fluid enters a straight pipe at 30 C and passes through a heat source, which it cools, to emerge 10 C warmer. In contrast the same pipe bent back upon itself, as in Fig. 29-9, *B,* would exchange heat between the outgoing and incoming fluid. The temperature of the fluid on entering and leaving would be the same as in *A,* but the heat source would not have been as effectively cooled, as shown by the temperature values at the bend.

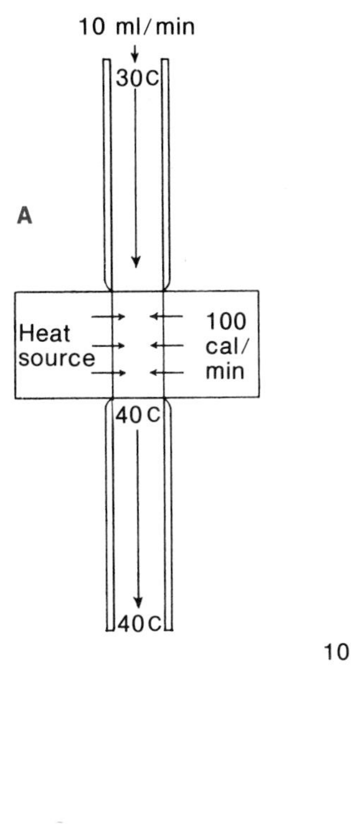

Fig. 29-9
Diagrams to illustrate the principle of countercurrent exchange. **A** shows heating of fluid in a straight pipe. **B** shows effect of bending pipe back upon itself. (From Berliner, Levinsky, Davidson, and Eden: Am. J. Med. **24**:730, 1958.)

The vasa recta (Fig. 29-7) function in an analogous way. Blood enters the capillary loop with a solute concentration of 300 milliosmols. Water diffuses out of the descending limb into the more concentrated interstitial fluid, while salt diffuses in and an osmotic gradient is established in the capillary between the cortical and medullary regions. In the ascending limb the diffusion of water and salt is reversed. Here water moves as a result of the favorable gradient created by the colloid osmotic pressure of the plasma. In effect, the salt tends to be recirculated in the medulla, thus preserving the latter's osmotic gradient, while the water previously reabsorbed by the collecting duct is carried away.

■ MICTURITION

The ureters, bladder, and urethra (Fig. 29-1) are lined with mucous membranes. These tubes are supplied with inner (longitudinal) and outer (circular) smooth muscle fibers. The ureters penetrate the wall of the bladder obliquely; this prevents the reflux of urine to the kidney. The bladder is a very strong muscular reservoir of almost a 1 pint capacity. The musculature of the bladder wall is composed of longitudinal and circular fibers; it is known as the detrusor muscle. The opening into the urethra is guarded by an internal vesical sphincter, composed of smooth muscle fibers, and by an external sphincter of skeletal muscle fibers. By these the urethra is kept closed except during urination.

The peristaltic contraction of the ureters forces the urine intermittently from the renal pelvis into the bladder. These peristaltic waves occur at intervals of from ten to twenty seconds. The muscles of the bladder exhibit a certain amount of tone. As the bladder fills, the pressure rises but slightly, due to the gradual relaxation of the muscles. When about 250 to 300 ml of urine have been collected and the pressure reaches about 180 mm H_2O, the sensory endings of the pelvic nerve in the wall of the bladder are stimulated, and there arises a sense of fullness and a desire to urinate. The impulses over the pelvic nerve are carried to a center (the vesical center) in the sacral cord. In a very young child the center discharges efferent impulses over the parasympathetic fibers in the pelvic nerves (or nervi erigentes, Fig. 11-2), which stimulate the detrusor muscle and inhibit the sphincters. In an infant this is purely a reflex action.

In the adult the same mechanism exists, but the impulses from the bladder also pass upward to the brain. The cerebrum normally has a controlling influence over the spinal center. The individual may either release the spinal center from this cortical inhibition and void the urine, or he is able, for a certain length of time, to reinforce the inhibition. By voluntary contraction of the abdominal muscles (with glottis closed and diaphragm fixed), external pressure may be applied to the bladder and the expulsion of the urine facilitated. We should call attention to this unusual situation of voluntary control being exercised over a process governed by the autonomic nervous system.

READINGS

Barger, A. C., and Herd, J. A.: Physiology in medicine: the renal circulation, New Eng. J. Med. **284**:482-490, 1971.

Davenport, H. W.: The ABC of acid-base chemistry, ed. 5, Chicago, 1969, University of Chicago Press.

Davis, J. O.: What signals the kidney to release renin? Circ. Res. **28**:301-306, 1971.

Earley, L. E., and Daugharty, T. M.: Medical progress: sodium metabolism, New Eng. J. Med. **281**:72-85, 1969.

Muntwyler, E.: Water and electrolyte metabolism and acid-base balance, St. Louis, 1968, The C. V. Mosby Co.

Pitts, R. F.: Physiology of the kidney and body fluids, ed. 2, Chicago, 1968, Year Book Medical Publishers, Inc.

Pitts, R. F.: Physiology in medicine: the role of ammonia production and excretion in regulation of acid-base balance, New Eng. J. Med. **284**:32-38, 1971.

30

REPRODUCTION

Reproduction of living things clearly differentiates them from inanimate bodies. A unicellular plant or animal duplicates its genetic material and divides into two equal parts which grow to the usual size of the full-grown organism *(asexual reproduction).* As the level of cellular organization increases, specialized cells, the reproductive cells, are found. The union of two of these cells is necessary for the continuation of the organism. In higher animals, including man, these cells are the male and female germ cells, or *gametes.* They are produced in specialized organs, the *gonads:* the two *testes* of the male and the two *ovaries* of the female. In addition to these primary sex organs, there are in man accessory sex organs which provide the necessary environment for the union of the reproductive cells and growth of the resulting organism. In the male these comprise the vas deferens, seminal vesicles, prostate, and penis; in the female, they are the Fallopian tubes (oviducts), uterus, vagina, and mammary glands.

■ MALE GENERATIVE ORGANS

■ Testes

The testes are two oval bodies placed in the scrotum (Fig. 30-1). Like most glands, the testis may be considered as composed of a large number of small tubes, the *seminiferous tubules,* which on uniting with each other form a long and much-convoluted tube. This tube, known as the *epididymis,* is closely applied to the posterior surface of the testis and gives rise to the *vas deferens* or the *seminal duct,* which constitutes the excretory duct of the testis. The vas deferens, enclosed with the arteries, veins, and nerves in the spermatic cord, passes out of the scrotum into the abdomen and reaches the posterior portion of the side of the bladder, as illustrated in Fig. 30-1. It then curves forward and downward, closely applied to the bladder, and opens into the *urethra* shortly after this leaves the bladder. In reaching the urethra, the two seminal ducts pass through the *prostate gland,* lying around the urethra close to the bladder.* The prostate discharges its secretion into the urethra. Near its opening into the urethra, the vas deferens is slightly dilated (the ampulla) and also gives off a long narrow pouch, the *seminal vesicle* or gland.

*For this reason, enlargement of the prostate (a condition frequently found in men over 50 years of age) may interfere with emptying the bladder and lead to frequent urination.

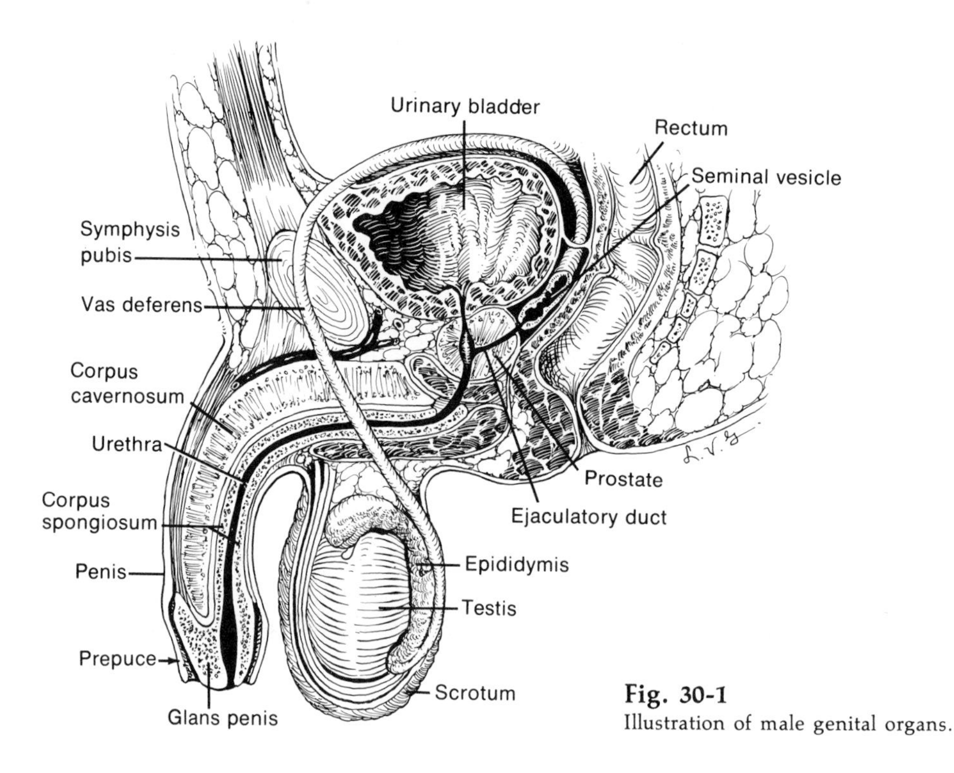

Fig. 30-1
Illustration of male genital organs.

Many parts of the seminiferous tubules and the vas deferens are supplied with cilia, and the walls of the last-named structure are well supplied with muscle tissue. The tubules in the testis are bound together by connective tissue, the stroma. In the stroma are found groups of specialized cells, the *interstitial cells,* or Leydig cells, which supply the internal secretion (androgens) of the testis.

Spermatozoa. ■ The epithelial cells that line the seminiferous tubules (known as the *germinal epithelium*) give rise, by a complex process of cell division, to the male germ cell, the *spermatozoon.* This constitutes the external secretion of the testis. The human spermatozoon (Fig. 30-2) consists of a head, middle piece, tail, and end-piece; the total length is about 0.05 mm. The head is formed almost altogether from nuclear material (DNA) of the epithelial cell that gave rise to the spermatozoon (spermatozoa; *sperma,* seed; *zoa,* animals). They are stored in the epididymis.

Spermatogenesis is influenced by some of the B vitamins. Vitamin E is designated as the fertility (or antisterility) vitamin, although this function has not been established for man. Lack of vitamin A causes keratinization of the epithelium in the seminiferous tubules and the ovaries. A diet low in proteins or merely lacking in the amino acid arginine decreases spermatogenesis.

Due to the pressure caused by their continual formation, the spermatozoa are constantly moved forward through the ducts; this motion is aided by the cilia of the tubes. The ducts, seminal vesicles, and prostate and other glands produce a fluid secretion in which the ejaculated spermatozoa are found; this complete fluid formed by these various parts is known as the

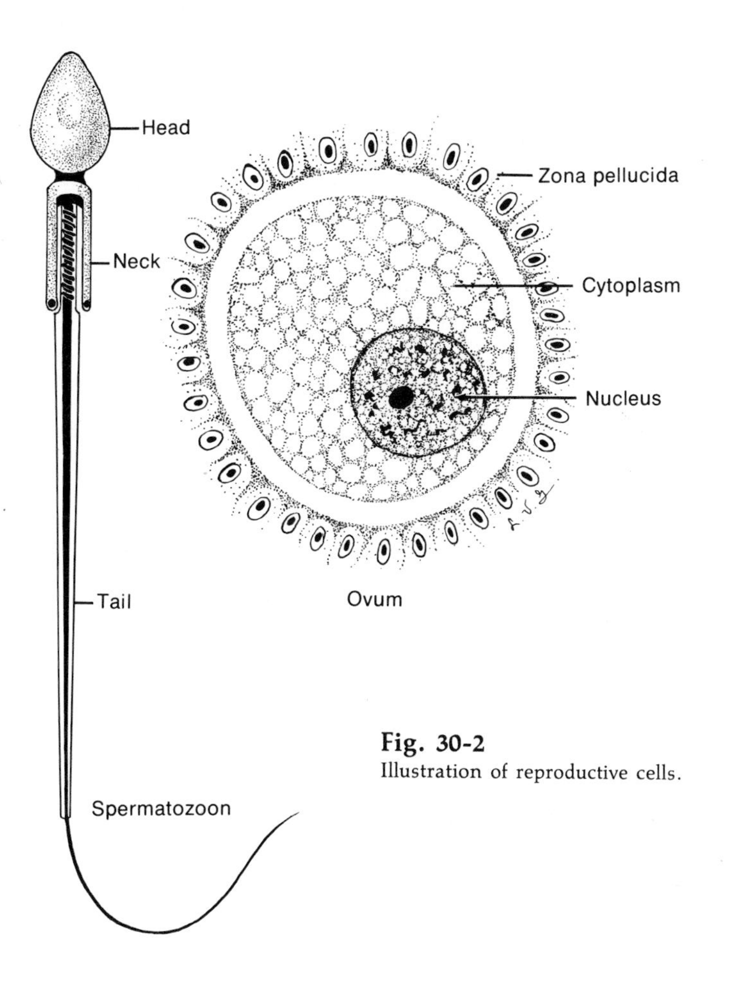

Fig. 30-2
Illustration of reproductive cells.

seminal fluid or *semen*. It contains a large amount of fructose, as nourishment for the sperm.

Various secretions are added from the male accessory glands as the sperm are transported from the testes. The secretions of the epididymis are high in potassium and glycerylphosphorylcholine, a potential energy source for the spermatozoa. The vas deferens (ductus deferens) is joined by the secretory duct from the seminal vesicles and together they form the ejaculatory duct. The secretions of the seminal vesicles have a high content of citric acid and fructose. Fructose is metabolized by the mitochondria in the middle piece of the sperm to provide energy. Sperm and the accumulated secretions are stored in the distal portion of the vas deferens, the terminal ampulla. As the ejaculatory ducts pass through the prostate the secretions of that organ are added. The prostatic secretion is slightly acidic and contains several strong proteolytic enzymes. The prostate gland contains a high concentration of zinc, and depletion of this metal results in damage to the glandular epithelium.

The various secretions of the male accessory organs serve to dilute the sperm, provide a suitable chemical environment, provide nutrient material, and increase the motility of the sperm.

The presence of small quantities of prostaglandins appears to be necessary to enhance sperm motility.

The complete fluid formed by these various parts is known as *semen*. By the peristaltic contractions of the vas deferens and the vesicles, the seminal fluid passes into the urethra, which courses through the penis. The penis is the copulatory organ by which the semen is ejaculated and deposited in the female vagina during *coitus* (insemination). This is brought about by the contraction of certain skeletal muscles. The number of spermatozoa produced is enormous; during a single ejaculation 300 million may be discharged. Discharged sperm live only one to two days in the female genital tract.

The penis is composed of cavernous or erectile tissues, consisting chiefly of connective tissue enclosing numerous spaces or venous sinuses. The filling of these spaces with blood from the greatly dilated penile arteries causes the erection of this organ. Emptying of the spaces into the veins during erection is prevented by the turgid condition of the sinuses or by the action of certain muscles. Arterial dilation may be brought about reflexly by the stimulation of the sensory nerves of the genitals; the center for this reflex lies in the lumbar part of the spinal cord, and the efferent nerves (vasodilator) are found in the first and second sacral nerves (pelvic nerves or nervi erigentes, Fig. 11-2). The center may also be influenced by mental states.

■ Puberty

Puberty is the period during which the gonads become mature and commence to function; interstitial cells begin to secrete. Puberty in the male is shown by an increase in the size of the testes and by the beginning of spermatogenesis. At this time the secondary sexual characteristics, such as growth of the beard and pubic hair and the deepening of the voice, also appear. This generally occurs at thirteen to fifteen years of age. Unconscious emissions of semen during sleep occur at intervals. The visible changes which occur during puberty are the result of an increased testicular secretion of testosterone and steroid secretion of the adrenals.

■ Hormones

No other field of physiology so beautifully demonstrates the absolute dependence for development and function of a system of organs upon the stimulating and controlling influence of a large number of hormones as that of the reproductive organs. The following hormones are involved.

Gonadotrophic hormones. ■ The previously noted changes occurring during puberty are initiated by the adenohypophyseal hormone FSH. These pituitary hormones are under the control of the hypothalamus (Chapters 10 and 28). FSH controls the multiplication of the cells of the seminiferous tubules and their maturation into spermatozoa. A second pituitary hormone, ICSH, which is identical with the luteinizing hormone (LH) in the female, stimulates the interstitial cells of the testis for the production of male sex hormones known as *androgens*. Both these functions of the testis fail in the absence of gonadotrophins. However, before the onset of puberty, the germinal epithelium must have developed sufficiently to be capable of responding to the stimulation by the pituitary hormones. The influence of the adrenal glands on the reproductive organs will be considered later.

Male gonadal hormones: testosterone. ■ The most important of the androgens or male hormones* is testosterone. It is a 17-hydroxy steroid synthesized by the Leydig cells from acetate or cholesterol. It stimulates the development of the genital tract and the development and maintenance of the accessory organs and secondary sexual characteristics. Its influence is not limited to the sex organs but is widespread throughout the body, as is shown by the results of castration.

*Several sex hormones, both male and female, have been discovered; collectively they are known as androgens and estrogens, respectively.

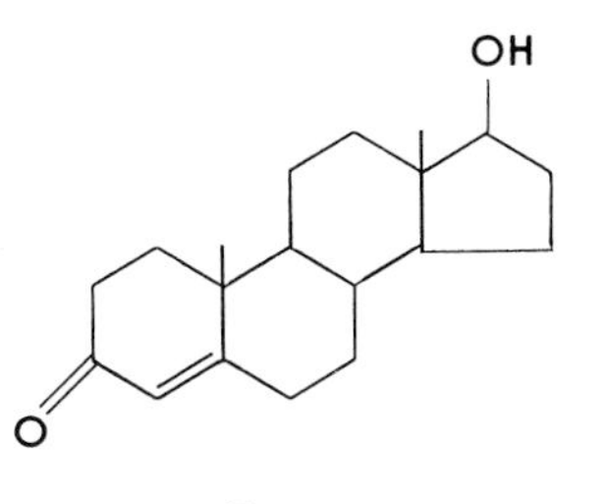

Testosterone

Gonadectomy or castration in the immature male animal or human being prevents further development of the secondary sex organs and characteristics; injection of testosterone restores them to the normal state. Castration also causes a marked fall in basal metabolic rate; this is due, at least in part, to lessened muscle activity, since the musculature of the body is less well developed and lacks tone (androgens have a protein anabolic effect). This also accounts for the greater storage of fat (e.g., the capon). In most instances there is delayed closing of the epiphyses (of the long bones).

In the adult animal, castration causes mental and emotional changes, for example, the loss of aggressiveness in an ox as compared with that of bull. In an adult man, the results of castration are less conspicuous than in animals; we may mention permanent sterility, limited atrophy of the secondary sex organs, and lessening, but not always a total loss, of the sex drive. Sex, as we shall see presently, is determined by the genes, and is, therefore, hereditary; but the degree of maleness is, at least in animals, largely hormonal. Differing from animals, the sexual life of man is controlled more by the higher brain centers, which govern mental and emotional states, and less by hormones.

Androgens are inactivated in the liver and are found in the urine of all males as conjugated water-soluble 17-ketosteroids. The small amount in the urine of young boys is perhaps formed in the adrenal cortex. At about puberty the amount increases. It is somewhat surprising to learn that the male sex hormone is also found in the urine of the female and vice versa. Evidently the gonads of either sex produce the hormones of both sexes, although predominantly those of its own sex.

FEMALE GENERATIVE ORGANS

Ovary

The ovary, like the testis, is a compound organ with dual function: the production of gametes, ova in this instance, and the elaboration and secretion of hormonal products that have important roles in growth, development, and maintenance of structures essential to the continuation of the species.

The relation of the female genital organs to the other organs of the pelvic cavity can be seen in Fig. 30-3. The ovaries, almond-shaped, are about $1^1/_4$ inches long. The body or stroma of the ovary (Fig. 30-4) is composed chiefly of connective tissue. In it are found the blood vessels. It is surrounded by a layer of cuboidal cells known as the *germinal epithelium.* During the growth of the ovary, the epithelium at various places dips into the body of the ovary, and by the growth of the connective tissue of the stroma a mass of epithelial cells becomes separated from the main layer. One cell of this mass gives rise to an immature ovum, or *oocyte;* the remaining cells form a layer surrounding the oocyte as a sac or follicle (Fig. 30-4). The ovaries of a young female are said to contain from 100,000 to 400,000 primordial follicles, most of them being present at birth. After puberty the follicles move further into the stroma. The surrounding cells continue to grow in number, and, by the formation of a liquid, *follicular fluid,* the cells immediately surrounding the ovum become separated from the more remote cells and give rise to what is known as the *Graafian follicle* (Fig. 30-4, *1* to *4*). As the follicle continues to grow in size, it moves nearer to the surface of the ovary. In the extreme right of Fig. 30-4 part of a well-developed Graafian follicle is represented, showing the layer of cells, *a,* lining the follicle, and the cells, *b,* surrounding the ovum, *c.* The immature ovum is a large cell

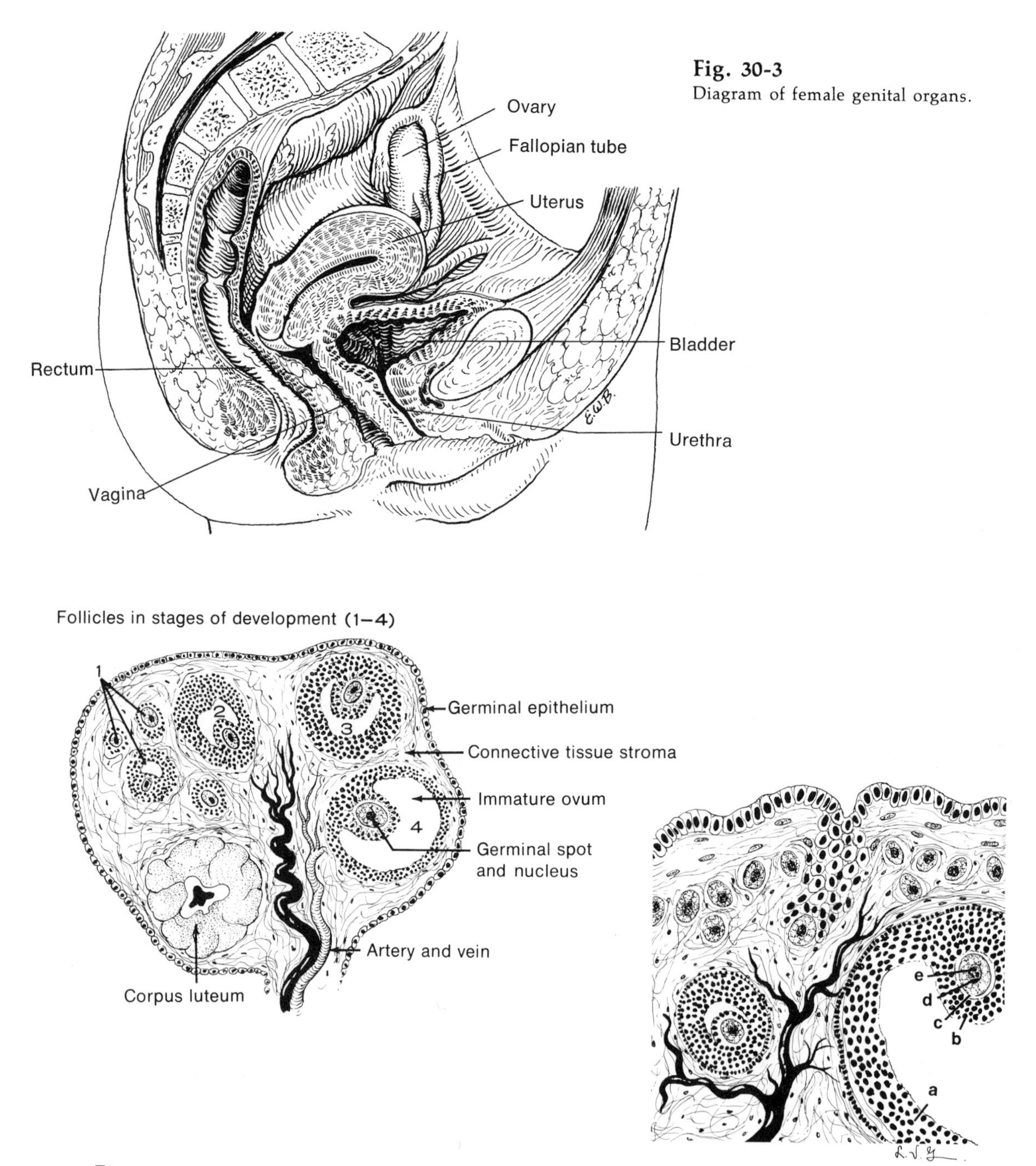

Fig. 30-3
Diagram of female genital organs.

Fig. 30-4
Section of an ovary on the left shows development of follicles, **1-4,** and corpus luteum. Section on the right shows an enlarged view of well-developed Graffian follicle.

containing a nucleus or germinal vesicle, *d*, with the enclosed nucleolus or germinal spot, *e*. After reaching a certain size, the Graafian follicle ruptures, and the ovum is liberated, a process known as *ovulation.* The ovum constitutes the external secretion of the ovary. Of the thousands of Graafian follicles, only 300 to 400 ripe ova are produced during the reproductive life of a woman; the other follicles degenerate and are reabsorbed. The human ovum measures about 0.25 mm in diameter.

Corpus luteum. ■ The cells, *a*, in Fig. 30-4, of the ruptured Graafian follicle give rise to a yellow mass, which encloses the blood clot formed by the blood escaping from the vessels of the ovary during the liberation of the ovum. This yellow mass, the *corpus luteum* (Fig. 30-4), is responsible for the secretion of progesterone, and unless pregnancy occurs it begins to regress and is replaced by fibrous tissue. If the ovum is fertilized, the corpus luteum increases in size and for the first three months continues to secrete progesterone, the hormone necessary for the maintenance of pregnancy.

■ Oviducts

The *oviducts*, or Fallopian tubes (Figs. 30-3 and 30-9), arise from the uterus and end at the ovary in a funnel-shaped end. This fimbriated end is brought close to the ovarian surface during ovulation. The oviducts are muscular and have a rhythmic contraction. This activity in conjunction with ciliary action are important in transporting the ovum to the uterus. Fertilization of the ovum by sperm occurs in the oviducts.

Prior to the discussion of further events concerned with reproduction, we shall consider the maturation of the germ cells.

■ MATURATION OF THE GERM CELLS

■ Spermatogenesis and oogenesis

In their final stage of formation, the spermatocyte, or immature sperm, and the oocyte undergo a cell division called *meiosis,* which differs radically from the mitotic cell division described in Chapter 2.

The maturation of spermatozoon and oocyte are diagrammed in Fig. 30-5 for an organism with a normal cell complement of two chromosomes.

Spermatogenesis. ■ When the germ cells, spermatogonia, in the germinal epithelium divide mitotically, one daughter cell remains as a spermatogonium and the other becomes a primary spermatocyte, *a* of Fig. 30-5. In the primary spermatocyte the nuclear membrane disappears, and the two chromosomes of the pair unite (synopsis) by twisting together. At this stage, replication of each chromosome occurs so that four chromatids, a tetrad, are present, *b*. Subsequent-

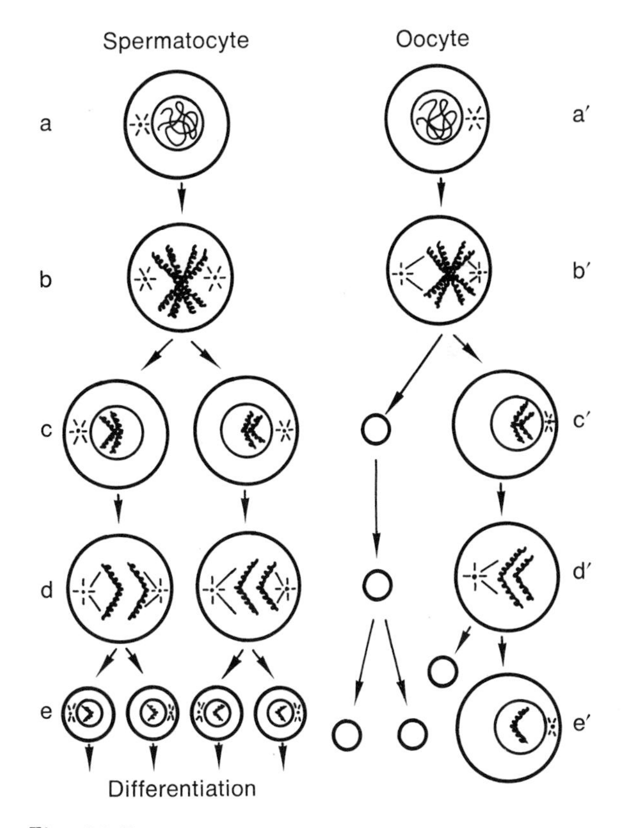

Fig. 30-5
Diagram illustrating maturation of spermatozoon and oocyte (spermatogenesis and oogenesis).

ly, the tetrad becomes aligned on the equator, the two original chromosomes separate, and two secondary spermatocytes, *c*, are formed. Each contains one half of the chromosome number found in normal somatic cells. Through this *reduction division* they have gone from the *diploid* to the *haploid* state. A succeeding mitotic *(equational)* division of the secondary spermatocyte, *d*, separates the chromatids and results in the formation of four *spermatids*. Gradual differentiation of the spermatid converts it into an active, motile spermatozoon.

Oogenesis. ■ The events in oogenesis are similar to those in spermatogenesis, except that in the reduction division of the primary oocyte, Fig. 30-5, *b'*, one daughter, the secondary oocyte, receives most of the cytoplasm. The other daughter becomes a polar body, *c'*. Again, during the equational division of the secondary oocyte, *d'*, the cytoplasm is unequally divided to leave a polar body and an *ootid, e'*, which requires no further differentiation. Polar bodies normally disintegrate.

The human spermatozoon and ovum *(gametes)* contain twenty-three chromosomes each. When the male and the female gametes unite, the fertilized ovum (*zygote*) has the original number of chromosomes, forty-six, one half of which originated from the male and one half from the female.

■ Genes

Each chromosome is composed of a large number of discrete submicroscopic bodies known as *genes* arranged in a definite linear order like a string of beads. These invisible chromosomal elements are the bearers of the hereditary characteristics in that by their chemical structure they control, in every detail, the development of the body from the zygote to its adult stature.

Each gene is known to occupy a definite position in its chromosome. In the longitudinal splitting of a chromosome during mitotic cell division, each gene reproduces itself in its own image and grows at the expense of the surrounding material.

From this discussion it must be evident that the frequently used expressions such as inherited blue eyes, musical talent, idiocy, or some other characteristic are not to be taken literally. Neither physiological nor psychological functions are inherited. All that an individual inherits from his forebearers, from Adam and Eve down to his father and mother, is a small lot of chemical compounds. Furnished with the proper environment, these are able to evolve a piece of physiochemical machinery. The precise structure of this machinery is determined (barring modifying environmental conditions) by the peculiar chemical composition of the genes or DNA handed down through countless generations of ancestors. "Each individual . . . is the product of the activity of these genes and heredity is the result of the shuffling of these genes in each generation." On this shuffling is based the value of inbreeding and outbreeding. The activity of the genes in determining the physical and chemical structure of the developing body may be modified by the environment of the ovum, embryo, or fetus.

■ Determination of sex

In the human oocyte and in the spermatocyte are two chromosomes that determine the sex of the offspring. In the oocyte these two sex chromosomes are similar and are known as the X chromosomes (Fig. 30-6). By the cell division (meiosis) occurring during maturation, there is, as we have seen, a reduction of the chromosome number; in consequence, each ovum contains but one X chromosome. One of the two sex chromosomes of the spermatocyte is similar to that present in the ovum and is, therefore, also called the X chromosome; the second sex chromosome is considerably shorter and is designated as the Y chromosome. As a result of the next cell division and reduction in chromosomes, one spermatozoon contains the X chromosome and the other the Y chromosome.

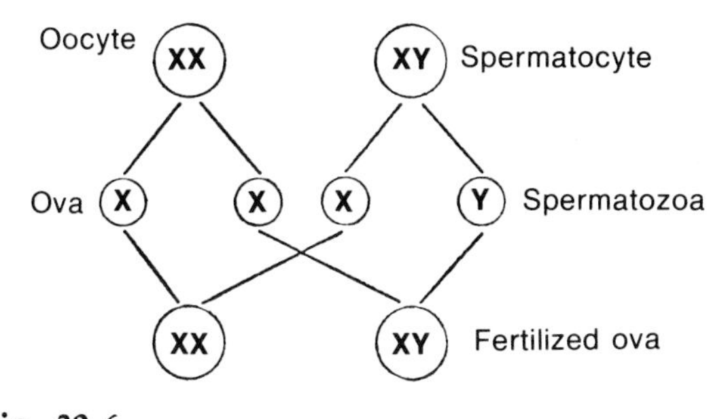

Fig. 30-6
Diagram to illustrate inheritance of sex.

By the subsequent union of an ovum with a spermatozoon containing an X chromosome, a cell with two X chromosomes, that is, a female cell, results. On the other hand, the union of an ovum with a spermatozoon possessing a Y chromosome gives rise to a cell containing one X and one Y chromosome or a male cell* (Fig. 30-6). The genetic sex of the offspring is fixed at the time of fertilization. Certain characteristics of the offspring are determined by genes located in the sex chromosomes; these are said to be *sex-linked characteristics,* such as color blindness and hemophilia.

■ Inheritance

The study of genetics is a discipline in its own right and the subject of entire textbooks. Here we can essay only to sketch in a few facets of the phenomenon of inheritance.

Alleles. ■ Except for the sex chromosomes, somatic chromosomes are called *autosomes.* In higher organisms the autosomes exist in pairs. The chromosomes of a given pair are known as *homologous chromosomes* and the genes they contain as homologous genes. However, the genes of homologous chromosomes may code for different messages; for instance, the genes of one chromosome of a homologous pair may specify pigmentation whereas the genes of the other chromosome specify a lack of pigmentation, albinism. When homologous genes transcribe different messages they are known as *alleles.*

Alleles interact to determine the characteristics or appearance (phenotype) of an organism. In some cases one gene predominates over the other. This is referred to as simple dominance. The interaction of some alleles results in partial dominance, or even in codominance. Furthermore, nonhomologous genes, on the same or different chromosomes, sometimes interact to complement or supplement one another and bring about multiple factor inheritance.

Simple dominance. Although pigmentation in the human in reality involves up to six gene pairs (multiple factor inheritance), we can illustrate simple dominance by assuming that a single allelic pair is responsible. The dominant gene of this pair permits pigmentation; the recessive gene causes albinism. We can designate the dominant gene as P and the recessive gene as p. As a result of reduction division the gametes of an individual will contain either P or p. The union of gametes from homozygous pigmented and albino individuals would yield four heterozygous progeny of the $P\,p$ genotype, each of whom would be phenotypically pigmented.

$$P\,P \times p\,p = P\,p, P\,p, P\,p, P\,p$$

Cross mating of two heterozygous individuals would yield three pigmented phenotypes and one albino phenotype. The ratio of genotypes would be: one of $P\,P$, two of $P\,p$, and one of $p\,p$.

$$P\,p \times P\,p = P\,P, P\,p, P\,p, p\,p$$

If a pigmented heterozygote is mated with an albino, the union produces equal numbers of pigmented and albino phenotypes and genotypes.

$$P\,p \times p\,p = P\,p, P\,p, p\,p, p\,p$$

In all combinations the pigmented phenotype occurs whenever the dominant gene appears in

*It is, therefore, the male germ cell and not the female germ cell that determines the sex of the offspring.

the somatic cells resulting from the union of gametes.

Partial dominance. When a phenotypic characteristic is expressed fully by an individual with homozygous alleles and only slightly by an individual with heterozygous alleles, partial dominance is responsible. Thus, in a certain type of inherited anemia (Cooley's) the three homologous genotypes can be *A A, A n,* and *n n*. The *n n* genotype is unaffected, and the *A n* individual shows only mild symptoms. The individual of the *A A* genotype is afflicted seriously, even mortally. Partial dominance is of significance because it will handicap the individual to a greater or lesser degree, if the partially dominant gene is a deleterious gene.

Codominance. The ABO human blood groups (Chapter 12) exhibit codominance. The three alleles can be combined in six ways: AA, AO, BB, BO, AB, and OO. Simple dominance between genes accounts for three blood groups: A, B, and O. The fourth blood group, AB, demonstrates codominance, each gene acting as if it were present without the other.

Sex-linked inheritance. In humans and many other species the sex chromosomes are unequal in their total gene content. The Y chromosome in man is physically smaller than the corresponding X chromosome and to large degree behaves as if it were devoid of genes. Consequently, many traits are attributable solely to the X chromosome. Altogether nearly sixty X-linked traits have been described as occurring in humans; most are produced by a recessive gene. Among these traits are red-green color blindness (Chapter 9), hemophilia, and the predilection to hemolytic anemia (Chapter 12). The recessive trait appears in males whose X chromosome carries the recessive gene. Females possessing the recessive gene on only one X chromosome are carriers, but do not exhibit the trait. Males cannot convey the recessive gene to their sons, since only the Y chromosome is transmitted to male issue; daughters of an afflicted male will be either carriers (if neither of the mother's X chromosomes carries the recessive gene) or exhibit the trait.

Premarital counseling is an important aspect of genetics, for it can, through the study of genetic traits in the families, offer guidance about the propriety of producing children.

The existence of an odd number of chromosomes in a cell is termed *aneuploidy.* The inequality may arise from either loss or gain of a chromosome due to an error during mitosis or meiosis. Aneuploidy may be either autosomal or associated with the sex chromosomes. Mongolism (Down's syndrome) is an example of autosomal aneuploidy where chromosome 21 is not paired but exists as a triplet. Since the incidence of mongoloid offspring increases with maternal age, it is thought that the age of the ovum may be critical in the production of this autosomal defect. The addition of an extra X chromosome to the normal male's XY complement (XXY) results in an abnormality known as Klinefelter's syndrome. This syndrome is associated with mental retardation, sterility, and tendencies toward female physical characteristics. The deletion of one X chromosome (XO) from the female produces Turner's syndrome, which is characterized by the absence of ovarian function and short stature. It should be noted that the abnormalities produced by aneuploidy, although not rare, are not frequent. Furthermore, modifications of the simple relationships described above occur.

■ SEX LIFE OF THE FEMALE

■ Puberty

In a young girl the Graafian follicles do not reach maturity; but at about age 10 to 14 years they begin to ripen and ova are then set free, as described previously. Even before the liberation of the first ovum, many changes, physical and psychic, take place in the individual. The accessory sex organs, uterus, vagina, and mammae, undergo a marked increase in growth due to pituitary gonadotrophin stimulation of ovarian estrogen secretion. The secondary sex characteristics make their appearance. These are the growth of pubic and axillary hair, the peculiar development of the skeleton (especially the enlargement of the pelvis), and deposition of fat on the hips, which causes the body to assume a more feminine contour. As in the male, pubescence in the female is accompanied by heterosexual (*hetero,* other) inclinations.

But not only is there an increased growth and development; sexual activity now begins with the appearance of the first menstruation. These uterine activities continue rhythmically (about every twenty-eight days) during the fertile period of life (which ends at about the age of from forty to fifty years) except during pregnancy. The cessation of the menses, known as the *menopause,* climacteric, or change of life, may be accompanied by serious physical and mental disturbances.

The gradual failure of ovarian function and decrease in estrogen levels to negligible values results in changes in the secondary sex characteristics. The muscle of the uterus is changed to fibrous tissue and shrinks; there is some instability in the vasomotor system during the menopause which results in "hot flashes." The glandular tissue of the breast atrophies although the size of the breasts may enlarge. Other effects of lowering of estrogen levels have been discussed in Chapters 24 and 28.

■ Menstruation and ovulation

The uterus is a hollow, thick-walled muscular organ (Fig. 30-3). It is lined with mucosa (the *endometrium*), which is well supplied with glands. The endometrium undergoes cyclic struc-

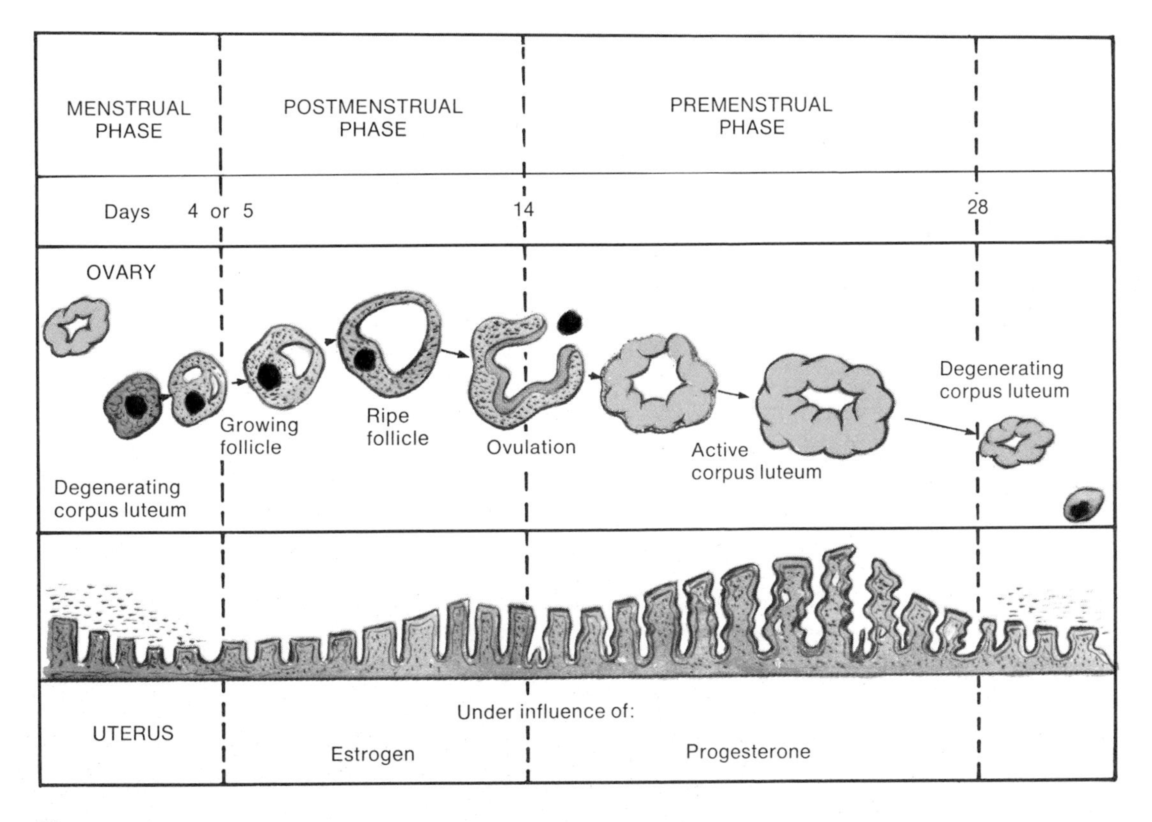

Fig. 30-7
Illustration of the events occuring in ovary and uterus during the menstrual cycle.

tural changes that can be considered as periodic preparation for fertilization of the ovum and pregnancy. This cycle in primates is termed the *menstrual cycle.* Events associated with this cycle in the uterus and ovary are depicted in Fig. 30-7. The menstrual cycle is conveniently dated from the first day of bleeding. During the four to five day period of the menses, the secretory endometrium undergoes degenerative changes so that a part of the mucosa disintegrates and is discharged with a quantity of blood. The amount of blood loss may vary from 20 to 200 ml and may be preceded or accompanied by severe pain, *dysmenorrhea.* Over the span of active reproductive life, a cumulative loss of 40 liters of blood may be experienced; this may explain, in part, why some women have a mild to moderate degree of hypochromic anemia. Loss of 150 ml of blood each month is equivalent to the loss of 75 mg of iron; therefore, it is essential that sufficient iron is present in the diet to replace the iron (Chapter 31).

After the flow has ceased, there is a gradual thickening of the endometrium (*proliferative* phase) due to the influence of *estrogens* from the ovary. At about the fourteenth day ovulation takes place, and *progestational* or *secretory* (phase) changes ensue. The endometrial glands become complicated and tortuous, and the submucosal layer becomes very vascular and edematous; these changes are induced by estrogens and *progesterone* from the corpus luteum. Ovulation can be ascertained by the difference between the basal body temperature of the preovulatory period and the postovulatory period; the latter is between 0.3 and 0.5 C higher. If fertilization is not accomplished within a period of a few days the ovum dies. The period of greatest probability of conception extends from about the thirteenth to seventeenth day of the cycle. However, this is a statistical fact, and there is wide variation from one individual to another. Fertilization leads to pregnancy; in the absence of fertilization, the secretory endometrium is maintained for a time, but on or about the twenty-eighth day menstruation recommences.

■ Hormones

Pituitary gonadotrophic hormones. ■ As in the male, the reproductive functions of the female are controlled by gonadotrophic hormones of the adenohypophysis. Removal of the pituitary in an immature animal or its lack of function in a child prevents the growth and development of the ovaries and, indirectly (because of the lack of ovarian hormones), of the accessory organs and sex characteristics. In an adult these organs atrophy and all reproductive functions cease.

Follicle-stimulating hormone (FSH). This hormone is responsible for the growth of the ovaries and the maturation of the ovarian follicle and its germinal cell. Without it, ova are never formed. It stimulates the secretion of the ovarian hormones, *estradiol* in particular. In the immature female the amount of FSH secreted is negligible, but at puberty it is increased; during pregnancy it is one hundred to two hundred times as abundant as in a nonpregnant woman.

Interstitial cell–stimulating (ICSH) or luteinizing hormone (LH). This hormone has three functions: (1) in conjunction with FSH, it aids in the ripening of the Graafian follicle; (2) it is responsible for ovulation and for the growth, development, and continued existence of the corpus luteum; and (3) it causes the secretion of ovarian hormones, *estrogen* (estradiol) and *progesterone* (progestin) (Fig. 30-8).

Ovarian hormones. ■ Under the influence of FSH and LH the ovaries and the Graafian follicles secrete hormones known as estrogens. The corpus luteum, during early pregnancy, and the placenta, through term, secrete both estrogens and progesterone. In addition, the placenta produces *chorionic gonadotrophin,* which prevents regression of the corpus luteum through the third month of pregnancy. The persistence of the functioning corpus luteum interrupts the menstrual cycle.

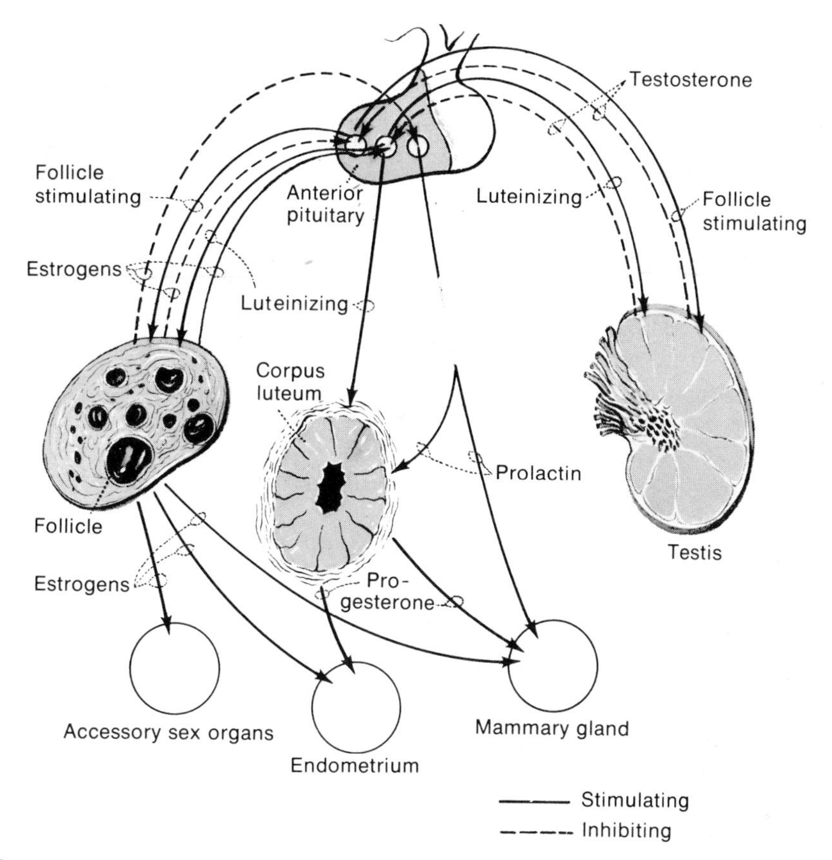

Fig. 30-8
Diagram showing influences of pituitary and ovarian hormones. Full lines indicate stimulation; broken lines indicate inhibition.

Estrogens. The principal hormone produced and secreted by thecal cells of ripening ovarian follicles in the humans is *estradiol. Estriol* appears to be the compound secreted by the placenta. Estrogens are 17-OH steroids synthesized from cholesterol, and the biochemical interrelations between estrogens, androgens, and hormones of the adrenal cortex are indeed complicated. A very brief outline of the relationships is shown here.

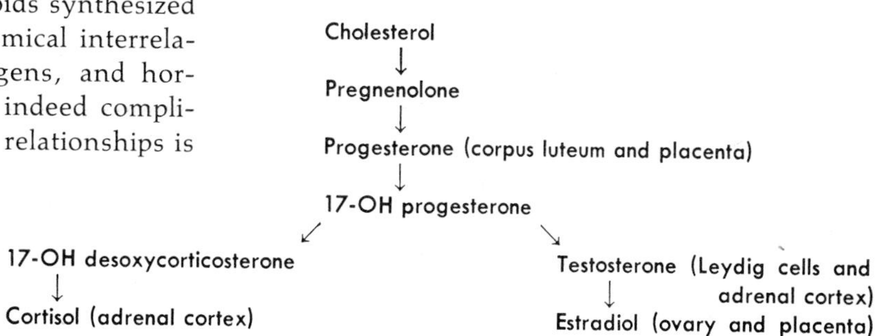

Estrogens accelerate the growth of accessory sex organs, prepare the uterus for reception of a fertilized ovum, and stimulate the growth of the ductile elements and enlargement of the mammary glands. They also induce the development of secondary sex characteristics. The feedback regulation of FSH secretion by estrogens (and progesterone) accounts for the cyclic nature of female reproductive activity. Although large amounts of estrogen inhibit FSH, small amounts are stimulatory, and moderate amounts induce LH secretion. High titers of progesterone inhibit LH secretion. During menopause no restraint is placed on the production of pituitary gonadotrophins, inasmuch as the ovaries have ceased to form estrogens and progesterone. Disturbances that frequently accompany the "change of life" have been attributed to the excessive amount of gonadotrophins. Administration of testosterone or estradiol gives relief.

Estrogens and testosterone are antagonistic. It is somewhat amazing that the adrenal cortex secretes, in both male and female, androgenic and estrogenic hormones; the androgens predominate. The object of this cortical activity is not clear. Abnormal adrenal activity (tumors) causes excessive secretion of adrenal androgens and masculinization of females. It may be relieved by removal of the tumor. Their effect on bone calcium is discussed in Chapter 24.

Estrogens, like all naturally occurring steroids, are inactivated by the liver, where they are conjugated with glucuronic acid or sulfuric acid. Conjugated steroids are water soluble and readily excreted in the urine.

Progesterone. In the development of the corpus luteum there is produced another ovarian hormone known as progesterone. We may note the following activities of progesterone:

1. By continuing the work of estrogens upon the uterus, it prepares the mucosa of this organ for the reception of the fertilized ovum and the maintenance of the embryo.
2. Progesterone inhibits the maturation of additional Graafian follicles and ovulation and thereby prevents the estrus cycle in lower animals and menstruation in women during pregnancy.
3. It also stimulates further growth of the mammary glands, provided these have previously been acted on by estrogens. We therefore see that in some respects these two ovarian hormones are synergic and, in other respects, antagonistic (Fig. 30-8). Progesterone is excreted in the urine as *pregnanediol.*

Chorionic gonadotrophins. ■ Human chorionic gonadotrophin (HCG) is a glycoprotein of about 100,000 molecular weight. It is secreted by the placenta as early as fourteen days after fertilization. HCG maintains and enlarges the corpus luteum and stimulates its secretion of estrogens and progesterone. After the third month, this function is not essential, and HCG secretion begins to decline. Estrogen and progesterone from the placenta itself continue to show increased secretion until parturition.

The presence of HCG in the urine is an almost unfailing *test for pregnancy.* Varied tests have been devised, but most are only modifications of the Aschheim-Zondek test, wherein a small quantity of urine from a pregnant (or suspect) woman is injected into a immature female mouse, rat, or rabbit. Chorionic gonadotrophin in the urine stimulates the recipient animal's reproductive organs and results in estrus (heat), with characteristic changes in the genital tract.

Hormones and the mammary glands. ■ The growth, development, and activity of the mammary glands are controlled by four hormones, each playing a specific part. At the onset of puberty the *estrogens* stimulate the growth of these glands; this, however, is limited almost entirely to the formation of connective tissue and the construction of a duct system. *Progesterone* by itself has no effect, but, given simultaneously with estrogens, it stimulates the development of secretory cells. Soon after parturition, the pituitary hormone *prolactin* starts the elaboration of milk;

oxytocin stimulates its release. Prolactin markedly accelerates the rate of glucose utilization and stimulates lipogenesis.

■ FERTILIZATION

Notwithstanding the opposing currents set up by the cilia of the uterus and of the oviducts, the spermatozoa penetrate by their motility from the vagina into the uterus and tubes.

Sexual intercourse. ■ Fertilization of the ovum by a spermatozoon is preceded by intercourse in which the erect penis of the male is introduced into the vagina of the female. The sexual response of the individual is determined both physiologically and psychologically. A variety of stimuli built up from the moment of birth through sexual maturity and beyond result in the development of conditioned reflex responses. It is possible that in man the psychological responses may completely dominate the physiological ones; therefore even though the genetic, endocrine, and gonadal characteristics of one sex are present, the sexual behavioral response may be of the opposite sex. All levels of the nervous system, as well as the hormones, are essential for directed sexual activity.

Sexual intercourse can be divided into four phases, although they overlap and cannot be sharply differentiated physiologically. *Excitement* in the male is evidenced by the erection of the penis with accompanying psychic tension. Engorgement of the erectile tissue of the penis with blood which causes erection is accomplished rapidly and is mediated by a reflex arc located in the sacral segments of the spinal cord. Outflow of blood from this organ is prevented by contraction of muscles located in the corpus cavernosum near the pubic area. In the female, the excitement phase also may be accompanied by erection of the clitoris and labia minora and by a transudation through the vaginal wall of a fluid which moistens this canal and the introitus. This period may last for a few minutes or hours and may increase and decrease during this time. There is a general heightening of muscular tone throughout the body.

The *plateau phase* is characterized by a general maintenance of the excitatory phase, accompanied by an enlargement of the female breasts and increased turgidity of vaginal structures that constitutes the development of an orgasmic platform. The rhythmic moving of the penis into and out of the vagina causes sensations of pleasure in both the male and the female. As these thrustings become more vigorous, *orgasm* is experienced. This is recognized in the male by the ejaculation of seminal fluid; it is brought about by contractions of the vasa deferentia, the seminal vesicles, and the bulbocavernosus which contracts vigorously. In the female a similar pattern occurs in which the orgasmic platform undergoes rhythmic contractions. Following *resolution* there is general feeling of physical well-being and mental lassitude. Orgasm in the female is not essential for fertilization and, indeed, does not always occur at every intercourse. During sexual intercourse there is an increase in respiratory rate (up to 40/min), pulse rate (100 to 170/min), and blood pressure (30 to 80 mm Hg systolic and 20 to 40 mm Hg diastolic). The durations of these changes are relatively short but may be of consequence in individuals suffering from respiratory or cardiac diseases.

Transport of sperm. ■ Sperm cannot survive for long periods in the acid vagina, although the large volume of seminal fluid does tend to neutralize this condition. Notwithstanding the opposing currents set up by the cilia of the uterus and the oviducts, the sperm rapidly progress through the uterus and into the Fallopian tubes. This transport may be assisted by prostaglandins present in the seminal fluid and by the high content of hyaluronidase which acts upon the mucoid substances present to make them less viscous.

Fertilization. ■ If a female gamete is present, one, and only one, sperm may enter it. The conjugation of the male and the female germ cell is known as *fertilization,* fecundation, or impregnation; normally this takes place in the Fallopian tubes.

When a sperm enters the ovum, it loses its tail. The nucleus of the spermatozoon and that of the ovum unite to form one single nucleus, the segmentation nucleus, which now possesses the number of chromosomes characteristic of the species, *half of these being of maternal and half of paternal origin.* In the subsequent divisions of the fertilized ovum in the usual manner of mitosis (Chapter 2), the chromosomes divide longitudinally, and each daughter cell thereby receives one half of the paternal and one half of the maternal chromosomes.

■ GESTATION OR PREGNANCY

During its passage from the Fallopian tube to the uterus, the zygote undergoes cell divisions and successively forms a structure of 2, 4, 8, 16, etc. cells. About three days after fertilization it arrives at the uterus. Meanwhile the uterine mu-

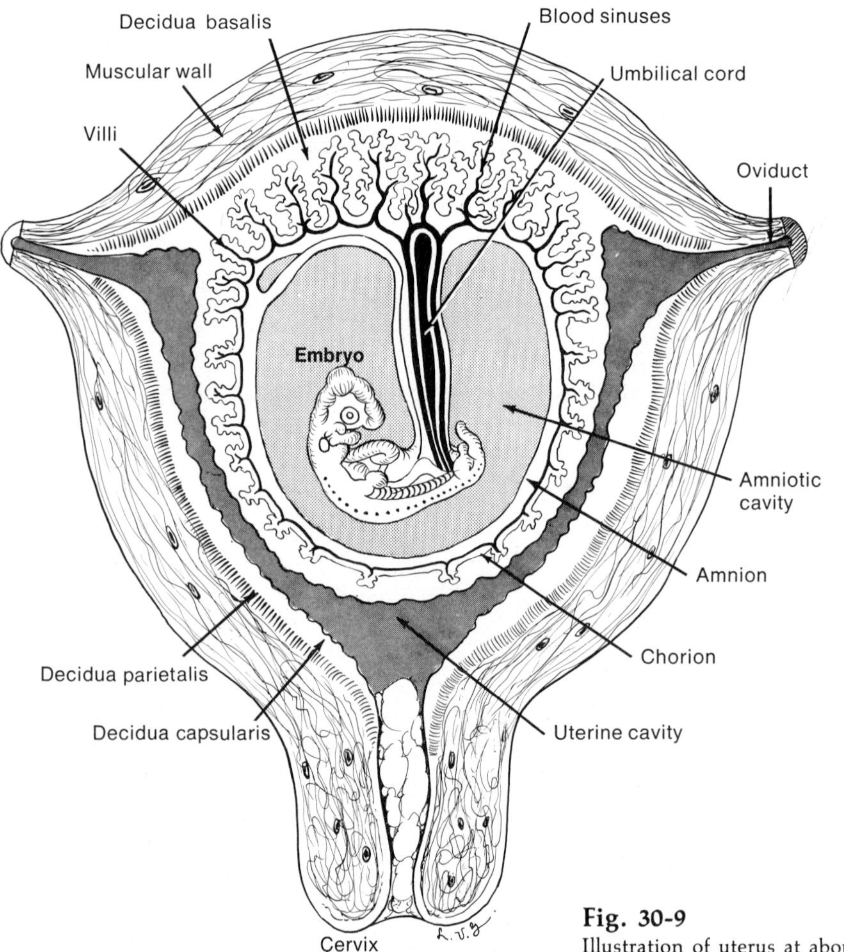

Fig. 30-9
Illustration of uterus at about seventh week of pregnancy.

cosa, under the influence of estrogens and progesterone, has undergone preparatory changes for the reception of the ovum. On coming in contact with the mucosa, the cells on the outside (trophoblasts) of the developing cell mass (blastocyst) begin to digest, by proteolytic enzymes, a limited area of the mucosa. This allows it to sink into the mucosa and to absorb nutrition (trophoblastic) until the placenta has been formed. By continued growth of the mucosa, the embryo becomes surrounded by two membranes, the *amnion* and the *chorion* (Fig. 30-9). The space between the embryo and the amnion, the amniotic cavity, is filled with amniotic fluid in which the embryo rests.

During pregnancy there occurs great growth of the walls of the uterus due to estrogens. The virgin uterus has a capacity of 2 or 3 ml; at full term its volume has increased to 5 to 7 liters; its weight increases about thirty times. The increase in size is due chiefly to an extremely great increase in the number and size of the smooth muscle fibers. That part of the uterine mucosa to which the embryo has become attached increases greatly in thickness and in it are developed large blood spaces into which the uterine arteries discharge blood; it is called the *decidua basalis.* The chorion (the outermost fetal membrane) fuses intimately with the decidua basalis and develops villi, which come to lie in the blood spaces (Fig. 30-9). This part of the mucosa (part of maternal and part of fetal origin) is designated as the *placenta.*

Two umbilical vessels leave the embryo at the umbilicus or navel and pass through the umbilical cord (Fig. 30-9) to break up into capillaries in the villi and thus lie in a pool of maternal blood. Here the exchange of materials between the maternal and fetal blood takes place. The placenta is therefore the respiratory, nutritive, and excretory organ for the embryo. Blood in the umbilical artery leaving the embryo is reduced blood, that in the umbilical vein is oxygenated blood. The fetal liver stores fat-soluble vitamins A and D; water-soluble vitamins are not stored. The fetus manufactures its own blood. The placenta bars nearly all bacteria that may be present in the maternal blood with a few exceptions, chief of which is the germ of syphilis *(Spirochaeta pallida);* congenital syphilis (not hereditary) is possible.

Maternal physiological adjustments. ■ The uterus and breasts increase during pregnancy, which increases the size of the vascular bed. This necessitates an increase in blood volume. At term there has been an increase in amount of hemoglobin and about a 50% increase in plasma volume; therefore, the content of Hb/100 ml is somewhat lower than in the nonpregnant state. During pregnancy there is an increase in cardiac output to about 5.5 liters/min. This is almost entirely the result of an increase in heart rate. Blood flow through the skin increases. Blood pressure appears to fall during the middle three months of pregnancy. Respiration rate does not usually increase during pregnancy but tidal volume (Chapter 17) does. Renal plasma flow increases early in pregnancy and glomerular filtration rate (GFR) is increased. Minor physiological adjustments occur in every tissue during pregnancy.

■ PARTURITION

After the full term of gestation—in the human being about two hundred and eighty days after the last menstruation—the fetus is expelled from the uterus. By the contractions of the uterine musculature (labor pains), pressure is exerted on the fetus; this causes the fetus to move toward the mouth of the uterus. This is aided, to some extent, by the contraction of the muscles of the abdominal wall, with the diaphragm fixed and the glottis closed. In consequence of this pressure and because of the relaxation of its circular muscle fibers, the mouth of the uterus dilates. When a sufficient amount of pressure has been generated, the membranes surrounding the fetus rupture and part of the amniotic fluid escapes. By the continued contraction of the muscles and by a further relaxing of the soft parts, the child is delivered through the mouth of the uterus and vagina.

After this, the contractions cease for a time. When they are resumed the fetal membranes and the placenta, being torn away from the wall of the uterus, are expelled as the afterbirth. This results in the escape of a certain amount of blood from the blood spaces of the placenta, but due to the strong contractions of the wall of the uterus, the ruptured vessels are mechanically occluded, and thus further hemorrhage usually is prevented.

As to why pregnancy terminates at a certain, fairly definite time is unknown; neither is it known what initiates the contractions of the uterine musculature. More than likely, it is not of neural origin, for it has been demonstrated that in a dog complete destruction of all the nerves supplied to the uterus does not interfere with pregnancy and parturition. Nevertheless, there is located in the lumbrosacral cord a center capable of influencing this activity, for it is well known that certain mental conditions and also the stimulation of sensory nerves may hasten parturition in a human being (e.g., miscarriage by fright). Whether oxytocin, a hormone of the posterior lobe of the pituitary (Chapter 28), is of any importance in normal parturition is doubtful.

The cause of the first breath of the newborn infant has been attributed to two sources: (1) to the increasing amount of carbon dioxide in the blood and (2) to cutaneous stimulation, both mechanical and thermal. Perhaps the latter is of greater importance.

■ MATERNAL DIET AND LACTATION

Maternal diet. ■ The embryo or fetus is a parasite living and thriving at the expense of the

Table 30-1. Average composition of human and cow's milk*

Kind of milk	*g/100 ml*			*mg/100 ml*			*Calories per 8 oz*
	Protein	*Fat*	*Lactose*	*Ca*	*P*	*Fe*	
Human's	1.2	3.8	7.0	33	15	0.15	168
Cow's	3.5	3.9	4.9	118	93	0.10	166

*From Church and Church: Food values of portions commonly used, ed. 9, Philadelphia, 1963, J. B. Lippincott Co.

Table 30-2. Average height-weight measurement of children from birth to five years of age*

	Height (inches)		*Weight (pounds)*	
Age (yr)	*Boys*	*Girls*	*Boys*	*Girls*
Birth	20.0	19.7	7.6	7.4
1	30.1	29.3	23.1	21.1
2	34.8	34.0	29.0	27.0
3	37.9	37.6	33.4	31.9
4	40.7	40.4	37.6	36.8
5	43.3	42.9	41.9	40.0

*Compiled by Howard V. Meredith, Child Welfare Research Station, State University of Iowa, 1960.

mother. In fact, this parasitic life continues during the lactation period. It is, therefore, necessary that the diet of the mother be adequate to feed both herself and the fetus. Lactation is generally associated with a marked spontaneous increase in food intake, as if the *appestat* of the hypothalamus were set at a higher level.

Among the materials especially needed by the fetus are proteins (for general tissue construction), calcium and phosphorus for bone formation, iron for making red blood cells, and the necessary vitamins. In general, the mother's diet should contain more protein of high physiological value. The process of ossification begins during the sixth week of pregnancy; at this time the intake of calcium and phosphorus should be increased; milk can be relied on. The extra iron can be furnished by green vegetables or fruits, or an inorganic iron preparation can be prescribed. A small increase in iodine may be necessary. Intake of the usual sources of vitamins should be more generous. The total calories must also be augmented after the third or fourth month of pregnancy by from 10% to 20% and even more during nursing. The fetus is a thorough-going parasite. It obtains the necessary iron and calcium even though this should cause the mother to become anemic and her bones (not her teeth) to lose a considerable part of their mineral constituents.

Lactation. ■ During pregnancy the large amount of estrogen secreted by the placenta in some way inhibits the formation of prolactin. But immediately after the expulsion of the placenta during parturition, the estrogen and progesterone content of the blood falls to a very low level and thereby allows the secretion of prolactin. After delivery, the mammary glands are able to secrete milk. Ablation of the pituitary in a lactating animal stops the flow of milk. Table 30-1 gives the average composition of human and cow's milk.

It is claimed that human milk is superior to cow's milk in the following respects: human milk is digested more economically, its proteins are of higher biological value, and its calcium is better utilized. Human milk is generally richer in vitamins A and D, but it contains less thiamine and riboflavin.

The energy value of milk varies but may be placed at about 20 Cal/oz. During the first few months the daily requirement is from 45 to 55 Cal/lb for the normal infant. The energy requirement gradually diminishes, so that by the end of the first year it is approximately 40 Cal/lb.

■ GROWTH

The average weight of a newborn infant is about 7.5 pounds; the average height is about 20 inches, but it must be remembered that the height-weight characteristics depend on the race and family type. Table 30-2 shows the average height and weight of children from birth to 5 years of age, inclusive. Both the height and the weight of female children are slightly less than those of males.

READINGS

Friedmann, T.: Prenatal diagnosis of genetic disease, Sci. Am. **225**(5):34-42, Nov. 1971.

Friedmann, T., and Roblin, R.: Gene therapy for human genetic disease? Science **175**:949-955, 1972.

Human reproduction, Science J. **6**(6):Jun. 1970.

Levine, L.: Biology of the gene, St. Louis, 1969, The C. V. Mosby Co., pp. 54-151.

Mittwoch, U.: Sex differences in cells, Sci. Am. **209**(1):54-62, Jul. 1963.

McKusick, V. A.: The mapping of human chromosomes, Sci. Am. **224**(4):104-113, Apr. 1971.

Odell, W. D., and Moyer, D. L.: Physiology of reproduction, St. Louis, 1971, The C. V. Mosby Co.

Rhodes, P.: Reproductive physiology for medical students, London, 1969, J. & A. Churchill Ltd.

Smith, C. A.: The first breath, Sci. Am. **209**(4):27-35, Oct. 1963.

31

NUTRITION

The subject of nutrition has been selected as the concluding chapter of this text to emphasize its importance to the entire functioning organism, human or animal. Few modifiable factors (except bacterial infections) influence the growth and development of the body, the maintenance of health, and the length of life more than the nature and the amount of the ingested food. The preservation of the constancy of the internal medium (homeostasis) on which the normal activity and the life of each cell depend is impossible without a supply of the necessary food ingredients. Failure to furnish these ingredients to the blood and tissue fluid constitutes malnutrition.

Malnutrition has recently become an area of concern in the United States, which possesses an adequate supply of food products for its population. Inadequate diets have been found to exist in all levels of cultural, social, economic, and ethnic groups. Malnutrition is increasingly prevalent in the world population. There is no subject matter of greater moment for the welfare of the human race than the subject of nutrition.

■ NUTRITIONAL PROBLEMS

Malnutrition. ■ There are several causes of malnutrition. Some appertain to the supply of available food; some are due to improper selection of food when available, the lack of a balanced diet; in other instances the cause lies within the individual. Lack of available food may be due to local conditions in the production, transportation, or marketing of food. The nature of the soil upon which food is raised may also come into consideration, for example, lack of iodine or of iron. The processing of foods (e.g., canning, milling, and pasteurization) for the market must also be mentioned; in the processing of food great improvements have been made in recent years by enriching certain foods with vitamins, iron, iodine, fluorine, and so on.

In investigating the physiologically most suitable foods and the quantities of these foods, rats have been extensively used. These animals are quite similar to man in their metabolism and in the nutrients required.

Food selection. ■ Offered a variety of foods, some animals have to a certain extent the power of *instinctive food selection.* One or two rather surprising examples may be given. The amount of carbohydrate consumed by a diabetic rat fell from the previous 65% to 30% (reckoned in terms of calories), whereas the amount of fat eaten increased from 15.2% to 38.8%. During pregnancy a rat by instinctive food selection takes up more calcium salts, and during lactation this uptake may be from thirty to forty times the amount otherwise consumed. After weaning, the calcium consumption gradually drops to normal. During a specific disturbance in the function of the adrenal glands, the excretion of NaCl is greatly increased (Chapter 28); rats thus afflicted develop an intense craving for this salt. However, in the lower animals this self-regulatory food consumption is limited, especially with respect to essential proteins.

In man instinctive food selection is practically nonexistent; were instincts a reliable guide in the selection of food, few individuals would suffer from deficiency diseases such as rickets, scurvy, or pellagra.

In selecting a poor diet, economic conditions are often contributing factors, for the most nourishing foods (fruits, vegetables, eggs, and dairy products) are expensive, especially when out of season; the energy foods (cereals, tubers, and dry legumes) are relatively cheap. A pound of food protein costs significantly more than a pound of carbohydrates. But ignorance plays perhaps an even larger part. Frequently a person selects his food (both in quantity and quality) not according to his needs but according to his prejudices—a cultivated like or dislike for certain foods.

Distribution of meals. ■ Most individuals have well-established habits as to the distribution of their meals. Although this may vary in different parts of the world, the plan of three meals per day seems to meet the nutritional requirements of most people.

Breakfast is the meal that is omitted most frequently. Research has shown that a substantial breakfast is conducive both to the feeling of well-being and to efficiency in the late morning hours. An adequate breakfast provides approximately one fourth of the total daily caloric requirement and protein allowance.

Undernutrition. ■ On a diet qualitatively satisfactory but lacking in quantity (a mild starvation), the body consumes itself. Energy and materials for repair must be provided. We do not have in mind merely the using up of fat but the wasting away of what is generally regarded as the vital organs—muscles, glands, and neural structures. All tissues and organs do not waste at the same rate and to the same extent; the more vital parts are spared as much as possible, as Table 31-1 shows.

Table 31-1. Relative loss in weight of fresh tissue or organ during starvation*

Tissue	*Percent loss*
Adipose	97
Spleen	67
Liver	54
Muscles	31
Blood	27
Kidneys	26
Skin	21
Intestine	18
Lungs	18
Pancreas	3
Heart	3
Nervous system	3

*From Voigt.

A loss of 20% to 25% of body weight (exclusive of fat) can be survived. It requires six months of well-managed nutrition to recover fully from a period of twenty-four weeks of semi-starvation that has caused such a 25% loss of body weight. The large loss sustained by the muscles, coupled with greater difficulty in mobilizing energy and a sluggish circulation, accounts for the severe muscular weakness. The nervous system also is greatly affected; there is loss of spontaneous activity and coordination.

Undernutrition of pathological origin may be due to anorexia, impaired digestion and absorption, or inability to utilize the food (as in diabetes and pernicious anemia).

Obesity. ■ In reality overweight also has become a significant nutritional problem in the United States. The storing of fat in adipose tissues sometimes becomes excessive, and the body weight and bulk increase beyond that which may be regarded as normal. It is then not only unsightly but also a constant threat to health.

Three causes of obesity may be listed: (1) overeating, (2) underexercising, and (3) so-called constitutional causes, meaning the underactivity or overactivity of internal-secreting glands (Chapter 28). Endocrine disturbances are not major causes of obesity.

In attempting to find an explanation for this common abnormality, scores of opinions have been offered. However, it must be obvious that, whatever the ultimate cause may be, *all forms of obesity are due to a larger caloric intake than caloric expenditure.*

The psychic factor is of great importance in obesity. Indeed, many observers maintain that *motivation by self-indulgence instead of by self-control is the most frequent cause of obesity.* We know "living dynamos" who literally run themselves lean; the phlegmatic type favors saving rather than expending. Frequently we hear obesity blamed on heredity. But as Newburgh remarked, "Body build is inherited; obesity is not." This so-called heredity frequently resolves itself into bad eating habits acquired by following the eating pattern of parents.

Results of obesity. A large amount of fat adds an unnecessary burden and may cause great harm.

1. In the performance of muscular work an obese person is handicapped. The extra work done in walking, stairclimbing, and so on puts more strain on the heart. Being 20% overweight adds 20% to the energy cost of such activities as walking and golfing. Since subcutaneous fat is a poor conductor of heat, the heat generated during muscular work is not lost so easily; this results in great discomfort and profuse sweating.

2. The extra work in carrying this useless weight of fat soon causes fatigue, and therefore obesity tends to reduce bodily activity; lessened activity may cause the deposition of more fat, establishing a vicious circle. The lack of activity ultimately leads to a deterioration of the functions of digestion, respiration, circulation, and elimination.

3. Obese persons are more susceptible to certain ills than are persons of normal weight. At the Mayo Clinic studies showed that 91% of diabetic patients were overweight, and almost 83% were more than 10% overweight before diabetes occurred. Mortality from heart and blood vessel diseases is much greater in the obese person; especially is this true for lesions of the coronary blood vessels. Liver cirrhosis is more frequently found in obese persons than in people of proper weight. Gallbladder disease is more frequently found in obese women. The overweight persons are more susceptible to common cold, dyspnea, bronchitis and pneumonia, postural defects, and osteoarthritis.

Control of body weight. ■ The correct body weight depends chiefly on height, body build, and sex. The average weights of normal men and women, as established by the Society of Actuaries, are given in Table 31-2. It must be remembered that the values given in the table are averages and not ideal weights. According to the actuaries' study, ideal weight is that associated with lowest mortality and is as much as twenty pounds less than the average.

Overweight is due to a greater caloric intake than output. Thus it is obvious that if weight is to be lost, the caloric intake must be adjusted so that it is less than the caloric output, and if body weight is to remain static, input and output must be equal.

It is common practice among those who wish to lose body weight to follow one of the many low-calorie diets, which are widely publicized, without consultation with a competent nutritionist. This approach to the weight loss problem is usually unsound because these diets, for the most part, are too low in calories and unbalanced as to the foodstuffs that they provide. The result of any low-calorie diet is a loss of poundage, to be sure, but the end result may be quite unsatis-

Table 31-2. Average weights of Americans*†

Height	Ages							
	15-16	17-19	20-24	25-29	30-39	40-49	50-59	60-69
				Men				
5′ 0″	98	113	122	128	131	134	136	133
2″	107	119	128	134	137	140	142	139
4″	117	127	136	141	145	148	149	146
6″	127	135	142	148	153	156	157	154
8″	137	143	149	155	161	165	166	163
10″	146	151	157	163	170	174	175	173
6′ 0″	154	160	166	172	179	183	185	183
2″	164	168	174	182	188	192	194	193
4″		176	181	190	199	203	205	204
				Women				
4′ 10″	97	99	102	107	115	122	125	127
5′ 0″	103	105	108	113	120	127	130	131
2″	111	113	115	119	126	133	136	137
4″	117	120	121	125	132	140	144	145
6″	125	127	129	133	139	147	152	153
8″	132	134	136	140	146	155	160	161
10″		142	144	148	154	164	169	
6′ 0″		152	154	158	164	174	180	

*The values given in this table are based on a study of 5 million persons by the Society of Actuaries.
†Courtesy Northwestern National Life Insurance Co., Minneapolis, Minn., and the Society of Actuaries, Chicago, Ill.

factory. The more recent tendency is to consider weight reduction as an individual problem and to prescribe the reducing diet accordingly. A long-term reducing diet regimen, rather than the short-term strenuous diets, is recommended.

The control of body weight is, in many instances, a difficult problem and can be solved best on the basis of sound nutritional advice.

Daily losses of materials and energy. ■ The materials and energy lost during the course of a day must be replenished by the intake of food. The amount of these losses varies from one individual to another and in the same individual with changing conditions. The average daily loss of carbon and nitrogen is 225 and 16 g, respectively; the energy loss is about 2,300 Cal.

The loss of energy can theoretically be supplied by any one of the three organic foodstuffs —fats, proteins, or carbohydrates. And to include in the diet at least a minimum of each of the three foodstuffs, various combinations can be made. In Table 31-3 a diet is suggested that meets all the losses of energy and materials.

■ PROTEINS

Amount of protein required. ■ The most advantageous apportioning of these foodstuffs has been the subject of considerable investigation. This applies especially to the daily allowance of proteins. Maintaining growth in the young and nitrogen balance in the adult are indications of an adequate supply of dietary protein. A diet frankly deficient in proteins may lead to anemia, edema, an inadequate supply of antibodies, and so on.

A diet rich in protein generally entails in-

Table 31-3. Contribution of foodstuffs to energy and material loss

	Materials (g)		*Energy*	
Amount of food	*Carbon*	*Nitrogen*	*Calories*	*Percent of total*
100 g protein	53	16	100 × 4.1 = 410	17.4
100 g fat	79	0	100 × 9.3 = 930	39.3
250 g carbohydrate	93	0	250 × 4.1 = 1,025	43.3
Total	225	16	2,365	

creased intestinal putrefaction, and always increased formation and excretion of nitrogenous waste products. Whether this increased activity by the liver and kidneys, respectively, works injury upon these organs is a moot point. Eskimos, living on a high-protein diet, are said to show no high incidence of renal disease.

Nitrogen balance. ■ When the building up and the destruction of proteins in the body are equal, the amount of nitrogen ingested (in the form of proteins) is equal to that excreted in the form of nitrogenous waste products. The animal or human being is then said to be in nitrogen balance. A healthy adult person maintains this balance.* Proteins being the only nitrogenous foodstuff, all nitrogenous waste products (e.g., urea and uric acid) are derived from them. The amount of these wastes excreted (chiefly by the kidneys) can, therefore, inform us of the amount of protein catabolized in the body.

Negative nitrogen balance. When the loss of body proteins exceeds its synthesis, the animal or human being is in negative nitrogen balance. Several conditions may lead to this. Most obvious is the failure to include a sufficient amount of adequate proteins in the diet. The lack of appetite (characteristic of many pathological conditions)† or disturbances in the digestion or absorption of ingested food may be causative factors. The cells may have lost their ability to reconstruct proteins at the usual rate. On the other hand, the destruction of body proteins may be increased, for some reason, beyond the normal constructive power of the tissue; such conditions obtain in fevers and severe burns.

Whatever the cause may be, the results of a severe or long-continued negative nitrogen balance may be disastrous. Let us note how some of the functions of the body thus far studied may be affected. With an insufficient supply of amino acids from its ingested food, the body indirectly consumes its own tissue and plasma proteins in an endeavor to meet the demands of the vital organs. This may lead to edema (Chapter 16). Among the depleted plasma proteins are the globulins. We have learned that the gamma globulins, in a modified form, constitute the antibodies by which we resist bacterial invasions and build up immunity. Therefore protein undernutrition increases the susceptibility to infectious diseases. The restoration of indispensable hemoglobin is impossible. The construction or renewal of certain hormones and enzymes also ceases. In negative nitrogen balance, no amount of nonnitrogenous food consumed by the individual is able to prevent the loss of body tissue; as a result, the body weight falls. This is seen readily in the case of a debilitating disease. In young children it stops growth.

Positive nitrogen balance. An excess of pro-

*Except a woman during pregnancy.

†An animal fed on a diet lacking one or more of the essential amino acids soon refuses to eat the food. This loss of appetite (anorexia) has also been reported in man.

tein construction over destruction leads to an increase in protoplasm. This is most evident during growth. It is for this reason that, in proportion to body weight (and at a certain age, absolutely), the diet of a growing child should be richer in protein than that of the parent. Increased protein construction occurs during convalescence after a wasting disease, necessitating a good diet high in protein. During pregnancy the increase in the size of the uterus and the growth of the fetus are associated with a large positive nitrogen balance. The healing of a wound and the growth of tumors (especially those of malignant type) also illustrate protein construction in excess of destruction. The increase in size of muscles during a prolonged period of strenuous work furnishes another example.

Levels of nitrogen balance. In a normal adult a positive nitrogen balance cannot be maintained except for a very brief length of time. Let us suppose that for some time, a daily allowance of 50 g of protein plus 400 g of nonprotein food sufficed to maintain nitrogen balance and also to meet all energy requirements. The amount of protein is now augmented to, for example, 75 g. After a few days of lag, during which a small amount of protein possibly may be retained, the individual catabolizes 75 g and the nitrogenous excretion in the urine increases. Nitrogen balance has again been established but now at a higher level. Nitrogen equilibrium can, therefore, be established at various levels, that is, with a larger or with a smaller intake of protein, provided a minimum amount is supplied.

In this manner more and more protein thus may be included with the food until the alimentary canal is no longer able to cope with it. All protein in excess of that required for repairing the actual loss of body protein is wasted except so far as the production of energy is concerned. As this function can be more economically performed by the nonnitrogenous foodstuffs, *carbohydrates and fats are protein sparers.* But this replacement has its limits; for no matter how large an amount of carbohydrate or fat the diet may contain, a certain irreducible minimum of tissue wear and tear takes place; this can be restored only by the eating of proteins.

The tendency of the body to be in nitrogen balance, irrespective of the amount of proteins eaten, causes the storage of proteins in the body to be very limited. The protein eaten in excess of that necessary to maintain nitrogen balance on a fairly low level is not stored but is transformed into carbohydrate and, indirectly, into fat. In our discussion of positive nitrogen balance exceptions to this statement have been noted.

Amino acid requirement. ■ Although we have been discussing the need for a diet with adequate amounts of protein, we recognize from our discussions in Chapters 2, 20, and 23 that various cells must be provided with an ample supply of amino acids to meet their various needs. Therefore, it is the amino acids which are the essential nutrient rather than the protein. Since the animal body has no mechanism to fix the element nitrogen, we are ultimately dependent on plants which are able to incorporate it and manufacture amino acids and proteins.

Essential and nonessential amino acids. Of the twenty-two amino acids known to be present in plant and animal tissue, only eight need to be supplied to the adult human body: leucine, lysine, tryptophan, valine, threonine, methionine, isoleucine, and phenylalanine. These are designated as *essential amino acids* because the body is unable to synthesize them at a rate sufficient to grow and maintain life. The lack of any single one of them results in malnutrition and finally in death. Unless all the amino acids required for the synthesis of particular protein are present simultaneously, no attempt is made at construction. Since little or no storage of free amino acids is present in the cell, *all essential amino acids* must be fed simultaneously. A *nonessential amino acid* not present in the food but needed by a cell for construction of proteins can be made from other components and a source of nitrogen (essential amino acids or amino acids in food protein) by *transamination* (Chapter 23). Protein as a source of the essential amino acids and as a source of nitrogen constitutes one of the three main ingredients in our diet. From this mixture of amino acids sent to the cells, millions of different types of proteins are constructed. The cell selects the proper amount of each amino acid it needs and fabricates from them proteins that are identical with the proteins characteristic of that particular cell.

It is possible to substitute intravenous injections of amino acids for dietary protein following serious surgical operations or in the case of extensive burns. This prevents the loss of body protein that may occur in these situations.

Daily allowances. ■ The daily allowance of protein, as shown in Table 31-12, varies with age and sex.

The values allowed by the Food and Nutrition Board are believed by some authorities to be exceedingly liberal. Most carefully conducted experiments have recently disclosed that, on a diet in which all the proteins were of vegetable origin, between 30 and 40 g of protein was sufficient for a man weighing 156 pounds. If meat were substituted, even a smaller amount sufficed.

The daily allowance of proteins for the average adult has been placed at from 0.67 to 1 g/kg of body weight. Children from one to three years of age require about 1.5 to 2.0 g/kg. A boy sixteen years of age needs more protein than his father. During pregnancy the allowance of protein should be increased by 40% to 50%, and during lactation by about 80%. Infants up to one year of age require a higher intake of protein per unit weight. Intake of protein should remain unchanged or possibly increased in elderly people (although the total caloric intake may be decreased). In hyperactivity of the thyroid gland (Chapter 28) and in fever and infections, the protein requirements are increased.

Nutritional value of food protein. ■ We may recall that of the twenty-two varieties of amino acids found in the human body, only eight are essential or indispensable; the remainder can be made by the tissues from the essential amino acids or other sources. A protein that does not contain all the essential amino acids necessary to maintain nitrogenous equilibrium or growth is said to be an *incomplete protein.* The smaller the amount of a protein required to keep an animal in nitrogen balance, the higher is its *physiological* or *biological value.* In this respect, the proteins of the whole egg rank highest of all (94). In descending order are milk, 85; liver and kidney, 77; heart, 74; muscle meat, 69; whole wheat, 67; potato, 67; rolled oats, 65; whole corn, 60; white flour, 52; and navy beans, 38.

Although some vegetable proteins may be inferior to those of animal origin, three exceptions to this must be noted. Some of the proteins of peanuts, soybeans, and cottonseed are complete. There is no appreciable difference between muscle meat of ox, hog, or sheep. Proteins can supplement each other. For example, one part of beef protein (biological value of 69) plus two parts of flour protein (55) have a combined value of 73. This also is true for combining proteins of wheat with those of skim milk.

Growth. The distinctive physiological value of the various proteins is nowhere better illustrated than in their influence on growth. Young rats fed, in addition to all the other necessary food materials, no other protein than casein grow normally, as shown by curve *I* in Fig. 31-1. Compare with this the growth curve *II* when the young rats depend on the protein gliadin for their amino acids. Although gliadin (abundantly found in wheat and rye) is capable of maintaining nitrogen equilibrium, it lacks a sufficient amount of the amino acid lysine to maintain growth. The response to zein (from corn), as shown by the curves in Fig. 31-1, is most interesting. From Table 23-1, it will be seen that this protein contains no lysine and very little tryptophan. With zein as the sole supply of amino acids, growth is impossible; in fact, the animal

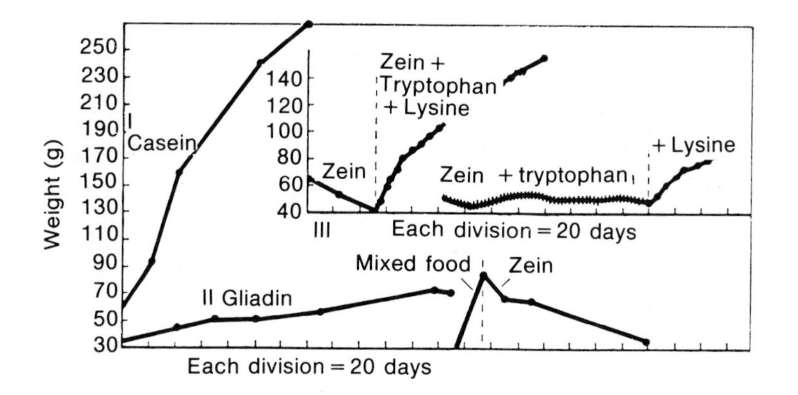

Fig. 31-1
Curves showing the effect of various proteins on growth of young rats, as indicated by changes in body weight. Curve **1** represents weight gain on casein diet and is to be regarded as normal. Both gliadin and zein are incomplete proteins. A diet of zein not only fails to maintain growth but also results in a loss of weight. As illustrated in the inset, the addition of two amino acids, tryptophan and lysine, restores curve to normal. The addition of these amino acids one at a time is not as effective. (After Mendel and Osborne.)

is unable to maintain nitrogen equilibrium and loses body weight. The addition of tryptophan to the zein prevents the loss in weight but does not promote growth; the addition of both lysine and tryptophan ensures normal growth. From this it is evident that these two amino acids do not play the same role in the animal body; tryptophan is necessary to maintain nitrogenous equilibrium and body weight; lysine is required for the building of new tissue, that is, for growth. Since one essential amino acid cannot substitute for any of the others, the various amino acids are specific in their function; they are not interchangeable. In the diet of an adult, 50% of the proteins should be of high biological value; for a young child this should be from 60% to 95%.

Kwashiorkor, a nutritional disease resulting from a diet deficient in protein, has recently come to prominence in many parts of the world. After weaning, children who subsist on a high-starch diet that includes a low quantity and quality of proteins are seriously affected; prevention of this disease is a world problem. Caloric intake in these children may be adequate but lack of protein leads to a failure to grow, mental apathy, edema, liver enlargement, anemia, and lack of ability to combat infections. *Marasmus* is used to describe children who have a protein deficiency combined with a caloric deficiency. Both of these conditions present great problems for the world population.

■ CARBOHYDRATES AND FATS

The yielding of energy by proteins is of secondary importance to their function as tissue builders. The bulk of the energy needed is supplied by the carbohydrates and fats. These two foodstuffs are indispensable for certain other functions, but the very limited amounts used for these purposes may be ignored in estimating the daily requirements. The energy requirements are determined almost exclusively by (1) the basal metabolic rate, (2) the amount of physical work performed, and (3) the external temperature.

An individual expending 2,400 Cal and maintaining nitrogen balance with 70 g of protein (287 Cal) derives 2,113 Cal from carbohydrates and fats. These two foodstuffs can largely replace each other in the ratio of their respective heat values, that is, $2^1/_4$ parts of carbohydrate to 1 part of fat. But neither the one nor the other should be entirely excluded from the diet. A large con-

Fig. 31-2
Effect of fat on growth of litter-mate rats is illustrated by these two animals. The rat on the left was maintained on a diet containing 20% lard; the rat on the right was maintained on a fat-free diet. (Courtesy Dr. George Burr; from Hawley and Maurer-Mast: The fundamentals of nutrition, Charles C Thomas, Publisher.)

sumption of sugars and starches may lessen the desire for foods rich in vitamins and minerals.

Carbohydrates are the cheapest obtainable source of muscle energy. The nervous system obtains its energy almost exclusively from carbohydrates. Fats and oils are less bulky and easily stored in concentrated form. They form the solvents for certain vitamins and supply the essential fatty acids. In practice, the partitioning of these two foodstuffs is largely a matter of personal preference and economic status. Perhaps the following may prove satisfactory for most people: 30% of total energy from fats, 12% from proteins, and 58% from carbohydrates.

The effect of a diet completely devoid of fat is shown in Fig. 31-2. Young rats cease to grow. This disastrous effect is due not to the want of olein, stearin, and palmitin but to the absence of higher unsaturated fatty acids, such as linoleic and linolenic acids. These are spoken of as *essential fatty acids* (Chapter 17). Being unable to manufacture them (similar to the essential amino acids), the animal is dependent for them on exogenous sources. Fortunately, they are widely distributed in such foods as lard, peanut oil, olive oil, soybeans, and egg yolk. Butter contains them to a very limited extent. In addition to being essential for growth, they are exceedingly important in maintaining a healthy skin. The amount of carbohydrate and lipids in the diet of Americans has been implicated as causative factors in the incidence of atherosclerosis. Insufficient evidence exists in the scientific literature to establish firm guidelines concerning the dietary factors involved in this disease. It would seem unwise with the information available to suggest any drastic changes in the American diet, although no harm should be done if the 30% fat content is adhered to, with the substitution of unsaturated fats for saturated and a reduction in intake of dietary cholesterol.

■ MINERALS

Seldom, if ever, is the diet deficient in sodium, potassium, magnesium, copper, sulfur, or chloride; these, therefore, do not need to be considered. But frequently there is an insufficient amount of iron, calcium, phosphorus, fluoride, or iodine.

Iron. ■ The total amount of iron in the body may be placed at 3 to 5 g. Over one half of this is found in hemoglobin, and the remainder is stored in myoglobin, liver, spleen, kidney, and bone marrow. It is only in the inorganic state that the body can absorb iron. The recommended daily allowance varies with age and is shown in Table 31-12. The very scanty supply of iron in milk cannot be relied on to furnish the liberal amount needed by the growing infant or child. The high concentration of hemoglobin in the

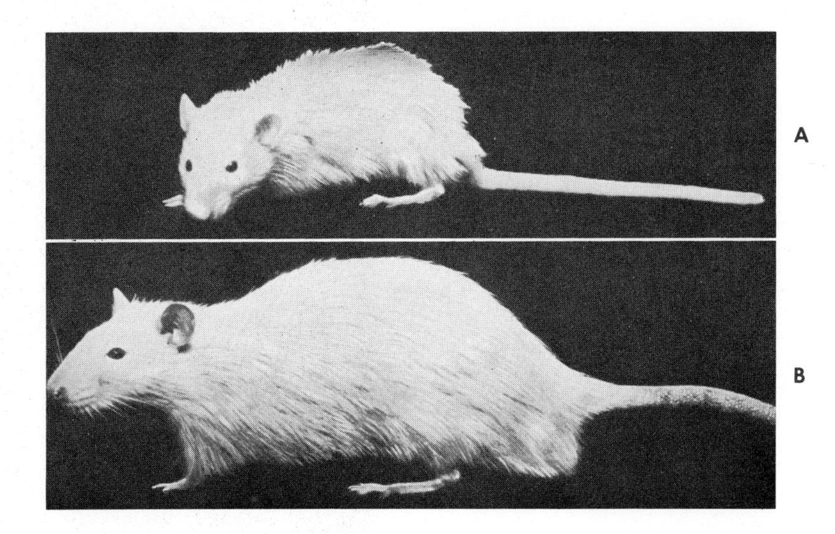

Fig. 31-3
Iron deficiency and growth in rats. **A,** 8-month-old rat on iron-deficient diet; weight, 109 g. **B,** $5^1/_2$-month-old rat on a diet with an adequate supply of iron; weight, 325 g. (From Bureau of Human Nutrition and Home Economics, U. S. department of Agriculture.)

blood at birth (from 22 to 23 g/100 ml) enables the newborn infant to tide over the shortage of iron during the suckling period of about six months. The retarding of growth in rats caused by an insufficient supply of iron is shown in Fig. 31-3. Iron deficiency may result not so much from a single, massive hemorrhage as from a constant, though mild, loss of blood.

Table 31-4 shows the iron content of a few common foods.

Calcium. ■ The role of calcium in bone and tooth development has been discussed (Chapter 24). The daily allowance is shown in Table 31-12. During pregnancy and lactation the amount should be increased. Foods with a liberal amount of calcium fall into two classes: the dairy products and the leafy vegetables. The calcium of milk is better absorbed than that of vegetables. Bread and meat are poor sources. Table 31-5 shows the calcium content of some common foods.

Phosphorus. ■ Phosphorus is found in every tissue of the body. It is important since in combination with calcium it gives strength and rigidity to the bones and teeth. The human diet is rarely inadequate in this element, especially if the recommended allowances of calcium and protein are met. In the diets of children the daily allowance should be at least equal to that of calcium. For adults the recommended allowance is one and one half times that of calcium. The phosphorus content of some common foods is shown in Table 31-6.

Fluoride. ■ The part played by fluoride in the development and preservation of teeth was discussed in Chapter 19.

Iodine. ■ Iodine, which is essential for thyroid hormone production, was considered in Chapter 28.

Trace elements. ■ Several other mineral elements (copper, cobalt, chromium, molybdenum, manganese, selenium, and zinc) are necessary for adequate human nutrition. These have been referred to as *trace elements* since they occur in the body in very small amounts (less than 0.005% of body weight), but this is not an indication that they are not nutritionally important. Many of them function as parts of enzyme systems. Seldom, if ever, do deficiencies exist of these elements in the diet, and excess intake of these elements can cause toxicity if ingested in large quantities.

VITAMINS
General considerations

In our previous discussions it was emphasized that nearly all body functions are made possible or are controlled by either neural or chemical means. Among these chemical agencies are enzymes, hormones, and vitamins. The first two of these are made in and by the body itself; vitamins must be supplied with our food or manufactured by bacterial action in the alimentary canal.

Some years ago (1912) it was found that on an artificially prepared diet, furnishing a sufficient number of calories and the required amount of protein (casein), carbohydrate (starch), fat (lard), water, and mineral substances, a young rat soon ceased to grow and, just as an adult animal on the same diet, sickened and died. These ill effects could be prevented by merely giving the animal 3 ml (about 1 teaspoonful) of milk a day (Fig. 31-4). Clearly, all the known foodstuffs then were not able to sustain growth and to maintain the health and life of higher organisms. A new class of foodstuffs had been discovered, and these new ingredients were called *vitamins*.

Deficiency diseases, avitaminosis. Diseases due to the lack of one or more indispensable food substances, such as amino acids, fatty acids, or vitamins, are known as deficiency diseases. A complete dietary lack of vitamins is called *avitaminosis* and may lead to a serious disease, such as scurvy, rickets, pellagra, night blindness, certain forms of anemia, and eye diseases. In the

Table 31-4. Iron content of some common foods*

Food	*Portion*	*Iron (mg)*
Beef liver, cooked	2 oz	4.4
Beef, pork, cooked	3 oz	2.6
Raisins	1/2 cup	2.6
Spinach	2/3 cup	2.4
Apricot, dried	1/2 cup	2.3
Chicken, cooked	3 oz	1.5
Egg	1 medium	1.3
Potato	1 medium	0.8
Cabbage, raw	1 cup	0.5
Milk, whole	1 cup	0.2

*From Church and Church: Food values of portions commonly used, ed. 9, Philadelphia, 1963, J. B. Lippincott Co. Also, Food—the year book of agriculture, Washington, D. C., 1959, U. S. Government Printing Office.

Table 31-5. Calcium content of some common foods*

Food	*Portion*	*Calcium (mg)*
Milk	1 cup	288
Cheese, American cheddar	1 oz	206
Mustard greens	2/3 cup	205
Ice cream	5 oz	100
Cottage cheese	1/4 cup	54
Egg	1 medium	26
Potato	1 medium	13

*From Church and Church: Food values of portions commonly used, ed. 9, Philadelphia, 1963, J. B. Lippincott Co. Also, Food—the year book of agriculture, Washington, D. C., 1959, U. S. Government Printing Office.

Table 31-6. Phosphorus content of some common foods*

Food	*Portion*	*Phosphorus (mg)*
Milk	1 cup	227
Liver, beef	2 oz	215
Beef (round), cooked	3 oz	191
Baked beans	1/2 cup	148
Cottage cheese	1/4 cup	106
Broccoli	2/3 cup	76
Orange	1 medium	36
Bread, white	1 slice	21

*From Church and Church: Food values of portions commonly used, ed. 9, Philadelphia, 1963, J. B. Lippincott Co. Also, Food—the year book of agriculture, Washington, D. C., 1959, U. S. Government Printing Office.

past, these and other deficiency diseases were more or less common; at present the complete lack of any one of the vitamins is seldom encountered. As a result the aforementioned diseases have almost disappeared from our country. Frequently, however, the diet does not contain the optimum quantity of one or more vitamins—*hypovitaminosis;* in consequence although the person may not be entirely incapacitated from carrying on his daily duties, his health and vitality are below par.

Interrelationship of vitamins. ■ The number of known vitamins is large and is still increasing. However only ten to twelve have been definitely proved to be needed by man. We shall limit our attention to these.

During the early history of vitamins it was thought that each performed just one definite function in the body and that its absence caused just one particular disease. For example: The lack of vitamin B_1 (thiamine) was said to be the sole cause of the disease beriberi, and B_1 was the specific and only cure for the disease. It is now known that the dietary lack of a certain vitamin may lead to two or more quite distinct disturbances and that in a pathological condition, such as beriberi, two or more vitamins may be concerned—*multiple deficiency.* Hence few generalized statements can be made as to the activity of any vitamin. The outline, as given in Table 31-7, may, however, be of some value, especially considering the bewildering array of body disturbances attributed to each vitamin.

A large number of vitamins have been obtained in pure form. Some have been produced synthetically; these are as effective as the natural vitamins. However, in using them, the individual is likely to neglect procuring an adequate diet and thereby suffer from other deficiencies.

Sources of vitamins. ■ Generally animals are dependent, directly or indirectly, for their vitamins on the vegetable world. Whatever the source, there is no uniform distribution of the various vitamins in the many different foods we

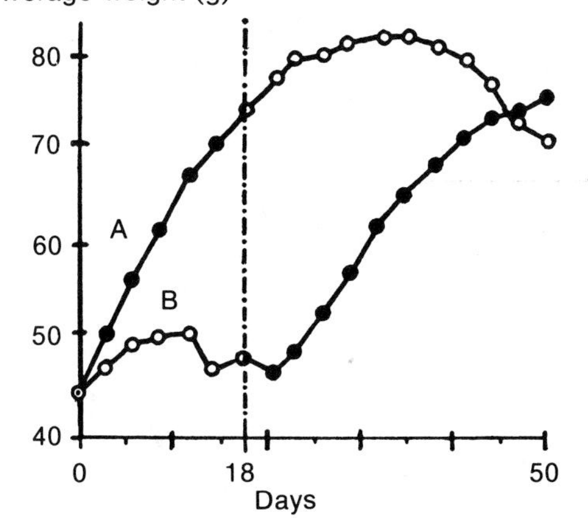

Fig. 31-4

Vitamin deficiency and its effect on growth in rats. **O**, Artificial diet sufficient in proteins, fats, carbohydrates, salt, and so on. **●**, Same diet to which 3 ml of milk was added daily. **A** illustrates effect on growth when vitamin source (milk) was removed at 18 days; **B** indicates restoration of growth when milk was added at 18 days.

Table 31-7. Effect of avitaminosis on structure and function

Avitaminosis	*Structure or process chiefly involved*	*Pathological condition, if severe*
A	Epithelial tissues	Night blindness, etc.
B_1, or thiamine	Neuromuscular	Beriberi
B_{12}	Erythropoiesis	Anemia
K	Formation of prothrombin	Noncoagulability of blood; bleeding
E	Reproduction (male and female)	Sterility
Niacin	Catalytic systems	Pellagra
C, or ascorbic acid	Intracellular material	Scorbutus (scurvy)
D	Bony structures	Rickets
B complex	Catalytic systems	Varied, decrease in general well-being

consume; to obtain an adequate supply, a wide selection of food is necessary. It is for this reason that people living on a restricted and monotonous diet are very likely to show hypovitaminosis. But to select a properly balanced diet, from a vitamin point of view, it is not necessary that the amount of each vitamin in each of the food materials consumed be known. In the so-called *protective foods* (dairy products, eggs, fruits, and vegetables) are found all the basic nutritional needs of the body. Fortunately, a diet balanced in all other respects is generally adequate as to vitamins.*

Daily allowances. ■ Both internal and external conditions determine the amount of the various vitamins required. Vitamin D is a necessity for the infant and adolescent; it is doubtful whether the adult needs it in his food. The construction of new tissue (as in growth and pregnancy) or the reconstruction of tissue (e.g., in convalescence and the healing of wounds) creates a greater demand for some vitamins. Whenever catabolism is accelerated, as during muscular work or exposure to cold, and in hyperthyroidism (Chapter 28), the intake of vitamins must be increased.

The amount of vitamins the body obtains depends not only on the amount found in the food but also on the thoroughness of their absorption; frequently this is reduced by intestinal disturbances. People who are limited in their choice of food (as those persons afflicted with such diseases as diabetes, peptic ulcers, and some forms of allergy) or those restricted in the total amount of food (calories) consumed daily may find some difficulty in procuring foods with a sufficient amount of all required vitamins.

Partaking of more vitamins than the usual optimum daily requirement has not been found to influence muscular activity favorably nor the resistance to or the recovery from fatigue (i.e., augmenting one's endurance). Although a few vitamins (e.g., vitamin A) are stored to some extent in the body (chiefly in the liver) and are gradually released as wanted, this is not true for some of the most needed vitamins; these must be supplied almost daily. Hence taking more than the optimum daily requirement is in most cases pure waste. The amount of vitamins required is expressed in milligrams, micrograms (0.001 mg), or International Units (IU) and has been established by the Food and Nutrition Board, National Research Council. Daily allowances are shown in Table 31-12.

*Many foods are rich in certain vitamins, but the amount of these foods eaten is so limited as to render them of little value (e.g., parsley contains large amounts of B_2 and C).

■ Vitamin A, antikeratinizing vitamin

Sources. ■ Since vitamin A is soluble in oils and fats and not in water, its concentration in most foods is very low. Vitamin A as such is found only in food of animal origin, chiefly in liver and milk (more in cream and butter) and in yolk of eggs. Fish liver oils and some body oils are exceedingly rich in vitamin A.* Our body derives its vitamin A chiefly from a pigment, called carotene, found in green-, red-, and orange-colored plants or parts of plants. Water-soluble carotene, the provitamin, is transformed into vitamin A, chiefly by the liver. The carotene content of fruits and vegetables is generally proportional to the intensity of the green, red, or orange color. It is practically absent in white (and blanched) vegetables; for example, the carotene content of the outer green leaves of lettuce is thirty times that of the inner leaves.

The amount of vitamin A is generally stated in terms of the International Unit; this unit is equivalent to 0.3 μg crystalline vitamin A. In Table 31-8 are given the amounts of vitamin A found in some common foods.

Stability. ■ Vitamin A is not affected by heat,

*One teaspoonful of cod-liver oil contains about 3,100 IU of vitamin A. One gram = 4,500,000 IU.

Table 31-8. Amount of vitamin A found in common foods*

Food	*Portion*	*Vitamin A (IU)*
Liver, beef, cooked	2 oz	26,340
Spinach, cooked	2/3 cup	14,132
Potatoes, sweet, cooked	1 medium	11,410
Carrots, raw	2/3 cup	8,800
Mustard greens, cooked	2/3 cup	6,700
Kale, cooked	2/3 cup	6,147
Broccoli, cooked	2/3 cup	3,400
Egg, whole	1 medium	550
Milk, whole	1 cup	390
Orange juice, fresh	1/2 cup	230
Apple, raw	1 medium	120

*From Church and Church: Food values of portions commonly used, ed. 9, Philadelphia, 1963, J. B. Lippincott Co. Also, Food—the year book of agriculture, Washington, D. C., 1959, U. S. Government Printing Office.

but it is destroyed by oxidation. Consequently long cooking in an open kettle entails some loss of this vitamin. The vitamin is altered by exposure to light, especially ultraviolet rays.

The vitamin content of milk and eggs fluctuates considerably with the amount of vitamin A or carotene in the feed of the cow or hen. During the pasturing season 1 quart of unfortified cow's milk contains approximately one fourth to one half of the amount of vitamin A required daily by a child. In the winter season, unless carotene is fed, the content of vitamin A in cow's milk is decreased by almost 50%. Grass and alfalfa are rich in carotene. Lard, meats of all kinds, poultry, flour, nuts, celery, grapefruit, turnips, and yeast are practically devoid of vitamin A.

Effect of deficiency. ■ Vitamin A is one of the many factors concerned with growth. In its absence growth is slowed down or ceases completely. The most readily detectable effect of a deficiency of vitamin A is upon the eye. When this substance is lacking, keratinized epithelium is substituted for normal epithelium. Keratin, a very insoluble protein, is formed in the tear glands and meibomian glands (a sebaceous or oil gland of the eyelid), causing the conjunctiva (the reddish membrane lining the eyelids) and the cornea to become dry and inflamed (Fig. 31-5). This condition, known as xerophthalmia, may lead to blindness, if severe.

As noted in Chapter 9, the inability to see in dim light, or night blindness, is frequently seen in humans when there is less retinol in the diet than the required minimum.

Dryness and scaliness of the skin are early manifestations of vitamin A deficiency. Involvement of the respiratory tract is frequent, and because of keratinization, the cilia are lost. This may lead to respiratory infections.

An insufficient supply of vitamin A in the young leads to a defective formation of tooth enamel, with a resulting exposure of the dentin.

■ Vitamin D, antirachitic vitamin

Sources. ■ Of all vitamins, this is the most limited in its distribution. It is probable that it is best to assume the diet contains no vitamin D. It is never found in natural food of plant origin. The most abundant source is the liver oil and the body oil of a number of fish, such as swordfish, salmon, halibut, mackerel, and cod.*

Egg yolk has some vitamin D; the amount depends on the irradiation of the hen and on the amount of vitamin in its food. The amount of vitamin D in milk is negligible unless the cows are fed irradiated yeast or vitamin D is added (fortified milk). When thus treated, milk contains 400 IU per quart, a quantity sufficient for an infant.

Stability. ■ Vitamin D is one of the most stable vitamins. It is not readily oxidized and is not destroyed by pasteurization or boiling.

Daily allowances. ■ The amount of vitamin D required varies with the extent of the calcium and

*To appreciate the very small amount of this vitamin necessary for preventing widespread and serious harm, we may state that a barrel of cod-liver oil contains about 0.25 g of vitamin D. Actually, cod is a poorer source than the other fish mentioned.

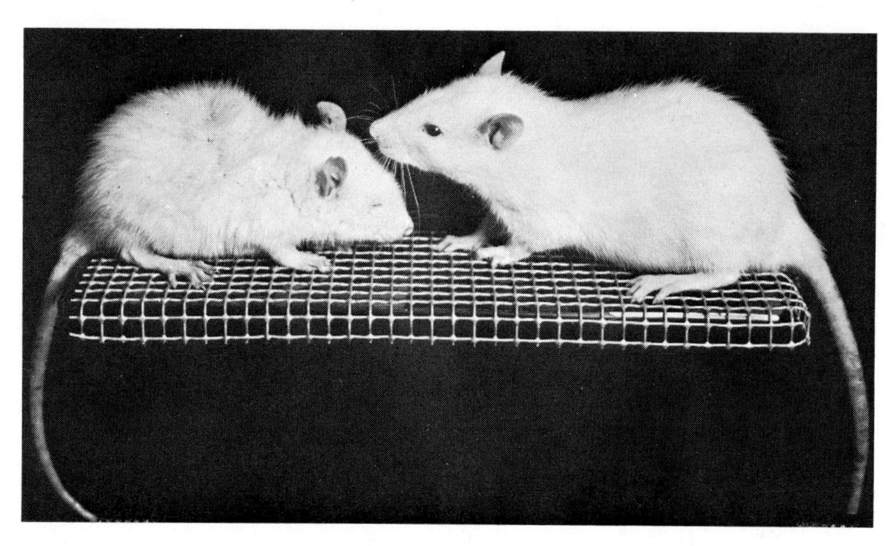

Fig. 31-5
Inhibition of growth in rat produced by restriction of vitamin A in diet. Animals, litter mates, were 21 days old at start of experiment, which was continued for 33 days. Animal at left received diet containing all nutritive substances except vitamin A; animal at right received adequate diet. Note xerophthalmia in rat deprived of vitamin A. (Courtesy The Upjohn Co., Kalamazoo, Mich.)

phosphorus metabolism. It is most needed during the first months of life. Its administration should begin at the end of the second week. During the last two months of pregnancy and still more during lactation, great demands are made on the calcium and phosphorus stores of the maternal body and, in consequence, large supplies of these minerals and vitamin D are needed in the diet (from 400 to 800 IU daily). It is doubtful whether an adult needs this vitamin in his food; the irradiation of the skin (Chapters 18 and 26) perhaps supplies a sufficient quantity.*

Effects of deficiency. ■ As discussed in Chapter 24, vitamin D is necessary for the proper formation of bones and teeth. In the adult a lack of this vitamin may cause osteomalacia. The bones are softened due to a change in the ratio of calcium and phosphorus, and various deformities ensue.

*The oil of the preen gland in birds is, by preening, distributed over the feathers; on irradiation by sunlight it gives rise to vitamin D. In preening, the vitamin is swallowed.

■ Vitamin E, tocopherol, or fertility vitamin

In 1920, it was found that rats fed with a sufficient amount of all of the vitamins known at that time became sterile. In the male the germ cells perish, the whole of the seminiferous tubules that form the spermatozoa (Chapter 30) being destroyed. In the female the ovaries remain normal, but the embryo dies a few days after fertilization. These results are avoided by feeding yeast or the germ of wheat. The beneficial effects are attributed to a vitamin called E or alpha-tocopherol. Vitamin E is thought to be concerned chiefly with cell maturation and differentiation. It is doubtful whether this vitamin normally influences the reproductive functions in man. Vitamin E, an oil-soluble vitamin, is found abundantly in vegetable oils (corn, cotton, soybean, and peanut) and in wheat germ. Recommended daily allowances of this vitamin are listed in Table 31-12.

■ Vitamin K and choline

Vitamin K, the antihemorrhagic vitamin, was discussed in Chapter 12. A synthetic product, menadione, is three times as effective as vitamin K.

Choline aids in the storage and mobilization of fats. In its absence the liver becomes loaded with fat. It also aids in the formation of phospholipids (Chapter 23) and is used in the construction of acetylcholine.

■ Vitamin B complex

Of the large number of water-soluble vitamins grouped together as the vitamin B complex, only three or four have been definitely shown to be of importance to man.

The B complex vitamins are of fundamental importance to life, being found in all living beings. They appear to be associated with the metabolic processes throughout the body. This is in agreement with the fact that the amount of these vitamins required is largely determined by the amount of energy expended in physical work and in maintaining the body temperature. Because the various B vitamins are found intermingled in many foods, it is difficult to differentiate positively between the disturbances created by the absence of each individual member of the group. Deficiency diseases attributed to a particular vitamin have in some instances been found to be multiple deficiencies (e.g., beriberi). Their absence may lead (in addition to the specific effects to be described presently) to a decrease in the general well-being of the individual, showing itself in a reduced work output, increased fatigue, and emotional disturbances (irritability and depression).

Thiamine, or vitamin B_1 ■

Sources and daily allowances. Of all deficiencies, that of thiamine is said to be the most common. Although it is found in a large variety of foods, it is not abundant in any. Polished rice is the chief cause of the widespread scourge of beriberi in the Far East.

Table 31-9. Amount of thiamine in some common foods*

Food	*Portion*	*Thiamine (mg)*
Wheat germ	1 tbsp	1.39
Ham, boiled	3 oz	0.86
Spaghetti, enriched	1 cup	0.83
Pork chop, fried	1 medium	0.59
Pecans, shelled	1/2 cup	0.34
Peas, green, canned	2/3 cup	0.27
Orange juice, fresh	1 cup	0.19
Milk, whole	1 pt	0.18
Figs, dried	2 small	0.16
Beans, lima, fresh, cooked	2/3 cup	0.15

*Food values from Church and Church: Food values of portions commonly used, ed. 9, Philadelphia, 1963, J. B. Lippincott Co.

Thiamine has been isolated in crystalline form and also prepared synthetically. The vitamin is not destroyed by cooking at 100 C for one hour, but the cooking water may contain from 10% to 60% of the vitamin originally present in the food. Thiamine values for some common foods are given in Table 31-9. Since this vitamin can be stored only to a limited extent, we are dependent on a daily supply. Recommended daily allowances for this vitamin are given in Table 31-12.

Effects of deficiency. A marked deficiency of vitamin B_1 in man is referred to as beriberi. This disease is characterized by a polyneuritis, muscular atrophy, edema, and cardiovascular changes. Although beriberi is a result of thiamine deficiency, there is usually a lack of other vitamins in addition. Vitamin B_1 is necessary for growth in young animals.

Nicotinic acid, nicotinamide (niacin) ■

Sources and daily allowances. Yeast and liver are excellent sources; tuna fish and peanuts are rich in niacin; lean meats, poultry, and enriched flour are well supplied. Milk, eggs, and most vegetables are poor sources. Table 31-10 gives the niacin content of some common foods.

Niacin is not destroyed by boiling, but it may dissolve extensively in the cooking water. A diet adequate in other respects is not likely to be lacking niacin. It is said that niacin is formed by bacterial action in the intestine. Daily allowances of this vitamin are found in Table 31-12.

Effects of deficiency. Deficiency of niacin gives rise to pellagra, a disease sometimes found during times of financial stress (especially during the winter months), when many individuals live largely on corn, molasses, and bacon—a poverty disease. Pellagra is characterized by want of strength and vitality, loss of appetite, indigestion, diarrhea, skin eruptions, pain, and in later stages mental disturbances. Formerly the cause of all these symptoms was attributed solely to the dietary deficiency of niacin. It is now known that the lack of thiamine, riboflavin, and perhaps other B vitamins is also involved; that is, pellagra is a multiple-deficiency disease.

It has been shown that a diet lacking the essential amino acid tryptophan has effects similar to those caused by niacin deficiency. Also, the amino acid can replace niacin in abolishing most of the pellagra symptoms. This is in agreement with the fact that corn is deficient in tryptophan and that pellagra largely afflicts individuals whose diet contains large amounts of unenriched corn products. Although eggs and milk are poor sources of niacin, they are excellent pellagra preventives due to their supply of tryptophan. This relationship between these two substances was solved by administering to an animal tryptophan labeled with radioactive ^{14}C and finding, in the urine, niacin with the characteristically labeled carbon. Tryptophan is the precursor of niacin. This view is supported by the fact that rats need niacin in their food only when tryptophan is lacking.

Riboflavin, or vitamin B_2. ■ In a modified form, known as flavin, riboflavin constitutes part of one of the hydrogen carriers discussed in Chapter 23. In conjunction with thiamine and niacin, it is, therefore, necessary for the oxidative processes in the tissues.

Sources. Riboflavin (or lactoflavin) in the form

Table 31-10. Niacin content of some common foods*

Food	*Portion*	*Niacin (mg)*
Chicken, roasted	6 oz	14.1
Halibut, cooked	4 oz	13.1
Liver, beef	2 oz	8.4
Heart, beef	3 oz	6.6
Ground beef, cooked	3 oz	4.1
Almonds, shelled	1/2 cup	3.3
Green peas, canned	2/3 cup	1.1
Orange juice, fresh	1 cup	0.6
Milk, whole	1 cup	0.3

*From Church and Church: Food values of portions commonly used, ed. 9, Philadelphia, 1963, J. B. Lippincott Co. Also, Food—the year book of agriculture, Washington, D. C., 1959, U. S. Government Printing Office.

Table 31-11. Riboflavin content of some common foods*

Food	*Portion*	*Riboflavin (mg)*
Liver, beef, cooked	2 oz	2.25
Kidney, beef, stew	3 oz	2.00
Brewer's yeast	1 tbsp	0.44
Milk, whole	1 cup	0.42
Turnip greens, cooked	2/3 cup	0.39
Ground beef, cooked	3 oz	0.16
Broccoli, cooked	2/3 cup	0.15
Orange juice, fresh	1 cup	0.06

*From Church and Church: Food values of portions commonly served, ed. 9, Philadelphia, 1963, J. B. Lippincott Co. Also, Food—the year book of agriculture, Washington, D. C., 1959, U. S. Government Printing Office.

Table 31-12. Recommended daily dietary allowances*

					Fat-soluble vitamins		Water soluble vitamins							Minerals				
Age (yr)	Wt (kg)	Ht (cm)	Cal	Protein (g)	A act (IU)	E act (IU)	Ascorbic acid (mg)	Folic acid (mg)	Niacin equiv (mg)	Ribo-flavin (mg)	Thia-mine (mg)	B_6 (mg)	B_{12} (μg)	Ca (g)	Phos (g)	I (μg)	Fe (mg)	Mg (mg)
Infants																		
0—1/6	4	55	kg × 120	kg × 2.2	1500	5	35	0.05	5	0.4	0.2	0.2	1.0	0.4	0.2	25	6	40
1/6—1/2	7	63	kg × 110	kg × 2.0	1500	5	35	0.05	7	0.5	0.4	0.3	1.5	0.5	0.4	40	10	60
1/2—1	9	72	kg × 100	kg × 1.8	1500	5	35	0.1	8	0.6	0.5	0.4	2.0	0.6	0.5	45	15	70
Children																		
1—2	12	81	1100	25	2000	10	40	0.1	8	0.6	0.6	0.5	2.0	0.7	0.7	55	15	100
2—3	14	91	1250	25	2000	10	40	0.2	8	0.7	0.6	0.6	2.5	0.8	0.8	60	15	150
3—4	16	100	1400	30	2500	10	40	0.2	9	0.8	0.7	0.7	3	0.8	0.8	70	10	200
4—6	19	110	1600	30	2500	10	40	0.2	11	0.9	0.8	0.9	4	0.8	0.8	80	10	200
6—8	23	121	2000	35	3500	15	40	0.2	13	1.1	1.0	1.0	4	0.9	0.9	100	10	250
8—10	28	131	2200	40	3500	15	40	0.3	15	1.2	1.1	1.2	5	1.0	1.0	110	10	250
Males																		
10—12	35	140	2500	45	4500	20	40	0.4	17	1.3	1.3	1.4	5	1.2	1.2	125	10	300
12—14	43	151	2700	50	5000	20	45	0.4	18	1.4	1.4	1.6	5	1.4	1.4	135	18	350
14—18	59	170	3000	60	5000	25	55	0.4	20	1.5	1.5	1.8	5	1.4	1.4	150	18	400
18—22	67	175	2800	60	5000	30	60	0.4	18	1.6	1.4	2.0	5	0.8	0.8	140	10	400
22—35	70	175	2800	65	5000	30	60	0.4	18	1.7	1.4	2.0	5	0.8	0.8	140	10	350
35—55	70	173	2600	65	5000	30	60	0.4	17	1.7	1.3	2.0	5	0.8	0.8	125	10	350
55—75+	70	171	2400	65	5000	30	60	0.4	14	1.7	1.2	2.0	6	0.8	0.8	110	10	350
Females																		
10—12	35	142	2250	50	4500	20	40	0.4	15	1.3	1.1	1.4	5	1.2	1.2	110	18	300
12—14	44	154	2300	50	5000	20	45	0.4	15	1.4	1.2	1.6	5	1.3	1.3	115	18	350
14—16	52	157	2400	55	5000	25	50	0.4	16	1.4	1.2	1.8	5	1.3	1.3	120	18	350
16—18	54	160	2300	55	5000	25	50	0.4	15	1.5	1.2	2.0	5	1.3	1.3	115	18	350
18—22	58	163	2000	55	5000	25	55	0.4	13	1.5	1.0	2.0	5	0.8	0.8	100	18	350
22—35	58	163	2000	55	5000	25	55	0.4	13	1.5	1.0	2.0	5	0.8	0.8	100	18	300
35—55	58	160	1850	55	5000	25	55	0.4	12	1.5	0.9	2.0	5	0.8	0.8	90	18	300
55—75+	58	157	1700	55	5000	25	55	0.4	10	1.5	0.9	2.0	6	0.8	0.8	80	10	300
Pregnancy			+200	65	6000	30	60	0.8	15	1.8	+0.1	2.5	8	+0.4	+0.4	125	18	450
Lactation			+1000	75	8000	30	60	0.5	20	2.0	+0.5	2.5	6	+0.5	+0.5	150	18	450

*Revised 1968 by the Food and Nutrition Board, National Academy of Science, National Research Council. Allowance levels are intended to cover individual variations of most normal persons living in the United States under usual environmental stresses.

of a yellowish green pigment was first studied in whey. Table 31-11 gives the riboflavin content of some common foods.

The vitamin is formed by bacterial action in the intestines. Riboflavin is not destroyed by heat, but exposure to light for one hour destroys 40% of riboflavin in milk.

Effects of deficiency. Riboflavin deficiency in man causes skin lesions, especially fissures in the corners of the mouth (cheilosis—ki-lo'sis). The cornea of the eye (normally an avascular structure) becomes bloodshot and may ulcerate. In lower animals cataracts are formed in the lens of the eye that, because of their opacity, cause blindness. Lack of riboflavin in the young arrests growth and finally results, in young and old, in death.

Pyridoxine—vitamin B_6. ■ Under vitamin B_6 are included *pyridoxine* and two other closely related compounds.

Sources and daily allowances. Excellent sources of vitamin B_6 are liver, eggs, fish, lettuce, celery, whole wheat, lemons, and milk. The daily requirement of vitamin B_6 for man has not been established. It has been suggested that 1 to 2 mg/day are sufficient.

Effects of deficiency. Vitamin B_6 seems to be needed by all animals thus far investigated, and a host of ill effects have been attributed to the lack of it. Its lack impairs growth in the young and causes loss of body weight in the adult animal. In some animals it leads to a form of anemia; skin lesions are universal. Special stress must be laid on the part this vitamin plays in the chemical reactions involving amino acids. A high protein diet demands a greater intake of vitamin B_6. However, in the descriptions of the disturbances caused by B_6 avitaminosis in human beings and in the accounts dealing with the value of this vitamin in treating human ills, there are so many contradictory statements and opinions that it would not profit us to go into this subject.

Pantothenic* acid. ■ Pantothenic acid is found in every living cell as a part of coenzyme A.

Sources. The richest sources of pantothenic acid are liver, eggs, kidney, milk, and fresh vegetables. An average daily diet in America contains about 10 mg of pantothenic acid, which is an adequate amount. No official daily allowance has been recommended for man.

Effects of deficiency. A disease due to the deficiency of pantothenic acid has not been found in man. In lower animals a deficiency can cause a dermatitis, graying of hair, atrophy of the adrenal cortex, and corneal changes.

Folic acid ■

Sources. Folic acid is widely distributed in nature, especially in the foliage of plants. It also occurs in meats such as liver and kidney.

Effects of deficiency. Absence of this vitamin in man results in megaloblastic anemia, glossitis, and gastrointestinal tract disturbances.

Biotin. ■ The usual diet offers a sufficient amount of biotin for man's need. Only a most unbalanced diet can evoke some of the symptoms of biotin deficiency (skin and tongue lesions). Good sources are kidney, liver, eggs, milk, and most fresh vegetables.

---

*From the Greek, meaning everywhere.

Vitamin B_{12} ■

Source. Liver is the chief source of vitamin B_{12}; it is present in milk, meat, eggs, and fish. It is manufactured by intestinal organisms under certain dietary conditions.

Effect of deficiency. Lack of vitamin B_{12} causes pernicious anemia in man. As noted in Chapter 20, the intrinsic factor produced by the gastric mucosa is necessary for the absorption of this vitamin, and deficiencies are a result of the lack of this factor rather than of a lack of the vitamin itself. Vitamin B_{12} may also be important for growth, since it has a growth-promoting property in some animals.

■ Vitamin C, ascorbic acid, or antiscorbutic vitamin

Sources. ■ In general, green vegetables (especially the growing parts of plants) are the best sources of vitamin C. Outstanding examples are broccoli, 118 mg per $^2/_3$ cup; turnip greens, 58 mg; kale, 37 mg; spinach, 36 mg; lemon juice, 81 mg; and orange juice, 81 mg. Meats, eggs, and cereals are poor sources. Cow's milk contains about 5 mg/pint, a quantity generally held insufficient for an infant. Ascorbic acid can be made synthetically.

Stability. ■ Ascorbic acid is destroyed by heating, salting, drying, and contact with air (storage). The effect of prolonged boiling is due to oxidation. For this reason commercially canned fruits, juices, and vegetables and those boiled in a pressure cooker (at 248 F) are superior to those boiled in an open kettle. The destruction of ascorbic acid is hastened by the presence of alkalies and heavy metals, especially copper (not aluminum). Acid fruits and juices, commercially canned, are almost as rich in vitamin C as the fresh fruits. As vitamin C is readily soluble in water, the vitamin content of the cooking water and of the liquid of canned fruits and vegetables is high.

Vitamin C deficiency. ■ Vitamin C deficiency causes a lack of tensile strength of bone and therefore leads to fractures. When the materials that bind the endothelial cells of the capillary wall are involved, capillary fragility is increased and hemorrhages result. In the absence of this vitamin, bone fractures do not mend and wounds do not heal properly. Its importance to adrenal cortical function was mentioned in Chapter 28.

READINGS

Brusis, O. A., and McGandy, R.: Nutrition and man's heart and blood vessels, Fed. Proc. **30**:1417-1420, 1971.

De Luca, H. F.: Recent advances in the metabolism and function of vitamin D, Fed. Proc. **28**:1678-1689, 1969.

Draper, H. H., and Csallany, A. S.: Metabolism and function of vitamin E, Fed. Proc. **28**:1690-1695, 1969.

Eichenwald, H. F., and Fry, P. C.: Nutrition and learning, Science **163**:644-648, 1969.

Guthrie, H. A.: Introductory nutrition, ed. 2, St. Louis, 1971, The C. V. Mosby Co.

Olson, J. A.: Metabolism and function of vitamin A, Fed. Proc. **28**:1670-1677, 1969.

Rivlin, R. S.: Riboflavin metabolism, New Eng. J. Med. **283**:463-472, 1970.

Silverstone, J. T.: Obesity, Sci. J. **6**(12):40-44, Dec. 1970.

Stadtman, T. C.: Vitamin B_{12}, Science **171**:859-867, 1971.

Suttie, J. W.: Control of clotting factor biosynthesis by Vitamin K, Fed. Proc. **28**:1696-1701, 1969.

Young, V. R., and Scrimshaw, N. S.: The physiology of starvation, Sci. Am. **225**(4):14-21, Oct. 1971.

GLOSSARY

PREFIXES, SUFFIXES, AND COMBINING FORMS

a- not, without (apnea)
ab- from, away (absorption)
ad- to, toward (adrenal glands)
ag-, af-, as- see *ad-*
-algia pain, complaint (neuralgia)
amylo- starch (amylopsin)
an- not, without (anemia)
ana- up (anabolism)
anti- opposite, opposed to (antitoxin)
apo- from, away (aponeurosis)
-ase termination denoting an enzyme (amylase)
auto- self (automatism)
bi- two, twice (biceps)
bili- pertaining to bile (bilirubin)
bio- life (biology)
calci- calcium, lime (calcification)
calor- heat (calorimeter)
cata- down (catabolism)
cerebro- pertaining to the large brain
chole- bile (cholecystokinin)
chrom- color (chromosome)
-cidal killing (bactericidal)
co-, com-, con-, cor- with, together
coll- glue (colloids)
contra- opposite (contralateral)
corpus- body (corpuscle)
-cyt- cell (leukocyte)
-dermic, -dermis skin (hypodermic)
di- two, twice (dichromatic)
dia- through, apart (diaphragm)
dis- negative (disinfect)
dys- bad (dyspepsia)
-ectomy to cut out (tonsillectomy)
em-, en-, endo- in, into (embolus)
-emia blood (anemia)
entero- intestine (enterokinase)
epi- on, above, upon (epidermis)
erythro- red (erythrocyte)
ex- out (expiration)
-fer- to carry (afferent)
-fract- break (refraction)
-gastric stomach (pneumogastric)
-gen- producing (glycogen)
-glosso- tongue (hypoglossal)
gluc- glucose, sugar (glucosuria)
-gnosis knowledge (diagnosis)
-gog or **-gogue** leading (secretogogue)
-graph to write (kymograph)
hemo- blood (hemorrhage)
hetero- other, different (heterogeneous)
hydro- water (hydrolytic)
hyper- over, above measure (hypertrophy)
hypo- under, less than (hypotonic)
in- in, into (insertion)
in- not, without (insufficiency)
inter- between, together (intercostal)
intra- within (intrathoracic)
-itis inflammation (tonsillitis)
-ject- to throw (injection)

juxta- next to (juxtaglomerular)
kin- to move (kinesthetic, kinetic)
-lac- milk (lactose)
leuco- or **leuko-** white (leukocyte)
-logy doctrine, science (physiology)
lympho- pertaining to lymph (lymphocyte)
-lysin, -lysis or **-lytic** dissolving, destruction (hemolysis)
macro- large (macrophages)
mal- bad (malnutrition)
-meter- measure (manometer)
micro- small (microorganism)
mole- mass, body (molecule)
mono- one (monosaccharide)
myo- muscle (myoglobin)
nephr- kidneys (nephritis)
neur- pertaining to nerves (neurasthenia)
nucleo- pertaining to nucleus (nucleoplasm)
-oid like (ameboid)
-ole small (bronchiole)
-oma swelling, tumor (sarcoma)
-opia sight (myopia)
-osis a condition (cyanosis); a process (phagocytosis)
oste- or **osteo-** bone (osteology)
ovi- or **ovo-** egg (oviduct)
para- near, by, beside (parathyroid)
patho- suffering, disease (pathology)
peri- around, near (pericardium)
phago- to eat (phagocyte)
-phil- loving (hemophilia)
-plasm- form (cytoplasm)
-plegia or **-plexy** stroke (apoplexy)
-pnea breathing (dyspnea)
pneumo- air, lungs (pneumonia)
poly- many (polysaccharide)
post- behind, after (postganglionic)
pre- before, in front of (precentral)
pro- before, giving rise to (proenzyme)
proprio- one's own (proprioceptive)
proto- first (protoplasm)
pseudo- false (pseudopod)
psycho- mind (psychology)
pulmo- lung (pulmonary)
re- back, again (regurgitation)
-renal kidney (adrenal)
rhino- nose (rhinology)
-rrhea flow (diarrhea)
-sarco- flesh, muscle (sarcoplasm)
-sclero- hard (sclera, sclerosis)
semi- half (semicircular)
-some- body (chromosome)
-sthenia strength (asthenia)
sub- under, below (subnormal)
thermo- heat (thermogenesis)
-thrombo- clot, coagulation (thrombin)
-tome or **-tomy** to cut (tonsillectomy)
trans- beyond, through (transudation)
trophic to turn (chemotrophic)
-ule small (saccule)
-uria pertaining to urine (glucosuria)
vaso- pertaining to blood vessels (vasodilation)

GLOSSARY

absorption the taking up of materials by the skin, mucous surfaces, or absorbent vessels.
acapnia a marked decrease of CO_2 in the blood.
acid-base balance the proper ratio of H and OH ions in the blood.
acidemia a relative increase in the H ions in the blood.
acidosis a condition in which the concentration of the bicarbonates in the blood is below normal.
actin a contractile protein found in muscle.
action current the electrical current generated when the active and inactive parts of a protoplasmic structure are connected by an outside circuit.
action potential a change in the electrical potential of an active cell or tissue.
active state the condition of a muscle immediately after excitation when it shows an increased resistance to stretch and capability of shortening or creating tension.
active transport the movement of materials against gradients by the expenditure of metabolic energy.
adenohypophysis the anterior lobe of the pituitary gland.
adequate stimulus that form of stimulus which is most efficient for a given structure.
adipose tissue a form of connective tissue in which fat is extensively deposited.
adolescence the period between puberty and maturity.
adrenal glands two glands of internal secretion lying just above the kidneys.
Adrenalin proprietary name for epinephrine.
adrenergic a nerve liberating or synapse activated by epinephrine.
adrenoglomerulotrophin a postulated hormone from the diencephalon which regulates aldosterone secretion by the adrenal cortex.
adsorption the adhesion of thin films of liquids, solids, or gases to the surfaces of solid bodies.
aerobic growing in air or oxygen.
afferent nerve a nerve which carries impulses to the central nervous system.
aldosterone the sodium-retaining hormone of the zona glomerulosa of the adrenal cortex.
alkali reserve the amount of bicarbonate in the blood available for neutralizing acid end products of metabolism.
alkalosis an increase in hydrogen ion accepting buffer sub-

stances of the plasma and decrease in hydrogen ion donor substances.

allergy or **allergia** excessive sensitivity toward certain substances which do not affect the majority of people.

alveolar air the air found in the end pockets of lungs.

alveolus one of the terminal air pockets of the lung.

amino acid an organic acid containing an NH_2 and COOH group and having both basic and acidic properties; the building stones of proteins.

amphoteric having opposite characteristics; capable of acting either as an acid or a base.

anabolism the building up of protoplasm from the simpler food molecules.

anaerobic living best or only without air.

anastomose to run together or to unite (as of two arteries).

androgen a male sex hormone.

anemia a lack of the proper number of red blood cells per cubic millimeter of blood or of the proper percentage of hemoglobin.

anesthesia the condition of total or partial loss of sensibility, especially to touch.

angiotensin a vasoconstrictor peptide derived from plasma globulin by the action of renin.

angstrom a unit of measure equal to 10^{-8} cm.

anode the positive pole of a battery, electropositive.

anorexia lack of appetite.

anoxemia the lack of the proper amount of oxygen in the blood.

antagonistic muscles muscles that oppose each other in their action.

antidiuretic hormone a product of posterior pituitary gland which, through its action on kidney, promotes the conservation of body water.

antiseptic having the power to prevent growth of bacteria; a drug having this power.

antrum a cavity or hollow space; the lower part of the stomach.

aorta the large artery springing from the left ventricle.

apnea the temporary cessation of breathing.

apoenzyme the protein portion of an enzyme.

appestat an area of the hypothalamus regulating appetite through two centers, the feeding and satiety centers.

arachnoid one of the meninges of brain and spinal cord.

areolar tissue a form of connective tissue in which the bundles of inelastic fibers and the elastic fibers form an open meshwork in the ground substance.

arteriole a very small artery.

arteriosclerosis a loss of elasticity or a hardening of an artery.

artery a vessel carrying blood from the heart.

assimilation the building of material into protoplasm; anabolism.

asthenia lack or loss of strength.

astigmatism a refractive error of the eye in which the various meridians of the cornea (or lens) do not have the same radius of curvature.

atrioventricular bundle a bundle of a peculiar tissue that conducts the impulse from the atrium to the ventricle; the bundle of His.

ataxia lack of muscle coordination.

atrium one of the cavities of the heart which receives the blood; auricle.

atrophy a wasting away of a tissue or organ due to a decrease in protoplasm.

atropine a poisonous alkaloid derived from the deadly nightshade *(Atropa)* and the seeds of the thorn apple.

auricle see *atrium.*

autolysis the disintegration of a cell or tissue by its own enzymes.

automaticity the property of a structure to initiate its own activity.

avitaminosis a disease due to lack of vitamins.

axis cylinder the axon of a nerve fiber.

axon one of the protoplasmic processes of a neuron; it carries the impulse away from the cell body.

basal ganglia collections of neurons (nuclei) at the base of the cerebrum consisting of caudate nucleus, putamen, globus pallidus, subthalamic nucleus, substantia nigra, and red nuclei.

basal metabolism (BMR) the minimum expenditure of energy compatible with life.

bilirubin a red pigment found in the bile.

biliverdin a greenish pigment derived from bilirubin.

biosynthesis the formation of a compound from its separate constituents by living organisms.

biuret test a test for native proteins and peptones.

bleeder an individual afflicted with hemophilia.

blind spot an area of the retina where the optic nerve leaves the eyeball; it is insensitive to light.

blood pressure the force which the blood exerts against the walls of the vessels or heart.

bouton a small terminal enlargement of an axon.

Bowman's capsule a saclike structure formed by the invaginated end of a uriniferous tubule of the kidneys.

brachial pertaining to the arm.

bronchus one of the two air tubes formed by the division of the trachea.

buffer substance a compound which has the power to combine with either acid or base, thus helping to maintain the acid-base balance.

bundle of His see *atrioventricular bundle.*

Calorie the unit of heat energy.

calorimeter apparatus for measuring the amount of heat produced by a chemical action or by an animal body.

afferent and efferent member of a reflex arc or two neurons with longer axons.

interstitial between cells; pertaining to the interspaces of a tissue.

interstitial cell–stimulating hormone (ICSH) a gonadotrophic hormone of the adenohypophysis which stimulates androgen secretion by the Leydig cells.

intrathoracic within the thorax.

intravascular within the blood vessels.

inulin a carbohydrate compound of 5,200 molecular weight used to estimate glomerular filtration rate since it is neither secreted nor reabsorbed by the tubules.

ionize the dissociation of a compound into ions, as in an electrolyte.

ions the products formed by the electrolytic dissociation of a molecule and carrying one or more positive or negative electrical charges. An electrically charged atom or group of atoms.

ipsilateral referring to structures on the same side of the body axis.

iris the pigmented diaphragm directly in front of the lens of the eye.

irradiated having been affected by electromagnetic waves, generally spoken of in connection with x-rays or the production of vitamin D.

irritability the power of protoplasm to respond to stimulation.

ischemia a temporary and local deficiency of blood.

islands of Langerhans isolated groups of alpha and beta cells in the pancreas concerned with the production of insulin and glucagon.

isosmotic having equal osmotic pressure.

isotonic having the same tension or pressure (e.g., osmotic pressure).

isotope one member of a group of atoms having the same number of protons but different numbers of neutrons in the nucleus so that the atomic masses are different, but the atomic numbers are the same.

jaundice the presence of bile in the blood.

juxtaglomerular apparatus the collection of modified smooth muscle cells in the media of the afferent arterioles at the entrance to the kidney glomeruli which produce renin.

karyokinesis cell division by mitosis.

keratin a highly insoluble, indigestible scleroprotein found in skin, nails, etc.

ketone a chemical compound formed by the oxidation of a secondary alcohol and characterized by the C-O group.

ketonemia a condition of excess ketones, largely acetoacetic acid, in the blood.

kilogram-meter a unit of mechanical work; equivalent to raising 1 kg to the height of 1 m.

kinase a proenzyme activator.

kinesthetic sensation the proprioceptive sensation by which we judge position and movement of our limbs.

lachrymal or **lacrimal** pertaining to tears.

lactase an enzyme splitting lactose into glucose and galactose.

lactation secretion of milk.

lacteals the lymphatic vessels of the intestine.

lactic acid an acid formed from carbohydrates, as in the souring of milk, having the formula $C_3H_6O_3$.

lactose milk sugar.

lacuna a little hollow space.

larynx voice box.

latent period the length of time elapsing between the stimulus and the response.

lecithin a complex fat containing phosphoric acid and a nitrogenous base known as choline.

lesion a pathological change in a tissue; a hurt or wound.

leukemia a pathological condition in which the number of white blood cells is greatly increased.

leukocyte a white blood cell.

ligament a white fibrous structure which binds, e.g., one bone to another.

limbic system the oldest part of the cerebral cortex and associated deep structures (amygdala, hippocampus, septal nuclei) concerned with olfaction, behavior, and emotions.

liminal the lowest level of perception or the threshold of excitation.

lipase a fat-splitting enzyme.

lipid fat and fatlike substances.

luteinizing hormone (LH) a gonadotrophic hormone of the adenohypophysis responsible for final maturation of follicular cells, ovulation, and corpus luteum formation.

lymph the fluid found in the lymph vessels and spaces.

lymphocyte a certain type of leukocyte.

lysosome an intracellular structure containing a large concentration of lytic enzymes.

macrophage a cell of the loose connective tissue having the power of phagocytosis; a resting wandering cell; histiocyte.

Malpighian body the beginning of the uriniferous tubule and composed of Bowman's capsule and the glomerulus.

maltase an enzyme that splits maltose into glucose.

maltose a disaccharide.

manometer an instrument for measuring pressure.

mastication chewing.

matrix intracellular ground substance.

maturation ripening.

Meibomian glands sebaceous glands of the eyelids.

meiosis the reduction division of germ cells which leaves

only half the normal complement of chromosomes in the mature sperm and ova.
Meissner's plexus a network of nerves in the submucosa of the small intestine.
melanophore-stimulating hormone a pituitary hormone of uncertain function, if any, in man which controls skin color in lower animals.
meninges the three membranes covering the central nervous system.
meningitis inflammation of the meninges.
menopause the cessation of the periodic menstruations; climacteric.
menstruation the periodic changes in the uterus during the sexual life of the female.
mesentery a fold of the peritoneum which connects the intestine to the posterior wall of the abdomen.
metabolism the sum total of the chemical changes occurring in the body; the combined anabolism and catabolism.
microelectrode a glass capillary or wire electrode with a tip diameter of less than 1 μ which can be inserted without undue damage into single cells for bioelectrical studies.
micron 0.001 mm.
mitochondria small granules or rod-shaped structures found in the cytoplasm of cells.
mitosis the normal division of growing cells which involves the complete reduplication of the chromosome complement.
mitral pertaining to the bicuspid valve.
molar solution a solution which contains in 1 liter as many grams of the solute as the molecular weight of the solute.
molecular solution a molar solution.
monosaccharides a class of carbohydrates of which the hexoses have the formula $C_6H_{12}O_6$.
motor areas the areas of the cerebral cortex which on stimulation give rise to muscular action.
mucin a glycoprotein found in saliva, etc.
mucosa a membrane lining the cavities and tubes communicating with the surface of the body and secreting a mucous fluid.
mucus a viscid, watery secretion.
myelin sheath the inner covering of a medullated nerve fiber.
myocardium the muscle tissue of the heart.
myofibril a subdivision of the muscle fiber found in the sarcoplasm.
myogenic originating in muscle tissue.
myoneural junction the motor end plates of a muscle; the structure between the muscle and its nerve.
myopia nearsightedness.
myosin a contractile protein found in muscles.
myxedema a pathological condition of the adult, associated with malfunction of the thyroid, in which the subcutaneous tissue becomes filled with a mucinlike material.

narcotic a drug that relieves pain and induces sleep; in large doses it causes stupor, coma, or death.
nerve a bundle of nerve fibers.
nerve cell a neuron.
nerve fiber a prolongation of the cell body of a neuron covered with one or more sheaths.
neurilemma the outermost covering of a nerve fiber.
neurogenic originating in nerve tissue.
neurohumor the chemical agent of transmission and excitation.
neurohypophysis the posterior lobe of the pituitary gland.
neuron or **neurone** a nerve cell.
niacin a vitamin chemically known as nicotinic acid.
nitrogenous equilibrium that condition of the body in which the amount of nitrogen ingested equals that excreted.
norepinephrine an amino acid derivative which constitutes the principal chemical transmitter substance at peripheral sympathetic (adrenergic) nerve terminals.
normal solution a solution containing one equivalent weight (weight in grams reacting with 1 g of H ion) per liter of solution.
nuclei a term used to denote collections of neurons in the central nervous system.
nucleoprotein a conjugated protein whose prosthetic group is a nucleic acid.
nucleus a highly differentiated body lying in the cytoplasm of a cell; the organ of cellular nutrition and reproduction.
nystagmus involuntary and abnormal oscillatory movements of the eyes.

obesity the state of being excessively fat.
occipital lobe the hindmost lobe of the cerebral cortex.
occlusion a form of indirect inhibition in the nervous system due to convergence.
olein one of the chemical fats found in the body fat and having a low melting point.
omentum a fold of the peritoneum connecting the stomach with the other abdominal viscera.
ontogeny the development of the individual.
oocyte the female germ cell.
organ a structure composed of two or more tissues and performing a special function.
ornithine cycle the biochemical pathway for the formation of urea from ammonia.
osmolarity an expression of solute concentration which takes into account the number of osmotically active particles in the solution, i.e., the product of the molar concentration and the number of particles per molecule.
osmosis the diffusion through a membrane of the solvent (e.g., water) from a lower to a more concentrated solution.
osmotic pressure the pressure generated by osmosis.
ossification the hardening of bone; calcification.

osteoblasts the specialized cells which produce a collagenous matrix for new bone formation.
osteoclasts the specialized cells which cause erosion and reabsorption of previously formed bone.
otoliths concretions found in the utricles and saccules of the inner ears.
ovary the female sex gland which forms the ovum.
ovulation the setting free of the ovum from the ovary.
oxygen one of the constituents of the air.
oxyhemoglobin hemoglobin united with oxygen.
oxyntic cells the acid-secreting cells of the stomach; the parietal cells.
oxytocin a hormone of the neurohypophysis whose principal action is on milk ejection in the lactating animal.

palmitin one of the chemical fats found in the body fat.
pancreas an abdominal organ secreting the pancreatic juice and containing the islands of Langerhans.
paralysis loss of motion or sensation in a part of the body.
parasympathetic nervous system that part of the autonomic nervous system in which the preganglionic fibers spring from the midbrain, medulla, or the sacral region of the cord.
parathormone the hormone of the parathyroid glands involved in calcium and phosphorus metabolism.
parathyroids four glands of internal secretion lying on or embedded in the thyroid gland.
paraventricular nuclei the specialized groups of neurons in the hypothalamus involved in control of neurohypophyseal secretion of oxytocin.
parietal pertaining to the walls of a cavity.
parturition childbirth.
patellar reflex the knee-jerk; the throwing forward of the leg due to the tapping of the patellar ligament.
pathogenic disease-producing.
pellagra a deficiency disease, claimed to be due to the lack of nicotinic acid.
pepsin the proteolytic enzyme of the gastric juice.
peptide a compound composed of two or more amino acids.
peptone a derived protein formed by digestion from native proteins.
pericardium the closed membranous sac surrounding the heart.
periosteum the fibrous membrane surrounding a bone.
peristalsis the wavelike muscular movement of a tube (e.g., intestine) by which the contents are moved forward.
peritoneum a serous membrane lining the abdominal cavity and enveloping the viscera.
peroxidase an enzyme which liberates active oxygen from a peroxide (e.g., H_2O_2).
pH the negative logarithm of the hydrogen ion concentration.
phagocyte a cell able to engulf solid particles.
phagocytosis the process by which a cell engulfs solid particles.
pharynx a pouchlike structure extending from the base of the skull to the level of the sixth cervical vertebra; into it open the mouth, posterior nares, Eustachian tubes, esophagus, and trachea.
phosphatide a complex fat.
phosphoglucomutase the enzyme which catalyzes the conversion of glucose-1-phosphate, the initial product of glycogen breakdown, to glucose-6-phosphate.
phosphoprotein a phosphorus-containing protein, e.g., casein.
photopia daylight, or bright-light, vision.
photosynthesis the process by which chlorophyll of plants produces organic compounds.
phylogeny the evolution of a race; history of ancestral development.
pia mater the innermost of the three meninges.
pineal gland a small body situated on the dorsal aspect of the brain.
pinocytosis a process of ingestion of molecular particles similar to that of phagocytosis but on a micro scale; cell drinking.
pituitary body an endocrine organ situated on the ventral aspect of the brain.
placenta a membranous structure which forms, with the umbilical cord, the connection between the uterine wall and the fetus and through which the exchange of material between mother and fetus takes place.
plasma the liquid portion of the blood.
plasma membrane the outer layer of the cytoplasm which is claimed to possess selective permeability.
plasmolysis the shrinkage of the protoplasm due to the withdrawal of liquid from the cell by a hypertonic solution.
plasticizer a substance which imparts softness or a viscous quality.
platelets thrombocytes; one of the three classes of blood cells.
pleura the serous membrane lining the chest cavity and covering the lungs.
plexus a network, especially of veins or nerves.
pneumogastric nerve the tenth cranial (vagus) nerve.
pneumograph an instrument for recording the respiratory movements.
pneumothorax the condition of having air in the thoracic cavity and outside of the lungs.
poikilothermic animals with a variable body temperature; cold-blooded animals.
polycythemia a condition of excess red blood cell population.
polydipsia the consumption of increased amount of water.
polypeptides simple proteinlike substances.

polysaccharides a class of carbohydrates having the general formula $(C_6H_{10}O_5)_n$.
polyuria the excretion of large volumes of dilute urine.
portal vein the vein conveying the venous blood from the small and large intestines, spleen, and stomach to the liver.
pregnanediol the product of progesterone conversion by the liver which is conjugated and excreted in the urine as a glucuronide.
presbyopia loss of accommodation due to hardening of the crystalline lens; the sight of old age.
pressor nerve an afferent nerve which on stimulation causes an increase in blood pressure by reflexly constricting the blood vessels.
proenzyme the antecedent of an enzyme.
progesterone a steroid hormone produced by the corpus luteum and placenta which causes proliferative changes in the endometrium and breasts.
prolactin a gonadotrophin of the adenohypophysis which, in primates, causes milk secretion from the breast after estrogen and progesterone priming.
prone lying with face down; opposite to supine.
propagation to spread, the passage of a bioelectrical potential along a membrane (e.g., nerve fiber)
proprioceptive impulses impulses received from the muscles, tendons, and joints.
prostate gland a gland of the reproductive system of the male, situated at the base of the bladder and surrounding the mouth of the urethra.
prosthetic group a small organic compound or ion firmly attached to enzyme protein that is essential to the enzyme's activity.
proteins nitrogenous compounds formed by the union of many amino acids.
proton the unit of positive electricity.
protoplasm the physical material exhibiting the properties of life.
protoplast a cell.
prothrombin the material from which thrombin is formed.
proximal nearest; opposite to distal.
pseudopod or **pseudopodium** a temporary projection of a portion of the protoplasm of a naked cell; an organ of locomotion.
psychic pertaining to the mind.
ptyalin a starch-splitting enzyme of the saliva.
puberty that period of life at which the young of either sex become capable of reproduction.
pulse a rhythmical dilation of the artery caused by the systolic output.
pulse pressure the difference between the systolic and diastolic pressures.
pupil the central aperture of the iris.
pylorus the part where stomach and small intestine meet.
pyramidal fibers the corticospinal nerve fibers, springing from cell bodies in the precentral convolution and extending into the spinal cord.

quantal summation an increase in contraction due to excitation of additional muscle fibers.

rachitis a disturbance in bone formation; rickets.
receptor an organ for the reception of stimuli; sense organ.
rectum the lowest part of the intestinal canal.
reflex or **reflex action** an action induced by the stimulation of a receptor and carried on without the intervention of the will.
reflex arc the pathway involved in a reflex, consisting of afferent and efferent nerves and interneurons.
refractory period the period of reduced irritability during the activity of a protoplasmic structure.
renal pertaining to the kidneys.
renal threshold the plasma concentration at which a substance first appears in the urine.
renin a hormone secreted by the kidney which causes an increase in blood pressure by its role in the production of angiotensin.
rennin a milk-clotting enzyme.
resonance an impressed vibration, the quality of voiced sound due to oral cavity structures; the sound elicited by percussion of the chest.
respiration the exchange of gases between an organism and its environment; the muscular movements by which the air in the lungs is exchanged.
resting potential the electrical potential difference between the extracellular and intracellular surfaces of a cell membrane.
reticular formation a diffuse aggregation of neurons in the brainstem.
retina the innermost coat of the eye which is stimulated by light.
rheobase the minimal strength of electrical current necessary to produce stimulation.
rhodopsin a pigment in the rods of the retina, necessary for scotopic or twilight vision.
ribonucleic acid (RNA) a high molecular weight compound found principally in the cytoplasm of cells.
ribose a 5-carbon sugar.
ribosomes the particles consisting of ribonucleic acid which occur in the cytoplasm of cells often in association with the endoplasmic reticulum.
rigor mortis the stiffening or hardening of the muscles soon after death.
rod one of the visual cells of the retina.

saccule a little sac; one of the divisions of the labyrinth of the inner ear; a sense organ of equilibrium.
sarcolemma the delicate sheath surrounding a muscle fiber.

sarcoplasm the more liquid portion of a muscle fiber in which the fibrils are embedded.
sarcoplasmic reticulum a network of fine tubules and vesicles found in muscle cells which is involved in the transmission of excitation from the surface of the cell to the contractile proteins.
sclera the outer, white coat of the eyeball.
scotopia the vision in very dim light; twilight vision.
scurvy a pathological condition due to lack of vitamin C.
sebaceous glands the oil glands of the skin.
sebum the oil secreted by the oil gland.
secretin a hormone formed in the wall of the small intestine which induces the secretion of pancreatic juice.
secretion the transfer of a substance from the blood onto a free surface by means of a gland; also the matter secreted.
secretogogue a substance that stimulates glandular activity.
semidecussation half-crossing, e.g., the crossing of the nerve fibers from the nasal half of each retina to the opposite side of the brain.
senescence growing old.
sensation a change in consciousness caused by the stimulation of a sensory surface.
serum the liquid part of coagulated blood.
sinoatrial node a mass of tissue lying in the wall of the right atrium and in which the impulse for the heartbeat originates.
sinus a cavity or recess, e.g., the cavities in the bones of the face.
sinus venosus the contractile structure of a frog's heart into which the systemic veins empty and which discharges its blood into the right atrium.
sinusoid relatively large space or tube in the circulatory system of the liver.
somatic pertaining to the body.
somatic nerves the nerves directly connected with the central nervous system.
somatotrophic hormone a secretion of the adenohypophysis which accelerates growth of the organism.
spasticity a paralysis characterized by tonic spasms of the affected muscles and increased tendon reflex response.
spatial summation an increase in response of a nerve due to an increase in the number of active presynaptic elements providing excitation.
spermatogenesis the stepwise production of spermatozoa from primitive germ cells, involving reduction in the number of chromosomes in man to twenty-three by meiosis.
sphincter a circular muscle surrounding and closing an opening, e.g., the pyloric sphincter and the sphincter of the iris.
sphygmomanometer an instrument for determining the amount of blood pressure.
spirometer an instrument for determining the amount of air respired.
spontaneity the property of initiating activity without external excitation.
squamous scalelike.
stasis a stoppage (especially of blood).
stearin one of the fats found in the body.
stenosis the constriction or narrowing of a channel or orifice.
stereognosis the recognition of the solidity and form of an object by means of the touch, pressure, and muscle senses.
sterols lipid compounds including cholesterol and its derivative steroid hormones, bile acids, and certain vitamins.
stimulus a change in the environment which modifies the activity of protoplasm.
stroma the framework of an organ.
substrate a substance on which an enzyme acts.
sudoriferous pertaining to sweat.
sulcus a groove or fissure.
supraoptic nuclei the specialized groups of neurons in the hypothalamus involved in control of neurohypophyseal secretion of vasopressin (antidiuretic hormone).
suprarenal glands the adrenal glands.
surface tension the net force existing at a free surface, e.g., of a liquid, due to unequal intermolecular forces about the surface molecules.
synapse or **synapsis** the junction between two neurons.
syncope fainting.
synergia the working together of two agents.
synovial fluid a form of lymph which bathes the synovial membrane covering the ends of a bone in a joint.
systole the contraction of the heart muscle.
systolic pressure the arterial pressure during the systole of the heart.
systolic sound the sound formed during the systole of the heart.

temporal summation an increase in response of a nerve or muscle (wave summation) due to an increased frequency of excitation.
tendon a connective tissue structure which binds a muscle to a bone.
testis (testes) the male reproductive gland; testicle.
tetanus a sustained contraction of a muscle produced by the fusion of twitches.
thalamus a mass of gray matter in the diencephalon.
thermogenesis the production of heat in the body.
thorax the chest.
threshold stimulus the liminal or minimal stimulus.
thrombin an agent involved in the coagulation of the blood.
thromboplastin one of the agents necessary in the formation of thrombin.
thrombosis the formation of a thrombus, or clot.
thrombus a clot formed in the vessels and adhering to the walls of the vessels.

thymus a gland believed by some to form an internal secretion.

thyroid gland a gland of internal secretion.

thyroxin a hormone produced by the thyroid gland.

tidal volume the amount of air inspired or expired during quiet respiration.

tissue an aggregate of similar cells.

tissue fluid the fluid found in the tissue spaces; lymph.

tonus a subdued continuous contraction of a muscle by which it resists stretching.

toxemia a condition in which the blood contains poisons formed either by the body itself or by microorganisms; blood poisoning.

toxin a poisonous compound of animal or vegetable origin.

trachea the windpipe leading from the pharynx to the bronchi.

transaminases the enzymes involved in the exchange of amine groups between keto acids and amino acids.

transducer a device which converts one form of energy into another.

transmembrane potential the electrical potential difference across a cell membrane.

transudate the material passing through the capillary wall from the blood into the surrounding spaces.

trauma a wound.

tricuspid valve the valve between the right atrium and ventricle.

trigeminal nerve the fifth cranial nerve supplying sensory fibers to the face, teeth, etc.

trophic pertaining to nutrition or nourishment.

tropomyosin a structural protein of muscle associated with contraction.

troponin a structural protein of muscle involved in regulation of contraction.

trypsin a proteolytic enzyme found in the intestine.

trypsinogen the material found in the pancreatic juice from which trypsin is formed.

tubular maximum (Tm) the maximum rate at which the kidney active transport mechanisms can transfer a particular solute into or out of the tubules; the term *transport maximum* is used also.

twitch a single contraction of a muscle.

tympanum the eardrum.

umbilicus the navel.

urea a nitrogenous waste produce found in the urine.

ureter the duct which conveys the urine from the kidney to the bladder.

urethra the duct by which the urine is voided.

uric acid a nitrogenous waste product found in the urine.

urinometer an instrument for determining the specific gravity of the urine.

urogenital pertaining to the urinary and genital organs.

uterus one of the secondary sex organs in mammals in which the egg and fetus develop previous to birth; the womb.

vagus the tenth cranial (pneumogastric) nerve which innervates the heart, bronchi, lungs, stomach, pancreas, small intestine, etc.

vas deferens the duct conveying the spermatozoa to the urethra.

vasoconstrictor nerve a nerve which causes constriction of a blood vessel.

vasodilator nerve a nerve which causes dilation of a blood vessel.

vasopressin an octapeptide secretion of the neurohypophysis important primarily for its antidiuretic activity.

vegetative nervous system the autonomic nervous system.

vena cava one of the large veins communicating with the right atrium.

venous blood blood returning from the tissues and containing more CO_2 and less oxygen than arterial blood.

ventral pertaining to the front or abdomen; opposite to dorsal.

venule a small vein.

vertigo dizziness; giddiness.

vesicle a small bladder or blister.

villus a minute structure of the intestinal mucosa projecting into the lumen of the intestines; an organ of absorption.

viscus (viscera) an organ in a body cavity, especially in the abdomen.

visual acuity the power of the visual apparatus to distinguish the detail of an object, such as the letters of a printed page.

vital capacity the amount of air that can be expired by the most forcible expiration after the deepest inspiration.

vitamin an organic compound necessary for normal metabolic function of the body.

xerophthalmia a pathological condition of the cornea of the eye due to the lack of vitamin A.

zein an incomplete protein found in corn.

zwitterion a dipolar compound; e.g., amino acids have a positively charged amino group and negatively charged carboxyl group; it is an amphoteric electrolyte.

zygote the single cell formed when an oocyte is fertilized by a sperm.

zymase an enzyme present in yeast.

zymogen the material from which an enzyme is formed.

INDEX

D

N

S

COMPARISON OF METRIC WITH ENGLISH MEASURES

■ Length

1 kilometer (km) = 1000 meters = 0.62 mile
1 meter (m) = 100 centimeters (cm) = 1000 millimeters = 39.37 inches
1 millimeter (mm) = 1/25 inch (approximately)
1 micron (μ) = 0.001 millimeter
1 angstrom (A) = 0.00001 micron = 1×10^{-7} millimeter
1 inch = 2.54 centimeters (approximately)

■ Weight

1 kilogram (kg) = 1000 grams = 2.2+ pounds or 35.27 ounces
1 gram (g) = 1000 milligrams
1 milligram (mg) = 1000 micrograms
1 microgram (μg) = 0.001 milligram = 1/28,000,000 ounce
1 pound (lb) = 453.6 grams
1 ounce (oz) = 28.35 grams

■ Volume

1 liter = 1000 milliliters (ml) = 1.05 liquid quarts
1 liquid quart = 0.9464 liter = 946.4 milliliters
1 fluidounce = 29.57 milliliters
1 cubic inch = 16.38 cubic centimeters

■ Energy

1 kilogram-meter (kg-m) = 7.25 foot-pounds
1 foot-pound (ft-lb) = 0.1381 kilogram-meter

■ Mechanical equivalent of heat

1 Calorie (Cal) = 1000 calories (cal)
1 Calorie = 426 kilogram-meters = 3087 foot-pounds
1 kilogram-meter = 0.00234 Calorie
1 gram-centimeter = 2.4×10^{-8} Calories

■ Temperature

To convert centigrade degrees into Fahrenheit, multiply by 9/5 and add 32.
To convert Fahrenheit degrees into centigrade, subtract 32 and multiply by 5/9.